D0081602

AgriScience

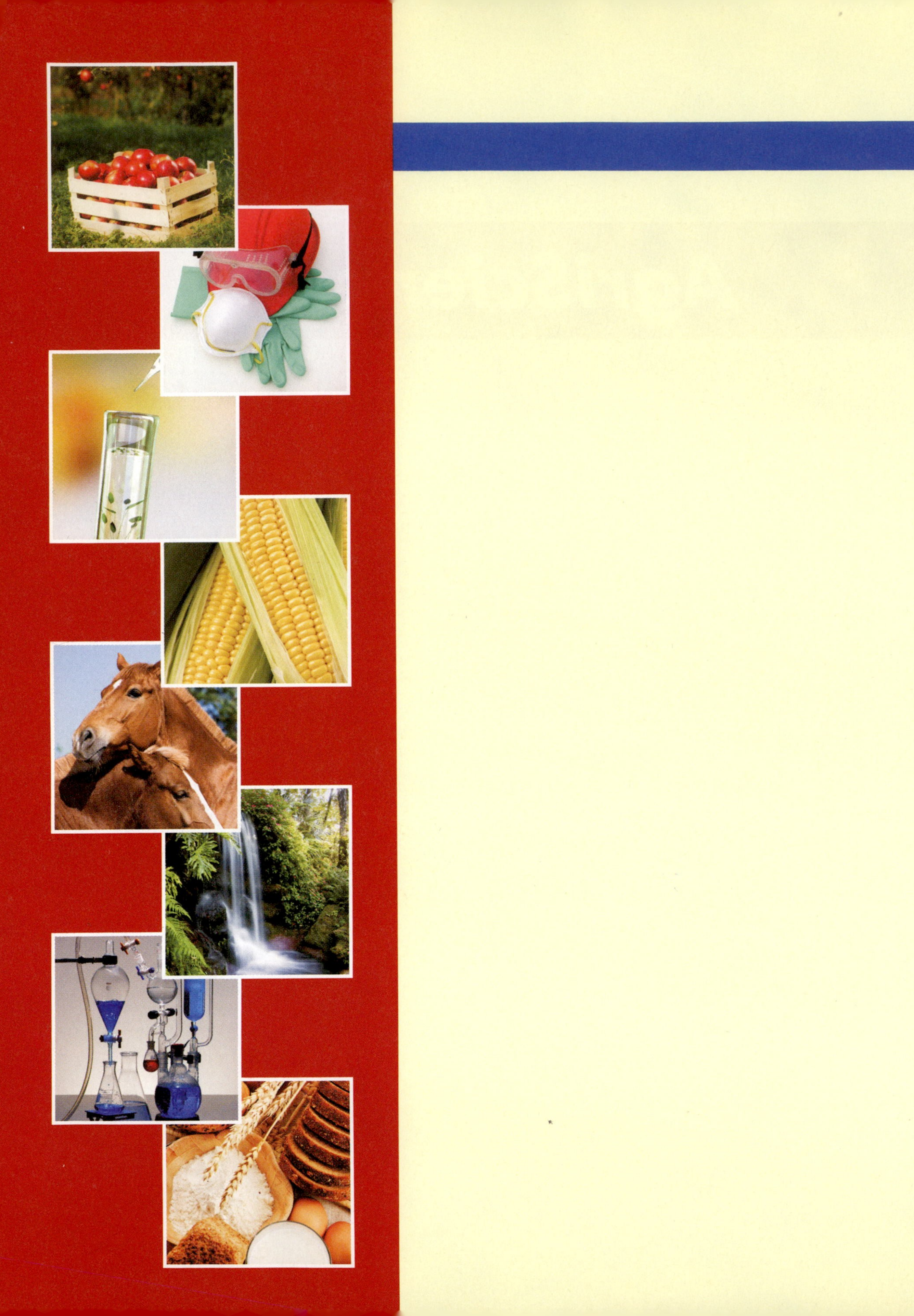

AGRISCIENCE

Fifth Edition

JASPER S. LEE
Agricultural Educator
Clarkesville, Georgia

DIANA L. TURNER
Science Educator
Starkville, Mississippi

Jasper S. Lee — Series Editor

Prentice Hall

Boston Columbus Indianapolis New York San Francisco Upper Saddle River
Amsterdam Cape Town Dubai London Madrid Milan Munich Paris Montreal Toronto
Delhi Mexico City Sao Paulo Sydney Hong Kong Seoul Singapore Taipei Tokyo

Editor in Chief: Vernon Anthony
Acquisitions Editor: William Lawrensen
Editorial Assistant: Lara Dimmick
Director of Marketing: David Gesell
Campaign Marketing Manager: Leigh Ann Sims
Curriculum Marketing Manager: Thomas Hayward
Marketing Assistant: Les Roberts
Associate Managing Editor: Alexandrina Benedicto Wolf
Senior Operations Supervisor: Pat Tonneman
Operations Specialist: Deidra Schwartz
Cover Designer: Keithley & Associates
Cover Art: Shutterstock—Background photo Varina and Jay Patel; top row left to right—Laboratory chemical reaction, Elemental Imaging; natural spring waterfall, Leighton Photography & Imaging; apples in orchard, Tomo Jesenicnik; bottom row left to right—Horses, Yegorius; corn on cob, crystalfoto; baking ingredients, Kiselev Andrey Valerevich
Product Development and Project Management: Emergent Learning, LLC
Composition: CAERT, Inc.
Printer/Binder: Courier Corporation
Cover Printer: Moore Langen, a Courier Company

Credits and acknowledgments borrowed from other sources and reproduced, with permission, in this textbook appear on appropriate pages within text.

10 9 8 7 6 5 4 3 2 1

Prentice Hall
an imprint of

PearsonSchool.com/careertech

ISBN-10: 0-13-509622-7
ISBN-13: 978-0-13-509622-2

Preface

Now is an exciting time in the evolving maturity of this book! Four previous editions have been extremely well received. ***AgriScience***, Fifth Edition, represents new student- and teacher-friendly approaches that blend science with the fundamentals of agriculture. Since agriculture is applied science, this is a highly useful and practical approach. Principles of science are presented and reinforced using agricultural applications. Of course, a broad interpretation of agriculture is used that includes crop and animal production as well as horticulture, forestry, marketing, and processing.

Teachers, students, scientists, agriculturists, and numerous others from throughout the United States have helped shape this Fifth Edition. Throughout, the goal of the authors was to acquire and use input that would aid in providing the best possible book for introductory science-based agriculture classes. First-hand observations of students and teachers in schools throughout the nation are the foundation on which many changes were made. Curriculum guides, lists of standards, blueprints, and other course materials from state education agencies were used.

The contents of ***AgriScience*** relate science information to useful applications in plant and animal production. Some things that sound complicated in the science classroom are used every day in the agriculture classroom. Practical applications result in academic connections that reinforce studies in related academic classes.

This edition has updated and expanded content in a number of areas. The contents of the book have been reshaped to incorporate the Agriculture, Food and Natural Resources (AFNR) Career Cluster and the eight career pathways that are included. The National Agriculture, Food and Natural Resources Career Cluster Content Standards were studied, interpreted, and applied to the book contents. Several areas of content have been expanded in this edition. Increased coverage is given to agricultural mechanics, including a new chapter on this topic. More information is included on student organizations and supervised experience. Content on leadership and personal development has been expanded. Areas have been added on entomology, wildlife, genetics, plant health, animal production and quality assurance, veterinary technology, animal restraint, land measurement and description, safety, integrated pest management, and the free enterprise economic system.

AgriScience is intended to be used in introductory science-based agriculture classes. It is the cornerstone of modern agriscience education. It is for people who are serious about the science and technology of agriculture.

Many exciting things (illustrations, examples, and content) are in this book to promote interest and learning. Go ahead and take a look for yourself! Enjoy and grow in agricultural knowledge.

DISCLAIMER CLAUSE/STATEMENT OF WARRANTY

The authors and the publisher are in no way responsible for the application or use of the chemicals mentioned or described herein. They make no warranties, express or implied, as to the accuracy or adequacy of any of the information presented about various chemicals in this book, nor do they guarantee the current status of registered uses of any of these chemicals with the U.S. Environmental Protection Agency. Also, by the omission, either unintentionally or from lack of space, of certain trade names and of some of the formulated products available to the public, the authors are not endorsing those companies whose brand names or products are listed.

About the Authors

Jasper S. Lee

Jasper S. Lee is an agricultural educator who has been a high school teacher, a university teacher educator, and a consultant in agricultural education. He is a native of Clinton, Mississippi, and grew up with farm and education interests. FFA activities were important during his high school years, and he held chapter, federation, and state offices. His education beyond high school was at Hinds Community College, Mississippi State University, and the University of Illinois. His former students are in leadership positions worldwide, with some serving as authors. He now lives in Georgia.

As the author or co-author of a number of textbooks, none is more exciting to him than ***AgriScience.*** Examples of other books used in agricultural education and applied science are *Introduction to Livestock & Companion Animals, Aquaculture, Environmental Science and Technology, Introduction to Plant & Soil Science and Technology, Wildlife Management Science & Technology,* and *Natural Resources.* In addition to authoring, his interests include curriculum development and agricultural photography. Lee welcomes your email contacts at jasperlee@windstream.net.

Diana L. Turner

Diana L. Turner has been a high school science teacher in Starkville, Mississippi, with major interests in chemistry, physics, advanced biology, and environmental science. Most recently, she has taught chemistry at East Mississippi Community College. Turner is a graduate of Mississippi State University and is widely recognized for excellence in science education, with an emphasis on human health care. She has directed an outstanding science program—one often recognized as the top program in the state. She has been named the Mississippi Secondary Teacher of the Year, inducted into the Education Hall of Fame by the Chamber of Commerce, and designated a STAR teacher. She is a strong advocate of quality science education. Her students often develop outstanding entries for science fair competition. Her former students are now in a wide range of graduate and professional schools and careers in science.

Turner has authored and assisted with several widely used textbooks and ancillaries. She is popular as a leader of inservice education workshops for agriculture and science teachers.

Acknowledgments

Many people have made important contributions to this and previous editions of ***AgriScience***. Students, teachers, scientists, and agriculturists throughout the United States are acknowledged for their help in shaping this book. Their guidance was used in outlining, writing, and illustrating the book. The result is an up-to-date book with outstanding student and teacher appeal. Individuals specifically acknowledged with the Fifth Edition are listed here.

Lindsay Drevlow, student, scholar, and scientist at Piedmont College, is acknowledged for her efforts in reviewing the previous edition of ***AgriScience*** and in providing suggestions on changes and improvements to keep the book current with science thought and practice. Further, she is acknowledged for preparing and carrying out scientific demonstrations and practices for study, photography, and inclusion in the book. Her role as a technical model and in arranging other technical models is also acknowledged.

Agriculture and science students and teachers at a number of high schools are acknowledged for their assistance with technical activities related to animals, plants, and other subjects for this edition. These include the following schools: Madera, Elk Grove, and Kingsburg, California; Edmond, Yukon, Moore, and Tecumseh, Oklahoma; Thief River Falls, Minnesota; Starkville Academy, Millsaps Career and Technology Center, and East Mississippi Community College in Mississippi; Piedmont College, Tallulah Falls, Malcolm Bridge, and Morgan County, Georgia; Orion and Oneida, Illinois; and Hancock (Charlottesville), Indiana. In addition, numerous teachers and students are acknowledged for their assistance with previous editions.

Students and former students at Piedmont College recognized here include: Hannah McKay, Mary Ann Alexander, J. J. McKay, Katie Wood, Bill Secor, Mike Santowski, Jenny Shane, Wesley Crow, Ryan Cobo, Samuel Coppage, Sarah Gardner, Whitney Matthews, Amanda Patrick, Sarah Simler, Katie Porter, Emily Brown, Heather Bardinelli, Cindy Williams, Tracy Westrom, Shae David, and Jenny Conner. These students are acknowledged for assistance with various stages of planning and illustrating this edition and for serving as technical models for photographs. The assistance of Tonia Shatzel, DVM, Georgia, is acknowledged with technical practices. Eric Watts of Pets Gone Wild, Georgia, is acknowledged for his assistance with animal photography.

Selected acknowledgments and citations are shown throughout the book for photographs and other materials.

Appreciation goes to the staff and associates of CAERT, Inc., for editing and designing the book. The individuals acknowledged are Dan Pentony, Ron McDaniel, Kim Romine, Rita Lange, Tim Shelton, Kristen Dallavis, and Jim Dowling. Special thanks goes to Peter McCarthy and Chris Katsaropoulos of Emergent Learning, LLC, for their enthusiasm in pursuing a revision of this book.

Jasper S. Lee
Diana L. Turner

Contents

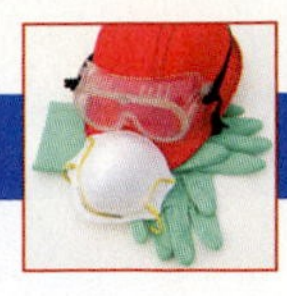

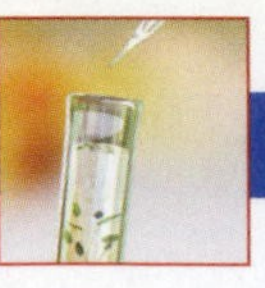

PART FOUR—Plant Science

PART FIVE—Animal Science

PART SIX—Earth Science and Technology

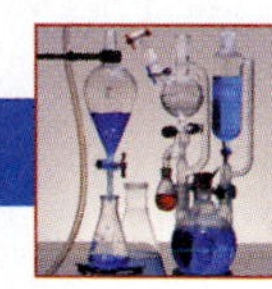

PART SEVEN—Physical Science and Technology

Part One

The World of AgriScience

CHAPTER 1

Science and Technology in Agriculture

OBJECTIVES

This chapter explains the meaning and importance of agriscience and related areas of technology. It has the following objectives:

1. Explain agriscience and technology.
2. Relate the meaning and importance of AFNR.
3. Identify career pathways in AFNR.
4. Explain the importance of human needs in AFNR.
5. Assess the role of consumers and their preferences.

TERMS

AFNR
agriscience
aquatic animals
career pathway
consumer
customs
demand
fiber
food
human need
nutrient
shelter
standard of living
technology
terrestrial animals

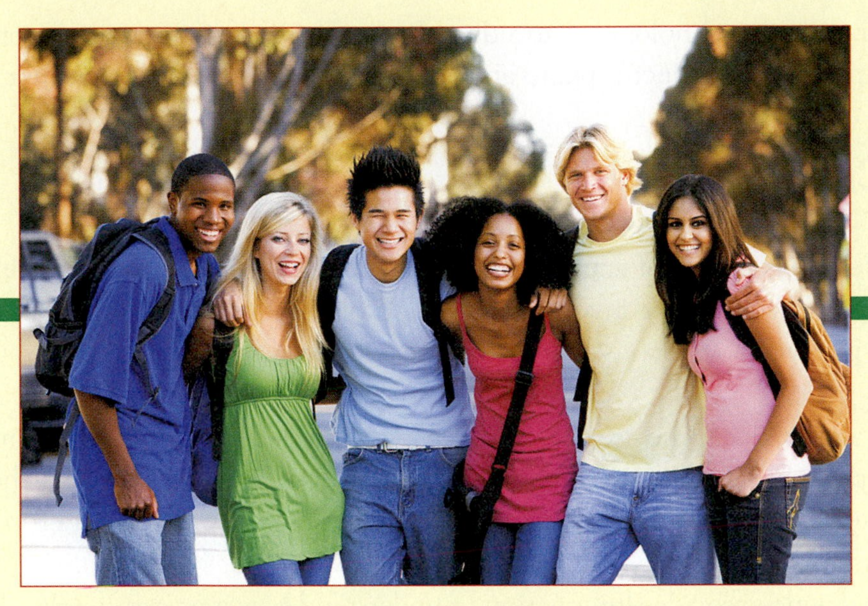

1–1. People have fun and enjoy life when their needs are met.

HOW DO people get the food, clothing, and other items needed in their lives? Most of the essential products result from efforts in producing plants and animals and their products. To do so, a sophisticated agricultural industry has emerged. This industry helps people have the products that they want in the forms that they want and where they want them.

Think about it a little more: Earth's population is more than 6 billion people. This large number requires massive amounts of food and clothing materials. The practices of yesteryear are no longer adequate. New approaches involving science and technology have emerged. Without science and technology, how we go about producing plants and animals would be much like it was 200 years ago. We would not produce nearly enough food and other materials!

Complex systems have emerged. These require knowledgeable people who can use science and technology in producing plant and animal products. People who are involved need good education and preparation in these areas. This vast industry may provide just what you want for a successful career!

AGRISCIENCE AND TECHNOLOGY ARE IMPORTANT

Agriscience and technology go together. Both contribute to meeting people's needs. Both are important in producing plants and animals for food, clothing, and other products.

AGRISCIENCE

Agriscience is the use of science principles in producing food, fiber, and shelter materials. Many areas of science are involved. The emphasis is on growing plants, raising animals, and getting products to the consumer. There are two broad areas of agriscience: applied and basic.

Applied agriscience is the use of knowledge in the production of plants and animals. Science principles help agriculture in many ways. Applied agriscience answers the question: "What should I do?" For example, knowing which chemical will solve a crop problem is applied agriscience.

Basic agriscience provides information to understand how a process works. The process by itself may be of little practical value, but it may be of great use in an applied agriscience practice. Basic agriscience answers the question: "Why and how does it do what it does?" For example, knowing why and how a plant reacts to a chemical is basic agriscience.

People who work in agriscience may be known as agriscientists. They have education in areas of both agriculture and science.

1–2. Agriscientists are harvesting greenhouse spinach leaves to investigate the presence of bacteria that could be a source of food contamination. (Courtesy, Agricultural Research Service, USDA)

1–3. A mechanical grain harvester uses high levels of technology. (Courtesy, AGCO)

TECHNOLOGY

Technology is the practical application of science. Some people view technology as the use of inventions. In agriculture, it may be known as agritechnology. Technology is used to increase the yield and quality of plant and animal products. It also reduces human labor requirements to produce crops.

With technology, the emphasis is on being practical. This means that we can put science to good use. Benefits can be gained from using technology.

Appropriate Technology

Technology improves the way things are done. However, good judgment enters into the decision about what to use. Not all technology should be used. Some just isn't right for a given situation.

Appropriate technology is that which best fits people's needs. Developed countries can use advanced technology. Lesser-developed countries cannot; more often, they benefit from using simple technology.

Large machinery may be well suited to agriculture in the United States and other developed countries. The level of agriculture allows good use to be made of the latest machines.

1–4. Careful testing is used to assess new technology. This shows a fungicide being tested on geraniums in a greenhouse. (A fungicide is a chemical that helps prevent diseases caused by tiny organisms called fungi.) (Courtesy, Agricultural Research Service, USDA)

In developing countries, the latest technology is often too advanced. The people aren't ready for it. They don't know how to use it or how to maintain it, and their farms are not set up for it.

Deciding when to use new technology is not easy. Government rules may delay the use of some technology. These rules are for the overall benefit of people.

Once new technology is ready, information about it must go to possible users. Educational programs will be needed. People can't adopt practices about which they are uninformed!

THE MEANING AND IMPORTANCE OF AFNR

AFNR is the abbreviation for Agriculture, Food, and Natural Resources. This term has emerged in the last decade to describe the broad area that we commonly refer to as agriculture. It is based on a way of organizing all careers created by the States' Career Clusters Initiative (SCCI).

All occupations are grouped into 1 of 16 career clusters. AFNR is the career cluster that addresses the production, marketing, and processing of agricultural commodities. It includes food and fiber products as well as wood, natural resources, horticulture, and other products of plant and/or animal origin.

AgriScience Connection

BECOMING A LEADER

All people have much potential. Some realize their potential; others don't. Students have many opportunities to develop and be successful.

Students in agriscience classes practice important skills in FFA. FFA is the national organization that provides exciting opportunities for students. Members have a lot of fun traveling, meeting other students, and being involved. FFA helps students learn in many ways. For example, this student is gaining leadership experience by presiding over a meeting at her school.

FFA has activities for students who live in cities as well as in rural areas. Talk to your agriscience teacher about FFA. Get involved in your local school. (Courtesy, Education Images)

The occupations in this career cluster focus on meeting basic human needs. Some people say that AFNR is the most important cluster. Certainly, it does deal with meeting the fundamental needs of human life and well-being.

The AFNR cluster includes production agriculture, often referred to as farming and ranching. It also includes the agricultural supplies and services used in farming and ranching as well as all the activities in marketing and otherwise getting the products to consumers in the desired forms. Horticulture, food processing, mechanical areas of agriculture, and natural resources and environmental science are in the cluster. All these are built on a solid base of science and technology.

1–5. Floral design requires knowledge of plants as well as artistic skills to create a pleasing arrangement. (Courtesy, Education Images)

You can learn more about the SCCI at **www.careerclusters.org**. Some exploration of your own may be quite useful. The SCCI materials are being carefully monitored to assure copyright protection and prevent unauthorized use.

CAREER PATHWAYS IN AFNR

All career clusters are organized into career pathways. A ***career pathway*** is a group of careers based on similarities of duties, subjects, and skills. A pathway often involves occupations with similar interests, though there is a great deal of variation. There are eight career pathways in the AFNR cluster. Each pathway has a number of occupational opportunities. Education and skill preparation are often needed to enter and advance in these occupations. Table 1–1 lists the AFNR pathways and briefly explains each.

Table 1–1. AFNR Career Pathways

Animal Systems—the production and care of animals and their products, such as those used for food, clothing, and companionship. Examples of occupations include poultry scientist, fish nutritionist, beef cattle producer, dairy farmer, veterinarian, small-animal breeder, and sheep rancher.

Plant Systems—the production and care of plants and their products, including those used for food, clothing, shelter, and beautification. Examples of occupations include wheat producer, tree surgeon, forester, agronomist, greenhouse grower, floral designer, cotton farmer, and grain inspector.

Power, Structural, and Technical Systems—the use of machinery, power, fuels, and other inputs for production in AFNR, including agricultural mechanics areas, such as wood construction, metals fabrication, and electricity. Examples of occupations include tractor mechanic, farm equipment operator, poultry house builder, agricultural engineer, remote-sensing specialist, and agricultural welder.

Biotechnology Systems—the use of organisms and their processes to achieve products and gain important benefits; genetics, molecular biology, and various technologies are used in the cluster. Examples of occupations include soybean geneticist, agricultural molecular biologist, plant biotechnology technician, animal disease researcher, and corn breeder.

Agribusiness Systems—the application of business principles and practices in AFNR; emphasis may be on nonfarm applications. Examples of occupations include agricultural economist, chemical sales representative, agricultural accountant, commodity broker, and farm loan officer.

Food Products and Processing Systems—the application of science and technology to develop, prepare, preserve, and otherwise improve human food materials and processes. Examples of occupations include milk sampler, meat grader, produce buyer, cannery manager, quality control specialist, food safety biologist, and frozen-food plant manager.

Natural Resource Systems—the use and protection of natural resources to assure sustainability in meeting future human needs. Examples of occupations include wildlife specialist, soil conservationist, ecologist, water quality technician, range manager, mine specialist, and fishing guide.

Environmental Service Systems—the use of technologies in assuring a quality environment, including providing for human needs and managing wastes. Examples of occupations include wastewater technician, solid-waste technician, landfill manager, recycler, hazardous materials specialist, toxicologist, and air quality control specialist.

Note: More information is available about AFNR and the other career clusters at **www.careerclusters.org/**.

1–6. The animal systems pathway includes the production and care of animals and their products. (Courtesy, Education Images)

AFNR AND HUMAN NEEDS

Ways of identifying and meeting human needs have been developed. These involve studying and determining the needs that are most important. Once the important needs are known, means are used to meet them. Many of these focus on the use of agriscience and the broad industry of agriculture—the industry created to meet human needs.

A ***human need*** is an essential element or component that supports human life. Food, fiber, and shelter are examples. Food provides the nutrition that helps the body to grow, to repair itself, and to reproduce. Fiber and shelter provide protection from the weather, dangerous animals, and other hazards of life. Food and fiber are carefully produced to meet human needs. A human also needs safety, health care, respect, self-esteem, and love to enjoy a full life.

FOOD

Food is the solid and liquid material humans and other living things consume. Some living things, such as plants, even have the ability to make their food. Food must provide the essential nutrients. A ***nutrient*** is a substance necessary for an organism to live and grow. Without sufficient nutrients, humans fail to grow properly, become diseased, and lack productivity. To some extent, animals compete with humans for limited nutrients.

1–7. Food product development often results in more nutritious foods, such as high-calcium carrots. (Courtesy, Agricultural Research Service, USDA)

Human Food Needs

Humans need food for four purposes: energy, growth and repair, good health, and body processes. Without proper food, these processes may not occur as they should. People, even as adults, may become ill and fail to thrive when they do not receive needed food. Adults who fail to have sufficient nutrients become diseased, age more quickly, and have shorter life spans.

Children who do not get proper food may get sick. They do not grow and develop as they should. Some will not become adults. Those who do may be unable to work, to raise a family, and to help other people.

People should strive for an appropriate diet each day. Such a diet should include nutrients in the right amounts. Fruits, vegetables, grains, milk, and meat are important sources of nutrients. Proper cooking also ensures that the nutrients are retained. The newest food guidelines stress the importance of fruits and vegetables along with sufficient exercise.

Sources of Food

Most food comes from two main sources: plants and animals. Nearly all the thousands of food products in a supermarket originated as plant or animal products on farms and ranches. A few, such as mushrooms and seaweed, may be from other organisms. Modern processing is used to change food materials into more desirable forms, such as wheat into flour for making bread.

Plant Sources. Many kinds of plants are used for food. These plants are carefully grown to provide high-quality food. Only about 600 species of the 250,000 species of plants are used for human food. An even smaller number provides most of the food. Wheat, rice, and corn—cereal grains—are the major human food items around the world.

1–8. Corn is widely used for human food and as feed for farm animals.

Table 1–2. Parts of Plants Used for Human Food

Parts	Plants
Leaves	Lettuce, cabbage, tea, and spinach; other parts are discarded.
Seeds	Beans, wheat, corn, rye, coffee, and nuts; other parts are discarded.
Roots	Potatoes, carrots, onions, and ginger; some of these are structures that grow on roots, such as the potato, which is a tuber.
Fruits	Apples, strawberries, pears, and oranges.
Flowers	Cauliflower and broccoli; the plants are discarded.
Stems	Celery, rhubarb, and asparagus; other parts of the plants are not eaten; some plants are destroyed when harvested.
Juice	Sugarcane (syrup and sugar); the stalk is discarded.
Multiple parts	Turnips; leaves and roots may be eaten.

Animal Sources. Humans eat foods from several animal sources. The meat of some animals is important. Other animals produce milk, eggs, and related products that we eat.

Both land and water animals are eaten. Those that grow on land are ***terrestrial animals***, while those that live in water are ***aquatic animals***. Chickens, hogs, sheep, and cattle are examples of terrestrial animals. Fish, shrimp, and clams are examples of aquatic animals.

Some 2 million species of animals live on Earth. Only about 50 species are used to any extent for human food. Of these, only four kinds are raised in large numbers: cattle, hogs, chickens, and sheep.

1–9. Catfish are cultured in carefully designed ponds, fed proper diets, and otherwise cared for on this farm. (Courtesy, Agricultural Research Service, USDA)

Table 1–3. Examples of Foods From Animal Sources

Foods	Animals
Meat	Muscles, organs, etc., of hogs, cattle, chickens, fish, horses, game animals, and others
Eggs	Chickens, fish, and a few other animals
Milk, cheese, ice cream	Cattle, goats, and sometimes other animals

Fish farming has increased. This is because of fewer wild fish and other species of aquatic animals in streams and oceans. We have developed new, efficient ways of raising aquatic animals. A new industry has emerged associated with aquatic animal production. It is known as aquaculture. Many people feel that a diet that includes fish promotes good health. Research has confirmed a relationship between the omega-3 fatty acids found in fish and good health.

FIBER (CLOTHING)

Fiber is primarily the material used to make clothing and shelter. Just as food is grown, natural fiber is grown also. Fiber is produced in three ways: by animals, by plants, and by manufacturing.

Animal Fibers

Animal fibers include wool, mohair, fur, and silk. They are used to make warm and attractive clothing and other fabric items, such as rugs and blankets.

Sheep produce wool. They naturally grow wool to protect their bodies from the weather. In the spring, the wool is removed from the sheep by shearing or other means. Cashmere and alpaca are variations of wool that are increasingly desired.

Furs are obtained from several animals, such as rabbits and minks. We raise some of these animals for their fur; we trap others in the wild. Besides fur, the hides from animals, such as cattle, may be used for making leather.

Silkworms produce silk when making cocoons. A cocoon is a case of tiny threads that protects the developing stage of the silkworm moth. Manufacturing silk cloth involves unwinding the cocoon. Silk is the strongest animal fiber.

Plant Fibers

Several fibers are produced from plants: cotton, flax, hemp, jute, and sisal.

Cotton is by far the most important fiber. It is used to make clothing, household goods (bath towels, bed linens, and furniture upholstery), and other products. Cotton has qualities that are highly desired, such as softness, absorbency, and durability.

Flax comes from the stems of flax plants. It is made into linen, which is a high-quality cloth used for clothing, table linen, and other items.

Hemp, jute, and sisal are coarse fibers that come from plants of the same names. All three are used to make twine, cords, and ropes. Sisal rugs are popular for their durability.

CAREER PROFILE

MOLECULAR BIOLOGIST

A molecular biologist studies properties of cells and other structures at the molecular level. A molecule is the smallest particle that retains the properties of a substance. It is composed of one or more atoms. Sophisticated equipment and methods are used to make the study possible. Most molecular biologists have specific areas of interest, such as plants, animals, medicine, or food.

Molecular biologists typically have advanced college degrees in biology, biochemistry, molecular biology, or a related area. Some may begin as animal or plant science majors in college and move into molecular areas for master's and doctoral degrees.

This shows a molecular biologist collecting a milk sample from a dairy cow to study infections in the cow. The milk is a body fluid that will exhibit qualities of the cow. (Courtesy, Agricultural Research Service, USDA)

In recent years, kenaf has emerged as a potential fiber crop. Research is underway to determine the possibilities of kenaf in making cloth, paper, and similar products.

1–10. Cotton is a popular natural fiber.

Manufactured and Mineral Fibers

Fibers produced by manufacturing are known as synthetic or human-made. Synthetic fibers are made in mills from various raw materials, such as petroleum. Some manufactured fibers are made from wood; others are made from chemicals. A substance in wood known as cellulose is the major wood fiber. Nylon and polyester are examples of synthetic fibers made from chemicals. Nylon fiber and yarn have great elasticity. Asbestos is a fiber made from minerals. It is no longer used because of health hazards.

1–11. Natural and synthetic fibers may be used to produce clothing with desired features.

SHELTER (FORESTRY AND WOOD PRODUCTS)

Humans need protection from the weather and other dangers. They use a shelter as defense against the elements (cold, heat, rain, wind, and snow) and against wild animals and other pests.

1–12. Lumber and plywood are forest products used in construction.

A ***shelter*** is a building used by humans for housing. The wood products used in constructing houses and apartments are produced by forestry.

Forestry is a large and specialized area of plant production. It involves native trees growing in the wild and cultured trees growing on tree farms. Tree farms grow improved kinds of trees. Scientific processes are used to get the trees to grow rapidly and economically. Some would say that forestry is a form of fiber production because wood is composed of fibers.

Trees are also important in the environment. Cities and real estate developers may plant and maintain trees for their beauty and cooling effects. The presence of trees can lower the temperature on the ground and make life more comfortable in warm climates. The use of trees in cities is known as urban forestry.

1–13. A knuckleboom loader is being used to move harvested logs before hauling them to a mill.

Table 1–4. Examples of Forest Products

Product	Description
Lumber	Lumber is made by sawing logs into boards; it is used to construct houses, furniture, and other products.
Plywood	Plywood is made by cutting logs into large, thin sheets of wood that are glued together in layers; strength and improved qualities over lumber for certain uses are advantages.
Veneer	Much like plywood, veneer is made by gluing a very thin layer of an expensive, beautiful wood over a lower-cost wood. (For example, a thin layer of valuable walnut wood can be glued over a thicker, less expensive wood. The result is a wood finish with the beauty of a fine wood but at a much lower cost.)
Composition board	Several types are made; this product is produced in sheets similar to plywood or in boards like lumber; wood chips or particles of wood are pressed and cemented together.
Paper	Paper is made from wood using a detailed chemical process; wood is cut into chips and "cooked" until it turns to pulp; pulp is rolled into thin sheets of paper; many products are made from paper.

PEOPLE MAKE CHOICES

People are consumers. They make choices about food, fiber, and shelter. Their choices are important and influence what is produced. People may have strong likes and dislikes, known as preferences. It helps to know about preferences.

Food and fiber are produced to satisfy human needs. If people don't want what is produced, there is no need to produce it. A goal of agriscience is to make food and fiber more appealing and readily available.

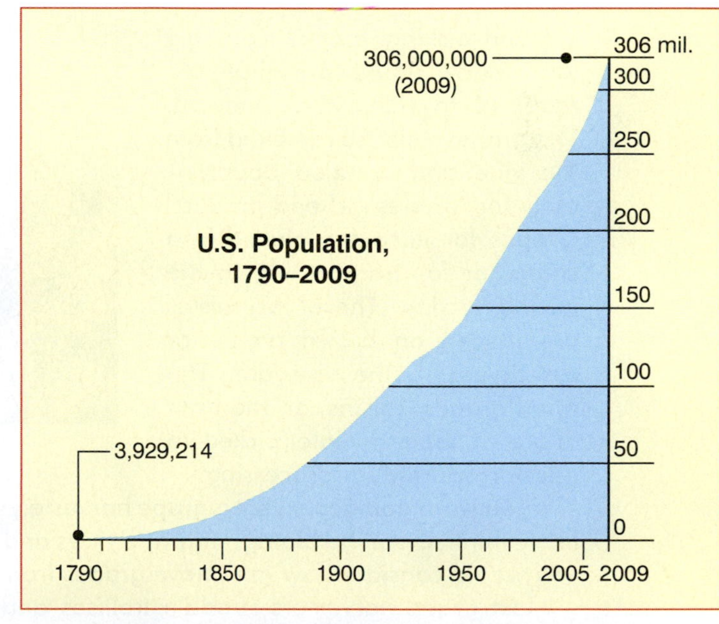

1–14. The population of the United States has increased rapidly since 1790. (The graph includes Puerto Rico.)

PRODUCING FOR THE CONSUMER

A ***consumer*** is a person who buys goods and services. All people are consumers! There are nearly 306 million consumers in the United States. World population is nearing 6.8 billion. Human numbers are projected to increase to 355 million in the United States and 8.9 billion in the world by the year 2050, according to the United Nations. Consumers eat food, wear clothing, and live in housing—needs that must be met. They also have opinions about what they will eat and wear and how they will be housed. These opinions dictate what is produced.

Consumer preferences can be powerful. Generally, consumers' cultural backgrounds, or ethnic heritage, will influence their choices about what they eat and wear. Knowing the cultural demographics (statistics) of an area can help producers meet consumer demands. A **demand** is a desire for a good or service and the ability to buy. Price is important; if it is too high, consumers won't buy at all, or they will limit the amount they buy.

The agricultural industry is concerned with meeting needs and demands. Plants and animals should be produced only if there is a demand and if a profit can be made. Producers should study the situation before beginning production. If no market exists, money should not be spent on production.

Technology Connection

GRAPE HARVESTER

Hand-picking grapes is a big job. Many hours of labor are required to harvest a vineyard. The grapes must be removed from the vine and handled appropriately for the desired end product. Grapes for juice are placed in a container for transporting to the juicing facility. Those for raisins are placed on brown paper for sun-drying in the vineyard. The dried grapes (raisins) on the rows of brown paper are later picked up and transported for processing.

If you were going to invent a grape harvester, what would you need to consider? You would likely think about the biology of grape vines and how they are grown in vineyards. Of course, you would consider how to remove grapes from the vines without destroying the grapes.

Most grape harvesters straddle trellises and use a bow-rod picking system that gently removes grapes with minimal vine damage. You would also consider how to design the equipment so that it would move through a vineyard. Grape producers would need to establish vineyards and use trellises and spacing that are uniform.

This image shows a mechanical grape harvester in a vineyard.

CONSUMER PREFERENCES

Many factors influence what people eat and what they use as clothing. People in various regions of the world are alike in their preferences. Standard of living, customs, and climate are the major factors in consumer choices.

Standard of Living

Standard of living is the level of choice about both essential and nonessential goods and services that people can make based on what they can afford. Income and education promote a higher standard of living. People use income from jobs to buy products. As income increases, food choices change. People with higher incomes eat more beef, seafood, and other higher-priced foods. As income grows, people often choose higher-priced clothing, housing, and other goods or services.

Education is related to the standard of living. People in nations with better education have higher incomes and more choices. In countries with an emphasis on technology, people may be interested in the background of their foods. They may choose foods based on the kind of technology used in production and processing.

Table 1–5. Per Person Yearly Consumption of Selected Foods in the United States

Commodity	Pounds Consumed
Red meats (beef, veal, lamb and mutton, and pork)	155.2[a]
Beef	93.8
Veal	0.5
Lamb and mutton	1.2
Pork	63.0
Poultry (chicken and turkey)	118.8[b]
Fish and shellfish (wild catch and farm raised)	16.5[c]
Eggs	32.3[d]
Milk	606.3[e]
Fruits, including grapes	269.6
Vegetables (fresh, frozen, and canned)	406.4
Potatoes (fresh, frozen, canned, chips, etc.)	69.3[f]
Rice, flour, and cereal products	191.3
Cocoa (chocolate equivalent)	5.2[g]
Coffee and tea	8.2
Tree nuts (pecans, almonds, coconuts, etc.)	3.9
Peanuts	8.4
Melons	27.7
Fats and oils (vegetable and animal sources)	84.5
	Gallons Consumed
Carbonated soft drinks (not an agricultural product)	50.6

Source: Economic Research Service, USDA, Accessed on January 20, 2009, at **www.ers.usda.gov/Data/FoodConsumption**.

[a] Dressed weight
[b] Ready-to-cook weight
[c] Edible weight and excluding game fish
[d] Retail store weight; all forms (converts to 250.6 eggs per capita)
[e] All dairy products (ice cream, cheese, butter, etc.) milk equivalent
[f] Farm weight
[g] Bean equivalent

Standard of living is often related to the livability in a country. Life expectancy, educational attainment, and adjusted real income are three important factors in establishing livability. Norway and Sweden are often considered most livable in terms of life expectancy,

1–15. Kinds of food, how food is prepared, and how it is eaten vary throughout the world.

education, and income. Sierra Leone and Burkina Faso are often considered least livable on these same criteria. (More information on livability is available at **www.infoplease.com**.)

Customs

Customs are established ways of doing things. People often grow up eating certain foods or observing certain social behaviors and continue to follow those habits as adults. Changes in customs occur slowly.

The number of people on Earth is increasing each year. Some scientists are alarmed that the resources of Earth may not be adequate to support an ever-increasing population. Each new individual requires food, clothing, and housing. New individuals also create wastes and place demands on the environment. Future generations may face challenges associated with meeting the needs of an increasing population. Some scientists project future populations worldwide and by continent and nation. You can access current population numbers at the World POPClock (**www.ibiblio.org/lunarbin/worldpop/index.html**). Information at this site is active and updated each second.

The earth is divided into seven continents. Six of them have many people. Antarctica has no native people. The continents and their populations are shown in Figure 1–16.

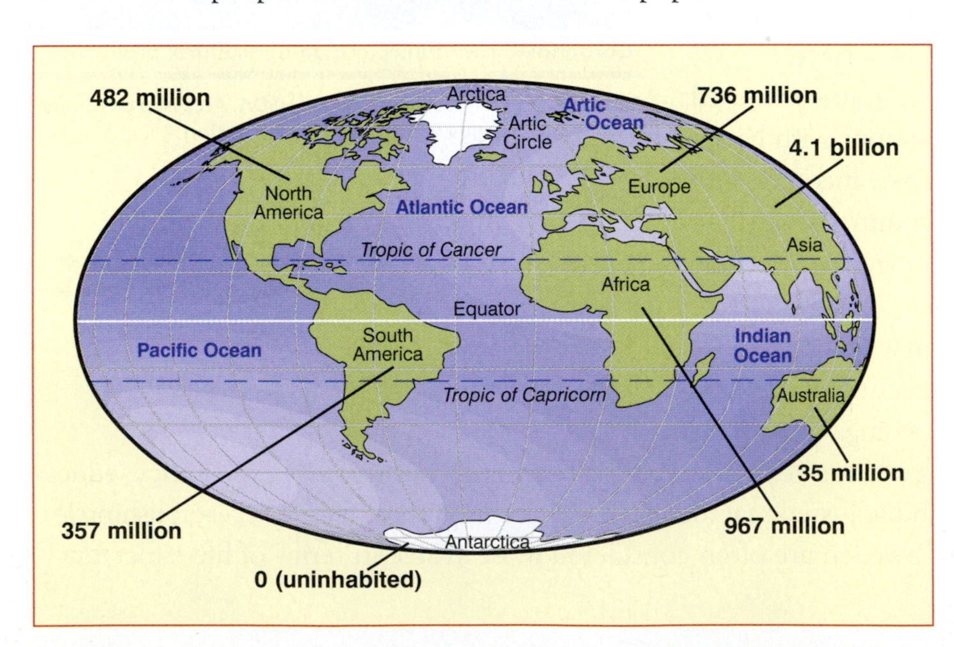

1–16. Population, climate, culture, and standard of living vary on the seven continents. World population is nearing 6.8 billion people.

Customs are identified with people from various continents and countries as well as within countries. In the United States, people in the North may eat more potatoes and wheat bread. Those in the South may eat more chicken, okra, and corn bread. People near rivers and oceans may eat more fish.

Continents of the world have certain customs. The customs are often related to climate and native food plants and animals. Asia is associated with rice, tea, and chicken. Australia and New Zealand are known for wheat, beef, mutton, and dairy products. Africa is known for corn, sorghum, cassava (a root crop), and sweet potatoes. Europe is associated with potatoes, hogs, cereal grains, and dairy products. South America is known for cassava, dry beans, corn, coffee, and potatoes.

Internet Topics | Search

Topics for Internet discovery and reporting are listed here. Select a topic and do an Internet search. Read and learn about the topic. Prepare a report on what you find.

1. Career cluster
2. Standard of living
3. Consumer preferences

The United States has much variety, which is related to the range of climates and the heritage of the people who settled the various areas. Vegetables, cotton, fish, and citrus crops are grown in the southern United States, while wheat, soybeans, corn, and cattle are raised in the Plains. Potatoes, wheat, and dairy products are important commodities from the northern United States. The West Coast contributes many different crops, particularly valuable fruits and vegetables.

Climate

The climate is the nature of the weather. Temperature, precipitation (rainfall), and direction and speed of wind are major factors in a climate. Of course, soil fertility and drainage and the "lay" of the land determine land use. Soil that has few nutrients won't produce much. Land that is too wet or too hilly cannot be farmed.

The foods commonly eaten by people are those that grow best in the region. People in cool climates eat more cereal grains, potatoes, and meat. Those in warmer climates eat more fruits and vegetables, such as papayas and bananas. Some areas of the world have poor climate conditions. Animals and plants will not grow well. The desert areas of Africa are a good example. There isn't enough rain to meet the needs of crops.

1–17. Agricultural practices vary, with animal power still used to produce crops in some places.

REVIEWING

MAIN IDEAS

Agriscience is the foundation on which modern food and fiber production is based. We are better producers when we understand and apply science principles. Technology is the application of science to gain useful products and services. We always need to assess technology to be sure it is appropriate for use in a given situation.

In the past decade, the broad area of agriculture has become known as AFNR—Agriculture, Food, and Natural Resources. This is a career cluster developed to depict agriculture more accurately from a modern perspective. Eight career pathways are in the AFNR cluster. The pathways focus on animals, plants, power and technical systems, biotechnology, agribusiness, natural resources, environmental service, and food products and processing.

The focus of the AFNR cluster is on meeting human needs. These are the essentials for quality human life. Food, clothing, and shelter are paramount. Without adequate food, human well-being is threatened. Most food materials are from two main sources: plants and animals. Likewise, most clothing materials are from plant and animal sources, though some synthetic fibers are used in making clothing. Shelter often involves wood products and natural resource materials.

As consumers, people make choices about what they consume. The choices represent strong preferences in agriculture. The goal is to satisfy human needs and preferences with products that are in most demand. Products with the greatest demand are more likely to be profitable to agricultural producers. Consumption is influenced by standard of living, customs, and climate.

QUESTIONS

Answer the following questions, using complete sentences and correct spelling.

1. What is agriscience? What is the distinction between applied and basic agriscience?
2. What is technology?
3. What is AFNR? Why is it an important career cluster?
4. What is a career pathway?
5. What are the career pathways in AFNR? Name one occupation in each pathway.
6. What is a human need? How do human needs relate to AFNR?
7. What are the three main kinds of human needs met through AFNR?
8. What are the two major sources of food? Briefly explain each source.
9. What is fiber? What are the sources of fiber?
10. What is forestry? How does it relate to meeting human needs?
11. What is the meaning of "consumer"? Approximately how many consumers are there in the United States? On Earth?
12. What three factors affect human preferences? Briefly explain each.

EVALUATING

Match the term with its correct definition.

a. aquatic
b. agriscience
c. demand
d. consumer
e. human need
f. standard of living
g. technology
h. terrestrial
i. AFNR
j. nutrient

_____1. The use of science principles in producing food, fiber, and shelter materials.

_____2. The application of science through inventions and other improvements.

_____3. A plant, animal, or other organism that lives in water.

_____4. The abbreviation for Agriculture, Food, and Natural Resources.

_____5. The desire for a good or service coupled with the ability to buy it.

_____6. A plant, animal, or other organism that lives on the land.

_____7. A substance necessary for an organism to live and grow.

_____8. A person who buys and/or uses goods and services.

_____9. The level of choice about goods and services that people can make based on what they can afford; often associated with livability.

_____10. An essential element or component that supports human life.

EXPLORING

1. Investigate the plant and animal crops produced in your local area (county or region). Write a short report on your findings.

2. Take a field trip to a local agricultural entity—farm, horticulture business, farm supplies store, or other business that uses agriscience. Interview an individual who works there. Determine the nature and the requirements of the work. Prepare an oral or written report on your findings.

3. Survey the products available in a local grocery store. Identify how the products have been packaged and kept from deteriorating. Select one product and investigate its background, including where it was produced, how it has been preserved, and how it is packaged. Prepare a brief oral report for the class on your findings.

CHAPTER 2

The Agricultural Industry

OBJECTIVES

This chapter covers the meaning and importance of the agricultural industry. It has the following objectives:

1. List and describe three major areas of the agricultural industry.
2. Discuss major events in the history of agriculture.
3. Relate the role of international trade.
4. Contrast international agricultural practices.
5. Explain the meaning and importance of biosecurity.
6. Describe the meaning and importance of research and development.
7. Explain the role of professional and commodity organizations.

TERMS

agricultural industry
agricultural services
agricultural supplies
agriculture
agronomy
animal science
aquaculture
biosecurity
commodity organization
country-of-origin labeling
domestication
export
fair trade
floriculture
food safety
food security
forestry
horticulture
import
international trade
landscaping
marketing
olericulture
ornamental horticulture
pomology
poultry science
processing
production agriculture
professional organization
research
research and development
tariff

2–1. Agricultural industry provides an abundance of products for consumers.

MEETING basic human needs is not so simple. Improved plants and animals, modern machinery, and scientific production techniques are used. The way our needs are met did not suddenly occur. It evolved over many years of scientific study, technology, and human effort.

The subsistence farms of a century ago have given way to modern commercial endeavors. People in agriculture no longer produce just for themselves. They produce for other people. What they produce is driven by market demand. Products must measure up to the demands of those who ultimately set the market: consumers.

Today, we are served by a large and diverse agricultural industry. High levels of technology are used to produce plants and animals and their products. Additional technologies are used to prepare the products into the desired forms and to get them to consumers. All of these activities involve using practices that ensure a profitable agriculture and healthful product.

AREAS OF THE AGRICULTURAL INDUSTRY

Agriculture is the science of growing crops and raising livestock that are used for food, fiber, and other purposes. Some people refer to this as farming or production agriculture. It is the foundation of the AFNR Cluster and is the major reason for the agricultural industry.

The ***agricultural industry*** is all the processes involved in getting the products of agriculture produced and made available to the consumer. It has three major areas: supplies and services, production agriculture, and marketing and processing. Other important areas include horticulture, forestry, food science, companion and service animals, and natural resources.

SUPPLIES AND SERVICES

Today, plant and animal production requires many things from off the farm. These inputs are known as supplies and services. ***Agricultural supplies*** are the inputs (e.g., materials and equipment) used to grow crops and to raise livestock. ***Agricultural services*** are the skills and knowledge of people who are highly trained to help with production. Veterinarians, soil conservation technicians, and insect specialists perform services. Machinery, equipment, seed, plants, chemicals, feed, and animal medicines represent major areas of supplies and services. Four groups of supplies and services are included here.

CAREER PROFILE

NUTRITIONIST

A nutritionist studies the food needs of people and animals. The work promotes eating the proper foods and helps people to be healthy and to grow properly. Some nutritionists are involved with research. This shows a nutritionist with a rat used in broccoli diet research.

Nutritionists need strong backgrounds in science. Many have college degrees in nutrition. Those nutritionists doing research often have doctoral degrees in nutrition or a related area. Begin preparing in high school by taking science, math, and agriculture classes.

Most jobs are with government agencies, large food companies, feed mills, and research stations. (Courtesy, Agricultural Research Service, USDA)

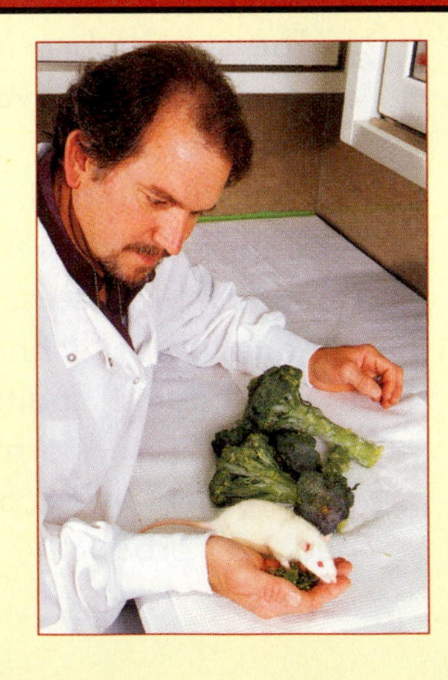

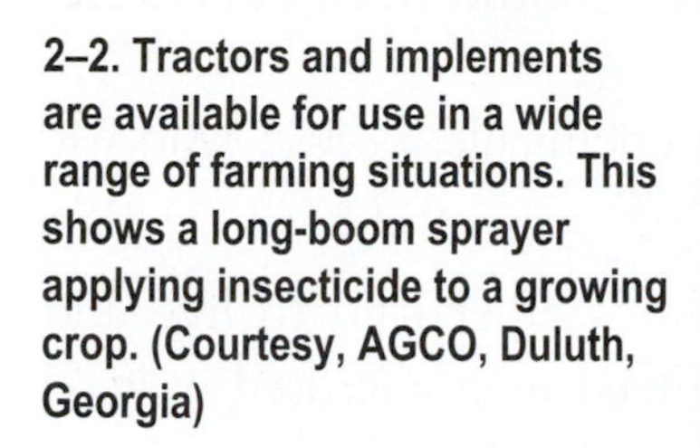

2–2. Tractors and implements are available for use in a wide range of farming situations. This shows a long-boom sprayer applying insecticide to a growing crop. (Courtesy, AGCO, Duluth, Georgia)

- **Machinery and equipment**—A large industry exists to provide the tractors and implements needed on farms and in forestry and horticulture. Agriscientists design equipment and test it in the field. Yet people are needed to manufacture, sell, and service the equipment. Much of today's equipment involves computers, GPS, and laser systems. Fuel and energy to power machinery and equipment are also included.
- **Seeds and plants**—Plant producers need seeds, bedding plants, and other living materials to grow. These items are provided by seed producers, plant growers, and others. Research stations, seed companies, and others develop improved types of plants. Seeds are produced from these plants. The seeds are cleaned and treated in a factory to make sure they are good. Proper bagging and storing keeps seeds in top-quality condition.

2–3. A display offers many different vegetable and ornamental plant seeds. (Courtesy, Education Images)

- **Chemicals**—Chemicals have many good uses in agriculture, and the chemicals are not bad if used properly. They help control weeds, diseases, and insect pests. Fertilizers are chemicals that help crops grow better. Growth regulators help plants produce desired products when needed. To some extent, all these are found in natural forms in nature. Without these chemicals, less food would be produced.

  Agriscientists are involved in developing and testing chemicals. People who use them often need to be certified.

- **Feed, health, and production supplies**—Animals require proper feed to live and grow. Producers may grow the feed needed or purchase feed manufactured by feed mills off the farm. In addition, animal medicines may be used to keep their livestock healthy. Without supplies and services, today's animal production would be much like it was 200 years ago.

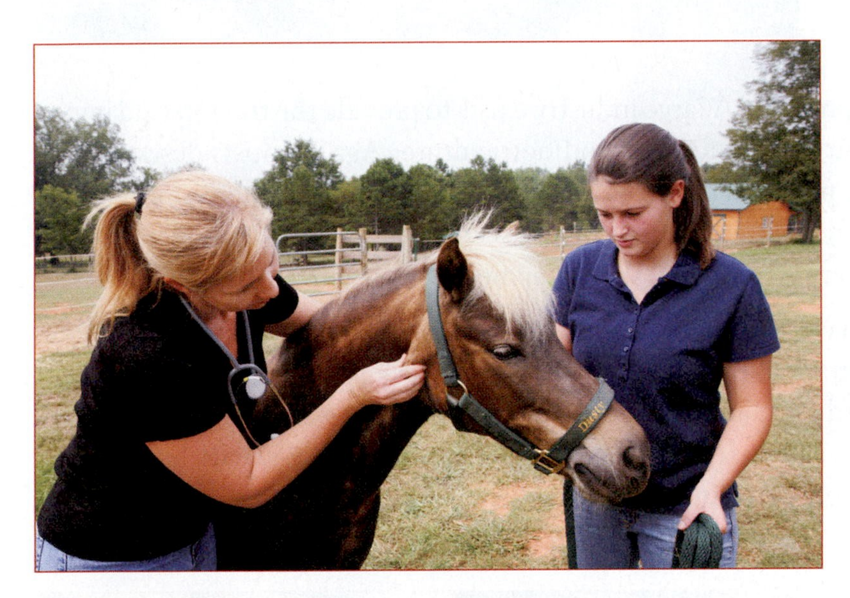

2–4. Veterinarians provide important services for animal producers. Here the lymph nodes of a horse are being examined. (Courtesy, Education Images)

PRODUCTION AGRICULTURE

Production agriculture is the farming part of the agricultural industry. It involves crops, livestock, and livestock products. The people who do the work are known as farmers or agricultural producers.

Today's farms are specialized. Most farms produce only one or two crops or products. For example, a dairy farm will likely produce only milk to sell. A corn farm may grow corn and nothing else. The corn is sold to a mill or a farmer who has livestock to feed. These are commercial farms because they are producing for a specific market.

Some individuals also include tree farming (forestry) as a part of production agriculture. Trees are farmed similar to other crops, but they take longer to grow and require different practices.

Table 2–1. Examples of Cultured Plants

Crop	Use
Rice	Grain
Cotton	Fiber and seed for oil
Wheat	Grain
Corn	Grain
Soybeans	Grain and hay
Sorghum	Grain and feed
Vegetables	Food; examples: tomatoes, potatoes, squash, snap beans, onions, lettuce, celery, and corn
Fruit	Food; examples: apples, oranges, grapefruit, and cherries
Berries	Food; examples: strawberries, blueberries, and raspberries
Nuts	Food; examples: pecans, walnuts, and almonds
Trees	Timber and ornamental use, such as holiday trees
Ornamentals (roses, petunias, daffodils, etc.)	Aesthetics and decoration
Turf (grasses and others, such as clover)	Animal feed, landscaping, and soil erosion management

Table 2–2. Examples of Agricultural Animals

Species	Product/Use
Hogs (swine)	Meat, known as pork
Sheep and goats	Meat (the meat of young sheep is called lamb; the meat of older sheep is called mutton) and wool
Cattle (beef and dairy breeds)	Meat, known as beef, and dairy (milk) products (meat of young calves is known as veal)
Poultry (chickens, turkeys, ducks, etc.)	Chickens for meat and eggs and turkeys for meat
Horses	Work and pleasure (meat in some countries; known as chevaline)
Fish and other aquatic species	Food and recreation; examples: catfish, trout, tilapia, hybrid stripped bass, oysters, prawns (shrimp), and crawfish
Dogs, cats, canaries, llamas, and others	Companionship (pets), research, and service

Production agriculture has come a long way from the subsistence farming of 200 years ago. Many farmers sell all that they grow and buy their food at the local supermarket. They do not even have gardens!

2–5. Modern crop planters promote efficiency in production agriculture. (Courtesy, AGCO Corporation)

Successful farmers apply agriscience. They understand science principles and use them to produce crops and livestock. Six areas of production agriculture are included here.

- **Agronomy—*Agronomy*** is the study of plants and how they relate to the soil. Most often, agronomy refers to field crops grown for food and fiber. New kinds of crops and better ways of growing crops are studied to aid in improving crop production and conserving natural resources. The emphasis in agronomy is on crops grown for human food and animal feed. These items are known as agronomic or field crops and produce large amounts of food and fiber. Wheat, corn, soybeans, canola, rice, cotton, grain sorghum, and sugar beets are examples of agronomic crops.

2–6. Row crops may need cultivation to manage weeds and to keep them from reducing crop yields and quality. (Courtesy, Case Corporation)

- **Horticulture**—***Horticulture*** is the science of growing vegetable, ornamental, fruit, and similar plants. It is sometimes listed separately from production agriculture. But the overall goal is to produce plants that meet consumer demand. Horticulture is sometimes further broken down into areas of its own.

Ornamental horticulture is the use of plants and other materials for decorative purposes. It is sometimes known as environmental horticulture. Within ornamental horticulture, floriculture and landscaping are major areas.

Floriculture is the production and use of flowers and plants. The plants may be grown in greenhouses, fields, or inside and outside of buildings for beauty. The flowers grown in greenhouses are used to decorate homes, businesses, and other places. Floral designers often use flowers in artistic arrangements. Local flower shops take orders and deliver floral arrangements to customers.

Internet Topics Search

Topics for Internet discovery and reporting are listed here. Select a topic and make an Internet search. Read and learn about the topic. Prepare a report on what you find.

1. American Farm Bureau
2. National Association of State Universities and Land-Grant Colleges
3. National Corn Growers Association

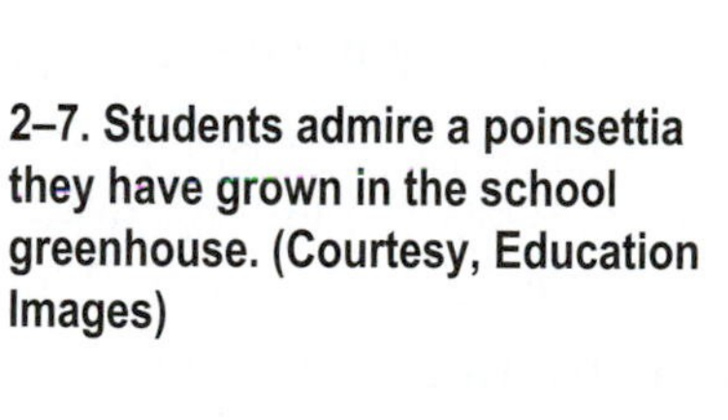

2–7. Students admire a poinsettia they have grown in the school greenhouse. (Courtesy, Education Images)

Landscaping is the use of plants to improve our outdoor environment. The plants are used to make buildings, parks, and other areas more attractive. Shrubs (small woody plants) are grown in nurseries. When shrubs reach the right size, they are moved and placed in landscapes.

Carefully prepared landscape plans describe the kind of shrub to use and where it should be placed. A good knowledge of plants is needed to make and follow these plans. If the characteristics of plants are unknown, the wrong shrubs might be chosen—different shrubs have different growth requirements.

Horticulture also includes two major food crop areas: olericulture and pomology.

2–8. Latex is being collected from a rubber tree.

2–9. A Boer goat is being trained for showing. (Courtesy, Education Images)

Olericulture is the science of growing vegetables, such as snap beans and tomatoes. While vegetables are grown in all areas of North America, some locations have climate advantages.

Pomology is the science of growing fruits and nuts. Pomology crops are found in many locations in North America. Common fruits are apples, blueberries, oranges, pears, and plums. Common nuts are pecans and walnuts.

- Forestry—***Forestry*** is the science of growing and using trees. Lumber, paper, and similar products are manufactured from harvested trees. These items are used to construct homes, barns, furniture, and other wooden structures. Additionally, trees provide syrup (maple), rubber (latex—ruber tree sap), rosin (pine), turpentine (pine), oil (tung, which is not edible), and decorations (holiday spruce).

  Native forests are those that grow naturally. Much of North America was once covered by trees. These native forests were cut as the land was settled. As a result, the amount of native forest has decreased. The trees in native forests grow without care and are often of mixed species.

  Farms where trees are produced have emerged in the past quarter century or so. These trees are planted, cultivated, and otherwise cared for to promote growth. The trees are established by planting (setting out) small seedlings of improved varieties. All of the trees on a farm are usually of the same species, such as pine or fir.

- Animal science—***Animal science*** is the area of agriscience dealing with the production of animals. The focus is on food animals, though some animals are used for fiber, recreation, companionship, and power. Cattle, hogs, and sheep are the major livestock used as food animals.

  Horses are used for pleasure and power. Outside the United States, some horses are used for food. Horseback riding, rodeoing, and cutting-horse competitions are examples of ways in which horses are used for recreation.

Other animals in agriscience include goats, bison (buffalo), and llamas. Animals used as companions or pets include dogs, cats, tropical fish, and various birds (e.g., canaries and parrots). Research and service animals are also included. Rabbits, rats, and llamas are among species that may be used in research. Examples of service animals include guard dogs and donkeys.

- **Poultry science**—***Poultry science*** is the study of birds used for food. It is sometimes included in animal science. Chickens are the most important poultry. Other kinds of poultry include turkeys, ducks, geese, pheasants, quail, guinea fowl, and the ratites (ostriches, rheas, and emus). Poultry is produced for meat, eggs, and other products. Therefore, poultry science is always seeking better ways of growing birds. Agriscientists work to develop improved chickens and turkeys. They determine the proper feed and environment needed for optimal growth.

2–10. White Leghorn chickens are used in egg production. (Courtesy, Agricultural Research Service, USDA)

- **Aquaculture**—***Aquaculture*** is the science of water farming. It has rapidly expanded since the mid-1900s. Aquaculture includes the production of aquatic animals, plants, and a few other organisms, such as algae.

  Animal aquacrops are more popular in the United States than plant aquacrops. Animal aquacrops include fish, such as trout and catfish; mollusks, such as oysters and clams; and crustaceans, such as shrimp and crawfish. Together, mollusks and crustaceans are known as shellfish. Frogs, eels, and alligators are other examples of animal aquacrops.

  Plant aquacrops include water chestnuts, watercress, and several ornamental plants, such as water hyacinths. Seaweed (certain species of algae) is popular as a human food in some places, particularly in Pacific island locations.

2–11. Market-size catfish are being harvested by seining. (Courtesy, Agricultural Research Service, USDA)

MARKETING AND PROCESSING

Marketing and processing are important in providing the kinds of products people want. People who buy and use products are known as consumers. Not many people buy directly from producers. Consumers often buy products that have been specially prepared, such as for the ease of cooking.

2–12. Peaches are being individually labeled in a processing facility. This automated equipment can individually label 8 to 10 peaches per second. The labels are transferred from the large reels to the peaches. (Courtesy, Education Images)

Marketing is the part of the agricultural industry that moves crops and livestock from the producer to the consumer. Many functions may be involved. A major function of marketing is processing.

Processing is the act of changing products, such as crops and livestock, into forms that people want. For example, the following activities might be involved in processing a fruit or vegetable:

1. The grower delivers the harvested crop to a packing shed.
2. At the packing shed, the crop is graded and packed into big crates for shipping to a processing plant.
3. In the processing plant, the crop is washed and otherwise prepared for canning or freezing.
4. As processing is finished, the produce is packaged and labeled for storage and delivery to supermarkets.
5. Specially equipped trucks haul the processed food to the supermarkets.

2–13. Harvested potatoes are being washed on a conveyor in a processing facility.

All steps are carried out to keep the food safe and secure. ***Food safety*** is the practices and procedures for keeping food safe to eat and preventing accidental contamination that would make it unsafe. The food is carefully handled to maintain nutrition and to prevent spoiling. Inspection ensures the consumer that the food is wholesome (fresh, healthy, and good to eat).

People also want food that is ready to cook and easy to use. They have less time to prepare meals. Marketing and processing take much of the work out of food preparation. This is why supermarkets have so many items. Some people like dinners that are ready to heat in a microwave oven. Other people like to spend a little time preparing original dishes. However, very few people would think of slaughtering their own hog or chicken. Times have changed, and so has agriculture.

A SHORT HISTORY OF THE AGRICULTURAL INDUSTRY

Humans have not always produced food, fiber, and shelter. Those living thousands of years ago searched and hunted for food. They were sometimes known as gatherers because they used what they could find. Much of each day was spent searching in forests and meadows for food.

Hunting was difficult because of a lack of weapons, so capturing large animals was nearly impossible. Snails, clams, oysters, and crawfish were among the first animals used for food. Wild berries, melons, beans, and other fruits and vegetables were also eaten. Early humans used leaves and animal skins for clothing. They sought shelter in caves and under big rocks or trees.

2–14. Wild bison once roamed the plains of the United States. They are now being domesticated and produced on a few ranches. (Courtesy, Agricultural Research Service, USDA)

DOMESTICATION

Domestication is taming, or controlling, wild plant and animal species and producing them for specific purposes. Domestication began in Asia about 7000 B.C. Donkeys, goats, cattle, and dogs were the first domesticated animals. Humans cared for and protected the animals. In addition, crops were being grown in China and Sweden in 1300 B.C. A few wealthy people had domesticated fish by 3000 B.C. Fish culture was well underway in China in 400 B.C. Today, cattle, swine, and many other domesticated animals are widely produced.

Domestication made it easier to have a dependable food supply. The animals and plants were nearby, so they could be easily harvested for use.

AGRICULTURE

Agriculture in America was still primitive as late as A.D. 1500. Much has changed since Christopher Columbus landed in 1492! Almost everyone lived a subsistence life. They produced for their needs on small acreage without machinery.

Early agriculture in America focused on growing crops that could supplement the foods that grew wild. In the late 1700s, 90 percent of the people lived “on the land.” Nearly everyone farmed, but most produced barely enough for themselves. However, a few people grew extra crops to sell to those who did not farm.

Humans provided most of the power and used hand hoes and shovels to till the soil. Later, animals were harnessed for power, so they pulled plows and other implements. It was the mid-1900s before mechanical power (tractors with engines) was widely used.

As developments made farming easier, farmers produced crops and livestock in increasing numbers. Fewer farm workers were needed, yet more food and fiber were raised.

Much of the agriculture technology developed in the United States is now used in other countries. By the year 2000, Americans had become dependent on foreign-produced food crops, such as off-season fresh fruits and vegetables. This makes it possible for supermarkets to have warm-season products in the winter months. It is anticipated that Americans will increasingly consume foreign-produced foods during the twenty-first century.

AGRISCIENCE AND TECHNOLOGY EMERGE

Many changes have occurred in the past 300 years. New insight into nature has helped improve crops and livestock. Inventions have allowed people to accomplish more work with less effort. With fewer people needed in farming, more people were available to work elsewhere. Factories began producing things for the farm as well as the home.

Inventions in agriculture changed farming. Many reduced the drudgery of agricultural work. These inventions led to agriscience and technology.

In the early 1900s, improved varieties of corn were developed in the United States. These varieties were known as hybrid corn. By the year 2000, nearly all corn planted in the United States was hybrid seed.

Table 2–3. Examples of Agricultural Inventions

Inventor	Description
Charles Townshend	Developed the first crop rotation system in the early 1700s in England. Legume crops that added nitrogen to the soil were rotated with others, such as grain.
Jethro Tull	Developed a seed planting machine in the early 1700s. Today's high-speed planters grew from this early effort.
Eli Whitney	Developed the cotton gin in 1793, an invention that greatly reduced the labor required to remove seeds from the cotton fiber. Large mills exist today thanks to Whitney's work.
Cyrus McCormick	Patented a harvesting machine (reaper) in 1834. This machine greatly reduced the amount of hand labor required to harvest grain.
John Deere	An Illinois blacksmith who developed the steel moldboard plow in 1837. This plow made it possible to till the heavy soils found in many parts of the United States.
Gregor Mendel	Discovered basic principles of heredity while working as a botanist in Austria in the mid-1800s. This work lead to improvements in crops and livestock.
Luther Burbank	Invented some 800 improved plants through his work in Massachusetts and California in the early 1900s. The white potato, Shasta daisy, Santa Rosa plum, and white blackberry are among his inventions.
Steve Stice	Successfully produced a cloned female beef animal (2002) that gave birth (2005) to a normal calf at the University of Georgia research facility.

2–15. The combine began as a reaper developed by Cyrus McCormick. Three combines are harvesting in one Illinois soybean field. (Courtesy, Deere and Company)

The mid-1900s had many changes. New chemicals were developed to control insects, diseases, and weeds. Some of these were controversial because of potential damage to the environment. More advanced equipment became common on farms.

In the late 1900s and early 2000s, genetic engineering and other advanced methods of improving crops and animals were being developed. Genetically enhanced crops were being planted. They safely increased yields and improved efficiency in pest management. Of course, some people who lacked knowledge of biotechnology questioned the safety of these products if used as food.

2–16. Keeping up-to-date is important in agriculture. Broadcast radio may provide agricultural news.

AGRICULTURAL INDUSTRY EMERGES

Agriscience has made new ways of producing food and fiber possible. With the use of technology, the agricultural industry has emerged.

Many changes occurred in agriculture in the 1900s. Each worker became much more productive, and farm activities began to occur off the farm. More people were needed to provide for the needs of farmers and to process what they produced.

More strides in the agricultural industry will be made in the twenty-first century. Science and technology will play a huge role in the developments. Opportunities will be great for young individuals who wish to pursue careers in these areas.

INTERNATIONAL TRADE

Agricultural commodities and the products manufactured from them are often traded by nations. A system that promotes such exchange has emerged. ***International trade*** is the buying and selling of goods by two or more nations. It is very important to the United States, as we import and export such products.

Trade among nations is said to have created a global economy. Nations can no longer exist in isolation and prosper. The ability to trade helps overcome disasters that may occur in one nation. For example, a drought may prevent crop growth. Another nation may have an abundant crop and, therefore, provide for the nation that had a failure.

An ***export*** is a good that is sold to another country. Many raw crop products are exported by the United States, such as wheat, cotton, and soybeans. The ability to export these commodities creates opportunities on farms in the United States. This exportation increases the economic well-being by providing jobs and producing valuable products for export.

2–17. Semi-trailer trucks deliver grain to an elevator for export. (Courtesy, U.S. Department of Agriculture)

An ***import*** is a good brought into another country. In practice, all exports become imports for the country that receives them. Examples of foods imported to the United States include honey, shrimp, coffee, frog legs, some fish species, fruits, and ornamental flowers. Some raw products exported by the United States are manufactured into useful goods that are then imported by the United States. An example is cotton. Bales of cotton are shipped overseas to be made into cloth, which is sewn into clothing, such as shirts, trousers, and socks.

In the United States, international agricultural trade is promoted by the Foreign Agricultural Service (FAS) in the U.S. Department of Agriculture. The FAS strives to open market access for agricultural commodities and build new markets. It also seeks to make U.S. products more competitive in world marketing.

Nations are sometimes reluctant to receive imports. To discourage imports, tariffs may be used. A ***tariff*** is a tax or fee placed on imports (or exports). Tariffs make the price of imported goods higher. Imports have contributed to economic problems in the United States. Low-cost labor in other nations allows manufactured goods to be produced at less cost. These goods may be imported into the United States. As a result, jobs are displaced, so citizens may lose their jobs and income. Some large stores now carry few American-made products. Government leaders try to determine the best approaches to use.

Fewer environmental regulations in some countries allow manufacturing without regard for the environment. Nations with environmental regulations may be at a disadvantage from an economic perspective. Regardless, some people believe that only products from nations that respect the environment should be purchased.

TRADE AGREEMENTS

Nations often hold meetings and cooperate to work out the details for orderly trade. The World Trade Organization (WTO) establishes procedures for orderly trade that often relate to tariffs and subsidies. A few nations subsidize farmers to keep costs of commodities low, allowing the commodities to be traded among nations.

The General Agreement on Tariffs and Trade (GATT) was developed by representatives of nations. GATT encourages trade and discourages trade restrictions. The general notion is that fair competition promotes trade.

The North American Free Trade Agreement (NAFTA) is an agreement between the United States, Canada, and Mexico to promote trade among the three nations. In the case of agriculture, the United States continues separate agreements with Canada and Mexico. Many economists feel that NAFTA has promoted agricultural trade that is beneficial to the United States.

AgriScience Connection

SOIL MOISTURE RESEARCH

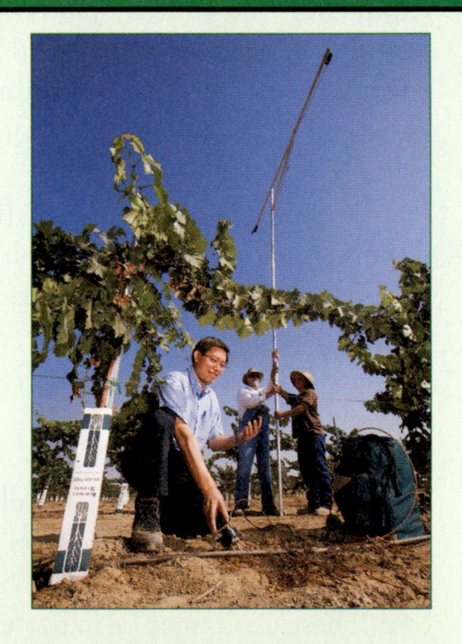

Crops need sufficient moisture to grow and to be productive. The moisture is obtained by plants from the soil through their roots. Understanding the processes that are involved helps producers gain greater efficiency in the use of water.

The amount of space plants occupy above ground is known as "canopy cover." Scientists have determined that canopy cover influences the amount of water plants use and need. This helps to conserve water supplies and to prevent excessive irrigation.

This scientist is using a probe to collect soil moisture data in a vineyard. A spectral camera is being used to record information for calibrating with satellite imagery. The research will allow growers using the information to instruct computer-managed irrigation systems in the amount of water to apply. Satellite imagery determines the canopy cover in the process. More information is available through the Agriculture Research Service Web site: www.ars.usda.gov/. (Courtesy, Agricultural Research Service, USDA)

TRADE IN PRACTICE

Americans want to be sure that they are consuming safe, quality products. They also want the producers to receive fair compensation for their work and to perform their work in a reasonable work environment. The U.S. commodities with the greatest volume exported are wheat, soybeans, rice, cotton, corn, and poultry.

Large volumes of agricultural commodities are involved in international trade. Major ports have facilities for loading ships to transport the commodities. Some facilities segregate products of genetically-altered varieties from those that have not been altered. An example is the soybean. Soybeans from varieties that have been altered are discriminated against by some nations in trade. The nations will not buy soybeans from varieties that have been genetically modified. Some people view this as unfair discrimination against producers who use modern, efficient technology.

Labeling imported products is an expectation of many American consumers. **Country-of-origin labeling** (COOL) is required on products imported by the United States. Such labels allow consumers to have information on the origin of products, including where products are manufactured. A provision of the 2002 Farm Bill provided for COOL on products in the United States. After many hearings, much debate, and a change in president, the U.S. Department of Agriculture announced that the final details of COOL would go into effect in March 2009. The regulations provide that commodities covered by COOL must be labeled at retail to indicate their country of origin. Covered items include beef, lamb, chicken, goat, pork, fish, shellfish, fresh and frozen fruits and vegetables, certain ground and tree nuts, and a few other products.

2–18. The label on this can provides country-of-origin information. (Courtesy, Education Images)

The safety and quality standards of some food products in international trade are established by Codex Alimentarious. The standards set rules for such things as pesticide tolerance levels in foods. The presence of lead, melamine, toxins, and other substances in imported foods can result in human disease when the products are consumed. Issues over food safety have resulted in some authorities suggesting that consumers only buy American-grown products.

Grading to ensure uniform quality is essential in trade. For example, all soybeans in trade should be uniform to promote pricing and exchange. Moisture content, the presence of trash, and uniformity are factors in grading. Insects, rat dung, and other impurities should not be present. Graders are present at ports to collect and test samples.

Fair trade is another notion among some people involved with international trade. ***Fair trade*** is a practice to make sure producers (farmers) receive fair compensation for their labor. It is intended to provide farmers with a fairer share of income. Some fair trade activities are through cooperatives. Coffee was among the first commodities involved with fair trade. Tea, sugar, and cocoa are also often included. An example is Equal Exchange, which involves the Presbyterian Church (U.S.A) as a partner. Most of the efforts, however, are with lesser commodities rather than major food commodities such as rice, corn, and wheat.

AGRICULTURAL PRACTICES VARY

The methods people use to obtain food and clothing vary considerably around the world. Developed nations—Canada, the United States, Japan, and those in Europe—have advanced food systems. In contrast, developing nations may have primitive agriculture. Little or no modern technology is used to produce plants and animals. This is changing as U.S. technology is increasingly being used by large corporations in these areas to produce items for export to the United States.

International agriculture deals with differences among nations. It includes production agriculture and needs for food, clothing, and shelter in different nations. International agriculture often focuses on helping people improve how they farm. Supplies and services may be needed. Marketing and processing are often included.

Nations with advanced food systems use science and technology. Meanwhile, nations with primitive food systems use few or no improved practices. The contrast between nations is great.

2–19. In some areas, rice may still be planted by hand.

Some areas of contrast in international agriculture are:

- **Mechanization**—Developed countries use tractors, power equipment, and low levels of hand labor. Site-specific practices involving global positioning and variable rates of application may be used. Countries that are developing use considerable hand labor and little mechanization. Animals may also be used for power.

- **Improved crops**—Developed countries use crops that have been carefully selected to produce high yields. New genetically enhanced crops may be more widely grown to help solve human nutrition problems. Yet developing countries may have few improved crops.
- **Commerce**—Developed countries have systems for buying and selling food and fiber. In contrast, developing countries may have poor transportation, few storage facilities, and no easy way to buy and sell.
- **Education of people**—People in the developed countries have higher levels of education. These countries have well-developed systems of agricultural education and know about agriscience. However, the developing countries generally have lower levels of education and very little education in agriculture. As a result, they know very little about agriscience.
- **Soil and climate**—Developed nations have soils and climates that support the production of food and fiber. Developing countries may have poor soils and extremely dry or wet climates.

2–20. A no-till grain drill plants without plowing the soil. (Courtesy, Case Corporation)

2–21. Papayas grow in tropical climates and are a popular food. (Courtesy, Agricultural Research Service, USDA)

BIOSECURITY

In recent years, national concern over the maintenance of a safe and sustaining food supply has led to the creation of the term and practices known as biosecurity. Some food safety issues have resulted from the failure of producers and processors, such as keeping a processing facility clean. An example is the salmonella food poisoning among consumers of peanut products from a manufacturing plant that failed to follow appropriate practices.

Biosecurity is the use of approaches to manage risk and ensure the production of disease-free animals and other products. The focus is to prevent the intentional damage of a biological product or damage caused by biological agents. For example, the introduction of pathogens (infectious causes of disease) into a herd or flock could cause major loss to the producer. It could also make the product unsafe for consumption.

RISKS

Biosecurity includes risks associated with food safety, plant and animal life, and environmental risks. It covers the introduction of plant pests, animal pests and diseases, and zoonosis. It also includes the inappropriate introduction and release of genetically-modified organisms as well as the management of invasive species. An invasive species is not native to an area and is capable of disrupting normal conditions where it is present.

At the producer level, disease prevention and control practices used in good production operations are usually sufficient. Every producer has a role in biosecurity. Of course, the major advent of the term and practices occurred following outbreaks of terrorism in the United States and other countries.

The goal of biosecurity is to produce farm products that are free of hazards and that promote the well-being of consumers. Producers follow production practices that ensure product quality. Processing plants and retail businesses also follow practices to assure food quality. Such practices should conform to accepted practices of food safety.

FOOD SECURITY

Food security is included here as part of biosecurity. ***Food security*** is keeping food safe by preventing its deliberate contamination with intent to cause harm or disruption. Food security focuses on terrorism that may occur in food supplies.

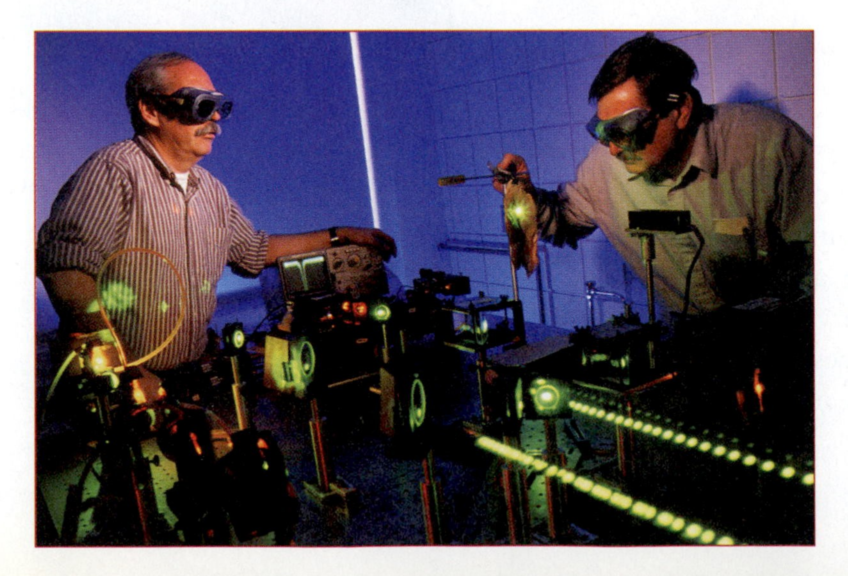

2–22. Researchers are studying the use of lasers and handheld scanners to detect food contamination. (Courtesy, Agricultural Research Service, USDA)

The U.S. government has attempted steps to guarantee food security. A broad range of practices and procedures for keeping food materials safe are followed, especially for foods brought into this nation from other countries. Risks may also arise in protecting domestic food supplies from terrorist action.

GENERAL APPROACHES

Appropriate practices can promote biosecurity. Approaches used in biosecurity include:

- Having gates, doors, fences, and other devices that restrict the entry of unauthorized people and vehicles in areas where plants and animals are kept.
- Following appropriate practices with new animals, such as isolation and observation.
- Protecting inputs from unauthorized contact during transportation and storage. An example is feed ingredients which should be secured for protection against contamination.
- Protecting food-processing facilities from unauthorized access.
- Protecting food containers from contact with unauthorized individuals.
- Traceability of food materials from the producer through all marketing channels to the consumer.

RESEARCH AND DEVELOPMENT

Research and development (R&D) is the process of creating and implementing new technology. Research is used to investigate and scientifically address issues and needs. Development is used to apply research findings and other information to invent new products or approaches.

R&D is carried out by government agencies, colleges and universities, and private agribusinesses and producers. A need is identified, and action is taken to fill the need. Sites of R&D sometimes work individually; other times the R&D sites work together. Most R&D carried out using public funds (tax dollars) is available to the public. Other R&D is proprietary, so it is carried out by an agribusiness or some other entity. Proprietary R&D is for the use of a private business that carried it out.

2–23. The Morrow Plots on the campus of the University of Illinois are the oldest agricultural research plots in the United States.

The hope of the business is that the R&D will be profitable. The possible opportunity for private businesses to gain profit from R&D is a driving force behind much of the work. Much new technology has resulted from private endeavors.

In many cases, a prototype or model is developed. This prototype undergoes considerable testing. If the tests prove it worthy, the product may be manufactured. Once manufactured, it is available for adoption.

RESEARCH

The vast growth and success of the agricultural industry in the United States are often attributed to research. Finding solutions to problems promotes agricultural productivity.

Research is the careful and diligent search for answers to questions. The questions may be based on problems faced by producers. The notion is that research can find better ways to produce plant, animal, and other products. Yields, quality, and production efficiency may be involved in the research.

Research is carried out by agricultural experiment stations, private agribusinesses, and others. The goal is to solve problems and to improve efficiency. In collaboration with land-grant universities, all states have agricultural experiment stations. These efforts were initiated by the Federal government through the states with the passage of the Hatch Act in 1887. Each state has one main station plus several branch stations. Approximately 13,000 scientists are employed at the research stations, which are usually associated with land-grant universities.

In the U.S. Department of Agriculture, the Agricultural Research Service (ARS) is the chief research arm of the Federal government. The mission of the ARS is to conduct research to develop and transfer solutions to agricultural problems. These reflect national

2–24. Demonstration plots are used so producers can compare growth and yields of different varieties. (Courtesy, Education Images)

priorities and provide solutions to problems. Currently, the ARS has 1,200 research projects within 22 national programs. Some 2,100 scientists are assisted by 6,000 technicians and others who support the research work.

GETTING THE WORD OUT

Once research has been carried out, its findings must be made known. Often, considerable research testing is needed to be certain of the findings. An experiment or demonstration may be repeated a number of times because the goal is to be absolutely sure that the results are consistent.

Experiment stations often host educational events to inform the public. These events are sometimes known as field days. Citizens, agricultural producers, and others are able to view research plots and hear scientific explanations. Bulletins, Web sites, and other means may be used to distribute the information.

In 1917, the Smith-Lever Act created an entity charged with providing education about agricultural research and related areas. The Extension Service (or Cooperative Extension Service as it is sometimes known) is an arm of the state land-grant colleges and universities. Local and regional offices have staff who carry out educational programs. We often think of these educators as County Agents, 4-H Agents, or Family and Consumer Science Agents.

Of course, agricultural education in the public schools also provides information based on agricultural research. This involves formal classroom instruction methods. More is presented about agricultural education in Chapter 3.

PROFESSIONAL AND COMMODITY ORGANIZATIONS

A number of organizations are important in the agricultural industry. These organizations serve useful roles for their members. Collectively, the members of these organizations have more power than individual members would have alone. Some organizations are for individual members while other organizations are for businesses or companies. Most organizations have some sort of dues structure to finance the operation. Organizations may focus on membership development, improvement of a commodity, political action to gain favorable laws, and promotion of a commodity. In this book, all organizations are grouped into one of two categories: professional or commodity.

A ***professional organization*** is typically developed around the career pursuits of its members. These organizations also may require a minimum level of education or experience for membership. Professional organizations usually promote the profession that is repre-

2–25. An exhibit at the National FFA Career Show is used to provide information about the American Association of Swine Veterinarians. (Courtesy, Education Images)

sented. Such organizations may provide continuing education for members. Some organizations may have a certification program that assures a minimum level of proficiency, such as the Certified Professional Agronomist Program (CPAg) of the American Society of Agronomy. Although they are not professional organizations, labor unions and similar organizations are included here because they promote the welfare of members through collective bargaining, education and training, and other benefits.

A ***commodity organization*** is developed around the production, marketing, or processing of a commodity or similar commodities. Some may be quite broad and represent a large sector of the agricultural industry, such as the American Farm Bureau. The American Farm Bureau professes that it is a voluntary organization promoting a unified voice for agriculture. Some organizations focus on one specific commodity, such as the American Soybean Association, which focuses on the production, marketing, and processing of soybeans. Organizations for animal producers may focus on particular species or breeds of animals. Some animal organizations promote breed purity and improvement. Other organizations provide education for members and member involvement, such as with livestock shows.

REVIEWING

MAIN IDEAS

The agricultural industry serves a major role in guaranteeing food and clothing for consumers. It also has a big role in forestry and housing. In addition, it includes horticulture, natural resources, environmental science, mechanical technology, and related areas. To some individuals, organizing the study of agriculture around the agricultural industry represents an approach that is realistic and practical.

Major areas of the agricultural industry are supplies and services used in production, production agriculture, and marketing and processing products. The supplies and services include the input used in growing plants and raising animals and their products. Production agriculture is farming and ranching. Marketing and processing is taking the produce of farms and moving it to the consumer in desired forms.

The agricultural industry as we know it today emerged over many centuries. Initially, people were hunters and gatherers. A first step toward agriculture was the domestication of certain species of plants and animals. As people began culturing crops, they studied what was needed to increase productivity. Agriculture began to emerge. Since the early 1700s, self-sufficient lifestyles have changed. Agriscience and technology gradually took on important roles in the advancement of agriculture. Crop and mechanical inventions had key roles. Later, commerce with crops lead to the agricultural industry. The focus shifted from subsistence to meeting market demand.

International trade is very important to agriculture today. Trade agreements among nations help facilitate trade. Major ports are involved in exporting and importing. Large volumes of commodities are involved in international trade. In the United States, country-of-origin labeling has moved from voluntary to mandatory for some products. Agricultural practices vary among nations, with some having advantages over others. Some cultures also use greater mechanization and improved crop practices. Biosecurity and food safety have achieved major status in the past decade.

Research and development have key roles in moving agriculture forward. These roles are needed to ensure that advances in knowledge and practice are able to provide for an increasing human population. Professional and commodity organizations also assume important roles in advancing agriculture.

QUESTIONS

Answer the following questions, using complete sentences and correct spelling.

1. What is the meaning of agricultural industry?
2. What are agricultural supplies and services? Why are they important?
3. What is production agriculture?
4. What are the major crops and animal species produced in the United States?
5. What is horticulture? What are four areas of horticulture?
6. What is forestry? What are three products from forestry?
7. What is the importance of marketing and processing?
8. What is food safety?
9. What is domestication? Why was it important in agriculture?
10. What is agriculture?
11. What is one example of an agricultural invention? Who made the invention, and why was it important?
12. What is international trade?
13. What is the distinction between imports and exports?
14. What is COOL?
15. How do agricultural practices vary among nations?
16. What is biosecurity? How is food security a part of biosecurity?
17. What is research and development? Why is it important?
18. What are two categories of organizations in the agricultural industry? Distinguish between the two.

EVALUATING

Match the term with its correct definition.

a. veterinary care
b. agronomy
c. ornamental horticulture
d. floriculture
e. pomology
f. olericulture
g. aquaculture
h. processing
i. export
j. biosecurity

_____1. The production and use of flowers and plants for their appeal.

_____2. The science of growing vegetable crops.

_____3. The changing of raw products into forms desired by consumers.

_____4. An example of an agricultural service.

_____5. The science of growing fruits and nuts.

_____6. The study of plants and how they relate to the soil.

_____7. The science of water farming.

_____8. The use of plants for decorative purposes.

_____9. The use of practices that manage risk and assure wholesome products.

_____10. A good sold to another country.

EXPLORING

1. Investigate the agricultural experiment in your state and the land-grant university with which it is associated. Determine the location of branch stations. Identify major areas of agricultural research. If possible, arrange to visit a nearby station and interview a research scientist about the nature of his or her research. Prepare a written report based on your findings.

2. Identify an agricultural organization in your local community. Determine the mission and overall program of work of the organization. Determine who can be members and inquire about the costs associated with membership. Investigate the benefits of membership. Interview a member of the group to determine the meaning and value of the organization to the individual. Give an oral report in class based on your findings.

3. Study the history of agriculture, horticulture, or another agricultural area in your local community. What products have been produced over the years? How has production changed? Prepare a report based on your findings.

AgriScience Education and You

CHAPTER 3

Education and Experience in Agriculture

OBJECTIVES

This chapter covers agricultural education and the importance and types of supervised experience. It has the following objectives:

1. Explain agricultural education and its three components.
2. Explain the purpose, benefits, and types of supervised experience.
3. Describe how to plan, manage, and advance in supervised experience.
4. Discuss the records needed with supervised experience.

TERMS

agricultural education
asset
balance sheet
budget
cash flow statement
depreciation
exploratory supervised experience
inventory
liability
owner equity
ownership supervised experience
placement supervised experience
planning
profit and loss statement
recordkeeping
research supervised experience
supervised experience
training agreement
training plan
training station

3–1. A teacher provides small group instruction for students enrolled in an agriscience class. (Courtesy, Education Images)

OVER A MILLION secondary school students take an agriscience class each year in the United States. Why do they do so? The reasons vary. Some do so to develop career skills. Others do so to be good consumers or to gain proficiency in useful areas, such as landscaping their home. Why are you enrolled in agriscience? No doubt, you have important reasons.

The number of students taking agriscience classes in high school has increased. There are a number of reasons for the popularity of agriscience. One reason for the increase has been the emphasis on a science-based approach to teaching agriculture. Other reasons include the opportunity to gain practical experience, develop personal skills, and enhance academic skills. Of course, many students enroll because of their budding career interests in the broad agricultural industry.

Agriscience study is available at every age level. Elementary school students may have projects about food, plants, or animals in science units. Middle school students may have short or year-long classes in agriscience. Many high schools offer specialized agriscience classes. Beyond high school, a variety of ways are available to study agriscience.

EDUCATION IN AGRICULTURE

3–2. A teacher is helping a student prepare a lamb for showing. (Courtesy, Education Images)

Agricultural education is instruction in agriculture and related subjects. Of course, instruction in the production of plants and animals is a major focus of agricultural education. The subjects are broad and reflect those found in the agricultural industry, such as those in horticulture, forestry, biotechnology, wildlife, veterinary science, and natural resources.

A science-based approach is often used in teaching agricultural education. The basics of the life sciences and physical sciences are taught. The classes are often known as agriscience.

LEVELS OF EDUCATION IN AGRICULTURE

Education in agriculture is available at every level of school. Goals may vary. If the goal is to prepare for a career in agriculture, instruction focuses on job skills. If the goal is to inform people about agriculture, a consumer approach is used.

3–3. A student is studying plant cells.

- **Elementary**—The goal at this level is to make students aware of agriculture. It may be taught as units in science, social studies, or other areas. Students may have projects, such as gardens. Other activities may involve field trips to farms, research stations, and agriculture museums. Ag-in-the-Classroom is a major national effort in elementary schools.
- **Middle and junior high school**—Middle and junior high school classes in agriculture may be in short blocks or

full-year classes. Some are integrated with technology or science education. The focus of these classes is on food, fiber, and natural resources. Students in these classes may be eligible for the FFA Discovery Degree.

- **High school**—Agriculture classes are available in more than 7,300 schools in the United States. Slightly over one million students are enrolled. Classes are planned to meet the needs and interests of students. In-depth classes may be offered in science-based agriculture. There may be specialized classes in horticulture, forestry, biotechnology, animal science, and other areas. Students usually have supervised experience and are active in FFA. Emphasis is on career skills and preparing for college.
- **Postsecondary**—Various institutions provide postsecondary education in agriculture. Many states have technical or two-year colleges. These schools have courses that lead to certificates or associate diplomas. Practical, job-oriented skills are taught. Four-year colleges and universities may offer baccalaureate degrees in agriculture. Many of these colleges have graduate and professional schools. Graduate education allows students to earn master's and doctoral degrees. Professional schools offer education in areas such as veterinary medicine and landscape architecture.
- **Continuing education**—Keeping up-to-date on technical topics is essential. This requires continuing education. It may be offered by colleges, sponsored by agribusinesses, or available through other sources. Associations sometimes have major roles in continuing education. An example is the American Society of Agronomy, which provides many workshops and other activities. Continuing education seminars and classes may be for a few hours in the evening or for longer periods. Field days and demonstrations are also used.

3–4. A teacher instructs students in an agriculture class. (Courtesy, Education Images)

INTEGRAL PARTS

Exciting ways of learning are used in high school agriscience classes. Students may use a range of modern materials. They may apply what they learn in the lab, at home, at a job, or in other ways. FFA helps students learn leadership and personal skills that are important in career success.

The Model

The three integral components of agricultural education classes are:

1. **Classroom and lab instruction**—This is the organized instruction by the teacher. Students often use textbooks as learning tools. Equipment and lab facilities are provided as needed for a particular area of instruction. For example, a class in small animal care would need a small animal lab with the equipment necessary to care for the well-being of animals.

2. **Supervised experience**—Supervised experience (SE) is the practical application of classroom and lab instruction. SE may occur during the school day, but it is usually after school hours. SE provides a good foundation for success in FFA. (In some schools, SE is known as SAE—supervised agricultural experience.)

AgriScience Connection

INTO PRINT

Supervised experience can lead to awards through the National FFA Organization. In fact, most all areas of supervised experience are suitable for FFA awards in one or more categories. For individuals who start a business, the AgriEntrepreneurship Award may be just what they are looking for.

This shows a student at her award-winning display at the National FFA Career Show. The student planned and established a publication design and layout business. She used programs available on the school's computer and at her home. She designed yearbooks for FFA chapters throughout her state using her B-ROD'S Graphic and Design Business. (Courtesy, Education Images)

3. **FFA**—This is the student organization portion of agricultural education. Time during a class period may be devoted to FFA. Most of the activities are outside regular class hours. Many FFA awards are built on success in supervised experience.

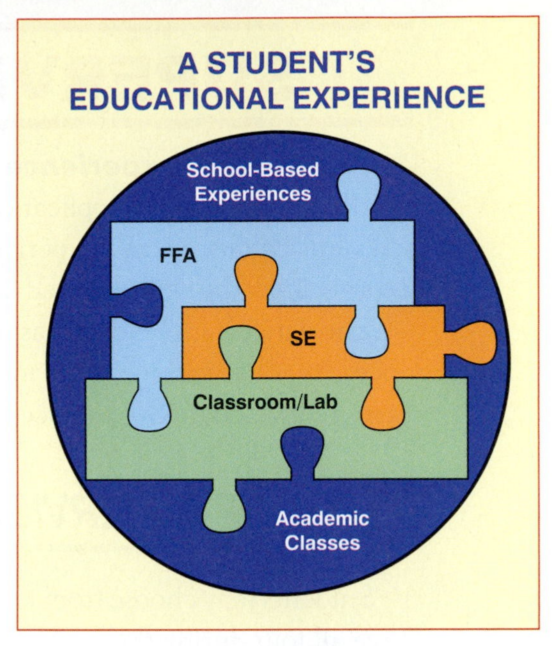

3–5. Relationships of SE and FFA to classroom and laboratory instruction in agricultural education.

All the parts overlap and promote learning. Success in one area usually requires emphasis in another. For example, achieving FFA awards may require good supervised experience.

The Teacher

Qualified teachers are needed in agriscience. They have an important job in helping students learn. In overseeing an FFA chapter, an agriculture teacher is known as an FFA advisor. Many schools have one teacher. Some schools have more than one teacher, with all helping advise FFA.

There are more than 11,000 secondary agriculture teachers in the United States. These individuals have completed university courses and other requirements to prepare them to be teachers. Agriculture teachers are hired by local school boards. Most teach in comprehensive high schools. Some teach in specialized career centers, academies, or charter schools. Some agriculture teachers are hired for the school year. Many are hired for a calendar year. Additional work days are often included because teaching agricultural education extends beyond the normal school day and year.

3–6. A university student who is preparing to be an agriculture teacher is planning, with her advisor, the courses to take. (Courtesy, Education Images)

SUPERVISED EXPERIENCE

Supervised experience (SE), often called supervised agricultural experience (SAE), is the planned application of skills learned in classes. The goal is to make learning relevant. Each student's experience is based on his or her interests and needs. Records are kept to show the skills learned and, in some cases, the income gained.

The agriculture teacher has a major role in helping plan appropriate supervised experience. An adult is responsible for direct supervision. Good supervision assures learning and helps reduce chances of an accident that could cause injury.

TYPES OF SUPERVISED EXPERIENCE

Students may choose from four major types of supervised experience. Some students may have all four during their school years. The types are:

- **Exploratory**—***Exploratory supervised experience*** is the foundation for other experiences. It is useful in learning firsthand about a wide range of agricultural areas. Exploratory experience helps students discover their interests and experience agricultural activities. Examples of exploratory supervised experience include shadowing (accompanying) a crop consultant for a day, spending a day in a small animal clinic, observing in a fish hatchery, and being with a rancher at work.

3–7. A student is reviewing plans with her teacher for research/experimentation supervised experience. (Courtesy, Education Images)

- **Research/experimentation**—This is a good type of supervised experience for students who have a great deal of interest in science. ***Research supervised experience*** is the planned investigation of a topic in food, fiber, natural resources, or a related area. The topic must be designed as a problem to be solved. Carefully planned procedures that follow the scientific method are used to investigate the problem. Many students use research experiences to prepare for school science fairs or the agriscience fair in FFA. Research problems are based on local situations. Solving water pollution, assessing tissue culture media, and studying methods of food processing to retain more nutrients are examples.

3–8. A student with ownership supervised experience cares for his lamb. (Courtesy, Education Images)

- Ownership—***Ownership supervised experience*** is the creation of an enterprise or business. Ownership experiences may be designed to meet unique market needs. Many students like to own animals, such as sheep or horses. Other students start lawn care businesses, greenhouses that produce flowers or bedding plants, and retail fruit and vegetable stands. Ownership supervised experiences should be carefully planned to develop important skills. Good records of the ownership activities are essential. Students with limited opportunities may use school labs or facilities at farms or agribusinesses for their ownership supervised experience.
- Placement—***Placement supervised experience*** is the assignment to a job related to school studies in agriculture. A wage may be paid, depending on the nature of the work. This type of SE is said to be work based. It takes place after school hours and must follow all federal, state, and local laws related to employment. Some restrictions apply to students based on their age. Careful records are kept of experiences

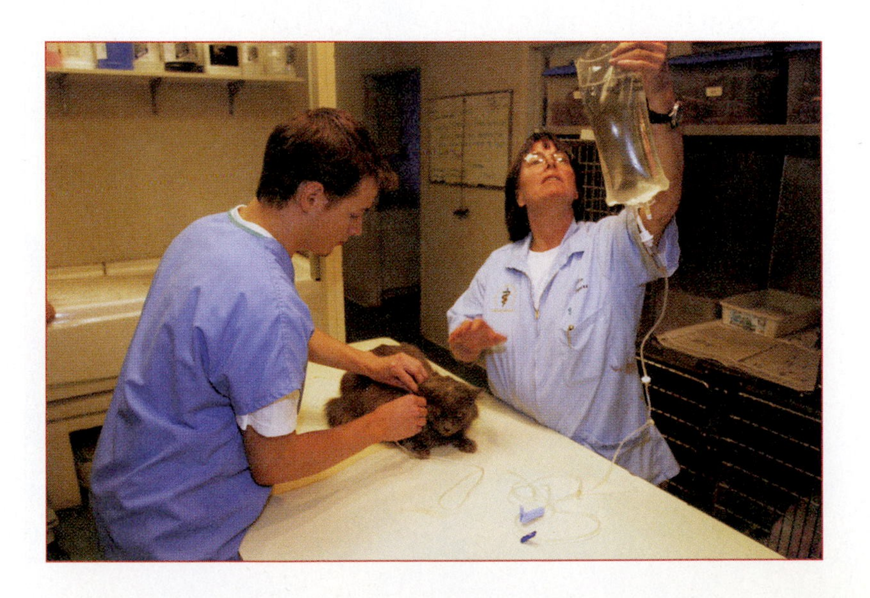

3–9. A student carries out his placement supervised experience developing skills in animal care under the direction of a veterinarian. (Courtesy, Education Images)

gained, hours worked, and other details. These records are important in advancing in FFA. Examples of placement are working at a flower shop, farm or ranch, fish hatchery, farm supplies store, or research facility.

- **Other**—Two other types of supervised experiences may be used. These are improvement and supplemental experiences. These two do not normally take the place of placement, ownership, or research/experimentation experiences. Improvement SE refers to learning activities that focus on improving the appearance or value of a home, school, or community. Most improvement activities are short-term and likely completed in a day or so. An example is installing landscaping around the home. Supplemental SE refers to specific agricultural skills developed outside of normal class time. The skills may not be related to the major supervised experience. Two examples are pruning a fruit tree or replacing a toggle switch.

WHERE IT IS CARRIED OUT

Supervised experience may be carried out in a variety of settings. A few examples are:

- **At school**—Some students carry out their supervised experience after regular class hours at school. They use school facilities under the direction of their agriculture teacher. This kind of experience is sometimes known as directed lab. The school facilities may include a project center where students can keep animals or a greenhouse where plants can be grown.
- **At home**—Some students have opportunities at their homes for supervised experience. Students whose families have land or operate an agribusiness, farm, or ranch

3–10. Schools may have facilities for supervised experience.

3–11. A student with placement supervised experience works for a landscape business.

often have good opportunities for gaining experience. This means that the student is supervised by his or her parents or another adult who is present.

- **At agribusinesses**—Some students work at garden centers, hardware stores, and other places relevant to their interests. The work must fit well into their supervised experience plans. Someone at the agribusiness is responsible for supervision, and the agriculture teacher will make regular visits to observe the student's work.
- **At agencies**—Students may work at government agencies or research stations. Opportunities vary from one location to another.

BENEFITS OF SUPERVISED EXPERIENCE

A major benefit of SE is the opportunity to gain practical experience in the world of work. Understanding work and how it is carried out gives meaning to education. This helps in getting and advancing in a job. Other benefits of supervised experience are listed in Table 3–1.

Table 3–1. Benefits of Supervised Experience

A few benefits are:
• Applying school learning in a job or problem situation
• Making classes more meaningful
• Learning about the "real world"
• Gaining practical experience
• Exploring career and education opportunities
• Working with other people
• Developing personal skills
• Earning money
• Understanding financial responsibility
• Advancing in FFA
• Developing self-confidence
• Developing communication skills
• Meeting other people

PLANNING AND MANAGING SUPERVISED EXPERIENCE

Good supervised experience requires planning, dedication, and recordkeeping. A student must do his or her part.

3–12. A student is individually planning supervised experience under the direction of her agriculture teacher. (Courtesy, Education Images)

PLANNING SUPERVISED EXPERIENCE

Supervised experience is carefully planned. Areas of possible interest are explored and discussed with the agriculture teacher. This helps students identify practical and useful experiences.

Planning is the process of identifying what is to be accomplished and how it is to be carried out. Plans provide direction to the experiences. Without plans, students may fail to achieve their potential. Begin planning by assessing yourself. Use Table 3–2 to help make an assessment.

Table 3–2. Assessing Personal Interests and Opportunities

Questions to think about while planning supervised experience:
1. What are my agriscience interests?
2. What would I like to explore?
3. What opportunities are available locally in my area of interest?
4. What can my family contribute to my supervised experience?
5. What is practical for me? (Consider opportunities, age, transportation, etc.)

Training Agreement

A ***training agreement*** is a written statement that lists the terms under which supervised experience is to be carried out. It includes the nature of the work and who will supervise the experiences. Training agreements specify a training station. A ***training station*** is the location where the supervised experience will be carried out. The student, his or her parents, the supervisor at the training station, and the agriculture teacher sign the training agreement. With placement, the work hours and wages, if any, will be specified.

> Internet Topics | Search
>
> Selected topics for Internet discovery and reporting are listed here. Begin by searching for each topic. Next, read and learn about it. Conclude by preparing a brief report on your findings.
>
> 1. Accountancy
> 2. Supervised agricultural experience
> 3. Financial recordkeeping

The details in a training agreement vary by local school district. In general, agreements specify the name of the student and school, who will be involved in the experience, responsibilities of those who are involved, and signatures of the student, parent, teacher, and others. (A sample training agreement form is presented in the Activity Manual.)

Table 3–3. Examples of Supervised Experience by Career Pathway

Pathway	Possible Supervised Experience (includes exploratory, placement, ownership, and research and experimentation examples)
Animal Systems	Raise dairy goats, develop a cow-calf operation, work at a dog kennel, produce feeder pigs, investigate nutrient content of various commercial feeds, work at a turkey farm, raise catfish in cages, operate a pet sitting business
Plant Systems	Produce organic vegetables, work on a crop farm, produce a corn crop, grow and sell mushrooms, operate a lawn care business, create and sell floral arrangements, investigate effects of environment on cut flowers, establish sod in a lawn
Power, Structural, and Technical Systems	Work at a local hardware store, construct compost bins, construct a utility building, build and sell birdhouse and feeders, investigate opportunities for alternative energy in the local area, steam clean and paint a used tractor, build a fence
Biotechnology Systems	Propagate plants from tissue, work in a biotechnology lab, investigate the effects of light intensity on GMO and non-GMO plants, graft trees, survey local citizen attitudes toward biotechnology

(Continued)

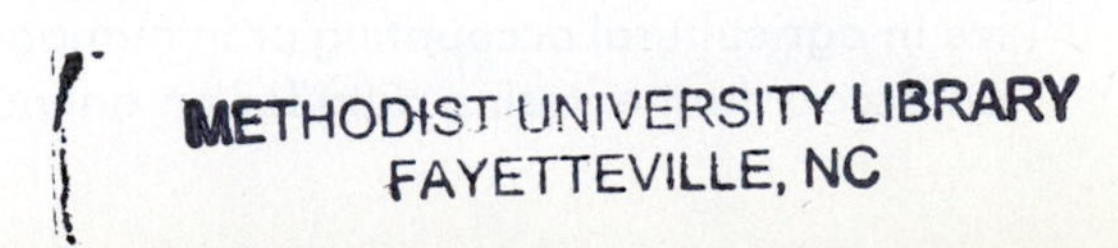

Table 3–3 (Continued)

Pathway	Possible Supervised Experience (includes exploratory, placement, ownership, and research and experimentation examples)
Agribusiness Systems	Work in a farm and garden center, write news releases about the local FFA chapter, operate a seed sales business, operate a roadside cooperative vegetable and fruit stand, develop a marketing plan for an agricultural commodity
Food Products and Processing Systems	Process cream-style corn in the school lab, investigate the effects of storage temperature on eggs, organize an FFA chapter fruit sale, work at a vegetable packing shed, investigate the effects of light on color and flavor of milk
Natural Resources Systems	Establish bird nesting boxes, work for a game management business, restore an eroded bank along a roadway, raise fish for stocking in streams, investigate the effects of moisture on feed in a bird feeder, grow lab cultures on water samples
Environmental Service Systems	Establish recycling bins on the school grounds, work for a water testing plant, investigate the biological and chemical characteristics of effluent from a local animal producer, job shadow a solid waste landfill manager

Note: These are a few examples of supervised experience opportunities. Use these to aid in identifying and planning supervised experience.

CAREER PROFILE

ACCOUNTANT

Accountants are individuals who are professionally prepared to do accounting. They know systems of accountancy and are skilled in recording, verifying, and reporting financial data. Most specialize in financial accounting. Management accounting and tax accounting are also widely used in the agricultural industry.

The nature of the work has changed over the years. It has moved from paper-based recording of information to computer-based data entry, reporting, and analysis. Most accountants hold baccalaureate degrees in accounting or a closely related area of business. More hold master's degrees, with some of these in business administration. Accountants may also strive for additional education to become a Certified Public Accountant (CPA). Good computer application skills are needed.

The work is typically in offices but may involve travel to the offices of clients. Some specialize in agricultural accounting or in management accounting with an agribusiness. This shows an accountant entering data for an online income tax report.

Training Plan

Many schools use training plans. A ***training plan*** is a list of the activities in supervised experience. The plan relates the activities to the areas studied in class. The supervisor at the training station will help develop the training plan. All activities should contribute to the overall goal for the experience. The training plan can also serve as a record of accomplishments.

SUPERVISED EXPERIENCE TRAINING PLAN

Student's Name ______________________ School ______________________

Objective/Career Goal ____

Training Station ______________________ Date Prepared ______________

Deadline for Completion ____

Check type of SE:

☐ Exploratory ☐ Research ☐ Ownership ☐ Placement

Activities/Skills/Competencies	Date Performed	Initials of Supervisor When Satisfactory

3–13. Sample training plan. (Add lines as needed to the form.)

Budgeting

A ***budget*** is a statement of anticipated income and expense for a given time or activity. It is a way of getting organized from the standpoint of money. Careful budgeting helps assess if an enterprise is likely to be profitable. If projected income is greater than costs, a profit may result. On the other hand, if costs appear greater than likely returns, a financial loss will likely result.

Several kinds of budgets may be used. With supervised experience, a start-up budget will be useful. Preparing a start-up budget helps identify costs and possible returns. The process of preparing a budget helps determine problems and opportunities. All data must be realistic and based on what is likely to occur. Of course, a start-up budget does not protect against risk.

ACADEMIC CONNECTION

MATHEMATICS

Relate your studies in mathematics class to recordkeeping with your supervised experience program. Summarize records and apply assessment skills to an analysis of your records. Revise supervised experience plans based on your findings.

Other kinds of budgets include personal or family budget, event budget (covers one activity such as a large farm show), corporate budget (use in large agribusinesses), and government budget (use by government agencies at the local, state, and Federal levels).

MANAGING SUPERVISED EXPERIENCE

In supervised experience, students learn the importance of being responsible—doing what is needed, when it is needed. For example, if animals are involved, they must be fed, watered, and otherwise cared for. If not, their well-being will be threatened.

Managing experience is the coordination of activities. It requires attention to detail and follow-through. Managing the activities might include ordering equipment for a project, caring for the equipment, and recording the cost and outcome of the project. There are so many learning activities in managing a supervised experience program!

Three areas in management are listed here:

- **Assume responsibility—**Supervised experience often involves activities that require you to be a responsible person. This includes paying attention to details, being on time, and doing the job right. Always follow safe practices. Live up to your commitments. Be on-task and never goof-off.
- **Expanding the program—**Supervised experience begins on a small scale. As you learn and gain experience, you will want to expand your program. How to expand varies with the kind of experience. For example, if you begin with one pig, expansion might involve increasing the number of pigs. Always think about how you can expand, grow, and become better. Think about how you can gain more skills and advance in what you are doing.
- **Keeping records—**Good records are essential. Always know what is expected of you and exceed the minimum. (The next section of the chapter addresses recordkeeping.)

RECORDKEEPING

Recordkeeping is the act of recording supervised experience activities. Special record books or computer programs may be used. These often fit well with FFA application forms. How records are kept often varies on a school-by-school basis. The records are used to prepare reports and assess progress.

Two kinds of records are typically kept: performance and financial.

PERFORMANCE RECORDS

Performance records are written accounts of the experiences of a student. In some cases, research logs or animal and plant care records may be kept. Animal care records may include when an animal was fed or watered, kind of feed and how much, and other details. Plant care records may include planting, fertilizing, and application of pesticides and irrigation. In other cases, records may be lists of the skills that were performed and when they were done.

Records of what has been accomplished and the skills that have been developed in supervised experience are very useful. These provide the information for advancing in FFA membership and related activities. Such records also provide an inventory of the skills that have been learned. Completing FFA awards forms requires information from records. In some cases, actual copies of records must be submitted with application forms.

3–14. A wireless computer is being used by a student to keep supervised experience records. Her teacher is providing instruction in the system. (Courtesy, Education Images)

FINANCIAL RECORDS

Simply, financial records are written lists of expenses and income. Expense is the money spent in carrying out the supervised experience. Income is the money gained from the experience activities. Financial records are much more detailed with the ownership type of supervised experience. With placement, the records are primarily of the income gained from

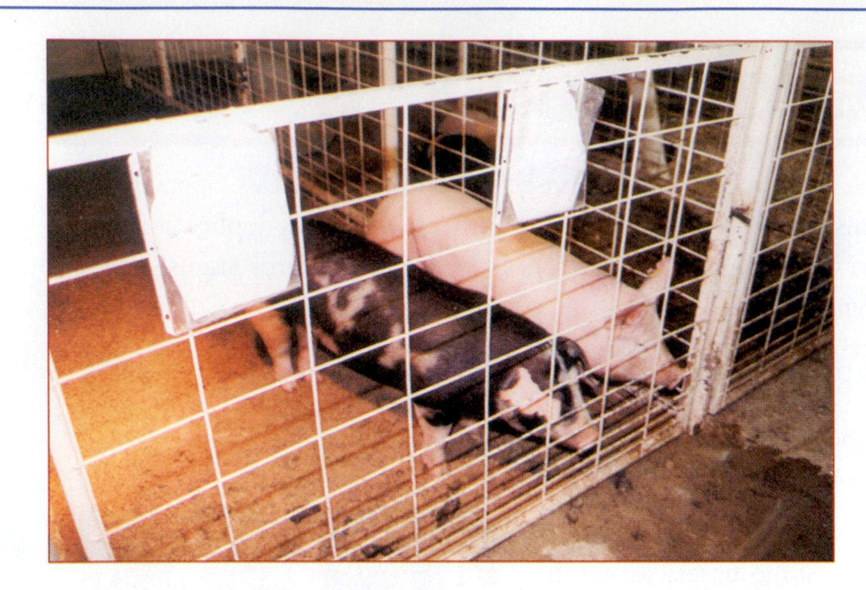

3–15. Sheets to keep a log of management activities are posted on the gates of pens.

work. At the end of a specific period, financial records are summarized to determine the total costs and income. This reveals whether a profit was made.

With supervised experience, records may be kept by hand in paper record books or with a computer using a program designed for such use. An example of a program is Excel software. Most supervised experience records involve simple systems. Large agribusinesses would be much more involved. Large agribusinesses and farms may use accountant services. Accountants are individuals who are professionally trained in recordkeeping, reporting, and analysis.

Records can be no better than the care and accuracy that are used in keeping them. Always record information promptly. Be sure that the information is entered accurately.

The financial picture of an ownership supervised experience program is determined through its sums of assets and liabilities. An ***asset*** is an item of value that is owned. When all assets are summed, the value of everything that is owned is known. An ownership supervised experience project is more than assets; liabilities are included. A ***liability*** is an item with an obligation, such as a loan to buy a tractor. The owner is responsible for repaying the loan.

With ownership SE, both assets and liabilities must be considered in order to have a clear picture of owner equity. ***Owner equity*** is the amount remaining after liabilities have been subtracted from assets. For example, the owner equity in a tractor valued at $50,000 but with a loan against it in the amount of $20,000 is $30,000. In summary, owner equity is the positive remainder after liabilities have been subtracted from assets.

Most records are summarized on an annual or fiscal year basis. Fiscal year is the 12-month period specified for tax or accounting purposes. Most of the time, a fiscal year is the same as a calendar year—January 1 through December 31. With supervised experience, school years or other fiscal years could be used.

Records are kept so that reports can be prepared. Reports summarize data in the records. The scope and detail of records vary with the size of an enterprise. The records for most beginning supervised experience ownership projects are quite simple. Larger projects and agribusinesses used more detailed recordkeeping, reporting, and analysis.

Financial Statements

Records are kept so that they can be summarized and reports prepared. The reports are analyzed to gain information about a business venture and make good decisions. Three kinds of business financial statements may be used:

- **Profit and loss statement**—The ***profit and loss statement*** shows how well an enterprise has performed to make a profit (or not make a profit). It is a summary of the revenues (income), costs, and expenses incurred during a specific period of time. The profit and loss statement is sometimes referred to as P&L.
- **Balance sheet**—The ***balance sheet*** is an itemized statement of the assets and liabilities of a business that gives its net worth at any moment of time. It shows the resources that were used in operating the business and its financial situation at any specific time.

Balance Sheet
Jason's Landscape Service
December 31, 2010

Assets		**Liabilities and Owner's Equity**	
Cash	$ 750.00	**Liabilities**	
Accounts Receivable	1,100.00	Notes Payable	$ 225.00
Tools and Equipment	1,650.00	Accounts Payable	110.00
		Total Liabilities	$ 335.00
		Owner's Equity	
		Stock (start-up investment)	$ 1,950.00
		Retained Earnings	1,215.00
		Total Owner's Equity	$ 3,165.00
Total	$ 3,500.00	Total	$ 3,500.00

3–16. Sample small business balance sheet.

- **Cash flow statement**—The ***cash flow statement*** provides information about the flow of cash into and out of a business. The money coming in is known as cash inflow; money going out is known as cash outflow. The cash flow statement may be divided into three segments: cash flow from operating activities, cash flow from investing activities, and cash flow from financing activities (loans and dividends).

Specific recordkeeping activities are described here. A number of business accounting concepts are involved. Some are explained in simple terms. Another resource will be needed for detailed information.

Inventory Records

Inventory is an itemized list of current assets—things owned that have value. They may be goods on hand, such as seed, feed, animals, equipment, and products to be sold. Inventory includes raw materials as well as manufactured products. Land and improvements, such as buildings, may also be inventoried. An inventory should list goods and materials held in stock by an agribusiness. Ownership supervised experience involves inventory.

Inventory is used to:

- Prepare owner equity statements
- Prepare income statements
- Obtain credit
- Determine insurance needs
- Manage taxation
- Aid in estate planning

Inventories may be done in two ways: physical and perpetual. A physical inventory is made by counting every item that is on hand. This is not a big task with a small business, but a large amount of time can be required with a large business. A perpetual inventory involves keeping records of inventory up-to-date based on sales and deliveries. Computer systems with bar coded merchandise facilitate perpetual inventorying.

Depreciation

Depreciation is an accounting process that spreads the cost of assets over the useful life of the asset. It is typically more than one year. The value of an asset is reduced due to age, wear, obsolescence, or other factors. For example, if you buy a new pickup truck for your

3–17. A round hay baler and tractor may be depreciated over several years. (Courtesy, AGCO, Duluth, Georgia)

business, it begins losing value due to age and use. It doesn't lose all of its value in one year. Properties vary in the number of years for depreciation. Buildings may depreciate out over 20 or 30 years. A farm tractor may depreciate over three or so years. Once fully depreciated, most things have a salvage value. This is the estimated value of something after it has been fully depreciated.

Methods used in calculating depreciation vary. The straight-line method involves dividing the lifespan of an asset into its cost (minus salvage value). Declining-balance methods are used to gain larger depreciation in the first few years of an asset's life. The assistance of an accountant in calculating depreciation is normally a good idea. The annual amounts of depreciation are important in income tax preparation.

REVIEWING

MAIN IDEAS

Agricultural education is offered in many schools at various levels. At the high school level, it includes three components: classroom and laboratory instruction, supervised experience, and FFA. Classroom and laboratory instruction is provided by a specially-prepared teacher. The teacher also serves as the FFA advisor.

Supervised experience is the practical application of classroom and lab instruction. It provides hands-on learning. Exploratory, research, placement, or ownership SE may be used. Good planning and recordkeeping are needed.

Recordkeeping is a process of recording supervised experience activities. Procedures vary. Some schools have students keep records in paper books. Others use computer programs. Regardless of how the records are kept, good information is essential. Performance records are written accounts of experiences of a student, including skills developed.

Financial records are written lists of expenses and income. Such records are simple when supervised experience is first begun. They become more detailed as the supervised experience expands and develops. Three kinds of business financial statements may be used: profit and loss, balance sheet, and cash flow. A record of inventory is also maintained.

QUESTIONS

Answer the following questions, using complete sentences and correct spelling.

1. What is agricultural education?
2. What are the three components of agricultural education classes?
3. What is supervised experience? What are the four main types?
4. What is a training agreement?
5. What is a training plan?
6. What are three areas of management in supervised experience programs? Briefly explain each.
7. What is recordkeeping?
8. What two main types of records are kept of supervised experience?
9. What is owner equity? Why is it important with ownership supervised experience?
10. What kinds of business financial statements may be used?
11. What is inventory? What two major ways are used in keeping inventory?
12. What is depreciation?

EVALUATING

Match each term with its correct definition.

a. budget
b. placement supervised experience
c. ownership supervised experience
d. depreciation
e. owner equity
f. asset
g. liability
h. balance sheet
i. profit and loss statement
j. inventory

_____1. Shows how well a business has performed to make a profit.
_____2. An itemized list of current assets.
_____3. A procedure for spreading the costs of assets over more than one year.
_____4. An itemized statement of the assets and liabilities of a business.
_____5. A statement of anticipated income and expense.
_____6. The type of experience in which a student works for another person.
_____7. The type of experience in which a student owns the components of the program.
_____8. The amount of cash remaining after liabilities have been subtracted from assets.
_____9. An item with an obligation attached.
_____10. An item of value that is owned.

EXPLORING

1. Good financial planning often involves preparing a personal budget. A personal budget is a list of anticipated income and expenses. You may work and earn money, or you may get an allowance. Regardless, using your money wisely is important. Prepare a personal budget that reflects the money you will have for one month and the ways you will spend it. Assess how well you are using the money you have. Here is a simple format to follow:

Budget					
Money Received			**Money Spent**		
Date	**Source**	**Amount**	**Date**	**Source**	**Amount**

2. Develop, implement, and expand supervised experience that is appropriate for you. Be sure to complete training agreement and training plan forms. Keep the appropriate records. Learn about career development events in FFA and incorporate these into your plans.

CHAPTER 4

Student Organizations

OBJECTIVES

This chapter explains the meaning, importance, and programs of student organizations in agricultural education. It has the following objectives:

1. Identify organizations for students enrolled in agricultural education.
2. Describe the purpose, objectives, and nature of FFA.
3. Trace the history of FFA.
4. Explain how FFA is organized.
5. Identify important factors in a successful FFA chapter.
6. Name and explain awards and events in FFA.
7. Identify membership degrees and requirements for advancing.

TERMS

active membership
ad hoc committee
alumni membership
association
career development events
ceremony
chapter
committee
leadership development events
membership degrees
mission
National FFA Advisor
National FFA Center
National FFA Foundation
Official FFA Manual
order of business
PAS
program of activities
standing committee

4–1. An FFA chapter president confers with the reporter at a meeting. (Courtesy, Education Images)

ONE OF THE special features of agricultural education is that it is more than regular classroom learning. It involves activities that extend beyond the classroom and into many areas of life. This is sometimes referred to as "whole person development."

Well-known for its important roles in the lives of young people, FFA helps students develop self-confidence, set goals, and aspire to achieve. It recognizes young people for excellence while they are still in high school and as they enter life beyond high school.

FFA allows for fun and achievements. Students can feel good about what they are doing. However, they must take the initiative to become involved because participation is the key to success.

STUDENT ORGANIZATIONS

Several organizations are available for students in agricultural education. They may be school-based or through organizations of the agricultural industry. Some organizations that primarily serve adults have junior membership, such as the American Junior Horticulture Association.

SCHOOL-BASED ORGANIZATIONS

FFA is the organization for students who are studying agriculture in secondary school. FFA activities help ensure student development. More than 500,000 students are FFA members because FFA is an integral part of the curriculum. Activities in FFA promote achievement in agricultural education. As such, FFA is more than a club. It is a co-curricular activity.

School clubs are of three main types: honorary organizations, interest groups, and curriculum-related clubs. Honorary organizations are for students who excel in an area and are invited to membership based on grades or other school achievements. Interest groups are for students with an interest in a particular area, such as stamp collecting, hiking, or dancing. Curriculum-related clubs are related to the curriculum and include science, history, and French clubs. Some schools have leadership and citizenship development clubs or groups that may or may not be honorary in nature.

Students in postsecondary agricultural education (beyond high school) may be members of various organizations, such as ***PAS*** (National Postsecondary Agricultural Student Organization). PAS is in many community colleges and technical schools. The organization hosts

4–2. Involvement helps students learn and develop personally. This shows a member at a University of Minnesota display at the National FFA Career Show. (Courtesy, Education Images)

a number of activities to reinforce learning and to create a desire to achieve. For additional information, refer to **www.nationalpas.org/**.

INDUSTRY-BASED ORGANIZATIONS

Industry-based organizations are associated with some part of the agricultural industry and may have an environmental or natural resource emphasis. These groups are off-shoots of adult organizations that support a particular area of the agricultural industry. They often focus on special interests and can be related to FFA activities and supervised experiences. These organizations are not typically associated with schools.

A few examples of industry-based organizations for students with interests in AFNR are listed here.

- **National Junior Horticultural Association (NJHA)**—This organization promotes and develops horticulture among youth. Leaders of the U.S. vegetable production industry founded it in 1935 to promote horticulture as an area of education and as a vocation among youth. For more information, refer to **www.njha.org/**.
- **National Junior Hereford Association (NJHA)**—This organization was formed to promote interest among youth in the Hereford breed of cattle, particularly to raise and train heifers and steers for showing. FFA members may also be members of this organization. For more information, refer to **www.jrhereford.org/**.

CAREER PROFILE

AGRICULTURE TEACHER

Agriculture teachers organize and provide education in a range of agriculture subjects. They use classrooms, labs, and equipment in teaching. Many of these teachers are active in community affairs and agriculture clubs.

Agriculture teachers must meet requirements to be teachers. They have college degrees in agricultural education or related areas. Many of these teachers have masters degrees, and some of them complete National Board for Professional Teaching Standards certification. Practical experience in agriculture work is beneficial. Begin in high school by taking agriculture classes and developing a strong background in science and communication.

Jobs for agriculture teachers are found in local schools. Nationally, a shortage of teachers has made it easy for teachers to obtain jobs. This photograph shows an agriscience teacher meeting with members of the FFA chapter officer team. (Courtesy, Education Images)

- **JAKES and Xtreme JAKES**—The National Wild Turkey Federation (NWTF) sponsors JAKES and Xtreme JAKES for youth members. JAKES (Juniors Acquiring Knowledge, Ethics, and Sportsmanship) is based on a jake turkey—the name given to a juvenile turkey, particularly a male. JAKES is for children (ages 12 and under); Xtreme JAKES is for youth (ages 13 to 17). This organization informs, educates, and involves youth with wildlife conservation. For more information, refer to **www.nwtf.org/jakes/**.
- **National Junior Angus Association (NJAA)**—This organization was formed in 1956 to encourage youth to become active with the Angus cattle breed. State, regional, and junior organizations are active throughout the United States. For more information, refer to **www.njaa.info/about.html/**.
- **Junior American Boer Goat Association (JABGA)**—This organization was formed to promote the raising and keeping of Boer goats by youth. Workshops, shows, and scholarships are among the activities of the JABGA. For more information, refer to **www.jabga.org/**.

4–3. Youth are participating in a JABGA show. (Courtesy, Education Images)

- **National Junior Swine Association (NJSA)**—This organization was formed in 2000 for people under the age of 21 who are interested in Duroc, Hampshire, Landrace, and Yorkshire breeds of hogs. It is sponsored by the National Swine Registry. For more information, refer to **www.nationalswine.com/njsa/njsa.html**.
- **American Junior Paint Horse Association (AjPHA)**—This organization is affiliated with the American Paint Horse Association. It was formed to promote Paint Horses among people under 18 years of age. These individuals are taught leadership, enjoyment of horses, and sportsmanship. For more information, refer to **www.ajpha.com/**.

PURPOSE AND NATURE OF FFA

FFA develops leadership and personal skills and promotes career success. Its various purposes and the manner in which it functions have emerged throughout its existence.

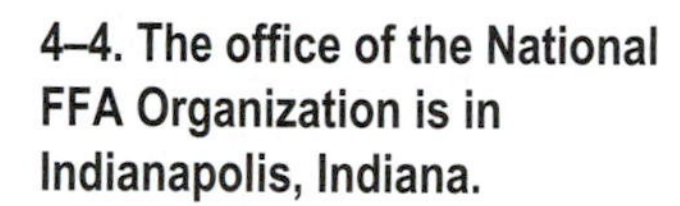

4–4. The office of the National FFA Organization is in Indianapolis, Indiana.

PURPOSE

FFA promotes learning and achievement. Emphasis is on skills for career success and personal development. These skills help build self-confidence, character, citizenship, and healthy lifestyles.

Each year the National FFA Organization produces an ***Official FFA Manual***. This book provides information about FFA. Every FFA member should have access to a manual.

The National FFA has a mission statement. A ***mission*** is the goal or task to be achieved. It reflects the purpose of the organization.

4–5. The *Official FFA Manual* provides information for FFA members. (Courtesy, Education Images)

Mission:

> *FFA makes a positive difference in the lives of students by developing their potential for premier leadership, personal growth, and career success through agricultural education.* (Source: *Official FFA Manual*)

OBJECTIVES OF FFA

The constitution of the National FFA Organization lists four major objectives. These objectives are achieved through cooperation among federal, state, and local educational agencies. The objectives are summarized as follows:

- To be an organized part of instructional programs in agricultural education
- To strengthen the confidence of students in themselves and their work by developing important personal attributes such as the following abilities: to assume responsibility and to communicate effectively
- To provide programs and activities that develop pride, leadership, character, scholarship, citizenship, patriotism, and wise financial management
- To encourage and recognize achievement in many areas, such as supervised experience and leadership development

More details on these objectives are available in Article II of the constitution, which is published in the *Official FFA Manual*.

AgriScience Connection

OFFICIAL FFA DRESS

Official dress is the uniform worn by FFA members. It may be required at FFA functions, such as a speaking contest, or as a delegate at a convention. Official dress provides identity and prestige for members.

The official dress of male and female members varies but both wear the official FFA jacket. Male members wear black slacks, a white shirt, an official FFA tie, black shoes and socks, and an official jacket zipped to the top. Female members wear a black skirt (or black slacks), a white blouse with an official FFA blue scarf, black shoes, and an official FFA jacket zipped to the top. (The FFA jacket is available from the National FFA organization.)

Members should always know the expected dress for FFA activities. Some events require official dress. Failing to wear official dress may result in disqualification. This photo shows an FFA member in official dress. (Courtesy, Education Images)

ACTIVE MEMBERSHIP AND BENEFITS

Internet Topics Search

Selected topics for Internet discovery and reporting are listed here. Begin by searching for each topic. Next, read and learn about it. Conclude by preparing a brief report on your findings.

1. National FFA Organization
2. American Farm Bureau
3. FFA jacket

Active membership is the FFA membership classification held by students (ages 12 to 21). Membership requires that a student be enrolled in an agricultural education program, attend meetings, work toward degree advancement, and pay dues. Other rules are listed in the National FFA Constitution. (See the *Official FFA Manual.*)

Enrollment in agricultural education is essential for membership in FFA. The area of interest in AFNR and the name of the classes being taken do not matter. All students are welcome to be members and are encouraged to advance in membership. Requirements for advancing are included later in this chapter.

FFA members gain many benefits. These vary with the local chapter and the nature of the agriculture classes. Overall, members are motivated and challenged by FFA activities.

Students give differing benefits of FFA membership. Some of the benefits are:

- Being involved in learning and doing
- Experiencing careers (tours, part-time work, etc.)
- Traveling to interesting places (camps, cities, etc.)
- Being recognized for work (cash, plaques, etc.)
- Getting to know new people (beyond your local school)
- Learning how to make good decisions (making choices)
- Being part of a team (relating to other people)
- Being self-confident (believing you can do the job)
- Developing a positive self-concept (feeling good about myself)
- Learning about money management (keeping records and making budgets)

ORGANIZATION DETAILS

- **Motto**—The motto of FFA reflects its purpose in promoting youth in agriculture. The motto is:

    Learning to do,
    Doing to learn,
    Earning to live,
    Living to serve.

- **Colors**—The official FFA colors are corn gold and national blue. These are used on banners, official FFA clothing, and published materials.
- **Emblem**—The FFA emblem has five symbols that depict the FFA history, mission, and future. The emblem is often on banners, jackets, plaques, and other official items.

  The trademark registered with the U.S. Patent Office protects the emblem. A federal law also protects the emblem to ensure that it is only used in official ways. The policies on the use of the emblem are in the *Official FFA Manual*.

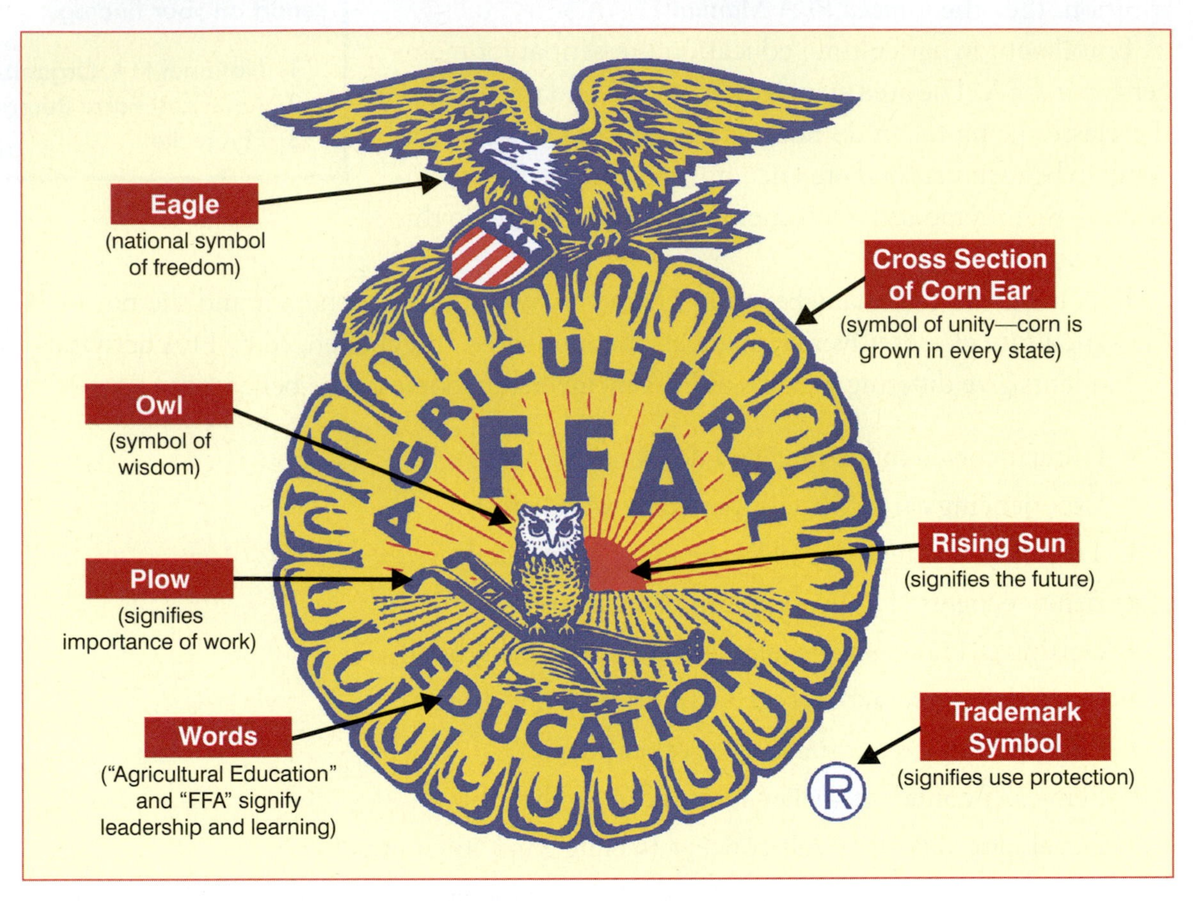

4–6. The FFA emblem and its meaning.

FFA HISTORY

FFA developed with agricultural education. In 1917, the U.S. Congress passed the Smith-Hughes Act to provide funds for agricultural education in public schools. Our leaders saw the need for a program to motivate and develop the personal skills of students.

THE BEGINNING

Henry Groseclose, Harry Sanders, Walter Newman, and Ed McGill of Virginia made the first major move. Their work resulted in the Future Farmers of Virginia club in 1926. Interest in such clubs grew in other states.

In 1928, students from 18 states gathered in Kansas City, Missouri, to form a new organization known as the Future Farmers of America (FFA). The U.S. Office of Education and state leaders provided adult leadership in agricultural education with an emphasis on preparing boys to farm.

Table 4–1. Major Events in the History of FFA

Year	Event
1917	Smith-Hughes Act passed Congress to provide funding to states for local schools to provide instruction in agriculture.
1926	Future Farmers of Virginia formed.
1928	Future Farmers of America established.
1950	U.S. Congress passed Public Law 740 granting a federal charter to FFA.
1952	FFA magazine began as *The National Future Farmer*.
1965	New Farmers of America (NFA) merged with FFA.
1969	Convention delegates voted to allow females as members.
1988	Convention delegates changed official name to National FFA Organization.
1989	Name of *The National Future Farmer* magazine changed to *FFA New Horizons*.
1998	National FFA Center dedicated in Indianapolis, Indiana. Public Law 740 replaced by Public Law 105-225.
2008	Launch of online ***FFA Nation***—a social network for members.

GROWTH

Within a few years after its founding, FFA had expanded to all states. New programs were developed. Membership grew, agricultural needs changed, and FFA adapted to the changes as best it could.

The first FFA office was in the U.S. Office of Education in Washington, D.C. A new building owned by FFA in Alexandria, Virginia, became headquarters at a dedication in 1959. The building was located on property that was a part of George Washington's Mount Vernon estate. In 1998, the main office was moved to Indianapolis, Indiana.

Initially, FFA was only for boys, but delegates at the 1969 National Convention voted to admit girls. This move, along with expanding agriculture classes, resulted in FFA being about more than farming. Interests expanded into a broad range of areas. In 1988, delegates

4–7. The Career Show at the National FFA Convention includes a display of the latest technology for viewing by FFA members. (Courtesy, Education Images)

at the national convention voted to change the name from Future Farmers of America to National FFA Organization.

The national convention is the major event of the year. It was held from 1928 until 1998 in Kansas City, Missouri. In 1999, the convention moved to Louisville, Kentucky. In 2007, the National FFA Convention was held in Indianapolis, Indiana. Every few years, bids are taken from cities that want to host the convention. It is a major event that attracts about 50,000 members and advisors each year.

HOW FFA IS ORGANIZED

FFA is organized into three levels: local, state, and national. State and national activities support local programs.

CHAPTERS

In schools, the local FFA is organized as a **chapter**. Each chapter is under the direction of an advisor. Large schools may have two or more advisors based on the number of agriculture teachers. Chapters qualify students for FFA membership, collect dues, elect officers, use committees, and carry out activities.

Within state guidelines, local school boards set the overall curriculum for schools and hire teachers. Local boards also support FFA by providing funds for the salaries of teachers, for facilities, and for instructional materials. In addition, the boards set policies that influence how a chapter is operated.

4–8. Officers are planning a chapter meeting. (Courtesy, Education Images)

Each chapter has a charter issued by the state-level office of FFA. Chapters carry out their work and report to the state office. Then the state reports to the national office.

ASSOCIATIONS

FFA at the state level is known as an **association**. Most associations are located in a state office of education. Many states have state advisors and executive secretaries. The state advisor is often in charge of curriculum and program leadership in agricultural education. The executive secretary handles details of FFA for the state. Large states with many FFA chapters may organize into districts or regions.

Associations hold conventions, elect officers, sponsor events, and support local chapters. Delegates from the local chapters attend the state conventions. State officers or others serve as delegates to the National FFA Convention, and associations make reports to the National FFA Organization office.

NATIONAL ORGANIZATION

An adult board of directors and the national student officers manage the National FFA. A staff is hired to carry out FFA work. Most staff members have offices at the **National FFA Center**. The facility is in Indianapolis and serves as the center of operations for the National FFA Organization.

Technology Connection

CAREER DEVELOPMENT EVENTS

FFA has a number of career development events (CDEs) that involve applying technology. Though each team member must be skilled, many of these CDEs require working together in small groups or teams to solve problems. Good knowledge of technology, along with skills in using it, is often beneficial in these events.

One example is the meats area. The Meats Evaluation and Technology Career Development Event promotes skill development for careers in the meat animal industry. In addition to taking a written exam, participants evaluate beef carcasses, identify meat cuts, and assess the quality of the meat. This shows students involved in such a CDE. (Courtesy, National FFA Organization)

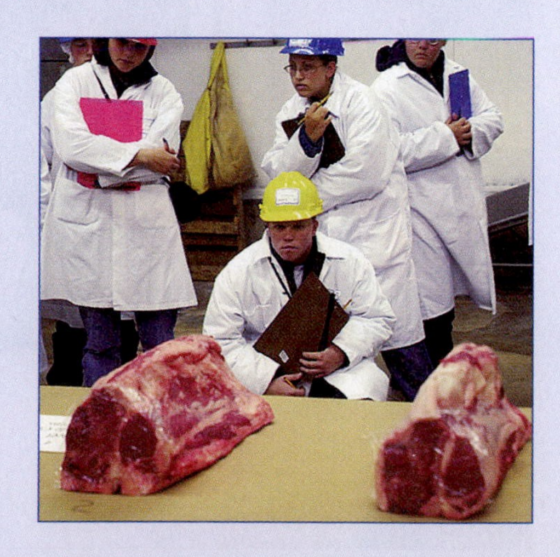

The **National FFA Advisor** is head of agricultural education in the U.S. Department of Education. An assistant serves as national FFA executive secretary. Both of these people have offices in Washington, D.C. A separate National FFA Headquarters office is in Alexandria, Virginia.

Six student officers are elected each year by delegates at the national convention, usually in October. The national officers are a president, secretary, and four vice presidents. New officers begin their terms following the convention. Their roles include managing FFA, meeting with business officials, traveling to states, and planning FFA activities.

The **National FFA Foundation** is the fundraising arm of the National FFA. The funds support FFA activities, such as providing awards for outstanding work to members. The Foundation has a Board of Trustees that oversees its work. Several million dollars are raised annually to support FFA.

The National FFA Archives are located in Indianapolis, Indiana, in the library on the campus of Indiana University Purdue University Indianapolis. They are housed in the special collections area and include a large amount of information on FFA. The Web site for gaining an overview of the contents of the archives is **www.ulib.iupui.edu/special/Ffa/Mss035.html**.

SUCCESSFUL LOCAL FFA CHAPTERS

The success of a chapter depends on the motivation of the students and teachers who are the members and advisors of FFA.

4–9. A committee of local FFA members is planning an activity. (Courtesy, Education Images)

BASICS

4–10. International understanding is a part of FFA activities.

Every student is important in the success of a chapter. All students enrolled in agriculture classes should be FFA members. Further, every student should carry out a planned, supervised experience program that is carefully documented to support FFA advancement.

A few basics in a strong chapter are:

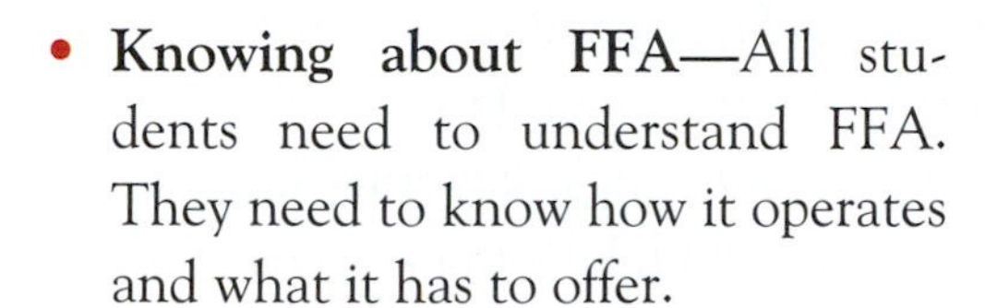

- **Knowing about FFA**—All students need to understand FFA. They need to know how it operates and what it has to offer.
- **Involving all members**—Each member of a chapter should have a worthy role. Committees are a big help in getting people involved. Committees plan their work and are accountable to the chapter.
- **Preparing a program of activities**—A good program of activities (POA) is needed. (More details are provided later in this chapter.)
- **Securing and managing finances**—Many activities require money. A budget is prepared to support the POA. A budget is a written statement of the anticipated income and expenses for a year. Fundraising activities may be used to raise needed money.
- **Electing good officers**—Having good officers is key. Each officer should be dedicated to FFA. Officers should learn their roles and take them seriously.
- **Having the needed equipment**—A chapter should have the meeting room symbols or banners for officer stations. One or more computers with programs for FFA records are needed.
- **Keeping records**—Good records of chapter activities are important. This includes financial records as well as minutes of meetings and member activities.
- **Meeting regularly**—Chapter meetings should be held on a regular schedule. Good programs are important. In some cases, meetings are held after school hours. Most are held during the school day in an activity period.
- **Gaining support**—Support is needed from school officials, parents, agribusinesses, and others. Former FFA members are usually quite interested in supporting the work of a chapter.

OFFICERS

The officers of a chapter form a leadership team. They may be elected near the end of a school year for the next year or at the beginning of the year. The officers include:

- **President**—The president is the lead officer. Duties include planning activities, presiding over meetings, appointing committees, and representing the chapter.
- **Vice president**—The vice president will assume the role of the president, if needed. The duties include leading the development of the program of activities and coordinating committee work.
- **Secretary**—The secretary helps activities go smoothly. Duties include preparing orders of business for meetings, keeping minutes, handling chapter mail, keeping records of meeting attendance, and serving as a strong officer team member.

4–11. A chapter secretary keeps minutes of meetings. (Courtesy, Education Images)

- **Treasurer**—The treasurer is responsible for keeping track of chapter money. This includes receiving, recording, and depositing funds and issuing receipts. A treasurer's book or computer records are kept.
- **Reporter**—This officer provides information to the public about chapter activities. The reporter may work with radio stations, newspapers, television stations, and cable channels.
- **Sentinel**—The sentinel is in charge of the meeting room. Duties include arranging officer symbols and other FFA materials as well as keeping the room neat and orderly.
- **Advisor**—The advisor is the adult teacher designated by the school to teach agriculture and oversee FFA.
- **Others**—In addition to those officers listed in the constitution, chapters may have a historian, parliamentarian, and chaplain.

MEETINGS AND CEREMONIES

Good meetings are important for leadership development as well as chapter development. Meetings provide opportunities to inform members, to plan activities, and to recognize members for their work. Banquets, assembly programs, and parents' nights are examples of special meetings. Some events are social activities, such as parties, trips, hikes, and campouts.

The meeting room should be comfortable and orderly. Officer stations should be set up, and each officer should be at the appropriate station.

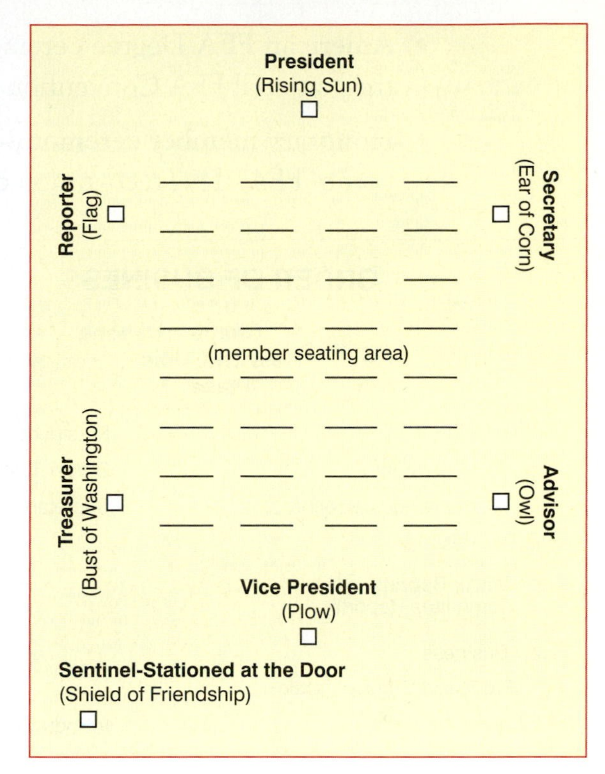

4–12. Sample room arrangement showing officer stations.

Ceremonies

A ***ceremony*** is a formal way to observe an accomplishment or carry out chapter activities. It increases the benefit and prestige of an event. Definite procedures are followed.

The people who are speaking at a ceremony should memorize and practice saying their parts loudly and clearly so everyone can hear. If the part has not been memorized, it should be read because reading it is preferable to delivering the part poorly. The words should be pronounced carefully, and appropriate FFA clothing should be worn. In addition, the speaker should stand using good posture. A poorly given ceremony detracts from the value of the recognition.

FFA ceremonies printed in the *Official FFA Manual* include:

- **Opening and closing ceremony—**This ceremony is used to begin and end chapter meetings. Most officers have parts in the ceremony that give some background on the nature of their duties.
- **Discovery Degree ceremony—**This ceremony is used to recognize students who are receiving the Discovery Degree.
- **Greenhand FFA Degree ceremony—**This ceremony is used to induct new members into FFA.
- **Chapter FFA Degree ceremony—**This ceremony is used to elevate Greenhands. It recognizes the accomplishments of those people who are receiving the Chapter FFA Degree.
- **State FFA Degree ceremony—**Typically used at the state FFA convention, this ceremony recognizes members who are receiving the State FFA Degree.

- **American FFA Degree ceremony**—The National FFA officers use this ceremony at the National FFA Convention to award the American FFA Degree.
- **Honorary member ceremony**—FFA members may honor nonmembers for their support of FFA. This recognition ceremony is used at the local level.
- **Ceremony for installing officers**—This ceremony is used to install new chapter officers.

ORDER OF BUSINESS

_______ Chapter Meeting
Date and Time
Place

Call to Order	President
Opening Ceremony	Officer Team
Minutes of Previous Meeting	Secretary
Old Business	
Officer Reports Committee Reports	
New Business	
Proposed Chapter Picnic	
Program	Introduced by President
"Agriculture in Our State" State Senator Bernardino Servier	Speaker
Announcements	
Closing Ceremony	Officer Team
Refreshments	

4–13. A sample order of business.

4–14. President presiding at a chapter meeting. (Courtesy, Education Images)

Meetings

FFA members have a variety of meetings. The most common is the local chapter meeting. The meeting may be during school or after school hours.

Officers should have an ***order of business***, which is a list of what is to be done in a meeting. A good order of business sequences items and ensures that nothing is overlooked. All officers should have a copy of the order of business and follow it. Members may receive copies. At a banquet, a printed program may be used for all members and guests.

All meetings should have clear goals of what is to be accomplished. Committee chairs should be ready to give succinct reports. The president should preside effectively. Guest speakers should be used to present informative programs. Speakers are selected based on their ability to relate to members. They should know details well in advance of a meeting. Details for speakers include when and where a meeting is being held and who will be attending. They should have a time limit for their presentation. After a program, always follow up with a written "thank you" to a presenter.

PROGRAM OF ACTIVITIES

A ***program of activities*** (POA) is an annual plan of the goals and procedures for an FFA chapter. All members should have a role in preparing a POA because all are involved in carrying it out.

A POA guides the work of an FFA chapter and may have three major divisions: student development, chapter development, and community development. Each division has committees that develop goals, ways of achieving the goals, and deadlines for doing so. Committees carry out the activities.

ACADEMIC CONNECTION

SPEECH

Relate your studies in speech class to participation in an FFA public speaking event. Apply what you learn in FFA events to enhance your speaking skills. Plan and deliver an FFA speech that is appropriate for your speech class.

Committees

A ***committee*** is a small group with specific duties. One member serves as the chair or organizer of the members. Other roles may be specified, such as secretary or vice chair. The number of committees and their duties vary with the number of members in the FFA chapter and the goals to be achieved.

Two kinds of committees may be used: standing and ad hoc. A ***standing committee*** has ongoing duties. Member terms are a year or more. Committees in POAs are usually standing committees. An ***ad hoc committee*** is appointed to carry out a specific duty.

4–15. An AgriEntrepreneurship winner finishing a display at the National Convention. (Courtesy, Education Images)

The committee is dissolved once it has completed its work. For example, ad hoc committees decorate the meeting room for a banquet, manage a fundraiser, or organize a chapter cookout.

Committees usually have names and functions—the banquet committee plans and handles the details of a banquet. The number of members varies, but groups of three to nine are best. Large committees are hard to manage, and small committees may not have enough people to do the work.

Effective committees are essential. Members share the work. They need to know the purpose and duties of the committee. Time must be available to meet.

Table 4–2. Standing Committees in an FFA Chapter

Division in POA	Committees and Brief Nature of Duties
Student Development	• Leadership—Promotes leadership development among members. • Healthy Lifestyles—Promote making positive choices that lead to good health and well-being and to the prevention of substance abuse. • Supervised Experience (or Supervised Agricultural Experience)—Promotes experiences that lead to education and FFA degree advancement. • Scholarship—Promotes good performance in school work among members and encourages lifelong learning. • Career Skills (or Agricultural Career Skills)—Promote information about careers among members.
Chapter Development	• Member Recruitment—Informs other students about agricultural education and FFA and promotes enrollment and membership. • Finance—Promotes good personal financial management among members. • Public Relations/Publicity—Provides information to local media and helps ensure a positive image among the general public. • Leadership—Promotes leadership development of members and all students; teamwork and cooperation are important. • Support—Coordinates with other groups, such as parents, alumni, and businesses, to gain support for the chapter.
Community Development	• Economic Development—Investigates and promotes businesses (including farms and entrepreneurship) in the local area. • Environment—Promotes maintaining a good environment and sustaining natural resources. • Human Resources—Promotes the development of human potential among members and citizens. • Citizenship—Promotes citizenship traits, including patriotism and involvement, among members and citizens. • Agricultural Awareness—Helps keep members and citizens informed about agriculture and related issues.

Adapted from: FFA Student Handbook; see the current *Official FFA Manual* for details.

Format

A good format for a POA helps. Several formats can be used. Most chapters prefer one that lists the goals, actions to achieve the goals, deadlines, fund allocations, and outcomes (results of the effort). Work is delegated to a committee and/or member(s).

PROGRAM OF ACTIVITIES

________________ FFA Chapter

________ Year

Division ____________ Committee Responsible ________________________

Goals	Steps/Actions	Date to Complete	Budget	Results

4–16. Example of a format for a program of activities. (A separate form is used for each committee. The number of lines on the form can be modified as needed.)

LIFE KNOWLEDGE

The National FFA Organization is moving forward with an initiative known as Life Knowledge. Classroom instruction and practice are used in developing essential leadership and career success skills. Life Knowledge includes learning in the areas of communication, teamwork, personal development, and leadership. Another focus is to prepare individuals for providing service to agriculture, careers, and communities. The name Life Knowledge connotes that the initiative is concerned with “real lessons for real life.”

FFA AWARDS AND EVENTS

FFA has many activities for its members. Some activities include awards that recognize hard work. Other activities involve travel, getting to know other students, and participating on teams. All are fun!

AWARDS

FFA has several kinds of awards. Some awards are for individuals, but other awards are for teams or the entire chapter. All are based on achievement and member performance.

Proficiency Awards

Students receive proficiency awards for their accomplishments in developing skills. Class and laboratory achievements are important, and supervised experience has a major role. Most proficiency awards are based on supervised experience.

Proficiency awards begin in the local chapter. The individual with the top award locally competes at the next level with members from other schools. A state winner competes nationally with other state winners. One individual is named the national proficiency award winner each year.

4–17. A member receiving an award on behalf of the local FFA chapter. (Courtesy, Education Images)

The National FFA Organization has about 50 proficiency awards. New awards are often added, and older ones may be changed to better serve the needs of members. The *Official FFA Manual* and the National FFA Web site (**www.ffa.org**) list the awards given each year. Leading businesses in the United States sponsor most of the proficiency awards.

Chapter

Chapter awards recognize local chapters that make outstanding accomplishments. The committee structure and the development of a program of activities are essential. The National FFA Organization publishes details on the chapter award program.

CAREER DEVELOPMENT EVENTS

Career development events (CDEs) are competitive activities that measure individuals and teams in the application of classroom-acquired knowledge. All CDEs are based on the agriculture curriculum. Events begin at the local level with intrachapter competition and team selection. Chapter teams advance through district, area, and state competition, with the winners traveling to the national finals.

Some examples of CDEs in which students can compete are agricultural mechanics, floriculture, forestry, horse evaluation, and poultry evaluation. In most CDEs, basic knowledge, a skill demonstration, and problem-solving ability are measured. For example, the Agron-

4–18. Members of a floriculture team are proud of their accomplishment: placing in the national FFA career development event. (Courtesy, Education Images)

omy CDE involves a 50-question written exam, seed and insect identification, and the development of a solution for a proposed problem.

The National FFA Organization publishes a list of CDEs in the *Official FFA Manual*. Details are available in bulletins from the National FFA Organization and at its Web site: **www.ffa.org**.

LEADERSHIP DEVELOPMENT EVENTS

Leadership development events (LDEs) are the competitive activities that measure the personal skills of FFA members in public speaking, communicating, and relating with people. State associations may design many of these events. Examples that lead to national competition include parliamentary procedure, prepared public speaking, creed speaking, a job interview, and an issues forum. Some LDEs are listed in the CDE information from the National FFA Organization.

OTHERS

Opportunities are great in FFA. No single listing is ever fully complete. Here are a few others:

- Star in Agribusiness Award
- Star Farmer Award
- National FFA Science Fair

- Star in Agriscience Award
- Scholarships
- Agri-Entrepreneurship Program
- Commodity Marketing Activity
- Food for America
- Partners in Active Learning Support (PALS)
- Partners for a Safer Community
- Washington Leadership Conference (WLC)
- Made for Excellence (MFE)
- International Programs

MEMBERSHIP DEGREES AND REQUIREMENTS

Within the active membership category are five levels (based on individual achievement) known as ***membership degrees***. Complete degree requirements are noted in the *Official FFA Manual*. The five degrees of active membership are:

4–19. A member receiving an FFA degree pin.

- **Discovery FFA Degree**—Students in grades 7 and 8 who are enrolled in an agriculture class may hold the Discovery FFA Degree. Knowing about an agricultural career and the local chapter's program of activities is also required.
- **Greenhand FFA Degree**—The Greenhand FFA Degree is for students entering high school. They must be enrolled in agriculture classes. Greenhands, as these members are known, must learn about FFA and have access to an *Official FFA Manual*.
- **Chapter FFA Degree**—The Chapter FFA Degree is for Greenhands who have completed 180 hours of agriculture instruction and have

supervised experience underway. To receive the degree, a member must have earned at least $150 or worked 45 hours after class time.

- **State FFA Degree**—To be eligible for this degree, two years of FFA membership and completion of at least two years of agriculture instruction are required. A minimum of $1,000 must have been earned or 300 hours should have been worked. Good records are needed. Only a small number of FFA members receive this degree.
- **American FFA Degree**—This is the highest degree. It is awarded at the National FFA Convention to a very few members. This degree is for students who have been out of high school at least a year before the convention at which it is awarded. A student must have been a member of FFA for at least 36 months and have completed at least three years of agriculture instruction. Earnings must be at least $7,500 or a combination of hours worked and invested money, as described in the *Official FFA Manual*.

In addition to active membership, the National FFA Organization has:

- **Alumni membership**—***Alumni membership*** is open to former members and others interested in FFA. Alumni dues are paid.
- **Collegiate membership**—Collegiate members are students in college pursuing degrees in agriculture, often to become agriculture teachers. Collegiate chapters are chartered much as local chapters.
- **Honorary membership**—People who have helped a chapter and its members may be honorary members. A majority vote of the chapter members is needed. There are no dues for honorary members.

REVIEWING

MAIN IDEAS

Several organizations are available for youth and students with interests in agriculture. These may be through the local school or the agricultural industry. FFA is the organization for students enrolled in agriculture classes in secondary school.

FFA promotes learning and achievement. Students are recognized for their accomplishments. The National FFA Organization has a formal structure of local, state, and national activities. The office of the National FFA Organization is in Indianapolis, Indiana.

Active FFA membership involves being a student in agriculture and paying dues. New members hold the Greenhand FFA Degree. All members strive to achieve chapter, state, and American FFA degrees. FFA has a motto, emblem, and official colors. All members are required to wear the FFA jacket or other appropriate attire at FFA events.

FFA activities have expanded with agricultural education. Career development events, leadership development events, and other programs and awards recognize the work of members.

Most members are involved in local chapters. Successful chapters help members develop their leadership, career, and personal skills. Chapter officers are important in the success of a chapter. Officers work with committees to establish and carry out a program of activities.

QUESTIONS

Answer the following questions, using complete sentences and correct spelling.

1. What is the purpose of FFA?
2. Who can be a member of FFA?
3. What are the degrees of active FFA membership? Explain each.
4. Why is the *Official FFA Manual* important to members?
5. Briefly trace the history of FFA in a few paragraphs.
6. What makes a strong FFA chapter? List any three basic features, and explain why they are important.
7. What officers are typically elected by a local chapter?
8. What is a program of activities? Why is a good program of activities important to a local chapter?
9. What are career development events?
10. What are the general requirements for the State FFA Degree?

EVALUATING

Match each term with its correct definition.

a. FFA advisor
b. standing committee
c. active membership
d. chapter
e. association
f. National FFA Foundation
g. National FFA Center
h. order of business
i. ceremony
j. committee

______1. A formal way of carrying out chapter activities or recognizing accomplishments.

______2. The agriculture teacher who oversees the FFA chapter.

______3. FFA organization at the local level.

______4. Located in Indianapolis, Indiana.

______5. The fund-raising arm of the National FFA.

______6. State-level FFA organization.

______7. A committee with ongoing duties.

______8. Membership classification for dues-paying students enrolled in agriculture classes.

______9. A small group with specific duties.

_____10. A written list of what is to be accomplished in a meeting.

EXPLORING

1. Use the Internet and explore the Web site for the National FFA Organization at **www.ffa.org**. Select an area of greatest interest to explore in detail. Prepare a report on your findings.

2. Review the constitution and bylaws of the National FFA Organization in the *Official FFA Manual*. Read the four objectives of FFA. Determine which objective you feel is most important. Prepare a short paper on your selection, and defend your choice.

3. Over the next year, arrange to attend FFA events. Record your participation in your supervised experience records. The events you will want to attend include chapter meetings, committee meetings, events beyond the local chapter, and the state FFA convention.

4. Investigate organizations other than FFA that may serve students or junior members. Several examples are included in the chapter.

CHAPTER 5

Leadership Development

OBJECTIVES

This chapter explains the meaning, importance, and development of leadership skills in agricultural education. It has the following objectives:

1. Identify important leadership qualities.
2. Discuss the role of communication in leadership.
3. Demonstrate speaking skills.
4. Demonstrate parliamentary procedure skills.
5. Demonstrate meeting organization and management skills.

TERMS

communication
extemporaneous speech
eye contact
impromptu speech
incidental motion
leadership
main motion
meeting
minutes
motivational speech
nonverbal communication
nonverbal cue
order of business
parliamentary procedure
prepared speech
presider
privileged motion
public speaking
quorum
recitation
subsidiary motion
verbal communication
voting

5–1. The ability to speak before a group is important for success in some careers.

WHY ARE some people able to get other people to do what needs to be done? Is it because they are bossy or domineering? Or, is it because they know how to get along with people and help others share in the work that is to be done? Anyway, these questions are about leaders and leadership.

Leadership is a relationship among people in which influence is used to meet individual or group goals. It is the ability to guide other people to achieve a desired outcome. This often involves organizing people, materials, and activities to accomplish goals.

Personal growth is usually a major part of leadership development. Leadership is far more than the act of "taking charge." It involves important ideas and values. Through your studies in agricultural education and FFA you will have the opportunity to develop leadership qualities.

LEADERSHIP QUALITIES

Both leaders and followers must be present for leadership to occur. A leader is a person who helps other individuals or groups achieve goals. The leader of a group is not always in an elected position. The leader role emerges by virtue of the individual's personal qualities. A follower is an individual who conforms to or accepts the ideas of leaders. They strive to assure that the goals of the leader are achieved. No one can be a leader without followers.

Leaders must have certain traits if they are to fulfill the role of leadership. Leadership traits can be taught and learned. They are useful in many different situations.

Some people are better at leadership traits than others. Understanding personal strengths and weaknesses is important for leaders. So is recognizing and valuing the talents and abilities of others to make the group strong.

Leadership traits and characteristics can be separated into four categories. A good balance of internal, technical, conceptual, and interpersonal traits makes it easier to be effective.

CAREER PROFILE

AGRICULTURAL SALESPERSON

An agricultural salesperson is responsible for selling agricultural supplies, services, or products. Selling is the process of promoting the sale of goods and services. Some selling involves considerable communication with the buyer. An example is a farm tractor salesperson. Other selling skills involve greeting customers, entering orders, taking payment, and thanking the customer. An example is the cashier at the check-out counter in a garden center.

Agricultural sales people need to have good communication skills. Some need to be able to assess customer needs and interests and match these with products. All need to be able to greet customers in a friendly, professional manner. Depending on what is sold, all salespersons need high school diplomas. Others need some college or a college degree in agribusiness or related area. The work is found where agricultural supplies and services are sold. More often, farming areas have more opportunities.

This photograph shows high school students in a floral class role playing making sales transactions. Role playing helps develop abilities for job performance. (Courtesy, Education Images)

INTERNAL TRAITS

An internal trait is a personal characteristic. Several internal traits are important. They are likely the most difficult traits to develop but are the most important. People need to have well-defined internal traits if they expect others to follow their lead. Several internal traits of effective leaders are:

- Know personal strengths and weaknesses
- Be creative in mind and action
- Be driven—internally motivated
- Be perceptive—pay attention to your environment
- Be moral
- Be hard working
- Be self-confident
- Be flexible
- Be responsible

TECHNICAL TRAITS

A technical trait is a "how to do it" skill. These traits are the easiest leadership qualities to master. Some technical traits are directly related to a specific task. For example, proficiency in mechanics skills is essential for the manager of an equipment repair shop. Other technical traits are more general and apply to many situations. Here is a list of some important technical traits:

- Follow directions
- Conduct a meeting—presiding skills
- Speak effectively in front of groups
- Lead discussions
- Organize events
- Communicate effectively

5–2. Participation in FFA events promotes leadership and human relations development. (Courtesy, Education Images)

CONCEPTUAL TRAITS

A conceptual trait is a "thinking" skill. Such traits may be a little more difficult to learn than technical traits. They are important abilities. Leaders need to be able to use their minds to evaluate situations and come up with new ideas. Several conceptual traits are:

- Think logically
- Analyze
- Anticipate problems
- See changes
- Recognize opportunities
- Combine ideas
- Solve group and individual problems
- Make good decisions

INTERPERSONAL TRAITS

An interpersonal trait is a skill that promotes getting along with people. These skills help leaders to work with others and develop good feelings within a group. Most of these skills are not difficult to learn. Many of them come from simply liking people and wanting to help them. Several interpersonal traits are:

5–3. This group has five people. Who is the leader? Why do you think so?

- Be trustworthy in all dealings
- Meet and greet people with ease
- Communicate clearly and effectively
- Listen effectively
- Have a positive attitude
- Accept other views even widely different from your own
- Understand the needs of others and their situations
- Respect people of other backgrounds and cultures

COMMUNICATIONS

Communication is the process of exchanging information. If information is not accurately exchanged, the process is not working properly. Fortunately, people can take steps to assure improved communication.

The communication process involves each of us every day. It is central to all social behavior. Humans cannot socially interact unless they communicate through the sharing of symbols. We do this when we talk with family, friends, and business associates. We write and read letters, send and receive e-mail and text messages, talk on the telephone, read the newspaper, and watch television to exchange information. Sometimes we give speeches to classmates, to adult groups, and at conventions.

VERBAL AND NONVERBAL

Communication may be verbal or nonverbal. Understanding these factors can promote your ability to communicate. They impact the messages you send as well as how you interpret the messages you receive.

Verbal

Verbal communication is communication that uses words to express ideas. The words may be spoken or written. They may be supported by nonverbal information. We must use care in selecting the words we use and how we use them.

5–4. With a small group, the use of a newsprint pad enhances the spoken word and organizes information. (Courtesy, Education Images)

Leaders need good communication skills. Without communication, there is no leadership. It is possible, however, to communicate without leading. In leadership roles, it is important that you communicate exactly what you want to communicate. Since people communicate daily, they assume that they communicate clearly and effectively. That is not always so. What is the communication process? How does it function?

5–5. No words are being spoken. What meaning is being communicated?

Nonverbal

Nonverbal communication is the exchanging of information without the use of words. As with other types of communication, nonverbal communication requires a sender, a receiver, a message, and a medium. Unlike other forms of communication, however, with nonverbal communication, the sender may not be aware that a message is being sent.

Have you ever had someone ask you "Have you had a bad day?" without your ever having said a word? Chances are, you were sending information about your emotional state through nonverbal cues.

A ***nonverbal cue*** is a signal we use to tell others our emotional state, our attitudes, and about ourselves in other ways. We also use them to give others feedback. When someone tells you a joke and you laugh, you are providing that person with feedback. Through the nonverbal cues of grins and laughter, you are letting the joke teller know your attitude about the joke. Of course, nonverbal cues vary with the culture, and cultural understanding is needed.

Specific cues in nonverbal communication include the following:

- **Eye contact**—This lets the individual with whom you are communicating know that you are paying attention. Eye contact also indicates self-confidence and honesty.
- **Facial expression**—This refers to how we use our eyes and face muscles to convey meaning. Examples include smiles, frowns, raised eyebrows, and wrinkled forehead.
- **Distance**—In a personal conversation, the distance between people is a factor in communication. A comfortable distance is normally at least 2 to 4 feet.
- **Tone of voice**—Several speech traits are included here. Loudness, pitch, and speaking rate are examples of factors in tone of voice.
- **Personal appearance**—Grooming, clothing, and other factors are part of personal appearance. These affect communication ability and how people feel about you.

- **Body movement**—Gestures, head shaking, finger pointing, and other movements of the body send signals to other people. Failing to use good posture and stand straight on both feet also send messages. Propping up is also a strong communicator.
- **Others**—Laughing, coughing, yawning, and clearing the throat are other nonverbal cues that impact communication.

THE PROCESS

The communication process has several parts that must function for communication to occur. In addition, the model includes a method to decide how effectively the information was exchanged.

- **Source**—The source is the sender or initiator of the attempt to communicate. Every effort in communication begins with a source. The sender must consider the receiver and choose appropriate means to assure that the receiver draws proper meaning from the message.
- **Message**—The message is the idea or information that is to be exchanged. The sender must prepare the message in a form for sending. This is known as encoding. It involves using some kind of code system, such as spoken language, written letters that form

AgriScience Connection

SPEAKING IN YOUR CAREER

Many careers require individuals to give public speeches. The speeches might be before small groups or large audiences. The purpose of speaking is often to inform people about a product or process or to motivate people to take an action, such as buy your product.

A key to success with speaking is planning and organizing what you are going to say and do. You will need to develop your topic and the appropriate support materials. You may also need to arrange a meeting room and for someone to take minutes. You will want to rehearse what you are saying and decide on the use of visuals.

This shows a person speaking before a large group. The dress and presence of the individual communicates competence, professionalism, and good communication.

words, drawings, photographs, gestures, or movements. The sender must select a system of codes that has meaning for the intended receiver of the message.

- **Channel**—The channel is the linkage between the sender and the individual for whom the message is intended. The kind of channel selected depends, to some extent, on the coding system. For example, speaking to an individual involves verbal communication in which the sender uses language that the receiver can interpret.
- **Receiver**—The receiver is the individual for whom the message is intended. The encoded message is decoded by the receiver in order to draw meaning from the symbols used by the sender of the message. Communication has occurred when the receiver gets the same meaning from the message as the sender intended.

5–6. Sometimes our concentration strays from the subject of a meeting—a blockage. We may not concentrate or yawn.

Blockages

Blockages sometimes occur. A communication blockage is interference with the communication process. Blockages are due to language, physical impairments, sounds, or other factors that prevent the exchange of information. A blockage in the communication process is known as noise. Noise includes sounds that interfere with oral communication and any other factors that cause the communication process to fail.

Anything that distracts the receiver is noise. Sometimes, not knowing a language is a blockage. An example would be trying to order from a menu in a French restaurant without knowing the language.

Visual communication blockages are also noise. Loud clothing, excessive jewelry, pictures, animals, and any visual object that distracts the receiver are noise. The sender should try to keep these away from a communication setting.

Perceptions

A perception is how people view or feel about something. It is based on what they have learned. Perceptions play powerful roles in how a message is encoded and decoded. A sender should select codes and channels based on how he or she perceives the information and the receiver. After receiving the message, the receiver uses experiences, values, and knowledge to decode the information. An individual's language, culture, education, and other attributes influence how a message is decoded.

PUBLIC SPEAKING SKILLS

Public speaking is a type of communication that uses oral methods of conveying information. Spoken words are enhanced with nonverbal symbols, such as gestures and visuals. Good preparation is needed to be an effective public speaker.

People who give speeches need to understand the communication process. They need to use the process in preparing and making their speeches. A good speech is focused on the audience—the listeners. The audience interprets the information. If members of the audience do not get the correct meaning from the speech, the speaker has not been successful.

Career success is often tied to speaking ability. People who can effectively use oral communication skills are more likely to advance up the career ladder. They are also more likely to hold positions of leadership and earn higher salaries.

KINDS OF SPEECHES

Public speaking is used to inform, motivate, and entertain. Being able to give a good speech is important for several reasons. It allows an individual the opportunity to emphasize strong ideas and ideals. What better way to express yourself publicly than through an organized and meaningful speech? Public speaking skills can be used as you give a prepared speech, oral report, illustrated talk, or demonstration, as you introduce a guest, or as you participate in a panel discussion.

Four kinds of speeches are commonly used: prepared speeches, extemporaneous speeches, impromptu speeches, and recitations.

Prepared Speeches

A ***prepared speech*** is a speech that is developed well ahead of time. Materials are gathered, and the speech is outlined, written, and practiced. With a prepared speech, knowing your purpose is important.

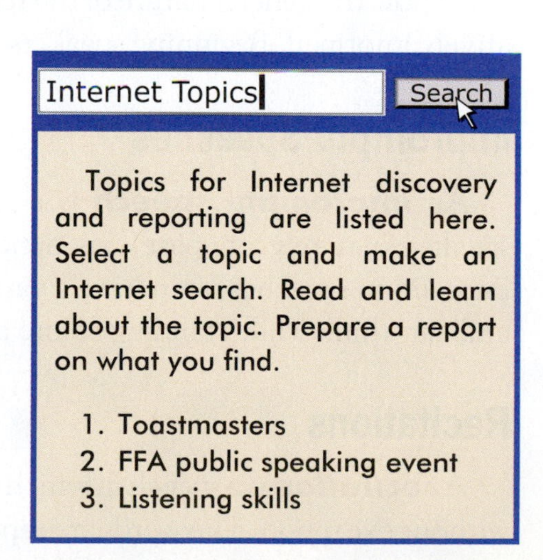

Topics for Internet discovery and reporting are listed here. Select a topic and make an Internet search. Read and learn about the topic. Prepare a report on what you find.

1. Toastmasters
2. FFA public speaking event
3. Listening skills

5–7. A speech can be outlined and written using computer word processing. (Courtesy, Education Images)

A popular kind of prepared speech is the informative speech, or oral technical report. An oral technical report is an informative speech that provides details of scientific or technical processes. Oral technical reports are often given at agricultural and scientific meetings. They demand careful attention to the accuracy of the information. Techniques are used to help the audience understand the information.

Another kind of prepared speech is the motivational speech. A ***motivational speech*** is a speech used to arouse people and encourage them to take a certain action. Speeches of this type are used to help people feel good about themselves and the world in which they live. Motivational speeches often use psychology in helping achieve their goals.

Some speeches are to entertain. Speeches that entertain often contain motivational or technical elements. These speeches are designed to help members of the audience relax and enjoy a few laughs.

Extemporaneous Speeches

An ***extemporaneous speech*** is a speech that is not prepared in detail in advance. Notes may be used. Limited time may be available to prepare the speech. This requires that the speaker be able to speak effectively without extensive prior preparation. The speaker may know the general nature of the topic and be able to gather information and become relatively informed. Beginning speakers are advised to give prepared speeches.

Impromptu Speeches

An ***impromptu speech*** is a speech that involves no preparation. Of course, the speaker can rely on past experiences and knowledge of a subject. People who give impromptu speeches often have a vast history of speaking in public. They know the subject and can think while standing before an audience.

Recitations

A ***recitation*** is a speech in which the words of a prepared statement are fully memorized without exception. The words are repeated from memory. The material is often prepared by someone other than the individual who recites it. Recitations may be used with poetry, ceremonies, and creeds. The National FFA Creed Speaking Career Development Event is an example.

5–8. Practicing helps develop public speaking skills. (Courtesy, Education Images)

BEING A GOOD PUBLIC SPEAKER

A good public speaker knows what to do and how to do it. Most good speakers follow basic guidelines. The following guidelines will help you in developing public speaking skills:

- **Be prepared**—Being prepared is essential. Plan your speech well in advance. Always give time to preparing, regardless of the audience, the subject, or the length of time for the speech. Good preparation helps you do a better job.
- **Be organized**—Being organized means that you plan and have all details in good order. Write an outline, and follow three simple steps:
  - Tell your listeners what you are going to tell them.
  - Tell them the message.
  - Summarize what you told them.
- **Use good voice**—Your audience needs to be able to hear you and understand what you are saying. Speak loud enough for all areas of a room to hear you. Choose your words based on your audience—what will they understand? Pronounce the words carefully. Vary your voice as it will help keep the audience's attention. You can use your voice to emphasize certain points. Varying your voice makes giving a public speech more fun.
- **Stay within the time limit**—Watch the clock! Adhering to the allotted time is extremely important. If a period of 10 minutes is allowed, plan your speech for that amount of time. When speaking, be sure to stay within the 10 minutes. It is important to say what needs to be said and not more. Extra small talk will only lose the attention of your audience.
- **Use a good introduction**—An attention-getting introduction alerts the listeners that you have something important to say. Use an introduction that is lively and one that

will catch the attention of your audience. You may want to use an appropriate anecdote. Humor in the right form and place is always appreciated by an audience.

- **Be enthusiastic**—Present yourself in an enthusiastic manner. You will hold your audience's attention longer. Enthusiasm drowns out nervousness.
- **Use proper facial expressions**—Your facial expression has an influence on your speech. Facial expressions can be used to emphasize certain points. A concerned expression can reinforce a serious point of your speech. Think positively and be positive; success will surely follow. Smiles always help the audience feel good about listening to you.
- **Use notes**—Keep important notes and the outline of your speech close by (usually on a speaker's stand). This will help you keep your train of thought on track. It will assure that you can refer to important facts without fumbling. You will find that the more you speak, the less you need to rely on notes and the outline.
- **Maintain eye contact**—***Eye contact*** is the technique of looking at the audience and into the faces of its members. It is important in public speaking. Eye contact helps get and keep the attention of the audience. It conveys a message of "I am talking with you and not at you." Look at all areas of the audience, but do not stare at any one individual.
- **Observe other speakers**—You can learn a great deal by watching and listening to other speakers. Pay special attention to their method of delivery.
- **Be knowledgeable**—Know your subject. Spend time reading and studying the subject. Have answers ready for typical questions. If you do not know the answer to a question, say so.

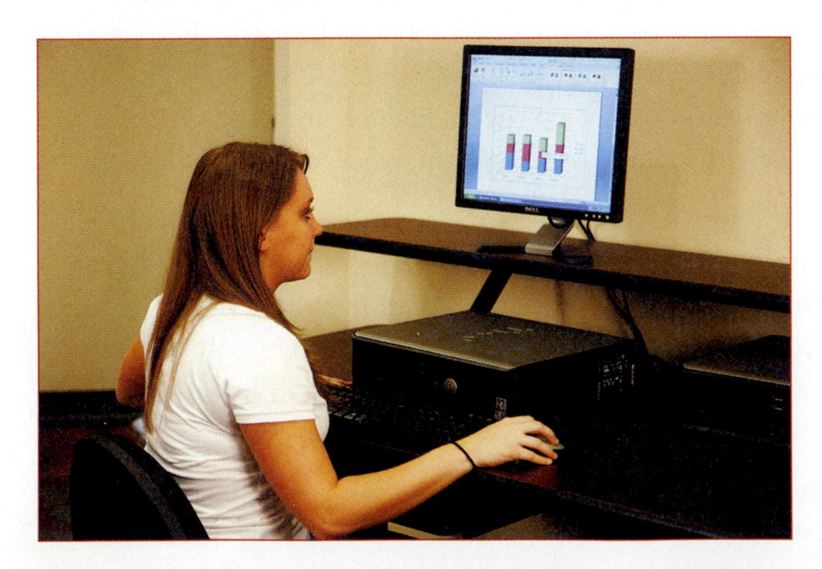

5–9. Preparing a bar graph to use in illustrating a speech. (Courtesy, Education Images)

PREPARING A SPEECH

Preparation is the key to becoming a good public speaker. Much time is needed to prepare—one or more hours for each minute you speak. Preparation includes developing the topic. In some cases, you can choose it. In other cases, the topic may be assigned to you. Develop the topic appropriate to the audience, including age, education, career, and socioeconomic status.

Gathering information is essential in preparing a good speech. Use books, magazines, the Internet, and other sources. Be sure that the information you use is accurate. With a written speech, you can have an authority on the topic review what you plan to include.

Once you have a topic and information, you will need to prepare an outline. It is the framework for your speech. The outline for a speech normally has three major parts: introduction, body, and ending/conclusion. The introduction should get the attention and interest of the audience. The body is the main part of the speech that gives the information you are covering. The ending/conclusion summarizes the information and motivates the audience to take a needed action.

Title: Recycling
Purpose: To convey the importance of recycling and encourage people to recycle

I. Introduction (importance of recycling)

II. Body
   A. First Main Point (everyone can recycle)
      1. Subpoint number one (at home)
      2. Subpoint number two (at business)
   B. Second Main Point (Items to recycle)
      1. Subpoint number one (when)
      2. Subpoint number two (where)
      3. Subpoint number three (how)

III. Conclusion (recycling needs to be done...)
   A. Summary of main points
   B. Action to take

5–10. Example of a simple outline for a short speech on recycling.

After the outline is ready, develop the speech. Gather and write the information to include for the introduction, body, and ending/conclusion. Using computer word processing is a big help. With some kinds of speeches you can include visuals (this depends on the ability to use visuals where your speech is given). Computer-based visuals such as PowerPoint are quite popular today. Good visuals are essential, as poor quality visuals detract from your speech.

DELIVERING A SPEECH

Practice giving your speech ahead of time. Determine if you will be using a podium or table. In some cases, a microphone and public address system will be used. A part of practice is getting your voice adjusted for the place where the speech will be given. Stand on both feet and use gestures in your practice. Look at the audience or the direction your audience will be. Video your practices and review them to identify ways you can improve. Get an authority on speaking to observe your practice and provide pointers on how to improve.

Here are some pointers on delivery:

- **Sincerity**—Be sincere in relating to your listeners. Reflect on who they are and their interests. Be positive about their attributes, their experiences, and the reasons they are present. Use a conversational approach.
- **Voice**—Speak with a voice that others can hear and understand. Pronounce words clearly. Word sentences so that they are grammatically correct. Avoid shouting. Vary voice volume and rate of speaking to help create and sustain interest.
- **Posture**—Stand erect, with hands at side. Avoid such nervous actions as shifting from one foot to the other and moving hands excessively. The way you handle your body is often as important as the words you speak.

5–11. Practice using gestures in rehearsing prepared speeches. (Courtesy, Education Images)

- **Eye contact**—Eye contact with the audience is very important. Look at the audience in general. Do not pick out one or two people and look at them all the time. Let your eyes move about the entire audience. Do not excessively look at notes, the podium, or other places.
- **Time**—Begin and end your speech on time. If there is a large clock in the room, occasionally glance at it to see how time is moving along and adjust your speech accordingly. Otherwise, place a watch on the podium or, in the case of a planned speech, note the actions of the timekeeper.

PARLIAMENTARY PROCEDURE SKILLS

Parliamentary procedure is a method of conducting meetings in an orderly manner. It is based on the rules first set by the British Parliament hundreds of years ago. All individuals are given the opportunity to participate in debating issues and making decisions.

Clubs or organizations will be more effective if parliamentary procedure is properly used. Parliamentary procedure provides rules for an organization to use in conducting its meetings. Everyone needs some skills in parliamentary procedure. Most people will find themselves participating in business or community meetings at some point. Parliamentary procedure is a democratic process. It is important to know and understand the basic rules.

The person in charge of running a meeting must know the fundamentals of parliamentary procedure. This includes knowing the rules and how to preside.

MOTIONS

Several kinds of motions are used. Motions can be classified into five categories: main motion, privileged motion, incidental motion, subsidiary motion, and unclassified motion.

- **Main motion**—A ***main motion*** is a motion that brings business before the assembly (group). A main motion requires a second, is debatable and amendable, and requires a majority vote. No main motion is valid unless it has been seconded, which shows that someone else also favors bringing the motion to the floor. If no second is provided, the motion dies for lack of a second. If the motion is seconded, the chair then states the motion for the group, and it is ready for consideration. Discussion takes place supporting and opposing the main motion. If no amendments are offered, the motion is voted on as proposed. Based on the vote, the motion either passes or fails.
- **Privileged motion**—A ***privileged motion*** is a motion not related to the main question. Privileged motions are used to help the meeting go smoothly. They keep the meeting on track.

5–12. Proper use of a gavel and a professional appearance are useful to a presider. (Courtesy, Education Images)

Table 5–1. Examples of Privileged Motions

- **Fix the time at which to adjourn—**This is a motion to set the time for another meeting to continue business of the current meeting. It requires a second, is not debatable, is amendable as to time, and requires a majority vote.
- **Adjourn—**This is a motion to close the meeting. It requires a second, is not debatable, is not amendable, and requires a majority vote.
- **Recess—**This is a motion to allow the group (assembly) to take a short break in the meeting. It requires a second, is not debatable, can be amended as to time, and requires a majority vote.
- **Raise a question of privilege—**This is a motion that allows a request relating to the rights and privileges of the assembly or any member to be brought up for immediate consideration. It does not require a second, is not debatable or amendable, and needs no vote. Action on the question of privilege is a decision of the presiding officer.
- **Call for the orders of the day—**This is a motion that allows a member to require the assembly to stay within the adopted agenda or order of business. It does not need a second and is not debatable or amendable. No vote is taken. However, the orders of the day may be suspended by a two-thirds vote so that an item not on the order of business can be considered.

Source: *Developing Leadership and Communication Skills*. Upper Saddle River, New Jersey: Prentice Hall Interstate, 2004.

- Incidental motions—An ***incidental motion*** is a motion used to provide proper and fair treatment to all members. Incidental motions have no particular precedence. They are usually handled as they arise during a meeting.

Table 5–2. Examples of Incidental Motions

- **Appeal—**To appeal is a motion that a member uses when he or she disagrees with the ruling made by the chair (presider). It allows the assembly (group) to have the final say on all matters of procedure within the organization. An appeal requires a second, is debatable only if the pending motion is debatable, is not amendable, and requires a majority vote to reverse the decision of the chair.
- **Division of the assembly—**This is a motion used when a member doubts the results of a vote and wants a standing vote for verification. It does not require a second and is not debatable or amendable. No vote is required—a single member can demand a division of the assembly.
- **Point of order—**This is a motion that a member can use when the rules of the assembly are not being followed. It does not require a second, is not debatable or amendable, and does not require a vote.
- **Object to the consideration of a question—**This is a motion used when it would be inappropriate for a motion even to be presented before the assembly. A second is not needed. The motion is not debatable or amendable. A two-thirds vote is required.
- **Division of the question—**This is a motion used to divide another motion into two or more parts. Its purpose is to allow each part to be discussed and voted on separately. It requires a second, is not debatable, can be amended, and requires a majority vote.

(Continued)

Table 5–2 (Continued)

- **Modify or withdraw a motion—**This is a request to modify a motion or withdraw it from consideration by the assembly. If the chair has not yet stated the motion, the maker may modify or withdraw it. No second is required. The matter is not subject to debate or amendment. No vote is necessary. If the chair has already stated the motion, however, the matter becomes more complex. For details, consult *Robert's Rules of Order*.
- **Suspend the rules—**This is a motion to set aside established rules because they interfere with what the assembly wants to do. It requires a second and is not debatable or amendable. A majority vote for standing rules (rules not related to parliamentary procedure) and a two-thirds vote for rules of order (rules related to parliamentary procedure) are required.
- **Close nominations—**This is a motion that directs nominations to be closed. A second is required. The motion is amendable but not debatable. A two-thirds vote is required to close nominations.
- **Reopen nominations—**This is a motion that directs nominations to be reopened. It requires a second, is amendable but not debatable. A majority vote is required.
- **Method of voting—**This is a motion used to direct a manner of voting on a motion by a method other than a voice vote or a division of the assembly. A second is required. The motion is amendable but not debatable and requires a majority vote.
- **Request for information—**This is a motion used to allow a member to obtain further information about the business being discussed. No second is required. The motion is not debatable or amendable. No vote is required. The chair answers the question or directs the question to a member of the assembly.
- **Parliamentary inquiry—**This is a motion used to get information about parliamentary procedure. No second is required. The motion is not debatable nor amendable. No vote is required. The chair provides his or her opinion or asks the group's parliamentarian for assistance. (A parliamentarian is a person who has considerable knowledge of parliamentary procedure.)

Source: *Developing Leadership and Communication Skills*. Upper Saddle River, New Jersey: Prentice Hall Interstate, 2004.

- Subsidiary motions—A ***subsidiary motion*** is a motion related to the main question. Subsidiary motions are applied to main motions to alter them, to dispose of them, or to stop debate. All subsidiary motions have precedence over main motions. Eight subsidiary motions are commonly used.

Table 5–3. Examples of Subsidiary Motions

- **Lay on the table—**This is a motion that allows the assembly to set the pending motion aside temporarily for more urgent business to be discussed. It requires a second and is not debatable or amendable. A majority vote is required for passage.
- **Call for the previous question—**This is a motion designed to close debate immediately and prevent any other subsidiary motion except Lay on the Table. It requires a second, is not debatable, is not amendable, and requires a two-thirds vote.
- **Limit or extend debate—**This is a motion to alter the amount of time for debate. It requires a second, is not debatable, is amendable, and requires a two-thirds vote.

(Continued)

Table 5–3 (Continued)

- **Postpone (also known as postpone definitely)—**This is a motion used to delay action on the pending motion to a definite day, to a specific meeting, or until a certain activity has been concluded. It requires a second and can be debated and amended. It requires a majority vote.
- **Refer—**This is a motion to direct the pending motion to a committee for further information or action. It requires a second, is debatable and amendable, and requires a majority vote.
- **Amend—**This is a motion used to adjust a main motion by changing its wording before a vote is taken. Three basic methods of changing a motion are by adding words, by striking out words, and by substituting words. An amendment requires a second and is debatable if the motion being amended is debatable. An amendment can be amended. To amend requires a majority vote.
- **Amend an amendment—**This is a motion used to adjust an amendment to a motion by changing its wording before a vote is taken. Three basic methods of changing a motion are by adding words, by striking out words, and by substituting words. An amendment to an amendment requires a second and is debatable if the amendment to which it applies is debatable. It cannot be amended. A majority vote is required.
- **Postpone indefinitely—**This motion provides extra time to consider the motion being debated. Such postponement may be used as an attempt to eliminate the motion entirely. The motion requires a second, is debatable, and is not amendable. It requires a majority vote.

Source: *Developing Leadership and Communication Skills*. Upper Saddle River, New Jersey: Prentice Hall Interstate, 2004.

- **Unclassified motions**—An unclassified motion is one that brings questions back before the assembly or group. Five examples of unclassified motions are summarized in Table 5–4.

Table 5–4. Common Unclassified Motions

- **Take from the table—**This is a motion used to bring back for further discussion a motion that was laid on the table. It requires a second, is not debatable, and is not amendable. It requires a majority vote.
- **Reconsider—**This is a motion that allows a majority of the assembly to bring back a motion that was already voted on for further consideration. It requires a second and is debatable only if the motion being reconsidered is debatable. It is not amendable and requires a majority vote. The motion can only be made by a member who voted on the prevailing side (the side that won the vote).
- **Reconsider and enter on the minutes—**This is a motion that allows a majority of the assembly to bring back a motion that was already voted on for further consideration. However, one important way in which it differs from a regular motion to reconsider is that it must be made on the same day that the motion to which it applies was voted on. It requires a second and is debatable if the motion being reconsidered is debatable. It is not amendable. A majority vote is required. The motion can only be made by a member who voted on the prevailing side (the side that won the vote).
- **Rescind—**This is a motion to cancel action taken by the assembly. It requires a second and is both debatable and amendable. Two-thirds vote is needed if notice of the motion to be proposed has not been given at the preceding meeting or in the call of the current meeting.
- **Ratify—**This is a motion that, if passed, will approve an action already taken by the organization that was not previously voted on. To ratify requires a second and is amendable and debatable. It requires a majority vote.

Source: *Developing Leadership and Communication Skills*. Upper Saddle River, New Jersey: Prentice Hall Interstate, 2004.

VOTING

Voting is how people make known their choices about issues or other matters. Voting is an integral part of parliamentary procedure.

Four common methods of voting are used:

- **Voice vote**—A voice vote is an oral response of yes or no to a motion when the vote is called for by the presider. A voice vote lacks precision because no count is made. The presider makes a judgment as to the side with the most votes.
- **Rising vote**—A rising vote is a vote in which people are asked to stand to indicate whether they support or oppose an issue. A modification of this is a show of hands. This is an easy method of voting but does not allow privacy when the votes are cast.
- **Secret ballot**—A ballot is a piece of paper on which a voter makes a mark or writes words showing his or her preferences. A ballot handled so the public does not know how any individual voted is a secret ballot.
- **Roll call vote**—A roll call vote is a method of voting in which the names of all members are read aloud and each individual indicates his or her choice. How a person votes is not a secret. Votes are recorded and could become a part of the records of an assembly.

5–13. Voting may be on paper with a secret ballot that is placed in a ballot box.

Most decisions are made by a simple majority of those who vote. A simple majority is one more than half of all votes cast. In some cases, a two-thirds or 60 percent vote may be required. In elections, runoffs are held to assure that the winner has at least a simple majority.

For an official meeting to take place, a quorum must be present. A ***quorum*** is the minimum number of members that must be present to carry out business legally. The bylaws of an association usually prescribe the quorum, such as half of all members.

MEETINGS

A ***meeting*** is the assembly of a group of people for a particular purpose. It is usually to conduct business or learn about some particular topic. Good meetings go about matters in an efficient and effective manner.

Meetings serve several important roles. Some roles relate to having a good club. Other roles involve serving the needs of members. The nature of a meeting is often based on the skills of the leader and the tradition established for the meeting.

Roles of meetings include

- Conducting the business of an association or organization.
- Providing educational opportunities for members.
- Helping members develop leadership and personal skills.
- Providing social activities or entertainment for members.
- Recognizing the outstanding work of members.
- Getting reports on events or activities that are being planned.

5–14. A good meeting appears to be underway here. The presider is informally leading the group. Minutes are being taken with a laptop computer.

HAVING A GOOD MEETING

Carefully planned and properly carried out meetings are necessary. A team of elected officers or a committee is usually in charge of an organization's meeting. The structure of meetings varies widely. Many young people prefer meetings with a definite structure. They also like for the structure to be flexible so individual needs and interests can be met.

The person who runs a meeting is the ***presider*** or chair. In many cases, this is the president of an organization. It is important for the presider to have good meeting skills, including knowledge of parliamentary procedure.

Here are some pointers on having good meetings:

- Schedule in advance and promote attendance
- Carefully plan the meeting well ahead of time; use an order of business
- Have an interesting program; meet the needs of members
- Choose a good location
- Follow the bylaws of an organization
- Carry out the meeting with good presider skills
- Use parliamentary procedure as appropriate
- Properly vote on matters of concern
- Allow all individuals equal opportunity to participate in discussion
- Begin and end a meeting on time

Technology Connection

USING VISUALS

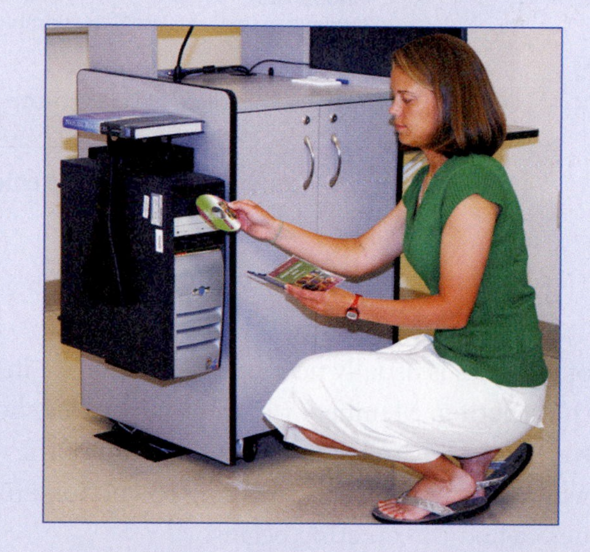

Effective speakers often integrate visuals or real things into their speeches. The use of presentation software is gaining in popularity. Some rooms and auditoriums where groups meet have specialized equipment to promote communication.

The speaker's stand or podium may have built-in computer capability. It projects visuals for the audience to see. The equipment may use CD-, DVD-, or flash drive-prepared images. It can also be connected to the Internet to download information or show real-time information in a speech.

This shows a speaker placing a CD into a computer for use in a presentation. The computer screen for the speaker is on the right of the podium. Posters, charts, and other materials that do not require electronic means could be used. (Courtesy, Education Images)

PREPARING THE ORDER OF BUSINESS

An ***order of business*** is a step-by-step plan for conducting a meeting. An order of business is sometimes known as an agenda. The order of business tends to vary with the organization and the traditions that have been established.

Order of Business
Meeting of FFA Chapter
Clinton High School
March 18, 2010
10:00 a.m.
Agriculture Classroom

Call to Order: President—Margie Oswego
- Opening Ceremony
- Welcome
- Announcements

Minutes of Previous Meeting: Secretary—James Gonzales

Reports:
- Treasurer—Angelica Loviza
- Social Committee—Jason Sloan

Old Business:
- Completing plans for making tour of horticulture industry

New Business:
- Selecting delegates for the State Convention
- Planning escargot sale

Program:
- Introduction—Jay Smith
- "Agriculture in Our State"—Honorable Rick Perry, Commission of Agriculture

Closing Ceremony
Adjournment: President—Margie Oswego

5–15. Sample order of business.

Prepare a written order of business well before a meeting takes place. The officers, program committee, or other responsible person usually prepare the order of business. An order of business is usually no more than one page in length. It contains a list of business items for the meeting. Members with responsibilities should get copies ahead of the meeting time so they can prepare for their roles. Copies of the order of business are often given to members when they arrive at the meeting room. Sometimes, the members approve the order of business when the meeting begins.

MINUTES

Minutes are the official written record of business in a meeting. The minutes are usually kept by the organization's secretary. If not, another person should do so. They may be handwritten on paper or kept on a computer. Only the major activities or decisions are usually recorded. Copies (usually on paper) of minutes are kept on file. Over the long-term, minutes reflect the history of a group.

The minutes of the previous meeting are acted on at the beginning of the next meeting. The order of business will indicate when minutes are to be acted on. The presider calls on the secretary to present the minutes. This may involve an individual reading the minutes, handing out written copies, or orally summarizing the events of the last meeting. Corrections, if any, are made, and the minutes are approved.

Minutes of a Regular Meeting

____
(Organization Name)

____
(Date)

Call to Order. The meeting was called to order at __________ (time) by the president, __________________________ (name).

Roll: ___________ (number of members present)

Minutes: Minutes of the previous meeting were read and approved.

Treasurer's Report: The treasurer reported a balance on hand of $__________ in the checking account and $___________ in savings.

Committee Reports: ____________________ (name) gave the following for the Recreation Committee: (Include a brief statement summarizing the report.)

Program: Example—Jane Smith presented slides on Australia.

Old Business: Example—Members were reminded to bring toys to the next meeting for the "Toys for Tots" campaign.

New Business: Example—A motion was made by Jill Olson that the organization sell Christmas wreaths. Sam Carter seconded the motion, and the motion carried.

Adjourn: The meeting adjourned at _____ (time).

____
Signed: President

____
Signed: Secretary

5–16. Sample minutes format.

REVIEWING

MAIN IDEAS

Leadership is the ability of an individual to influence others to meet certain goals or perform activities leading to a desired outcome. Leaders usually have qualities that lift them into a leadership role. The traits of these individuals are internal, technical, conceptual, and interpersonal.

Communication is an important part of the lives of people. In the process of exchanging information, people may use verbal or nonverbal means. Communication involves a sender, message, channel, and receiver. Blockages, known as noise, sometimes occur. Effective communication is most likely to occur if the sender selects means of communication based on the background of the intended receiver. In short, know your audience!

Public speaking is the process of using oral means of conveying information. Visuals and real things are sometimes used to enhance the oral part of a speech. A prepared speech is one that is developed well ahead of time, practiced, and delivered with passion. Extemporaneous speeches are not prepared in detail ahead of time though some preparation is used. Impromptu speeches involve no advance preparation other than general knowledge of a subject. Characteristics of good speakers include preparing in detail ahead of time, getting organized, using a quality voice, and showing enthusiasm.

Parliamentary procedure is a way of conducting meetings in an orderly manner. The rights of all people are respected. Various kinds of motions are used, with the main motion being most important. Groups typically make decisions by voting.

Meetings are important in our culture. A good meeting helps promote the overall well-being of an organization. A detailed order of business should be prepared ahead of time. Presiders should follow good practices in running a meeting, such as begin and end on time, be business-like, and respect the opinions of all members.

QUESTIONS

Answer the following questions, using complete sentences and correct spelling.

1. What is leadership?
2. What are four general qualities or traits of leaders? Briefly explain each.
3. What is communication?
4. What are the components of the communication process? Briefly explain each.
5. What is nonverbal communication?
6. What is public speaking?
7. What kinds of speeches may be used? Distinguish among each.
8. What are the important guidelines to help a person become a good public speaker? List and briefly explain any two.

9. What should be considered in delivering a public speech? Name and explain any two pointers on delivery.
10. What is parliamentary procedure? Why is it important?
11. What is a main motion?
12. What is voting? How are votes taken?
13. What is an order of business?
14. What are meeting minutes?

EVALUATING

Match the term with its correct definition.

a. communication
b. leadership
c. eye contact
d. verbal communication
e. nonverbal communication
f. minutes
g. parliamentary procedure
h. voting
i. presider
j. order of business

______1. The individual who runs a meeting.
______2. How groups of people often make decisions.
______3. The exchange of information.
______4. A written step-by-step plan for conducting a meeting.
______5. Relationships among people in which one influences others to achieve desired goals.
______6. A written record of the business carried out in a meeting.
______7. Looking into the faces and eyes of an audience.
______8. A method of conducting meetings in an orderly manner.
______9. Communication that uses words to express ideas.
_____10. Exchanging information without the use of words.

EXPLORING

1. Prepare a speech for delivery in class or other event. Follow the regulations of the FFA-prepared speaking event. Video record your delivery and observe your performance. Develop a plan to improve on your speaking ability.

2. Develop an order of business for the meeting of a school club or other organization. Be sure to follow the example presented in the chapter.

3. Participate on a parliamentary procedure team. Follow the regulations of the FFA parliamentary procedure event. Video record the performance of your team and critique how well you did. Prepare a plan to improve on your parliamentary abilities.

CHAPTER 6

Job Skills and Safety

OBJECTIVES

This chapter covers important skills needed for success in a career as well as those related to safety at work. It has the following objectives:

1. Identify needed job skills for your success.
2. Explain and develop appropriate interpersonal skills.
3. Demonstrate appropriate citizenship.
4. Explain the role of computer technology in agriculture.
5. Locate, assess, and use agriscience information.
6. Describe and follow appropriate safety practices.

TERMS

citizenship
computer
digital safety
employer expectations
information
first aid
MSDS
patriotism
personal appearance
PPE
program
safety
social behavior

6–1. A veterinarian must have specific job skills. What do you think they are?

EVERYONE wants to be successful. No one wants to fail. By developing and following important personal and job skills, anyone can prepare to be successful.

What is success? It is the progressive achievement of worthy goals. Being successful begins with having goals. Then people must take the needed action to achieve the goals. Although some goals can be accomplished in a few days or weeks; other goals require many years.

By being smart, people can position themselves to be successful. They can develop and apply appropriate skills that will help them move forward. Success includes relationships with people, so this chapter is a good place to begin.

SKILLS FOR CAREER SUCCESS

Career success is an important measure of progress. Employers hire people to help reach the goals of their business, agency, or institution. They want productive employees who have good work ethics. Interpersonal and job skills are also important.

EXPECTATIONS

Employers have expectations of the people they hire. ***Employer expectations*** are behaviors employers seek and reward in their employees. Some expectations require considerable skill. Other expectations relate to being at work on time and focusing on the work to be completed.

Here are a few general expectations of employers:

- **Acceptance of responsibility**—Being responsible includes caring for tools, equipment, and facilities as well as doing what you say you will do. If you make a mistake, admit it. Correct errors. Do not waste the employer's resources or your time. Do not abuse equipment and facilities.
- **Ability to communicate**—The ability to receive and share information is important. Listen to what others say, especially supervisors. Speak and write using good language skills.
- **Job proficiency**—People need skills and knowledge to do a good job. Some training may be provided by the employer, but most people need to have basic skills. A good general education is always important.
- **Ability to be a team player**—Most jobs involve groups of people working together. People must get along and assume their share of the work. In many cases, doing more than the minimum promotes good team relationships. Consider the other workers. Help ensure a safe, productive work environment.

6–2. Harvesting soybeans with a combine is a big responsibility. (Courtesy, Case Corporation)

- **Professionalism**—Dress and act professionally. An appropriate appearance is always important.
- **Honesty**—Being honest is essential to long-term success. Tell the truth. Admit mistakes. Take steps never to repeat a mistake.
- **Skills in satisfying customers**—Most jobs involve some dealings with people who are not employees. Having a positive, helpful, and friendly approach always pays dividends. In selling, remember that "the customer is always right." That is true even if the situation is flawed. Practice good interpersonal skills.

6–3. The ability to get along with people is essential for career success.

PERSONAL TRAITS

Personal traits can assist people with job success. Many things are involved at work. Productivity and the success of a business often depend on the personal traits of the employees.

A few examples of personal traits are:

- **Loyalty**—Employers like employees who are loyal. They want employees to say good things about their work and their employers.
- **Wholesome lifestyle**—Have an approach to life that makes you a good worker. What you do after work hours can affect your ability at work.
- **Willingness to learn new things**—Work demands change. People must learn new job skills. Always be willing to learn. Be anxious to do a better job for your employer.
- **Acceptance of others**—All people can make contributions. Never be biased toward others. Skin color, gender, language, and other differences should never be the basis of prejudice.
- **Respect for others**—All people deserve respect. Differences of opinion must be settled in constructive ways. Violence is never acceptable at home or away from home. Physically abusing anyone—friend, coworker, spouse, child, parent, or sibling—is unacceptable.
- **Sense of humor**—A sense of humor is the ability of a person to see the amusing side of situations. Often it helps to be able to laugh at mistakes and then take actions to correct them.

INTERPERSONAL SKILLS

How people relate to each other is important in their personal lives and in their careers. Good interpersonal skills enhance technical job skills. They are often the difference between advancing in a career and failing to move ahead.

SOCIAL BEHAVIOR

Social behavior is the way in which we relate to the people around us. It includes our manners and mannerisms. Social behavior involves following guidelines of conduct. It is a part of our culture and may change with time.

Several principles to guide social behavior are:

- Be courteous. Know when to say "Please," "Thank you," "Excuse me," and "I'm sorry."
- Respect the property of others.
- Respect the rights of others.
- Properly greet other people. Know how to shake hands. Look other people in the eyes.
- Use appropriate grammar.
- Be caring. Help others when they need help.
- Follow up on commitments. Do what you say you will do.
- Have empathy (demonstrate understanding by identifying with the feelings of another).

6–4. Personal appearance is enhanced by a smile, a neat hair style, and well-cared-for teeth.

PERSONAL APPEARANCE

Your ***personal appearance*** is what other people see when they look at you. We have no control over some of our personal features. We do, however, have a great deal of control over many things that make up our personal appearance.

Some features are genetic. They are our heritage, and we can be proud of them. The color of our eyes is an example. No eye color is better than another. Another example is hair color. All natural hair colors are equally impressive. (Using dyes to alter hair color is a matter of personal preference.)

Other features of personal appearance can be controlled and are important in shaping how people view us, including how we dress and groom ourselves. Attire and grooming make the all-important first impression. Our personal appearance projects how we want other people to see us. It also influences how others perceive our values and lifestyles.

Here are several guidelines for personal appearance:

- **Clothing**—Select appropriate clothing; do not overdress or underdress. Clothing should be conservative and stylish and should fit properly. Fad clothing should be avoided. Keep clothing clean, in good repair, and properly pressed. Shoes should be clean, and dress shoes should be polished.
- **Personal hygiene**—Keep clean and free of odor. Daily baths and the use of deodorant are important. Oral hygiene (brushing your teeth) helps in appearance, reduces bad breath, and prevents tooth decay.
- **Grooming**—Keep hair at an appropriate length. Avoid extreme colors and styles because these fads don't appeal to employers or customers! Be neat!
- **Jewelry**—Keep jewelry simple and suitable. Body piercings and tattoos are quite distracting and inappropriate in most job settings. Tattoos are difficult and costly to remove and are typically on your skin the rest of your life.

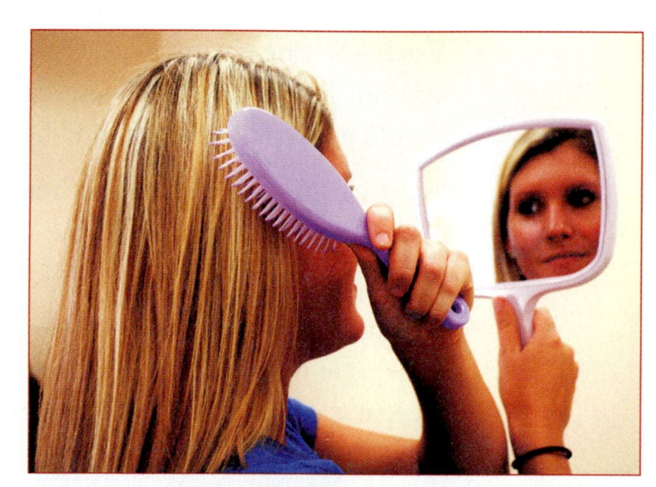

6–5. Brushing helps maintain hair and promotes an attractive appearance. (Courtesy, Education Images)

- **Fitness**—Get regular exercise, and eat properly. Avoid substances that damage the body. Get plenty of sleep and rest. Regular medical and dental checkups are important to ensure fitness and good health.
- **Demeanor**—Be happy, and let it show. Be friendly. Avoid holding grudges and talking negatively about someone.

CITIZENSHIP

Citizenship is your conduct as a member of the population of the United States. Your citizenship means that you are a legal resident of the country and are entitled to specific rights. Being a citizen requires us to carry out certain duties and meet certain responsibilities.

The laws of the United States provide for a citizen's right to vote, to own property, and to enjoy freedom of speech, religion, and the press. Each of these rights has certain conditions that protect people and the overall well-being of the nation.

People who are good citizens have desired traits. Good citizens:

- **Abide by the law**—Laws are followed in all regards. Laws govern. Some examples are paying taxes, not violating the rights of others, and operating motor vehicles.
- **Support charity**—Needy people benefit from the help of others. Good citizens support legitimate charities.
- **Vote**—Voting is a privilege. Citizens can vote at age 18, provided they have registered and not lost the right to vote because of a criminal conviction. Before voting, they should gather information and become informed to make the best choice.
- **Support the well-being of others**—Citizens participate in decision making for their communities. They look at the overall well-being of a community rather than their personal interests.
- **Are patriotic**—***Patriotism*** is admiration for and loyalty to one's country. Citizens take pride in their country. Good citizens pledge allegiance to the flag, ascribe to service, and demonstrate enthusiasm for their nation. Patriotism is also shown by voting

AgriScience Connection

TEAMS AT WORK

People usually don't work alone. They work with other people. Each person has specific duties. Those duties, however, are part of a larger "picture" of what is needed for a team to accomplish the work.

A team is a group of two or more people working together to achieve a goal or mission. A goal or mission is what the team is to accomplish. It provides direction to the work of the group. Without a direction, the team members would wander aimlessly in doing their work. The goal would not be accomplished as a team.

Everyone needs to be able to work as a team member. Getting along with others on the team is essential. Each person must do their share, or a little more, of the work. When working together, teams accomplish more than each individual would accomplish working alone.

in elections, serving in the armed forces, and participating in a democracy to improve the nation.

- **Take pride in their communities, jobs, and homes**—Good citizens appreciate the role of a quality environment in promoting well-being.
- **Contribute**—Being a productive worker, good student in school, or other socially useful person contributes to the overall betterment of society.

COMPUTER TECHNOLOGY IN AGRICULTURE

Computer technology has changed how many functions in AFNR are carried out. This requires individuals to use computer technology in their work and daily lives. For example, many engines and mechanical devices are operated by using computer technology.

HARDWARE AND SOFTWARE

The use of computer technology involves two major products: hardware and software. Hardware is the equipment—keyboard, processor, and screen. It is often referred to as a computer. Some computers are relatively invisible in our daily lives, such as those that control engines and robotic operations.

A ***computer*** is a machine that manipulates data. Software is the program that runs the equipment to perform the desired tasks. The first computers of the mid-1900s were very

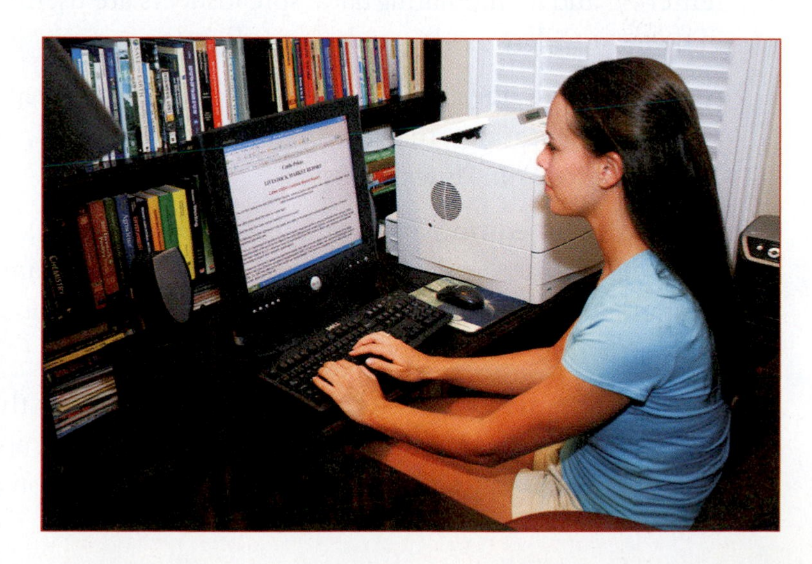

6–6. Modern computer technology promotes efficiency and profitability in agriculture. (Courtesy, Education Images)

ACADEMIC CONNECTION

COMMUNICATION

Use your studies in communication class to increase your understanding of nonverbal communication. Assess the effectiveness of your personal nonverbal communication. Devise a strategy to improve this area of your interpersonal skills.

large and expensive. They used vacuum tubes and other space-consuming components. In the mid-1980s, the use of transistors allowed computers to be smaller and, hence, the personal computer came into being. This computer could sit on desks and tables. Improved circuitry allowed even smaller computers (laptops) to be made. In the early 2000s, smartphones and similar hand-held devices allowed computers to be quite portable. The miniature devices that perform functions are known as microprocessors. With all computers, a source of electrical energy is needed for operation. Data are stored on a hard drive in the computer or may be loaded into a flash drive, CD, DVD, or another transportable device, such as an external hard drive.

A ***program*** is a set of instructions that makes it possible for a computer to store and execute lists of instructions. You probably use several different programs when you use a computer for school work or to manage your supervised experience. Some programs will perform a wide range of functions; other programs perform only one or two things.

Programs widely used in AFNR include the following:

- **Word processing**—Word processing is using software that aids in preparing reports and messages; some applications include desktop publishing; examples are WordPerfect and Word.
- **Spreadsheets**—A spreadsheet is an application that simulates paper with cells for entering and manipulating data; spreadsheets are useful in preparing financial reports. The most widely used spreadsheet is Excel.
- **Presentation**—Presentation software aids in the preparation and delivery of illustrated presentations that integrate various graphics, such as line art, images, and words; PowerPoint is widely used.
- **Photography editing and graphics**—This software aids in improving images made with digital photography; Adobe Photoshop is an example. Specialized graphics programs are available, such as those used in preparing landscape plans or the design of a new structure.
- **Information systems (Internet and email)**—This is the software that connects and aids in using various World Wide Web and email applications; examples include Internet Explorer and Microsoft Outlook. Information systems is sometimes lumped together with telecommunications.

Beyond these, many specialized kinds of software may be used. Some software facilitates the design of Web sites, the maintenance of business records, the billing of customers, and the inventory process through bar codes, electronic chips, and other means.

APPLICATIONS IN AFNR

Computer technology is increasingly used in AFNR to focus on gaining greater efficiency and profitability from a farm, agribusiness, or other venture. Four uses are listed here:

- **Keep records**—Inventory, financial, performance, and other records can be useful. They often involve using spreadsheet programs that will summarize data. Summaries of the information may be viewed on the screen or printed on paper. Students may use record-keeping programs with their supervised experience and FFA participation records.
- **Prepare reports**—Word-processing programs are used to prepare reports. In some cases, spreadsheet programs provide reports of financial information. Desktop publishing may be used to prepare flyers and brochures to promote products or events. Reports may also include tax forms and other materials for government and association needs. Financial and performance activities of FFA and supervised experiences can also be produced.
- **Exchange information**—The Internet is widely viewed as an information source. Reports prepared with word processing and spreadsheets may be exchanged in electronic or paper forms. Popular social networking is also a part of the information exchange. Many agriculture students use FFA Nation, which is a social network espe-

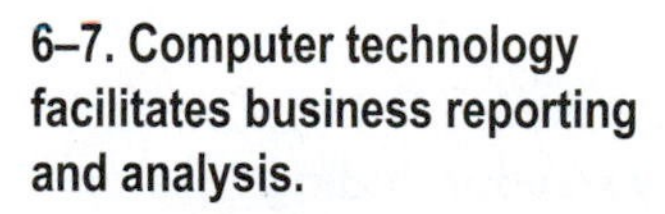

6–7. Computer technology facilitates business reporting and analysis.

cially for members. The World Wide Web and email are widely used today in agriculture to receive and send information.

- **Marketing**—Computer technology aids with a wide range of marketing activities. Sometimes it is used with a Web site to promote the sale of products. Other times it is used to produce promotional materials or to prepare advertising layout copy. It is also used to receive reports of marketing information.

DIGITAL SAFETY

Digital safety is using computer applications to reduce risk and to promote safety. It is sometimes known as online safety and Internet safety. It is, however, greater than these and includes reports and other information in computers or saved on flash drives or CDs. Promoting digital safety is increasingly important.

Here are a few digital safety rules to follow:

- Use only safe, protected sites. Establish a list of favorite or safe sites related to your needs and interests in AFNR.

Technology Connection

JOB FUN

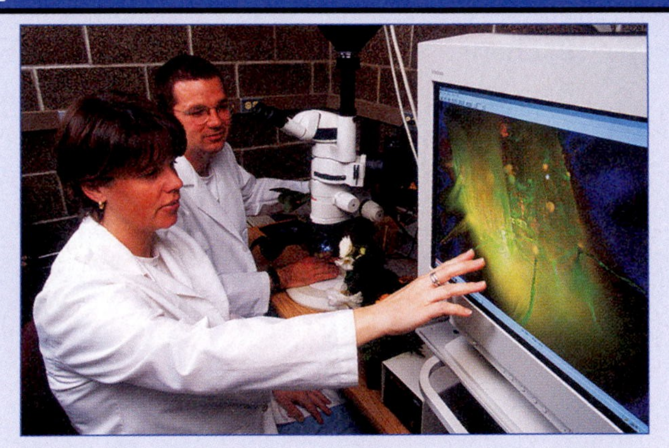

Do you think people like their jobs? Many do; some don't. As you consider career options, be sure to assess your likes and dislikes. Consider what you enjoy. You want to have fun with your work. You also want to be competent and achieve important success goals.

How would you like it if your work involved using high-level technology? Scopes, probes, electronics, cameras, monitors, and other devices might be used. If so, you will need to have the necessary skills to use them and take pride in your work to make sure that the items are used properly. You will want accuracy so your findings are on target. Of course, most beginners experience a time of training or of learning on the job.

An example of a job in agriscience is shown here. A greatly enlarged fungus is viewed on a screen using a microscope and camera. This *Botrytis cinerea* fungus forms a mold on plant leaves. In the research, the investigators found that the fungus can move from one tiny leaf hair to another without ever touching the leaf. This type of movement reduces the effectiveness of pesticides applied to the surface of leaves. (Courtesy, Agricultural Research Service, USDA)

- Use blocks on your computer that deny access to certain sites.
- Do not provide personal information (including pictures) in chat rooms or most other situations. Sometimes unscrupulous people will take your information and use it for themselves. This is a part of identity theft. (Phishing is a criminal activity whereby a Web site tries to obtain personal information, such as a social security or credit card number.)
- Disable Web cookies to prevent information about the user from being sent to the server. (Some Web sites require Web cookies for access, so this may need to be turned on occasionally.)

Internet Topics | Search

Selected topics for Internet discovery and reporting are listed here. Begin by searching for each topic. Next, read and learn about it. Conclude by preparing a brief report on your findings.

1. Voting in public elections
2. Laboratory safety
3. Personal appearance

- Use firewall software to prevent access by hackers—unauthorized users—into your computer.
- Report harassing activities, including cyber bullying and cyber stalking.
- Avoid hate, sexually-oriented, and other inappropriate sites.

Laws apply to digital safety. The Children's Only Privacy and Protection Act (COPPA) was enacted by the U.S. Congress in 1998. The overall purpose of the act was to protect youth from the collection and use of inappropriate information.

AGRISCIENCE INFORMATION

Information is knowledge or news acquired in any manner. People need good information to make the best possible decisions. Locating, assessing, and using information are important in agriscience.

LOCATING INFORMATION

Several sources of information are widely used in agriscience. When looking for information, choose the source by format and by provider.

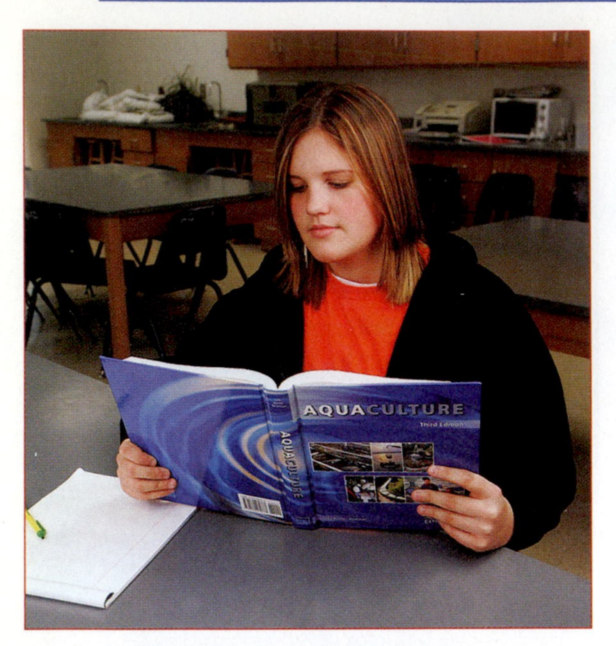

6–8. Well-written books are good sources of information.

Format

The most common sources of information by format are:

- **Books**—Books provide basic information about a subject. Most books have been carefully written and reviewed to verify accuracy. Most students in school use textbooks. A textbook is carefully organized and written to help students learn. Research has shown that students learn more in classes where textbooks are used. They make higher test scores than those in classes where textbooks are not used, including classes where computers are the main tool for learning.
- **Bulletins**—A bulletin is a small publication that often reports the findings of research. Bulletins are good sources of information, but several are needed. Readers must often draw their own conclusions from the research.
- **Magazines**—Magazines are published weekly, monthly, or on some other schedule. Many magazines address agriscience and related areas. Some are easier for the nonscientist to read and understand than others.
- **Newspapers**—Newspapers are published daily or weekly. Many have a wide range of news stories. Some trade papers relate specifically to agriculture.
- **Manuals**—Manuals are publications that accompany products to provide details about them. Manuals may also include assembly and use instructions.
- **Electronic media**—Electronic sources of information include the Internet and disks. The Internet has many sites that offer agricultural information. CDs and DVDs are often used to produce published information and are often easily used on a personal computer, with Adobe Acrobat Reader.

6–9. Wireless connectivity to the Internet enables people to find information in a wide range of locations.

Provider

Providers are the businesses or agencies that produce forms of information. Examples of providers are:

- **Publishers**—Commercial publishers produce books, magazines, and newspapers. Many publishers provide information that is useful in agriculture, horticulture, forestry, food science, aquaculture, and related areas. Some publishers also produce information in electronic forms.
- **Universities**—Many universities have information offices that produce bulletins and other materials. These offices often operate in association with research departments or the Cooperative Extension Service. Most maintain Web sites where information can also be obtained.
- **Industries**—Suppliers of feed, seed, fertilizer, equipment, and other inputs used in agriculture often produce bulletins and manuals. These materials may deal with how to use specific products.
- **Associations**—Associations may produce materials with helpful information. For instance, the Irrigation Association produces magazines, bulletins, and books.

CAREER PROFILE

SMALL ANIMAL VETERINARIAN

A small animal veterinarian provides for the health needs of small animals. They may have private practices in veterinary medical clinics. The animals are rabbits, birds, dogs, cats, iguanas, snakes, and others. The work involves greeting owners, handling animals, examining animals, diagnosing problems, providing medications, doing surgery, and supervising assistants.

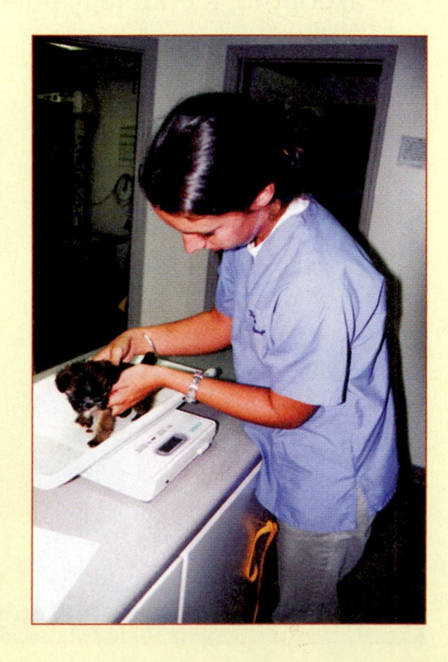

Small animal veterinarians have degrees in veterinary medicine. They receive extra education and experience with small animals. Before entering veterinary medical school, most have degrees or strong backgrounds in animal science, zoology, or a related area. Courses in science, agriculture, and business are important in high school.

Jobs for small animal veterinarians are in clinics near cities where people own animals. They may have their own clinics or work in a partnership with other veterinarians. Some work with pet food factories, pharmaceutical companies, and government agencies.

ASSESSING INFORMATION

Information should be assessed to see if it is appropriate for your needs. Using incorrect or out-of-date information may result in making poor decisions. A few factors to consider in assessing information are:

- **Source**—Is the source reliable? Does the source use reviewers or follow other procedures to verify technical accuracy?
- **Date**—How long ago was the information prepared? Up-to-date information is essential.
- **Illustrations**—Is the material illustrated with photographs and line drawings? Are the illustrations attractive and accurate?
- **Safety**—Does the material promote safe practices?
- **Availability**—Is the information readily available at a reasonable cost?

USING INFORMATION

For information to be beneficial, it must be used properly. Read and consider how the information applies to your situation.

- **Adopting technology**—Information can assist you in deciding if adopting technology will be beneficial.

6–10. Educational requirements and career information is available from a teacher or an advisor. (Courtesy, Education Images)

- **Determining when to use technology**—When technology is used is as important as adopting it. If technology is used at the wrong time, you may have poor results. For example, if a pest control practice is used too early or too late, there will be little benefit.
- **Choosing what to produce**—Good information should readily indicate the advantages and disadvantages of a crop, animal, or other product.
- **Implementing a practice**—Information should give details on how to go about getting started. It may focus on how something is done—instructions on how to use a chemical or assemble machinery.

SAFETY

Safety is the condition of being free of harm and danger. It affects you and those around you. Being safe applies to school settings and job settings. Removing hazards, preventing accidents, and having a productive learning and working environment are part of safety. Safety applies in many areas of our lives. As a result, everyone should "think safety" by anticipating dangers and taking steps to avoid them.

Routinely follow good safety practices in the school lab and in daily living. This will prepare you to practice safety on the job. Employers expect workers to have developed basic safety skills that can be refined in the workplace.

- ★ Identify potential hazards.
- ★ Learn practices that reduce risks of hazards.
- ★ Always follow safe practices.
- ★ Read and follow instructions.
- ★ Keep equipment in good condition.
- ★ Remove hazards.
- ★ Alert other people to hazards.
- ★ Never take unnecessary risk.
- ★ Always use personal protection equipment.
- ★ Know the steps to take in case of an accident.
- ★ Have a telephone nearby and emergency numbers posted.
- ★ Place well-stocked first-aid kits near potential hazards.

6–11. General safety rules.

SAFETY IN THE LAB

The labs at school and work may present hazards. Good practices help reduce the likelihood of injury from potential hazards, such as chemicals, hot and cold materials, equipment, and various situations.

Material safety data sheets (***MSDS***) accompany most chemical products and provide information about the safe use of chemicals and the steps to take in case of an accident. Labs often have notebooks of MSDS materials readily available or post the sheets in the work area.

6–12. A notebook of MSDS materials is conveniently placed in a wall-mounted bracket. (Courtesy, Education Images)

Posters and other symbols may be used to alert people to hazards. This is true in a laboratory environment as well as on equipment and supplies that may be used. Always study the symbols and know the appropriate safety practices.

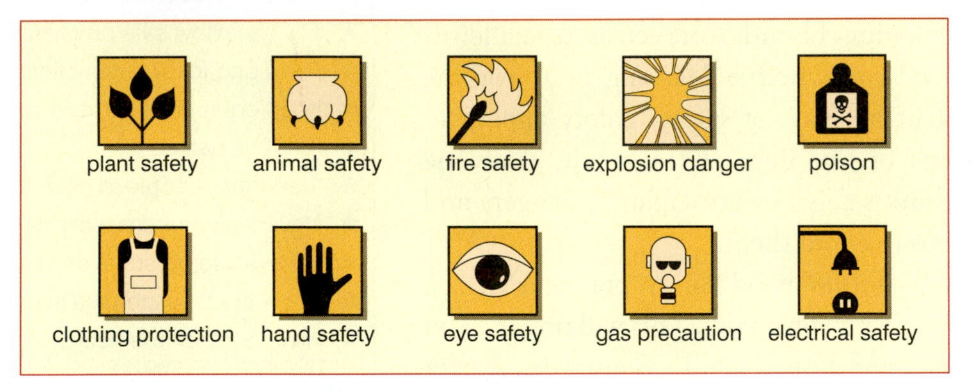

6–13. Safety icons and posters may be used to identify safety hazards.

Personal protective equipment (***PPE***) is used to protect people from injury. The equipment is designed to protect the eyes, hearing, respiratory system, skin, and overall body. Safety goggles and glasses are particularly important in any lab activity. Properly fitted goggles are best. PPE should always be used in a lab.

Hazards in a lab may come from several sources:

- **Chemicals**—Some chemicals are very hazardous. Know how to use all chemicals properly, and always follow safety procedures.
- **Organisms**—Living plants, animals, and other organisms can cause disease and injury. Know the characteristics of organisms and the hazards they pose. Always respect the well-being of animals. Discarded tissues should be disposed of correctly.

6–14. Proper eye and hearing protection are essential in many lab and work situations. (Courtesy, Education Images)

6–15. Knowing how to carry an animal properly promotes personal safety as well as animal safety. (Courtesy, Education Images)

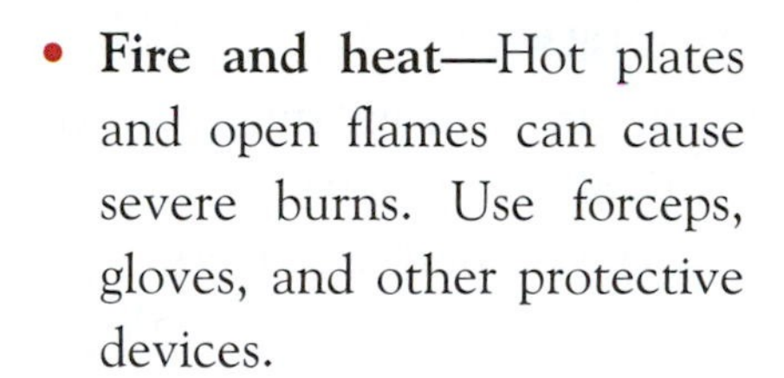

- **Fire and heat**—Hot plates and open flames can cause severe burns. Use forceps, gloves, and other protective devices.
- **Equipment**—Lab equipment ranges from hand tools to motorized devices. Know how to operate equipment properly before attempting to use it.
- **Sharps**—Sharps are objects that can cause cuts and penetrate the skin. These include knives and pins as well as broken test tubes, microscope slides, and various equipment. Every lab should have an appropriate way of disposing of sharps. Wearing the proper gloves can help prevent injury.

6–16. Examples of personal protective equipment. (Note: The construction materials of PPE will vary with the hazards involved. For example, steel-toed leather shoes, not rubber boots, are used in welding.)

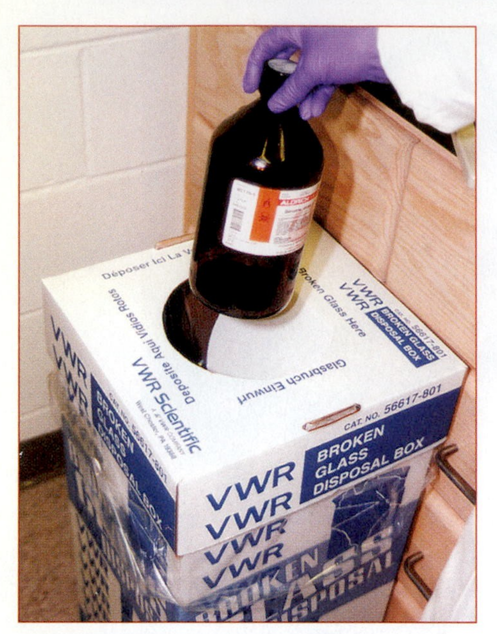

6–17. Proper disposal of broken glass, used materials, and empty containers is essential.

SAFETY AT WORK

Safety hazards are often present at work. Some of these are the same as those in labs. Other hazards may be found depending on the nature of the work.

Hazards at work may include the following:

- **Hand tools**—Common hand tools include hammers, wrenches, and measuring devices. Always use hand tools properly. Be sure the tools are in good condition. Never use a tool for a job it wasn't designed to do. Wear eye, hearing, skin, and respiratory system protection.
- **Power tools**—Power tools vary with the job. Examples of power tools include drills, saws, grinders, trimmers, and edgers. Know how to use power tools properly. Know the hazards associated with them. Take steps to prevent injury, such as using eye and hearing PPE.
- **Equipment**—Equipment ranges from that used on a farm to that used in horticulture, forestry, and agribusiness. Be sure a tractor, a mower, or other equipment is in good condition. Know how to use the equipment properly to avoid injuries. Never go too fast. Avoid unstable areas and steep hills or banks.
- **Engines and motors**—Engines and motors often produce a lot of power and heat. Know how to use them. Pay particular attention to the locations of moving parts. Keep engines and motors properly serviced and in good condition. Always follow safety practices. Safety shields should always be in place.
- **Electricity**—Electricity can cause shock, resulting in injury or death. Use electricity properly. Be sure all installations have been properly made. A ground fault circuit interrupter (GFCI) should be on all lines to power tools. A GFCI protects against electrical shock.
- **Water**—Some jobs are around water. Fish farming, boating, and similar work may involve water hazards. Never take unnecessary risks. Use safety rings, life jackets, and other devices as appropriate.

Increasingly, producers and others involved with plant and animal products need to follow practices that promote safe, quality products. This involves taking steps to prevent contamination and to keep the products safe. Many producers follow quality assurance programs to promote the production of safe products. This is sometimes said to be biosecurity.

SAFETY IN DAILY LIVING

The world in which we live has many hazards. Use common sense, and follow safety practices to reduce the likelihood of injury.

Homes

Our homes have electrical and gas appliances, motors and tools, and household and garden chemicals. Practicing safety is important. Everyone should know the safe procedures for handling and disposing of hazardous materials. Help protect children and others—pick up trash around homes. Remove junk and broken-down cars.

Some homes have pets, such as dogs and cats. These animals can pose hazards. Some animals can become aggressive and cause injury or death. Hundreds of people are seriously injured or killed each year in the United States by dogs. Any pet that attacks a person should be confined, and a veterinarian should be called to determine if the animal has rabies. The animal may have to be euthanized (put to death). Know the local laws about pets, and follow them. Never let pets roam about. Respect other people; clean up after your pets. Keep down pet noise, such as annoying barking.

Highways and Driving

An important part of growing up in the United States is learning to operate a motor vehicle. Driver training is needed, and proper licensing is required in all states. However, safety depends on the care and skill of the operator. Driving a car or truck is a major responsibility. Always follow regulations; never violate laws.

Here are a few safety suggestions:

- Know the features of the vehicle.
- Keep the motor vehicle in good condition.
- Wear a seatbelt and use other appropriate safety devices.
- Obey signs that provide information to drivers. (For example, a stop sign means "STOP.")
- Avoid fast starts and sudden braking.
- Keep the proper distance from other vehicles and from structures.

6–18. Always fasten the seatbelt on any moving vehicle.

- Have passengers follow safety rules.
- Anticipate the actions of other drivers.
- Avoid substances that impair your ability to drive.
- Avoid distractions, such as telephones, radios, and eating.
- Leave in plenty of time to reach your destination. Do not rush.
- Take driver training, and know the laws of the road.

Recreation

Recreation is important to many people. Fishing, hunting, skiing, and similar activities can be hazardous. Take steps to reduce the risks. Complete appropriate safety education. Comply with laws and have the appropriate license as required.

Always know the hazards associated with recreation. Anticipate hazards, and take steps to eliminate them. This makes the environment you are in safer for you and for those around you. Use equipment properly. Have safety devices in good working order and nearby.

FIRST AID

If an injury occurs, it is important to know how to respond properly. Therefore, it is a good idea to take a first-aid course to be prepared. ***First aid*** is providing initial care to an individual who has had an injury or is ill.

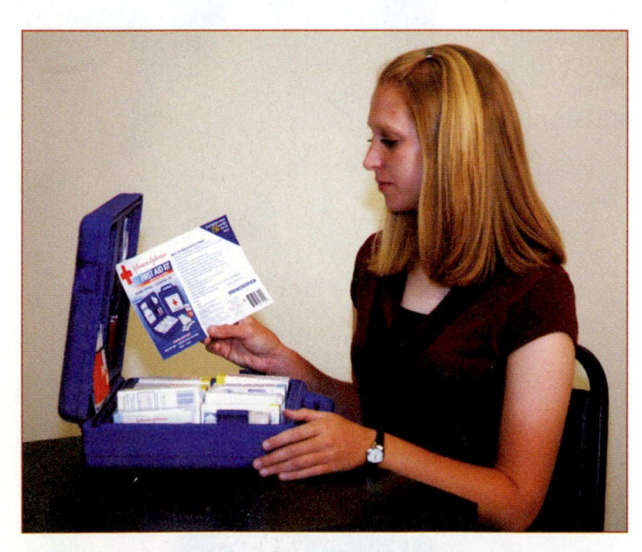

6–19. Become familiar with the contents of a first-aid kit. (Courtesy, Education Images)

Depending on the severity of the injury, it is essential to make the best possible decision. With an injury that appears major, it is best to call for professional help immediately. In most places, dialing 9-1-1 will connect the person to law enforcement and emergency medical services. However, it may be necessary to assist someone until emergency medical service arrives. If bleeding is occurring, pressure should be applied directly on the wound. If the person is not breathing, cardiopulmonary resuscitation (CPR) should be performed.

With a minor injury, such as a small cut or skin abrasion, you can likely take the needed steps. Always have access to a well-stocked first-aid kit. This kit should have a range of medical supplies.

In general, here are five steps with a minor wound:

1. Stop the bleeding.
2. Cleanse the wound.
3. Apply antiseptic.
4. Cover the wound.
5. Secure the cover.

If a wound doesn't heal within a few days, seek medical help. The key to dealing with small wounds is to keep them clean and to protect them from infection.

REVIEWING

MAIN IDEAS

People want to be successful in their careers and personal lives. Interpersonal skills are needed for career success. These skills include social behavior and personal appearance. These skills help in relating to other people and help establish the way you want others to see you. Personal traits and citizenship skills may be involved.

Computer technology is playing more important roles each day. Hand-held devices aid in sharing and gaining information. Computer software assists in record-keeping, communication, and education.

In planning and carrying out careers, good information is needed. The information may be about occupations, educational needs, and technical details needed to complete our work safely and efficiently.

Safety is important in all work. Preventing injury and loss is beneficial to the employee and the employer. Students can begin by practicing safety in the school lab. Using personal protective equipment is a major step. Safety at work involves many of the same hazards as in the school lab. Safety in daily living includes actions in our homes, in the operation of motor vehicles, and in recreation. First aid may sometimes be needed. Have a kit available and know how to respond.

QUESTIONS

Answer the following questions, using complete sentences and correct spelling.

1. What important personal skills are needed for career success? List any four.
2. How are computers used in the AFNR career cluster? Name and briefly discuss three ways they are used.
3. What are three applications of computer technology in agriculture? Briefly explain each.
4. What principles guide our social behavior? (List any four.)
5. Why is personal appearance important?

6. What areas are included in personal appearance?
7. Why is practicing safety important?
8. What safety hazards might be present in a school lab?
9. What safety hazards may be found at work?
10. What safety suggestions should be followed in operating motor vehicles?
11. What are the major expectations of employers?
12. What are the important citizenship traits?

EVALUATING

Match each term with its correct definition.

a. digital safety
b. program
c. safety
d. citizenship
e. information
f. social behavior
g. personal appearance
h. patriotism
i. first aid
j. computer

______1. Your conduct as a member of the population of the United States.
______2. Using computer applications to reduce risk.
______3. What other people see when they look at a person.
______4. How people relate to other people with manners and mannerisms.
______5. Recorded knowledge about a subject.
______6. Instructions a computer uses in operation; sometimes known as software.
______7. Admiration for and loyalty to one's country.
______8. The condition of being free of harm and danger.
______9. A machine that manipulates data.
_____10. Providing initial care to a person who is ill or has been injured.

EXPLORING

1. Tour the agriscience or science labs at your school. Identify potential sources of danger, and describe actions that have been taken to reduce risks. Where hazards have not been dealt with, suggest ways of reducing risks.

2. Select eye PPE suitable for use in the lab of your school. Practice properly wearing the PPE. Make the needed adjustments. Keep the PPE clean. Store it properly when not in use.

3. List agricultural employers in your community, including farms, agribusinesses, government agencies, associations, and others. Select one of the employers, and interview an employee regarding the nature of the work. Develop a written list of questions ahead of time. Use the list as a guide in conducting the interview.

Part Three

The Science of Agriculture

CHAPTER 7

Science Relationships

OBJECTIVES

This chapter identifies and explains science areas of agriculture and use of the scientific method in agriscience. It has the following objectives:

1. Identify and describe areas of science in agriscience.
2. Identify new and emerging areas of technology.
3. Explain the scientific method and its use in agriscience.

TERMS

biotechnology
botany
computational science
experiment
genetic engineering
geology
laser
life science
mathematics
physical science
precision farming
radiation
remote sensing
science
scientific method
traceability
zoology

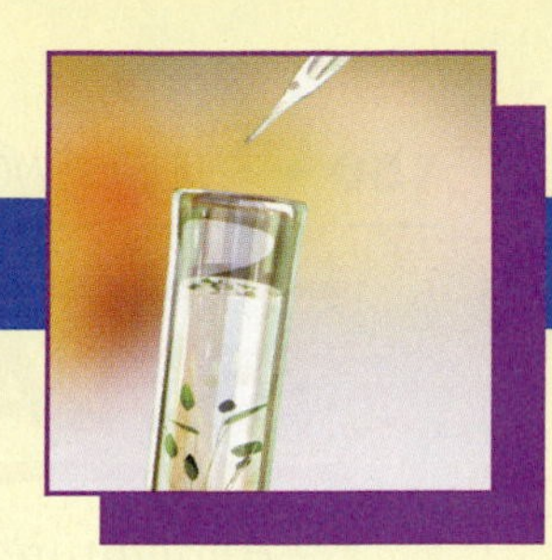

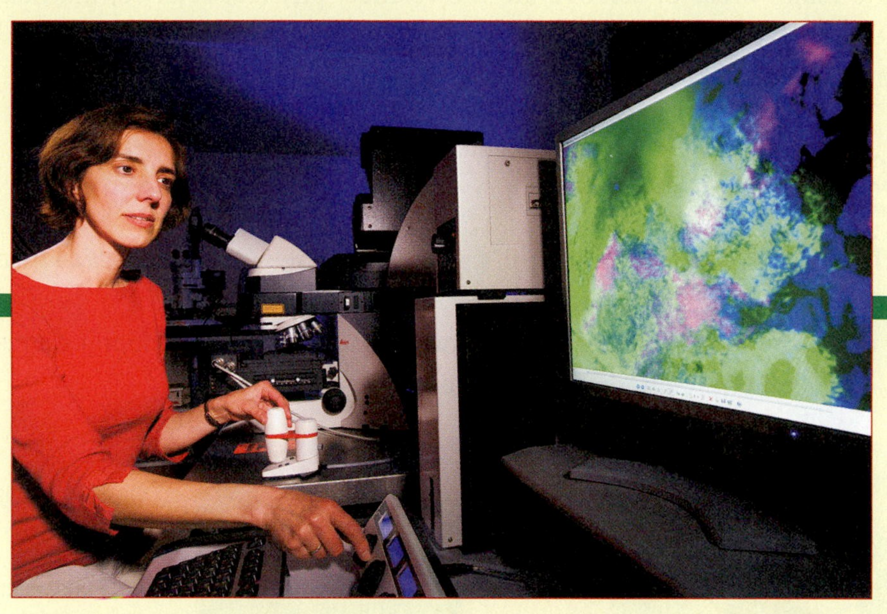

7–1. A microbiologist is using a confocal laser scanning microscope to study *Salmonella* on cilantro leaves. (Courtesy, Agricultural Research Service, USDA)

SCIENCE is knowledge of the world around us. It uses general truths or laws and involves specific methods of study. New ways of producing food and fiber and meeting human needs come about through science. Science is an exciting and informative area of learning.

Observing an event and recording what is seen is important in science. The event must be carefully carried out, and good measurements made. For example, a new growth chemical may be used on a crop. The crop is observed for change. The observations are compared with those of other fields of the same crop but where the chemical was not used.

Today, we use science to learn about our world. We often want to know whether something actually exists and how to solve problems; we want the facts! To help get the facts, special ways of looking at things are used. These give answers to our questions.

RELATIONSHIPS WITH SCIENCE

Agriculture has close relationships with science. It is sometimes said to be "science in action." The major areas of science are life science, physical science, mathematics, and social science. Each of these areas has many divisions. Differences between the areas are not always clearly defined. Most areas of science overlap.

LIFE SCIENCE

Life science is the study of living things. It is sometimes called biology. The major areas of life science are botany and zoology.

Botany and zoology have several areas that overlap. These overlapping areas apply to both plants and animals. Genetics is an example.

7–2. Plant scientists are studying high-lycopene watermelons. (Lycopene is the red pigment that, when eaten, reduces in humans the risk of certain diseases, such as cancer.) (Courtesy, Agricultural Research Service, USDA)

Botany

Botany is the science of plants. Plants have the unique ability to convert energy from the sun into food. Plants use the food themselves, and animals use it when they eat plants. Humans use this "captured energy from the sun" (in the form plant and animal products), such as the oil and meal from soybeans and the delicious and nutritious fruit from tomato plants.

Agriscientists who specialize in agronomy, horticulture, forestry, and related areas study ways to help plants grow better. They study what plants need for growth and how to provide for these needs. When plants grow more efficiently and productively, more valuable products are produced.

Table 7–1. Overlapping Life Science Disciplines

Area	Subject
Morphology	Internal and external structure of living things; important in understanding plants, animals, and other organisms.
Anatomy	Arrangement of the parts of plants and animals.
Physiology	How living things function, including life processes and what happens when they fail.
Biochemistry	Chemical processes in plants and animals.
Genetics	How plants and animals pass characteristics to their offspring; often used to improve plants and animals.
Molecular biology	Organization of living matter; important in understanding disease, how feed is used, etc. As molecular methods have emerged, new understanding of genetics and the classification of organisms has been gained.

Zoology

Zoology is the science of animals. Many kinds of animals are important in agriscience. Some animals are beneficial; others cause problems.

Beneficial animals provide products that people need. Food, clothing, medicines, soap, and other products may come from animals. Agriscientists study ways to improve animals and make them grow better.

Animals can cause harm in several ways. Small animals, such as insects, may damage crops. Large animals, such as wolves, can attack and destroy sheep or cattle. Harmful animals are often called pests.

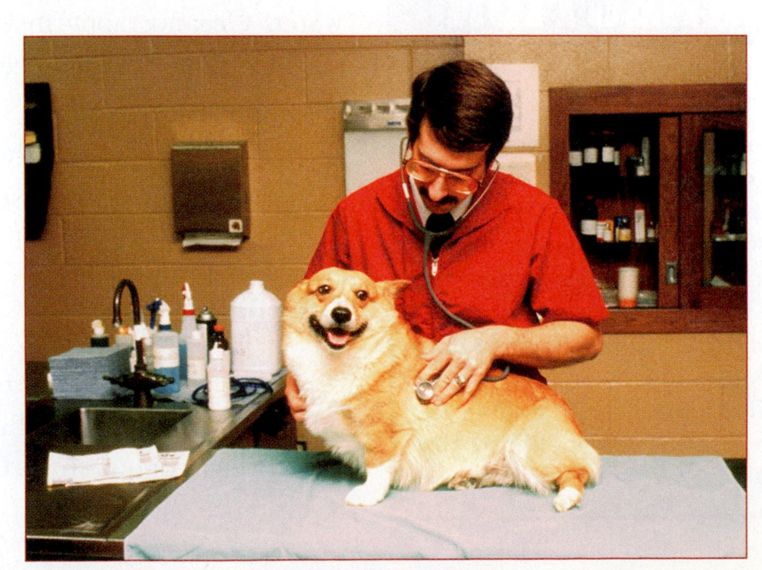

7–3. In-depth knowledge is needed in treating a dog. (Courtesy, College of Veterinary Medicine, Texas A&M University)

PHYSICAL SCIENCE

Physical science is the study of nonliving things. Areas of physical science used in agriscience are earth science, chemistry, and physics.

7–4. A hydrologist has collected a sample of water for testing.

Earth Science

Earth science is the study of the environment in which plants and animals grow. This includes soil, water, and the atmosphere. Branches of earth science include geology, meteorology, hydrology, oceanography, and astronomy.

Geology is the study of the earth's composition, structure, and history. Included is movement in the earth by earthquakes and volcanic eruptions. Study of the soil is a part of geology that is important in agriscience.

Meteorology is the study of the earth's atmosphere. It includes the weather. Weather information will influence many activities, such as cutting hay, harvesting wheat, picking cotton, plowing land, protecting crops from freezing, and using irrigation.

Hydrology and oceanography involve studying water. We use both in agriculture. Hydrology is concerned with water on land. It deals with the movement of water on and through the ground and with ways to keep good-quality water. Oceanography involves studying the water in the oceans. Changes in ocean water have affected our environment.

Astronomy is the study of celestial bodies, such as the planets, the sun, and the stars.

ACADEMIC CONNECTION

BIOLOGY

As the study of living things, biology has a great deal of application in agriculture. Relate your study of biology to AFNR. Think about how what you learn in biology class helps with success in producing plant and animal products. You may wish to do a science fair project on one of the areas. The project might qualify you for the agriscience competition of the National FFA Organization.

Chemistry

The science of chemistry is concerned with the makeup of matter. Matter is anything that occupies space and has mass (weight or quantity). Chemists often refer to matter as a substance.

A substance that cannot be broken down chemically into different substances is called an element. Iron is an example. If a substance can be separated into two or more elements, it is a compound. Salt (NaCl) is an example. Salt can be broken into two elements—sodium (Na) and chlorine (Cl).

Table 7–2. Examples of Chemistry in AgriScience

Example	Description/Importance
Fertilizers	Compounds used to enrich the soil. Chemistry tests are used to tell which nutrients are present and needed in the soil; tests allow selection of the best fertilizer.
Pesticides	Compounds used to control insects, weeds, and other pests. Some have elaborate chemical structures.
Rust	A chemical process that occurs when iron chemically combines with oxygen in the presence of moist air. Iron, common in agricultural equipment, will rust if not protected. Painting iron protects it from moisture and prevents rust.

Physics

Physics is the study of the physical nature of objects. Areas of physics include heat, light, electricity, and mechanics.

Mechanics is the area that gets the most attention in agriscience. It deals with objects and their motions. Most agricultural machinery and equipment apply many principles of physics. A large machine may be made up of many small, simple machines.

7–5. An engineer is investigating the working of large gears.

Table 7–3. Examples of Physics in AgriScience

Example	Description/Importance
Loading ramps	Sloping ramps that make it easy to lift heavy objects. A loading ramp uses a simple machine known as an inclined plane.
Wheels and axles	Devices used to move equipment from one place to another and to transfer power. Power is the amount of work done. (Think of all the tractors and equipment we wouldn't have if there were no wheels and axles!)

MATHEMATICS

Mathematics is the science of numbers. With mathematics, people can make exact statements and predictions. Systems for measuring, calculating, and inferring have been developed. Computers have changed and expanded the use of mathematics in AFNR.

Mathematics includes arithmetic, algebra, geometry, trigonometry, calculus, probability and statistics, and other more advanced areas. Many arithmetic functions are used in AFNR, such as adding, subtracting, multiplying, and dividing. Geometry is used in shapes, angles, and various functions in building construction, earth construction, landscaping, and the like. Note: The appendixes of this book present useful conversions and formulas.

7–6. A tape measure is being used to determine the width of a board. (Courtesy, Education Images)

Measurements

Measurement is the process of assigning a number to an attribute. We use it to determine the number of units in something. The qualities of an object or phenomenon can often be measured or quantified. Scales of measurement are used. Some scales are fairly simple to use, such as a tape measure or weighing scale.

Table 7–4. Examples of Measurements in AFNR

Kind of Measurement	What It Means
Linear	The distance between two points; used to make other measurements, such as area and volume; performed with a linear measuring device, such as a tape or a ruler; also known as length.
Area	The measurement of surfaces; linear measurements are made before area is calculated; reported in square units, such as square feet (ft^2).
Volume	The total size of an object or container; the amount of space something takes or holds.
Weight	The heaviness of something; related to gravity; determined using a weighing machine.

The system of measurement varies and creates some blockages in communication. In the United States, the customary (English) system of measurement is used. In other nations, the metric system (SI or International System) is used. Formulas can help convert from one system of measurement to the other.

Table 7–5. Examples of Customary and Metric System Measures and Equivalents

Customary	Metric	Equivalents
Volume:		
Quart (qt)	Liter (L)	1 qt = 1.057 L
Gallon (gal)	Liter (L)	1 gal = 0.2642 L
Weight:		
Ounce (oz)	Gram (g)	1 oz = 28.3495 g
Pound (lb)	Kilogram (kg)	1 lb = 0.45359 kg
Linear:		
Inch (in.)	Centimeter (cm)	1 in. = 2.54 cm
Foot (ft)	Centimeter (cm)	1 ft = 30.5 cm
Yard (yd)	Meter (m)	1 yd = 0.91 m (1 m = 39.37 in.)

Note: Some rounding has been done with the equivalents. Refer to the appendixes of this book and other sources for more details.

Two examples using the customary (English) system of measurement are:

- **Weight**—A ton of feed is 2,000 pounds to both buyer and seller. It is a uniform measure on which both can agree. Mathematics allows for exact and uniform measurements. In the metric system, weight is measured in grams, with larger amounts measured in kilograms.

7–7. Large scales are being used to weigh this grain truck as it arrives at an elevator. (Courtesy, Education Images)

- **Area**—An acre of land is always the same size: 43,560 square feet. When people say "acre," they know the exact area of land. There is no disagreement. In the metric system, land area is measured in hectares.

Statistics

Statistics is a branch of mathematics that deals with the collection, analysis, interpretation, and presentation of data. We use statistics is many ways. Agriscientists use statistical methods to draw meaning and forecasts from the observations they have made. Statistics helps people to make good decisions.

In agriculture, statistical methods can be used in two main ways: to describe and to infer. Descriptive statistics summarize data. They tell the reader what was found in a study. Examples of descriptive statistics include count (n), mean (average), mode (most frequent score or value), and median (midpoint of a distribution—half the values are above a particular score and half are below). Inferential statistics allow us to draw inferences about a population or process. With inferential statistics we can project the attributes of a sample to the population from which it was drawn.

Probability is the chance or likelihood that an event will occur. It helps in drawing conclusions from data about a potential event occurring. If you flip a coin, what is the probability that it will be heads? If a player made 90 percent of her free throws in the first 10 basketball games of the season, what percent will she likely make in the eleventh game? Probability

Technology Connection

LAB RESEARCH

Research in a laboratory involves many precise treatments, measurements, and observations. Keeping careful notes of all activities is essential. When were treatments begun? What changes occurred? When did the changes occur?

Accuracy and precision are most important. Without these, the findings of an experiment may lack validity and reliability. Valid results mean that the reported findings are what occurred. Reliable results mean that the experiment can be repeated with the same findings resulting. If validity and reliability are missing, the report of an experiment is of no value.

This shows a scientist making observations using a microscope. The observations are being recorded with a pen on paper. A computer keyboard is also present, meaning that the observations are entered into a computer database.

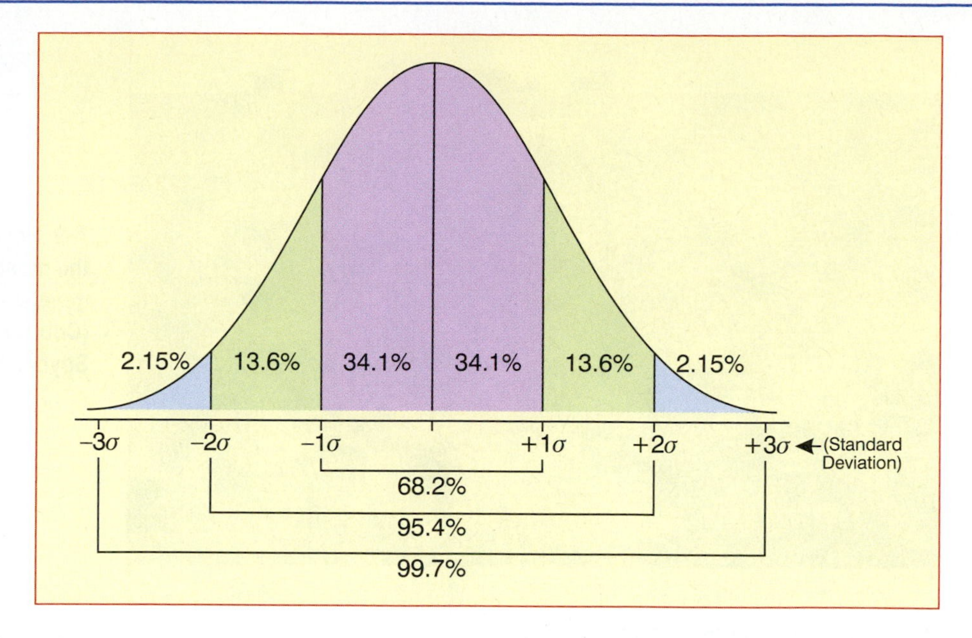

7–8. A normal-distribution curve (bell-shaped curve). The curve is vertically lined into standard deviations. One standard deviation near the center of the baseline contains 34.1 percent of the population.

has a number of underlying mathematical formulas and theories, including randomness. Probability is often used with statistical analysis of data.

Normal distribution is another concept associated with statistical methods. The standard normal distribution forms a bell-shaped curve. The attributes of populations or events usually conform to a normal distribution. The statistical method of standard deviation can be used to determine how well an observation fits within the bell-shaped curve.

More information can be found in statistics books or at online sites.

SOCIAL SCIENCE

Social science deals with human society. It is also known as behavioral science. Anthropology, psychology, and sociology are areas of social science. Many traits of human behavior are studied, including people's interest in foods, clothing, and shelter.

Some social science areas have direct relationships with agriscience. Consumer behavior, food consumption patterns, and attitudes toward new technologies involve social science. Understanding these relationships helps explain why people respond as they do.

NEW TECHNOLOGY

Exciting developments are being made in technology. New ways of producing food, fiber, and shelter materials will result. Several emerging technologies are described here.

7–9. Soybeans are among the most widely planted genetically altered crops. (Courtesy, American Soybean Association)

BIOTECHNOLOGY

Biotechnology is the application of science in the development of new products or processes that involve living organisms. The work is often in laboratories, where conditions can be controlled. Some biotechnology has been around for a long time. More advanced forms are now being used.

New uses of biotechnology hold much promise. These deal with increasing production and protecting the environment. New vaccines are being developed to protect animals. Methods to get more lean meat from animals result in improved efficiency. "Waste-eating" bacteria are protecting the environment.

7–10. Plum-pox resistance was developed by adding one gene from a plum-pox-resistant variety of French plum. This pox is caused by a virus that can destroy a plum crop. (Courtesy, Agricultural Research Service, USDA)

GENETIC ENGINEERING

Genetic engineering is an advanced form of biotechnology. It involves changing the nature of living organisms by "cutting and pasting" genes. The changes are intended to be very small and only of desired traits.

Genetic engineering has already produced important new traits. Plants and animals with these traits are known as genetically enhanced organisms (GEOs). Crops have been developed that resist insect damage and spoilage. Some crops are being developed that kill the weeds around them.

Genes that provide natural benefits are used. An example involves black walnut trees. These trees have a natural substance that kills certain plants that grow near them. The identification of the gene in black walnut trees that has this trait is the first step. Once the gene is isolated, it can be moved to other plants. The "other" (genetically enhanced) plants will kill weeds that grow near their roots. Chemical weed killers will not be needed.

ANIMAL IDENTIFICATION AND TRACEABILITY

Animal producers are being involved in several ways to promote animal identification and tracking. A National Animal Identification System (NAIS) is being implemented for horses, cattle, hogs, and other species. In some cases, the animals are individually identified, with beef cattle being an example. In others, they are identified by group or lot, as with chickens and other smaller animals. Individual animals may be identified with eartags, radio frequency identification (RFID) eartags, and injectable transponders. Each animal receives an Animal Identification Number (AIN).

CAREER PROFILE

BEEF CATTLE SCIENTIST

A beef cattle scientist studies the production of beef animals. This includes animal well-being, nutrition, efficiency, reproduction, and food product quality and safety. The work may be to identify threats to animal well-being, solve problems they cause, and maintain a good environment. The work typically involves being around cattle, managing and corralling cattle, and dealing with variable dispositions among animals.

Beef cattle scientists have college degrees in animal science or a closely related area. Most have master's degrees and doctorates. Practical experience and a fondness of beef cattle are beneficial. In high school, take courses in science and agriculture to begin preparing to be a beef cattle scientist.

Jobs for beef cattle scientists are with universities, research stations, government agencies, and large ranches. This photograph shows a beef cattle scientist (left) working with a food technologist to obtain microbe samples from the cow's hide. (Courtesy, Agricultural Research Service, USDA)

7–11. An eartag has information that identifies this beef animal. (Courtesy, Education Images)

Animal tracing is also being implemented. ***Traceability*** is the identification of an animal or lot of animals so that it can be traced back to the producer. This is important in the event of a disease outbreak or other situation. The origin and movement of an animal can be determined.

Individual producers may register the premises where animals are kept and produced. In early 2009, more than a half million premises in the United States where animals were kept or produced had been registered through the NAIS. Registration is voluntary but will likely become increasingly important.

PRECISION FARMING

Precision farming is the use of cropping practices that improve yields based on the needs of the land. Precision farming is also known as site-specific agriculture.

Several technologies are used. A global positioning system (GPS) tracks location of equipment. Monitors collect yield data or other information. A computer keeps track of all the information. Pest infestation, soil fertility and compaction, and other characteristics are studied. Computer-controlled equipment applies the needed fertilizer and other inputs to grow a crop. Records of crop yields are kept for the areas.

7–12. Precision farming uses several technologies. This combine is equipped with a global positioning system and a grain-flow sensor to gather yield information. (Courtesy, Agricultural Research Service, USDA)

REMOTE SENSING

7–13. Aerial photographs give details about land. Can you find a stream? A road? (Courtesy, USDA)

Remote sensing is the gathering and recording of data from a great distance. Most remote sensors are on satellites some 500 miles above the earth. Some methods used are highly accurate.

Landsat is the term applied to United States satellites that make photographs of the earth and plot the earth's resources. The photographs are used to make maps that provide meaningful information. Not only is the earth mapped, but sea-floor elevations are also mapped. Photographs taken from airplanes have been used for several years as sources of information. Remote sensing in agriscience is beneficial in forecasting the weather, locating natural resources, detecting crop disease, and protecting the environment.

LASER TECHNOLOGY

A ***laser*** is a device that produces a type of light known as coherent light. Lasers are sometimes called electric eyes. An intense, narrow beam of light is produced that is useful in many ways.

The following are examples of laser technology in agriscience:

- **Land forming**—Laser equipment can be used in forming land to a certain contour. Rice fields, terraces, and fish ponds are also designed with lasers. The line of light does

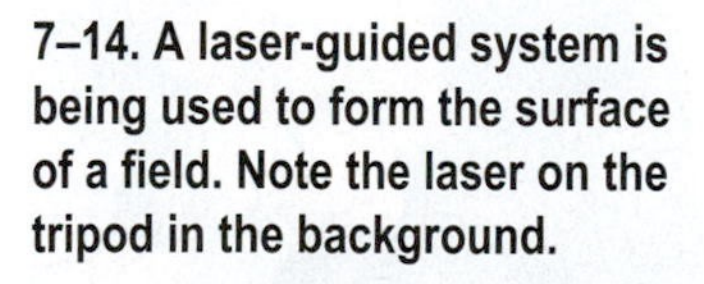

7–14. A laser-guided system is being used to form the surface of a field. Note the laser on the tripod in the background.

not bend and cause unwanted changes in elevation when moving the earth. The laser beam guides the tractor as it lays out the land.

- **Monitoring**—Laser beams can be used to "watch" processes in manufacturing plants and other facilities. The light will detect changes in movement or other things indicating a problem.
- **Scanning**—Laser beams scan bar codes on merchandise. Supermarkets use scanners that list items and prices. The scanners are hooked to a computer that has all the prices.

COMPUTATIONAL SCIENCE

Computational science is the use of computers to solve problems in mathematics and science. Supercomputers with considerable capacity are needed. Computational science is used in agriscience in designing facilities, developing products, and modeling the earth's resources.

RADIATION

Radiation is a form of energy that travels as waves. Radiation may be natural or artificial. Life depends on some kinds of radiation, such as that from the sun. Nuclear plants use radioactive materials to produce radiation power. Some people are afraid of the potential danger of radiation.

AgriScience Connection

KEEPING CORN SAFE

Corn and other grains may contain a very toxic fungus known as *Asperigillus flavus*. This fungus causes aflatoxin in grain. Aflatoxin is a natural poison and possible cause of disease.

A biochemist is using a photoacoustic infrared sensor to evaluate corn for fungal contaminants. The procedure has potential for use at grain elevators and food processing plants. When perfected, the procedure can be used to reduce the natural toxins in food. (Courtesy, Agricultural Research Service, USDA)

Electromagnetic radiation, an artificial type, has many uses in agriscience. Foods can be exposed to radiation to prevent spoilage. For example, meat products can be irradiated to kill bacteria that may be on or in them.

Radiation can be used to control insects in food. Wheat, corn, and other cereal grains often have weevils, worms, or other pests. By irradiating grain, the pests are destroyed and the food is kept wholesome.

THE SCIENTIFIC METHOD

Agriscience uses the ***scientific method***, which is an organized way of asking questions and seeking answers. This is also known as scientific problem solving. By practicing the scientific method, you are "thinking like a scientist."

Questions often come up in agriscience. These may be the "what" or "why" of a plant or animal. All problem solving involves answering questions. Thinking like an agriscientist involves precise thinking. It must be exact and free of error. The experiences of others are valuable in solving problems, especially new problems. You may find the scientific method useful in creating a science fair project or solving a problem in your daily life or supervised experience.

STEPS OF THE SCIENTIFIC METHOD

The scientific method has several important steps. Sometimes the order or number of the steps may change, but a logical process is always used.

Step One: Identify the Problem

A problem can't be solved until it is defined. To help with this, most problems are first stated as questions.

Sometimes, people see only symptoms of a problem. More exploration is needed to identify the problem accurately. If hogs refuse to eat, failure to eat is a symptom; it is not the problem. Studying a symptom will not result in solving a problem.

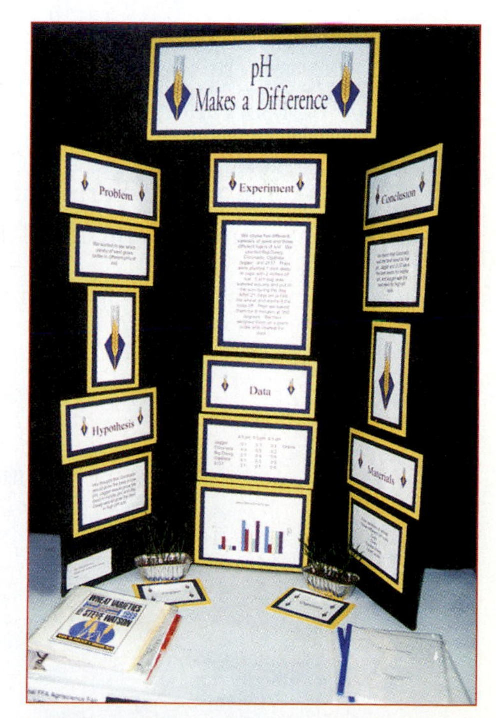

7–15. A student's display on studying the effects of pH on plants.

Step Two: Get Information

Get the facts about a problem. Sometimes, this is called data collection. The facts are what is observed about the prob-

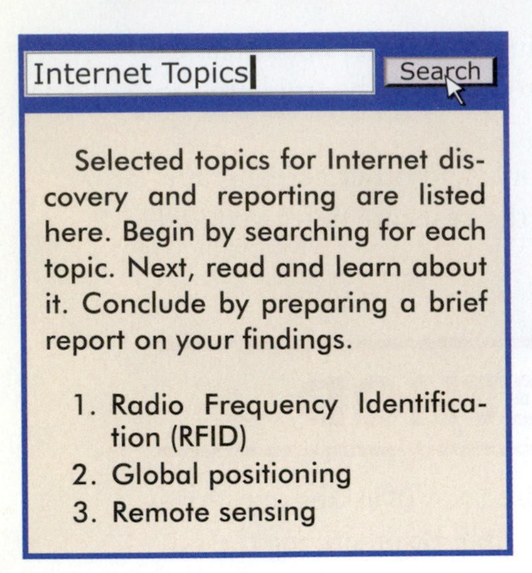
Internet Topics
Search
Selected topics for Internet discovery and reporting are listed here. Begin by searching for each topic. Next, read and learn about it. Conclude by preparing a brief report on your findings.
1. Radio Frequency Identification (RFID)
2. Global positioning
3. Remote sensing

lem. Everything people see, hear, touch, taste, or smell can be a part of the observation.

Careful measurements of observations are essential. Several kinds of measurements may be used. In agriscience, measurements are often made of increased growth, quality, or amount produced.

Step Three: Suggest an Answer

Agriscientists suggest answers to problems. These suggestions are known as hypotheses. Hypotheses are statements about the problem that can be tested. Hypotheses may be proven true or false. The findings are important either way. A sample hypothesis is: The Coronado variety of wheat will grow better in soil with low pH than other common varieties.

Step Four: Experiment

This step uses an experiment to prove each hypothesis. An ***experiment*** is a trial or test. All conditions are controlled except the one being studied. The condition that is changed is known as the variable. In some experiments, several variables may be used. The part of an experiment that is not changed is the control, sometimes called the controlled variable.

Observations are made. Changes are carefully measured. Inaccuracy results in poor-quality experiments. The solution to a problem is no better than the accuracy of the information.

Agriscience experiments frequently use control and trial plots. For example, experiments to measure the effects of increased fertilizer on yield include plots that do not get the

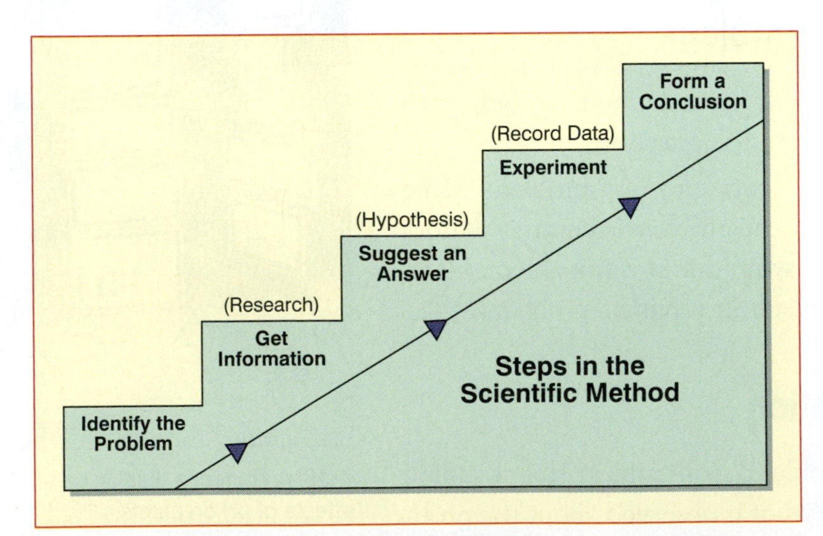
Form a Conclusion
(Record Data)
Experiment
(Hypothesis)
Suggest an Answer
(Research)
Get Information
Identify the Problem
Steps in the Scientific Method

7–16. The five-step scientific method.

increased fertilizer as well as those that do. All plots should be identical. If not, the test that was carried out might not cause differences.

Step Five: Form a Conclusion

A conclusion is a judgment formed after an experiment. It involves a process called inferring. The data are carefully looked at for meaning. Facts are used to infer traits to the problem being solved.

Sometimes conclusions need more study. This is known as replication (repeating). Most new agriscience developments need replication. This assures that the findings can be trusted.

Additional Step

Some scientists include an additional step in the scientific method: report the results. Sharing the findings of an experiment is important and an obligation of a scientist. Findings may be shared in research reports, in magazine articles, on Web sites, in exhibits, through broadcast media, and in other ways.

7–17. An experiment is underway to test the use of various washes in removing microbes from spinach. (Courtesy, Agricultural Research Service, USDA)

PRACTICAL USE

The scientific method is used with everyday problems. A few examples are listed here. You can probably think of many others.

- If a tractor engine stops running, the mechanic can use the scientific method to get it going again.
- If a plant shows signs of disease, the horticulturist can use the scientific method to decide how to control the condition.
- If the fish in a pond refuse to eat, the aquaculturist can use the scientific method to solve the problem.
- If flowers wilt for some unknown reason, the floriculturist can use the scientific method to find an answer.

REVIEWING

MAIN IDEAS

All areas of science are applied in agriculture. The major areas of science are life science, physical science, mathematics, and social science.

Life science is also known as biology. It has two major areas: botany (plant science) and zoology (animal science). Many plants, animals, and other organisms are produced in agriculture. These require knowledge of life science.

Physical science includes earth science (the study of soil, water, and the atmosphere), chemistry, and physics. These are applied in many aspects of agriculture. Mechanical technology in agriculture uses a number of physical science applications.

Agriscience uses the scientific method in which experiments are made to test hypotheses. Research is the careful and diligent search for answers to problems. The scientific method is the primary tool used in research. The five major steps are: identify the problem, get information, suggest an answer, experiment, and form a conclusion. In addition, communicating the findings to others is an important step.

New ways of producing plants and animals using technology sometimes raise issues. A few people may have concern about biotechnology. They want foods that are safe to eat. They want production methods that don't damage the environment and that protect resources for future generations. They also want the rights of people protected.

QUESTIONS

Answer the following questions, using complete sentences and correct spelling.

1. What is science? Why is it important?
2. What are the four major areas of science? Define each.
3. What are the two major areas of life science? Why are these areas important in agriscience?
4. What are the areas of physical science? How does each relate to agriscience?
5. What is mathematics?
6. What two systems of measurement are used?
7. What is statistics?
8. What is the scientific method? What steps are used? Briefly explain each step.
9. What is a normal distribution?
10. What are the new areas of technology? Briefly explain each. Give examples of how each can be used in agriscience.

EVALUATING

Match each term with its correct definition.

a. bell curve
b. measurement
c. research
d. botany
e. zoology
f. biology
g. descriptive statistics
h. linear measurement
i. remote sensing
j. scientific method

_____1. The science of plants.

_____2. The shape of a line on a graph that depicts population attributes.

_____3. The process of assigning numbers to an attribute.

_____4. The distance between two points.

_____5. A five-step process of getting answers to questions.

_____6. The careful search for answers to problems.

_____7. Collecting information from a distance.

_____8. Used to summarize data.

_____9. Means the same as life science.

_____10. The science of animals.

EXPLORING

1. List the areas of science taught at your school. Interview the teachers of the classes; ask what subjects they cover and how those subjects relate to agriscience. Prepare a short report on your findings.

2. Do a science fair project. Conduct an experiment related to agriscience, and enter it in the local science fair. Also, enter your project in the agriscience fair.

3. Identify an agriscience problem (question) in your local community. Design a process for answering the question using the scientific method.

CHAPTER 8

Natural Resources and the Environment

OBJECTIVES

This chapter covers important areas of natural resources and the environment, including responsible enjoyment of nature. It has the following objectives:

1. Define natural resources, environmental science, and sustainable agriculture.
2. Describe the role of ecosystems.
3. List and describe examples of natural resources.
4. Identify important wildlife species.
5. Describe responsible use of wildlife.
6. Discuss the meaning of pollution, and identify its sources.
7. Explain methods of agricultural waste disposal, including composting.

TERMS

abiotic factors
biodegradable
biosphere
biotic factors
birding
composting
ecosystem
effluent
environmental science
erosion
fishing
food chain
fuel
habitat
hunting
landfill
natural resource
niche
nonrenewable natural resource
particulate
pollution
recycling
renewable natural resource
soil
sustainable agriculture
symbiosis
toxin
waste
water cycle
wildlife
wildlife management

8–1. Earth's natural resources provide many opportunities for enjoyment, such as kayaking in a white-water stream.

PLANET EARTH has many resources that support life. Understanding and using them wisely will help assure adequate resources for future generations.

Fortunately, scientists have found ways we can responsibly enjoy our natural resources. We can do so while protecting the environment and ensuring that our resources will last indefinitely. Each of us has a responsibility to be a good steward of the environment.

Agriculture uses sustainable practices. These respect our resources and help assure that some will be available many years from now. Yes, in AFNR, we act responsibly when it comes to natural resources and the environment.

EARTH'S RESOURCES

Three major areas are included: natural resources, environmental science, and sustainable agriculture. These all very prominently relate to and promote wise use of Earth's resources.

NATURAL RESOURCES

A ***natural resource*** is something found in nature that supports life, provides fuel, or is used by humans in other ways. There are a number of important natural resources. Some natural resources can be renewed; others cannot.

A ***renewable natural resource*** is a resource that can be replaced, but a long time may be needed for replacement to occur. Renewable resources include soil, water, air, and wildlife. These also relate to a quality living environment for people.

A ***nonrenewable natural resource*** is a resource that cannot be replaced. Fossil fuels (coal and petroleum) and minerals (gold and iron) are examples. While they cannot be replaced, nonrenewable resources can often be reused or recycled. We should use these resources carefully to assure that they are available for many years.

8–2. Strip cropping is used in this field to conserve soil and water.

ENVIRONMENTAL SCIENCE

Environmental science is the study of using and protecting the resources around us. Humans use resources to produce goods and make life better. We need to make sure that a good environment is available in the future.

The emphasis in environmental science is on conserving and improving natural resources. These resources include air, soil, water, wildlife, native plants, and minerals.

Disposing of wastes is a major area in environmental science. How we deal with solid wastes, hazardous wastes, and waste water poses big challenges in some locations.

8–3. Air samples are being collected as part of an environmental study in Iowa. (Courtesy, Agricultural Research Service, USDA)

SUSTAINABLE AGRICULTURE

Sustainable agriculture is the use of practices that will maintain our long-term and indefinite ability to produce plants and animals to meet human needs. The practices used represent a combination of approaches. No one practice stands alone.

With sustainable agriculture, producers maintain the soil and manage water to protect and prevent loss or damage. Pest management involves carefully selecting practices to protect the environment and using control measures only when the pest population is a definite threat. Some producers use precision technologies. These allow the collection of information at harvest and the varying of inputs the next year based on yields of the previous year.

ECOSYSTEMS

8–4. Rare and exotic animals are kept for people to enjoy. The giraffe is the tallest four-legged animal.

Plants and animals live in places that meet their needs. The area of Earth that supports life is the ***biosphere***. The biosphere extends from the earth's surface into the surrounding atmosphere and the oceans. Within the biosphere, various ecosystems exist.

All the parts of a particular environment form an ***ecosystem***. Some parts are living; others are nonliving. The living things strive for balance in the system. However, balance is never fully achieved. Living things

8–5. Fish live in an aquatic ecosystem.

8–6. What biotic and abiotic factors do you see in this mountain stream?

depend on nonliving things, such as water and sunlight. The parts of an ecosystem can be divided into two factors—biotic and abiotic.

BIOTIC FACTORS

Biotic factors are all the living parts of an ecosystem. Most ecosystems have many living things, or biotic factors. These include plants, animals, and other organisms. Different types of living things compete for space, water, food, and other resources. Plants and animals compete with their own kind and with other species. For example, deer compete with their own kind and with others for food plants.

Biotic factors interact with each other. Some relate well even though many kinds of living things may be present. They form a dependent relationship that benefits at least one of them. This relationship is known as ***symbiosis***.

Some animals do not relate well. Wolves may kill and eat sheep. This is the act of predation. Animals that kill are known as predators; those killed are known as prey.

ABIOTIC FACTORS

Abiotic factors are all the nonliving parts of an ecosystem. These include water, temperature, sunlight, rocks and soil, the lay of the land, and the available space. Abiotic factors interact with each other and with the biotic factors.

Abiotic factors determine which plants and animals can live in an ecosystem. Plant growth is essential for animals to live. Biotic factors cannot exist without the right abiotic factors.

HOW ECOSYSTEMS WORK

A meadow with a stream flowing through it has a working ecosystem. All the biotic and abiotic factors interact.

The system may include water in the stream, fish and other life in the water, fertile soil by the stream, plants growing on the surrounding land, insects in the plants, and birds in the air above.

Sometimes, people damage an ecosystem. Polluting the water, trashing the grounds, or killing all of a certain kind of plant or animal causes problems. This is true of anything that changes a natural ecosystem. Maintaining ecosystems requires understanding how they work.

Habitat

The area where a plant or animal lives in natural conditions is its ***habitat***. Different living things have different habitats. Understanding a habitat is important in helping an organism survive and grow.

Many things make up a habitat: the climate (temperature, rain, etc.), the lay of the land, the nature of the soil and water, and other plants and animals in the area.

Changes in a habitat affect an entire ecosystem. For example, a lack of rain results in fewer plants. Fewer plants result in a lack of food for animals. The animals may starve or move to another location searching for food.

Agricultural practices may sometimes affect a wildlife habitat. For example, the habitat of some animals may be changed or eliminated by plowing the land. To solve this, strips of land around fields are not plowed. They are left to serve as a habitat.

8–7. Hare prefer habitat with certain features, such as food, water, and shelter. (Hare (*Lepus timidus*) vary from rabbits in habitat preferences, ear size, and nesting.)

Niche

A ***niche*** is the function or role of a living thing within its habitat. This relates to food, shelter, raising of young, and other aspects of life.

To illustrate, beavers may live in a stream that passes through a wooded area. They cut down small trees, pull them into the water, and make dams that hold water year-round. Beavers live in and around the water pool and will defend the area against the intrusion of other beavers.

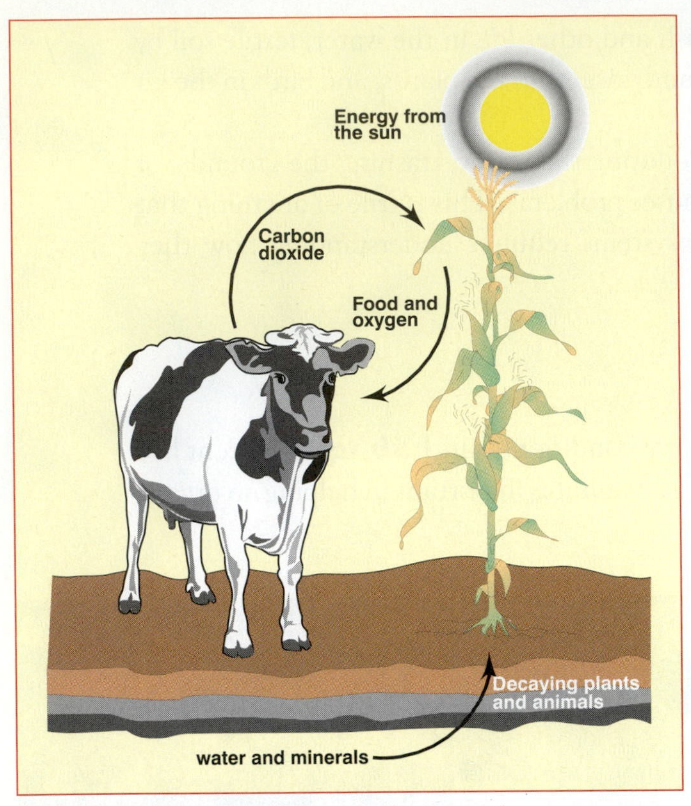

8–8. A simple food chain: Plants grow using sunlight and soil nutrients. A cow eats the plants and returns manure and tissues to the soil. Future plants use the nutrients from the decomposed manure and tissues for growth.

Food Chain

A ***food chain*** is the ranking of species into successive levels, where each feeds on the one below. It begins with very basic foods and advances upward to more complex foods. The highest level of feeding is by members of a higher level on those of the levels below, particularly, the level just below.

A food chain begins with plants using energy from the sun to make food, which results in growth of the plants. Animals eat the plants, grow, and die or are eaten by animals that later die. The remains of plants and animals rot and return nutrients to the soil. The cycle begins again as new plants use the nutrients for growth.

All living things participate in the food chain. Some are tiny and can be seen only when enlarged. Others are quite large and cultured by humans.

AgriScience Connection

ENJOYING OUR WORLD

People seek ways of having fun and enjoying life. Many like outdoor activities. They are looking for good places to relax and renew themselves. Camping, boating, picnicking, and being in fresh air are important.

While we are enjoying our world, we also need to keep it in good condition. We must strive to prevent pollution. We must properly dispose of wastes. We should work to be sure future users have good places for fun. Never leave debris behind after camping, hiking, hunting, or fishing.

MAJOR NATURAL RESOURCES AND RENEWABILITY

Natural resources are classified into two groups: renewable and nonrenewable. Practices are used to sustain both. *Sustain* means that the resources will last indefinitely. They will be available for future generations.

RENEWABLE NATURAL RESOURCES

Renewable natural resources are those that can be replaced. When used, more can be created. Our actions can encourage the renewal of natural resources. The important renewable resources are soil, water, wildlife and fish, forests, and air.

Soil

Soil is the outer layer of the earth's surface. It consists of minerals, organic matter (rotting plants and animals), moisture, and small amounts of other materials. Soil supports plant life. Essential nutrients must be in the soil for plants to live and grow.

Topsoil is the thin layer of the most fertile soil. It is where seeds are planted and crops grow. Grass, trees, and other plants have roots in the topsoil. (Of course, big trees with deep roots go below the topsoil!) Many years are required for a 1-inch layer of topsoil to be formed. Topsoil is made from rotting plants and animal remains.

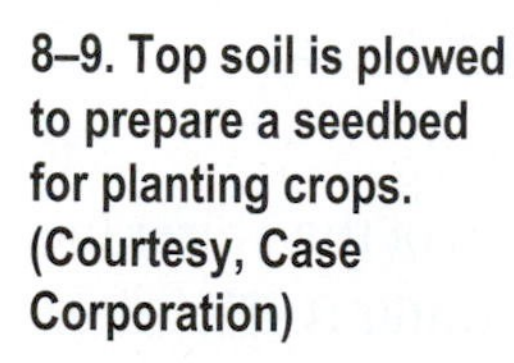

8–9. Top soil is plowed to prepare a seedbed for planting crops. (Courtesy, Case Corporation)

Since topsoil is plowed, it is also easily lost. Water and wind can wash and blow the loose particles away. Washing or wearing away of soil is known as **erosion**. Preventing erosion helps sustain the soil.

Erosion is often most severe on land where the topsoil has been disturbed. Fields and construction sites for highways, houses, and other buildings are major erosion sites. A big rain can wash away more topsoil than can be formed in 20 or more years. Conservation measures are used to preserve the soil and promote formation of new topsoil.

8–10. Soil erosion has formed a deep gully on plowed land that was not wisely managed.

Water

Water is likely the most important natural resource. Crops require water for growth. Animals need water to live. Humans need water for drinking, bathing, and other uses. Industries use water in many ways.

Water is a resource that appears in three forms: solid, liquid, and vapor. It is the only natural resource in three forms within the temperature range of the earth. The solid form is ice. Water becomes ice at 32°F (0°C). Water changes to vapor at 212°F (100°C). As the only planet to have liquid water, Earth is sometimes known as the "water planet."

More than 70 percent of the earth's surface is covered with water. Much of this water has too much salt for most agricultural uses. About 97 percent of the earth's water is salt water. Of the 3 percent said to be fresh water, 2 percent is frozen in glaciers and ice caps at the North and South Poles. This leaves only 1 percent of the water for human, agricultural, and other uses.

Agricultural production is closely tied to water. Some places get too much rain; others get virtually none. In the United States, some mountains in Hawaii may get more than 500 inches a year (too much for crops). Desert areas of Nevada may get only 7 inches a year (too little for crops).

All the water on Earth forms the hydrosphere. This includes water in streams, lakes, oceans, underground, and other places. It includes liquid, vapor, and solid (ice).

The circulation of water in the hydrosphere forms the ***water cycle*** (also known as the hydrologic cycle). In the cycle, water is continuously moving.

The water cycle works like this: Water evaporates into the atmosphere. As a vapor, it rises high into the air. As it rises, it begins to cool. The vapor forms tiny droplets in clouds. When the droplets get larger, they become too heavy to stay suspended in the clouds and begin falling to Earth as rain or other forms of moisture. Some water soaks into the soil; some of it runs into streams, lakes, and oceans. The cycle is repeated.

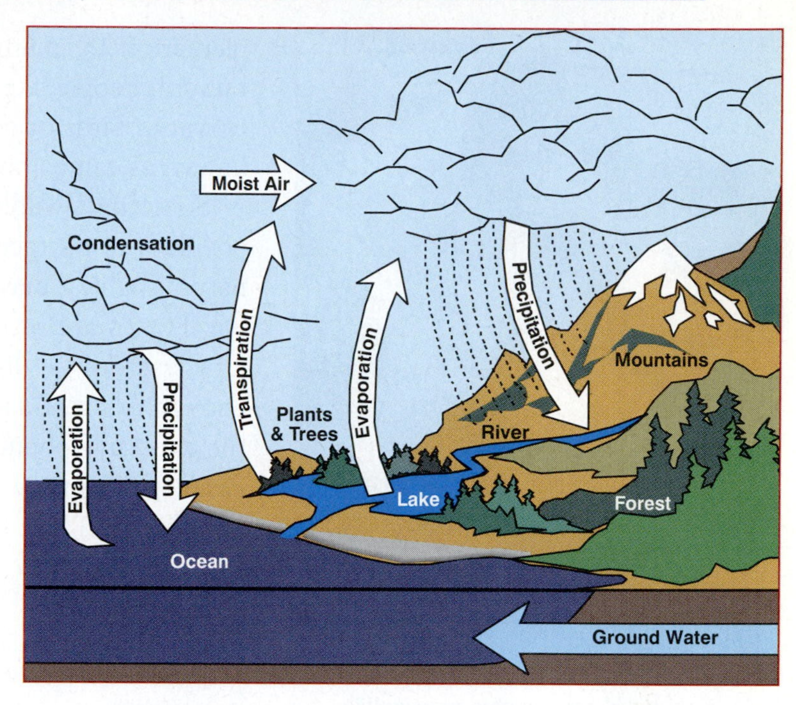

8–11. The water cycle.

Water can be renewed and conserved. The first step is to use no more water than necessary. Wasting water adds to the water shortage. Reducing runoff after a rain helps conserve water. Ponds, terraces, crop residue (stems, leaves, etc.), and other means hold water so it can soak into the ground.

Waste water can be treated for reuse. Sewage treatment plants can reclaim water for irrigating golf courses and other nonfood production purposes. Water from fish farms can be used to irrigate crops. Water from factories can be treated and used for fish culture and other purposes.

Wildlife and Fish

Wildlife and fish are plants and animals that haven't been domesticated. People often think of wildlife as the deer or rabbits they hunt or the fish they catch. These provide

8–12. Earth's natural resources include beautiful and useful wildlife, such as this scarlet ibis.

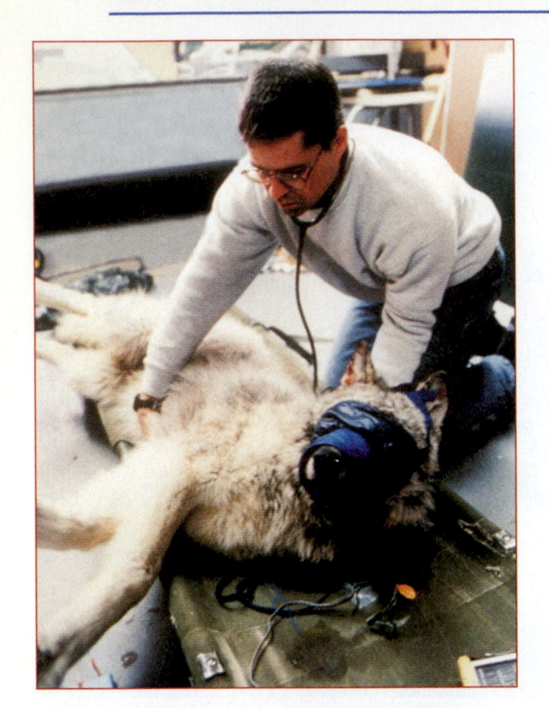

8–13. A captured and sedated gray wolf is being examined before being moved to an area where wolves are no longer found. The goal is to have a wolf population again in the area. (Courtesy, U.S. Fish and Wildlife Service)

recreation for many people. Not all wildlife and fish are hunted. People like to see the beauty of nature. Some like to watch birds or go hiking. Animals that are hunted are known as game animals.

Sometimes wildlife may have too much protection. The population may get so large that it causes problems. The natural habitat may not provide enough food for the wildlife. They may go to new places to get food. Sometimes they attack crops or livestock. Other times they don't get enough to eat and may get sick and die. A good example is the excessive population of bison in Yellowstone National Park. (More information about wildlife is presented later in this chapter.)

Forests

Native forests once covered much of North America. In the 1600s, about half the United States was in forests. As our nation developed, forests were cut. Some wood was used to build shelter and factories. Other wood was burned to make room for crops.

Today, many areas have more trees than in 1700! Tree farms are used in much the same way as crop or livestock

Technology Connection

LASERS FIND DUST

Pollution in the air is more readily visible with special equipment known as LiDAR. The equipment uses laser beams to make tiny particles in the air more visible to special equipment. Computers analyze reflected light in measuring air pollution. For example, using LiDAR at night helps detect dust particles in the air from grain mills and harvesting. This will allow improved ways of reducing emissions from these important operations.

LiDAR is the short name for "Light Detection and Ranging." The technology has been used in other areas, such as studying clouds in the air and pollution from oil refineries. It is now being applied to agricultural situations.

This shows nighttime use of LiDAR by USDA scientists in Iowa to assess air pollution in agricultural areas. (Courtesy, Agricultural Research Service, USDA)

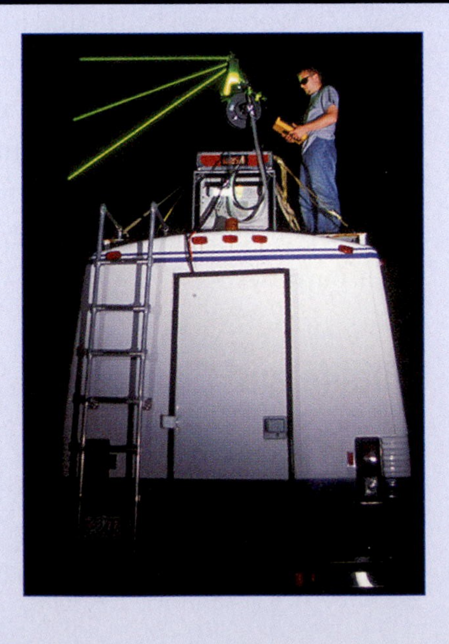

farms. Improved species of trees are planted. Approved practices are used in growing them. Unwanted fires are prevented and put out. Careful harvest practices are used.

Trees help the environment. They provide habitat for wildlife and conserve soil and water. Trees help maintain air quality by producing oxygen.

Trees add beauty and provide recreation. Many people admire the giant sequoia trees of California, the tall pine trees of Mississippi, and the many-rooted banyan trees of Hawaii.

Most commercial forestry operations use practices that sustain the forests. People need wood products as well as a healthy environment. Large companies need wood if they are to stay in business.

8–14. Giant sequoia trees are impressive.

Air

Air is Planet Earth's atmosphere. It is evenly distributed over the surface. All living things need good air.

Air is made up of gases, moisture droplets, and small solid particles known as ***particulate***. The major gases are nitrogen and oxygen, which together make up 99 percent of the air. Living things need oxygen for life processes. The air they get needs to be free of unwanted particulate.

Air quality is often measured as dust count. Measuring the number of particles in a cubic inch of air establishes an air quality index. Over large cities, 1 cubic inch of air may contain 5 million particles. Air from areas far removed from city life, such as over oceans, may have as few as 15,000 particles per cubic inch. The normal level is about 100,000 particles per cubic inch.

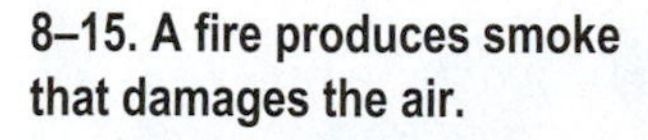

8–15. A fire produces smoke that damages the air.

Living things need air that is as free as possible of dust, smoke, and other particulate. Fires, factories, and engines release substances into the air. Incinerators (places where trash is burned) may release dangerous materials. For example, batteries in flashlights are made using cadmium, a heavy metal that can damage human life. If batteries are burned, the cadmium is released into the air.

The goal is to avoid releasing anything that damages air quality. Devices have been installed on engines to reduce emissions (wastes into the air). Factories use procedures to "clean" the air before it is released. Open burning of waste products has been regulated.

NONRENEWABLE NATURAL RESOURCES

Some resources can't be renewed. Once used, they are gone. These nonrenewable natural resources are sometimes said to be "exhaustible"; they cannot be replaced when used.

Nonrenewable resources must be used wisely. There won't be any more! Conserving them is often easy. The major nonrenewable resources are those that provide fuel and certain minerals.

Fuel

Fuel is any material that provides energy. People use fuel to heat and cool their homes, cook food, power engines, and make electricity.

CAREER PROFILE

FORESTER

A forester works to improve trees, forest productivity, and related areas, such as game and minerals. The work may be outside managing forests, planting trees, measuring trees, or preventing fires. In this photo, a forester is using a clinometer to measure the heights of trees.

Most foresters have college degrees in forestry. Some have master's degrees and doctorates. Begin preparation in high school with science and agriculture classes. Practical experience in forestry is very useful.

Jobs are with government agencies, forestry companies, colleges, and research stations. A few foresters are private consultants.

8–16. Petroleum is pumped from reserves deep in the ground.

8–17. A small solar panel collects energy from the sun and conserves fuels.

The major fuels are from natural sources, such as coal, oil, and natural gas. Artificial fuels are sometimes made from corn and other products. These fuels are alcohol based. Gasohol is a blend of gasoline and alcohol.

There are three major kinds of natural fuels: solid (coal, peat, and wood), liquid (petroleum), and gas (natural gas, propane, and butane). More than 95 percent of all energy produced is from one of these.

Minerals

Minerals occur on Earth naturally. Iron, zinc, copper, lead, and magnesium are examples. They are used in making many products.

Minerals are often dug (mined) from the earth and refined before a product is made. For example, iron must be smelted (melted) and cast (made into a certain shape). High temperatures melt iron. Other minerals may be mixed with iron to give it better qualities. For example, stainless steel is made by adding 12 percent chromium (another mineral) to molten iron. Iron is brittle (not bendable), while steel has more flexibility.

Many minerals are reused, and this cuts down on the rate of depletion. All of us are familiar with ugly automobile junk yards. The old cars can be melted and made into new products.

WILDLIFE MANAGEMENT AND SPECIES

Wildlife includes all animals, plants, and other living organisms that have not been domesticated. A few species, such as elk, are both wild and domesticated. Once domesticated, some individuals in a species revert to living wild. These formerly domesticated individuals that again live wild are known as feral.

WILDLIFE MANAGEMENT

Wildlife management is the manipulating of wildlife systems to achieve desired goals. For example, steps may be taken to increase the population of some species. In other cases, populations may be controlled to prevent excessive numbers of a species.

Uses of wildlife are in two major groups: consumptive and nonconsumptive. Consumptive uses involve harvesting or taking wildlife. Hunters and fishers are consumptive of wildlife. The species hunted for food and sport are known as game. Nonconsumptive uses involve enjoying wildlife without harvesting or taking it. These include bird watching, enjoying spring flowers, or other activities.

8–18. Deer are widely found throughout the United States.

Some wildlife species become extinct, endangered, or threatened. Extinct means that a species is no longer living on Earth. Endangered refers to populations of species that are near extinction. Threatened means that a species is likely to become endangered if not protected. Over-harvesting, disease, lack of habitat, inability to adapt to changes, and other conditions lead to extinction.

Habitat refers to the physical area where a species is found. Species vary in habitat preferences. Rabbits, deer, and woodpeckers may all exist in the same forest, but their habitats vary.

The practices used to protect and improve game wildlife are known as game management. Game management begins with understanding habitat needs. The needed habitat can be improved. In some cases, plants or animals are moved. Only species suited to a habitat should be moved into it. Streams and lakes are frequently restocked with fish. Young fish (fry) are raised to a few inches long in a hatchery and released. Trout, salmon, and catfish are three examples. Many locations have laws that protect wildlife and fish. Over-fishing can remove all the fish so no new ones can grow. Over-hunting can kill all the game. Game laws are designed to protect animal species from being killed off.

ACADEMIC CONNECTION

ENVIRONMENTAL SCIENCE

Relate your study of natural resources in agriscience to your study of environmental science. Investigate habitats in your area and the species found in them. Determine the food chain followed in each of these habitats by the species you identify.

WILDLIFE SPECIES

People most commonly think of animals when the word *wildlife* is mentioned. After some thought, people realize that any organism that has not been domesticated is wildlife. This includes many plant species as well as fungi and others.

Some wildlife live on land, others in water. Those that live on land are terrestrial wildlife, with deer and squirrels being examples. Those that live in water are aquatic wildlife, with fish and beavers being examples.

Animal Wildlife

Many species of animal wildlife are popular. Some are used as game. All contribute to the overall diversity and ecological balance of nature. The first five groups listed here are vertebrates—that is, they have backbones.

8–19. The American bald eagle, popular with bird watchers, has increased in population in recent years.

- **Mammals**—Mammals are animals whose young are nourished by milk from their mothers. Mammals maintain a body temperature. Most live on land, but some are found in water. Examples are deer, bear, rabbits, and raccoons.
- **Birds**—Birds lay eggs (which are incubated and hatch) and normally have the ability to fly. Common examples are sparrows, ducks, robins, and pelicans.
- **Reptiles**—Reptiles are animals with scale-covered skin and body temperature regulated by their envi-

8–20. The American alligator is a large reptile.

ronment. Some have legs; others do not. Examples are alligators, turtles, lizards, and snakes.

- **Amphibians**—These species typically live on land but reproduce in water. Some stages of their lives may be in water. Examples of amphibians are frogs, toads, newts, and salamanders.

8–21. The red-eyed tree frog is an amphibian.

- **Fish**—Fish include a large number of species. They live in water (either fresh water or salt water) and have the ability to swim about. Their bodies are covered with scales or skin. Most have gills to acquire oxygen from water. Examples are perch, catfish, trout, tuna, and salmon.

- **Invertebrates**—A large number of wildlife species are invertebrates—that is, they do not have backbones. They play valuable roles in nature and for human purposes. Examples are insects, crabs, lobsters, worms, and spiders.

8–22. The opened shell of a mussel illustrates overall body structure.

Plants and Other Species

Many wildlife plants are found on Earth. Some grow quite large; others remain small. Some are known for the beauty of their flowers, such as the dogwood and rhododendron. Others provide valuable products, such as wood from pine trees and sap (for syrup) from sugar maple trees. Still others are best known for providing habitat for animal wildlife, such as hickory trees with nuts desired by squirrels.

Besides plants, fungi are interesting and important wildlife. Mushrooms may be hunted for food. Some fungi are aquatic and grow in salt water. These seaweeds are used as human food and to make other products.

Other species, such as bacteria and algae, grow wild and perform useful roles in the environment. Fallen trees decay because of action by such organisms. Garbage is converted into degraded forms by the action of bacteria.

RESPONSIBLE USE OF WILDLIFE

Nearly everyone enjoys wildlife. Fishing, hunting, birding, walking in a forest, and hiking on mountain trails to see the wonders of nature can be refreshing. People may establish and maintain wildlife areas, including parks and preserves. Fishing and hunting are the most widely used methods of taking wildlife. Trapping is also sometimes used.

FISHING

Fishing is the harvesting or capturing of fish and related species. Some fish are harvested for commercial purposes—they are processed and sold for human or animal food. Increasingly, cultured fish are used commercially. These are produced in aquaculture. Natu-

8–23. A good stringer of trout.

ral populations in streams, lakes, and oceans have been overfished and, in some cases, almost fully depleted.

Sport fishing has long been a favorite pastime. It is capturing fish for relaxation, enjoyment, and food. With sport fishing, the fish may be released back into the stream or lake or taken for food. Artificial fishing lakes are becoming more popular. Small ponds may be stocked with fish and open to the public. The fish are fed and otherwise kept healthy. New fish are added to replace those that are caught. Some lakes charge a fee for the fish that are caught.

Common freshwater species in sport fishing are sunfish, bluegill, crappie, catfish, walleye, pike, bass, and muskie. Common saltwater species are snapper, pompano, flounder, bluefish, marlin, and tuna. In addition, crabs, oysters, mussels, and clams may be taken as a sport.

HUNTING

Hunting is the harvesting or taking of wildlife (game) for recreation or food. Today, emphasis is on enjoyment of the hunting experience. Years ago, hunting was an important source of food. Guns are used in hunting to kill the harvest. Traps and other devices are sometimes used.

Laws protect wildlife and specify when, how, and by whom wildlife can be hunted. Always follow the laws. A hunter must usually have a license and, in some cases, a special permit, such as a duck stamp. Only harvest legal species in season and during shooting

8–24. A hunter awaits waterfowl game.

hours. Strictly adhere to bag limits (the number that can be taken). Use appropriate hunting gear, including firearms (guns). Always follow hunter safety practices—guns are dangerous! Before you hunt, be sure to take advantage of appropriate hunter education.

Hunting preserves are becoming more common. Preserves are needed because much of the wild game has been killed. Preserves protect wildlife and regulate how it can be taken.

Examples of game mammal species are deer, caribou, moose, black bear, pronghorn, rabbit, and squirrel. Waterfowl include several species of ducks, such as mallard and wood. Upland bird species that are hunted include turkey, dove, bobwhite, ring-necked pheasant, and ruffed grouse.

BIRDING

Birding is the hobby of watching birds. It may include identifying species and determining habitat needs, establishing habitats, and otherwise studying birds. Some birds are tagged or micro-chipped to study their migration and patterns of movement. Birding is a nonconsumptive use of wildlife. Always respect birds and their habitats. Avoid frightening birds or destroying their habitats.

Birding may include providing nesting boxes and feed. Nesting boxes are intended to provide safe places for birds to nest and hatch young. The boxes are constructed for specific species and to keep pests out.

Some bird populations are too great for naturally available food to support their well-being. Feeders are used to provide appropriate food materials, such as thistle seed for finches and safflower kernels for cardinals. Migrating birds often benefit from feeding. They travel long distances and need the energy from feed to enable them to continue their flights.

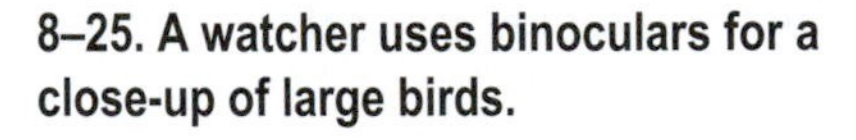

8–25. A watcher uses binoculars for a close-up of large birds.

Birding may involve a wide range of species. Waterfowl, such as ducks, egrets, and cranes, may be observed near lakes and marshes. Robins, sparrows, finches, cardinals, and humming birds are popular in other areas.

POLLUTION

Pollution is the result of substances that damage our environment. When something is polluted, it is unclean. Who wants to breathe dirty air or drink bad water? Air, water, soil, and natural beauty are damaged by pollution. Pollutants (materials that pollute) can be serious. Examples are acid rain, radiation released by nuclear accidents, noise, and odors.

There are two types of pollution: point source and nonpoint source. Point source pollution is pollution from a readily identifiable source, such as a factory. Nonpoint source pollution is pollution from many sources that may not be individually identifiable. An example of nonpoint source pollution is runoff that may contain chemicals from farmland or parking lots. Stopping one such source does not stop the pollution.

Among the many sources of pollution are toxins, soil loss, waste products, discharged water, and junk and litter.

TOXINS

A ***toxin*** is a substance that contains poison or has the potential to poison plants and animals. Toxins can ruin the air, water, and soil. Even helpful products can be poisonous when not used properly.

The effects of toxins are classified as acute or chronic. Acute effects appear immediately after exposure. They are associated with short-term exposure to high amounts of a toxin. The effects are often severe for a short time and then are gone.

Chronic effects are long-term and may not appear until some time after first exposure. Exposure to toxins over a long period is often associated with human cancer, birth defects, and related problems.

Toxins in the air, water, and soil can create problems in workplaces and schools. Many agricultural workers use chemicals or are present when others use them.

Internet Topics | Search

Selected topics for Internet discovery and reporting are listed here. Begin by searching for each topic. Next, read and learn about it. Conclude by preparing a brief report on your findings.

1. Food chain
2. Composting
3. Sustainable agriculture

Table 8–1. Steps in Preventing Agricultural Pollution

Regulations	Know regulations when using chemicals; get the needed training.
Follow rules	Read instructions when using chemicals; never fail to follow them.
Storage	Properly handle and store chemicals and the equipment used to apply them; buy only the amount needed, and protect the containers from damage and leaking.
Waste control	Never let water used to wash equipment flow into streams, ponds, or other places where it may cause damage.
Dispose of containers	Properly dispose of chemical bags, cans, plastic jugs, or other containers; some companies take back containers; never throw containers into streams, woods, pastures, or fields.
Safety	Practice personal safety; dress properly; use gloves, hats, eye protection, and air masks (respirators) as needed; wash the skin if dangerous materials get on it; know how to handle an emergency in case of an accident.

The Environmental Protection Agency (EPA) has guidelines known as the Worker Protection Standards for Agricultural Pesticides (WPSAP). Following these guidelines will reduce illness or injury from pesticides. Workers must have information on the dangers of the pesticides they are applying. Protective clothing must be used. Training in safe pesticide use and handling emergencies is essential.

SOIL LOSS

Wind and water can blow or wash soil away. Besides the loss of nutrients, soil particles get into the air, streams, and lakes, causing pollution. Air may appear hazy. Water may have a muddy (brown) color.

Living things in polluted water may be damaged. Natural processes in the water are disrupted. Oxygen levels may be too low to support aquatic life. Use of the water by industries or by cities may not be possible. The soil particles float down streams to the ocean, where they damage the environment for sea life.

WASTE PRODUCTS

Waste is the solid and semisolid material that results from the activity of people and animals. There are many types of wastes: household (from our homes), institutional (from schools, hospitals, research facilities, and prisons), industrial (from manufacturing and processing facilities), construction (from building and demolition sites), and agricultural (from livestock and crop production and harvest). Of high concern is the manure from animal production.

8–26. A feedlot surface has been treated with urease inhibitor to block ammonia production and curb bad odors. (Courtesy, Agricultural Research Service, USDA)

The primary household wastes created by people are sewage, paper, glass, aluminum and tin cans, food wastes, and yard wastes, such as grass clippings and leaves. The average person in the United States creates more than 1,300 pounds (590 kg) of waste per year!

Industrial wastes vary with the resources used and products made. Paper mills, food processing plants, steel mills, and chemical plants create industrial wastes.

Agriculture creates some waste products—used engine oil, worn-out tires, and even dead animals. Many of these produce toxic substances if not disposed of properly.

DISCHARGED WATER

Factories may use water in making products. Farms use water for irrigation, fish culture, barn cleaning, and other purposes. After use, some water is discharged. This water is known as ***effluent***.

Effluent can pollute streams, lakes, and oceans. In the mid-1900s, many streams had water dumped into them that killed natural life. The water was toxic. Laws to prevent the release of dangerous effluent were passed. Today, some streams and lakes are clearing up and losing toxins.

JUNK AND LITTER

People discard all sorts of things: vehicles and vehicle parts, furniture, appliances, food and beverage containers, and construction materials. Farms may have empty seed, feed, or chemical containers; worn-out batteries; old equipment; and barns in poor condition, with scraps of lumber and roofing lying around. All these pollute!

Litter on roadsides is a problem. Not only is litter unsightly and a potential health hazard, but it is also costly! Consider the money spent by city, county, state, and national programs to remove litter. Everyone can help keep our earth safe, attractive, and free of litter.

Old machinery and junk are eyesores. They detract from the natural beauty and value of the land. Old machinery and junk are often dangerous. There is danger from stepping on a nail or being hit by a falling piece of rusty machinery. Livestock can be injured.

Many things can be reused. Reuse makes the environment more attractive and saves natural resources.

8–27. Junk pollutes the environment and wastes nonrenewable natural resources. Discarded items can be recycled. Cleaning up would definitely make the scenery better!

AGRICULTURAL POLLUTION

Agricultural pollution is any pollution caused by agricultural activities. It can occur on farms, in greenhouses, at processing plants, or at many other places.

Natural resources are threatened every time the soil is plowed, an animal is raised, and other work takes place. Knowing about pollution helps prevent problems. Most people in agriculture take care to prevent pollution. They strive to maintain a good environment.

Table 8–2. Possible Sources of Agricultural Pollution

Sources	Description/Examples
Introducing pests	Pests can be brought in on seed, plants, fruit, and other products; for example, papaya shipments from Hawaii to the U.S. mainland are inspected.
Exotic plants or animals	Non-native plants or animals may escape or be released and cause damage; for example, fish and plants kept in aquariums cause damage if released.
Chemicals	Agricultural chemicals can cause damage if used improperly.
Waste water	Excess water from irrigation, fish ponds, food processing plants, and other places can damage natural lakes and streams; treat water before release to prevent damage to lakes and streams.
New life forms	Non-natural forms of plants and animals could be threats if not properly tested and used; some people fear a plant or animal of this type will "get loose" and cause problems.

DISPOSING OF WASTES

Waste disposal is the process of getting rid of wastes. Our way of life creates much waste. Think about the debris in a theater after a movie or in a stadium after a ball game. Agricultural activity creates wastes. One example is the stalks that remain in a field after a crop has been harvested. Another example is the manure excreted by pigs on a hog farm. A third example is the waste from a vegetable cannery that includes leaves and stems that have been removed from the harvested crop and the waste water from washing the vegetables.

Table 8–3. Forms of Wastes

Forms	Description/Examples
Gases	Some gases are very hazardous; gases include carbon dioxide (given off by animals), carbon monoxide (produced by auto and tractor engines), and methane (given off as plant and animal matter rots).
Liquids	Water runoff and effluent are two important examples; city sewage treatment facilities handle most residential and business liquid wastes.
Solids	Paper, plastic, food scraps, cans, manure, grass clippings, dead animals, and the unused stems and leaves of plants are common; some smell bad, while others promote the growth of rats and roaches and harbor diseases; ugly to see on roads, in parks, and other places.

METHODS USED

Properly disposing of wastes is a big job. People often don't realize how much waste they create and the cost of properly disposing of it. Four methods of waste disposal are hauling to landfills, incinerating, recycling, and composting.

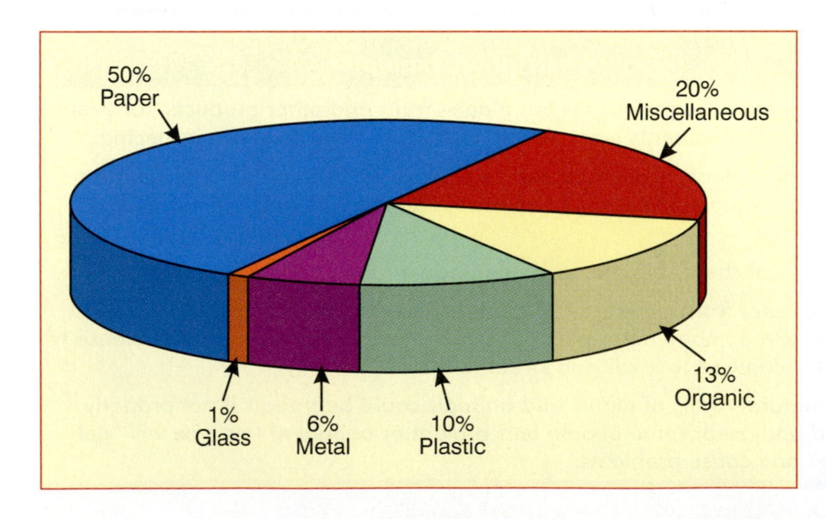

8–28. This pie graph shows percentages of various wastes, by volume, produced by households in the United States. (The "Organic" group contains materials that readily degrade, such as food scraps. The "Miscellaneous" group includes discarded furniture, old tires, and similar wastes.)

Hauling to Landfills

A ***landfill*** is a large earthen pit for waste disposal. It is designed to protect the environment from pollution. About 80 percent of the waste in the United States is disposed of in landfills.

Landfills are located on carefully selected sites. A landfill is made by digging a large hole in the earth. The ground should have a high clay content to keep wastes from seeping into the underground water. A landfill may have a large plastic liner inside the earthen pit. The wastes are dumped on top of the liner or clay bottom, packed with a bulldozer, and covered with earth. The goal is to promote decay of the materials.

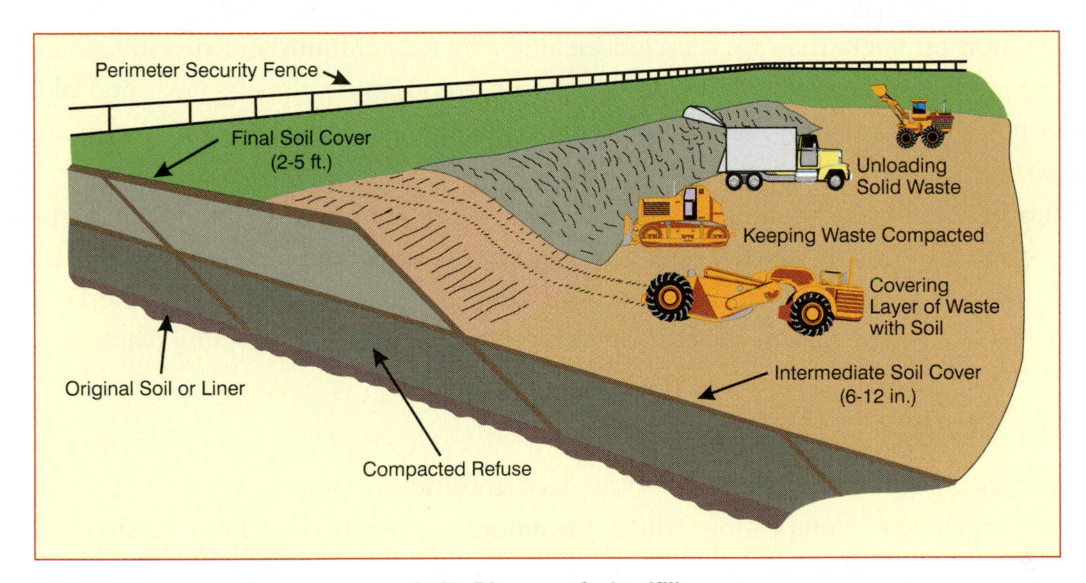

8–29. Diagram of a landfill.

Rotting (decomposition) takes a while. Some materials decompose quickly; others break down slowly. Since landfills are packed and little air is present, materials are slow to rot. Further, some materials, such as glass and plastics, do not decay very fast. This means that we are likely to find ourselves with a massive amount of waste buried in landfills.

The term ***biodegradable*** describes materials that can be broken down by bacteria. Some things, such as newspapers, are biodegradable, while other things, such as glass, are not. New ways of manufacturing products promote degrading. For example, adding cornstarch to plastic speeds degrading.

Incinerating

Some materials are burned in special structures known as incinerators. Smoke containing gases and particulate may be released into the air. These substances may settle to the earth and harm plants, animals, and structures, such as the paint on farm buildings.

Most incinerators are equipped to prevent releasing harmful smoke into the air. Pollution control devices catch harmful materials. We incinerate only 9 percent of our waste in the United States.

Recycling

Recycling is the act of recovering and reusing materials instead of throwing them away. It conserves raw materials and reduces pollution. Use of recycling has increased in recent years because of concern over protecting the environment. For example, the hamburger you had today might have been in a bag created by recycling! (Look for the symbol that it was made of recycled paper.)

Common products that are recycled include paper, aluminum and tin cans, and glass. Other recyclable items include motor oil, auto batteries, auto tires, plastic, and old cars. Empty pesticide containers may be recycled. Just about anything can be recycled in one way or another.

Many places have recycling programs to help reduce pollution. Drop-off centers may be set up where people can take their trash. Some communities have curbside pickup of materials for recycling. Paper, plastic, glass, and cans must usually be sorted. Even different kinds of plastics need to be sorted. Mixed materials are big problems at recycling plants.

Composting

Composting is the promotion of the decomposition of organic matter. The resulting material is compost. Composting reduces the amounts of wastes that must be disposed of in other ways. It is a natural process that occurs on the floors of forests.

8–30. Compost bins have been constructed so that oxygen can get into the materials to speed decomposition.

Paper, sawdust, leaves, stems, food materials, animal wastes, and small dead animals, such as birds, can be composted. Poultry farmers may use composting to dispose of chickens that die on the farm. Human wastes (feces) should not be put in a home compost pile. Sewage treatment plants may use composting with wastes.

Compost piles or bins promote the action of microbes and other organisms on the materials. The microbes include bacteria, fungi, and actinomycetes—organisms between bacteria and fungi. Small animals, such as worms, insects, and mites, speed composting.

A mixture of materials decomposes faster. Different materials have varying nutrients for microbes to use. Oxygen is needed for composting to occur. Watering the materials and, in some cases, adding partially decomposed animal manure promotes bacterial action. Materials in smaller sizes compost faster. Activity in a compost pile may raise the temperature of the material to 100°F (37.8°C) or more. Compost bins should allow oxygen to enter the material. Usually a compost bin rests on the ground and uses the ground for its bottom.

Compost from home systems is used as fertilizer or mulch. It improves the soil. Nutrients in the materials that are composted are returned to the soil.

REVIEWING

MAIN IDEAS

Earth has natural resources that are important to human life. These resources are a part of the environment, or all the factors that affect living things.

Plants and animals live in the biosphere (area of the earth that supports life). Within the biosphere, there are ecosystems. An ecosystem has biotic (living) and abiotic (nonliving) factors.

The area where a plant or animal lives in natural conditions is its habitat. Habitats vary for different plants and animals. A niche is the role of a living thing within its habitat. Plants and animals use different foods and are part of the food chain.

Some natural resources are renewable; others aren't. The major renewable resources are soil, water, wildlife and fish, forests, and air. Once used, nonrenewable resources are gone. Examples of nonrenewable resources are petroleum and minerals. Sustaining natural resources is using them in such a way that they remain for future generations.

Pollution degrades the environment. Factories, homes, and farms may release toxic substances. Using materials according to recommendations will reduce threats. Properly disposing of wastes reduces pollution. Hauling wastes to landfills, incinerating, recycling, and composting are approved means of disposal. Never dump trash and garbage into streams, along roadsides, or in open areas.

QUESTIONS

Answer the following questions, using complete sentences and correct spelling.

1. What is the environment? What are natural resources? How are they related?
2. What is the biosphere?
3. What is an ecosystem? Distinguish between factors in an ecosystem.
4. How does an ecosystem work? Describe habitat, niche, and food chain as parts of the ecosystem.
5. What is sustaining natural resources?
6. Distinguish between renewable and nonrenewable natural resources.
7. What are the most important renewable natural resources?
8. What is the importance of the following renewable natural resources: soil, water, wildlife and fish, forests, and air?
9. What are the major kinds of nonrenewable natural resources?
10. What is pollution?
11. What are five important sources of pollution? Explain each.
12. Name three kinds of wastes.
13. What methods of waste disposal are used? Briefly describe each.
14. What are possible sources of agricultural pollution?
15. What can people do to reduce agricultural pollution?
16. Distinguish between consumptive and nonconsumptive uses of wildlife.
17. What are three uses of wildlife? How are these carried out responsibly?

EVALUATING

Match each term with its correct definition.

a. biosphere
b. sustainable agriculture
c. ecosystem
d. habitat
e. biodegradable
f. soil
g. water cycle
h. game management
i. effluent
j. composting

______1. Materials that can be broken down by the action of bacteria.

______2. The area where an organism lives in natural conditions.

______3. The practices used to maintain our ability to produce food, fiber, and shelter.

______4. Promoting the decomposition of organic materials.

______5. The area of Earth that supports life.

______6. Used water discharged by factories and farms.

______7. The outer layer of the earth's surface.

______8. Circulation of water in the hydrosphere.

______9. Practices to protect and improve wildlife.

_____10. All the parts of a particular environment, including living and nonliving things.

EXPLORING

1. Investigate the ways of disposing of wastes in your community. Contact the local solid waste disposal office, landfill, or government office for information. Prepare a report on your findings. In your report, assess how well these methods are working and offer recommendations for future actions.

2. How do people recycle in your community? Identify the different locations for depositing wastes to be recycled. Determine how the materials are sorted and prepared for recycling.

3. Make a survey of the water used by agriculture in your community. Determine sources of the water and its major uses, such as for drinking by livestock, cleaning barns, irrigating crops, and raising fish.

4. Study the wildlife found in your local area. List the species, and indicate if the species are considered game. Determine if there are laws about hunting these animals. Talk to a local game conservation officer (game warden) to get the details. Prepare a report of your findings. Include a poster that has pictures of the major wildlife species.

CHAPTER 9

Organisms and Life

OBJECTIVES

This chapter focuses on the common characteristics of all living things. Important differences are also included. It has the following objectives:

1. Explain important characteristics of organisms.
2. Explain the meaning of life span and list its stages.
3. Name and discuss the life processes of living organisms.
4. Describe the structural bases of living organisms.
5. Identify cell growth processes.

TERMS

autotrophs
biogenesis
carnivore
cell
cell division
cell membrane
cell specialization
cell structure
cell theory
chemosynthesis
circulation
circulatory system
cytoplasm
decomposer
detrivore
digestion
diploid cell
eukaryote
fertilization
food web
growth
haploid cell
herbivore
heterotroph
life span
living condition
locomotion
meiosis
mitosis
multicellular organism
nucleus
nucleolus
omnivore
organ
organism
organ system
ovum
phloem
photosynthesis
producers
prokaryote
repair
reproduction
respiration
sperm
tissue
unicellular organism
vascular system
xylem

9–1. The upper layers of a tropical rain forest (canopy) often have new living organisms.

HOW MANY KINDS of living things are there on Earth? Scientists have identified 1.5 million species, but many more living things exist and have not been studied.

New species of living things are found in remote places like in tropical forests. Some species are large enough to be easily seen. Other species are extremely small, so we must use microscopes to find them. Once they are located, scientists attempt to classify and name them.

In classifying new organisms, scientists study how they are similar or different than existing known types (e.g., How many petals are on the flower? How many toes are on the animal? What about its genetic material?) If the organism does not match any known categories, it may be new.

Certain structures and processes are present in all living things—plants, animals, and other species. Scientists study the ways in which living things are alike and different to group them, and then they study their characteristics. Agriscience deals with the study of similarities and differences between species that affect our daily lives.

CHARACTERISTICS OF ORGANISMS

9–2. The red-eyed tree frog on a branch in Costa Rica is an example of a unique organism found in the wild.

Any living thing is an **organism**. For instance, trees, dogs, humans, bacteria, and fungi are organisms. No matter what the size and shape, life-supporting processes occur. When these processes stop, life ends. In agriscience, the emphasis is on using living things or their products to meet the needs of humans.

Knowing about life processes helps in growing plants and animals. Applying basic principles learned about the structures inside plants and animals helps explain why plants and animals respond the way they do to their environment.

Organisms have important characteristics. They (1) are chemically and structurally unique, (2) need energy, (3) are capable of growth and reproduction, (4) respond to their environment, and (5) have a life span.

9–3. Cacao bean pods are harvested and used to make chocolate.

ORGANISMS ARE UNIQUE

Organisms are made up of carefully organized substances. The way in which these substances are grouped and their compositions determine the species of the organism. Plants reflect a range of ways in which substances are organized, but animals reflect a different pattern of arrangement. Even among a group of plants or animals, many differences exist.

CELL THEORY

Many scientists throughout the years have made discoveries that gave us information about cells. For example, the development of the microscope was a major step in enabling people like Leeuwenhoek to be the first person to see living organisms in a drop of water. Robert Hook studied cork under the microscope and called the small chambers that composed the cork cells. This discovery indicated that the **cell** is the basic unit of all living things.

A summary of all the information found in the discoveries by scientists is called the ***cell theory***, and it states the following:

1. All living things are made of cells.
2. Cells are the basic units of structure and function in living things.
3. New cells are produced from existing cells.

Some organisms, such as cattle and trees, are made up of billions of cells; they are multicellular organisms. Other organisms consist of only a few cells. Many species of one-celled or unicellular organisms exist.

BIOGENESIS

Biogenesis is the theory that life comes from life. Nonliving things cannot produce life. Only living things can produce new life, and the conditions must be right for life to occur.

Technology Connection

VIEWING TINY ORGANISMS

Some organisms are quite small. They are visible only by using high-powered magnification. Other organisms can be seen with a hand lens or a less powerful microscope. Even the smallest organism carries out life processes that allows it to remain in the living condition.

Tiny organisms are sometimes pests. They are so small that we cannot see them until they are collected, taken into a laboratory, and viewed with magnification equipment. A good example is the spider mite (about the size of the period at the end of this sentence). Spider mites are pests on many kinds of plants, ranging from cotton to roses.

Scientists have found that spider mites use their mouthparts to suck juices from plant cells. This image of a two-spotted mite feeding on a rose leaf was made by a scanning electron micrograph. The mite is shown at about 200 times its actual size. (Courtesy, Agricultural Research Service, USDA)

Scientists use synthetic biology when trying to create lifelike processes in a laboratory from nonliving chemicals. The work has not yet produced a living organism. However, there have been advancements in producing organisms from a single living cell, called cloning. The cloning process may find wide use in medical, scientific, and agricultural research, but it has raised serious ethical issues.

ORGANISMS NEED ENERGY

Cell activity requires energy to carry out life processes. Energy must be available, and the organism must be able to use it. Most energy comes from the sun in one way or another. Of all the sun's energy that reaches Earth's surface, less than 1 percent is used by living things. Three types of organisms exist based on the way in which they receive food and use energy.

A limited number of organisms obtain energy from a source other than sunlight. Some types of organisms get energy that is stored in inorganic chemical compounds. For example, mineral water found in hot springs and undersea volcano vents is loaded with chemical energy. If an organism uses chemical energy to produce sugars or starches, the process is called ***chemosynthesis***. This process is performed by several types of bacteria.

Only plants, some algae, and certain bacteria can capture energy from sunlight or chemicals and use that energy to produce food. These organisms are called ***autotrophs***, and they use energy from the environment to assemble simple compounds into more complex ones. Because autotrophs make their own food, they are called ***producers***. Some autotrophs harness the sun's energy through a process called ***photosynthesis***. These autotrophs use light energy to power chemical reactions that change carbon dioxide and water into oxygen and energy-rich compounds such as sugars like glucose. This chemical process is important for the production of oxygen that adds to the atmosphere and for removing carbon dioxide.

Many organisms—including animals, fungi, and many bacteria—cannot use energy directly from the sun or from chemicals like autotrophs do. Instead, they receive energy from other organisms. If an organism relies on other organisms for its energy and food supply, it is called a ***heterotroph*** or consumer.

There are several types of heterotrophs. A ***herbivore*** is an organism that gets energy by

Light from the sun

Carbon dioxide + Water $\xrightarrow[\text{Chlorophyll}]{\text{Light energy}}$ Glucose + Oxygen

9–4. Plants produce sugar and oxygen by photosynthesis.

eating only plants. Examples of herbivores are cows, caterpillars, and deer. A **carnivore** eats the flesh of animals, with examples being snakes, dogs, and owls. Humans, bears, and crows are called omnivores. An **omnivore** is an animal that feeds on plants and animals. A **detrivore** feeds on nonliving plant and animal remains—dead matter. Examples include mites, earthworms, and snails. Another heterotroph group is the decomposers. A **decomposer** breaks down organic matter. Bacteria and fungi are examples of decomposers.

FOOD WEB
Meat Eaters (carnivores)
Eat Both (omnivores)
Plant Eaters (herbivores)
Green Plants (producers)
Sun Soil
Resources
Water Air

9–5. Food webs show the feeding relationships among organisms in an ecosystem.

The energy in an ecosystem moves along a one-way path. The energy begins at the sun or with inorganic compounds and is used by autotrophs (producers) and then goes to heterotrophs (consumers). The relationship between producers and consumers results in feeding networks called food chains. A food chain is a series of steps in which organisms transfer energy by eating or being eaten. These feeding relationships among the various organisms in an ecosystem form a network of interactions called a **food web**.

ORGANISMS GROW AND REPRODUCE

Organisms will not last long if they do not grow. This does not mean that they always increase in size. New cells are needed to replace those that wear out in a mature organism.

Growth is the process by which an organism increases in size by adding cells or by cells becoming larger. This differs from replacement, in which new cells replace those that have been destroyed. Mature organisms must constantly replace older cells. When cell replacement slows, the organism declines. The activities in cells are similar during growth and replacement.

9–6. New life sprouting from a seed—a coconut. (This coconut washed ashore and sprouted. Note the roots underneath.)

Reproduction is the process by which new organisms are produced. This results in new organisms of the same species that resemble the originals. Some organisms may not reproduce, but most are capable of reproduction at some stage of life. Without reproduction, a species would cease to exist.

The growth and reproduction of crops and livestock are important because we need their products. Skill is necessary to grow and reproduce crops and livestock successfully.

ORGANISMS RESPOND TO THEIR ENVIRONMENT

The surroundings of an organism produce the many factors that make up its environment. Organisms grow best in environments in which they are adapted. In agriscience, the goal is to provide the best possible environment for a plant or animal.

Many organisms have specific environmental needs. Rice, for example, grows best in a warm climate and needs much water. Winter wheat, on the other hand, is cold tolerant and can be planted in the fall for harvest in the late spring.

Organisms can adjust to small changes in their environment. Yet large changes will result in loss of growth and production and possibly death.

Plants sometimes need help adapting, so breeding practices are used to develop adapted varieties. For instance, several varieties of "dwarf" fruit trees have been developed. This means that the tree size remains small, making it easier to harvest the fruit.

LIFE SPAN

The entire length of life is known as the ***life span***. All organisms have a fairly definite life span. Some organisms live longer than others, but the average life span of humans in the

9–7. Adult butterflies, such as this tiger swallowtail, may live only a few days.

United States is about 77 years. Cattle have life spans of 12 to 15 years. Some plants live just one growing season, and insects may live only a few days.

THE STAGES

Regardless of the length of a life span, organisms go through five stages:

1. **Beginning**—All organisms have a beginning point that varies by species, depending on how the species reproduces.
2. **Growth**—In this stage, an organism grows rapidly. The growth stage occurs when an organism is young and immature. The rate of growth declines as maturity is reached.

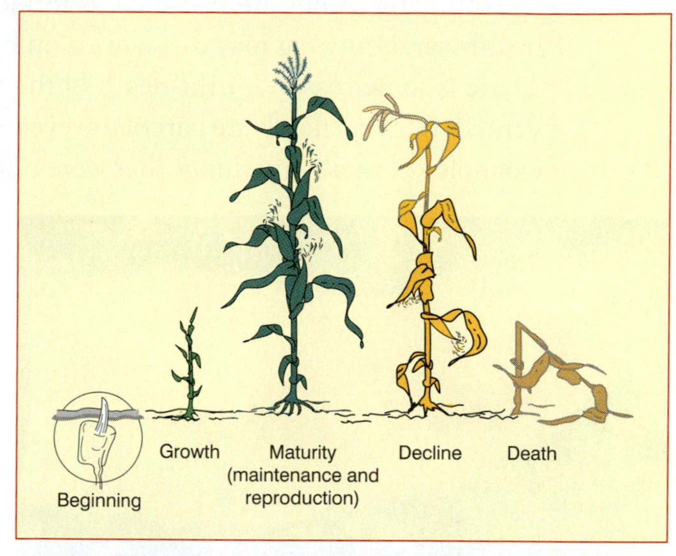

9–8. The life stages of a corn plant are shown here: beginning, growth, maturity, decline, and death.

3. **Maturity**—At this stage, the organism is said to be fully developed. Growth stops, except for the repair and replacement of cells. The organism can reproduce in this life stage.
4. **Decline**—This stage follows maturity. The organism loses its ability to maintain itself. Cells are no longer repaired and replaced. Decline is associated with aging.
5. **Death**—Death occurs when the organism cannot replenish itself. Protoplasm stops carrying out chemical activity, and the organism no longer functions.

Plants and animals experience these life stages. With a horse, for example, new life begins when a sperm from a male fertilizes the female's egg. The fetus (developing embryo or unborn horse) grows rapidly in the mare's (female's) reproductive tract. After birth, the foal (baby horse) continues to grow. The rate of growth slows as the young horse reaches maturity in two to three years. The mature stage may last 10 to 12 years. Decline begins when the horse starts to lose teeth, develops bone problems, and otherwise deteriorates. Death follows decline.

WHEN LIFE STOPS

All things may be classified as living or nonliving. Life ends when an organism's system stops functioning. The activity in the cytoplasm of the cells no longer results in life. Animals and plants may be die if they are not handled properly. Growing crops and raising animals depends on encouraging life processes.

Nonliving things are classified as formerly living or never living. A tree that was once healthy and growing may die. This is an example of a living thing that has ceased to live. There is some reason for the death of the tree. Sometimes the death could have been prevented. Life and death are part of the cycle of nature. Minerals, such as clay and iron ore, are examples of nonliving things that were never alive.

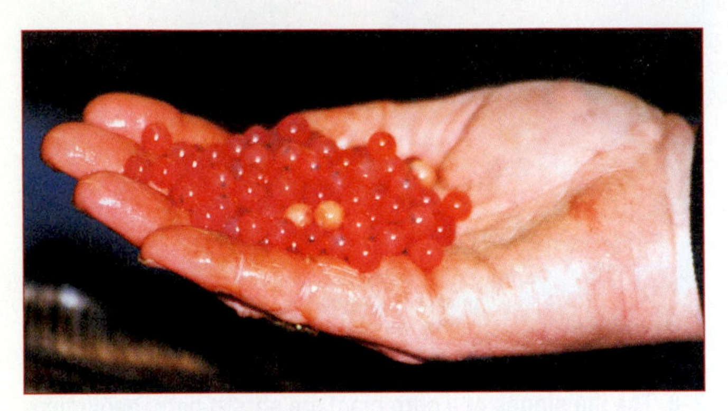

9–9. Sixty days are needed for fertile salmon eggs to hatch. Dark red eggs with eyes (black dots) have developing embryos. Light-color eggs are not developing; they are dead or were never fertilized.

In some cases, nonliving things have to be carefully examined to determine if they were ever alive. Limestone rock may contain fossils, such as the shells of once-living sea animals. Many fuels, such as coal and oil, are now nonliving; however, the fuels contain materials that were once alive.

The application of science in agriculture helps us understand how to provide for plants and animals. Controlling disease, furnishing nutrients, and providing protection from harm help plants and animals grow.

LIFE PROCESSES OF ORGANISMS

All organisms carry out life processes. The ***living condition*** is the total of all the life processes. Eight life processes compose the living condition. They are (1) getting and using food, (2) movement, (3) circulation, (4) respiration, (5) growth and repair, (6) secretion, (7) sensation, and (8) reproduction. All living organisms, both plants and animals, experience these processes in one form or another.

OBTAINING AND USING FOOD

All organisms must have food. Without it, life processes stop. Food provides energy needed for activity. The life processes of a living organism never stop, so the need for food does not stop. Food is used for growth, repair, and movement within an organism, even while the organism is resting. Food also provides minerals, vitamins, and other substances needed for life processes to occur. Some organisms require food in ready-to-use forms, but other organisms can take in nutrients and manufacture food.

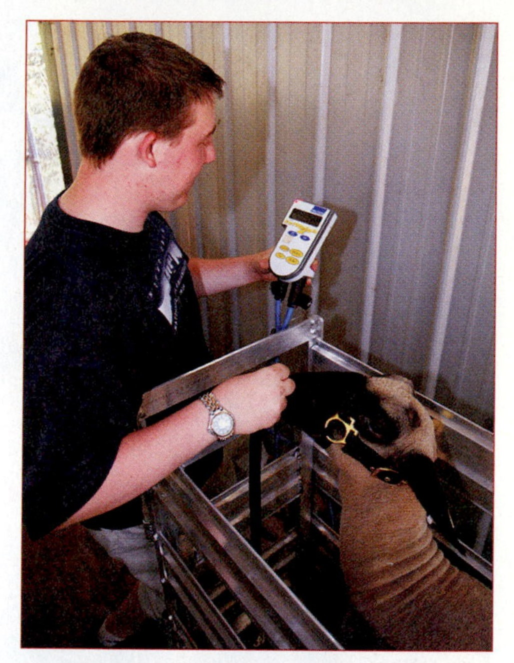

9–10. A lamb is being weighed to assess the rate of gain and adequacy of feed provided. (Courtesy, Education Images)

Animals

With animals, getting and using food can be broken into four processes:

- **Ingestion**—Ingestion is the act of taking in food. We may think of it as eating.
- **Digestion**—***Digestion*** is the process that changes food into simpler forms that can be absorbed. Digestion takes place in the digestive system. Various chemical processes occur that allow the nutrients to be absorbed.
- **Absorption**—Absorption also takes place in the digestive tract. It is the passage of nutrients into the bloodstream. Once the food has been digested and nutrients absorbed, certain wastes remain in the digestive system.
- **Elimination**—Elimination is the process of expelling wastes. It is also known as excretion. Liquids (urine) and solids (feces) are expelled.

Animals vary in their ability to use foods. Some animals need foods in ready-to-use forms because their digestive systems do not easily break down foods, but other digestive systems can convert lower-quality foods into more nutritious forms. These animals have more complex stomachs with more than one compartment.

Plants

Through photosynthesis, plants take in nutrients and convert them to food (a simple sugar). Most nutrients enter the plant through its roots. The types of nutrients needed vary with the species of plant, and some species require more nutrients than others.

9–11. Roots of plants, like on this bulb, absorb nutrients for growth.

9–12. Desert plants require nutrients to grow. (Annual growth of this saguaro cactus is shown.)

Culturing plants requires knowledge of the nutrients they need. Young plants need nitrogen for fast growth. Phosphorus and potassium are also important nutrients. Calcium, iron, magnesium, zinc, and others are required in smaller amounts. Plants must have water and sunlight to manufacture food from the nutrients. (More information on plant nutrition is in Chapter 15.)

MOVEMENT

Animals and plants move in different ways. Some movement is needed within an organism for life processes to occur. Internal movements are part of using food and carrying out life processes. Some organisms are stationary in their environment; they do not move about. Other organisms move from place to place. Energy is required for all movement.

Animals

Locomotion is the ability to move from one place to another. Some animals must move to find food and to protect themselves from danger. A complex system of nerves, bones, and muscles is necessary for movement.

9–13. Fin and body muscles produce movement so a fish swims.

Animals vary in their method of and need for locomotion. Dairy cattle may graze in the pasture and walk to the barn for milking and additional feeding. Oysters move very little; they ingest food by filtering it from the water that passes through their shells.

9–14. Snails move by extending and contracting the muscles in their foot (at bottom of shell touching the hand in this image). (Courtesy, Education Images)

Plants

Movement in plants is usually related to growth and other life processes. Plant roots probe into the soil, reaching for nutrients and anchoring the plant. The nutrients move through the roots and stems to the leaves. Stems and leaves grow and change positions to use energy from the sun. Flower buds form and open, and seeds develop in various ways, depending on the plant species. An example of plant movement can be observed in a houseplant on a windowsill; the leaves of the plant grow toward the sunlight, which is known as phototropism.

CAREER PROFILE

MICROBIOLOGIST

A microbiologist is a scientist in the field of microbiology. The work is with organisms and structures that are microscopic—quite small—such as algae, bacteria, yeasts, and viruses. A microbiologist may specialize in a particular area with, for example, a specialist in bacteria being known as a bacteriologist. Instruments are used to enlarge the organisms and structures being studied.

Microbiologists need college degrees in microbiology, biology, or a related area of science. Many have master's degrees and doctorates in microbiology. Begin in high school by taking biology, advanced biology, and agriculture classes that emphasize science.

Jobs for microbiologists are often with colleges, health facilities, research organizations, biotechnology companies, and agricultural seed producers. This shows a microbiologist and a research associate examining a cloned gene related to paratuberculosis (commonly called Johne's disease [pronounced "yo-knees"]). (Courtesy, Agricultural Research Service, USDA)

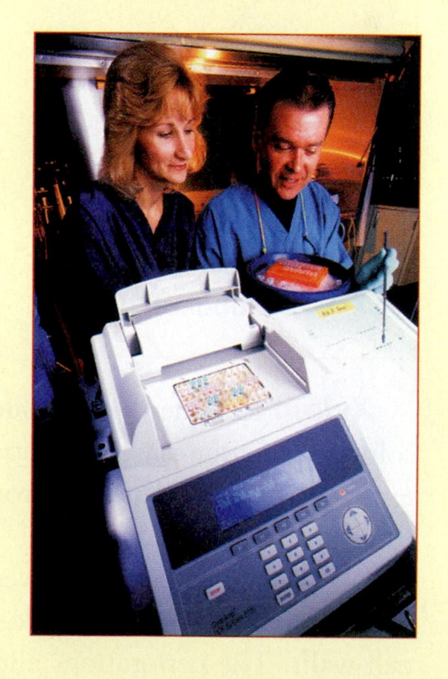

CIRCULATION

Circulation is the movement of food nutrients, digested food, and other materials within an organism. The way in which circulation occurs varies between animals and plants.

Animals

In most animals, circulation is carried out by a complex blood-pumping system—the ***circulatory system***. The heart is a hollow muscular organ. It contracts (squeezes) and then releases, which moves blood through the vessels. Heart movement is sometimes referred to as a heartbeat. The blood carries nutrients, oxygen, and other materials to all parts of the body. As it flows, the blood picks up cell wastes and carbon dioxide. These wastes are carried to organs where they are eliminated.

Injuries to animals can result in loss of blood or bleeding. If too much blood is lost, animals may be weakened or die.

Plants

The circulatory system of plants is known as the ***vascular system***. It consists of two kinds of tissue: xylem and phloem.

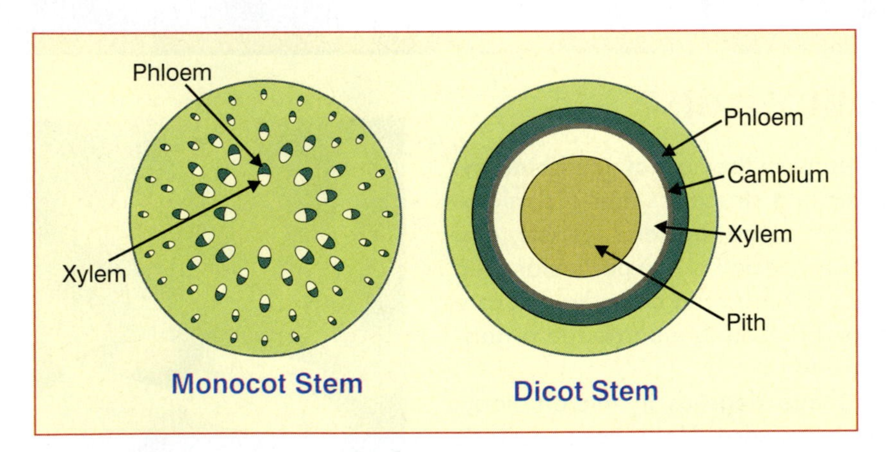

9–15. Stem cross sections show vascular structures.

Xylem is tissue that conducts water and nutrients throughout the plant. This watery solution is frequently called sap. Xylem is found in roots, stems, and leaves. Water typically enters the roots and moves through the xylem to the leaves. In woody plants, xylem consists of dead cell walls.

Phloem is tissue that moves food from where it is manufactured (usually in the leaves) to other parts of the plant. Phloem cells are living cells with perforations (small holes) in the cell walls. The perforations allow nutrients in the cytoplasm to connect from one cell to another. This connection forms a continuous system to move food (sugar and protein) down the stem.

Plants can "bleed" just as animals do. Cuts or breaks in stems will result in sap flowing out of the plant. When a plant loses sap, the effect is similar to an animal losing blood. Some cultural practices cause cuts or breaks. When plants are pruned (unwanted stems cut away), sap is lost through the open cuts. A good example is the pruning of grape vines late in the winter; excessive sap is lost resulting in reduced grape production.

9–16. Newly started plants in hanging baskets are receiving water to promote growth. (Courtesy, Education Images)

RESPIRATION

Respiration is the process an organism uses to provide its cells with oxygen so energy can be released from digested food. Respiration occurs all the time in living cells. Animals and plants need oxygen to use food.

Animals

In animals, the blood carries nutrients, various compounds, and oxygen throughout the body. Oxidation is a chemical process that occurs between oxygen and digested food. This process produces chemical energy and carbon dioxide. Carbon dioxide is a byproduct that must be removed from an organism.

Animals have respiratory structures to obtain oxygen and release carbon dioxide. Cattle, hogs, and chickens, for example, have lungs. Air is inhaled through nostrils and temporarily held in the lungs. Blood vessels in the lungs remove oxygen and release carbon dioxide.

Some animals do not have lungs. Fish have gills that remove oxygen from the water and release carbon dioxide. As a result, fish farmers must monitor water and add oxygen if it becomes low. Insects have tiny openings (called spiracles) in their bodies. Air enters through the spiracles and goes throughout the body in a network of tubes known as tracheae.

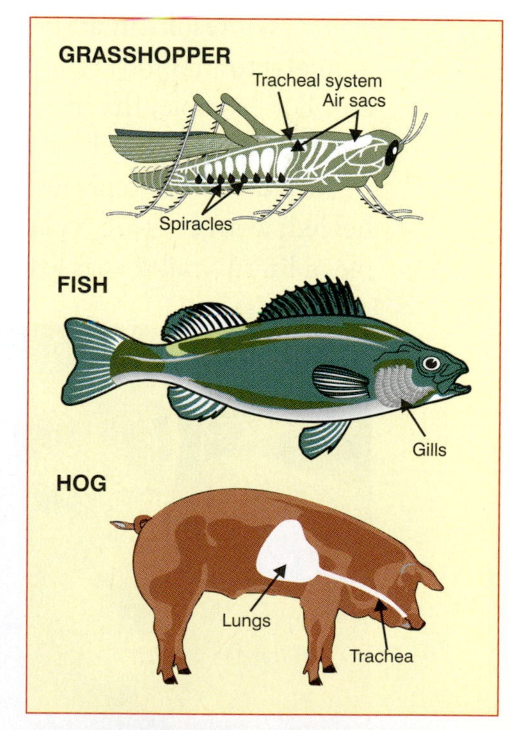

9–17. Respiratory structures vary in the grasshopper, fish, and hog.

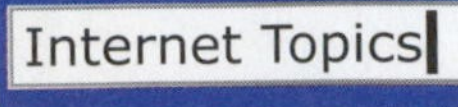

Selected topics for Internet discovery and reporting are listed here. Begin by searching for each topic. Next, read and learn about it. Conclude by preparing a brief report on your findings.

1. Biogenesis
2. Regeneration
3. Micrograph

Plants

Unlike animals, plants can use carbon dioxide in a chemical process to produce oxygen. Carbon dioxide and water are used during photosynthesis to manufacture food and oxygen. The food is a sugar compound known as glucose. Just like animals, plants use oxygen to react with the glucose to produce chemical energy for cell functions.

Plants use oxygen day and night in respiration. The glucose reacts with the oxygen resulting in the production of carbon dioxide and water. The carbon dioxide is emitted through the stoma (a tiny pore) in a leaf or stem. The stoma may also give off water.

GROWTH AND REPAIR

Organisms grow and/or repair themselves from the beginning of life until death. The process of growth and repair is known as assimilation and involves changing food substances into new living material. Since many foods are living or were formerly living, the food substances are reorganized to form new cells. The rate of growth and repair is rapid in young organisms, slows during maturity, and virtually stops during decline.

Growth results in an increase in the size of the organism. Obviously, young plants and animals grow rapidly. Mature forms (adults) also grow, but the nature of the growth is different. For example, the hooves on mature horses continue to grow. Mature animals may not change in size to much extent, but the growth of specific areas continues.

Repair is the replacement of the worn or damaged parts of an organism. Injuries are healed; worn cells are replaced; and in some animals, parts regrow or regenerate. For example, a lizard's tail that is lost in an accident may regenerate.

9–18. A lizard can regenerate its tail if the tail is broken off.

SECRETION

An organism depends on the production and availability of certain chemical substances for life that are known as secretions. Specialized structures known as glands produce secretions needed for a wide range of functions. Secretions are fluids carried from a gland by a tube or duct. All internal and external body fluids, except urine, are secretions.

Saliva is a good example of a secretion in an animal. Saliva softens and lubricates food so it can be swallowed. It also begins the digestive process by breaking down certain foods, such as starch. Other examples include tears (secreted by tear glands to cleanse and protect the eye), bile (secreted by the liver into the first part of the small intestine to aid in the breakdown of fat), and gastric juice (secreted by gastric glands in the wall of the stomach to aid in digesting food).

SENSATION

A sensation triggers a response in an organism to its environment. Complex systems of nerves communicate within an organism. Detecting changes in the environment is known as sensitivity. Plants and animals respond to sensation.

9–19. Barbels ("whiskers") on a catfish are sense organs. They help the fish maintain its position in the water and find food. (Courtesy, Education Images)

Animals

Animal responses to sensation are more obvious than plant responses. Larger animals have five types of receptors (senses): touch, taste, smell, sight, and hearing. The senses are receptive to stimuli—things that excite or cause a response. Most stimuli cause reactions from the organism. For example, if a cow brushes up against an electric fence, the stimulus shocks the animal and causes it to move away and stay in the pasture. As another example, when a lamb sees a person bringing feed, it will head to the feed trough.

Plants

Plants are particularly sensitive to light. Leaves may turn and stems may grow toward sources of light. Plants are also sensitive to heat, lack of water, and other environmental conditions.

REPRODUCTION

Reproduction varies between plants and animals and between species. In general, sex cells from the parents are contributed so a new plant or animal will form. This new individual will reach maturity and will likely reproduce.

Reproduction is not necessary for an organism to stay alive. However, it is necessary for new organisms to develop. Each plant or animal develops reproductive organs or reproductive structures in the growth stage. In the mature stage, these organs can create new organisms.

The two forms of reproduction are sexual and asexual. In nature, most plants and animals reproduce through sexual means. Life begins when a ***sperm*** (male sex cell) unites with an ***ovum*** (mature female sex cell; egg). The union of sperm and ovum is known as ***fertilization***. The process of fertilization varies between plants and animals and within each group. Sexual reproduction involves two parents. (Later chapters will include sexual reproduction of plants and animals.)

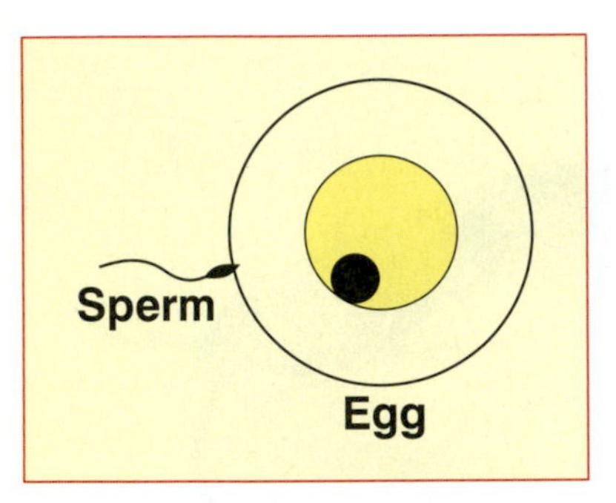

9–20. Sexual reproduction involves the union of a sperm and an egg.

Asexual reproduction does not involve the union of cells. Only one parent is needed. It is typical with plants. Existing organisms are divided into one or more organisms independent of each other. The offspring are identical to the parent. Asexual reproduction is most common in lower life forms and is used in producing plants with desirable characteristics. The producers of fruits and nuts often apply asexual reproduction.

Various forms of grafting or cuttings are used to create new plants. For example, a pecan bud from a tree that has the desired qualities can be grafted (joined) to a tree that is not desirable. The grafted part will grow and produce the desired pecans. Apple, peach, plum, pear, and other fruit trees are often grafted as well.

9–21. Sugarcane reproduces sexually or asexually. Stalks have flowers and bear seed. Most sugarcane is reproduced by planting sections of stalk. Buds at the joints in the stalk will grow into new plants.

Cuttings are used in reproduction. In the process, part of a stem is placed in a growing medium. Roots develop on the cutting, and a new plant will grow. Many ornamental plants are reproduced by cuttings.

STRUCTURAL BASES OF LIFE

Cells are the structural bases of life. Cells bond together and specialize to form tissues, systems, and organisms.

CELL STRUCTURE

Cells appear to be similar in plants, animals, and other organisms, but close examination reveals the differences. A microscope is used to see the differences in structure. **Cell structure** is the general pattern of organization and relationship of parts within a cell.

Cells have three major structural parts: cell membrane, nucleus, and cytoplasm. Each part has specific functions.

- **Cell membrane**—The cell membrane is a thin, flexible barrier that surrounds the cell and controls the movement of materials into and out of the cell. The membrane consists of several types of protein and fatty substances. The arrangement of the structure of the cell membrane varies in plants and animals.

  Scientists divide cells into two categories: prokaryotes and eucaryotes. The cells of a **eukaryote** have a nucleus; the cells of a **prokaryote** do not. Prokaryotic cells do not have a separate membrane enclosing the DNA or other internal parts that would form the nucleus. Prokaryotic cells carry out all activities associated with life, even though they are relatively simple organisms. Bacteria are examples of prokaryotes. Eukaryotic cells have at least one membrane that encloses the nucleus. They usually contain specialized structures that carry out higher functions than found in the prokaryotes. Some eukaryotes live solitary lives as single-celled organisms, but many are large, multicellular organisms. All plants, animals, fungi, and many microorganisms, are eukaryote.

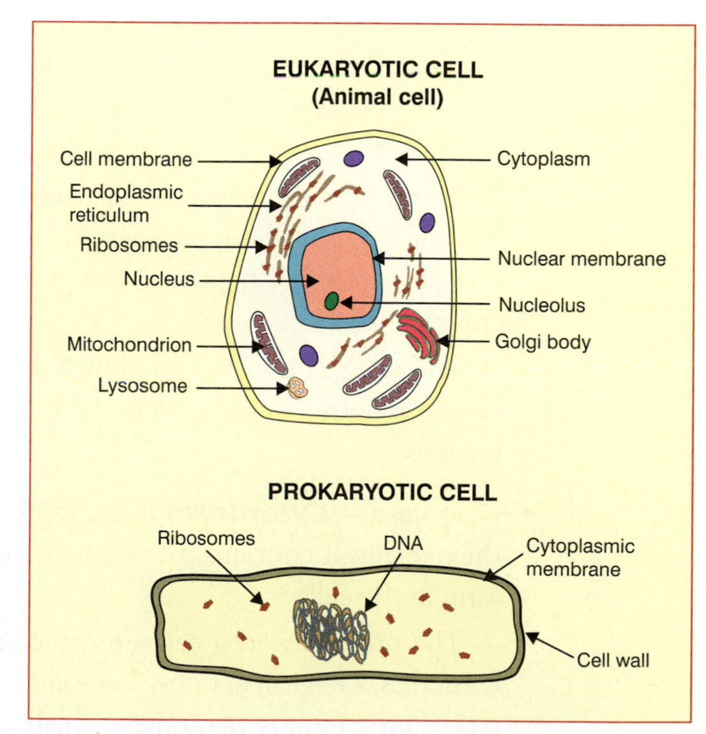

9–22. A comparison of eukaryotic and prokaryotic cell structure is shown here.

9–23. Since most cells are quite small, a microscope is often required for viewing them. (Courtesy, Education Images)

In plant cells, a wall outside the cell membrane provides support and protects the cell. The cell wall contains materials that make our foods appealing and tasty. Pectin is an example. It is found in fruits and berries. Pectin is important in preparing jelly and preserves. Cooking fruit releases pectin so jelly is formed as it cools.

Animal cells do not have walls. To provide body structure, many animals have skeletons that give rigidity and body shape. Because their cell membranes are soft and pliable, animals can move about more easily than plants.

- **Nucleus**—Usually near the center of a cell, the ***nucleus*** is the storehouse of genetic information. The nucleus controls most cell processes and contains the hereditary information of DNA (deoxyribonucleic acid). The DNA holds the coded instructions for making proteins and other important molecules. The granular material visible within the nucleus is called chromatin. It is made of DNA bound to protein. Generally, chromatin is spread throughout the nucleus. When a cell divides, however, chromatin condenses to form chromosomes.

Most of the time, the nucleus also contains a dense region called the ***nucleolus***. The nucleolus is where the assembly of ribosomes begins. Ribosomes aid in the production of proteins. (Chapter 11 has more information on nucleus contents, particularly chromosomes.)

The nucleus is surrounded by a double-membrane layer called the nuclear envelope. It is dotted by thousands of pores that allow material to move into and out of the nucleus.

- **Cytoplasm**—***Cytoplasm*** is the thick, semi-fluid material inside the cell but outside the nucleus. It contains the organelles (little organs) that have specialized jobs to perform in the cell.

The organelles in a cell are the ribosomes, endoplasmic reticulum, mitochondria, lysosomes, Golgi apparatus, lysosomes, chloroplasts, and vacuoles. The endoplasmic reticulum contains ribosomes, which control protein production. Mitochondria use energy from food to make high-energy compounds that the cell can use to power growth, development, and movement. They also control the amounts of calcium, water, and other substances in a cell. Chloroplasts, important for photosynthesis in plants, are found only in plant cells and contain a green pigment known as chlorophyll. Minor structures called plastids contain pigments that may create red, yellow, or other colors in certain plants. These plastids make the beautiful foliage seen on many plants in the fall!

MULTICELLULAR ORGANISMS

Organisms may consist of one or more cells. Organisms, such as cattle and corn, are multicellular organisms. A ***multicellular organism*** is comprised of many cells. The cells in these organisms are organized to form more complex structures, known as tissues, organs, and organ systems.

Cell Specialization

Cell specialization is the development of a cell for a particular purpose or function. The grouping of specialized cells forms unique structures like tissue or organs. If all cells were alike, higher level organisms could not exist!

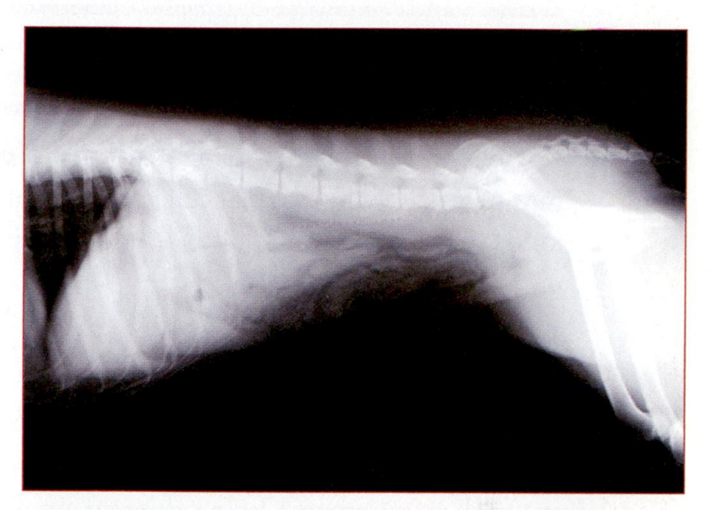

9–24. An X-ray of the skeleton system of a female dog shows the combination of tissues and bones to form an organ system.

Tissues

A ***tissue*** is a group of cells that are alike in structure and activity. The cells in tissues are specialized, so the cells have unique functions to support specific kinds of activities. In animals, tissues include muscles, nerves, and bones. In plants, tissues include melon rinds, potato skins, and flower petals.

Tissues have specific jobs in an organism. For example, the cells in muscle tissues provide motion. These specialized cells do not perform other jobs.

9–25. The eye is a specialized organ that sends sensory information to the brain. (This is a closeup of a cat's eye.)

Organs

An ***organ*** is a collection of tissues that work together to perform a specific function. Although the tissues may differ in the jobs they do, each contributes to the overall function of an organ. Examples of organs in animals are the skin, heart, lungs, stomach, and liver. In plants, organs include roots, leaves, and stems.

Organ Systems

An ***organ system*** is several organs that work together to perform an activity. These compose the major systems of the bodies of many animals. For example, the digestive system is made up of several organs: mouth, stomach, small intestine, large intestine, and others. Organ systems are typically found in complex animals.

An example of an organ system in a plant is the vascular system.

CELL GROWTH PROCESSES

Cells grow through division. ***Cell division*** is the duplication process of one cell splitting into new cells. The original cell is the parent cell. The new cells are daughter cells. The two kinds of cell division are mitosis and meiosis.

MITOSIS

Mitosis is cell division for growth and repair. Each parent cell produces two daughter cells with identical genetic material. The number of cells in an organism increases by mitosis, with the outcome being a larger organism or the replacement of damaged cells. In

AgriScience Connection

GROWING CROPS BEGINS WITH IMPROVED SEED

Successful crop production requires good seed. Better yields are obtained from varieties of crops that have been improved. The improvements are made by controlled breeding and genetic manipulation.

Commercial seed companies sell seeds that have been carefully improved and are guaranteed to grow. A crop can be no better than the seeds that are planted. Using poor-quality seeds wastes time, land, fertilizer, and other inputs. (Courtesy, Terra)

agriscience, the emphasis is on helping organisms grow. Nutrients are provided to plants and animals so mitosis can occur quickly and more frequently.

Mitosis is a sequential process of cell division. It is the phase of the cell cycle when one cell becomes two cells but maintains the same chromosome number as the original parent (***diploid cells***). The process is repeated over and over during the life of an organism.

Mitosis occurs in four steps:

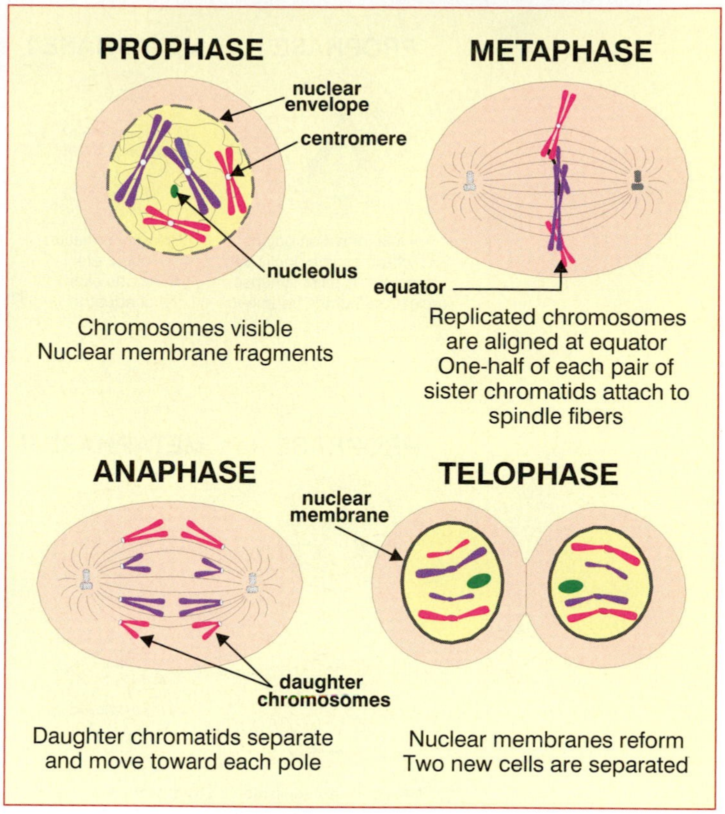

- **Prophase**—The prophase involves the development of chromosomes. Replication of DNA occurs at the end of the interphase. This step follows a time of rest for the cell between nuclear divisions.
- **Metaphase**—During the metaphase, the chromosomes move to the middle of the cell along an invisible line called the equator.
- **Anaphase**—In this step, the chromosomes separate and move to opposite sides of the cell pulled by shortening spindle fibers.
- **Telophase**—In telophase, the chromosomes undergo additional maturity, the nucleus reforms, and a membrane appears between each mass of chromosomes and divides the cytoplasm. The two separate cells can repeat the process.

9–26. Mitosis is a four-step process of one cell becoming two cells.

MEIOSIS

Meiosis is cell division in the sexual reproduction of organisms—plants and animals. Each parent cell produces four daughter cells. Each daughter cell has half the number of chromosomes of the parent cell. Fertilization restores the number of chromosomes. Meiosis is sometimes known as reduction division.

The cells produced by meiosis are called gametes. With animals, male gametes are known as spermatozoa (sperm) or microgametes. Female gametes are known as ova or macro-

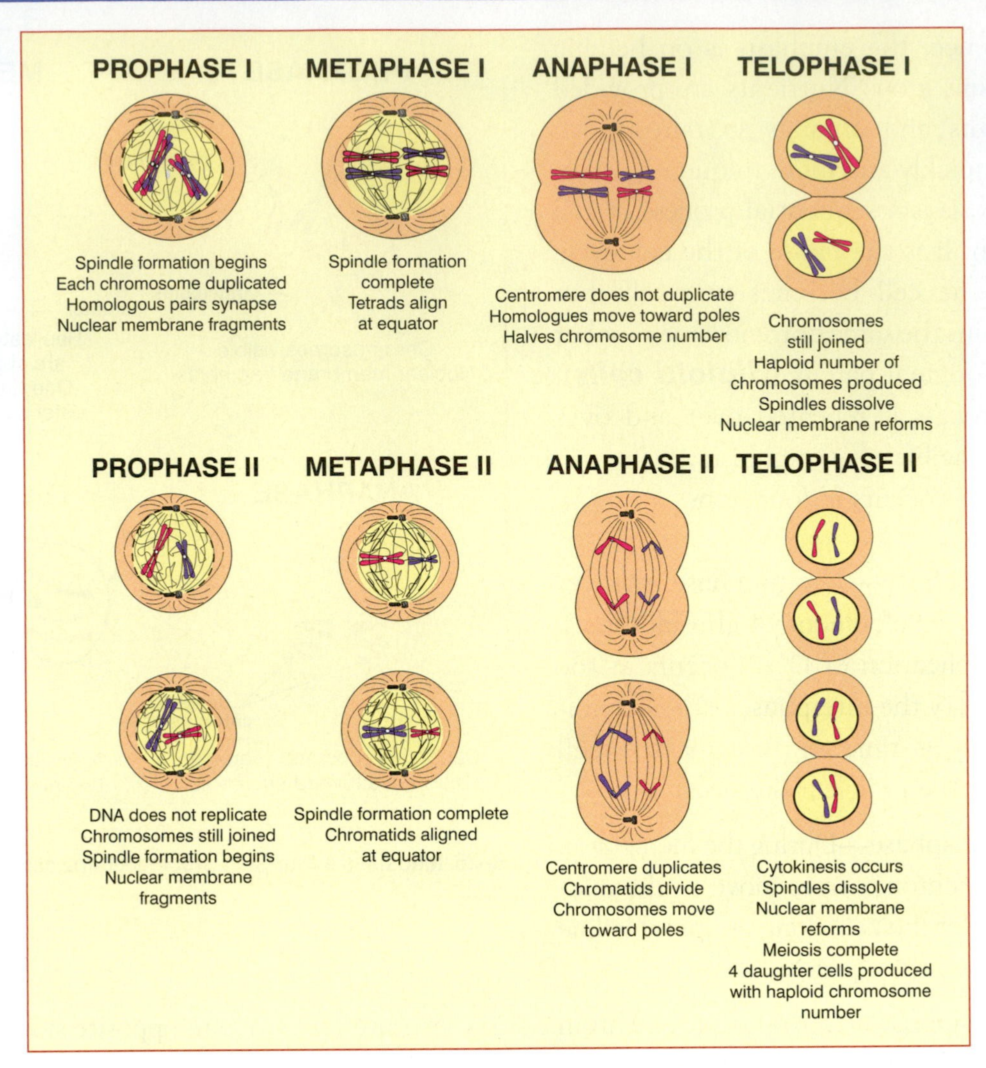

9–27. Divisions and phases of meiosis.

gametes. Sperm are produced in the testicles of males, and ova are produced in the ovaries of females.

A cell produced by meiosis is a ***haploid cell***. It has single chromosomes rather than homologous (matching) pairs. When a sperm unites with an ovum, they form a new individual having diploid cells. Diploid cells have similar complements of chromosomes.

Meiosis I:

- **Prophase I**—DNA is replicated at the end of the interphase before the prophase begins. The formation of threadlike spindle fibers that will attach to the chromosomes begins. The homologous chromosome pairs (the ones that are alike) are located together. Synapse or crossing-over of the chromosomes can occur in this phase, giving a different combination of genes. The nuclear membrane and nucleolus will be dismantled by the end of this phase.

- **Metaphase I**—The tetrads (group of four chromosomes that are alike) align at the equator with spindle fibers attached from the centromere (the point of attachment between the two chromatids) to the poles.
- **Anaphase I**—The centromere does not duplicate. The paired chromosomes are pulled to the poles by the shortening spindle fibers, which halves the chromosome number.
- **Telophase I**—Chromosomes are still joined with half the original chromosome number (haploid number). The nucleolus and nuclear membrane reform. Two daughter cells are produced with one-half the original chromosome number.

ACADEMIC CONNECTION

BIOLOGY

Relate your study of cells, life span and processes, body structures, and genetics in agriscience to your studies in biology class. Compare the processes of agricultural animals with those of animals you study in biology. Identify similarities and differences between various species. Use laboratory equipment to view cells and to identify structures within the cells.

Meiosis II:

- **Prophase II**—DNA does not replicate before the phase begins. Spindle formation begins.
- **Metaphase II**—The chromatids align at the equator of the cell with spindle fibers attached from the centromere to the poles.
- **Anaphase II**—Centromere duplicates, which divides the chromosomes with each half pulled by the spindle fibers to the poles.
- **Telophase II**—A cell membrane forms between the two developing nuclei. The nucleolus and nuclear membrane reform. Cytokinesis (division of the cytoplasm) occurs. Four daughter cells are produced and contain the haploid number of chromosomes.

In the process of producing sperm from the parent cell (spermatogenesis), four spermatozoa (sperm) are produced from each parent cell. In oogenesis (the process of making eggs), the parent cell produces one ovum (egg) and three polar bodies that do not participate in reproduction. The spermatozoa and ova are the cells of sexual reproduction. When an ovum with a haploid number of chromosomes and a haploid sperm join together, a new individual with a diploid number of chromosomes for a species is produced.

REVIEWING

MAIN IDEAS

Scientists have made many discoveries and produced equipment, like microscopes, that have given us the information we know about cells. The cell theory states the main ideas about cells that are a fundamental concept of living things.

Agriscience is concerned with promoting life processes. The efficient production of plants and animals depends on being aware of the stages in an organism's life span and on understanding the fundamental life processes. The life processes of organisms are (1) getting and using food, (2) movement, (3) circulation, (4) respiration, (5) growth and repair, (6) secretion, (7) sensation, and (8) reproduction.

All living things are made up of cells. A cell is much like a combination of building blocks. Each cell has a cell membrane, a nucleus, and a cytoplasm. As cells specialize, they are grouped into tissues, organs, and organ systems.

Life forms reproduce themselves by sexual and asexual means. Mitosis is a process by which organisms grow and repair cells. Meiosis is the process by which sex cells are produced. Parents pass certain traits to offspring.

QUESTIONS

Answer the following questions, using complete sentences and correct spelling.

1. What is an organism?
2. Name and describe five important characteristics of organisms.
3. Name the three parts of the cell theory.
4. What is the difference between living and nonliving things?
5. Describe the difference between autotrophs and heterotrophs.
6. Name the five types of heterotrophs, and give examples of each.
7. Name and define the five stages of life.
8. How do organisms respond to their environment? List examples of plants and animals and their preferred environment.
9. What are the life processes of living organisms? List and explain each.
10. What is a cell? What are the major parts of a cell, and what are their functions?
11. What is cell specialization?
12. How are tissues, organs, and organ systems related?
13. Name and explain the two kinds of cell division.

EVALUATING

Match each term with its correct definition.

a. respiration
b. mitosis
c. organ
d. tissue
e. cell specialization
f. food chain
g. meiosis
h. life span
i. growth
j. biogenesis
k. eukaryote
l. producer
m. prokaryote
n. nucleolus

______1. The theory that life comes from life.
______2. The process of providing cells with oxygen.
______3. An increase in the size of an organism.
______4. Division of cells for growth and repair.
______5. Division of cells for reproduction.
______6. The development of a cell for a particular purpose or function.
______7. A collection of tissues that work together to perform a specific function.
______8. A dense region of nucleus where assembly of ribosomes begins.
______9. The entire length of life.
_____10. A group of cells alike in structure and activity.
_____11. Cells that do not have a separate membrane enclosing DNA of nucleus.
_____12. Autotrophs that make their own food.
_____13. Series of steps by which organisms transfer energy by eating or being eaten.
_____14. Cells with an organized nucleus surrounded by a nuclear membrane.

EXPLORING

1. Make a survey of your local community to learn the kinds of plants and animals that are produced. Interview different producers to determine if improved varieties or species are being used. Prepare a report on your interviews.

2. Visit a local farm supply store or garden center. Study the labels on seed to find the improved varieties. Prepare a report on your observations.

3. Review seed catalogs for varieties that have been improved. If possible, determine if the varieties were improved by selection, inbreeding, hybridization, or genetic manipulation.

4. Arrange to tour a research facility where crops and livestock are studied. Determine the nature of the research. Prepare a report on your findings, and show a relationship to the content of this chapter.

CHAPTER 10

Classifying and Naming Living Things

OBJECTIVES

This chapter provides an introduction to classifying and naming living things. It has the following objectives:

1. Describe the classification of living things.
2. Explain scientific names, and match scientific names with species.
3. Name and discuss the three domains into which organisms are placed, and give examples within each.

TERMS

Archaea
Bacteria
binomial nomenclature
breed
cladogram
common name
cultivar
dicot
domain
Eukarya
monocot
phytoplankton
scientific name
species
taxonomy
variety
virus

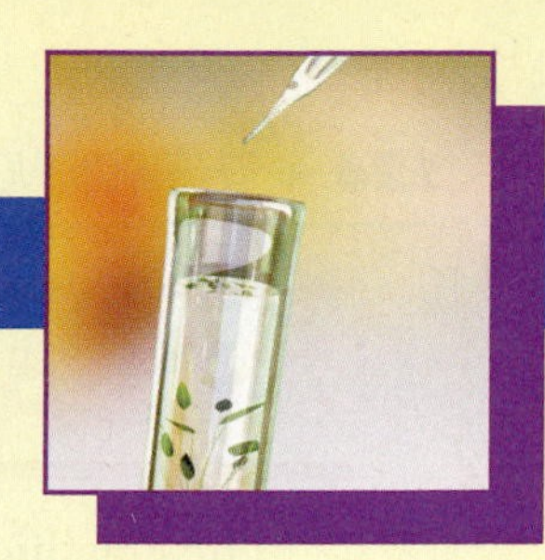

10–1. Common names may vary, but scientific names are uniformly used. (Courtesy, Education Images)

FINDING unique ways of identifying individuals shows up in every area of our lives. People are often identified with numbers, such as the number of a player on a sports team or a student number. Adults are often asked to show a driver's license or a passport to prove identity. Many times we are identified by regions where we live or countries from which our ancestors moved.

On a personal level, we are identified by the names our parents gave us. Most of us have three names: first, middle, and last. These names often reflect our heritage, with the last name placing us in our family.

Just as we have names that indicate who we are, all living things have scientific names. These names are used to uniquely identify living things by their characteristics and show how they relate to each other. Scientific names allow us to describe and compare living things more easily and more accurately.

CLASSIFYING LIVING THINGS

The classification of living things began more than 2,000 years ago in Greece in an effort to show how living things relate to each other. Aristotle developed a simple system of placing plants and animals in separate groups. Around 1735 Linnaeus developed two divisions called kingdoms. Linnaeus also developed a system called binomial nomenclature, which gave living things two-part names and is still used today. This science of classifying living things is called ***taxonomy*** or systematics.

10–2. The tag on this plant gives both its common name and its scientific name.

SCIENTIFIC SYSTEM

Through the years there have been revisions to the classification system in an attempt to organize all living things into groups based on similarities and differences between them. The Linnaean system of classification of living things lumped organisms together based on presumed structures of common origin. The more structures the organisms had in common, the closer they were considered to be in historical development. An example would be that all animals were placed into the same kingdom. All animals that were bears were placed into a group together. Eventually this system was expanded into seven taxonomic categories, ranging from largest to smallest: kingdom, phylum, class, order, family, genus, and species.

Around 1969 Whittaker devised a five-kingdom system that has been widely used since. In recent years there has been much discussion about a six-kingdom system. In the six-kingdom system, archaebacteria (Archaea) were separated from bacteria to form the sixth kingdom. The other five kingdoms remained the same.

Table 10–1. The Process of Change and Development in the Classification System Over Time

Author	Linnaeus	Haeckel	Chatton	Whittaker	Woese, et al.	Woese, et al.
Year	1735	1866	1938	1969	1977	1990
System	2 kingdoms	3 kingdoms	2 empires	5 kingdoms	6 kingdoms	3 domains
	Bacteria & Archaea not included	Bacteria & Archaea classified within Kingdom Protista	Prokaryotes (no organized nucleus)	Monera (includes Bacteria & Archaea)	Eubacteria	Bacteria
					Archaebacteria	Archaea
	Vegetablia	Protista	Eukaryotes (organized nucleus)	Protista	Protista	Eukarya (includes Kingdoms: Protista Fungi Plantae Animalia)
				Fungi	Fungi	
	Animalia	Plantae		Plantae	Plantae	
		Animalia		Animalia	Animalia	

Around 1990 a system based on three domains was introduced by Woese and his associates. A domain is a way of grouping living things that represents lines of evolutionary descent, not just physical similarities. The method uses a system of cladograms. A ***cladogram*** is a diagram that shows development of structures within groups (clads). The cladograms help in understanding how one organism branched from another over time.

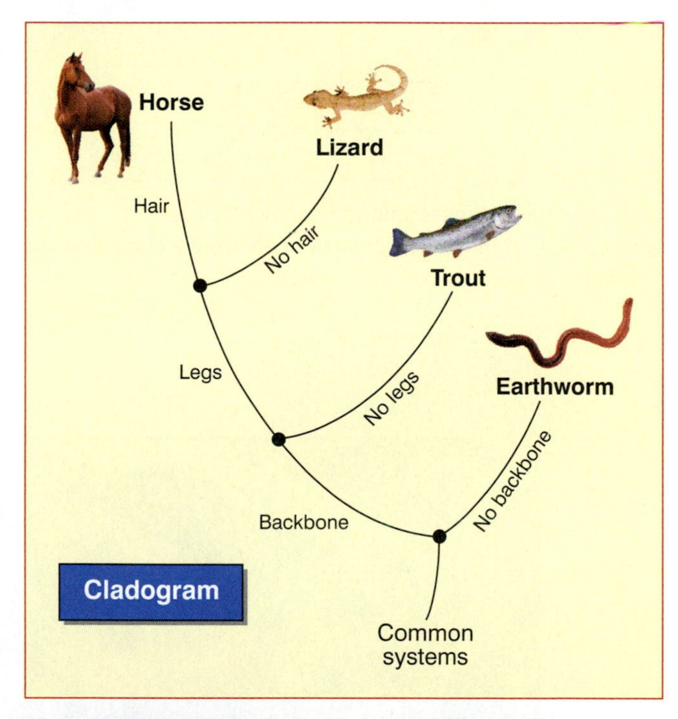

10–3. A cladogram shows how characteristics of these animals have separated them as they developed over time.

The domain system includes eight divisions or stages in the classification system. These divisions show the relationships and differences between various organisms. The modern classification system includes:

Domain
Kingdom
Phylum or Division
Class
Order
Family
Genus
Species

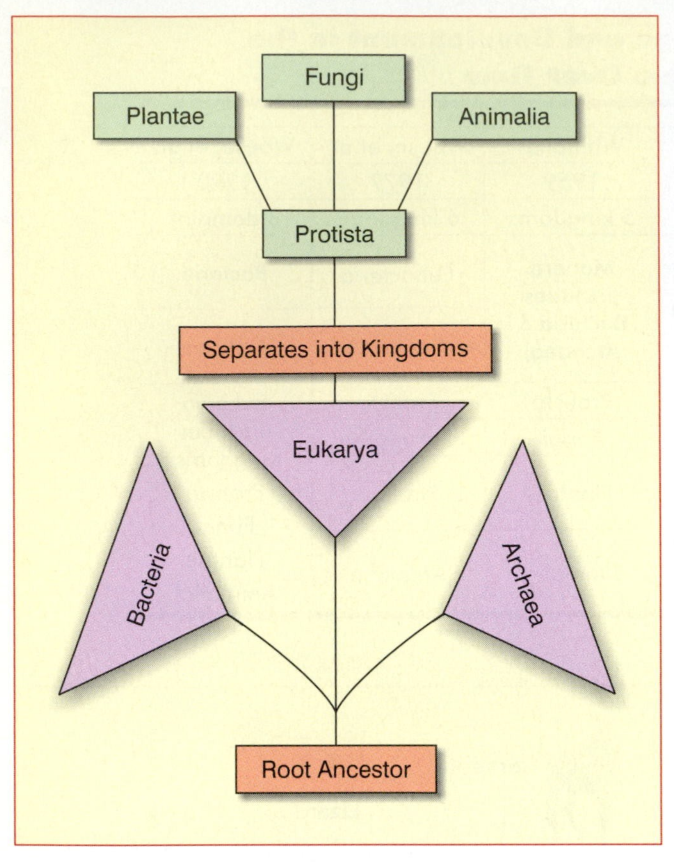

10–4. A simple representation of the evolutionary development of the three domains of life from a common-root ancestor.

With this eight-division system, ***domain*** is the broadest group into which organisms are classified. The common features are more specific with each move to the next division. The kingdom, which is followed by phylum (or division), class, order, family, genus, and species, is more specific than the domain. ***Species*** is the most specific division of classification. All organisms in a species are very similar. They are more similar than those grouped together in the higher divisions, such as genus and family.

ADDITIONAL DIFFERENCES

Organisms may be further divided into subphyla or superfamilies. Agriscientists often work with varieties, cultivars, and breeds. These are used to classify species further.

A ***variety*** is a group of related organisms within a species. The difference is not enough to merit another species. Producers often plant certain varieties of crops, such as improved varieties of rice or wheat.

10–5. People everywhere use the same scientific name for papaya: *Carica papaya*.

10–6. Agricultural research scientists are producing new varieties of rice that will be more nutritious. (Courtesy, Agricultural Research Service, USDA)

10–7. The Brahman breed of cattle has definite identifying characteristics. (Courtesy, American Brahman Breeders Association)

A ***cultivar*** is a crop variety that is cultivated and retains its features when reproduced. Many crop varieties have been developed that retain their characteristics from one generation to the next. The seeds of hybrid crops do not retain their features when reproduced. They are often said not to "breed true."

A ***breed*** is a group of animals of the same species that have definite identifying characteristics and a common origin. Breeds of beef and dairy cattle, hogs, sheep, goats, and horses

10–8. The Chow is a distinct dog breed that may be black or red.

are important in livestock production. For example, common breeds of beef cattle are Angus, Hereford, Santa Gertrudis, and Charolais. Registry associations record information about animal breeds to assure breed qualities are maintained.

Breeds of companion animals are also important. We are familiar with dog breeds, such as Border Collie and Golden Retriever, and with cat breeds, such as Burmese and Maine Coon. Competitive events—dog shows, for example—may feature only selected breeds. The American Kennel Club (AKC) maintains a registry of purebred dogs and otherwise promotes dog events. The Cat Fanciers' Association (CFA) maintains a breed registry, promotes cat activities, and provides information on cat care.

10–9. The Odd-Eyed White Persian is an interesting pedigreed cat.

We also know that certain breeds have specific uses. For example, Beagle dogs are good hunters, and German Shepherds are good guard dogs. The word *pedigree* is sometimes used in relation to dogs, cats, and other animals that have a pure ancestry of the breed.

CAREER PROFILE

BIOLOGIST

A biologist studies living things. Biologists may investigate the structure and physiology of organisms. Some biologists do research to answer questions about living things and life processes. This shows a biologist examining the leaves of inoculated safflower plants.

Biologists need college degrees in biology. Many specialize in particular areas, such as zoology (animals), botany (plants), and insects (entomology). Some biologists have master's degrees and doctorates. If you want to become a biologist, begin now by taking science and agriculture courses in high school.

Jobs for biologists are found in a wide range of places. The nature of biologists' work depends on their specialization. Jobs are with colleges, research stations, private companies, and other employers. Some biologists are consultants and work for themselves. (Courtesy, Agricultural Research Service, USDA)

SCIENTIFIC NAMES

Taxonomy is used in scientific names. A scientific name is the two-word Latin name, which is written in italics or underlined. Every known organism has a two-part ***scientific name***. The name identifies the organism in much the same way as our names identify us.

A plant or animal's ***common name*** is the name used by non-scientists. Common names often vary for the same species from one location to another. The common name of the ornamental aster is the Greek word meaning "star."

10–10. In this family, all the members have the same last name. Each family member has a first name and a middle name to identify him or her specifically. These names are their common names. The scientific name for all humans is *Homo sapiens*.

BINOMIAL NOMENCLATURE

Scientists call the system of assigning the two-part scientific names ***binomial nomenclature***. The scientific name of every living thing is made up of two names. The first name is the genus, and the second name is the species. After the first use, the genus name may be abbreviated. For instance, *Bos taurus* might be written as *B. taurus*.

The scientific names of common agriculture species are presented in Table 10–2.

Importance of Scientific Classification

Three reasons scientific names and classification are important are:

- People use the same name for an organism worldwide. For example, a housefly has the scientific name *Musca domestica*. There are many kinds of flies. Using the scientific name ensures that everyone will be talking about the same species of fly.

Table 10–2. Common and Scientific Names of Selected Species

Common Name	Scientific Name (genus and species)	Common Name	Scientific Name (genus and species)
TERRESTRIAL ANIMALS			
Cat (domestic)	*Felis catus*	Goat (domestic)	*Capra hircus*
Cattle (those with humps)	*Bos indicus*	Hog	*Sus scrufa*
Cattle (those from Europe)	*Bos taurus*	Horse	*Equus caballus*
Chicken	*Gallus domesticus*	Sheep	*Ovis aries*
Dog (domestic)	*Canis familiaria*	Turkey	*Meleagris gallopavo*
Duck	*Anas domestica*		
AQUATIC ANIMALS			
Bullfrog	*Rana catesbiana*	Oyster (American)	*Crassostrea virginica*
Channel catfish	*Ictalurus punctatus*	Rainbow trout	*Salmo gairdneri*
Common carp	*Cyprinus carpio*	Shrimp (brown)	*Penaeus aztecus*
Crawfish	*Procambarus clarkii*	Tilapia (blue)	*Tilapia aurea*
Goldfish	*Carassius auratus*		
FOOD AND FEED CROPS			
Corn	*Zea mays*	Oats	*Avena sativa*
Cotton (upland, as grown in North America)	*Gossypium hirsutum*	Rice	*Oryza sativa*
Cotton (long staple, as grown in Egypt and island locations)	*Gossypium barbadense*	Soybean	*Glycine max*
		Sugarbeet	*Beta vulgaris*
		Wheat	*Triticum aestivum*
ORNAMENTAL PLANTS			
Astor	*Astor novae-angliae*	Ginger	*Zingiber officinale*
Carnation	*Dianthus caryophyllus*	Petunia	*Petunia × hybrida**
Gardenia	*Gardenia jasminoides*	Tulip	*Tulip gesneriana*
Geranium	*Pelargonium graveolens*	Zinnia	*Zinnia elegans*
TREES			
Cedar:		Oak:	
eastern red	*Juniperus virginiana*	live	*Quercus virginiana*
western red	*Thaja plicata*	Rocky Mountain white	*Quercus utahensis*
Fir:		Pine:	
Douglas	*Pseudotsuga taxifolia*	loblolly	*Pinus taeda*
Maple:		longleaf	*Pinus palustris*
sugar	*Acer saccharum*	northern white	*Pinus strobus*
		Spruce:	
		blue	*Picea pungens*

(Continued)

Table 10–2 (continued)

Common Name	Scientific Name (genus and species)	Common Name	Scientific Name (genus and species)
FOOD PLANTS			
Apple	*Malus domestica*	Lima bean	*Phaseolus limensis*
Asparagus	*Asparagus officinalis*	Orange	*Citrus sinensis*
Banana	*Musa paradisiaca*	Papaya	*Carica papaya*
Blueberry	*Vaccinium virgatum*	Pineapple	*Ananas comosus*
Cabbage	*Brassica oleracea*	Strawberry	*Frageria chiloensis*
Cauliflower	*Brassica oleracea*	Sugarcane	*Saccharum officinarum*
Coffee	*Coffee arabica*	Tomato	*Lycopersicon esculentum*

*This scientific name designates a hybrid cross. Most ornamental plants have crosses as well as numerous varieties.

- Relationships between organisms are evident in their names—all cattle have the same genus name of *Bos*. The species name will vary with the location where the breed originated and whether it has a hump.
- Differences between organisms are obvious from their family names. For example, corn, oats, and many other plants are in the grass family, Gramineae. Strawberries and apples are in the rose family, Rosaceae. By knowing the family name, differences and similarities in organisms are easy to determine.

10–11. In the same family (Rosaceae), the strawberry (*Frageria chiloensis*) and the apple (*Malus domesticus*) are much alike yet different. What are the similarities and differences?

Importance in Agriculture

In agriculture, plants and animals go by their common names. A rancher hauls a load of beef cattle to market—not a load of *Bos taurus*! In fact, a cattle buyer would be surprised if the rancher said that the *Bos taurus* were for sale.

10–12. A hybrid striped bass is obtained by crossing the male white bass (*Morone chrysops*) with the female striped bass (*Morone saxatilis*).

When are scientific names used? They are used when agriscientists need to communicate accurately about plants and animals. Since many people in agriculture use science, they need to know about scientific names.

Scientific names are universal. An organism has the same scientific name in all parts of the world. *Zea mays* is known as corn in North America and maize in other parts of the world.

- **Cultural practices**—Producers of plants and animals often rely on information prepared for worldwide use. Products used to kill weeds or treat sick animals may use scientific names. The products, as well as the plants or animals they are to be used on, may have scientific names.

  Labels or instructions that use scientific names have little meaning unless the reader knows about the names. People need only to remember a few names. For the others, they need to know how to find the scientific names.

10–13. Many species of plants can be identified by studying leaf shape and other plant characteristics.

- **Research**—Scientists nearly always use scientific names when doing research. They do so because other scientists worldwide are interested in the results. Research findings are reported using scientific names, unless the study is for a local area, such as the Shenandoah Valley of Virginia. People in this area would typically use the same common names. There would be no confusion about the name or the organism involved.

  When agriscientists talk with the public, they may use the common name followed by the scientific name. In some places, only the common name may be used. This is because everyone would know the exact organism being discussed.

10–14. The large leaf size of the Princess Tree of China (*Paulownia tomentosa*) and other features help in identifying it. (Leaves of the tree range from 24 to 36 inches in diameter.) (Courtesy, Education Images)

THE DOMAINS

All living organisms can be classified into one of three domains. A domain is the first division in the scientific classification system. The domains are Bacteria, Archaea, and Eukarya.

In addition to the indicated domains, there is a pre-cellular group known as viruses. The viruses are examples of the difficulties we face in trying to place all organisms into specific groups. Some organisms don't have an exact fit. A ***virus*** is a tiny particle that is not a complete organism but that contains genetic material and protein, similar to a living organism. As incomplete cells, viruses cannot reproduce, and most scientists do not consider them to be alive. They invade other cells and to use the enzymes and other materials of the hosts. Scientists believe that the first viruses probably came from living material and have continued to evolve over billions of years. Viruses are associated with many disease problems of plants and animals. Viruses were discovered several years ago, when electron microscopes were developed. These microscopes enlarge objects that are too small for regular microscopes. Viruses are not placed in any domain.

DOMAIN BACTERIA

The domain ***Bacteria*** is made up of one-celled organisms that have cell membranes, cell walls, and cytoplasm but don't have nuclei (prokaryotes). The domain Bacteria corresponds to the kingdom Eubacteria. The bacteria are very diverse, ranging from free-living soil organisms to deadly parasites. Some make food by photosynthesis, while others do not.

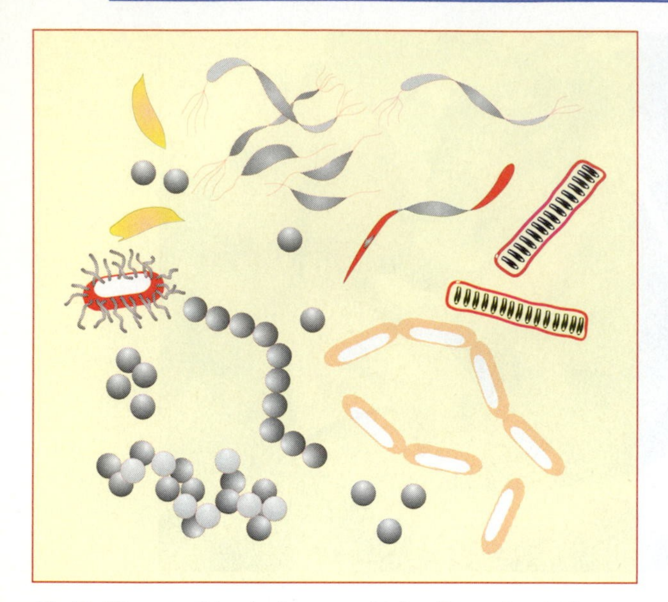

10–15. Shapes of bacteria vary widely. (Some bacteria appear round, while others may appear rod-shaped, spiraled, or as tiny filaments. Some form colonies. Some have flagella for movement.)

Some need oxygen to survive, while others are killed by oxygen. Particular types of bacteria help break down the nutrients in dead matter and the atmosphere. This allows other organisms to use the nutrients. Bacteria are useful but also cause disease.

Many types of bacteria are all around us. One common type of bacteria is known as blue-green bacteria. These bacteria are sometimes mistakenly called algae. Blue-green bacteria grow in water, such as ponds and streams. They may grow in large numbers and make the water appear green in color. The growth of blue-green bacteria is important in aquaculture ponds. Too much growth (known as bloom) and the death of the bacteria can cause water problems. Most commonly the blue-green bacteria tie up oxygen and cause fish to die.

One of the most common bacteria is *Escherichia coli*, an organism found in the intestines of humans. The *E. coli* live on unabsorbed nutrients and help the body obtain vitamins.

Technology Connection

KNOWING AND DOING

A person who knows the environment required for a plant can usually establish an artificial environment for it. Greenhouses are often used to protect tender plants from cold weather and to promote growth. Warm-season plants may be grown in a greenhouse during cold weather under the proper conditions. These conditions protect the plants from frost and provide nutrients needed for growth.

Warm-season plants are often started in a controlled environment and then transplanted outside. Seeds are planted. Germination occurs under the right conditions. The tiny plants shown here will be grown for a couple of months in the greenhouse. This will allow the plants to be set outside after the weather is warm. Some people say that this gets a "jump on summer." Such plants reach maturity earlier in the summer than those planted from seed after the danger of frost has passed.

A pesky problem in health-care facilities is MRSA. MRSA stands for "methicillin-resistant *Staphylococcus aureus*." It is commonly called staph. We all have staph bacteria on our skin. Occasionally, the bacteria get into the body and cause infections, such as pimples, boils, pneumonia, and blood infections. Some *S. aureus* are resistant to methicillin. Methicillin is an antibiotic commonly used to treat staph infections. *S. aureus* may form colonies on the body without causing disease. Sanitation is important in stopping the spread of staph infections. If you are around a person with an MRSA infection, wear protective gloves and clothing, wash hands after contact, and disinfect affected areas. Avoiding skin-to-skin contact helps prevent the spread of staph.

The following are a few examples of ways bacteria affect or are used by living things:

- **Cause decay**—Bacteria act on materials to cause decay. This involves breaking down dead plants and animals into forms that can be reused. Bacteria are essential in natural recycling. The organic matter in soil results from the action of bacteria on leaves and twigs.
- **Used in food production**—Bacterial action helps produce many important food products. Cheese, vinegar, and yogurt are three examples. Bacteria are needed to change the original form of the food into a new product. Of course, only selected species of bacteria can be used for these purposes.
- **Used in medicine production**—Certain species of bacteria are used to make antibiotics, artificial insulin, and other medicinal products.
- **Serve as food**—Bacteria in ocean, lake, and stream water may provide food for animals. Several small aquatic organisms, including oysters and shrimp, feed on bacteria in the water.
- **Clean water**—Some forms of bacteria help to keep water clean. They convert materials in water into useful forms or forms that are not dangerous. In fish ponds, bacteria convert a substance known as ammonia into a form that is not harmful. These bacteria are added to filter systems in fish tanks to help keep the water clean.

10–16. Making Swiss cheese involves using bacteria. Action of the bacteria creates big bubbles of gas as the cheese is being made. The bubbles appear as holes in cheese slices.

- **Provide nutrients**—In rice fields, bacteria convert nitrogen from the air into a form that the rice plants can use. This saves the farmer the cost of buying fertilizer.

DOMAIN ARCHAEA

The domain **Archaea** is composed of organisms that look very similar to Bacteria but are genetically more like Eukarya (the third domain). Like Bacteria, they are small, lack nuclei, and have cell walls. However, the cell walls of Archaea have a different chemical composition than those of Bacteria. The arrangement of the genetic material makes Archaea more like Eukarya than Bacteria. Scientists look at this gene arrangement to reason that Archaea may be the ancestors of Eukarya.

Many Archaea live in extremely harsh environments. One group produces methane gas in an oxygen-free environment, such as thick mud or the digestive tracts of animals. Other Archaea live in extremely salty environments, such as Utah's Great Salt Lake, or in hot springs, where the temperature can get near the boiling point of water.

DOMAIN EUKARYA

The domain **Eukarya** consists of all organisms that have nuclei. Most organisms also contain cell membranes, cytoplasm, and specialized structures called organelles. The domain Eukarya is organized into four kingdoms: Protista, Fungi, Plantae, and Animalia.

Kingdom Protista

The protists, as the organisms in the kingdom Protista are known, are one-celled. Some form collections of cells that may resemble plants or animals. Their cell structure is more advanced than that of the Bacteria or Archaea. More than 65,000 organisms have been classified into the kingdom Protista.

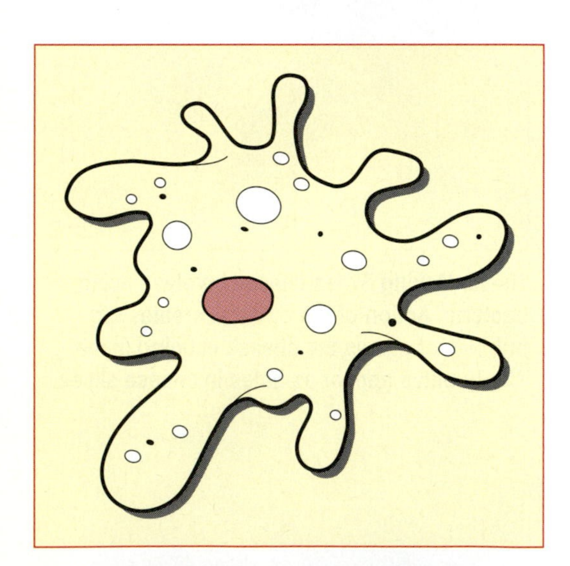

10–17. An ameba moving by extending part of the cell in one direction, with the remainder of the cell flowing into it.

The phyla are:

- Phylum: **Sarcodina**—The members of this phylum include the amebas. These organisms have cell membranes that are flexible. An ameba moves when part of its cell reaches out, somewhat like a foot, and the rest of the cell flows into it.
- Phylum: **Ciliophora**—Organisms in this phylum have cilia, or tiny hairs, that help them move about. The most common Ciliophora, or ciliate, is the "slipper-shaped" paramecium. Ciliates are found in most fresh and salt water. There may be many different ciliates in a stream or lake near you.

- Phylum: **Zoomastigina**—Every organism in this phylum has a long, whiplike structure known as a flagellum, which it uses to move about. The tiny euglena is an example. Zoomastigina often cause disease, such as human sleeping sickness. Insects carry the organisms. When the insects bite, they transfer the organisms into their victims' blood, where they continue to live. They also cause amebic dysentery. A symptom is severe diarrhea. These organisms are most common in areas with poor sanitation but are occasionally found in other areas.
- Phylum: **Sporozoa**—Organisms in this phylum have no means of moving about. They are carried in the bodies of host animals. Sporozoa cause diseases in humans and other animals. Coccidiosis, a deadly disease in poultry, is caused by sporozoans.
- Phyla: **Chlorophyta, Phaeophyta, Rhodophyta, Chrysophyta, and Pyrrophyta**—These are the five phyla of algae. Unlike other protists, algae can make their own food. Algae are plantlike organisms that live as single cells or in groups. Some scientists formerly put red, brown, and green algae into the plant kingdom. This is because they make food and live as plants.

  Colonial algae are groups of alga cells that act as a unit. They share functions needed for life. Filamentous algae grow in stringlike forms or rows of cells. Thalloid algae have cells that divide in all directions. They do not form definite structures. The thallus forms are known as seaweed.

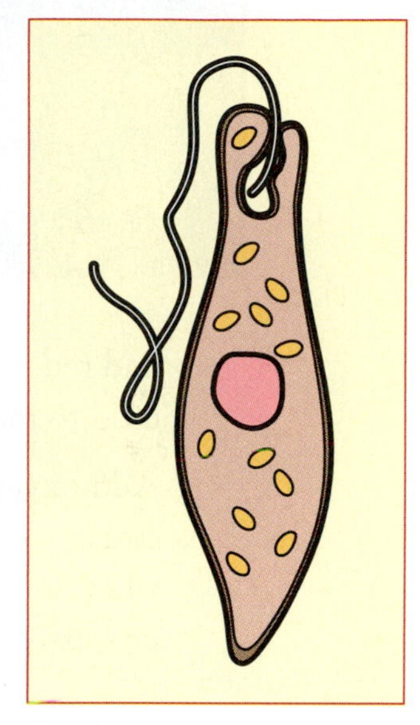

10–18. An enlarged euglena shows a flagellum.

Here are some examples of ways protists affect life or are used:

- **Cause disease**—Protists cause disease in humans and other organisms. A common example is malaria. It is transferred to humans with the bite of the anopheles mosquito. In the human bloodstream, the Protista genus known as *Plasmodium*, a form of sporozoan, produces a toxin. The toxin causes chills and fever.
- **Polish or grind stones**—Diatoms, which are protists, have extremely hard cell walls. They are mined and used to make materials to polish or grind metal and other materials.
- **Provide food**—Brown, red, and green algae are cultivated for food in some parts of the world. They are often known as seaweed. The brown form is called wakame in Japan and kelp in Europe and North America. Red algae are known as laver in Europe and North America and as nori in Japan. Green algae are not grown as much as the brown

10–19. Nori seaweed is cultured in some locations. (Courtesy, Mike Guiry, National University of Ireland)

and red. Of course, alga is not a popular food in North America. It is, however, often used to make food additives, some of which are included in low-fat hamburger.

- **Add oxygen to water**—Forms of algae produce oxygen in the water of fish ponds and streams. They are called ***phytoplankton***. The oxygen is released into the water when the algae make food. Fish take the oxygen from the water and use it in their life processes.

10–20. Mushrooms are fungi that are frequently used as food.

Kingdom Fungi

The kingdom Fungi includes yeasts, mildews, and mushrooms. Fungi are more complex than bacteria, archaeans, and protists. About 100,000 species have been identified, with several used as sources of an antibiotic known as penicillin. Four phyla are listed:

- Phylum: **Zygomycota**—This group includes black bread mold, often found on bread, fruit, and other food products.
- Phylum: **Ascomycota**—Yeasts, truffles (food that grows below the surface of the soil), and *Penicillium notatum*, which is used to make penicillin, are included in this group of fungi.
- Phylum: **Basidiomycota**—This group includes mushrooms.

- Phylum: **Deuteromycota**—This group includes imperfect fungi, which cause ringworm but are also a source of penicillin.

Following are examples of ways fungi affect life or are used:

- **Cause food spoilage**—Some fungi, in the form of mold, attack bread and other foods, causing them to be unfit to eat. Mold is easily seen on bread. Fungi may also attack and ruin animal feed. Proper storage can slow the growth of mold.

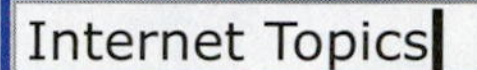

Selected topics for Internet discovery and reporting are listed here. Begin by searching for each topic. Next, read and learn about it. Conclude by preparing a brief report on your findings.

1. American Kennel Club
2. Phytoplankton
3. Seaweed

- **Make food products better**—Yeast is used as a leaven in baking bread. A leaven causes fermentation in the bread. Small amounts of yeast organisms are added to the mix as it is being prepared for baking. Warmth causes the yeast to grow. When the bread is baked, the holes left in it are due to gas bubbles caused by the fermentation.
- **Used as food**—Mushrooms and truffles are desired as food. Mushrooms are preferred in the United States, Canada, Japan, and other countries. Truffles are preferred in areas of Europe. (Since truffles grow under the soil, trained hogs and dogs are used to sniff out and dig up the mature truffles.)
- **Used in the production of medicine**—Penicillin is an important medicine made from the *Penicillium* genus of fungi. It is used to treat many kinds of disease in humans, livestock, pets, and exotic animals.
- **Cause plant disease**—Some plants are subject to mildew. This is a disease easily seen as a white coating on the affected plants. Control involves spraying the plants with fungicides (fungus killers).
- **Damage property**—Some species of fungi grow on surfaces during damp weather. The stain or off-color on the trim and walls of houses is mildew. It must be removed before another coat of paint is applied. Many people use a chlorine and water solution to wash it off. Mildew is a problem in warm, humid climates.
- **Cause human and animal disease**—Some fungi are parasites. They live in or on other plants and animals. Ringworm is a common form that attacks humans, pets, and other animals. Athlete's foot is also a fungus.

10–21. A puffball is a fungus. The inside of a mature puffball is filled with tiny spores that can be released into the air. Also known as devil's snuffbox, the puffball is in the Lycoperdaceae family.

10–22. The banana (*Musa paradisiaca*) is a large tropical plant. This shows a reddish bud with petals that open to reveal developing bananas.

Kingdom Plantae

The kingdom Plantae is a large, important kingdom. Plants provide food, fiber, and shelter for humans. They provide feed for livestock, poultry, and other animals. Some 350,000 plant species have been identified.

Plants are made of many cells and make their own food by photosynthesis. Plants cannot move from place to place. They have cell walls that contain cellulose. Some plants are quite large, such as the sequoia trees in California. One sequoia named General Sherman is the earth's largest living organism. Other plants are very small, such as mosses and ferns.

The kingdom Plantae has eight phyla or divisions, as follows:

- Phylum: **Bryophyta**—Mosses and liverworts are the simplest land plants. They are generally found in damp, shady places, such as in woods or near springs.
- Phylum: **Lycophyta**—Club mosses make up this phylum.
- Phylum: **Spenophyta**—This group is composed of horsetails.

AgriScience Connection

LEARNING IN AN ARBORETUM

An arboretum is a type of outdoor botanical garden where trees, shrubs, and other plants grow in a natural setting. Some people refer to an arboretum as a plant identification laboratory. People go to the garden to study the plants and learn proper identification.

Most specimens in an arboretum are labeled. This helps people learn the common and scientific names. In some cases, other information is given, such as where the plants originated.

Many universities have arboretums that also include trial sections for new varieties of plants. Colorful flowering plants are often included in horticulture gardens.

- Phylum: **Pterophyta**—Many species of ferns were thought to exist in prehistoric times. Today a small number of species remain.
- Phylum: **Coniferophyta**—Conifers are trees or shrubs that bear cones, such as pine trees. Most remain green the year round; therefore, they are often called evergreens. Conifers are used in many ways, such as in making paper, lumber, plywood, and other wood products. Some produce rosin, which is used to make turpentine and other products.
- Phylum: **Cycadophyta**—This phylum contains palmlike plants that grow in tropical areas.
- Phylum: **Ginkgophyta**—Ginkgo plants grow fan-shaped leaves. Some are used for landscaping. Ginkgoes were common when dinosaurs were alive, but the phylum now has only one species, *Ginkgo biloba*. This plant may be one of the oldest seed plant species alive today.
- Phylum: **Anthophyta**—Flowering plants make up this phylum.

10–23. Leaves from the *Ginkgo biloba* tree.

The flowering plants phylum (Anthophyta) is the most important in agriscience. This phylum is divided into two classes:

- Class: **Dicotyledonae (dicots)**—A ***dicot*** is a plant with two seed leaves. Most fruits, flowers, trees, and vegetables are dicots.
- Class: **Monocotyledonae (monocots)**—A ***monocot*** is a plant with one seed leaf. Examples are onions, corn, wheat, oats, and sorghum.

10–24. Neatly labeled plants at the San Antonio Botanical Gardens help people learn to identify plants.

10–25. Cultural practices are used to improve yields and assure a quality tomato crop.

Here are some ways plants are used:

- **Used as food**—Many plants are used as human food. These include fruit, such as the apple; grain, such as wheat; leafy vegetables, such as lettuce; roots, such as the carrot and beet; and flowers, such as the cauliflower.
- **Made into clothing**—Clothing is made from the fibers of cotton, flax, and a few other plants. Cotton is by far the largest provider of clothing material. Many acres of cropland in the United States are devoted to cotton production.
- **Provide shelter**—Trees provide lumber and other materials used to build houses and to make furniture and other things we need. Some trees naturally grow on the land. Other trees are planted as tree farms. The role of tree farming has increased tremendously in the last 50 years. Today, more trees are planted each year than are harvested for wood products. Many tree farms are planted to improved varieties that grow more rapidly and produce greater amounts of quality wood.
- **Made into paper**—A number of different trees are used to make paper. Other plants, such as kenaf, are also being grown for making paper. (Kenaf is a kind of hibiscus, *Hibiscus cannabinus*, and is also called Indian hemp.)

More information about plants will be presented later in this book.

Kingdom Animalia

The kingdom Animalia is made up of organisms that (1) have many cells, (2) can move about, and (3) get their food by ingestion (eating). The cells of animals do not have cell walls. Animals can be found in nearly every part of the planet.

More than a million species of animals are known to exist. These range from spiders, butterflies, and ants to elephants, horses, cattle, and chickens. Each year, new species are being identified for the first time. These include some that live deep in the ocean.

Animals are divided into 14 phyla, as follows:

- Phylum: **Porifera**—This group is composed of the sponges that grow in the sea.

10–26. Colorful red finger sponge and brown tube sponges on Belize reef.

- Phylum: **Cnidaria**—This phylum is made up of the jellyfish, coral, and other sea animals.
- Phylum: **Platyhelminthes**—This phylum comprises the flatworms.
- Phylum: **Nematoda**—The roundworms of this phylum are both useful and problem causing in agriculture. Nematodes in the soil destroy crop roots. Roundworms in livestock cause animals to grow poorly.
- Phylum: **Rotifera**—This group contains tiny round or wormlike animals found in the soil and in stagnant water.
- Phylum: **Bryozoa**—The animals in this phylum resemble moss but aren't plants.
- Phylum: **Brachiopoda**—Organisms in this phylum are referred to as the lampshells.
- Phylum: **Phoronida**—Tube worms make up this phylum.
- Phylum: **Annelida**—The segmented worms in this phylum include earthworms and leeches.
- Phylum: **Mollusca**—The animals in this phylum are soft-bodied and nonsegmented. Some have hard shells, such as the oyster and the clam. Examples of those without shells are slugs and squid.

ACADEMIC CONNECTION

BIOLOGY

Relate naming and classification in agriscience to the study of taxonomy and nomenclature in biology or advanced biology class. Use what you learn in agriscience to support your study of biology.

10–27. Sheep and dogs are in the phylum Chordata and class Mammalia, but that is where the similarities end. Dogs are in the order Carnivora (meat eaters with claws). Sheep are in the order Artiodactyla (even-toed hoofed animals).

- Phylum: **Arthropoda**—The arthropods have bodies divided into segments and hard outer-body covers known as external skeletons. The legs are jointed. Examples are lobsters, shrimp, crawfish, insects, mites, and ticks.
- Phylum: **Echinodermata**—The starfish and other sea animals make up this group.
- Phylum: **Hemichordata**—Animals known as acorn worms are included in this phylum.
- Phylum: **Chordata**—This is by far the most important phylum of animals. The subphylum of vertebrates is widely grown in agriculture. Vertebrates are animals with internal skeletons and backbones. Examples are horses, pigs, cattle, sheep, birds, and most fish.

Table 10–3. Taxonomy of Common Animals

	Hogs	Cattle	Sheep	Horses
Kingdom	Animalia	Animalia	Animalia	Animalia
Phylum	Chordata	Chordata	Chordata	Chordata
Class	Mammalia	Mammalia	Mammalia	Mammalia
Order	Artiodactyla	Artiodactyla	Artiodactyla	Artiodactyla
Family	Suidae	Bovidae	Bovidae	Equidae
Genus	*Sus*	*Bos*	*Ovis*	*Equus*
Species	*scrofa*	*taurus** or *indicus***	*aries*	*caballus*

*Breeds of European origin.
**Breeds with humps.

Here are some ways animals are used:

- **Used as food**—Many animals provide food for humans. Some are raised, such as hogs and chickens; others are captured in the wild, such as lobster and fish. Today, with the emergence of aquaculture (water farming), more fish are being grown on farms. Various animal products are eaten, including meat, eggs, and milk.
- **Used to make clothing and jewelry**—Animals provide fur, wool, leather, pearls, and other products used to make clothing and jewelry.
- **Byproducts used for the production of medicine**—Various medicines to treat human disease come from animals. Insulin is made from cattle byproducts and is used to treat diabetes. The intestines make sutures (threads) used to repair wounds or cuts in the human body. New methods allow transplanting from one species to another, such as the valves in the heart of a pig into a human.
- **Used as sources of power**—Though the use of animals as sources of power has declined, animals are still used for this purpose. Some countries rely heavily on animal power for agriculture. Other countries use animals in unique situations where they are best suited to provide power, such as to harvest maple sap in the United States. Animals used for power include the ox, horse, and water buffalo.
- **Used for recreation**—Many animals are used for recreation and companionship. These uses vary from horses and dogs that race to hunting hounds, homing pigeons, and the pets many people keep.

More information about animals will be presented later in this book.

10–28. Animals bring joy to human life! (A young girl is hugging her puppy.)

REVIEWING

MAIN IDEAS

Classification systems are used to show relationships and differences between organisms. The system used today is based on eight divisions.

Domain
Kingdom
Phylum or Division
Class
Order
Family
Genus
Species

In addition, breeds, varieties, and cultivars may be used. These are based on variations in species that retain their unique characteristics when the animals or plants reproduce.

Scientific classification is important because it helps people in all parts of the world communicate. Everyone looks at organisms the same way. Taxonomy is the science of the classification of organisms into groups.

Scientific names are based on scientific classification. The system of giving every organism a two-part scientific name is called binomial nomenclature. The name is made up of the genus and the species. Scientific names are usually in Latin and are either written in italics or underlined.

Plants, animals, and other organisms have important roles in human life. They provide food, clothing, housing, and many other benefits. They also perform important roles in the environment and help maintain a balance in nature.

QUESTIONS

Answer the following questions, using complete sentences and correct spelling.

1. What is taxonomy?
2. What are the eight divisions in the modern classification system? What is the sequence of the divisions from the broadest group to the most specific group?
3. What three domains are used to classify organisms? Briefly describe each domain.
4. How do members of the four kingdoms of the domain Eukarya provide for the needs of people? Give two examples for each kingdom.
5. What are the scientific names for the following animals: horse, sheep, goat, hog, chicken, humped cattle, and turkey?
6. What are the scientific names for the following crop plants: wheat, corn, cotton, and soybeans?

7. What are the scientific names for the following trees: live oak, sweetgum, and shortleaf pine?
8. Why is the use of scientific names important?

EVALUATING

Match each term with its correct definition.

a. taxonomy
b. species
c. kingdoms
d. cultivar
e. binomial nomenclature
f. domain
g. breed
h. Archaea
i. variety
j. cladogram

_____1. Animals of the same species that have definite identifying traits.
_____2. The system of giving two-part scientific names to organisms.
_____3. The classification of living things.
_____4. The most specific scientific name of an organism.
_____5. A group of related organisms of the same species.
_____6. A cultivated crop variety that retains its features when reproduced.
_____7. The name of the main divisions of the domain Eukarya.
_____8. The name of a group of organisms that live in harsh environments.
_____9. The broadest classification of all living things.
_____10. A diagram that shows lines of evolutionary descent in organisms.

EXPLORING

1. List the common names of the major trees that grow in your community. Determine the scientific names of the trees by using reference materials or the assistance of a forester. Collect leaves, fruit, and other identifying materials from the trees. Prepare a display of your findings. (An alternative is to use a digital camera and make a photographic presentation using an electronic program, such as PowerPoint®.)

2. Review the labels on weed killers (herbicides) or insect killers (insecticides), and list the scientific and common names that are found. You may need to go to a garden center or farm supplies store to find labels. Prepare a written report on your findings.

3. List the common pets (companion animals) in your community. Use reference materials or the assistance of an agriculture teacher to determine their scientific names. Study the classification of the pets to determine their relationships and differences.

4. Using a world atlas, geography book, or other references, determine the common names used in different countries for the following: *Oryza sativa*, *Capra hircus*, and *Zea mays*. (In North America, these are known, respectively, as rice, goat, and corn.)

CHAPTER 11

Genetics and Biotechnology

OBJECTIVES

This chapter covers the fundamentals of genetics and biotechnology. It includes information on genetic engineering. It has the following objectives:

1. Discuss the role of heredity and genetics.
2. Relate genetic approaches used in improving organisms.
3. Discuss the meaning and use of nucleic acid science.
4. Describe organismic biotechnology methods.
5. Describe molecular biotechnology methods.

TERMS

alleles
apomexis
asepsis
chromosome
cloning
controlled breeding
DNA
DNA isolation
DNA profiling
dominant trait
double helix
embryo splitting
embryo transfer
gender selection
gene
gene splicing
gene transfer
genetic code
genetic engineering
genetic manipulation
genetically modified organism
genetics
genome
genotype
heredity
heterozygous
homozygous
molecular biotechnology
nucleic acid
organismic biotechnology
phenotype
plant tissue culture
polymerase chain reaction
Punnett square
recessive trait
recombinant DNA
RNA
stem cell
superovulation
transgenic organism
ultrasonics

11–1. Three litter-mate kittens have a very similar appearance. What is the role of heredity and genetics in their appearance?

GENETICS and biotechnology are important areas of study in today's agricultural industry. Scientists use knowledge of genetics to improve the efficiency and productivity of agricultural products.

Biotechnology is used to develop new or better plants and animals. Sometimes the nature of an organism is changed. Other times new substances are produced. Biotechnology is used with plants and animals as well as with other organisms, such as bacteria and fungi.

People hold different views of biotechnology. If scientists have a broad view, biotechnology is concerned with all the ways organisms are improved and used to make products. Other scientists view biotechnology as a useful tool in creating improved plants and animals.

People need accurate information about biotechnology to be able to weigh the advantages and disadvantages of the processes involved. They should not make judgments without good information.

HEREDITY AND GENETICS

Heredity is the transmission of parental traits to offspring. New generations have traits of their parents. Yet heredity varies with asexual and sexual reproduction.

11–2. The arrangement of genes in the chromosomes determines whether the genetic code makes a butterfly, a hand, or a plant.

In asexual reproduction (involving only one parent), the offspring are identical to the parent. Plants reproduced by cuttings or buds will have the same traits as the stock from which the cuttings or buds were taken. A rose cutting from a bush that has red blossoms will produce red blossoms. This trait consistency is important in agriscience because it guarantees the offspring traits before reproduction.

Heredity is more complicated in sexual reproduction because the offspring is the product of two parents and will have traits of both parents or traits that are not obvious in either parent. These differences are known as variations and arise because of new combinations of genetic factors. The characteristics of a new organism occur by chance though they reflect those of its parents. The information in the genetic material of an organism's parents plays a major role.

AgriScience Connection

GETTING ORGANIZED

Organizing the work area promotes good outcomes with experiments and demonstrations. Take time to read instructions and to obtain the needed equipment and supplies before beginning work. Have a place to set equipment and supplies once you are finished with them. Keep the work area neat, organized, and safe.

This shows preparation to extract DNA from strawberries. A kit of recommended equipment and materials is being used. The protocol (procedure) in the instructions should be followed precisely. If not, the experiment may not provide accurate results in a timely manner. (Courtesy, Education Images)

GENETICS

Genetics is the science of heredity and variation in living plants, animals, and other organisms. It is an attempt by scientists to explain why offspring possess certain traits and not others. Several genetic methods are used to improve plants and animals. Each method applies fundamentals of heredity.

DNA is the substance in chromosomes that makes up genes. In general, a **gene** is a unit of hereditary information consisting of a specific sequence in DNA. Genes are the determinants of heredity. The **genetic code** is the specific order or arrangement of DNA within the genes. An understanding of DNA is essential in using genetic code information. Some genes are dominant and show up more often; other genes are recessive and show up less often. The new organism will be a combination of the two sets of genes.

Internet Topics | Search

Selected topics for Internet discovery and reporting are listed here. Begin by searching for each topic. Next, read and learn about it. Conclude by preparing a brief report on your findings.

1. Genetic map
2. Forensic DNA fingerprinting
3. Stem cell

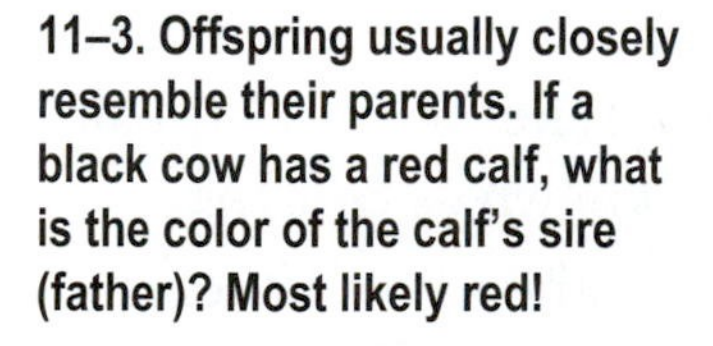

11–3. Offspring usually closely resemble their parents. If a black cow has a red calf, what is the color of the calf's sire (father)? Most likely red!

GENOME

The total genetic makeup of an organism is its **genome**. All the chromosomes in a cell form a chromosomal set. Chromosomes serve as the genetic link between parents and their offspring. Every new body cell formed in an organism has an identical genome. All the genomes in an organism are copies of each other.

When organisms sexually reproduce, the genome of the offspring is a combination of its parents. All the cells formed in an offspring have identical genomes.

Chromosomes

A ***chromosome*** is a threadlike structure inside a cell nucleus that contains genetic material and protein. The genetic material is made of DNA (deoxyribonucleic acid). Chromosomes are found in pairs, and the offspring of sexually reproduced organisms receive pairs of chromosomes from their parents. Since parents vary in traits, offspring manifest a combination of these traits. Some are expressed (visible), and some are not.

The cells of organisms have several chromosome pairs. For instance, corn cells have 10 pairs, mice cells have 20 pairs, and humans have 23 pairs of chromosomes in each cell. Cattle and horses have 30 pairs of chromosomes, swine have 20 pairs, and sheep have 27 pairs. The number of chromosomes determines what organism is produced.

A cell with all of the chromosomes characteristic of a species is a diploid cell. The chromosomes are arranged in pairs in the nucleus. These identical pairs are known as homologous chromosomes. All the chromosome pairs in a particular individual are identical except for the sex chromosome pair, which differs between males and females.

11–4. A giraffe has a definite phenotype. Can you describe it?

Genes

A gene is a part or segment of a chromosome that contains hereditary traits. Each chromosome has many genes. Since chromosomes are in pairs, genes are also in pairs. The different forms of genes are known as ***alleles***.

The alleles may be different or the same. When they are the same in structure, the organism is said to be ***homozygous***. When they are different, the organism is said to be ***heterozygous***.

Heredity Type

Two groups of characteristics are associated with type: genotype and phenotype.

The genes for a trait, represented by a combination of letters, are known as ***genotype***. This makeup is typically not readily visible but can be determined through DNA procedures. Genotype is a link to offspring and the expression of traits.

The appearance of an organism because of its genotype is known as **phenotype**. Offspring, as products of their parents, have both genotype and phenotype heredity. Phenotype refers to the traits of a plant or animal. In agriscience, this is important because of the value associated with products. An example is the "muscling" of a meat animal. The preference is for an animal phenotype with high productivity of valued meat products, such as large hams and large loins.

If the parents have similar genes, the genotype and phenotype of the offspring are much like those of the parents. There is little variation; therefore, the offspring are homozygous. If the parents' genes are different, the offspring could have genotypes and phenotypes different from the parents and be heterozygous (different gene types). Systems of breeding make use of similar and different gene types. Outcrossing is used to obtain heterozygous offspring. Homozygous offspring have like genes. Inbreeding is used to develop homozygosity.

Trait Expression

Geneticists often place traits in two groups: dominant and recessive. The traits are based on alleles in the genotype of an organism. An allele is the alternate form of a gene. Some are quite similar and are known as homozygous for a trait. Others involve alleles that are different and are known as heterozygous. (The terms homozygosity and heterozygosity are sometimes used to reflect these traits.)

ANIMAL GENOTYPES AND PHENOTYPES

Animal	Dominant Trait	Recessive Trait	Possible Genotypes	Resulting Phenotypes
Dog	L	l	LL Ll ll	long hair long hair short hair
Guinea pig	B	b	BB Bb bb	black hair black hair white hair
Cat	W	w	WW Ww ww	all white hair all white hair mixed hair color
Cow	P	p	PP Pp pp	hornless hornless horns

11–5. Dominant and recessive alleles illustrate how genotypes are expressed in phenotype.

A ***dominant trait*** is one that covers or masks the alleles for a recessive trait. For example, the trait of being polled is dominant over the trait of having horns in cattle. A ***recessive trait*** is one that is masked by a dominant trait. In the example of polled cattle, the presence of horns is a recessive trait.

Stated another way, recessive traits are covered up by dominant alleles. Recessive traits can be expressed if both parents have the recessive alleles. This happens when solid black animals that have alleles of white faces are mated. Their offspring could have a white face, but not always.

An example of dominant and recessive genes is the presence of horns on cattle. When a purebred polled animal is mated with a horned cow, the offspring will be polled. Not all polled cattle are genetically purebred polled. Some of them carry a recessive gene for horns. When mated with an animal with horns, the offspring could develop horns.

The Red Angus offers another example of recessive traits. Angus cattle have been black for hundreds of years. Yet, a red-colored Angus is occasionally born; one out of every 400 Angus calves is born red. Mating two red animals will always produce red offspring because it is a recessive trait.

Trait Prediction

With classical genetics, trait prediction is based on probability or the chance that a trait will occur. This means that mating animals of specific traits does not guarantee that the offspring will have these traits. Scientists commonly use two methods of trait prediction: Punnett square and chi square. In this book, only Punnett square is illustrated.

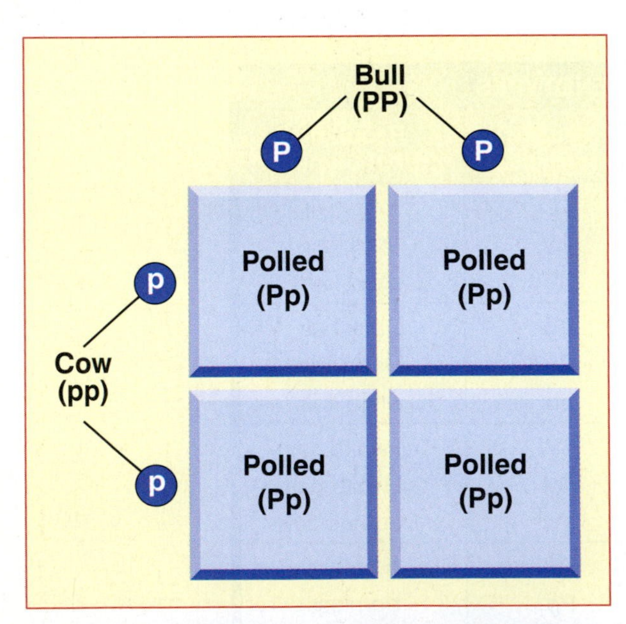

11–6. Sample Punnett square for pollness prediction in cattle. (The bull is homozygous [PP] for the dominant trait of being polled; the cow is heterozygous [pp] for the recessive trait of horns.)

The ***Punnett square*** is a diagram used to predict the outcome of breeding. It was developed by Reginald C. Punnett to determine the probability of an offspring having a particular genotype. It represents every possible combination of alleles from each parent. Traits are classified as dominant or recessive. Dominant alleles are noted as capital letters and recessive alleles as lower case letters. Parents are often noted as maternal (mother) and paternal (father). Of course, maternal and paternal terms do not apply to plants, but two plants are used.

A common example of trait prediction is the presence of horns versus the lack of horns (known as poll) in cattle. If a bull that is homozygously dominant for being polled mates with a cow that is heterozygously recessive for pollness, the offspring will all be polled.

HISTORICAL PERSPECTIVES

No one individual is more associated with heredity and genetics than Gregor Johan Mendel (1822–1884). The work of this Austrian botanist was with garden peas and lead to the science of genetics. Mendel bred and crossbred thousands of pea plants. He concluded that plant traits are randomly passed to offspring through hereditary elements in genes. Today, the work of Mendel is often referred to as Mendelian or classical genetics.

New methods in understanding genetics began to occur in the mid-1900s. James D. Watson and Francis Crick determined the structure of DNA in 1953. This development and the work of other scientists led to the emergence of molecular genetics later in the 1900s. In the early 2000s, molecular genetics provided new understanding of genetics and its applications.

The information in an organism that directs its functions is its genetic code. As a result, an organism is what it is because of its genetic code. Cattle have one genetic code, horses another, and corn plants still another. Understanding this code is essential in molecular biotechnology.

CAREER PROFILE

MOLECULAR BIOLOGIST

A molecular biologist develops organisms or processes with useful functions. The work may involve genomes and genetic codes, recombinant DNA, vectors, field or greenhouse duties, and other areas. Most work as team members with a specific organism or process.

Molecular biologists have considerable training in science, such as genetics, microbiology, and physiology. Most have college degrees in biology, agriculture, or a related area plus a master's degree and a doctorate. Experience in lab work is essential. Begin by taking biology, chemistry, physics, agriscience, and computer classes in high school.

Jobs are with biotechnology companies, research stations, universities, and government agencies. This photo shows molecular biologists using bovine (cattle) gene sequences in studying cattle genetics. (Courtesy, Agricultural Research Service, USDA)

The genetic codes of a number of organisms have been sequenced. Rice was the first crop to have its genes mapped, with the work completed in 2001. By knowing the genetic code, it is easier to modify the genes in a rice plant to gain the desired quality.

IMPROVING ORGANISMS

Overall, the approaches in agriscience are used to improve plants and animals and can be clustered into two main areas: controlled breeding and genetic manipulation.

11–7. Potato plants are being hand pollinated in controlled breeding. (Courtesy, Agricultural Research Service, USDA)

CONTROLLED BREEDING

Controlled breeding is the selective act of mating plants and animals to achieve desired traits in the offspring. For example, ranchers want cattle that grow faster, resist disease, and provide more beef. Wheat growers want varieties that produce more grain without increased fertilizer and other inputs.

Controlled breeding is a natural method of improving plants and animals. Three kinds of controlled breeding are used.

- **Selection**—Selection is the act of choosing parents with the desired traits. These parents are bred with the intent of increasing the desired qualities in the offspring. A good example is the milk production of dairy cattle. Dairy producers want cows to produce 20,000 pounds, or more, of milk per year. The offspring of cows that produce large amounts of milk are selected and mated. This mating should produce offspring with an even higher milk-production capacity.
- **Inbreeding**—Used with plants and animals, inbreeding is the mating of the offspring of the same parents to each other. The purpose is to produce a new generation without introducing new genes. For instance, a plant with desired traits can be used to produce other plants with the same traits. The plant is protected so pollen from other plants cannot reach it. Most of the new plants will be like the parent. No new genes are a part of the heredity. However, inbreeding also has problems. The weak or undesirable traits

are also reproduced. In some cases, undesirable traits result in problems that make the desired traits impractical.

- **Hybridization**—Hybridization is the crossing of two different, but closely related, plants or animals. Hybridization combines the traits of different parent strains. Strains are plants or animals that have a common heredity. With hybridization, the strains vary considerably. When bred, certain traits predominate in the offspring. Hybridization has been widely used. Examples include hybrid corn, hybrid striped bass, and mules. Yet hybridization may create some problems. Many hybrids do not reproduce; if they do reproduce, the offspring do not have the desired traits. Regardless, the use of hybrids has improved agriculture.

GENETIC MANIPULATION

Genetic manipulation is the use of artificial ways of obtaining desired traits in plants and animals. Hereditary traits are moved from one species to another. In effect, new life forms are created. Gene splicing is used.

Gene splicing is transferring genes from one plant or animal to another. The species may be unrelated to each other. In some cases, genes have been exchanged between plants and animals. For example, the luminescent (light) trait of a firefly has been spliced into tobacco plants. As a result, the tobacco plants that were produced emitted a low level of light. Though of little practical value, these efforts supply new information about genetic manipulation. Several crops resistant to pesticides and pests have been developed using genetic manipulation, such as soybeans, corn, and cotton.

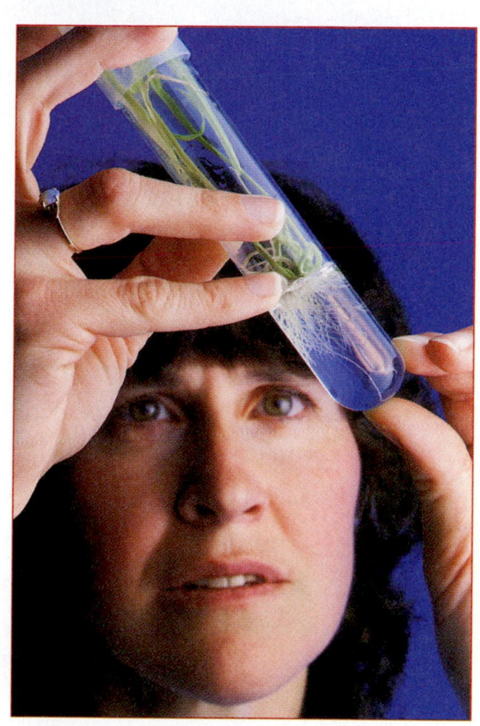

11–8. Genetically engineered wheat plants are being developed for resistance to the disease *Fusarium*. (Courtesy, Agricultural Research Service, USDA)

Gene splicing is used to gain desired traits in organisms and is sometimes referred to as **genetic engineering**. It is used to alter the natural arrangement of genes to receive desired traits. Genes of one type can be removed, and genes with desired traits can be added. For example, gene splicing was used to develop cotton plants resistant to certain chemicals. This made the crop easier to grow.

Gene splicing creates new life forms. Some people are concerned about what they consider the unnatural creation of life. They fear that these forms may have harmful effects on the environment and quality of life. Such fears appear to be not well founded and based on a lack of information.

11–9. Cells in a laboratory environment.

STEM CELLS

Probably no area of biotechnology holds more potential than that of stem cells. A **stem cell** is a cell with the ability to divide for indefinite periods in the proper medium and to give rise to specialized cells, such as those of the heart or brain. Stem cells retain the ability to form various specialized cells that have deteriorated due to disease. With humans, stem cells may be able to halt or cure Type 1 diabetes, Parkinson's, and heart failure. They are thought to have high potential in cancer treatment and to reverse the aging of humans. Scientists believe research may lead to ways of programming stem cells to perform any desired function or specialization.

Stem cells are classified by source: adult and embryonic. An adult stem cell is one that is undifferentiated in tissue. Such cells are from adult organisms and are found among differentiated cells in specific tissues. They can usually be used to form only one specific kind of cell. For example, stem cells from bone marrow can generate bone, cartilage, fat, and fibrous connective tissue. With humans, adult stem cells are now used to treat more than 100 diseases and conditions.

Embryonic stem cells are those of a tiny, developing embryo. These cells form after conception and are undifferentiated. Cells differentiate when they become specialized to form tissues and organs. Embryonic stem cells are obtained from the inner mass of cells of a blastocyst (an embryo of about 150 cells). Embryonic stem cells naturally multiply and specialize to form the tissues, organs, and systems of the body as an embryo develops. Once available, these cells can be indefinitely cultured in the laboratory *in vitro* (in glass). Embryonic stem cells are felt to have the greatest potential for advancing life.

Three types of stem cells are totipotent, pluripotent, and multipotent. A totipotent stem cell can grow into a complete organism—just one cell is needed. A pluripotent stem cell can differentiate but cannot grow into a complete organism. A multipotent stem cell can become only a certain type of cell, such as a blood cell.

The cells obtained from adult and embryonic stem cells can be used in what is known as cell-based therapy, which is the use of stem cells to repair damaged or depleted cells or tissues. Such use can likely be applied among all animal species. Additional research is needed. Some of the processes of cell-based therapy may occur naturally… think of the lizard that regenerates a tail if it is lost.

NUCLEIC ACID SCIENCE

Nucleic acid is the substance in cells that directs all cellular structures and activities. The molecules of nucleic acid are known as polymers. These polymers contain subunits known as monomers. As such, the monomers are sometimes referred to as the building blocks of polymers and, hence, the nucleic acid. The major functional molecule is protein, which is responsible for the cellular structures and activities of a cell.

Two types of nucleic acid have been identified: DNA and RNA.

DNA

DNA (deoxyribonucleic acid) is the molecule that codes the genetic information of all living things. Knowledge of DNA and its structure makes it possible to alter the genetic material in an organism. Information can be removed or added to the DNA.

DNA is a long polymer of repeating units known as deoxyribonucleotides. Each DNA molecule is made of two strands. The name, ***double helix***, refers to this two-stranded structure. Each strand has many alternating units of sugar and phosphorus. The two strands are held together by hydrogen bonds between base pairs.

A deoxyribonucleotide has three parts: a phosphate group, a sugar, and nitrogen-containing bases. The sugar is known as deoxyribose. The four bases are adenine (A), guanine (G), thymine (T), and cytosine (C). Genetic information is stored in the nitrogen-containing bases. Alternating sugar and phosphate parts compose the backbone of the DNA.

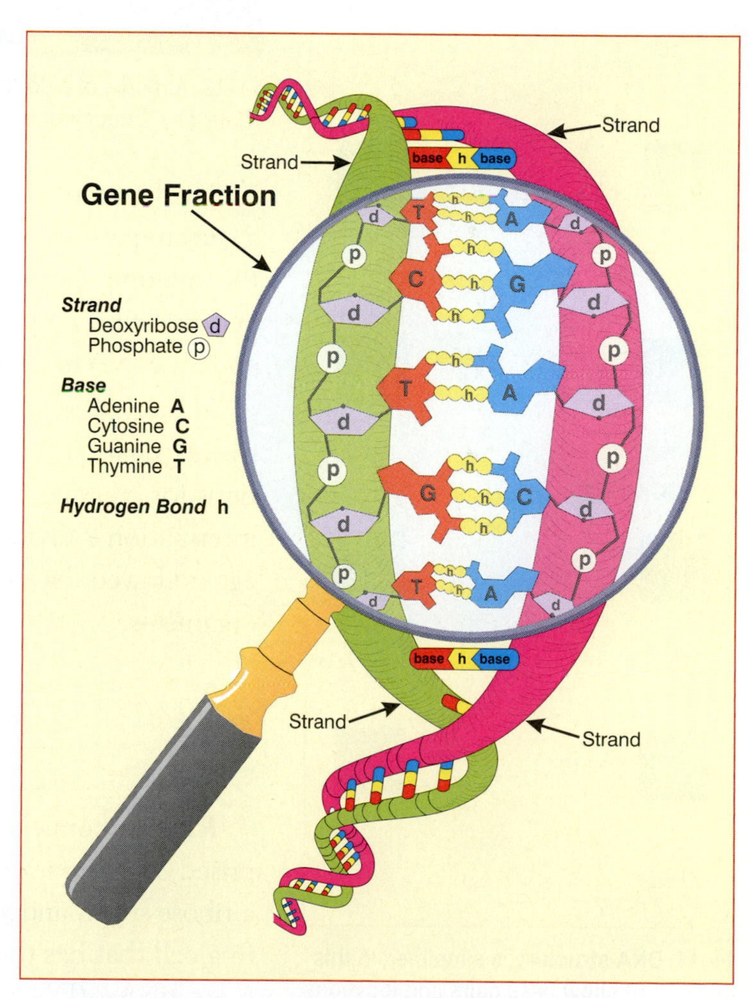

11–10. Representation of a DNA double-helix molecule.

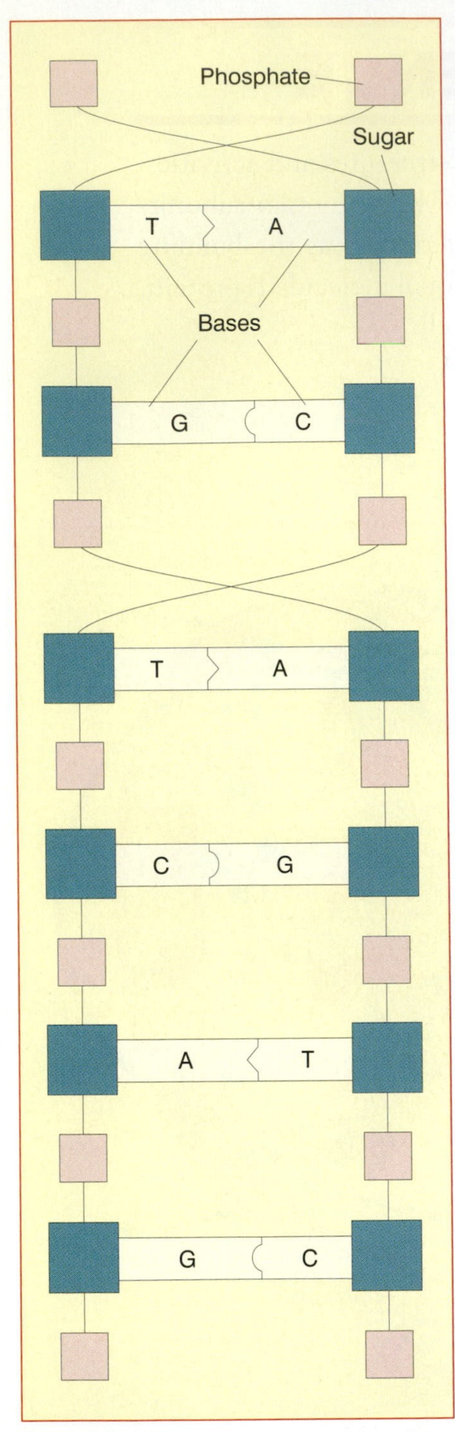

11–11. DNA structure is simplified in this drawing to show base pairs and locations of A, G, C, and T.

11–12. A model of a double helix is being investigated for DNA structure. (Courtesy, Education Images)

The bases in a DNA molecule always arrange themselves in a certain pattern. The thymine always pairs with adenine, and the guanine always pairs with cytosine. This provides the ladder-like appearance of a DNA double helix. This arrangement and the DNA molecule are of major interest in biotechnology.

In reproduction, DNA replication occurs. Because DNA is the physical carrier of inheritance, it contains all of the instructions for creating and maintaining a new living organism. The information is divided into genes. The union of a sperm and an egg followed by cell division results in unique, individual organisms.

RNA

RNA (ribonucleic acid) is quite similar to DNA. It is comprised of a long strand of nucleotide units with a nitrogen base, a ribose sugar, and a phosphate. RNA is usually a single strand in a cell that has the base uracil rather than thymine and contains ribose (DNA contains deoxyribose). RNA consists of short segments rather than long double helixes of DNA. Double-stranded RNA is in some cells and life forms, such as viruses.

RNA is transcribed from DNA by enzymes known as RNA polymerase. Other enzymes may also be involved. RNA is essential for the synthesis of proteins. Scientists continue to investigate RNA to learn more about its nature and role.

Several kinds of RNA are found. Some kinds are involved in translation; other kinds are involved in the regulation of cellular processes and RNA processing. A few examples are listed here.

- **Translation**—Several kinds are involved in translation. Messenger RNA (mRNA) carries information from DNA to the ribosome where protein translation occurs in a cell. Transfer RNA (tRNA) moves specific amino acid to a polypeptide chain. Ribosomal RNA (rRNA) serves as a catalyst (promoter) of protein synthesis and combines with protein to form a ribosome in the cytoplasm of a cell. Transfer-messenger RNA (tmRNA) is found in some species and tags protein for encoding.
- **Regulation**—Several RNAs are involved in regulating gene expression and other processes. The microRNAs (mRNA) breaksdown mRNA, block its translation, or accelerate its degradation. Other RNAs are involved in regulation.
- **Processing**—Processing is the modification of other RNA by specialized RNA. An example is small nuclear RNA (snRNA), which may alter nucleotides.

11–13. The genetic code of a plant that removes heavy metal from soil is being studied. (Courtesy, Agricultural Research Service, USDA)

11–14. An automated DNA sequencer is being used to gain a detailed genetic analysis of an unidentified microbe. (Courtesy, Agricultural Research Service, USDA)

PROTEINS

Proteins are responsible for functions and development in a cell. All living things must have proteins (amino acids), and several are present.

DNA codes for the synthesis (manufacture) of proteins. The process is known as transcription. In the process, messenger RNA codes for protein synthesis. The mRNA carries DNA information from the nucleus to the ribosomes. When mRNA reaches the ribosome, translation begins. Translation is the process of a cell manufacturing protein.

TOOLS OF NUCLEIC ACID SCIENCE

Scientists have developed procedures for the identification of DNA codes and for using this information in productive ways. The DNA code of any organism (animal, plant, bacterium, or other) can be analyzed and organized into a genome. This allows scientists to draw meaning from DNA and to make comparisons that are useful in improving life, preventing disease, and gaining useful products.

DNA Isolation

DNA isolation is the process of extracting and separating DNA from other materials in a cell. Different procedures can be used to isolate "clean" DNA. The procedures vary with

Technology Connection

DNA MODEL

We know that DNA is the molecule in a chromosome that furnishes the genes with information for the development of the individual. DNA has the hereditary traits of the organism. Each DNA molecule is a long, two-strand polymer of repeating DNA units. The structure is held together by hydrogen bonds. Genetic information is stored in the four bases: adenine (A), guanine (G), thymine (T), and cytosine (C). When cells divide, the genetic material is replicated so each daughter cell contains identical copies of the DNA.

The structure of DNA can sound and appear rather complex. With the help of a plastic DNA model kit, you can build your own DNA molecule. Various colors and shapes are used to represent the phosphates, sugars, and nitrogen-containing bases. These students have assembled a DNA double helix using a kit.

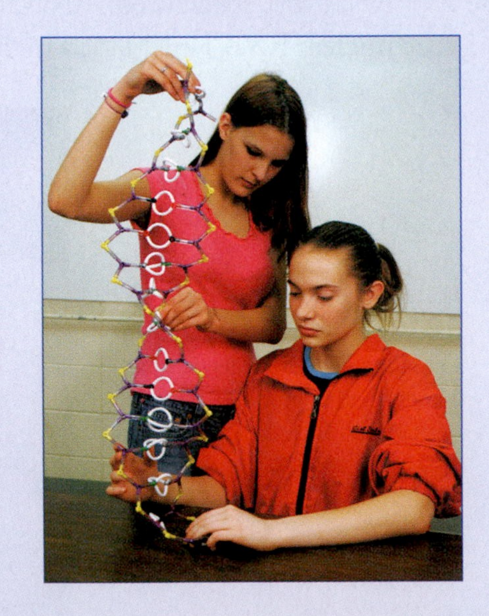

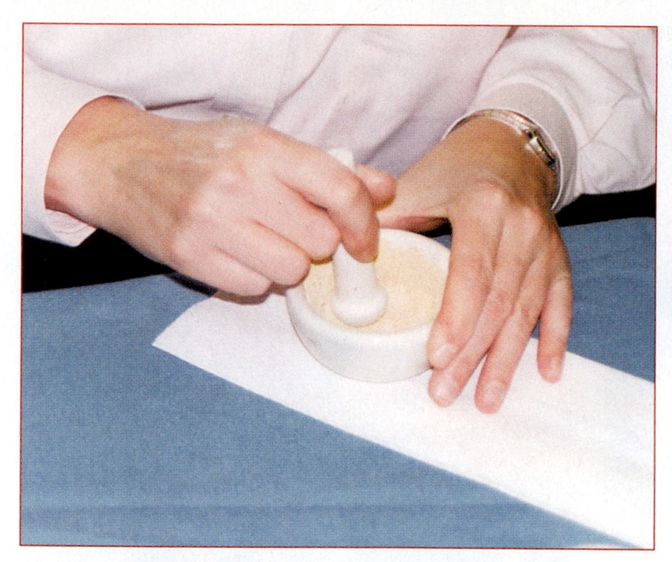

11–15. A mortar and pestle are being used to crush wheat germ for DNA extraction.

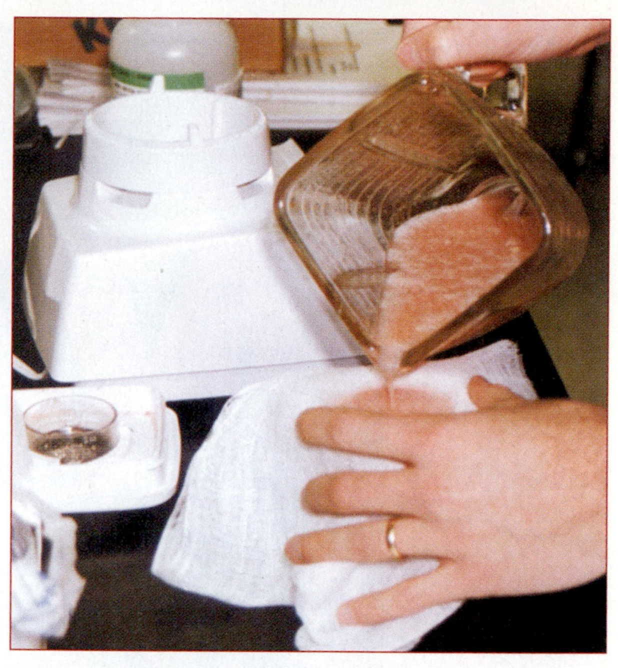

11–16. The blended calf thymus solution is strained through a cheesecloth to remove clumps and gristle. (Low-speed centrifuging may help settle the nuclei to the bottom of a test tube and leave the cellular material in suspension.)

the cells/tissues being used. Wheat germ, strawberries, onions, bovine thymus glands, and other tissues are used in school labs. It is likely best to use cells that have large nuclei. Most of these procedures follow the same basic steps.

1. Break open the cell wall and/or membrane. This is done with liquid nitrogen and/or grinding. This step varies with the source of the DNA. A mortar and pestle may be used to crack and grind seed cells, such as wheat germ. A food blender may be used with animal tissues, such as a thymus gland. Some soft plant cells can be broken open by thoroughly mashing, such as strawberries in a plastic bag.
2. Digest the cellular components. This is often accomplished with a detergent. With wheat germ, a small amount of detergent solution is mixed with the ground powder. The wheat germ solution is heated to 60°C for 10 minutes. With the thymus solution, a small amount is

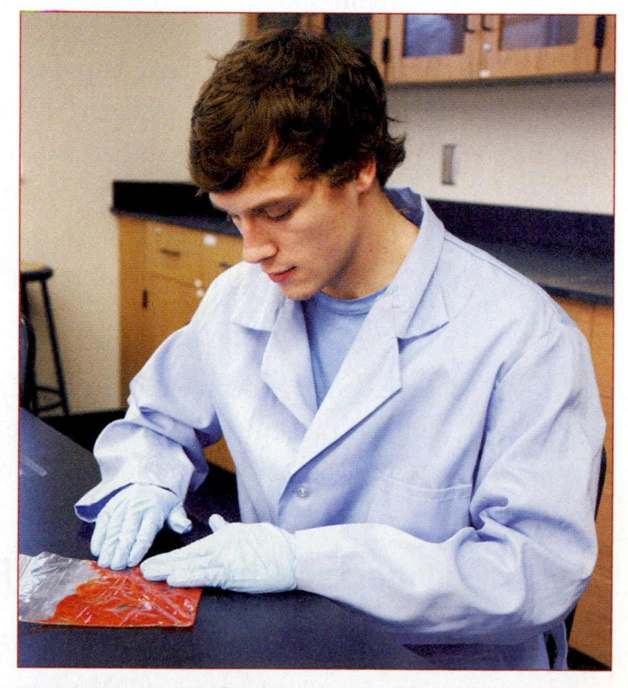

11–17. Ripe strawberries may be finely mashed in a clear plastic bag. (Courtesy, Education Images)

11–18. Adding cold alcohol to precipitate DNA. (Courtesy, Education Images)

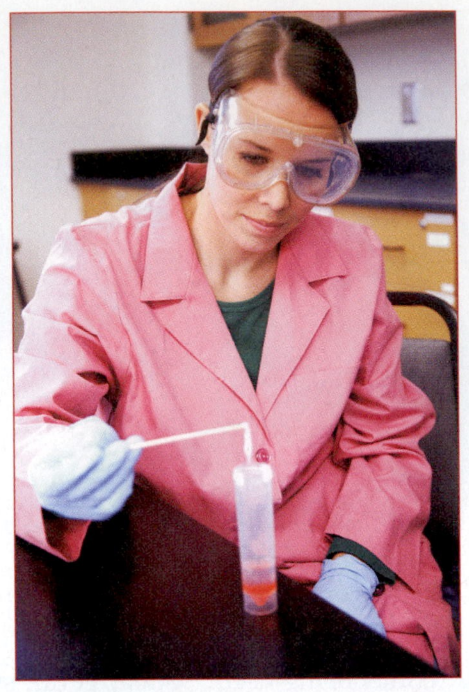

11–19. DNA is being lifted on a wood stir rod. (Courtesy, Education Images)

put into a small test tube. EDTA solution is added to weaken membranes and inactivate DNA-digesting enzymes. With strawberries, no heat or EDTA are needed.

3. Separate the polar compounds. This involves using a detergent solution, such as sodium dodecyl sulfate (SDS). The detergent dissolves the lipids (fat) in the nuclear membranes. Salt, such as NaCl, may be added to the solution. Gently stirring the solution with a slow circular motion promotes the process. Procedures may vary slightly depending on the DNA source. Some procedures involve using chloroform or ethanol; other procedures are fairly easily completed without these materials.

4. Extract and precipitate the DNA. Remove the top aqueous layer with a pipette, and place it into cold absolute alcohol. The white mass of DNA that precipitates may be "spooled" or collected onto a glass or wood stir rod for observation.

Polymerase Chain Reaction (PCR)

A ***polymerase chain reaction*** is a procedure that replicates pieces of DNA for viewing. A target section of DNA for a particular gene is amplified. In some cases, only very small amounts of DNA will work satisfactorily. Small amounts of blood, tissue, or another

DNA source can be used. With a crime scene, for example, PCR can be used to gain a DNA profile for comparison to that of suspected criminals.

Various procedures are used with PCR. One procedure involves using a thermocycler. This machine alters the temperature at each step of the process. Heating, cooling, and extension are used to build a new DNA strand.

11–20. Using a thermocycler to obtain PCR information. (Courtesy, Agricultural Research Service, USDA)

Electrophoresis

Electrophoresis is a process used to separate DNA and RNA molecules based on size. The process is widely used to visualize DNA products. It provides a sequence of DNA fragments. Two approaches are listed here: capillary electrophoresis and gel electrophoresis.

Capillary electrophoresis involves the use of an electrical field that separates the DNA fragments in a glass test tube. Florescent dyes are used to enhance the ability to view the fragments.

Gel electrophoresis uses an electrical field in agar to perform the separation. Dye is used to make it easier to see the molecules. Because DNA is negatively charged, it moves away from a negative electrode and toward a positive electrode. The smaller fragments move faster.

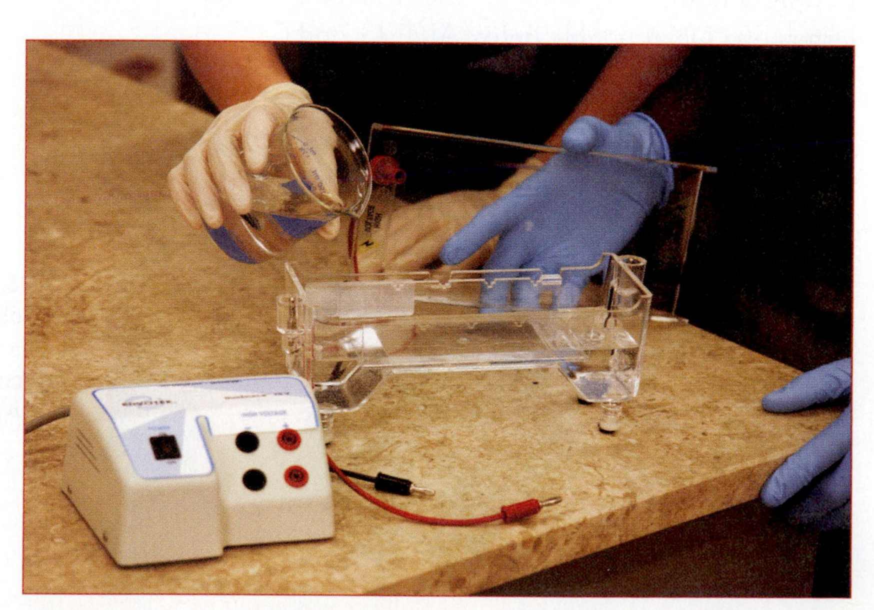

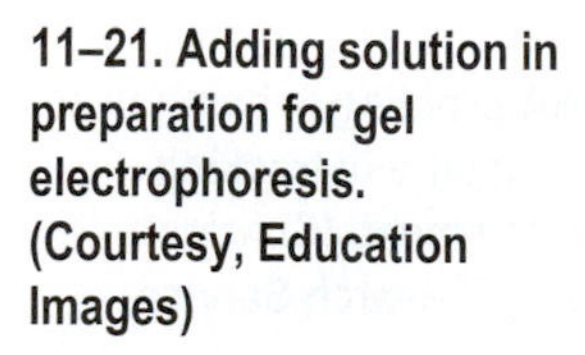

11–21. Adding solution in preparation for gel electrophoresis. (Courtesy, Education Images)

11–22. Shorter, easier-to-harvest apple trees have been developed using DNA fingerprinting to discover molecular markers associated with dwarfing genes. Note the taller trees in the background. (Courtesy, Agricultural Research Service, USDA)

DNA PROFILING

The tools of nucleic acid science can be used in DNA profiling, which is also sometimes known as DNA typing and DNA fingerprinting.

DNA profiling is a process to identify individuals of a species based on DNA profiles. The DNA sequences in individuals of the same species are 99.9 percent alike. However, the 0.1 percent variation is useful in identifying individuals, plants, animals, bacteria, or other species.

The DNA profiling process uses repeat sequences in a DNA sequence. The only difference between individuals of the same species is the base pair order. A base pair is a connection or bonding of two nucleotides in DNA or RNA strands. The bases involve four code letters of a DNA double helix: A, T, G, and C.

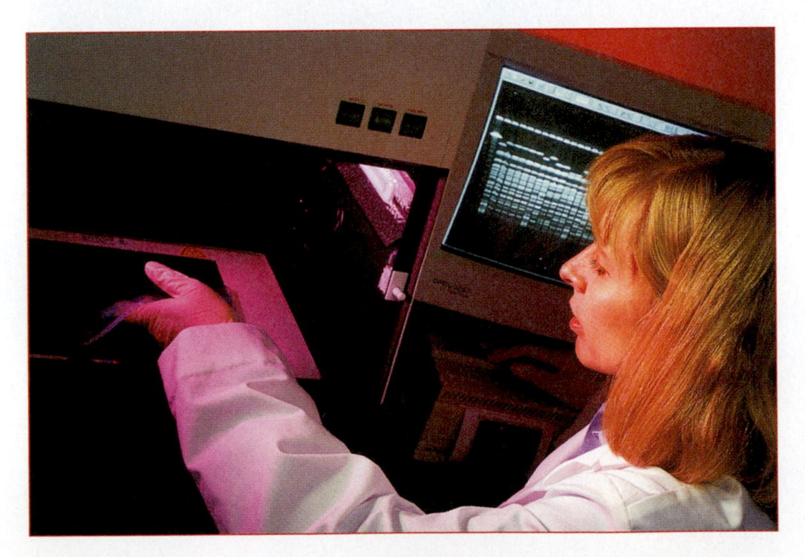

11–23. DNA profiling is being used to isolate bacteria in tracking contaminated meat. (Courtesy, Agricultural Research Service, USDA)

No two individuals have identical DNA sequences. Identical twins may be an exception, though slight differences may be found.

The profiling process applies to all species. Several methods of profiling can be used. The method widely used today is known as STR (short tandem repeats). STR locations on DNA can be targeted for sequence and amplified using PCR.

ORGANISMIC BIOTECHNOLOGY

Organisms are complete living things. **Organismic biotechnology** is the branch of biotechnology that deals with intact or complete organisms, but the natural processes of organisms may be altered in some way, such as making a tissue culture of a begonia. The objective is to help organisms live and grow better as well as perform useful processes. Working to improve plant or animal growth using nutrients, or by controlling hazards, is important. Organismic biotechnology is most widely used in agriscience.

Organismic biotechnology deals with helping organisms live and grow better. It takes an organism as it is and makes improvements. The genetic material in the organism is not artificially modified. Many of the practices carried out in producing plants and animals are included.

ASEPSIS

Work in biotechnology is often carried out in aseptic environments. **Asepsis** is the condition of being free of disease-causing germs; it is also the technique used to achieve the aseptic condition. A work area may be free of all germs (sterile), or it may be one that tolerates a few germs.

Three levels of asepsis may be used:

- **Sterilization**—Sterilization is the removal or destruction of all germs from work surfaces and equipment. High temperatures and chemicals (sterilants) are used when the work that needs to be done will not tolerate the presence of any germs. Instruments are sterilized in a steam environment with an autoclave or pressure cooker.

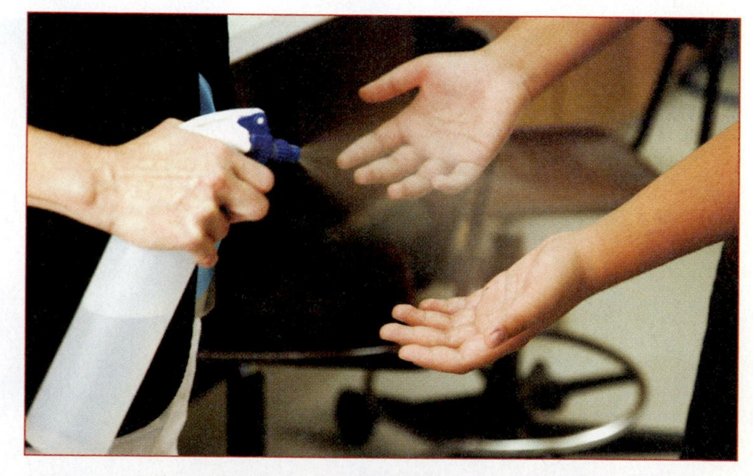

11–24. Hands and the work area in a flow hood should be sprayed with a bleach solution or other substance to promote asepsis. (Courtesy, Education Images)

11–25. An autoclave is used to sterilize instruments.

- **Disinfection**—Disinfection is the act of removing or lowering the germ population. The area is not sterile, but it has a low germ population that is not likely to cause a problem with the biotechnology work. Disinfectants, such as 70 percent ethanol alcohol or bleach, are used. Work may be confined to a protected area, such as within a laminar flow hood.
- **Sanitation**—Sanitation is the practice of lowering the germ contamination to an acceptable level. Some germs may be tolerated. Sanitation is used where sterile environments are not needed. Bleach solutions, 70 percent ethanol alcohol, Lysol®, and other products may be used to sanitize a work area.

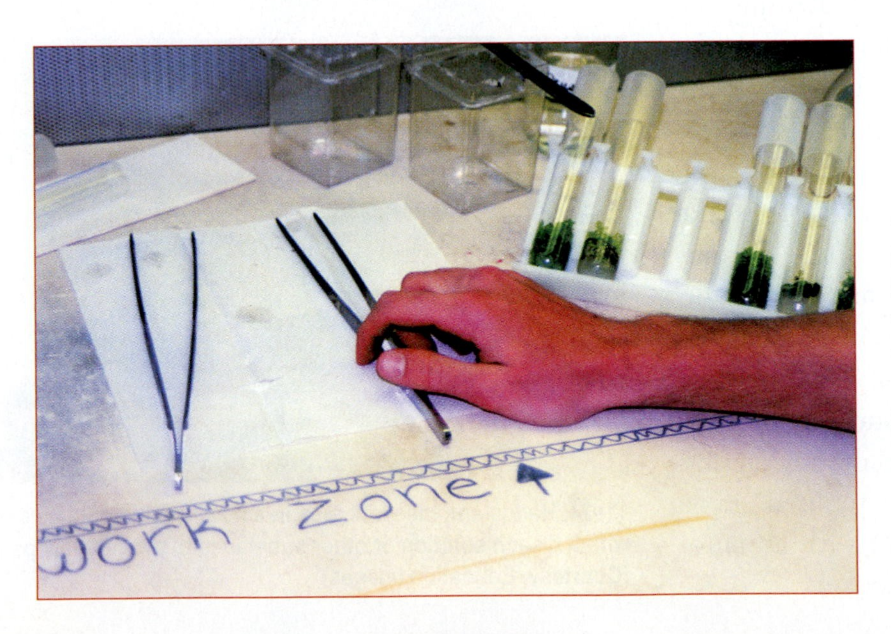

11–26. The work zone has been marked in this flow hood. All instruments as well as hands and materials should be sprayed with a bleach solution. (Courtesy, Education Images.)

CLONING

Cloning is the process of asexually reproducing organisms. No union of male and female germ cells occurs. The new organisms are identical to their parents. Genetic information in the organism has not changed. Three examples of cloning are plant tissue culture, embryo splitting, and apomixis.

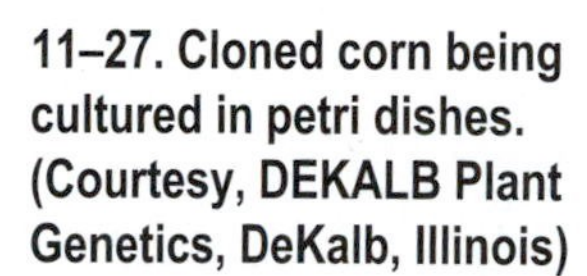

11–27. Cloned corn being cultured in petri dishes. (Courtesy, DEKALB Plant Genetics, DeKalb, Illinois)

- **Plant tissue culture**—***Plant tissue culture*** is the propagation of plants using single cells or small groups of cells. The cells are removed and placed in a solution of nutrients—vitamins, plant hormones, and sugar. The temperature and photoperiod are controlled. Tissue culturing does not change or modify the cell structure of the organism. It is simply a method of reproducing identical plants known as clones. Tissue culture is used to propagate carnations, ferns, orchids, and many others. Special equipment and some training are needed.

11–28. Checking the growth of a tissue culture in a test tube. (Courtesy, Education Images)

- **Embryo splitting**—An embryo of a few cells can be split into two parts. Each part develops into a complete organism. With cattle, ***embryo splitting*** is the asexual reproduction process in which an embryo is removed from a cow seven days after conception, cut in half, and each half is placed in the uterus of a cow known as a recipient. The halves grow to full term and produce two identical calves. The genetics of the recipient cows are not transferred to the calves. The cow providing the embryo and the bull providing the sperm give the genetic information to the new animals. Embryo splitting is expensive and requires skill. It is used to increase the number of offspring of valuable animals.

- **Apomixis**—***Apomixis*** is the asexual reproduction of plants using unfertilized seed. All the plants produced are identical to the female plant. They are clones of the female. Nearly 300 species of plants can be reproduced by apomixis, including corn, wheat, strawberries, and beets. Although it is a natural phenomena, apomixis is used to develop improved varieties of some crops.

GREATER FERTILITY

The reproductive capacity of superior animals can be increased. Outstanding cattle, hogs, and horses are more valuable when they produce more than the normal number of offspring.

- **Superovulation**—Most females naturally produce a small number of eggs during estrus. ***Superovulation*** is an increase in the female's egg production. For example, a cow normally produces one egg (ovum) at a time. By injecting a cow with an ovulation-inducing hormone, such as gonadotropin, the cow may release 8 to 20 eggs (ova) during estrus (heat). The eggs are flushed from the reproductive tract, fertilized in vitro

11–29. Adding antibiotics to bovine semen protects sperm. (Courtesy, American Breeders Service, DeForest, Wisconsin)

(in glass), and transplanted into the womb of a recipient cow. A cow that provides ova is known as a donor.

Another approach to increasing an animal's productivity is to inseminate the donor cow to fertilize the eggs. Seven days after fertilization, the developing embryos are flushed from the donor. Flushing is the use of a saline solution (water with some salt) in the uterus of the cow to remove embryos. The embryos are "washed out" with the solution. They are very small—no bigger than the tip of a small match. The embryos are immediately put into recipient cows to develop into full-term calves. In some cases, the embryos are frozen for later use. Embryos can be frozen in straws to make future implantation easier. Surgically removing the embryos has also been done, but it is more expensive and likely to cause infection.

11–30. Laparoscopy is being used to artificially inseminate a ewe. (An incision is made into the reproductive organs and a metal tube with a lens guides the veterinarian.) (Courtesy, Elite Genetics, Waukon, Iowa)

- **Embryo transfer—*Embryo transfer*** is the removal of an embryo from its mother (donor) and the placement of the embryo in a recipient. Embryo transfer may be used with or without superovulation. With cattle, an embryo could be flushed from a donor cow and put into a recipient cow to allow the donor cow to begin estrus. Most cows will begin estrus 21 to 60 days after the embryo is removed. This allows a valuable cow to produce more offspring. Recipient cows are less valuable, but they are good mothers.

MORE PRODUCTION

Several practices are used to increase production, such as fertilization. Fertilizer gives plants more nutrients, thereby improving cell growth. This changes the normal growth of the plant—growth that would occur without the fertilizer.

- **Milk hormones**—A product named bovine somatotropin (bST) can be given to cows to increase milk production. This hormone naturally occurs in cows and is present in all milk. Injecting a cow with additional bST causes her system to become more productive. Of course, feed must be increased if more milk is to be produced. Research has shown that cows produce 25 percent more milk when given bST.
- **Meat hormones**—The pituitary glands of hogs produce porcine somatotropin (pST). This hormone regulates growth. Injections of pST cause hogs to produce more muscle cells, which results in the valuable cuts of pork being larger than they would be otherwise. This hormone also decreases body fat on a hog. Larger pork chops and hams give the producer a higher price for the animal when it is sold. Therefore, daily injections of pST are usually given.
- **Growth implants**—Implants are small pellets placed under the skin of animals behind an ear. They are primarily used to promote growth. The U.S. Food and Drug Administration has approved the use of several implants. The materials used in implants help animals make more efficient use of feed. Some implants contain hormones; other implants are substances derived from corn molds. The implants slowly release materials over several weeks.

ULTRASONICS

Ultrasonics is a science that uses high-frequency sound waves (ultrasound), which people cannot hear, to obtain an image of an organ or other feature. Most ultrasound is produced by an ultrasonic transducer that converts electric energy into ultrasonic waves. Waves are sent and received. Instruments show the presence of objects and may be used to indicate movement. Three examples are covered in more detail.

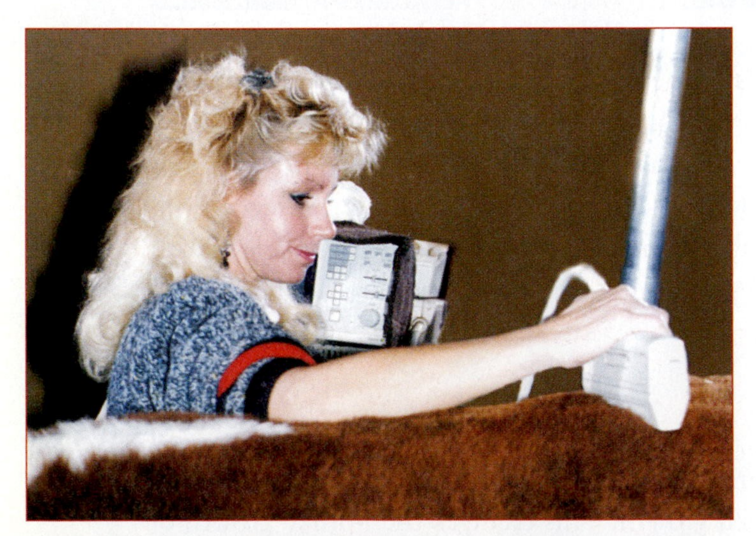

11–31. Ultrasound is being used to determine loin area on a cow. (Courtesy, American Polled Hereford Association)

- **Product quality**—Ultrasonics can be used to assess the amount of fat on animals. Though not widely used on market animals, it provides the opportunity to assess the thickness of fat and muscle. It is more widely used in selecting breeding stock. Ultrasound may be used on animals in livestock shows to assess their quality.

- **Pregnancy assessment**—Ultrasound can be used to determine embryo and fetus development in a pregnant female. It has been used on cattle and other more-valuable animals, such as registered breeding stock. Ultrasound develops an image of the unborn animal on a screen similar to a television.
- **Locate fish**—An adaptation of ultrasonics, known as a sonar device, is used to locate objects under water, such as wild fish in the ocean. The equipment sends out sound waves through the water. The presence of fish will result in a sound wave being sent back, similar to an echo. Sonar will indicate the depth of the fish in the water.

ANIMAL NUTRITION AND FEEDS

Many approaches have been used to improve the growth of animals. These include treating feedstuffs before they are fed, as well as intervening in the digestive process.

- **Ammoniating hay**—Low-quality grass hay may be treated with anhydrous ammonia to make it more nutritious for cattle. The hay is usually treated in a specially designed ammoniation chamber. Research using ammonia on Bermuda grass hay has shown that the protein content is increased up to 3.5 percent. The digestibility of protein in the hay may increase up to 14 percent.
- **Digestibility testing**—Digestibility studies may use direct intervention in the digestive systems of cattle. One procedure is to create an opening into a part of the digestive system. Surgical procedures are used to make a fistula (opening in the hollow of the stomach). A cannula (tube) is placed in the fistula. The cannula is made of rubber or glass and has a cover to close the opening. A cannula is usually 3 to 4 inches in diameter.

11–32. A steer fitted with a cannula is used in feed digestibility tests. The steer is in the squeeze for a few moments at a time for research.

11–33. Cattle in this research are fitted with sensors that allow the animal to eat at only one place. The amount of feed ingested can be studied in relation to growth rate.

Feed can be taken from, or put into, the digestive system through the cannula. Sometimes, feed is placed in the system using a mesh cloth bag attached to a string that is fastened to the cover of the cannula. This allows easy retrieval of the partially digested feed sample.

A veterinarian makes the fistula using careful surgical procedures. The animal experiences no discomfort from the cannula when it is properly placed.

- **Controlled feeding**—Improving nutrition and feeding require careful study of animal feed consumption and growth. Cattle are sometimes fitted with sensors that allow the animal to eat only feed prepared for it. The kind and amount of feed can be varied and compared with the rate of growth of the animal. The animal can receive feed only at the feeding station that the sensor will open.

11–34. A growth chamber that controls the environment of the plants is used to study plant growth.

PREDICTING THE FUTURE

Biotechnology methods can be used to predict future life. How will a warmer Earth with more carbon dioxide affect life?

- **Plant growth chambers**—A growth chamber is a structure that creates a controlled environment in which to study changes in the growth of plants. Regular measurements are made of the plants. With a large plant, such as a tree, only part of the plant may be in the chamber. Growth in the regulated environment can be compared with growth on the outside.
- **Computer simulations**—Computer simulation is the use of computer methods to depict or represent a situation. It is used in crop production, landscaping, and other areas. Cropping models

have been developed. A cropping model is a computer simulation that imitates or provides a likeness of a crop production situation. Information on weather, soil moisture, and stage of growth of the crop is used to make decisions about what to do with the crop. Should an insecticide be used? Should irrigation be used? These models help producers choose the right practices at the right times.

A computer simulation on erosion can be used in preventing soil loss. Known as SLOSS (a short form of "Soil LOSS"), the model helps estimate average soil loss under varying conditions, such as the kind of crop, land, and climate.

ACADEMIC CONNECTION

ADVANCED BIOLOGY

Relate your study of genetics and biotechnology to the study of advanced biology. Explore the human genome and how it may be useful in promoting human health. Use your learning in agriscience to support your mastery in advanced biology.

GENDER PRESELECTION

Gender selection is determining the sex of offspring before birth or hatching. With some animals, it can be achieved by sorting sperm. The technology allows animal producers to predetermine the gender (sex) of the animals that are produced. Several processes of sperm sorting have been used, and some are more successful than others.

The process begins with semen collected from a male of the chosen animal species. The methods used separate sperm based on the content of their DNA. For example, Y-chromosome sperm produce male offspring; X-chromosome sperm produce female offspring and have 4 percent more DNA than Y-chromosome sperm. Newer methods using fluorescent

11–35. A heifer calf newly born to a cow inseminated with sorted X-chromosome sperm. (Courtesy, Agricultural Research Service, USDA)

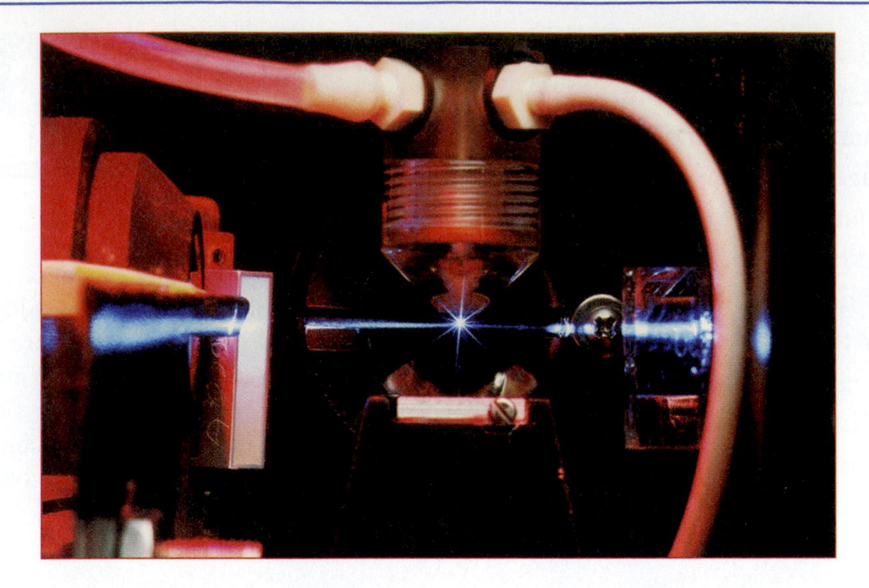

11–36. A sperm sorter that uses fluorescent dye and a laser to separate X and Y sperm. (Courtesy, Agricultural Research Service, USDA)

dye have 90 to 100 percent accuracy in the sorting process. The amount of dye that clings to a sperm is based on the amount of DNA in the sperm. More than 500 animals of predetermined sex have been produced using the fluorescent dye method.

OTHER EXAMPLES

Almost every practice used to produce plants and animals has some biotechnology implication. Examples include:

- Using growth regulators to get plants to grow or to inhibit growth, such as using gibberellic acid on plant cuttings to get them to root, controls the end result.
- Forcing plants to grow at times of the year when they would not normally grow, such as tomatoes in a greenhouse, increases availability.
- Using animal wastes as feed reduces disposal problems and provides economical feed, such as feeding chicken litter to cattle.

MOLECULAR BIOTECHNOLOGY

Molecular biotechnology is the branch of biotechnology that deals with changing the structure and parts of cells. It is often known as genetic engineering.

Plants or animals produce offspring with many of the same characteristics as their parents. Within each species, each animal is different. This difference is known as variability. Molecular biotechnology uses variability and other principles of genetics.

All life forms can be reduced to molecular levels. This means that cells—the building blocks of organisms—and their parts can be divided into smaller pieces, known as molecules. A molecule is the smallest particle of a substance. When a molecule is divided, it is broken down into two or more atoms. An atom is the smallest unit of an element. Elements (substances that chemists study) cannot be broken into other substances.

All things (living and nonliving) on Earth are made of chemical elements. Most of the things in our daily lives are made from no more than 20 elements. More things are made of carbon, hydrogen, and oxygen than any other elements.

The processes used in molecular biotechnology are complex. Molecular biotechnology begins with cells and the materials that are in them, including the information in cells and their organization. A background in genetics and molecular biology is essential.

GENETIC ENGINEERING

Genetic engineering changes the genetic information in a cell. Sections of DNA may be cut out and new sections inserted. Since chromosomes contain thousands of genes, finding the right gene to remove or add requires considerable searching. Genetic engineering is also known as gene manipulation. Moving a gene from one organism to another is ***gene transfer***.

Recombinant DNA

Recombinant DNA is the result of gene transfer in which a tiny amount of DNA is cut from one chromosome and moved to another. The DNA of two different organisms is combined.

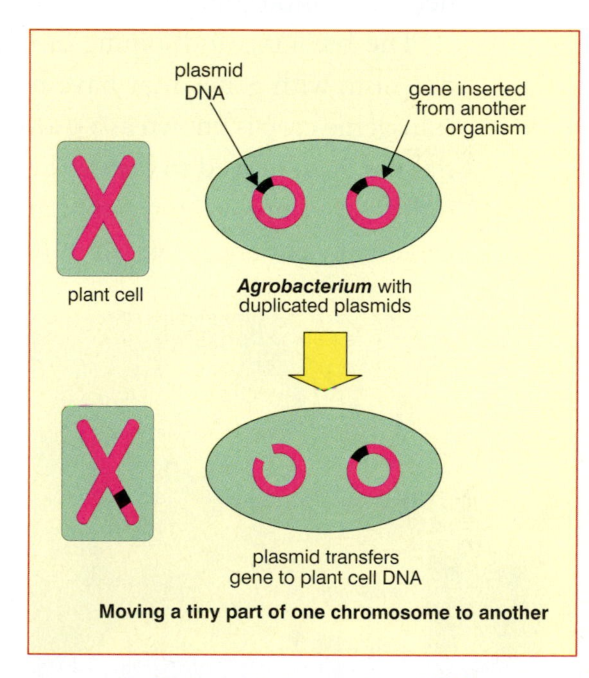

11–37. Simplified recombinant DNA.

All cells have membranes. In addition, plant cells have walls that give them rigidity. The membrane and wall are intended to protect the cell. Invading the nucleus of the cell with new DNA is not easy. DNA is a rather large part of a cell. Several approaches have been used, including mixing the DNA in a solution and microinjection (injecting with a very small instrument). With both methods, many cells die. Most of those that survive may not have the new DNA in their nucleus. Success is often slow.

Vectors. A vector is a carrier of new DNA into a cell. Yeast cells and the bacterium *Escherichia coli* are often used as vectors. *E. coli* bacteria are commonly found in the intestinal tract of humans and other vertebrates. Vectors are cells that are easy to work with and lend themselves to being carriers.

Breaking and Joining. Ligation is the act of uniting or attaching two DNA fragments. It includes cutting the DNA in the cell that is to be transformed and joining it with the new DNA. Ligation is often done in vitro—usually in a test tube.

Cells that receive new material are known as transformed cells. The process of changing the cell is known as transformation. Cells that provide the new material are known as donor cells.

Getting the new DNA to join the cut DNA is a process known as cleavage. Several laboratory procedures have been used in this process. Refinements of these procedures are needed. Various substances are being used to try to more efficiently unite strands of DNA.

The breaking and joining of DNA results in a ***transgenic organism***, which is an organism with genes that have been artificially altered. Molecular methods were used. A transgenic crop is known as a ***genetically modified organism*** (GMO). Genetic material has been moved to create the desired organism. When the productivity of a GMO has resulted in improvements that are useful in agriculture production, the GMO is called a genetically enhanced organism (GEO).

11–38. A microprojectile unit is used to shoot tiny particles coated with DNA into cells during genetic engineering. (Courtesy, University of Illinois)

Genetic Engineering in AgriScience

Genetic engineering is changing how food and fiber are produced. A few examples follow.

- **Herbicide-resistant plants**—Weeds and crop plants are often much alike. Even when selective herbicides are used, they kill both the weed and crop plants. Recombinant DNA is used to move the gene from a plant that is resistant to a herbicide into a plant that is not resistant. The resulting plant is not killed by the herbicide. Research by the Monsanto Company, among others, has produced herbicide-resistant plants, such as Roundup Ready Soybeans.
- **Insect-resistant plants**—Genetic engineering is used to develop plants with natural resistance to insects. Insecticides are not needed. Considerable progress has been made with cotton. Similar procedures are being used on food crops, including potatoes and tomatoes.

11–39. Bags of genetically improved potatoes at an Idaho Farm. (Courtesy, Ashton Hi-Tech Seed Company, Ashton, Idaho)

Cotton has been modified to contain a gene that resists damage by bollworms. The gene is from Bacillus thuringinesis (a bacterium) and is commonly called the Bt gene. Transgenic Bt cotton will control the worm. This saves applying pesticides and protects the environment from the residues.

The strong interest in insect-resistant plants is due to insecticide- resistant insects. More than 500 insects have developed resistance to some kind of insecticide, including roaches, mosquitoes, and flies.

- **Disease-resistant plants**—Plants that resist disease should be more productive and less costly to grow. Further, the quality of the produce should be improved. No chemicals to control diseases will need to be sprayed on the crops. Major efforts have been made to control viral diseases in tomatoes and tobacco.

- **Transgenic animals**—A process known as embryo microinjection is used to introduce growth hormone genes into livestock embryos. This process enhances the quantity of lean meat to body fat. Animals with more lean meat bring higher prices.

- **Frost protection**—Crops lost to frost cost producers millions of dollars each year. Frost-resistant plants would prevent these losses.

  Some bacteria are useful in protecting crops from frost damage. The bacteria have a protein that prevents the formation of ice crystals. These bacteria can be used on crops when below-freezing temperatures are occurring.

  An experimental tomato has been created that resists frost. A gene was moved from an arctic flounder (a fish) into a tomato plant. The gene produces a substance similar to antifreeze that keeps frost away.

11–40. Scientists use fast protein liquid chromatography to purify bovine recombinant CD14. (Courtesy, Agricultural Research Service, USDA)

- **Longer storage life**—Many fresh foods can be stored only a few days before they begin to go bad (rot). Efforts in molecular biotechnology have focused on improving the shelf life of products. The tomato is a good example.
- **Udder health**—Cows with healthy udders are the most efficient milk producers in a dairy herd. Medical treatments for mastitis have often been costly, and a cow with the condition was removed from production while her udder was infected. Now, a bioengineered gene has been placed in a dairy animal that will code for a soluble protein known as CD14, which sensitizes the lining of the mammary glands so that it attacks the bacteria that cause inflammation.
- **New animal products**—Immunochemistry animal farms raise animals for the purpose of making medical products for humans. The major goal is to produce immunizing agents. Pharmaceuticals produced by animals are sold in much the same way as other animal products.

11–41. Annie, a bioengineered purebred Jersey heifer, has modified genes that resist mastitis. (Courtesy, Agricultural Research Service, USDA)

REVIEWING

MAIN IDEAS

Heredity is the transmission of parental traits to offspring. New generations have traits of their parents. Genetics is the science of heredity and variation in living plants, animals, and other organisms. It is an attempt to explain why offspring possess certain traits and not others. The Punnett square is widely used in agriculture in predicting traits of offspring.

Biotechnology is the use of biological processes to change plants and animals or the products they produce. At the organism level, it helps them grow better and produce more. At the molecular level, it changes the genetics of plants or animals. Stem cells are emerging as major opportunities for preventing disease and promoting overall health of humans and other organisms.

Many organismic biotechnological practices are used in producing plants and animals, including cloning, greater fertility, more production, and predicting the future. Molecular biotechnology changes genetic information in cells. Knowledge of genetics, genomes, chromosomes, DNA, and genes is needed.

DNA is a nucleic acid that gives instructions and patterns to the cell. DNA has a double-helix (twisted ladder-like) shape and is the only known chemical that replicates itself. DNA molecules make up the chemical factors that determine hereditary traits (genes). The genes carry individual traits like gender (male or female) or eye color.

Another nucleic acid in the cell is RNA (ribonucleic acid). It conveys information for producing proteins. Several kinds of RNA are found, based on their roles in the cell.

Genetic engineering is molecular biotechnology. It uses recombinant DNA procedures. With gene splicing, a gene can be removed and another inserted in the DNA of an organism.

Major advances have been made in the use of genetic engineering. Widespread use now exists in some products. Major agricultural crops, horticultural and forestry plants, and animals (such as cattle, swine, and poultry) have been studied for genetic engineering applications.

QUESTIONS

Answer the following questions, using complete sentences and correct spelling.

1. What is heredity? Genetics?
2. What is a genome?
3. Distinguish between dominant and recessive genetic traits.
4. What is biotechnology?
5. Name and distinguish between the two major areas of biotechnology.
6. What are the contents of a genome?
7. Distinguish between genotype and phenotype.
8. What is the Punnett square?

9. Distinguish between homozygosity and heterozygosity.
10. What is genetic engineering?
11. What is recombinant DNA? Relate the process of gene splicing to vectors and breaking and joining.
12. What areas of agriscience have developments underway using genetic engineering?

EVALUATING

Match each term with its correct definition.

a. genome
b. recombinant DNA
c. vector
d. genotype
e. phenotype
f. double helix
g. cloning
h. superovulation
i. embryo transfer
j. asepsis

_____1. The process of asexually reproducing organisms.
_____2. Increasing the number of eggs released during estrus.
_____3. Moving an embryo from a donor to a recipient.
_____4. The total genetic makeup of an animal.
_____5. The genes for a trait, represented by a combination of letters.
_____6. The appearance of an organism because of its genotype.
_____7. The structure that DNA forms.
_____8. The result of combining DNA from two different organisms.
_____9. A carrier of new DNA into a cell; sometimes used in genetic engineering.
_____10. The condition of being free of disease-causing germs.

EXPLORING

1. Clone a plant using tissue culture. Have your agriscience or biology teacher help you. Possible plants for cloning include carnations, carrots, and African violets. Use references on tissue culturing. Several weeks will be required. Prepare a report on your experience.

2. Write a short paper on the use of embryo splitting in beef cattle. Use magazines, books, and authorities on the subject to obtain information.

3. Biotechnology raises issues and concerns with some people. Take a survey of students to receive their opinions about the use of biotechnology. Your class may wish to work as a group on this project. Carefully develop the questions you want to ask. Collect and compile the answers.

4. Use electrophoresis to study DNA. Follow the appropriate protocol in your investigation. Prepare a report on your findings.

Part Four

Plant Science

CHAPTER 12

Plant Structure and Growth

OBJECTIVES

This chapter focuses on the structure and growth of plants. Uses of plants are also included. It has the following objectives:

1. Describe how plants are adapted to climate.
2. Explain plant life cycles.
3. Identify the major vegetative parts of plants, and discuss their functions.
4. Explain the meaning and kinds of tropisms.
5. Identify useful plants.

TERMS

aerial stem
annual
biennial
bulb
compound leaf
corm
fibrous root system
geotropism
growing season
herbaceous stem
irrigation
leaf
life cycle
lodge
osmosis
perennial
photoperiod
phototropism
primary root
rhizome
root
root hair
root cap
secondary root
simple leaf
stem
subterranean stem
taproot
thigmotropism
transpiration
tropism
tuber
woody stem

12–1. Working with plants is enjoyable as a hobby or as a career.

JUST ABOUT ALL living organisms depend on plants in one way or another. People use plants to help meet their needs—food, clothing, and shelter. Some plants are used as ornamentals, such as the flower in a homecoming corsage. Animals used by humans feed on plant materials. For example, the cows in the pasture are eating plants!

Of the 350,000 plant species on Earth, approximately 1,000 are used by humans for food. Of these, a few are dominant. The cereal grains, such as wheat and rice, are major human foods.

The most important plants in agriscience are those that have flowers and bear seed (Anthophyta phylum). They are the advanced forms of plants and have specialized tissues and organs; they are vascular plants. These plants are important because of their product uses.

Knowing the basics of plant science helps in growing plants. Understanding what to do to make plants grow reduces the chance of famine and improves life for humans and animals.

CLIMATE ADAPTATIONS OF PLANTS

Plants vary in their growth characteristics. The variations result in different plant species being suited to particular environments. Some plants require more water than others and will grow in wet environments. Other plants take longer to grow and cannot tolerate cold weather. As a result, climate is the major factor in determining the suitability of a plant species.

12–2. Ornamental plants must be adapted to the climate where they are grown. (Courtesy, Agricultural Research Service, USDA)

CLIMATE FACTORS

Climate is the long-term weather in an area. Major climate factors are temperature and precipitation (rain, snow, etc.). The length of the seasons and the number of hours of daylight are important to some plants, but the climate is considered in selecting a crop variety. For example, some varieties of sweet corn need 120 days or more to mature; other varieties mature in 90 days or less.

Temperature

The length of the growing season determines the suitability of a particular plant species for an area. A ***growing season*** is the period after the last frost in the spring and before the first frost in the fall. It is measured as the number of frost-free days. In North America,

12–3. At your home, grow a grape variety that is suited to conditions. (Courtesy, Education Images)

the growing season is longer in the southern United States and shorter in the northern United States and Canada.

Some plants are cool-season crops, but other plants must be grown in warm weather. Plants must be adapted to the climate to be grown successfully. Therefore, the growing season of a plant must be known.

Warm-season crops need a warm growing season because cold weather is not tolerated. Examples of warm-season crops are bananas, papaya, oranges, tomatoes, cotton, corn, and soybeans.

AgriScience Connection

MACHINES ADAPTED TO PLANTS

Mechanical harvesting requires specially designed machinery. The machinery must select the desirable parts of plants and remove the wastes. A good example is the potato. As tubers, potatoes grow in the soil on the roots of potato plants. Getting potatoes out of the ground is not easy!

Machines that harvest potatoes must remove them from the soil, sort out stems and trash, and move the potatoes to a truck or container. All of this must be done so the potatoes are not damaged and so no potatoes are left unharvested in the soil. The above photograph shows a high-tech potato harvester on a Minnesota farm. (Courtesy, R. D. Offutt Farms, Fargo, North Dakota)

12–4. Selected food and agricultural plants.

The life cycle of cool-season plants in the southern United States begins in the fall and ends in the summer. Cool-season plants cannot tolerate warm weather, so most cool-season crops are planted in the fall, except in climates that are cool in the summer. Examples of cool-season crops include wheat, oats, and some vegetables.

Precipitation

Plants vary in the amount of water they need. Some plants require much more water than other plants. Crops need to be grown in areas where they are best suited to the available water, unless irrigation will be used to add water. If crops do not receive enough water, they will not grow. Sometimes they die without producing anything.

Irrigation is the artificial application of water to meet plant growth needs. Moisture levels in the soil are assessed to determine when to use irrigation. Then water is typically applied to the soil.

12–5. A drip-tube irrigation system is used with hanging baskets in this greenhouse. (Courtesy, Education Images)

Light

Plants vary in the amount of light they need. Cool-season crops will begin to die when the days become longer. In contrast, warm-season crops die when the days become shorter, even if the temperature is still acceptable.

Light is often referred to as the photoperiod. The ***photoperiod*** is the length of the light period in a day. Light is most readily used by the leaves of plants. Broad, flat leaves provide good surfaces for light collection.

Long-day plants change from the growth to the reproductive stage as the day length changes. For example, the mustard plant, a cool-season crop, "goes to seed" when the days get longer in the spring. It quits growing leaves and grows flowers and seed.

The day length has no influence on day-neutral plants. They may produce flowers and grow regardless of the day length.

Short-day plants change from the growth stage to the flowering stage as the length of light exposure decreases. Plants that grow in the summer and bloom in the fall are short-day plants.

12–6. A plant physiologist is studying nutrients in the leaves of sugar beets. (Courtesy, Agricultural Research Service, USDA)

Artificial light can be used to extend day length. In fact, greenhouse growers use light to "trick" plants into a condition of longer days.

PLANT ADAPTATIONS

Plants can usually be grown successfully in an environment to which they are adapted. Water, temperature, and the length of the growing season are factors. Before trying to grow a crop, its requirements must be studied in detail.

Many plants are sensitive to weather changes and die when changes are too sudden or too drastic. Other plants adjust to weather changes and live for years.

Plant growth structures can be used to create an artificial climate suitable for plants. Greenhouses and hotbeds are two examples. A typical use of growth structures is to create warmer climates in winter to grow plants or to protect plants from frost damage.

PLANT LIFE CYCLES

The ***life cycle*** of a plant is the sequence of changes the plant goes through from its inception to reproductive maturity. The life cycle includes the beginning, growth, and maturity stages of the life span.

Life cycles may last for one growing season or for many years. Growing seasons may be only a few months long and are related to the climate. Plants require certain climates to grow.

Plants are classified into three groups based on life cycle: annuals, biennials, and perennials.

ANNUALS

An ***annual*** is a plant species that completes its life cycle in one growing season. Annuals grow from a seed, mature, and reproduce or produce seed. Some annuals go through a life cycle in a few weeks; other annuals take months. Radishes and mustard plants have a life cycle of six to eight weeks. Corn and wheat may require three months or more. Many cultured crops are annuals. They may be divided into those that grow in the winter and those that grow in the summer.

12–7. The collard is grown as an annual but will live more than one year under the right conditions. (Courtesy, Education Images)

Winter Annuals

Winter annuals are planted in the fall, grow in the winter, and mature in the spring. They die after they have produced seeds, when summer arrives.

Climate influences the life cycle of winter annuals. In the southern and Midwestern parts of North America, wheat is a winter annual. In the northern areas, wheat is a summer crop. Wheat cannot survive the winter in very cold weather. It also cannot live in hot weather. Other examples of winter annuals include oats, rye grass, vetch, and clover.

12–8. The oat is grown as a winter annual in warm climates and as a summer annual in cooler climates.

Summer Annuals

Summer annuals are planted in the spring. They grow, mature, and produce seeds. Then they die in the fall.

Summer annuals are usually sensitive to cold weather and are killed by frost, so their life cycle is related to the frost-free time of the year. The frost-free time begins after the last frost in the spring and ends at the time of the

12–9. Soybeans are summer annuals. These young beans will produce a crop that can be harvested in the fall.

first frost in the fall. Examples of summer annuals include tomatoes, beans, corn, squash, and rice.

BIENNIALS

A ***biennial*** is a plant that lives two seasons. During the first season, the plant grows from its seed and reaches maturity. Food is made and stored in the plant for the second season. During the second year, the biennial grows a little more and reproduces or makes seeds. After the seeds are produced, the plant dies. Examples of biennials include cabbage, beets, some clovers, and carrots. Biennials are not grown as widely as annuals or perennials.

12–10. Cabbage is a biennial.

PERENNIALS

A ***perennial*** is a plant that lives for more than two seasons. Some perennials live for many seasons and may be hundreds or thousands of years old. Bristlecone pine trees in California may be 4,000 years old! As a result, perennials are often large plants.

Perennials may grow only vegetative parts (leaves, stems, and roots) the first years. Flowers and seeds are produced later. The way a perennial develops varies with the plant species. For example, an oak tree may grow for years before it produces acorns (seed).

Perennials adjust differently to changing seasons. This further classifies them as deciduous or evergreen.

Deciduous plants lose their leaves in the winter, resulting in leaves to rake in the fall. Examples of deciduous plants are some oak trees, sweet gum trees, and sycamore trees.

Evergreens retain their leaves year round. Examples include pine trees, hollies, junipers, and boxwood shrubs.

12–11. Some species of deciduous trees have vibrant fall colors. Can you distinguish the evergreens from the deciduous trees? (Hint: The deciduous trees are changing color; the evergreens are not.) (Courtesy, Education Images)

CHANGING WHEN PLANTS GROW

Some plants are not what they appear. Their natural life cycles differ from how they are grown in agriculture.

Crops may be planted and harvested without the plant completing its life cycle. The carrot is a good example. It is planted and harvested in the first season, after the enlarged root has developed. Except when producing seeds, carrots are treated as annuals.

Cotton is another example. It is planted in the spring and harvested in the fall. Cold weather kills the plant, but if it is grown in a greenhouse, a cotton plant can live for years. Cotton is a perennial, but it is grown as an annual.

Agriscience is used to grow plants at times when they do not grow naturally. It involves studying how plants grow and the conditions needed for growth. Growing structures can be used to provide the necessary conditions.

12–12. Cotton is farmed as an annual but can live for years if protected from cold weather. Young cotton plants are growing in a field where no tillage has been used. (Courtesy, Education Images)

12–13. A pansy plant started in a greenhouse can be placed in a flower bed.

12–14. A commercial greenhouse facility for producing bedding plants.

Greenhouses are common structures in which to grow plants. They are made of plastic or glass placed over pipe or wooden frames. Heating or cooling makes the temperature right for the plants. The plants are carefully watered; diseases and insects are controlled; and fertilizer is used as needed.

The photoperiod may be lengthened or shortened in the greenhouse. Using artificial light a few hours at night can lengthen the photoperiod. Also, covering the greenhouse with cloth or other material to keep light out can shorten the photoperiod.

Examples of environmental control to produce plants:

- Tomatoes are warm-season plants. However, they can be grown in the middle of winter in greenhouses where the climate is controlled. The grower must provide the best possible climate for the tomato in the greenhouse.

12–15. A greenhouse structure at the San Antonio Botanical Garden is used to create tropical climates for some plant species.

12–16. Heated beds may be used to start plants. These sweet potato beds promote plant growth while the weather is cold.

- Bedding plants are annuals, such as petunias and marigolds, used in landscaping. These plants are often sensitive to cold weather. Growing structures can be used for a head start on the growing season. This helps to supply flowering plants earlier in the season. For example, petunia plants started in a greenhouse when the weather is cold are transplanted to flower beds after the weather warms.

Other ways of controlling when plants grow may be used. Some methods involve genetic engineering to alter the plants. Other methods involve applying substances to the plants to protect them from bad weather. Many of these methods are now used on a small scale or experimentally.

VEGETATIVE PARTS OF PLANTS

Plants have important parts that carry out specialized functions in their life cycle. There are two major kinds of parts: vegetative and reproductive. The vegetative parts are covered in this chapter. The reproductive parts are covered in Chapter 9.

Plants have three main vegetative parts: leaves, stems, and roots. These work together as organs to help a plant carry out life processes.

LEAVES

A ***leaf*** is an organ on a plant stem that is usually green. Some leaves are large, thin, and flat. Other leaves are long and narrow, like a needle. Regardless of the shape and color, leaves are alike in many ways.

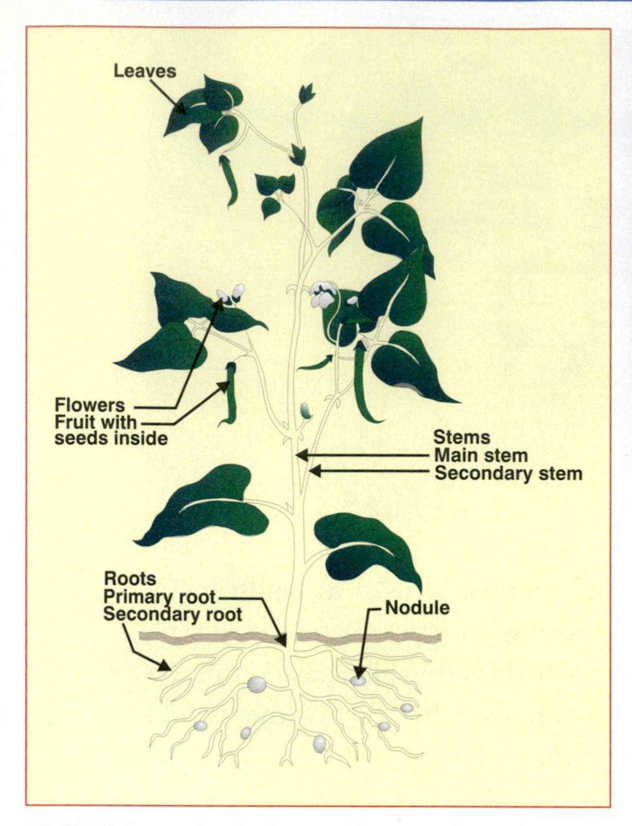

12–17. Major parts of a legume plant are shown here. Legume plants form nodules of nitrogen on their roots.

12–18. Some plants, such as the fiddle leaf fig, have large leaves with prominent veination.

Leaves make food for the plant through a process known as photosynthesis. During photosynthesis, leaves convert water and other materials absorbed by the roots into food. Oxygen is given off in the process. Photosynthesis will be covered more completely in Chapter 14.

Parts of Leaves

Most leaves contain several parts that work together. The leaves of dicots (two cotyledons) and monocots (one cotyledon) vary in arrangement.

The major parts of a dicot leaf are:

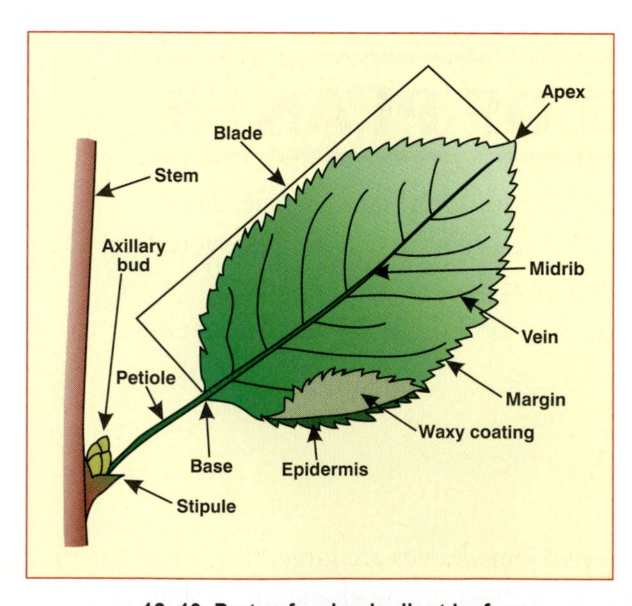

12–19. Parts of a simple dicot leaf.

- **Blade**—The blade is the large, flat part where most of the photosynthesis occurs. Some leaves are not large and flat (such as the pine needle), but they function as a blade.
- **Veins**—The blade has veins that distribute water in a leaf. Tiny tubes (vascular bundles) form the framework of a leaf. The arrangement of the veins is known as venation. Veins

in monocots (grass family) are parallel (run side by side) while those in dicots have net or branching venation.

- **Epidermis**—The outer surface of a leaf is the epidermis. This surrounds and supports structures inside the leaf.
- **Stomata**—The stomata are tiny holes or openings in the epidermis. They are the openings through which water vapor, oxygen, and carbon dioxide pass into and out of leaves.

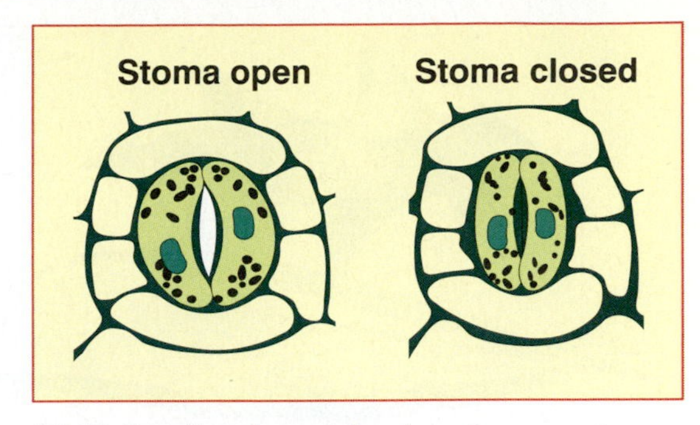

12–20. Greatly enlarged drawing of open and closed stomata.

- **Internal cells**—Various kinds of spongy and tall cells fill the space between the two layers of epidermis and veins. These cells carry out much of the photosynthesis.
- **Waxy coating**—The epidermis is coated with a waxy layer to protect the leaf. It prevents excessive evaporation of water.
- **Petiole**—The leaf is attached to the stem by the petiole.
- **Stipule**—Stipules are small structures that may grow where the petiole joins the stem.

Monocot leaves are long, narrow, and thin with a point at one end. They are attached to the stem by a sheath, which is a structure that wraps around the stem and holds the leaf in place. The middle vein of a monocot leaf is the midrib. It is strong and can support the entire leaf, such as on sugar cane and corn plants.

Technology Connection

FUTURE GROWTH

The climate on Earth is changing in ways that may influence plant growth. A few of the changes include global warming, increased CO_2 levels in the atmosphere, and increased ozone. Scientists use various means of investigation.

One approach is to grow plants in enclosed structures where the change being studied can be artificially produced. Such investigations are made in greenhouses and other specially designed facilities. While the study is under way, careful observation is made of the growing plants.

Findings about global warming indicate that crops grown for seed may be more damaged by an increase of a few degrees than plants grown for forage. This image shows a scientist measuring the growth of rice to determine the effect of elevated temperature on grain yield. (Courtesy, Agricultural Research Service, USDA)

12–21. Differences in dicot and monocot leaves are easy to see. The leaf of a red oak tree (dicot) has net venation. The grass (monocot) leaf has parallel veins.

12–22. Differences in simple and compound leaves are easy to see. The locust tree leaf (top) has a compound leaf. The bottom leaf is a simple leaf from a magnolia tree.

Types of Leaves

Leaves are simple or compound. A ***simple leaf*** has only one blade, while a ***compound leaf*** is divided into two or more leaflets. Plants with simple leaves include corn, oak, and elm trees as well as wheat. Plants with compound leaves include clovers, roses, and locust trees.

Shapes of leaves vary from those with smooth edges to those with lobed edges and needles. The leaf type and shape are important in identifying plants. Many agriscientists can look at a leaf and instantly know the plant type. Leaves also reveal the scientific classification of the plant. For example, all grasses have simple leaves with parallel venation.

Leaf Attachment

Leaves are attached to stems in certain ways. The leaves of plants of the same species are attached in a specific and consistent pattern. Three basic patterns are:

- **Alternate**—One leaf is located at each node on a stem. (A node is the place on

12–23. Three types of leaf arrangement are opposite, alternate, and whorled.

a stem where leaves develop.) Grass plants have alternate leaf arrangement. The adjacent leaves point in different directions.

- **Opposite**—Two leaves are attached at a node opposite each other.
- **Whorled**—Three or more leaves are attached at each node.

STEMS

A ***stem*** is the structure that supports the leaves. Stems vary in shape and size. Some stems are large and strong, such as a tree trunk. Other stems are short and small, such as lettuce.

Stems:

- Support the leaves and hold them so they receive as much sunlight as possible.
- Support flowers, fruit, and other structures grown by a plant (including thorns on some species).
- Transport water and other materials between the leaves and the roots.
- Produce new living tissue; the tip of each stem has a bud. When the bud grows, the plant gets taller.
- Store food and water. The potato, for example, is a special kind of stem that grows in the ground. It stores food manufactured by the leaves.

CAREER PROFILE

GRAZINGLAND SCIENTIST

A grazingland scientist studies grazinglands and solves problems associated with grazingland. The scientists use remote sensing, land-based equipment, and other means to assess conditions and how they affect plants. This photo shows a grazingland scientist using a yoke with sensors to collect and record information about plants and soil moisture. (Courtesy, Agricultural Research Service, USDA)

Grazingland scientists have college degrees in agronomy, turf, or a related area. Most of these scientists have a master's degree and a doctorate in the field. In high school, take biology, agriculture, and academic classes to begin preparing.

Jobs are found at research stations and government agencies, with owners of large land tracts, and with other people who are involved with grazingland.

Kinds of Stems

The kinds of stems vary in two ways: location and structure.

Stem Location. There are two stem locations: aerial and subterranean. An ***aerial stem*** grows above the ground. They are the plant parts that are typically called stems. A ***subterranean stem*** grows below the ground. Many subterranean stems are quite important.

12–24. A potato is a subterranean stem known as a tuber. This shows "eyes" beginning to grow. An eye with a small piece of the potato can be planted to start a new potato plant.

Examples of subterranean stems are:

- Tuber—A ***tuber*** is an enlarged underground stem. The potato is a common example. Careful examination of a potato reveals buds, which are often called eyes. These eyes can be cut off and used to grow new potato plants. Potatoes store food for the plant. We generally enjoy potatoes baked, fried, or mashed.
- Bulb—A ***bulb*** is an underground stem that has layers resembling leaves. An onion is a good example. Tulips, jonquils, and other ornamental plants grow from bulbs. Slicing a bulb in half reveals layers of leaflike structure. The bulb stores food for the plant.
- Corm—A ***corm*** is an underground stem that is similar to a bulb, but it has thinner leaves and a thicker stem. Examples of plants with corms are taro, gladiolus, and garlic.

12–25. An onion is a bulb. Note that the outer layer of leaves is dry and protects the inner, moist leaves.

12–26. The Chinese Water Chestnut is a corm.

- **Rhizome**—A ***rhizome*** is a long, underground stem that sends up shoots to start new plants. Some pesky weeds have rhizomes, such as Johnson grass. Even when cut apart, a rhizome can grow into another plant.

Stem Structure. Stems are classified as woody or herbaceous (nonwoody). The classification refers to the amount of woody material in the stem.

Woody stems are the stems of trees, shrubs, and many crop plants, such as cotton. A ***woody stem*** has rigid tissue known as xylem. The main stem of a tree is the trunk. Tree trunks have many cells with strong walls, making the trunks ideal for lumber. Shrubs and other woody plants have more stems than trees and are usually shorter than trees.

12–27. A hollow tree trunk is weak and of no value for lumber.

A ***herbaceous stem*** is a nonwoody stem that is soft and green and that contains only a small amount of xylem. Many flowering, vegetable, and crop plants, such as corn, have herbaceous stems.

Parts of Stems

Stem parts vary by the kind of plant and the woodiness of the stem. Stems with a lot of wood have a more complicated structure.

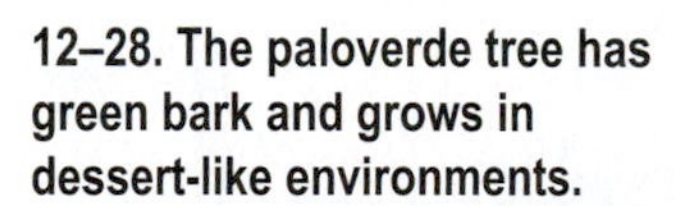

12–28. The paloverde tree has green bark and grows in dessert-like environments.

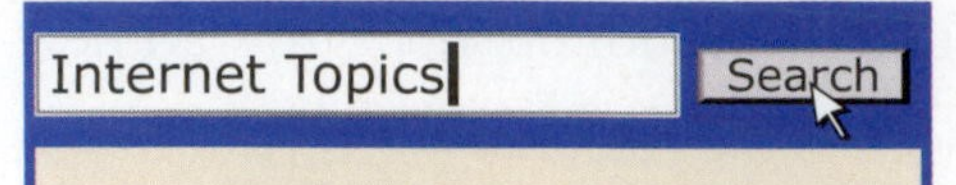

Selected topics for Internet discovery and reporting are listed here. Begin by searching for each topic. Next, read and learn about it. Conclude by preparing a brief report on your findings.

1. Chinese water chestnut
2. Growing season
3. Paloverde tree

All stems are divided into nodes. Leaves grow at the nodes, and the space between nodes is known as the internode. The end of each stem has a terminal bud where growth in length occurs. Terminal buds on trees usually show a spurt of growth at the start of every growing season.

The inside of a stem is designed to transport substances. Water and other nutrients travel from the roots to the leaves. Food is moved from the leaves to the roots and other parts of the plant.

The cross section of a woody stem has the following parts:

- **Bark**—The bark is the outer skin that protects the stem from injury and holds it together.
- **Cortex**—The cortex is the primary tissue in a stem. It is located between the bark and phloem.
- **Phloem**—The phloem transports sugar (manufactured food) throughout a plant.

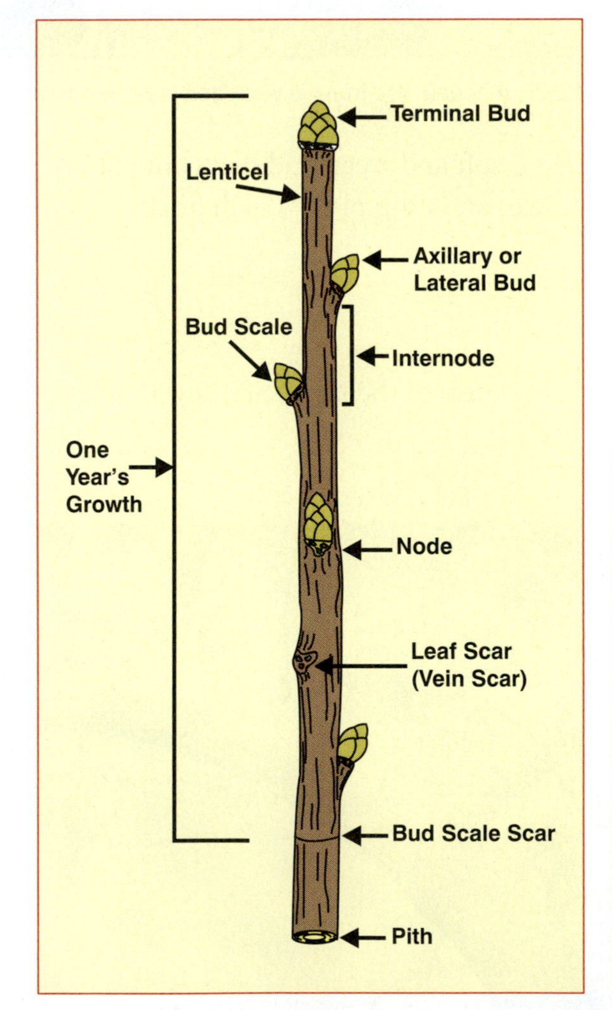

12–29. Typical parts of a stem.

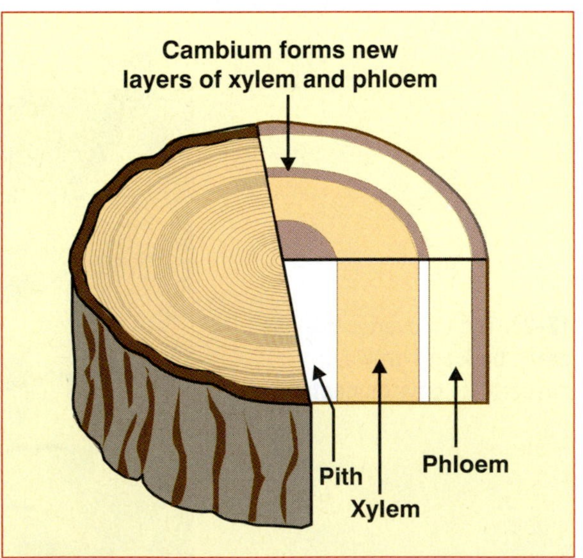

12–30. The structure of a cross section of a woody stem.

- **Cambium**—The cambium is where growth occurs. Both phloem and xylem cells grow at the cambium.
- **Xylem**—The xylem conducts absorbed water and other substances from the roots to the leaves for food production.
- **Pith**—The pith is the middle part of a stem. It stores moisture and food.

Agriscientists refer to the combined xylem and phloem as the vascular system of plants. Vascular tissue consists of vessels and tubes that can convey fluids, known as sap.

12–31. Bamboo, a member of the grass family, has stems that are hollow, woody, and jointed. (Courtesy, Education Images)

ROOTS

A ***root*** is the part of a plant that usually grows underground. Roots have three important functions:

- **Anchor the plant**—Roots hold a plant in place and keep it from falling onto the ground.
- **Take in water and minerals**—Roots absorb water and minerals from the soil. These materials are transported to the leaves for use in making food.
- **Store food**—Food made in the leaves may be moved to the roots for storage. Plants with large root structures, such as carrots and beets, store the most food.

Kinds of Root Systems

Two major kinds of root systems are: tap and fibrous.

A ***taproot*** is one large root that grows downward. It may have a number of branches known as secondary roots. Pine and pecan trees, carrots, and beets have taproots. The taproots of trees are woody, while other taproots (such as those of the carrot) are fleshy. Some weeds have well-developed taproots, which makes them difficult to kill. A good example is the dandelion.

12–32. Two underground parts of plants: the carrot is a taproot, and the jicama is a tuber.

12–33. Corn plants have prop roots to help hold the stalk upright.

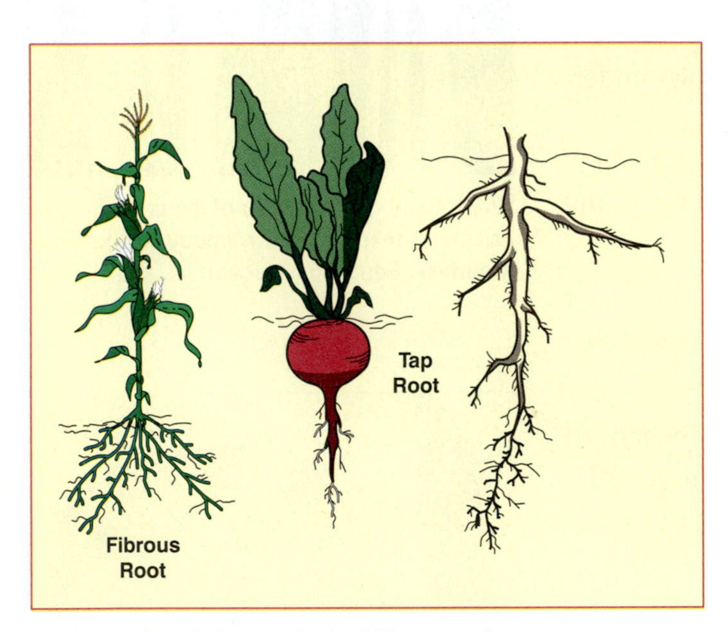

12–34. Comparison of fibrous and taproots.

A ***fibrous root system*** is made of many small roots that spread out in the soil. Plants in the grass family, such as wheat and corn, have fibrous root systems. Other plants have fibrous root systems, such as the soybean, tomato, and most ornamental bedding plants.

Some plants have modified root systems for special needs. Corn has prop roots that grow out of the stem above the soil. When the plant becomes tall, prop roots (adventitious roots) help to keep it from falling over. Poison ivy and other vines, some lawn and pasture grasses, and strawberries also grow adventitious roots. These roots anchor the plant and grow new plants. The roots on ivy anchor it to trees, buildings, and other structures.

Parts of a Root System

Root systems help the plant live and grow. They are connected to the lower end of the stem. Water and minerals are absorbed through the roots and are passed to the stem.

Root systems have a primary root, secondary roots, and root hairs. The ***primary root*** is the first root to grow from the seed. Roots that branch from the primary root are known as

secondary roots. Both primary and secondary roots have root hairs. A ***root hair*** is a small, hairlike growth that helps anchor a root. Every plant has many root hairs that may form a large mass in the root system. Root hairs absorb water and nutrients that are passed to the secondary and primary roots.

The tip of the primary root has a ***root cap*** that protects the root as it grows into the soil.

On the inside, roots have vascular tissue. This is the tissue that conducts substances in the roots. It connects with the vascular tissue in the stem.

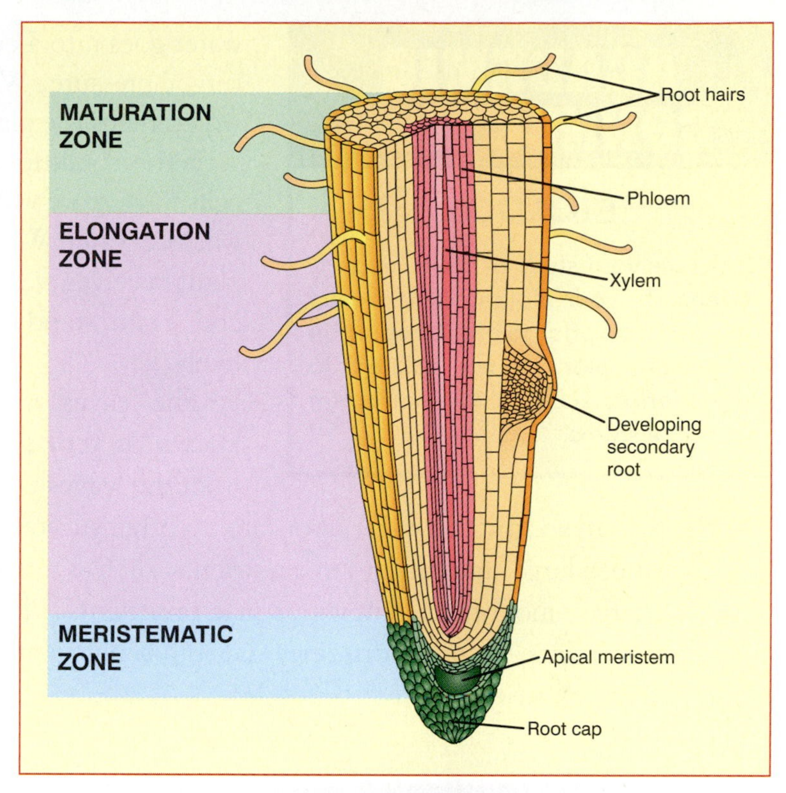

12–35. A primary root tip.

How Roots Absorb Water

Water and nutrients are primarily absorbed by the tiny root hairs. These root hairs have thin coverings that allow water to enter through a process known as osmosis.

Osmosis is the movement of liquid from a greater concentration to a lesser concentration. The greater concentration of water is usually in the soil. Plant roots have higher levels of certain nutrients than the surrounding soil and water, but they have a lower concentration of water. This "draws" the water to the nutrient. Water moves into roots until the level of concentration is equal inside and outside of the roots. Of course, dry conditions in the soil may not provide sufficient water for the plant.

After the roots absorb the water, it is passed from cell to cell until it reaches the xylem. Upon reaching the xylem tubes in the roots, the water moves up the stem and to the leaves. The petiole takes the water from the stem to the leaf veins, which distribute it throughout the leaf.

Plant cells swell (get bigger) as they fill with water. They do not swell beyond the limits of the cell walls, however. As

12–36. Root systems of bean plants grown in a water solution (known as hydroponics) are quite massive. (Courtesy, Agricultural Research Service, USDA)

ACADEMIC CONNECTION

BIOLOGY

Relate your study of plants in agri-science to your biology class. Select an area and offer to give a report on important plant topics related to agriscience, such as thigmotropism or aerial stems.

water goes into a cell, pressure is formed. This is known as turgor pressure. When filled with water, the cell is rigid, which helps the plant and leaves to stand erect.

If the conditions outside a plant are dry and the plant cells have more water than a plant's surroundings, the cells will lose water. When this occurs, the plant wilts. If a wilted plant receives water soon, it will absorb the needed water and again stand erect. Some decrease in production is likely.

The leaves of plants give off water through a process known as ***transpiration***. This creates an upward pull from the leaves through the stem to the roots.

In addition to water, roots absorb other substances known as inorganic nutrients. Plants need large amounts of some nutrients, such as nitrogen and potassium. Nutrients needed in large amounts are known as macronutrients. Plants also need smaller amounts of other nutrients, such as iron, zinc, and sulfur. Nutrients needed in small amounts are known as micronutrients.

HOW PLANTS MOVE

Most plants do not have the ability to move about on their own (locomotion). Their roots anchor them in place. Some movement is essential, however, if a plant is to receive sunlight and respond to its environment. Of course, some "in-place" movement occurs as plants grow.

12–37. The upward growth of these young almond trees is phototropism. (Courtesy, Education Images)

Plant movement is known as ***tropism***. Conditions in the environment of plants cause them to move. These conditions are known as stimuli. Tropisms are named for the various stimuli that influence plants.

12–38. Plants grow toward light.

KINDS OF TROPISM

The most common tropisms are:

- Phototropism—Plants respond to light. ***Phototropism*** is the growth or turning of a plant in the direction of light. A plant that sits a long time on the window sill bends toward the window or light. If the plant is turned around, the stems will again bend toward the window, if given enough time. Phototropism is why stems grow upward.
- Geotropism—Plants respond to gravity. ***Geotropism*** is the downward growth of roots. Other plant structures, such as tubers, also grow down.
- Thigmotropism—***Thigmotropism*** is the growth of plants over and around solid objects, such as fence wire attached to a tree. Over time, the tree will grow around the wire so it ends up inside the tree. This damages the tree and makes it dangerous to saw or cut where the wire is located.

12–39. A tree perched on a rock has grown around the hard surface—thigmotropism.

HOW TROPISMS HELP AGRICULTURE

Tropisms help crop plants grow properly. Sometimes plants get twisted or knocked over by a storm or an animal. Tropisms help plants regain their normal position.

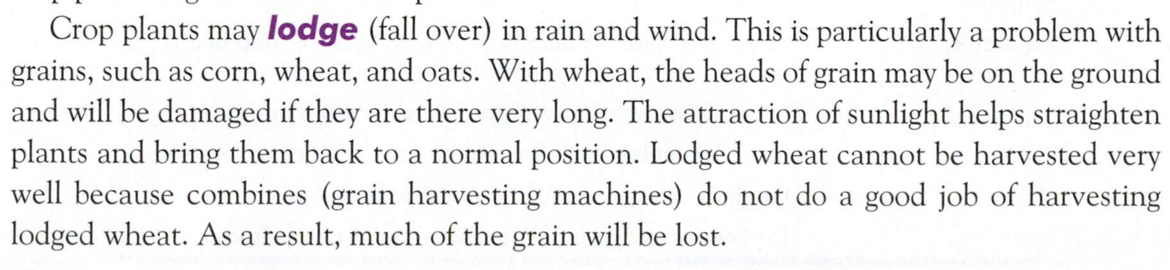

Crop plants may ***lodge*** (fall over) in rain and wind. This is particularly a problem with grains, such as corn, wheat, and oats. With wheat, the heads of grain may be on the ground and will be damaged if they are there very long. The attraction of sunlight helps straighten plants and bring them back to a normal position. Lodged wheat cannot be harvested very well because combines (grain harvesting machines) do not do a good job of harvesting lodged wheat. As a result, much of the grain will be lost.

Tropisms result in plants growing upright and producing desirable flowers. A flower on a long, straight stem requires appropriate sunlight. This is extremely important to producers of ornamental flowering plants.

Tropisms cause seeds to sprout and grow normally. Seeds that are planted upside down will grow as well as those planted upright. Geotropism attracts roots downward as they grow from a seed. Phototropism attracts stems, buds, and leaves upward. Regardless of how a seed is turned, the plant grows correctly.

Many of the ways crops are grown and harvested are based on the plants being in a normal growing position all of the time. Tropisms help to keep them there.

USEFUL PLANTS

All plants have an important and useful role in their natural ecosystems. Other plants have been domesticated and are produced under carefully managed conditions. Plants can be placed in three groups for purposes here: agricultural, forestry, and ornamental.

AGRICULTURAL PLANTS

Agricultural plants are grown on farms to gain desired products to meet human and other needs, such as animal feed. Grain, forage, and fiber crops are the most widely grown. Table 12–1 lists examples of major crops in the United States and the uses of the produced items.

Table 12–1. Examples of Agricultural Crops

Crop or Plant	Use
Rice	Grain
Cotton	Fiber and seed for oil
Wheat	Grain
Corn	Grain
Soybeans	Grain and hay
Sorghum	Grain and feed
Vegetables	Food; examples: tomatoes, potatoes, squash, snap beans, onions, lettuce, celery, and corn
Fruit	Food; examples: apples, oranges, grapefruit, and cherries
Berries	Food; examples: strawberries, blueberries, and raspberries
Nuts	Food; examples: pecans, walnuts, and almonds
Trees	Timber and ornamental use, such as Christmas trees

12–40. Examples of agricultural plants.

FORESTRY PLANTS

Forestry plants are used for the wood, oil, nuts, and other products. Many of these plants are produced on tree farms; yet native forests continue to be major sources of these products. The species of trees include those that produce softwood and hardwood. Table 12–2 lists examples.

Table 12–2. Examples of Forestry Plants

Crop or Plant	Use
Oaks (several species such as red oak, white oak, and pin oak)	Lumber, furniture, flooring
Gums (several species such as black gum and sweetgum)	Lumber, pulp for paper
Poplars (several species such as the yellow poplar)	Lumber, veneer, and pulp
Maples (several species such as sugar maple and red maple)	Furniture, syrup (sap), sugar, and cabinets
Pines (several species such as long leaf, white, loblolly, and short leaf)	Lumber, plywood, particle board, and pulp
Spruces and firs (several species)	Lumber, pulp, and plywood

ORNAMENTAL PLANTS

Ornamental plants are those grown for cut flowers as well as those in landscapes such as bedding plants, shrubs, and turf or sod. Some trees may also be included as ornamental plants. Table 12–3 lists a few examples.

Table 12–3. Examples of Ornamental Plants

Plant	Use or Product
Roses	Cut flowers and landscape gardening
Geraniums	Hanging baskets, potted plants, and bedding plants in landscapes
Petunias	Bedding plants and containers, such as hanging baskets and potted plants
Daffodils	Bedding plants, containers, and cut flowers
Blue grasses and fescues	Turf or sod, and forage (forage is non-ornamental)
Boxwoods	Landscaping
Azaleas	Landscaping
Hollies	Landscaping (bedding and accent plant materials)

Note: These examples are only a few of the many species and uses of ornamental plants.

REVIEWING

MAIN IDEAS

Plants are essential to life on Earth. They provide food, fiber, and shelter for humans. They also provide food for other animals, including cattle, hogs, chickens, fish, and pets.

A life cycle is the length of a plant's life. Plants are classified based on life cycle: annuals (live one year and die), biennials (live two years and die), and perennials (live more than two years and may live hundreds of years, depending on the plant). Cultural practices make it possible to grow plants at times of the year that are not natural for them. For example, greenhouses are used to produce summer annuals in the winter.

Plants have three main vegetative parts: leaves, stems, and roots. All three are needed for a plant to grow. Leaves are usually thin, flat organs that make the food for the plant. They are positioned above the ground to receive sunlight. Leaves may be simple (one blade) or compound (several blades or leaflets).

Stems support the leaves and connect the leaves with the roots of the plant. Stems conduct water and nutrients from the roots to the leaves and manufactured food from the leaves to the roots. Stems grow above and below the ground. Specialized kinds of stems below the ground include tubers, bulbs, corms, and rhizomes.

By osmosis, roots absorb water and nutrients from the soil. Roots are connected with the stem so the water and nutrients flow upward. Roots anchor plants and store food for the plant. Many of the roots that store food, such as carrots and beets, are eaten by humans.

Tropisms influence the direction in which plants grow. Geotropism is the attraction of certain parts of plants to grow toward and into the ground. Phototropism is the attraction of plants to light.

Plants have many uses and provide an array of agricultural, forestry, and ornamental plant products.

QUESTIONS

Answer the following questions, using complete sentences and correct spelling.

1. What are the factors in the adaptation of plants to climate?
2. What are the three ways of classifying plants based on life cycle? Distinguish between the life cycles.
3. How can the life cycles of plants be altered?
4. What is the major role of leaves?
5. What are the major parts of a leaf? (Sketch a leaf and label its parts.)

6. What are the types of leaves? Give an example of each.
7. How are leaves attached to plants? What patterns are found?
8. What is the role of the stem?
9. What are the two major kinds of stems? Distinguish between them.
10. What are the parts of a stem? (Sketch a cross section of a woody stem, and label the parts.)
11. What does the root system do for a plant?
12. What are the two major kinds of root systems? Distinguish between them.
13. What are the parts of a root system? What are the functions of these parts?
14. What is tropism? Name three kinds, and explain how they affect plants.

EVALUATING

Match each term with its correct definition.

a. perennial
b. growing season
c. annual
d. biennial
e. root cap
f. root hair
g. transpiration
h. tropism
i. osmosis
j. bulb

______1. An underground stem with layers that resemble leaves.
______2. The tip of a primary root.
______3. The movement of water from a greater concentration to a lesser concentration.
______4. The process by which leaves give off water.
______5. Plant movement.
______6. Small, hairlike growths that help anchor the root.
______7. A plant that lives for two or more years.
______8. A plant that completes its life cycle in two years.
______9. The period between the last frost in the spring and the first frost in the fall.
_____10. A plant that completes its life cycle in one growing season.

EXPLORING

1. Determine the climate where you live. Select five crops that are adapted to the climate. Use reference materials to help with this activity. Prepare a report that uses electronic presentation software and that includes digital images of the five crops.

2. List the important crops and/or ornamental plants in your area. Classify them as annuals, biennials, and perennials. Prepare a report on your study.

3. Tour a greenhouse growing facility in your area. Determine the names of the plants being grown. Study how the environment is regulated to help the plants grow.

4. Collect the leaves of the common plants in your area. Classify the leaves as simple or compound. Use reference materials, and identify the common and scientific names of the plants.

CHAPTER 13

Reproducing Plants

OBJECTIVES

This chapter provides information on reproducing plants. It has the following objectives:

1. Explain how plants reproduce.
2. Identify and explain the kinds and parts of seeds.
3. Explain the types and functions of flowers.
4. Describe germination and the conditions needed for it to occur.
5. Explain the use of vegetative propagation.
6. Explain the importance of seed variety and quality.

TERMS

asexual reproduction
budding
clone
complete flowers
cutting
dry fruit
embryo
endosperm
fleshy fruit
flower
fruit
germination
grafting
imperfect flower
incomplete flowers
layering
ovule
percent germination
perfect flower
planting depth
pollen
pollination
propagation
scion
seed
seed certification
seed coat
seed germ
sexual reproduction
tissue culture
vegetative propagation
viability

13–1. Wheat is a major grain crop. (This field is ready for harvest.)

HOW DO WE get more plants? With crop culture, plant reproduction is promoted. It is an essential part of growing plants. The process occurs naturally, but growers can also regulate plant reproduction to gain the desired products.

Seeds are planted regularly, and new crops are grown every year. Tree seeds sprout and grow during spring and summer. It is a continuous cycle of new life.

Reproducing plants is a major area of agriscience. In nature, plants reproduce on their own. With agricultural crops, the process is promoted so crops reproduce when and where desired and in a way that provides products in a timely manner.

Understanding plant reproduction helps us give plants the proper care to yield large crops. Knowing how plants reproduce is a skill in improving plants. Many improvements in crops result from processes that control reproduction in some way.

WAYS PLANTS REPRODUCE

Most plants reproduce naturally to keep their species alive. Many plants reproduce by seeds; others reproduce vegetatively with runners, sprouts, or other means. Reproduction is an important part of nature. New ways of reproducing plants make better plants possible.

PROPAGATION

Propagation is the reproduction or increase in the number of plants. Propagation includes natural plant reproduction as well as other means of obtaining more plants. Plants are propagated by sexual and asexual means.

Sexual Reproduction

Sexual reproduction is the use of seeds to propagate plants. Flowers are important in this process because they contain the reproductive organs that produce sex cells.

Two kinds of sex cells are produced: female and male. The female sex cell is known as the egg. The ***ovule***, which eventually becomes a seed, is the small part in an ovary where the female sex cell, or egg, is produced.

13–2. Butterflies and other insects transfer pollen when gathering nectar.

Pollen is the powdery substance that contains sperm—the male sex cell. The sperm mature during pollination. Through the process of fertilization, new seeds are formed.

A seed contains the tiny embryo of a new plant. An ***embryo*** (***seed germ***) is an immature plant that will grow into a mature plant if planted properly.

Sexual reproduction is the creation of a genetically new individual. Many combinations are possible, depending on the parent plants involved. Scientists may produce improved plants in ways that nature could not.

Asexual Reproduction

Asexual reproduction is the duplication or reproduction of a plant by using the vegetative parts of a parent plant. There is no union of sperm and egg. The new plant is a clone of its parent. Plant structures involved with asexual reproduction include bulbs, corms,

13–3. Removing a cutting for asexual propagation. (Courtesy, Education Images)

13–4. Garlic may be reproduced by removing parts known as cloves from the bulb. This shows a clove being separated from the bulb. Cloves are planted similar to other bulb and root structures and grow into new garlic plants. (Courtesy, Education Images)

tubers, stolons, and rhizomes. Asexual reproduction is widely used with crops and ornamental plants.

Some asexual reproduction occurs in nature, but most of it involves the work of agriscientists. In nature, some plants send out runners or develop shoots that grow into new plants. A good example is the strawberry. Mature strawberry plants grow runners (stolons) that are a few inches long. New plants are formed along the runners, and each runner may start several new plants. The stolon can be cut between the plants so each plant lives on its own.

Asexual reproduction results in an offspring that is a genetic duplicate—a **clone**. In the example of the strawberry, the plant that develops from the runner is genetically identical to the parent plant. As a result, the berries of the new plant will be very similar to those of the parent.

Asexual reproduction ensures that the new plant will be like the old one. This is important in fruit and nut trees, ornamental plants, and many others. For example, native pecan trees produce

13–5. Iris plants are propagated by dividing the rhizomes (underground structures).

very small pecans. Occasionally, a tree will produce larger pecans. When this occurs, grafting (a procedure in which "pieces" of the tree are cut off and grown on another plant) is used to grow new trees. The new trees will produce pecans just like the old tree. If the tree were reproduced by planting the large pecan as a seed, the resulting pecans would likely be small and not like those of the parent. Orchards with many trees may be just one tree growing separately many times!

13–6. Plants are being started in a greenhouse for setting outside after the weather is warm. (Courtesy, Education Images)

USING PROPAGATION

Growing plants requires the use of propagation. People in agriscience study the best propagation methods and use them. New plants fail to develop if the wrong practices are followed.

When properly planted, a seed will sprout (germinate) and produce a new plant. If planted incorrectly, seeds with living embryos will not grow. Knowing when and how to plant seeds is important. After the plants come up, the care (culture) needed for them to grow must be provided.

Knowing how and when to use asexual reproduction requires skills. Care is needed to make sure the new plant lives.

CAREER PROFILE

PLANT BREEDER

A plant breeder develops new varieties of plants. The goal is to gain desired plant qualities. Plant breeders study plants, control pollination, artificially pollinate, and use advanced methods, such as genetic engineering.

Plant breeders need a strong foundation in biological science. Many have college degrees in botany, agronomy, or plant physiology. Most have master's degrees and doctorates. Preparation can begin in high school by taking biology, agriculture, and other science courses.

Jobs are at colleges, research stations, private seed companies, government agencies, and other places where plant breeding is carried out. This shows a plant breeder pollinating Snake River wheatgrass in Utah. (Courtesy, Agricultural Research Service, USDA)

SEEDS

A ***seed*** is a container of new plant life that is formed in the ovaries of flowers. All seeds are designed to allow new life to grow and become established as plants. Seeds ensure continuing life and provide food and other products. Therefore, plants need to produce many seeds because they are used to reproduce plants, and they are major sources of food products, such as wheat grain used in milling flour.

KINDS OF SEED

Seeds may be classified as monocot or dicot. The major difference in the seeds is evident in the plants that grow.

- **Monocot**—A monocot (monocotyledon) is a plant that has seeds with one seed leaf (cotyledon). The embryo in the seed will have one leaf. As the embryo grows, leaves develop with parallel venation. Corn, wheat, rice, and all other grasses are monocots.
- **Dicot**—A dicot (dicotyledon) is a plant that has seeds with two cotyledons. The embryo has two seed leaves. In addition, the leaves of dicots have net venation. Examples are tomatoes, beans, petunias, and most non-grass plants. More dicot than monocot plants exist. Scientists estimate that there are 185,000 species of dicots, as compared to 90,000 species of monocots.

PARTS OF A SEED

As a container of life, a seed must perform certain functions. It must keep the living embryo viable (alive) and provide food for it to grow.

The structure of dicot and monocot seeds varies, but the same basic parts are present.

13–7. Two views of a corn kernel (monocot) and a lima bean (dicot) show outside differences in the seeds.

Dicot Seeds

Dicot seeds vary with the species of plant. Bean seeds are often flat, while other seeds (e.g., radish seeds) may be round.

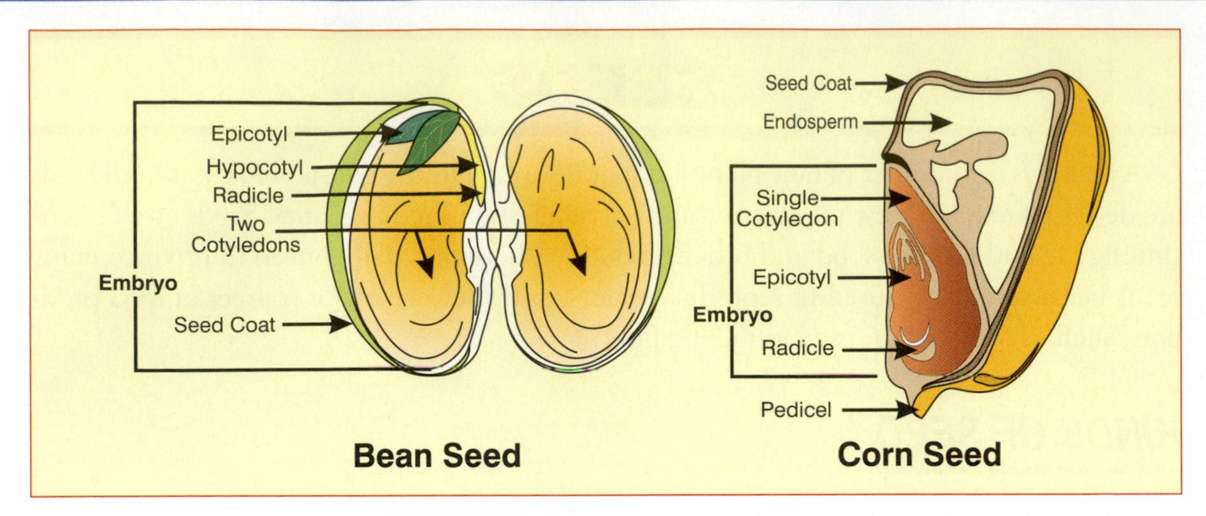

13–8. The major parts of dicot and monocot seeds are shown here using a drawing of a bean and corn kernel.

A bean seed grows in a pod with several other seeds. The pod (bean fruit) develops where the flower grew on the plant. At maturity, the pod opens, and the beans fall out. Each bean seed can start a new plant.

The outside (external) parts of a bean seed are:

- **Seed coat**—The seed coat (testa) is the outer covering of a seed. It protects the fragile embryo from injury and holds the seed together. The coat must let moisture enter when a seed is planted so it can sprout (germinate).
- **Hilum**—The hilum is the place where the seed was attached in the pod. It is sometimes known as the seed scar.
- **Micropyle**—The micropyle is the tiny opening near the seed scar where the pollen entered the ovule to form the seed.

The inside (internal) parts of a bean seed are:

- **Cotyledons**—A bean seed has two large cotyledons. They break apart, forming two symmetrical (alike) halves of the bean. The cotyledons are usually fleshy in form because they contain the food for the developing embryo. The food is high in protein and oil. This is why beans are important crops!
- **Radicle**—The radicle is the seed part that becomes the root system of the plant. It is also known as the embryonic root.
- **Hypocotyl**—The hypocotyl is the seed part that connects the radicle and the cotyledons. It is shaped like a stem and connects the embryo with the food supply in the cotyledons.

- **Epicotyl**—The epicotyl forms the stem of the plant. It is attached to the hypocotyl on one end and has tiny, undeveloped leaves (embryonic leaves) on the other end.
- **Plumule**—The plumule is the part of a seed that develops the above-ground part of the plant. It includes the epicotyl and embryonic leaves in a seed.

13–9. Pea seeds show sprouts—a sign of viability and good germination.

Monocot Seeds

Monocot seeds contain an embryo and a food supply, just as dicot seeds do. The differences in the seeds can be learned by examining a corn seed.

A corn grain (kernel) is a fruit, much as the entire pod of a bean plant is a fruit. Under proper conditions, a corn seed will sprout and grow.

The external parts of a corn seed are:

- **Seed coat**—The ***seed coat*** protects the embryo and holds the seed parts together.
- **Hilum**—The hilum, or seed scar, is the place where the corn grain was attached to the cob. A corn cob is a long, rounded structure formed inside the corn ear. The ear is the part of the corn plant that holds the seed (grain). The ear is covered with husks (shucks) that protect the developing seeds. Ears form from the fertilized female flowers of the corn plant (stalk).
- **Silk scar**—The silk scar is on the opposite side of the grain from the hilum. The silk scar is where the silk (pollen tube) in the female flower was attached to the ovule. Silks extend outside the developing corn ear in the female flower to receive the pollen from the male flower.

The internal parts of a corn seed are:

- **Cotyledon**—The cotyledon in a corn seed does not store food as it does in a bean seed. It absorbs food from the endosperm and moves it to the embryo.
- **Radicle**—The radicle becomes the root system. It grows into the soil as a seed germinates and the plant develops.

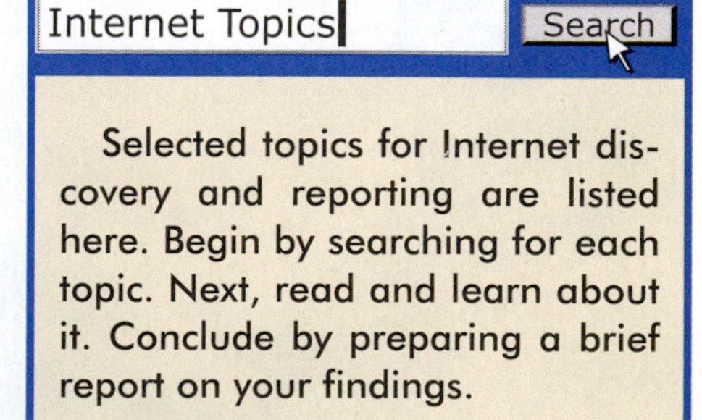

Selected topics for Internet discovery and reporting are listed here. Begin by searching for each topic. Next, read and learn about it. Conclude by preparing a brief report on your findings.

1. Fern
2. Grafting
3. Seed certification

- **Hypocotyl**—The hypocotyl connects the radicle with the food source. It does not grow in monocots as it does in dicots.
- **Epicotyl**—As with dicots, the epicotyl forms the stem of the corn plant.
- **Plumule**—The plumule is the part of a corn seed that develops into leaves on the growing plant. In a seed, it is only one leaf.
- **Endosperm**—The ***endosperm*** is the tissue where food is stored for the embryo to use when a seed sprouts. It is also the part of the grain that provides much of the nutrition in food and feed made from corn.

FLOWERS

Seeds are borne (formed by and in) by flowers. A ***flower*** is the reproductive part of flowering plants. Nearly all plants of importance in agriscience have flowers. Although flowers vary in shape, color, and fragrance, they are all involved in the sexual reproduction of plants. They ensure reproduction!

13–10. Flowering ornamental cherry trees are among the first trees to blossom at the end of winter... a sign of spring. Attractive blossoms have aesthetic appeal and serve several useful roles. (Courtesy, Education Images)

TYPES OF FLOWERS

Flowers are categorized in two ways: complete or incomplete and perfect or imperfect.

13–11. The flower of the tulip poplar (*Liriodendron tulipifera*) is among the most appealing flowers found in a forest. (Courtesy, Education Images)

Complete or Incomplete

Complete flowers are flowers that have all four principal parts, or whorls:

- **Sepals**—The sepals are the outer parts of the flower. They are usually green but may be brightly colored. An example is the tulip. Sepals cover the bud and protect it as it develops. In open flowers, the sepals are found at the base of the petals. Sepals support the petals and help hold them open.

13–12. Food plant flowers. (Note: The flower on a pecan is known as a catkin.)

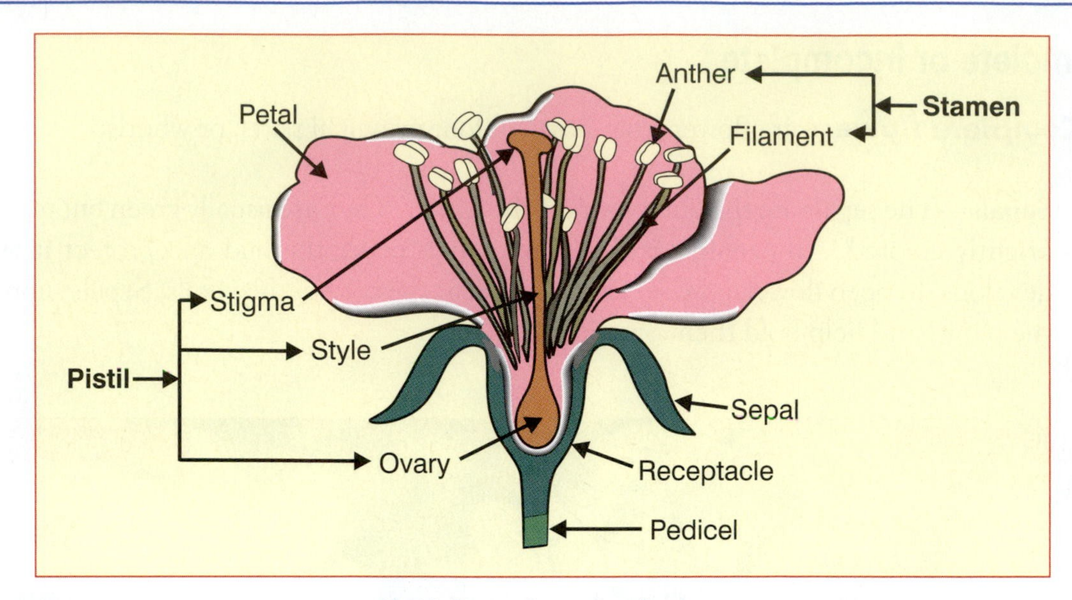

13–13. Parts of a perfect flower.

- **Petals**—Petals are located just inside the sepals. They are often brightly colored and showy. They attract insects that help with the fertilization process by carrying pollen. All the petals on a plant form the corolla.

Technology Connection

CARROT COLORS

Carrots can be found in a rainbow of colors. Though the orange carrot is predominant in the United States, carrots are available in several colors. It has been reported that purple and yellow carrots were eaten 1,000 years ago in Afghanistan and 700 years ago in Europe.

Carrot color is more than eye appeal, according to USDA researchers in Wisconsin. Color (pigment) is an indicator of the nutrient content that has unique properties in promoting human health. Carrots with red pigment are high in lycopene, which is also prominent in red tomatoes. Lycopene helps guard against cancer and heart disease. Those with purple pigment are high in anthocyanin, which serves as an antioxidant. Yellow carrots have xanthophyll, which promotes eye health.

Maybe our green salads need to be made with a rainbow of carrots. We might enjoy better overall health! (Courtesy, Agricultural Research Service, USDA)

- **Stamens**—The stamens are the male reproductive parts of the flower. They are located inside the ring of petals. Each stamen has an enlarged anther at the end that produces pollen. The stalk-like structure that supports the anther is the filament.
- **Pistil**—The pistil is the female reproductive part of the flower. It is located in the center of a flower. The outer end of the pistil has an opening known as the stigma. The tube-like structure that connects the stigma with the ovary is the style. The ovary is the enlarged base where the ovules (egg cells) are located. Fertilization occurs when pollen grains come into contact with the stigma and are moved through the style to the ovules. The ovules develop into seed after being fertilized.

Incomplete flowers are flowers that do not have all four principal parts. Some flowers do not have sepals and petals, such as the flowers on wheat and oats. Fertilization can occur without sepals and petals. However, flowers must have ovules and anthers for seed to form. In some cases, the male and female flower parts are separate on plants. Both are needed for a seed to be formed.

Perfect or Imperfect

A ***perfect flower*** is a flower that has both the stamens and the pistil. In contrast, an ***imperfect flower*** is a flower that lacks either the stamens or the pistil. Some imperfect flowers have the male sex organs; other imperfect flowers are female.

Plants may have both male and female imperfect flowers on them. When they do, they are known as monoecious. Corn is a monoecious plant. On corn, the male flower is the tassel at the top of the plant. The tassel produces pollen. The female flower forms where the ear grows on the plant. The female corn flower is a small shoot with silks at the end. Usually, one to three shoots develop on a corn stalk where leaves join the stalk, but only those in the middle of the stalk are involved. Other monoecious plants are squash, melons, and pumpkins.

13–14. Corn tassels (male flower) produce pollen that falls onto silks and shoots (female flower) where fertilization occurs.

Flowers on these plants are easily observed as male or female. The female bud and flower has a small structure that resembles a fruit. The male flower does not.

Some species of plants are dioecious—some of the plants have male flowers and others have female flowers. Male plants do not bear fruit. This is why dioecious plants are often propagated by vegetative means. People who grow crops must have plants that bear fruit! Strawberries and asparagus are examples of dioecious plants. To be sure that a strawberry plant will produce berries, it is propagated vegetatively by using stolons (runners).

13–15. The flower on a pine tree will produce an abundance of pollen.

13–16. Gingko trees are dioecious. Only the female trees have fruit. Unfortunately, the fruit has an undesirable odor. Select male gingko trees for a landscape around your home.

Combinations

Flowers can be combinations of complete and incomplete and perfect and imperfect. For example, wheat is incomplete and perfect, and cotton is complete and perfect.

POLLINATION

Pollination is the transfer of pollen from an anther to the stigma of a flower of the same species. Wind, insects, birds, and other means may move pollen. Insects and birds distribute pollen because it sticks to their bodies. As they travel from flower to flower, some pollen rubs off. In plant breeding, agriscientists may move the pollen to guarantee that the plant is properly fertilized.

Some plants have attractive, brightly colored flowers. This is nature's way of attracting insects to help with pollination.

Flowers may be cross-pollinated or self-pollinated. Cross-pollination is the transfer of pollen from an anther on one plant to the stigma on another. Self-pollination is the transfer of pollen from one flower to another flower on the same plant.

A tassel on a corn plant may make as many as 15,000,000 pollen grains. Only one grain is needed for the formation of a kernel in an ear of corn. In a field of corn, the plants may self-pollinate and cross-pollinate. Research has found that no more than 5 percent of corn kernels are self-pollinated. Most of the kernels on a corn plant are the result of different parent plants.

FERTILIZATION

Fertilization is the union of the pollen cell with the ovule (egg). After the pollen grain lands on the stigma, it grows a long tube cell through the style toward the ovule. A pollen grain develops two sperm. One of the sperm unites with the ovule in the ovary and forms an embryo. The other sperm forms tissue, known as endosperm, in the embryo sac. This is sometimes referred to as double fertilization, which is unique to flowering plants.

SEED FORMATION

Fertilization initiates the growth of fruit and seed. A flower that is not fertilized will not produce seed.

A **fruit** is a matured ovary. In some cases, the ovary grows very large and has seed inside. Examples include watermelons, peppers, and tomatoes.

Once an ovary has been fertilized, the flower is no longer useful. It may dry up and fall to the ground. For example, the flower on a squash plant may last only one or two days. It quickly dries up and may stick on the end of the developing squash.

AgriScience Connection

SUGAR FROM A BEET

We like to eat sweetened foods, such as desserts. To have sweet foods, we must have sweeteners. Sweeteners are from several sources, with cane and beet sugar being widely used. About 30 percent of Earth's sugar comes from beets. A little more than a million acres of sugar beets are planted each year in the United States.

The sugar beet (*Beta vulgaris*) is a hardy biennial plant. In its first year of growth, a sugar beet plant grown from seed will produce a large root (beet) weighing 2–5 pounds and containing 15–20 percent sucrose (sugar). The beets are produced as row crops in large fields. They are mechanically harvested in the fall and are transported for processing.

This shows a nearly mature sugar beet being checked for growth and conformation. You will note a fairly large amount of soil is adhering to the beet. A genetically-modified sugar beet that is smooth and resists soil adherence was approved for planting in 2008. (Courtesy, Agricultural Research Service, USDA)

The seeds formed are products of their parents. All parts of a seed must develop if the seed is to be planted. Seeds must remain on the parent plant until mature. If fruit is pulled too soon, the immature seeds may not grow.

13–17. The apple is a fleshy fruit with seed in the middle.

KINDS OF FRUIT

Fruit varies with the species of plant. In general, all fruit is fleshy or dry.

Fleshy fruit is fruit that has large fibrous structures surrounding the seed. Examples include apples, pears, blackberries, oranges, and grapes.

Dry fruit is fruit that develops as a pod or in a hull. Pod fruits grow in structures that have lines or places where the fruit has definite seams. They can usually be divided into halves. Beans, peas, peanuts, and cotton have pods. Hull fruits do not have definite seams in the shell of the fruit. Hull fruits include pecans, acorns, corn, oats, wheat, elm tree seeds, and the dandelion.

The fruit that develops from the ovary of a plant is important for food and other uses. A goal is to produce more and larger fruit.

13–18. Corn is a dry, hull fruit. This ear of dry shoe peg corn illustrates the arrangement of kernels on the cob.

13–19. The snap bean is a dry fruit. The beans form in pods with definite seams.

13–20. Examples of fruit.

SEED DISPERSAL

Seeds need to be dispersed (scattered) over the land to grow new plants. In nature, seeds are scattered in many ways. Birds may remove a plum from a tree and carry it a few hundred yards before the seed is dropped. Wild animals may carry fruit a longer distance before it is eaten and the seeds are discarded. Some seeds may float in the air.

As plants were domesticated, humans developed systems of dispersing and planting seeds. The methods used were to grow and produce large crops.

SEED GERMINATION

Germination is the sprouting of a seed that will grow into a new plant. A seed must be mature to germinate. Some seeds will germinate soon after full size is reached in the fruit. Other seeds go through a time of rest, known as dormancy. With proper storage, seeds can be kept viable for years.

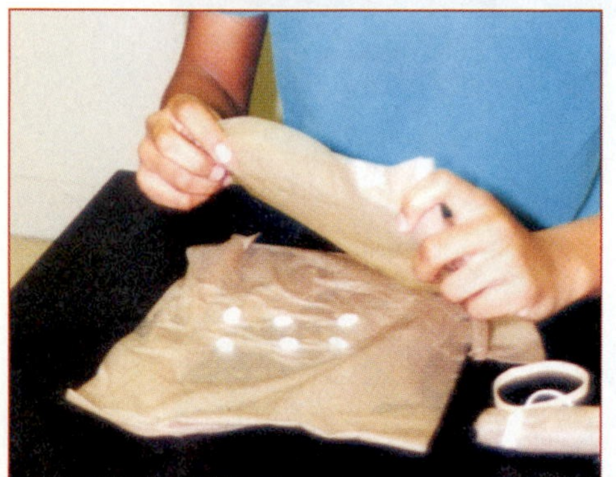

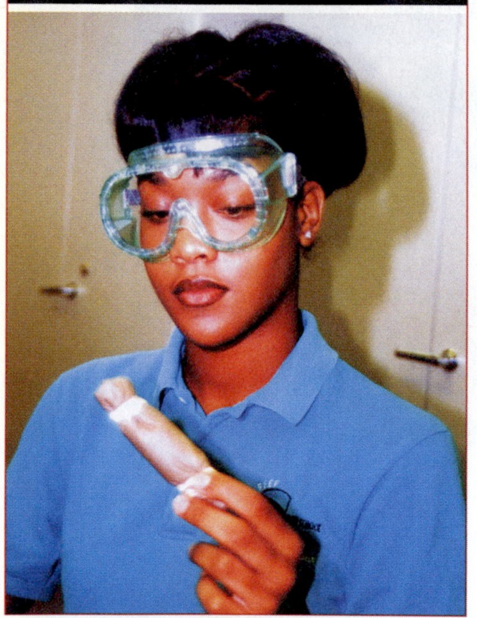

13–21. A seed sample is being placed (top) and wrapped (bottom) in moistened paper. After the sample has been in a warm place for several days, sprouts will be counted and percentages calculated for the lot from which the seeds were selected. (This is known as the "rag doll test.")

VIABILITY

Viability is the ability of a seed to sprout and grow. When a seed is viable, it germinates, grows rapidly, and has the potential to produce a good yield.

To germinate, a seed must be in an environment that is appropriate for the species. Seeds of different species sprout under different conditions. Some seeds need warmer temperatures than others. Some need more moisture and must be protected from sunlight.

Seed Testing

Seeds should be tested for viability. Seeds that are sold have labels indicating the percent of seeds that will sprout and grow—***percent germination***. If a label says that the seeds have 90 percent germination, 90 out of every 100 seeds should sprout and grow. Of course, the seeds must be properly planted!

Germination testing is done by sprouting a sample of seeds. For example, 100 seeds may be taken from a sack of corn and artificially sprouted. This may involve wrapping the seeds in moist paper towels and keeping them at a warm temperature, such as 85°F, for several days. After a few days, the number of seeds with sprouts is counted and divided by 100 (the number of seeds in the test). If very few seeds sprout, the seeds have low viability.

Percent germination is important because it influences the number of seeds to plant. If 20,000 plants are to grow on each acre of land, more than 20,000 seeds must be planted. If percent germination is 90 percent, 10 percent more seeds should be planted. In this case, the number of seeds to plant per acre is 22,223 (22,223 × 0.90 = 20,000). Of course, the seeds must be properly planted, and the weather must be good.

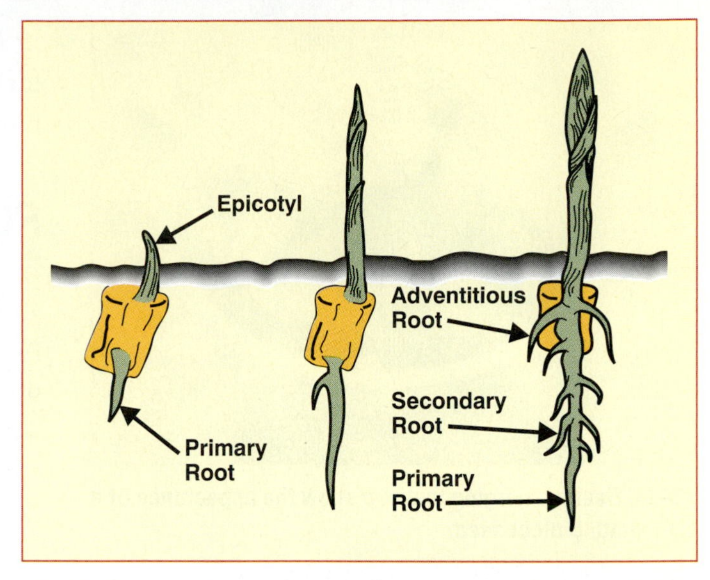

13–22. The sprouting and emergence of corn—a monocot plant.

Viability is related to the amount of food stored in the seeds and how the seeds are handled before planting. Seeds that are fully developed as opposed to shriveled usually have enough food. Seeds should be stored in cool, dry places and protected from injury by insects, rodents, and rough handling. Dropping a bag of seeds injures them internally and may cause them not to grow.

Germination Needs

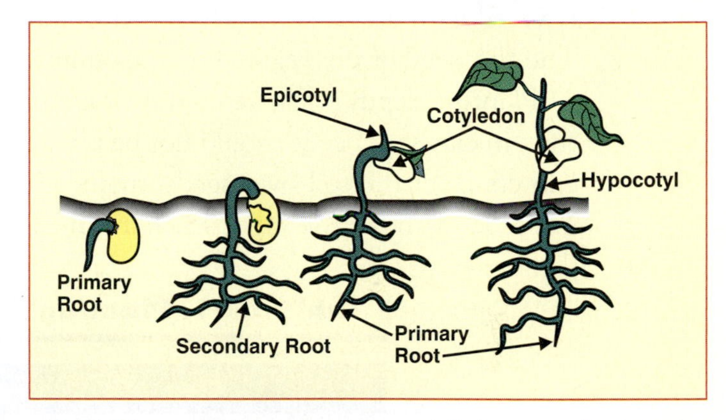

13–23. The sprouting and emergence of a bean—a dicot plant.

Seeds are planted in a medium that promotes germination and growth. Water and other nutrients should be present. Soil is the most widely used growing medium.

Seeds grown in greenhouses and other structures may be planted in "artificial" media. The media may be made of natural materials, such as vermiculite, peat, and sphagnum moss. Soil is sometimes mixed with media to form the best medium for a species.

Most seeds require a range of three conditions for germination: moisture, temperature, and oxygen. Some seeds require light, but most of those used in agriculture do not. Light can damage some emerging sprouts.

Moisture is needed for germination. Seeds usually absorb considerable water when planted. The water softens the seed coat and causes the seeds to swell. Too much moisture can damage the seeds. Bacteria, fungi, and other organisms can attack the seeds. Wet seeds will rot and will not sprout, particularly in cold or cool weather.

Proper temperature promotes germination. Seeds will not sprout if the soil is too cool. Some species require warmer soil than others. Corn needs soil that is at least 50°F, but many crops need a soil temperature of 70°F or higher. Seeds planted in greenhouses may be placed in a growing medium that is heated to promote germination.

13–24. Beans emerging in a cup show the appearance of a germinating dicot seed.

Oxygen is needed for the embryo to live and for the sprout to grow. Soil that is too wet may not have much oxygen present.

PLANTING DEPTH

Planting is covering seeds with soil or other media. The layer of soil incubates the seeds. Conditions should encourage the seeds to sprout. Temperature, moisture, and available oxygen vary by depth in the soil. ***Planting depth*** is the depth at which seeds are planted.

Seeds of different species have different planting depths. A general rule is to plant a seed at a depth eight times the thickness of the seed at its thinnest place. Very small seeds may not be covered with any soil. Merely applying water will be sufficient for the seeds to germinate. The tiny seed of the begonia is an example.

Planting depth is also related to soil type. Seeds are planted a little deeper in sandy soil than in clay soil. Seeds should not be planted quite as deep if rain is expected. A heavy rain can compact the soil over seeds, making it hard for new plants to push through the soil. Table 13–1 shows the suggested planting depths for some crops.

Table 13–1. Planting Depth of Selected Crops*

Crop	Planting Depth (Inches)
Beet	1
Carrot	½
Corn	2
Cucumber	1
Peas	2
Potato	4
Radish	½
Snap beans	2
Spinach	1
Tomato	½
Watermelon	11
Wheat	11

*Seed should be planted a little deeper in sandy, dry soil and not quite as deep in heavy, moist soil.

VEGETATIVE PROPAGATION

Vegetative propagation is a method of reproduction in which a plant part is grown into another plant. It is cloning that uses leaves, stems, or roots to grow the new plant. In some cases, parts of desired plants are transferred to less desirable plants. Regardless, the genetic makeup of the new plant is identical to its one parent. Opportunities for genetic improvement do not exist with vegetative propagation. Consequently, selecting only superior plants for vegetative propagation results in the desired products.

13–25. A Douglas Fir has been grafted to form a decorative shape at a Washington tree nursery.

ADVANTAGES OF VEGETATIVE PROPAGATION

Vegetative propagation helps ensure quality crops and ornamental plants. Two major benefits are:

- **True traits of parents**—Seeds produce plants that may vary from the parents. This is known as not breeding true. Variation occurs when sexual reproduction is used. Many fruit trees, such as apple trees, have been carefully selected for large, delicious fruit. If a seed is planted, the plant that grows might not produce the same high-quality fruit, but vegetative propagation ensures that the new plant is identical to its parent.
- **No seed**—Some plants do not produce seeds. The seedless grape is an example. Other plants are vegetatively propagated because they have seeds that are very small or difficult to get.

DISADVANTAGES OF VEGETATIVE PROPAGATION

Some disadvantages of vegetative propagation are:

- **Lack of genetic improvement**—Without the introduction of improved genetics, as would occur with sexual reproduction, the new plants will have the same genetics as the parent.

13–26. Ferns reproduce both sexually and asexually. This shows fern leaves with dark structures on the undersides. Spores are formed in these structures, which is asexual reproduction. (Courtesy, Education Images)

13–27. A sweet potato has sprouted and has grown a sizeable top. (Courtesy, Education Images)

- **Lack of genetic diversity**—Asexual reproduction does not introduce new genetic material to a plant species or variety. The traits are those of the one parent. No opportunity exists for expansion of the genetic heredity.
- **Retained weaknesses**—If the parent plant has a genetic fault, it will be transferred to the new plant.

13–28. "Eye" cut from a potato tuber that is ready to plant.

METHODS OF VEGETATIVE PROPAGATION

Vegetative propagation varies with the plant species. Some plants are easy to propagate with buds, stems, or other parts; other plants are difficult to propogate. Some plants cannot be vegetatively propagated with success.

Plant parts located above and below the ground can be used in vegetative propagation.

Below-Ground Parts

Some plants grow stems and root structures below the ground that are used for vegetative propagation.

Plant parts with buds may be cut into pieces for propagation. For example, a potato can be cut into sections with buds (known as eyes) and planted. Each eye will develop into a potato plant. Sometimes small potatoes are planted whole.

With some species, the parts are used without cutting. Bulbs and corms are two examples. Some plants grow a number of bulbs or corms. This is sometimes called multiplying. The parent plant can be dug and the new bulbs pulled apart and planted separately. The gladiola is a good example. Glads have corms. In propagation, corms are dug, separated, and planted where desired.

Above-Ground Parts

Many plants are propagated using buds, stems, and other structures that grow above the ground. The major methods are layering; using cuttings, buds, and grafts; and using tissue culture.

Layering. ***Layering*** is a method of propagation in which roots grow from the nodes on the stem of a plant while the stem is still attached to the parent plant. Layering is accomplished in several ways.

13–29. Air layering is being used to propagate a plant. After several weeks of being in a moist growing medium, wrapped in plastic, and covered with foil, roots have developed. These photographs show the wrapping being removed, the roots being observed, and the new plant being cut from the parent and set in a container.

Soil may be pulled up around a stem to get it to grow roots. A gardenia may be layered by using a low limb on a gardenia plant. The limb is bent down to the ground and partially covered with soil. The covered part will grow roots in a few weeks. After roots have developed, the limb can be cut from the parent plant. The limb will have enough roots to grow on its own in another location.

Air layering is placing a growing medium around a stem that includes a bud or a node. Sometimes a cut is made in the stem, and it is covered with a medium known as sphagnum moss. Aluminum foil or plastic may be wrapped around the moss to hold moisture. After roots develop, the covering is removed, and the top part is cut off just below where the plant was layered. This new plant is set out to grow on its own. It is free of support from the parent plant.

Cuttings. A ***cutting*** is a short section of plant stem used for propagation. The section of stem is cut from the parent plant and is put in a growing medium to sprout roots. Once the roots have developed, the plant can be set out and grown to maturity. Many plants can be grown from cuttings. Roses, begonias, pineapples, and coleus are examples.

13–30. ***Camellia japonica*** **cuttings are beginning to develop roots and grow into new plants. (Courtesy, Education Images)**

13–31. A cutting is being set in a basket. (The medium needs to be initially saturated with water to promote rooting.) (Courtesy, Education Images)

Budding. ***Budding*** is a method of propagation in which a bud is taken from one plant and moved to another plant. The bud is removed from a parent that has desired traits. It is placed on another plant (stock) to grow. If the layers of the stem match, the bud will grow.

13–32. The cleft graft is a popular way of grafting young trees.

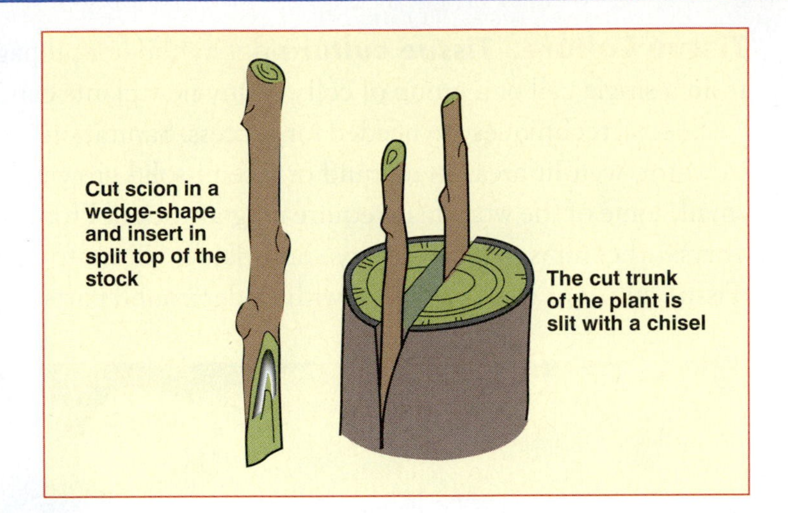
Cut scion in a wedge-shape and insert in split top of the stock
The cut trunk of the plant is slit with a chisel

Of course, buds can only be placed on stocks that are compatible. Examples of plants that can be propagated by budding include fruit trees, roses, and nut trees.

Grafting. ***Grafting*** is a method of propagation in which a section of a stem (known as a ***scion***) of one plant is placed onto another plant. The scion and stock must be compatible.

In the grafting process, a scion is attached so it naturally grows on the stock. The part that grows from the scion will have the traits of its parent. Any growth on the stock will have the traits of the stock. With pecans, small sections of stem from the desired pecan tree are placed on small trees grown from seed (seedlings). The seedlings would not produce large pecans, but they are excellent stocks for the scions. Grafting requires matching the cambium (growing part) of the stock and the scion. If done properly, the two parts will grow together as one plant.

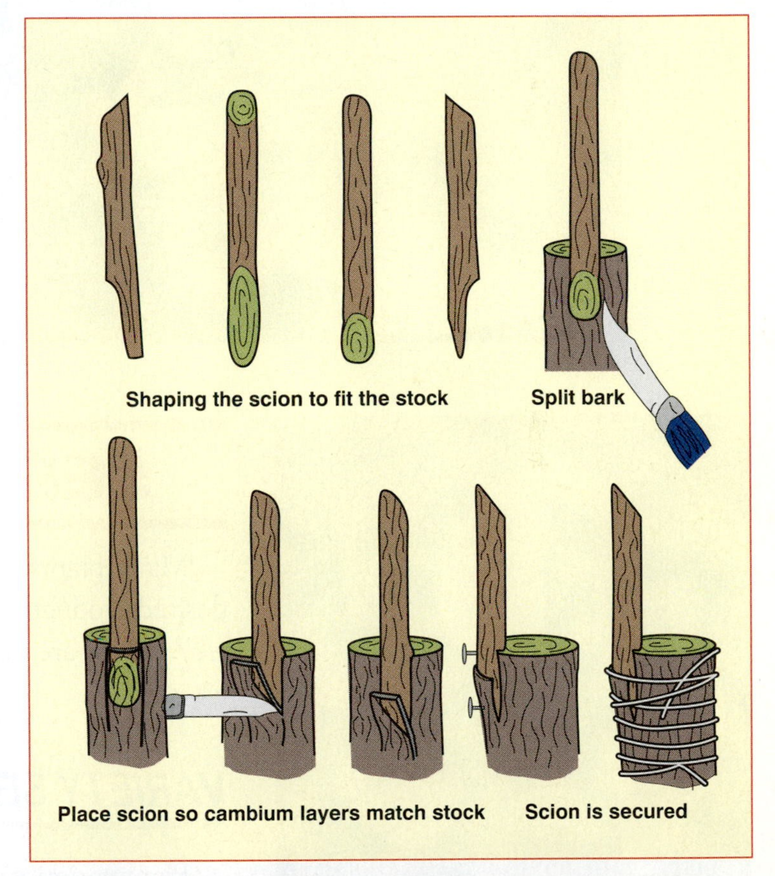
Shaping the scion to fit the stock
Split bark
Place scion so cambium layers match stock
Scion is secured

13–33. The process of making a bark graft involves several steps. The scion is prepared by shaping it to fit the stock, the stock bark is split, the scion is placed so that cambium layers match with the stock, and the scion is secured in position to begin growing.

Tissue Culture. ***Tissue culture*** is a method of propagation in which a plant is grown from a single cell or a group of cells. Many new plants can be grown from one parent.

Special techniques are needed for success. Sanitation is essential. The cultures are placed in warm, well-lit areas in a liquid or a semi-solid growing medium. Since cultures are very small, some of the work may require magnification. Producers of ornamental plants, such as chrysanthemums and carnations, find tissue culture to be a good way to propagate plants. Tissue culture can also be used with underground parts.

13–34. Tissue culturing is being done in a sterile environment to prevent contamination. (Courtesy, Education Images)

13–35. Cotton seeds have lint attached after ginning. The seeds are often chemically cleaned to remove the lint. Seeds with lint do not work well in planting equipment.

SEED SELECTION

Many plants have been selected and improved to gain desired products. The goal is to have plant varieties that are well suited, are productive, and meet the needs of consumers.

VARIETY SELECTION

Plant varieties are suited to a particular location or a particular use, so it is best to obtain information on the varieties to plant. For example, varieties of corn that grow well in Montana might not do well in Texas.

Varieties suited to the climate should be selected. Most local seed stores sell only the varieties recommended for the area. Ordering seed from companies that are located far away might result in varieties that are not suited to the area. If unsure of the variety to purchase, contact an agronomist at a college or a research station.

Seed Certification

Seed certification is a method of classification that ensures that the seed is of good quality. Seed certification is primarily used with major agricultural crops, such as wheat, soybeans, and corn.

There are four classes of certified seed:

- **Breeders seed**—This seed is used to breed or maintain a crop variety. Only plant breeders usually have breeders seed, as it is not sold to the public. The seed is placed in a container labeled with a white tag and the words "Breeders Seed."
- **Foundation seed**—This seed is produced from breeders seed. The seed is carefully grown to maintain the genetic traits established by the breeder. The container for this seed is usually labeled with a white tag and the words "Foundation Seed." The tag is issued by the seed certification association.
- **Registered seed**—This seed is produced from foundation seed. It is carefully grown for purity of the variety. The container has a purple tag that provides information about the seed.
- **Certified seed**—This is the most widely used class of seed. It is produced from registered seed. Seed producers follow high standards to ensure good seed. A purple label is used for certified seed.

ACADEMIC CONNECTION

BOTANY

Relate the content of this chapter to your study of botany. Place the emphasis on plant sexual and asexual reproduction. Identify five native plants in your area, and investigate the method of reproduction that occurs in nature. Prepare a brief report on your findings.

13–36. Quality seeds are pure and free of weed seeds, trash, and disease. The seed coats have no cracks or breaks. (Courtesy, American Soybean Association, St. Louis, Missouri)

QUALITY SEEDS

Good seeds, when planted properly, come up, grow, and become the intended plants. Avoid using seeds of mixed varieties. Also, avoid those that are cracked or diseased or that otherwise lack the ability to grow and produce desired crops. Always consider seed purity.

Some people grow plants and save their own seeds. Other people buy the seeds they plant. In either case, quality is a must.

Traits to look for in good seeds are:

- **Purity**—Seeds should be pure (contain only the seeds of the intended variety). Seeds that have other varieties of seeds mixed in are impure.
- **Contamination**—Seeds should be free of trash, such as pieces of leaves, pods, and stalks; wood pieces; and stones. Most seeds are cleaned to remove trash.
- **Percent germination**—Seeds should come up and grow. Only seed with 90 to 100 percent germination should be selected.

APM PERENNIAL RYEGRASS

LOT # B29-3-36APM

	PURITY	GERMINATION	ORIGIN
PURE SEED	97.20%	90%	OR.
OTHER CROP SEED	2.36%		
INERT MATTER	0.44%		
WEED SEED	0.00%		

NOXIOUS WEED SEEDS: NONE

TEST DATE 05-09

NET WT. 50 LBS.

PMA 185

MEDALIST AMERICA
1490 INDUSTRIAL WAY S.W., ALBANY, OR., 97321

NOTICE READ CAREFULLY BEFORE OPENING

1. NOTICE OF REQUIRED ARBITRATION
Under the laws of some states (including Idaho), arbitration is required as a pre-condition of maintaining certain legal actions, counterclaims or defenses against a seller of seed. The buyer must file a complaint along with the filing fee within such time as to permit inspection of the crops, plants or trees and notify seller of complaint by certified mail. Information about this requirement, where applicable, may be obtained from a State's chief agricultural official.
2. NOTICE OF EXCLUSION OF WARRANTIES AND LIMITATION OF DAMAGES AND REMEDY.
The labeler warrants that this seed conforms to the label description, as required by federal and state seed laws. WE MAKE NO OTHER WARRANTIES, EXPRESS OR IMPLIED, OF MARKETABILITY, FITNESS FOR A PARTICULAR PURPOSE, OR OTHERWISE CONCERNING THE PERFORMANCE OF THIS SEED.
The liability for damages for any cause including, but not limited to, breach of contract or breach of warranty or negligence with respect to this sale of seed is limited to a refund of the purchase price of this seed. This remedy is exclusive. IN NO EVENT SHALL THE LABELER BE LIABLE FOR ANY INCIDENTAL OR CONSEQUENTIAL DAMAGES, INCLUDING LOSS OF PROFITS.

13–37. Sample seed label.

- **Uniformity**—Seeds should be of uniform size and shape. Small seeds or those that are not shaped properly may not germinate. Buying them is a waste of money.
- **Damage**—Seed coats should not be cracked or broken. Seeds that have been damaged may not come up. Rough handling in harvesting, cleaning, and moving seeds causes damage. Broken seeds are a waste of money.
- **Disease**—Seeds should be free of disease. Seeds that contain disease will not produce well and can infect an entire field with a disease.
- **Treatment**—Some seeds are treated to prevent disease. These seeds should be carefully handled in planting and should never be fed to animals because the chemicals could injure them.
- **Name of grower and/or dealer**—Most seed growers and dealers are honest and reputable. They produce only quality seeds. Avoid using seeds from unknown sources or sources known to be dishonest.

REVIEWING

MAIN IDEAS

Plants are propagated (reproduced) sexually and asexually. Sexual propagation is the union of a male and a female sex cell. Asexual propagation is the use of the vegetative parts of a plant to grow another plant.

Sexual reproduction involves the formation of flowers, pollination, fertilization, fruit and seed growth, and proper planting of the seeds. Flowers have male and female parts. In some cases, separate flowers are male and female. Plants produced by sexual means have genetic traits of both parents. Pollen is the male sex cell produced by the anther in a flower. Pollination involves the pollen contacting the stigma—female part of the flower. The pollen develops sperm that fertilize ovules in the ovary of the flower. The fertilized ovary develops into fruit that contains seeds. Seeds are containers with living plants inside.

Germination is the sprouting of a seed. A seed must have the right conditions to germinate: moisture, temperature, and oxygen. Most seeds are planted in growing media, usually soil, that provide the needed germination environment.

Asexual propagation results in a plant identical to its parent. The new plant is a clone. Many plants can be produced from one parent plant. Both above-ground and below-ground plant parts are used in asexual propagation, depending on the species. Asexual propagation is done in several ways: layering, cuttings, budding, grafting, and tissue culturing. All methods produce a clone of the original plant.

The success of a crop can be no better than the seeds used. Growers should select seeds of the best possible quality. Seeds should be pure, be free of disease and damage, and have a high percent germination. It is best to purchase seeds only from reputable sources. All seeds should be labeled with information about their quality.

QUESTIONS

Answer the following questions, using complete sentences and correct spelling.

1. What is the purpose of plant propagation?
2. Distinguish between the two major ways in which plants are propagated. Give advantages and disadvantages of each.
3. What is a seed? Sketch a seed, and label its parts.
4. What is the purpose of flowers?
5. What are the parts of a complete flower? Sketch a flower, and label its parts.
6. What is pollination? How does it occur?
7. What is fertilization?
8. How is seed formed?
9. What kinds of fruit are produced by plants?
10. What is germination? Why is it important?
11. What does a seed need to germinate?
12. Why is vegetative propagation used?
13. What parts of plants can be vegetatively propagated?
14. What methods of vegetative propagation are used?
15. Why should quality seed be used? What determines seed quality?

EVALUATING

Match each term with its correct definition.

a. propagation
b. seed
c. flower
d. fruit
e. pollination
f. germination
g. viability
h. percent germination
i. planting depth
j. layering

______1. The depth at which seeds are planted.

______2. The reproduction or increase in the number of plants.

______3. A method of propagating in which roots grow from the stem of a plant while the stem is still attached to the parent plant.

______4. A container of new plant life.

______5. The reproductive part of a flowering plant.

______6. The transfer of pollen from an anther to a stigma.

______7. A matured plant ovary that has been fertilized to produce seed.

______8. The sprouting of a seed.

______9. The percent of seeds that will sprout and grow.

______10. The ability of a seed to sprout and grow.

EXPLORING

1. Visit a place where seeds are sold (e.g., farm supplies store, garden center, or the display rack in a supermarket). Read the labels on the seed containers. Determine the stated percent germination, purity, and other characteristics.

2. Conduct a germination test of seeds. Use corn, beans, peas, or another crop with large seeds. Place the counted number of seeds (usually 25) on a moist paper towel on a tray and cover it with three or four layers of moist towel. Place the towels with the seeds in a warm location. Keep the towels moist. Check the seeds in three to five days and note any swelling or sprouting. The amount of time varies with the species of plant, but a week is usually long enough for common seeds. After a week, count the number of seeds that sprouted, and use that number to obtain a percent of germination for your seed test.

3. Use the eye on a potato to grow another potato plant. The eye should be planted 2 to 4 inches deep in soil and kept moist and warm. The potato eye will begin to grow in a few days and should come up in two to three weeks.

4. Arrange to tour a seed processing facility. Note how the seeds are cleaned and protected from injury. Prepare a report on your observations.

CHAPTER 14

Plant Growth and Culture

OBJECTIVES

This chapter covers how plants grow. It has the following objectives:

1. Describe how plants grow and the conditions needed for growth.
2. Explain why plants grow.
3. Explain photosynthesis and why it is important.
4. Explain transpiration and why it is important.
5. Name the nutrients plants need, and describe how plants get them.
6. Describe nitrogen fixation in legumes.
7. Explain the use of fertilizer.
8. Identify general practices in field crop production.
9. Discuss the use of plant growth structures for specialty crops.

TERMS

annual growth ring
auxin
chlorophyll
cultural practice
fertilizer
fertilizer analysis
gibberellic acid
glucose
glycolysis
greenhouse
heartwood
legume
major elements
mulch
nitrogen
no-till cropping
phosphorus
photosynthesis
phytohormone
potassium
primary growth
sapwood
secondary growth
soil analysis
tissue analysis
trace elements
wilting

14–1. Watering is essential in producing plants in a greenhouse. (Courtesy, Education Images)

PRODUCING PLANTS when, where, and how we want them requires more skills than some people think. We need to know how plants grow and what can be done to make them grow better. Agriscience can supply the answers.

Plants grow wild in nature; the forest is filled with them! The plants in native forests grow without much attention to spacing, to the season, or to how fast they are growing. But people are impatient; most people cannot wait for nature to take care of matters. Producers want crops to grow quickly and bear abundantly. They also want top-quality products.

How do we know what to grow and how to grow it? The kinds of plant products produced and how the products are manufactured reflect consumer demand. How we grow a plant requires that we understand plant growth and that we understand how to meet plant needs in production.

HOW PLANTS GROW

The tiny new plant that emerges from a seed usually gets off to a good start. It survives on the food stored in the seed until it can establish its roots and send leaves through the soil's surface into the air to receive sunlight. Once the plant reaches this stage, it can make its own food.

14–2. Gaining a high volume of quality plant product is essential. This field of alfalfa is being cut and windrowed for use as forage. (Courtesy, AGCO, Duluth, Georgia)

14–3. This tiny oak seedling still has the acorn attached.

Plants are composed of many tiny cells. Growth is the increase in the number of cells in a plant. The size, shape, and function of cells vary. The cells in a plant all work together to form tissues and organs. This makes it possible for plants to live and to grow.

KINDS OF GROWTH

Plants have two kinds of growth: primary and secondary. Both are important when growing plants that will produce good crops.

Primary Growth

Primary growth is the increase in the length of a plant—the plant becomes taller. Stems and roots get longer, and leaves and flowers grow. Primary growth is sometimes known as linear growth. Growth is the result of cell division, which is the process (meiosis or mitosis) by which a cell divides.

New cells grow at the tips of stems and roots in the apical meristem. The meristem is the plant tissue capable of cell division. The buds on the ends of stems are made of meristem tissue. Spurts of meristem activity occur in the spring. Cells elongate and divide rapidly. The stem tips of trees may grow several inches in a few days.

Primary growth is important when plants need to grow rapidly and bear fruit or to be used for products.

14–4. A bud on a chestnut tree will grow rapidly and will increase the size of the tree several inches per year.

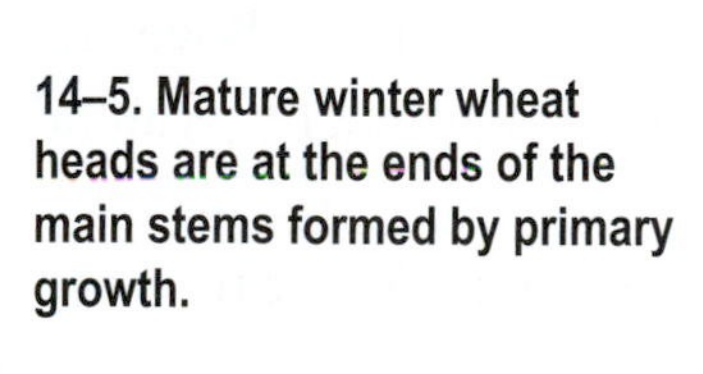

14–5. Mature winter wheat heads are at the ends of the main stems formed by primary growth.

Secondary Growth

Secondary growth is the increase in the diameter of stems and roots. Tree trunks increase in size due to secondary growth. Secondary growth is quite useful in producing logs, pulpwood, and other products.

New cells grow in meristem tissue known as lateral meristem. Much of this growth is in the cambium layer of a plant. Some of these cells become xylem; others become phloem. All plants tend to grow alike. Of course, annuals do all of their growing in one year.

14–6. The annual growth rings are easy to see. How old is the tree? (Count the number of dark circles.)

In trees, the phloem is near the bark, and the cambium lies between the phloem and the xylem. The xylem is toward the inside of the tree and is often called wood. Each year another layer of cells is added. The cells form rings that are easy to see when a tree is cut across.

An ***annual growth ring*** is the result of the secondary growth in a plant's lateral meristem. Typically, one ring is the xylem a tree grows in a year. The width of rings tells how rapidly the tree was growing. Narrow rings indicate that the tree was growing slowly. Wide rings indicate fast growth. Under good conditions, pine trees may have layers of growth that are one-half inch or more each year.

To estimate the age of a tree, count the number of rings. A tree usually grows one ring per year; therefore, a tree with 25 rings would be approximately 25 years old. To take a sample of the annual growth rings, use an increment borer. If a tree has fallen, make a cut across the trunk to count the annual rings. Remember that the tree most likely grew several years before reaching the height where the count was made. If the count was made 5 feet above the ground, the tree was most likely at least five years old when it reached that height. Annual rings are typically good indicators of age.

Rings have light and dark colors. The light-colored rings were grown in the spring when growth was rapid and cells were larger. The darker color is a sign of smaller cells that grew in the summer.

14–7. A sample of annual rings is being removed to determine the rate of growth and age of the tree. The tool used to obtain the sample is an increment borer. The tree will soon heal where the sample was taken. (Courtesy, Education Images)

The innermost xylem of a tree is the ***heartwood***. In pine trees, the heartwood has rosin and resists decay better than the sapwood. You can readily see the heartwood of a cross section of a log. Heartwood has qualities that help it resist decay and promote long-term weathering. Older heartwood may be more susceptible to fire.

Sapwood is the xylem that is two to four years old. The color of sapwood is lighter than heartwood and may even be white. Sapwood lumber has a low resistance to rot. Therefore, it should only be used to build structures that are protected from the weather. Heartwood lumber resists rot much better.

14–8. This wood sample was taken with an increment borer (shown in Figure 14–7). The width of the rings shows the rate of growth. The number of rings is the age of the tree.

GROWTH PROCESSES

As organisms, plants carry out many complex processes that make it possible for cells to grow and divide. These processes keep the plant healthy and growing.

The processes include absorbing water and nutrients, moving water and other materials in the vascular system of the plant (circulation), making food through the process of photosynthesis, converting food to tissue by the assimilation process, taking in and giving off carbon dioxide and oxygen through respiration, and adapting to the environment through transpiration and reproduction.

Plants are indeterminate in their growth, so they grow throughout their lifetimes. Animals are determinant in their growth; they reach a certain size at maturity. Before a new animal is born or hatched, we have an image of the mature size of the new animal in the adult form.

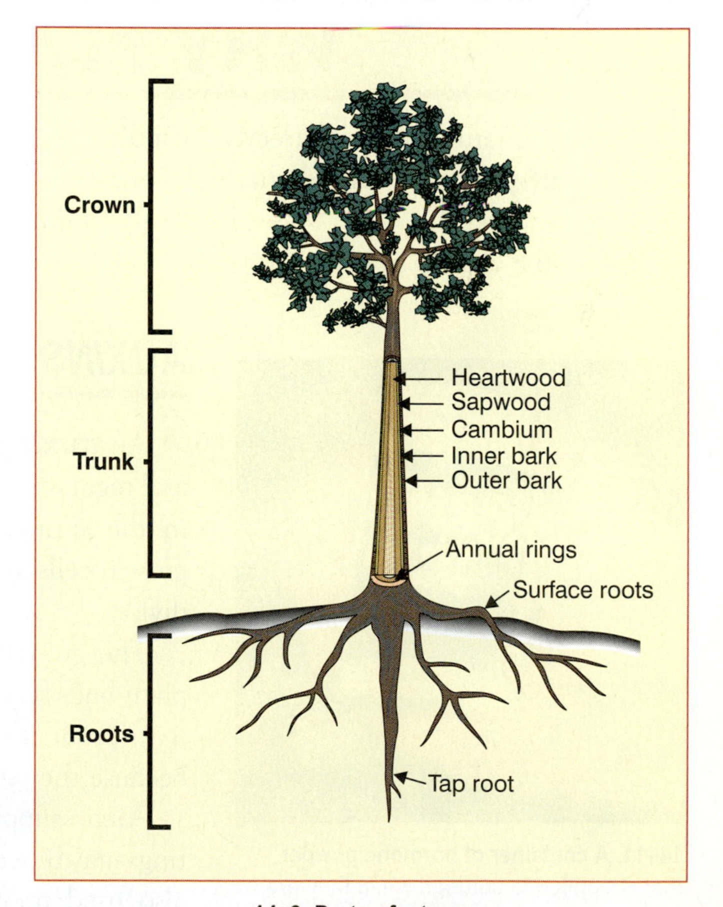

14–9. Parts of a tree.

14–10. Observing the growth of a tree in a container. (Courtesy, Education Images)

WHY PLANTS GROW

Plant growth is directly, or indirectly, caused by phytohormones or plant hormones. A **phytohormone** is a naturally occurring hormone that activates or regulates plant growth and development. Plants need several phytohormones. Even very small amounts can result in a change in plant growth.

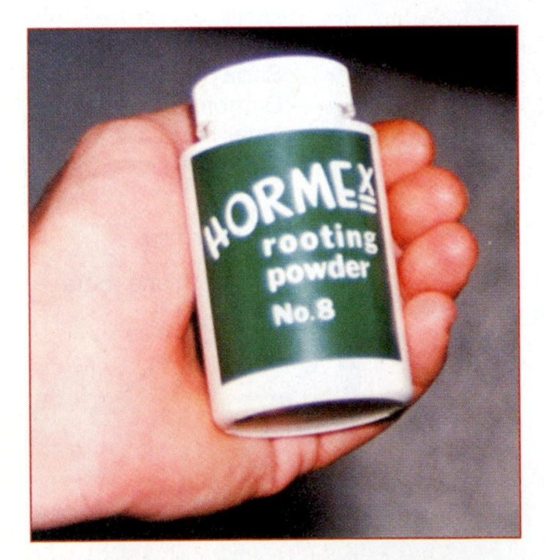

14–11. A container of hormone powder that is applied to cuttings when they are being rooted.

AUXINS

An **auxin** is a plant hormone that controls stem growth and regulates fruit development. It causes plant growth spurts in the spring. Auxin promotes the elongation of primary growth cells in stems and roots and causes cells to stretch and divide.

The growth cycle of plants is caused by auxin. Swelling plant buds send auxin to the cambium in the spring and cause it to produce xylem cells. In the fall, leaves fall from trees because they stop making auxin.

Auxin supplements are sometimes used. For example, cuttings may be treated with auxin to encourage rooting. Auxin is also used to regulate fruit setting on plants. Auxin-based materials are used to produce weed killers, known as herbicides. These are made from artificial auxins.

GIBBERELLINS

Gibberellins, or **gibberellic acid**, is a plant hormone that induces stem cell elongation and cell division.

Gibberellins are as important in plant growth as auxins. They affect stem and leaf growth, fruit development, flowering, cell division, and other plant activities. Gibberellins are closely associated with organ enlargement and stem, fruit, and leaf expansion.

Gibberellins cause some species of plants to flower. Spraying gibberellic acid on some plants will cause them to produce flowers at times of the year when they would not do so naturally. This use is important to growers of ornamental flowers.

CYTOKININS

Cytokinins are hormones that promote cell division and other functions in plants, as do the auxins and gibberellins.

AgriScience Connection

WHEN NATURE RUNS LOW

Sometimes the soil does not have the moisture needed by plants to grow. Producers help nature out with irrigation—the artificial application of water to the soil. Irrigation provides needed moisture.

How plants are irrigated varies with the kind of plant and how it is grown. Some fields are sprinkled, others are flooded, and in others each plant has its own water emitter—drip irrigation. The later uses water more efficiently because only the area immediately around the roots is irrigated.

Four methods of irrigation are shown here: center-pivot sprinkler, furrow flood, individual emitter, and sprinkler irrigation. Variations in these methods are used with different crops.

Cytokinins regulate cell expansion in roots, leaves, and stems. They are often used when asexually propagating plants, particularly with tissue culture. The use of cytokinins causes the cells to grow more rapidly.

Cytokinins are used to encourage the development of buds and flowers on ornamental plants. For example, buds form more quickly when cytokinins are put on an African violet.

14–12. Strawberry plants growing in a hanging basket are flowering and have young berries. (Courtesy, Education Images)

ETHYLENE

Ethylene is a hormone that is a gas. It causes potato tubers to sprout, pineapples to flower, fruit to ripen, and plants to shed their leaves. It also regulates the height of small grains, such as oats and wheat.

As a gas, ethylene has been difficult to use artificially because the plants had to be in an enclosed place where the gas could not escape. However, new methods of producing ethylene have been developed. Some newer approaches involve spraying plants with liquids that contain ethylene.

ABSCISIC ACID

Abscisic acid is named for the effect it has on plants: dropping of leaves and fruit. Leaves and fruit fall from a plant because a layer of cells develops over the stem where the leaf or fruit is attached. This layer of cells is the absciss layer.

This hormone helps regulate plant functions and promotes hardiness. Hardy plants can withstand changes in their environment and can tolerate colder weather. Hardiness also refers to how easily a plant can be moved. Bedding plants may be damaged by the sun if suddenly moved from a protective greenhouse to a hot field. Plants that are not hardy are known as tender plants. Abscisic acid helps plants develop hardiness. Gradually exposing plants to a change develops hardiness.

PHOTOSYNTHESIS

Photosynthesis is the process plants use to make food. Although it is a fairly complex process, photosynthesis is easily carried on by plants under the right conditions. Sugar is the major product of photosynthesis; it provides energy for plant functions.

THE PROCESS

14–13. Photosynthesis in the leaves of a coffee plant provides food for the plant to live and grow and to develop coffee berries. (The large leaves are good solar energy collectors!)

Photosynthesis has two major phases: energy gathering and sugar making. Energy is gathered from sunlight by the ***chlorophyll***—a green-colored substance in the leaves of the plant. The light energy is changed to chemical energy. Sugar is produced when the chemical energy rearranges the combinations of carbon dioxide and water in the plant.

The sugar made by photosynthesis is known as glucose, or simple sugar. Bonding more than one glucose sugar together may make other kinds of sugars. An example is sucrose (table sugar). Some of the sugar also changes to starch.

What Plants Need

Photosynthesis occurs only in plants, and the plants must be in an environment that supports the process.

Plants need the following for photosynthesis:

- **Chlorophyll**—A green-colored substance in leaves, chlorophyll is located in bundles known as chloroplasts. Sugar is made in the chloroplasts.
- **Sunlight**—Leaves will turn toward the sun to receive more light. Chlorophyll transforms the light energy of the sun into chemical energy.

14–14. Young container-grown plants have been spaced apart to allow good conditions for photosynthesis.

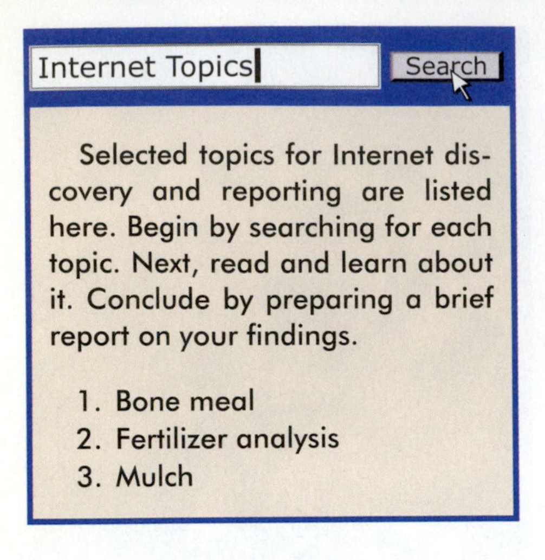

Internet Topics
Search
Selected topics for Internet discovery and reporting are listed here. Begin by searching for each topic. Next, read and learn about it. Conclude by preparing a brief report on your findings.
1. Bone meal
2. Fertilizer analysis
3. Mulch

- **Carbon dioxide**—Plants receive carbon dioxide (CO_2) from the air through stomata. Carbon dioxide is a major substance from which glucose is made. The carbon dioxide is split apart by the chemical energy in the plant.
- **Water**—Water (H_2O) is also an ingredient needed for plants to make sugar. Water is absorbed by the roots and travels to the leaves through the vascular system. Chemical energy splits water apart and partially combines it with the elements of carbon dioxide. Water also serves other roles, such as transporting materials throughout the plant.

The Equation

Photosynthesis is expressed as an equation that shows the splitting apart and recombining of ingredients. The ingredients begin as molecules and end up as different kinds of molecules. A molecule is the smallest-sized particle that a substance can have. A molecule is often made of two or more elements. For example, water (H_2O) contains two elements (the basic units of matter) of hydrogen (H) and one of oxygen (O).

The equation for photosynthesis is:

$$\text{light} + 6CO_2 + 6H_2O \xrightarrow{\text{chlorophyll}} C_6H_{12}O_6 + 6O_2$$

The equation can be explained as follows:

> Six molecules of carbon dioxide (CO_2) and six molecules of water (H_2O), plus energy from the sun, are combined. Sugar (glucose) is formed; oxygen (O_2) is released. The sugar ($C_6H_{12}O_6$) remains in the plant as food. Six molecules of oxygen (O_2) are emitted into the air through the stomata.

In the equation, chlorophyll is a catalyst (causes the process to occur). The chlorophyll is not changed in the process. It remains to continue photosynthesis.

THE PRODUCT

Photosynthesis produces two products: oxygen (O_2) and ***glucose*** (sugar) ($C_6H_{12}O_6$). Plants use or dispose of these products.

Most of the oxygen is released into the air. It replenishes the oxygen used by animals, the combustion of engines, and other activities. The plants also use some of the oxygen.

Sugar is transported to various parts of the plant and stored for future use. The form of the sugar may change. Much of it is stored as starch in the stems and roots. For example, the potato tuber is filled with starch made by the plant from sugar. Other examples are fruit and seed. Both have a lot of sugar and starch that are stored to help the young embryos live.

Plants must grow and carry out photosynthesis to produce the desired products, such as fruit and seed. Culturing plants involves trying to provide the best possible conditions for photosynthesis.

RESPIRATION

Respiration is the process of plants using the products of photosynthesis. It occurs in living cells when energy is "burned" (oxidized). The energy comes from the sugar made by photosynthesis. Waste products (carbon dioxide and water) are given off.

Plants use energy to grow and to reproduce. Energy is released through a process known as **glycolysis**. Glucose is broken down into pyruvic acid in the cytoplasm of cells. Glycolysis without oxygen is fermentation.

Respiration occurs 24 hours a day. Unlike photosynthesis, which occurs mainly in light, respiration takes place in the dark. Plants use energy all of the time.

Technology Connection

GENES AND BEANS

Soybeans are the second-largest cash crop in the United States. More than $1 billion in soybean value are lost each year due to the pest known as the soybean cyst nematode. Scientists are determining how to reduce losses.

A few soybean plants are naturally resistant to the pest. If that trait of resistance could be identified and bred or engineered into all soybeans, losses would be eliminated. Scientists are now using new marker technology to study the DNA of soybeans to identify the genes associated with natural resistance of the soybean cyst nematode. DNA stretches with simple sequence repeats have been identified in an effort to flag the whereabouts of desirable genes and their alleles. But, the work is not over as the ability to produce all-resistant soybeans has not been developed.

This shows a scientist studying the partial DNA fingerprints of 24 soybean varieties. The goal is to identify signposts associated with resistance genes. (Courtesy, Agricultural Research Service, USDA)

14–15. Lettuce in the Salinas Valley of California grows rapidly under the proper conditions.

TRANSPIRATION

Transpiration is the loss of water from a plant. Most water is lost through the leaves. Transpiration is sometimes called water evaporation. A tree may transpire as much as 80 gallons of water in one day! If the loss to transpiration is not replaced, a plant may suffer and wilt. As a result, it may fail to grow and produce as desired.

Plants must have enough water. It is used to make food, transport nutrients and food for storage, and help the plant fit into its environment.

Most of the force that moves water up the plant stems is due to transpiration. Plant roots must be able to absorb more water from the soil. The process is known as transpirational pull or turgor pressure. In addition, roots appear to have some upward pressure on the water. This is evident when a plant is cut and water comes out of the cut.

14–16. A drop of water has formed on this cut stem. Pressure within the plant is forcing the water out.

Weather affects transpiration. Warm weather results in more transpiration. Also, transpiration increases in windy weather and when the relative humidity is low. Some of these conditions can be controlled in greenhouses.

Water exits leaves through stomata. The stomata open to allow water out when the surrounding air does not provide the needed environment. If the air is humid and not too warm, the stomata do not let out much water.

Too much water loss damages plants. Every species of plant has different needs. For instance, plants that grow in deserts are adapted to the hot, dry climate. Yet plants that grow in swamps are best suited there. Even when plants are being grown in their best environment, excessive water loss can occur.

14–17. Plants in a greenhouse must receive adequate water. (Courtesy, Education Images)

WATER SHORTAGE

Shortages of water affect plants in different ways. The first sign may be **wilting**—a plant condition in which the stems and leaves are no longer rigid and erect. It may be due to excessive transpiration, low soil moisture, disease, or other factors.

Plants wilt for several reasons:

- **Dry soil**—Plants with shallow roots are especially hard hit by dry soil that results from lack of rain, failure to provide proper water, or other reasons. It is relatively easy to take care of the water needs of a few plants in a greenhouse or lawn area. Adding water to the soil (irrigation) will help the plants receive needed water.
- **Damage to root system**—Plant root systems may be damaged by plowing, transplanting (moving), or other methods. For example, deep plowing or digging near a plant can cut the roots, and damaged roots cannot absorb water as usual. All plants should be

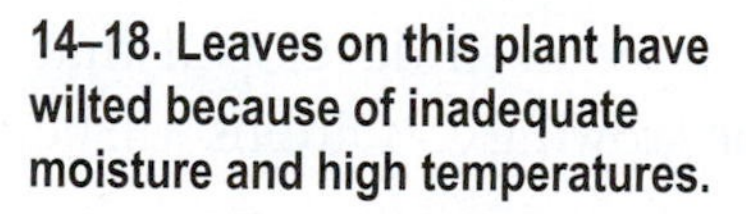

14–18. Leaves on this plant have wilted because of inadequate moisture and high temperatures.

14–19. Tillage practices should not damage plant roots. (Courtesy, Case Corporation)

given water when transplanted. When transplanting shrubs, the tops are usually cut back (pruned) so there are fewer leaves for the roots to provide with water. All plants have a balance of roots, stems, and leaves. When one is damaged, the others suffer.

- **Disease**—Plant wilting may be due to disease because disease may damage the ability of the plant to absorb water and carry on life processes. Some diseases live in the soil and interfere with the work of the roots.
- **Too much water**—The roots of plants need oxygen. When soil is too wet for a time, the oxygen is driven out. Roots suffer from oxygen depletion and cannot use moisture.
- **Too much fertilizer**—Plants that receive too much fertilizer have their roots burned. This is especially true if the fertilizer is applied too close to the roots. Concentrated fertilizer can draw water out of the roots! Plants should not be fertilized when the soil is dry.

HELPING PLANTS

To reduce plant stress through water loss:

- **Avoid damaging roots**—Plowing around plants can damage the root system. Do not plow (cultivate) too close to plants or too deeply. Improper cultivation does more harm than good. Cultivation is used to control weeds and to create a soil mulch on the land. If not controlled, weeds damage crops. They compete for water, nutrients, and space.
- **Plant drought-tolerant species**—Some species of plants have been developed that require less moisture. These species should be selected if the growing conditions will be dry.
- **Practice fallowing**—This method is used in places with small amounts of precipitation, such as western parts of North America. Crops are grown on the land every other year. The year following the growth of the crop, the land is kept clean of all plants. This allows the soil to store water for the next crop.

- **Use a no-till culture**—***No-till cropping*** is the practice of growing crops without plowing the soil. Water loss is greater in plowed soil. Plowing exposes the soil to the sun and wind, which dry up the moisture. No-till cropping also reduces erosion. The dead crop vegetation stays on the top of the soil and holds it in place. The use of no-till has increased in recent years. Minimum-till is also used. This is partially breaking land to plant a crop.
- **Mulch**—A ***mulch*** is a cover on soil to hold in the moisture. It also limits the growth of weeds. Common mulches are straw, tree bark and leaves, sheets of plastic, sawdust, and paper. Plants are set in the soil through the mulch. Mulches are used in flower beds and around shrubs. Freshly plowed soil also serves as a mulch. Fields where previous crop remains cover the ground have a natural mulch.

14–20. Pine needle mulch has been used with these pansies to reduce moisture loss, prevent erosion, and control weeds.

- **Terrace**—The land can be formed into ridges or embankments to hold back water. These structures, known as terraces, keep rainfall from running off quickly. Slowing the rate of runoff allows more water to soak into the soil.
- **Irrigate**—Irrigating is adding water to the land. Some crops, such as rice, require much water. Other crops, such as wheat, require less water. Water can be pumped from wells or reservoirs and put on the fields. Reconditioned water from sewage plants is sometimes used to irrigate areas that are not used to grow food crops. Of course, irrigation may deplete the supply of fresh water in the earth, reservoirs, or streams. Crops are irrigated by sprinkling, flooding, or dripping the water around the plants.

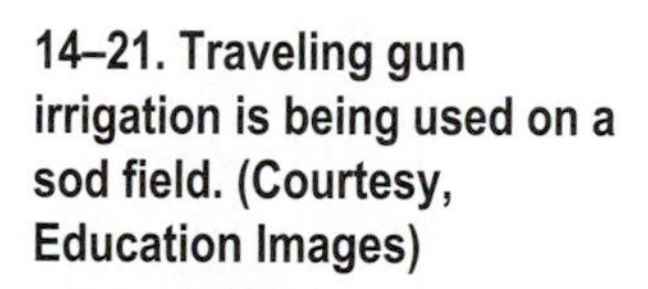

14–21. Traveling gun irrigation is being used on a sod field. (Courtesy, Education Images)

PLANT NUTRIENTS

Plants need nutrients for growth. Nutrients are the minerals (substances) in the soil that are typically absorbed with the water through the roots. The color, size, and productivity of plants are related to nutrient supply. Some nutrients are needed in large amounts; others are needed in small amounts. Nutrients are often stated as chemical elements.

Plant stems and other parts are analyzed to determine the contents. More than 95 percent of the total dry weight of plant material is composed of carbon, hydrogen, and oxygen. The remainder (approximately 5 percent) is made of other elements.

ESSENTIAL ELEMENTS

Elements needed in large amounts are ***major elements*** or macronutrients. The elements needed in small amounts are ***trace elements*** or micronutrients. Both major and trace elements are needed.

Seventeen elements are essential for plant growth. Plants use three other elements, but research has yet to determine their importance. Table 14–1 lists the essential elements and their symbols.

14–22. The label on a fertilizer container should specify nutrient content and other information. (Courtesy, Education Images)

Table 14–1. Essential Elements for Plant Growth

Macronutrients	Symbol	Micronutrients	Symbol
Carbon	C	Iron	Fe
Hydrogen	H	Zinc	Zn
Oxygen	O	Copper	Cu
Nitrogen	N	Boron	B
Potassium	K	Molybdenum	Mo
Phosphorus	P	Manganese	Mn
Calcium	Ca	Chlorine	Cl
Magnesium	Mg	Cobalt	Co
Sulfur	S	Vanadium	V
Sodium	Na	Silicon	Si

The three major elements are nitrogen, phosphorus, and potassium. These are needed in large amounts.

Nitrogen

Nitrogen is the most important element in the growth of plants. Nitrogen makes leaves healthy and green. Dark-green, fast-growing plants, such as corn, require large quantities of nitrogen.

Unfortunately, nitrogen is the most frequently deficient element. Even though the air is 78 percent nitrogen gas, plants cannot use it. Bacteria, algae, or industrial processes must convert the nitrogen into other forms.

As a vital part of cell protoplasm, nitrogen is found in chlorophyll and other plant substances. Plants that do not receive enough nitrogen are stunted and have pale green or yellow leaves. Plants that get too much nitrogen are too green and grow weak, succulent (juicy) foliage.

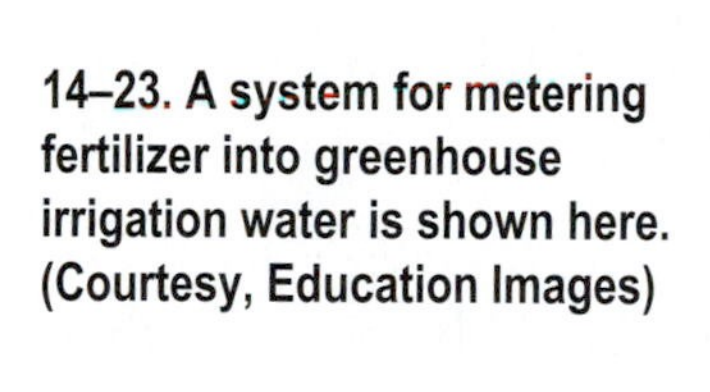

14–23. A system for metering fertilizer into greenhouse irrigation water is shown here. (Courtesy, Education Images)

Sources of Nitrogen. Natural processes in the soil produce nitrogen, and it is added as a chemical fertilizer. Plants absorb nitrogen in nitrate (NO_3^-) or ammonium (NH_4^+) forms.

Decomposing organic matter produces nitrogen. Bacteria form nitrogen when they act on stems, leaves, and other materials in the soil. The nitrogen becomes available when the bacteria die.

Crops may receive commercial fertilizers. The most effective fertilizer sources of nitrogen are:

- **Ammonium nitrate (NH_4NO_3)**—34 percent N; solid, round, white prills about 1/8 inch (0.32 cm) in diameter
- **Anhydrous ammonia (NH_3)**—82 percent N; a liquid under pressure

14–24. Ammonium nitrate is 34 percent nitrogen—an excellent source of nitrogen for many crops. It is manufactured as prills—small round particles.

- **Urea [$(NH_2)_2CO$]**—45 percent N; applied as a solid or liquid, with the liquid suitable for spraying on the leaves
- **Sodium nitrate ($NaNO_3$)**—16 percent N; originally mined in Chile, now manufactured

Since nitrogen is readily lost from the soil, the amount applied should be no more than necessary. Recommended levels of fertilizer use have been established for many crops. The levels vary with the climate and the crop variety. Corn, for example, can benefit from 120 pounds or more of nitrogen fertilizer per acre, depending on the plant population (number of corn stalks). Smaller amounts are sometimes applied because applying more than the crop will use is a waste.

Phosphorus

Phosphorus is a nutrient that plants need to store and transfer energy and to grow. Plants without enough phosphorus do not grow well. They appear stunted, and the older leaves fall off. Some leaves may turn red or purple. Phosphorus is also needed for seed germination. It stimulates flowering and aids the plant in seed formation.

Unlike nitrogen, phosphorus is not plentiful in the environment. The soil may contain small to large amounts. Most phosphorus is in the phosphate form (PO_4^{-3}).

Phosphorus is often applied as superphosphate or mixed with other fertilizer elements. Superphosphate is made by treating rock phosphate with sulfuric acid. The phosphorus content in superphosphate is about 9 percent.

Potassium

14–25. Tomato size is related to the potassium level in the soil. (Courtesy, Potash and Phosphate Institute)

Potassium is a nutrient needed by plants for photosynthesis, moving sugar, and other functions. The opening and closing of stomata are regulated by potassium. Plants need potassium to be healthy and tolerant of changes in weather. Seed and grain need potassium to develop properly.

Potassium deficiency shows in the older leaves first. For instance, the edges of the leaves lose their green color and may die. Dead leaf tips are often signs of potassium deficiency. Plants that do not have enough potassium will have small seed and fruit.

Potassium is usually added to soil as potash (K_2O) fertilizer. Muriate of potash (KCl) is mined in North America, Europe, and other places as a source of potash.

OTHER ESSENTIAL ELEMENTS

Healthy plant growth depends on other elements in smaller amounts. These elements include calcium, sulfur, magnesium, iron, boron, zinc, manganese, copper, molybdenum, chlorine, and sodium. Of course, carbon, oxygen, and hydrogen are available from the atmosphere.

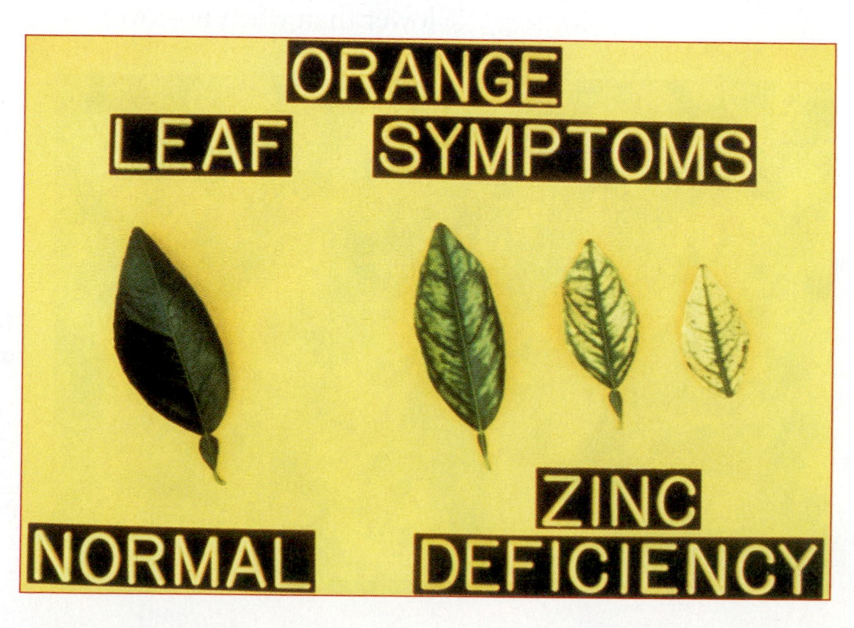

14–26. A normal leaf from an orange tree is compared with three leaves of increasing zinc deficiency. (Courtesy, Potash and Phosphate Institute)

LEGUMES

A ***legume*** is a plant that stores nitrogen from the air in nodules that grow on its roots. This process is fixation (nitrogen fixation). The roots of a legume may have small, rough, and rounded attachments about the size of a pea. If you pull up a plant such as clover, fescue, or bean growing in soft soil, you will likely see the nodules on the roots.

Certain kinds of bacteria must be present for fixation to occur. The *Rhizobium* bacteria are the most important. If the soil does not contain these bacteria, fixation will not occur.

ACADEMIC CONNECTION

CHEMISTRY

Relate the content of this chapter to your study of chemistry. Plant growth and the use of fertilizer to provide additional nutrients involve chemical processes. Visit a local farm or garden supplies store, and observe the labels on fertilizer containers to determine chemical analysis. Prepare a brief report on your findings.

Legume seeds are sometimes inoculated with the nitrogen-fixing bacteria before they are planted. Packages of inoculant are available at farm supplies stores. The bacteria must be alive and capable of fixing nitrogen.

Legumes are plants that have seed pods. Soybeans, peas, vetch, and locust trees are examples of legumes.

When growing legume crops, it is seldom necessary to apply nitrogen to the field. Sometimes legumes are grown with other crops and provide nitrogen for them also. Many so-called cover crops (often grown on land in the winter to protect it from erosion) are legumes that add nitrogen. Crops grown following the cover crops in the summer may not need nitrogen fertilizer. If they do, the requirement is lower than when no cover crops are grown.

14–27. Vetch is a legume that is often planted as a cover crop.

USING FERTILIZER

Producing plants often requires the addition of fertilizer. **Fertilizer** is any substance used to provide plant nutrients.

KNOWING WHAT IS NEEDED

Crops use nutrients to grow and bear fruit. Good crop yields require nutrients. However, growing crops removes nutrients from the soil. As a result, the nutrients must be replaced if future crops are to do well.

The kind and amount of fertilizer to use depend on several major factors: the requirements of the crop that will be grown; the nutrients available in the soil; and soil pH (if the soil is acidic or basic).

Table 14–2. Amounts of Selected Nutrients Used by Certain Crops

Crop	Pounds of Nutrients Used			
	Amount of Yield*	Nitrogen (N)	Phosphorus (P_2O_5)	Potassium (K_2O)
Cabbage	20 tons	130	35	130
Corn	100 bu	90	35	25
Cotton (seed and lint)	1,500 lb	40	20	15
Potatoes	400 bu	80	30	150
Soybeans	40 bu	150	35	55
Sugarcane	30 tons	96	54	270
Tomatoes	15 tons	90	30	120
Wheat	40 bu	50	25	15

*Yield is based on the production from one acre of land. Many producers get higher yields. To do so, they must add more nutrients to the soil because plants cannot produce without the needed nutrients!

Crop Needs

Crops have varying nutrient needs. For example, some crops require more nitrogen than others; some crops require other elements in greater amounts.

Different crops remove different kinds and amounts of nutrients from the soil. For instance, a yield of 100 bushels of corn requires 90 pounds of nitrogen, 35 pounds of phosphorus, 25 pounds of potassium, and 6 pounds of calcium. Table 14–2 shows the amounts of nutrients required by selected crops.

Once the nutrient requirement is known, the amount to apply can be determined. This varies with the climate, nature of the soil, and other factors.

14–28. Soil analysis results can be no better than the sample. Proper procedures should be followed in collecting soil samples.

14–29. A soil analysis kit can be used in nutrient analysis.

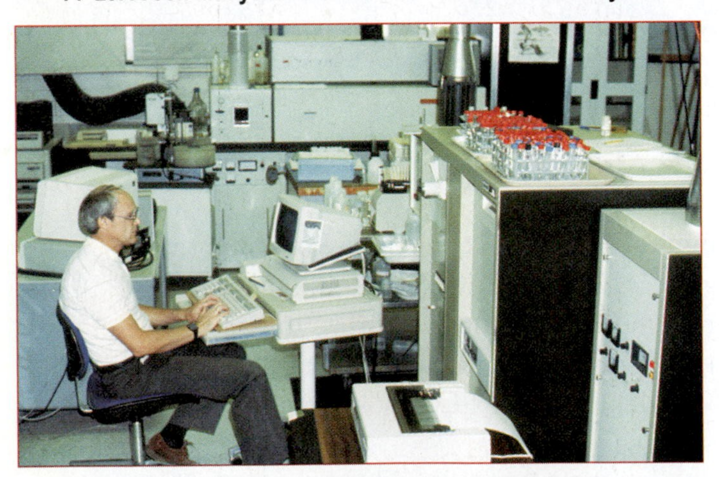

14–30. The soil testing lab at the University of Minnesota. (Courtesy, Potash and Phosphate Institute)

Soil Analysis

Soil analysis is performing a chemical analysis of soil to determine the amounts of nutrients that are present. This information is used to determine what to add to the soil for a crop.

Soils change. Some nutrients wash away when it rains. Other nutrients are used by crops. Therefore, soil should be tested annually or at least every three years.

Steps in soil testing are:

- **Sampling**—Sampling is collecting portions of soil for analysis. The samples must accurately represent the soil in a field. They must be collected from several places in the field at a depth of 4 to 6 inches. Clean tools and containers must be used. Otherwise, the samples will be contaminated, and the results of the analysis will not be based on the soil that is actually in the field. Bad tests are no better than no tests!
- **Sample analysis**—Soil samples are chemically analyzed to determine the nutrients present. Samples are often sent to soil-testing laboratories that have the equipment to do a good job. Some people test their own soil using various test kits and laboratory procedures. These kits can be accurate if used properly. Based on the analysis, recommendations can be made regarding the fertilizer to use on a field for a given crop.

- **Interpreting results**—Once test results are known, the nutrient content of the soil is compared to the needs of the crops to reveal how much fertilizer to add. For example, if a crop requires 100 pounds of nitrogen per acre and the soil only contains 10 pounds, 90 pounds must be added. If the right amount is not added, the crop yield will be low.

Tissue Analysis

Tissue analysis is the testing of samples of plant tissue. The analysis provides information about nutrient deficiencies in the plant. These deficiencies reflect the content of the soil in which the plant is growing. Fertilizer can be used to add needed elements.

Tissue analysis normally involves testing samples of leaves. The green leaves are dried, ground to a powder, and dissolved in a solution. The solution is tested for the presence of particular elements. Tissue samples must be taken and tested carefully. They should be representative of the crop and not contaminated with other substances.

Results of tissue testing can be used to determine the kind and amount of fertilizer to use. Soil and tissue analysis findings may be used in determining fertilizer needs.

TYPES OF FERTILIZER

Many kinds and types of fertilizer are available. In general, fertilizer is dry or liquid, although other types are available.

Dry Fertilizer

Dry fertilizer comes in a pellet or granular form. The small particles are about the size of a BB. Dry fertilizer has been used for many years and is often sold in bags or in bulk (unbagged). Bags add to the cost of fertilizer, so large quantities are bought in bulk. Homeowners, and others who do not have the equipment to handle large amounts, purchase dry fertilizer in bags.

Liquid Fertilizer

Liquid fertilizer is dissolved in water and applied by spraying on the soil. Its use has increased in recent years. Special tanks with pumps and spray rigs are needed to apply this type of fertilizer.

14–31. Special tanks are used to store and transport anhydrous ammonia.

14–32. Fertilizer is made in easy-to-use forms for potted plants. Nutrients in this spike are slowly dissolved by moisture in the growing medium.

Other Forms

Fertilizer is sometimes applied in other forms, such as gases or solid materials. Regardless of the form, proper handling and application are essential.

Anhydrous ammonia, a source of nitrogen, is a liquid that becomes a gas in the atmosphere. It is held in pressure tanks and is injected into the soil. Any anhydrous ammonia released into the air quickly becomes a vapor and is lost. Anhydrous ammonia has been widely used, but it is decreasing in importance.

Solid and spike forms of fertilizer are used in special situations. Some fertilizer may be made into nail-like spikes that are pushed into the soil. These are used in decorative potted plants and in other places. The fertilizer slowly dissolves and is released into the soil.

Fertilizer and other chemicals are sometimes combined. Lawn care products may include pellets of fertilizer and herbicide (weed killer) mixed together.

CAREER PROFILE

AGRONOMIST

An agronomist studies crops, soils, and related areas. Agronomists often specialize in an area such as cotton, corn, or wheat. Some work with turf, forage crops, and rangelands. Other agronomists focus on soils, fertility, and irrigation.

Agronomists need college degrees in agronomy or a closely related area. Many agronomists have master's degrees and doctorates in agronomy. Begin preparing to be an agronomist in high school by taking biology, chemistry, agriculture, and related courses. Practical experience in crop production is very useful.

Jobs for agronomists are with large farms, agribusinesses dealing with crops and fertilizers, colleges, research stations, and government agencies. Some agronomists are private consultants. This photo shows an agronomist investigating a condition in a rice crop. (Courtesy, Animal Plant Health Inspection Service, USDA)

FERTILIZER ANALYSIS

Most fertilizer is made of several different materials that are blended. The materials give it an analysis or grade. ***Fertilizer analysis*** is a list of the amounts of nutrients in a fertilizer.

A label on a fertilizer container usually has three important numbers as well as other information. These numbers are always in the same sequence and have the same meaning.

- **First number**—The first number is the percent of nitrogen in the fertilizer.
- **Second number**—The second number gives the percent of the anhydride of phosphoric acid (P_2O_5) in the fertilizer. Phosphorus is not reported in the elemental form, or P.
- **Third number**—This number represents the percent of potash (K_2O) in the fertilizer. Potassium is not reported in the elemental form, or K.

Together, the three numbers give the overall analysis of the major nutrients in the fertilizer. For example:

- **0-20-20**—This means that the fertilizer contains no nitrogen, 20 percent P_2O_5, and 20 percent K_2O. A 50-pound bag of fertilizer would have 10 pounds of P_2O_5 and 10 pounds of K_2O.
- **9-12-12**—This means that the fertilizer has 9 percent N, 12 percent P_2O_5, and 12 percent K_2O. A 50-pound bag would contain 4.5, 6, and 6 pounds, respectively, of each of the three nutrients.

Other information on the label will tell about trace elements and availability of the nutrients.

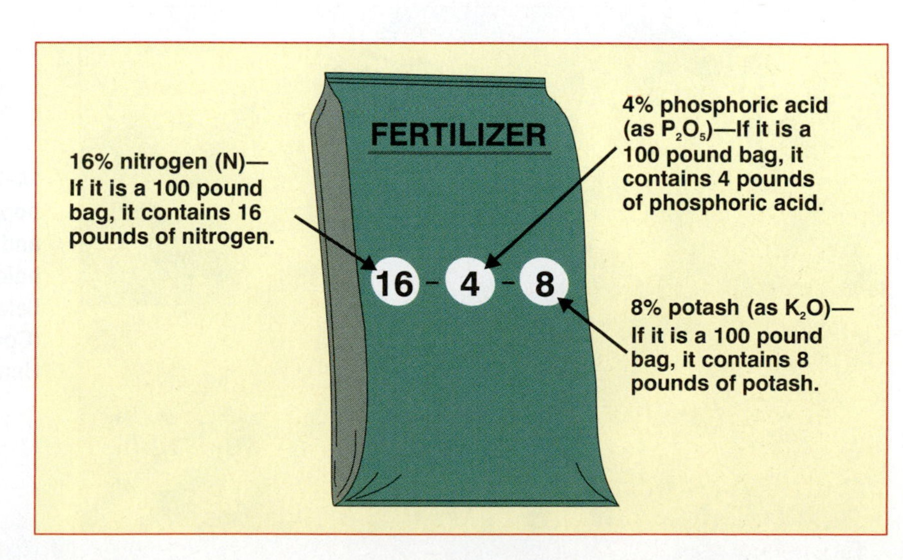
FERTILIZER
16% nitrogen (N)—If it is a 100 pound bag, it contains 16 pounds of nitrogen.
4% phosphoric acid (as P_2O_5)—If it is a 100 pound bag, it contains 4 pounds of phosphoric acid.
16 - 4 - 8
8% potash (as K_2O)—If it is a 100 pound bag, it contains 8 pounds of potash.

14–33. Interpreting fertilizer analysis usually involves three numbers, as shown above.

FERTILIZER PLACEMENT

Fertilizer must be placed where plants can get it. In most cases, plant roots absorb nutrients in the fertilizer from the soil. Proper placement means that the fertilizer is close enough for the roots to receive it. Yet putting fertilizer too close to the roots results in damage. Some dry fertilizer can also damage leaves and stems if it lands on them.

Pre-Planting Application

Fertilizer may be applied to a field before the crop is planted. Solid-types may be spread or broadcast evenly over the soil. The field may be tilled with a disk harrow or "do-all" (a large tiller pulled by a tractor) to mix the fertilizer into the soil.

Planting Application

Fertilizer may be applied at the time of planting. Some planter equipment may have fertilizer applicators attached. Fertilizer is often placed in narrow bands 4 to 6 inches from the seed. Never place fertilizer in the same drill (small trench) as the seed because fertilizer can burn sprouts and keep seed from coming up.

Post-Planting Application

Fertilizer may be applied to crops after they are up and growing. This is known as top-dressing or side-dressing, depending on the location of the fertilizer. Pastures, wheat, oats,

14–34. The three large hoppers hold dry fertilizer, and the six small hoppers hold seed. One pass over a field plants and fertilizes. (Courtesy, Mississippi State University)

and other crops are often top-dressed. The fertilizer is applied over the ground where the crop is growing. Rain dissolves the fertilizer, and the water soaks it into the soil.

Side-dressing is the application of fertilizer along the sides of rows of growing crops. Corn, cotton, and many vegetable crops are side-dressed. The fertilizer is placed six inches from the root area. Sometimes a crop is cultivated to mix the fertilizer into the soil. Other times, the fertilizer is placed on top of the soil and carried into the soil by rain or irrigation water. Care should be used to avoid damaging plant roots, stems, and leaves when side-dressing.

14–35. Small lawn areas may best be fertilized with a carefully calibrated push-type spreader. (Courtesy, Education Images)

HANDLING FERTILIZER

Fertilizer is a chemical compound. Proper handling is needed to prevent damage to people, property, and the environment. Fertilizer bags and other information accompanying fertilizer will provide important details. Because some kinds/forms of fertilizer are explosive, attention to safety is essential. One example is ammonium nitrate. This fertilizer is a commonly used explosive. When improperly handled, it can explode when being stored, transported, and applied to the soil.

Here are a few rules to follow when using fertilizer:

- Use only the recommended kind and amount. Using the wrong kind or more than is needed wastes fertilizer and pollutes the environment. Excess fertilizer may wash into streams.
- Apply fertilizer only where it is needed. Wasting fertilizer on the ground at the ends of rows costs money and does not grow a crop.

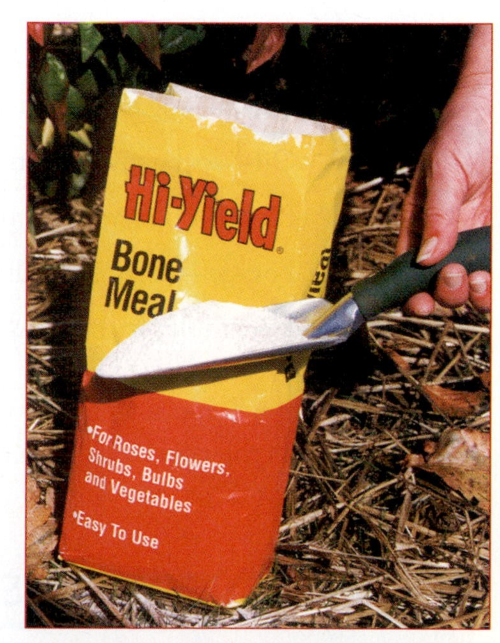

14–36. Bone meal is a fertilizer used when planting some kinds of bulbs.

- Store fertilizer in dry places to prevent nutrient loss. Use care when handling bags of fertilizer to avoid bursting or cutting them. Because some fertilizer ingredients are explosive, keep fire away from storage areas. Fertilizer should not be stored near animal feed or water.
- Clean applicator equipment after use. Fertilizer will chemically react with metal and other materials. Equipment can be damaged and may fail to work properly when next used.
- Wear protective clothing, and wash after handling or using fertilizer. Some fertilizer can burn the skin.

Remember, fertilizer is a chemical. It can cause reactions that are dangerous and damaging.

FIELD CROP PRODUCTION

The vast majority of major food and fiber crops are grown in large fields. These crops are carefully managed to ensure efficient and profitable yields. Examples of field crops are as follows:

- **Grain**—corn, wheat, rice, rye, sorghum, oats, and barley
- **Sugar**—sugarcane, sugar beets, and sweet sorghum
- **Oil**—soybeans, cotton, sunflower, canola, peanuts, and safflower
- **Fiber**—cotton, flax, and kenaf
- **Forage**—corn, sorghum, millet, legumes, and grasses
- **Vegetable, fruits, and nuts**—carrots, lettuce, broccoli, apples, oranges, and walnuts

Note: Some authorities do not include legumes and grasses for forage as field crops. Also, some of the vegetable, fruit, and nut crops would not be field crops as they may be produced in orchards or vineyards.

FIELD CROP CONSIDERATIONS

Before beginning to produce a crop, study its needs as related to the characteristics of the land available for its production. Specialized equipment and other technology may be needed, depending on the crop. Consider the following:

- **Climate**—Climate is the average of the weather conditions over many years. Since crops vary in climate suitability, always get background information. If the production will be in a warm climate, choose a crop that is suited to warm weather. If production will be in a cool or cold climate, choose a crop suited to that climate.
- **Soil**—The nature of the soil must be within the requirements of the crop to be grown. Some crops require sandy soil; other crops need loamy soil. Lay of the land will also need to be considered, including slope and flooding possibilities.
- **Moisture**—All crops require moisture for growth, but some require much more than others. Moisture needs for wheat production are less than for rice or cotton production. The moisture may be in the soil and naturally available through precipitation. In some cases, irrigation may be used; if so, the available water should be sufficient for needs.
- **Market**—Once a crop is produced, the producer must be able to sell it at a price that results in a profit. Never produce a crop for which there is no market. Careful study may be needed to identify niche crops where markets can be developed or expanded.
- **Technology**—Some crops are more sophisticated to produce than others. The technology is greater. A producer must be able to afford the equipment and have the ability to use it or hire a competent equipment operator.
- **Skills and preferences**—A producer needs the skills and knowledge to produce a crop. Since conditions are continually changing, a producer needs to keep current by attending field days, seminars, and other educational events. Further, the crop should be one that the producer will gain reasonable satisfaction from producing.

FIELD CROP PRODUCTION PRACTICES

Overall, all field crops have certain general production requirements that are sometimes referred to as cultural practices. A ***cultural practice*** is an activity or treatment needed by a crop to grow and be productive. Cultural practices are the procedures used in producing a crop. Without appropriate practices, yields and efficiency will be lower than they should be.

Here are several general cultural practices:

- **Variety selection**—The variety chosen should be one that is adapted to the climate and growing conditions. It may also have other features that are desired in the market or that promote its production. GMO (genetically modified) varieties may be preferred, but market clearance may be needed before production. Growers producing under contract may be required to plant specific varieties, such as sugar beets or soybeans.

- **Planting or seeding**—Planting involves placing seeds in the soil to promote germination and plant growth. Some seeds can be planted using high-speed planting equipment. Proper seeding rate, adjustment, and operation are essential to gain the desired plant population. All seeding should be at the proper time of the year to maximize yields by preventing losses due to unexpected frost.

14–37. Seedbed preparation may involve breaking the soil. (Courtesy, AGCO, Duluth, Georgia)

- **Seedbed preparation**—A seedbed is the medium or soil in which seeds are placed. How the seedbed is prepared varies by crop and cultural practices. Some seedbeds are prepared into finely tilled rows. Other crops involve planting without any tillage but with the application of a herbicide to "burn-down" weeds that may be growing on the field. Specialized no-till planters will be needed. Seeds should be planted at the proper depth and spacing for plant growth.

- **Fertilization**—Fertilization involves the application of fertilizer (nutrients) to the soil to promote plant growth. Some fertilizer is applied pre-planting while other fertilizer is applied as a top-dress or side-dress application after the plants are growing. A few forms may be sprayed on the foliage of plants. Liquid, solid, or other forms of fertilizer may be used. Specialized equipment will be needed to apply the fertilizer. Always base applications on the results of a soil analysis as related to the requirements of the crop being produced.

- **Pest management**—Many crop plants have pests that may damage plant growth efficiency or the product that is produced. Pests include weeds, insects, diseases, and others. How the pests are managed depends on the nature of the crop and the potential loss caused by the pest. Having a few pests in a field may not merit the use of pest control practices. However, once the number has reached a certain level, such practices will likely be needed. (More information is presented on pest management in Chapter 15.)

- **Harvesting**—Once a crop has been grown, it must the harvested. Specialized equipment and practices may be needed. Harvesting varies with the kind of crop and the product use. For example, corn for grain is harvested to ensure clean, quality kernels are efficiently obtained, while corn for silage involves chopping the entire above-ground portion of the plant.

14–38. Modern farms may use seeding equipment to ensure proper seeding rates and placement of seeds. (Courtesy, Education Images)

- **Post-harvest handling**—Post-harvest handling relates to how a crop is transported, stored, and otherwise managed after it has been harvested. Always use practices that prevent loss and damage and that retain the quality of the crop. In some cases, such as grain, drying may be needed. In other cases, such as cotton, the product will be ginned to separate seeds and lint and to remove trash and excess moisture.

STRUCTURES AND CROPS

Some crops are produced in specialized structures that may be used to "start" plants before setting them in the field or to produce the entire crop. Such crops are typically of higher value than those grown in fields. The structures allow production at times of the year when plants could not tolerate the weather or other conditions in outside fields.

Examples of crops that may be produced in special structures are:

- **Flowering plants**—Flowering plants include plants in pots or baskets as well as those that may be cut; examples are impatiens, Easter lillies, petunias, chrysanthemums, African violets, and daffodils.
- **Foliage plants**—Plants in pots or baskets may be used; examples are poinsettia, banana plants, caladiums, ferns, croton, corn plants, and spider plants.
- **Bedding plants**—These plants are raised from seed to a size for setting in flower beds, gardens, and fields when the weather is appropriate; examples are ornamentals (e.g., marigolds, petunias, and pansies) and vegetable plants (e.g., tomatoes, eggplants, and pepper plants).

14–39. A teacher and students are preparing to enter a freestanding greenhouse. (Courtesy, Education Images)

- **Food crops**—Large greenhouses may be used to produce crops out of season; examples are tomatoes, cucumbers, and beans.

Special structures include greenhouses, shade houses, hot beds, and cold frames. The greenhouse is most widely used.

GREENHOUSES

A **greenhouse** is a specially-constructed, enclosed building where plants can be cultured. Typical construction involves glass or plastic materials over or supported by a frame of wood, metal, brick, or concrete. Glass and plastic are useful in helping regulate the environment for plant production. The covering allows heat and sun rays to enter and prevents the loss of artificial heat in cold weather. Greenhouses may have tables or benches for plants as well as irrigation systems and built-in electrical lighting.

The environment in greenhouses is regulated to provide a good cultural situation for the plants that are produced. Environmental conditions that may be managed include temperature, light, moisture, and air quality. Automated systems may be installed for each of these needs.

Temperature may be controlled with fans, cooling pads, and heaters, depending on the outside air temperature and the needs of the crops. Thermostats may be used to establish and maintain the desired temperature.

Two major factors related to light management are intensity and duration. Intensity is the "brightness" of the light. Some greenhouses are covered with a fabric material that reduces the amount of light that reaches crops. This is done to "force" crops so they are ready when desired. Light duration refers to the number of hours of light in a greenhouse.

On short days, artificial light can be used to lengthen the number of hours of light each day to gain the desired plant performance.

14–40. Greenhouse-grown Easter lilies are nearing market readiness.

Moisture is regulated through irrigation and air humidity management. Some crops are irrigated using an automated system set on timers. Other plants are hand watered as needed. Moisture needs vary with the kind of plant and the desired goal. For example, rooting plant cuttings may require more moisture than gaining the desired growth of seedlings.

Air quality includes the gaseous composition of the air as well as humidity. Dry air (low humidity) places stress on plants, but excess humidity may promote plant disease. Since plants need carbon dioxide gas from the air for photosynthesis and rapidly-growing plants use carbon dioxide, ventilation may be needed to guarantee that adequate carbon dioxide is present. Processes are available to add carbon dioxide to the air, which is known as carbon dioxide fertilization.

Greenhouses may also have benches and other facilities for work areas. Media storage, fertilizer storage, pesticides, tools, and other small equipment may be stored in a greenhouse.

REVIEWING

MAIN IDEAS

Understanding and providing what plants need for growth is the goal of crop producers. Plants have two kinds of growth: primary and secondary. Primary growth is the increase in the length of a plant. It occurs at the buds on the ends of stems in tissue known as meristem. Secondary growth is the increase in the diameter of stems; they get thicker. Layers of cells expand the stem size.

Several hormones are important in plant growth. They influence the way in which plants grow and the stages of growth. Producers may apply some hormones to encourage plants to grow in different ways.

Photosynthesis is the process plants use to make food. Plants use energy from sunlight, carbon dioxide from the air, and water from the soil to make a simple sugar known as glucose. Chlorophyll is needed for the process to occur. The sugar is moved to other locations in the plant and may change form, such as to starch. Some of the sugar is stored; the rest of the sugar is used by the plant in respiration, which is the process an organism uses to provide its cells with oxygen so energy can be released from digested food.

Transpiration is the release of water by a plant. This occurs as the plant adjusts to its environment. Wilting is a sign of a lack of water. Plants need water so they are not damaged and so productivity is not lost.

Plants need nutrients to grow. Twenty nutrients are needed, three in large amounts: nitrogen, phosphorus, and potassium. Plants can receive some elements from the air, such as carbon dioxide and oxygen. Other elements come from the soil as water or nutrients in the water.

Fertilizer is used to increase the nutrients available to plants. Most fertilizer is applied to the soil. Soil testing and tissue analysis are used to determine the kind of fertilizer that should be used. As with all chemicals, fertilizer should be used carefully. It can damage animals, crops, people, equipment, and the environment.

Field crops and those plants produced in special structures require appropriate cultural practices to issue efficient growth and profitable production.

QUESTIONS

Answer the following questions, using complete sentences and correct spelling.

1. How do plants grow?
2. What kinds of growth occur in plants? Distinguish between the two.
3. What kinds of hormones are found in plants? What does each do for the plant?
4. What is photosynthesis? What are the major phases of the process?
5. How is photosynthesis carried out? Write the equation.
6. What is respiration?
7. What is transpiration?
8. Why do plants wilt? What can be done to prevent wilting?
9. What are the essential nutrients for plants? List their names and chemical symbols.
10. What is the importance of nitrogen? What are the sources of nitrogen?
11. What is a legume?
12. What is the importance of phosphorus and potassium?
13. How is the appropriate fertilizer determined?
14. What types of fertilizer are used?
15. What does "fertilizer analysis" mean? Explain the following label on a fertilizer bag: 5-10-5.
16. How is fertilizer placed?
17. What are the important rules to follow in handling fertilizer?
18. What general practices are used in producing field crops?
19. What should be considered in making decisions about field crop production?
20. What is a greenhouse?

EVALUATING

Match each term with its correct definition.

a. annual growth ring
b. primary growth
c. wilting
d. mulch
e. major elements
f. legume
g. fertilizer
h. soil analysis
i. tissue analysis
j. fertilizer analysis

______1. Analyzing samples of plant tissue to determine nutrient needs.

______2. A cover on soil that holds moisture and prevents weed growth.

______3. Nutrients that plants need in large amounts.

______4. The result of secondary growth in a plant's lateral meristem.

______5. The nutrients in fertilizer.

______6. Any substance used to provide plant nutrients.

______7. A plant that stores nitrogen from the air in nodules that grow on its roots.

______8. The increase in the length of a plant stem.

______9. A condition in which stem and leaves are no longer rigid and erect.

_____10. Chemical analysis to determine the nutrients in soil.

EXPLORING

1. Visit a local farm or garden supplies store. Ask to see the labels on fertilizer bags. Study the labels for their content. List the different kinds of fertilizer. Prepare a report on your findings.

2. Analyze the soil at your home or on the school grounds. Take the sample and send it to a lab, or use a testing kit and do it yourself. Your teacher will need to assist with this activity.

3. Design an experiment in the school laboratory or at your home. Use corn or bean plants. Vary the kinds of fertilizer you use. Note the differences in the color of the leaves and the rate of plant growth.

4. Arrange to tour a tissue analysis laboratory. Take a sample of plant leaves to be analyzed. Prepare a report on your observations.

5. Tour a greenhouse to observe the plants being grown and the techniques used in their culture. Prepare a report on your findings.

CHAPTER 15

Healthy Plants

OBJECTIVES

This chapter contains information on plant pests, pest management, pest prevention, and methods used. It has the following objectives:

1. Explain plant health and how pests cause damage.
2. Discuss the major kinds of pests.
3. Explain the role of entomology in plant health.
4. Identify conditions that promote damage to plants by pests.
5. Describe the meaning and use of integrated pest management.
6. List safety practices in using pesticides.

TERMS

bactericide
beneficial insect
biological pest management
chewing insects
contact poison
cultural pest management
entomology
environmental disease
fumigant
fungicide
genetic pest management
herbicide
insect pest
insecticide
integrated pest management
mechanical pest management
metamorphosis
nematocide
nematode
pesticide
plant disease
plant health
plant pest
protectant
rodent
roguing
seed-borne pest
stomach poison
sucking insects
surfactant
systemic poison
therapeutant
translocated herbicide
trap cropping
weed

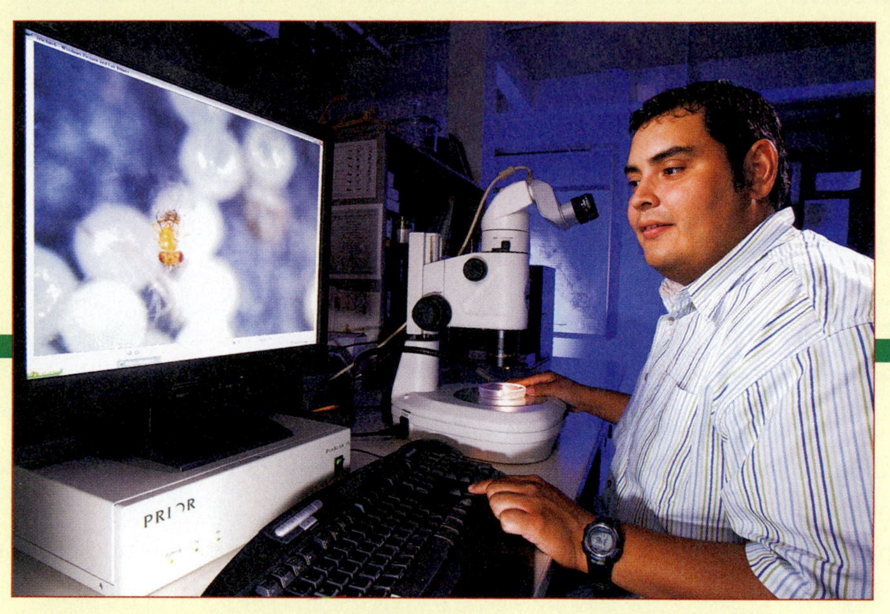

15–1. Using a dissecting microscope and videography, a scientist is viewing a tiny *Trichogramma* wasp as it is parasitizing insect eggs. (The wasp is visible near the middle of video screen.) (Courtesy, Agricultural Research Service, USDA)

HEALTHY PLANTS flourish; diseased plants do not. A healthy plant is one that is free of disease and growing properly. Healthy plants will produce large amounts of desired products, such as grain, flowers, and wood.

Plants that are not heathy have pests or disease. They do not grow well and have the appearance of being unhealthy, such as yellow leaves, stunted growth, and damaged parts. Entire crops can be destroyed by pests and diseases. Fortunately, most pests can be managed and losses reduced or eliminated.

Plant pests often affect people and cause big problems. An example is the potato famine of nearly 200 years ago in Ireland. In the 1840s, the potato crop in Ireland was lost several years in a row because of blight. More than a million people starved to death. Thousands of others left Ireland for new places—many for the United States—because they were faced with starvation. Another example of plant pest damage is the corn leaf blight that struck in the United States in 1970. The result was 40 million fewer bushels of corn that year. The loss had a huge impact on human food and livestock feed.

PLANT HEALTH

Plant health is the condition of a plant related to growth and productivity. The goal is to have plants that are in good health rather than poor health. Healthy plants grow rapidly and produce an abundance of quality products. These are plants that will help a producer realize a profit from plant production.

Plants that are free of disease and damage by pests will grow and make efficient use of nutrients. Plant health is related to proper nutrients and water, freedom from disease and pests, and growth in a good environment.

PLANT HEALTH DAMAGE

Plant health may be degraded by a number of pests. The damage interferes with plant growth and reproduction. When this happens, the products that plants provide are damaged and/or the amount produced is less.

The kind of damage varies with the pest and the plant. The presence of pests does not bother some plants, while the same pests may severely damage other plants.

Some ways pests damage plants:

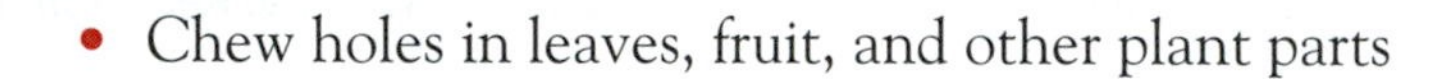

- Chew holes in leaves, fruit, and other plant parts
- Attack the vascular system, causing damage that may go unseen for a long time
- Attack the fruit, causing damage or a total loss
- Contaminate a product merely by being present, such as a worm in an apple
- Rob plants of food by sucking out the juices
- Damage land by digging holes and building mounds
- Compete for space in which to grow, as is characteristic of weeds

15–2. Worms have destroyed the productivity of this corn.

DAMAGE AND THE PRODUCER

People who produce plants are affected when pests attack crops. The damage reduces income and increases production costs.

Some ways plant pests affect producers:

- Reduce yields by lowering a plant's productivity
- Decrease crop quality by damaging fruit, leaves, and other valuable plant parts
- Increase production costs because of the need to buy pesticides, equipment, and other inputs to control pests
- Restrict marketing by quarantines

DAMAGE AND THE CONSUMER

Damage caused by plant pests affects consumers. A consumer is an individual who uses a product. Anything that influences production of a product will have an impact on consumers.

Some ways plant pests affect consumers:

- Higher costs for food, clothing, and other products
- Lower quality in available goods because of pest damage
- Possible product contamination from materials used to control pests
- Possible environmental contamination or damage from materials applied to diseased plants, such as insecticides

15–3. A corn ear with a worm is not appealing to consumers.

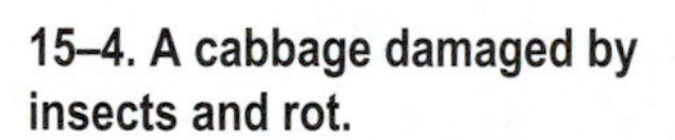

15–4. A cabbage damaged by insects and rot.

KINDS OF PLANT PESTS

Many different things can damage plants. Five categories of pests are insects, nematodes, diseases, weeds, and rodents and other animals.

15–5. The boll weevil (*Anthonomus grandis*), a major pest of cotton, is nearing eradication, with many cotton-growing areas of the United States identified as boll weevil free. (Courtesy, National Cotton Council of America, Memphis, Tennessee)

INSECTS

An ***insect pest*** is a small, six-legged animal that damages plants in one way or another. Some insects eat the leaves, stem, fruit, or roots of a plant; others use the plant for raising their young. The young develop and feed on the plant, causing damage. Several other tiny animals are included with insects, even though they are not really insects. Examples include spiders, millipedes, and centipedes. (More information is presented on insects in another section of this chapter.)

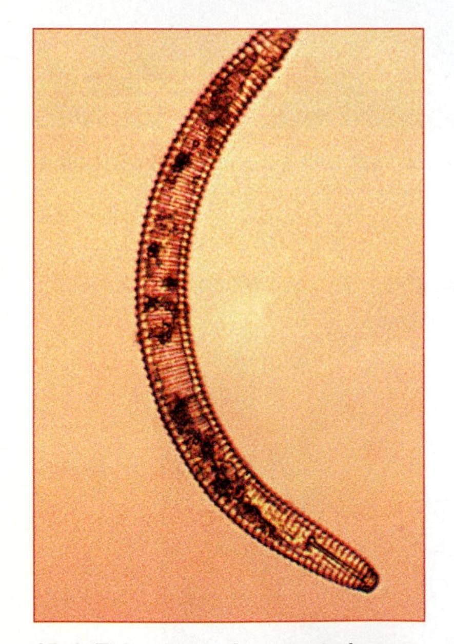

15–6. This photo of a nematode was made using a microscope to enlarge the tiny organism. (Courtesy, Frank Killebrew, Plant Pathologist, Mississippi State University)

NEMATODES

A ***nematode*** is a tiny wormlike organism that lives in the soil and attacks the roots and stems of plants. The damage may go unnoticed for a while. Nematodes sometimes cause knots on the roots of plants. Nematodes are not insects, though they are sometimes listed as insects.

Identification

Mature nematodes are about 1 millimeter long. Their identification often requires laboratory study by a trained agriscientist. Nematodes are often tentatively identified by the damage they cause. Roots, stems, buds, and other plant parts show damage in different ways depending on the kind of nematode.

Life Stages and Damage

Nematodes go through several life stages. Adults lay eggs that hatch into nymphs. The nymphs molt (shed old skin and grow new skin) four times.

Adult nematodes have spearlike structures that pierce plants and suck out juice. Nematodes may carry plant disease from one plant to another, such as ring-spot virus in apples and peaches. Nematodes attack many crops, including corn, soybeans, cotton, and vegetables.

DISEASES

A ***plant disease*** is an abnormal condition in a living plant. The plant often appears stunted or does not grow properly. A disease may interrupt plant growth, reproduction, or other functions and even result in plant death. Many different living organisms and nonliving substances cause diseases.

15–7. The area without clover has a fungus disease called brown spot. The spot will spread and destroy more clover if not controlled.

Identification

A disease does not just happen; it results from a cause. Many things can trigger a disease. Careful laboratory study is often needed to identify the disease accurately.

Diseases are visible by the kind of damage (signs) we see. Signs of disease include (1) rotting plant parts, particularly the fruit; (2) leaves turning yellow or having an unnatural color; (3) plants wilting; (4) plants having twisted leaves or stems; (5) buds, flowers, or fruit not developing or falling off; and (6) dead plants. Various combinations of these may also be found.

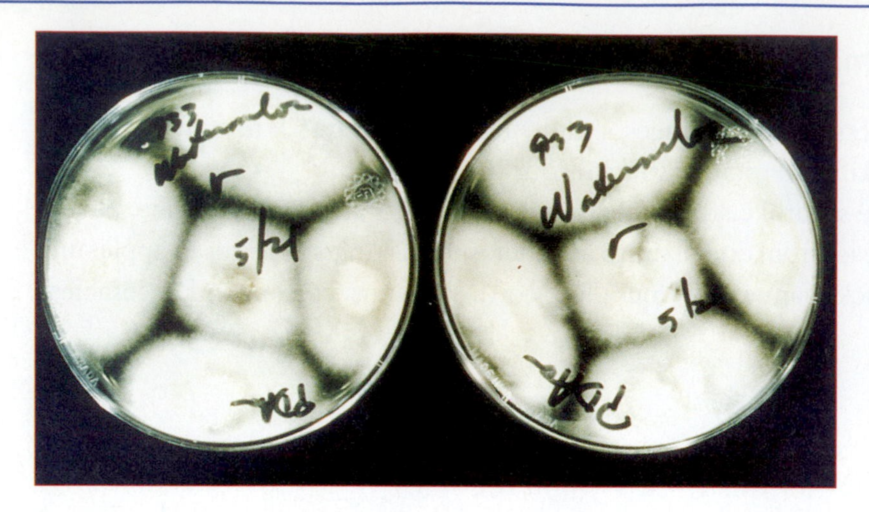

15–8. Magnified *Fusarium* spores from a watermelon plant were identified by growing plant tissue in a petri dish. (Courtesy, Jackie Mullen, Auburn University)

Types of Disease

Two major types of disease are environmental and parasitic.

An ***environmental disease*** is a plant disease caused by factors in the environment that are not "right" for the plant. Since plants vary, their tolerance of environments also varies. Some plants can live in a condition in which others will die.

Examples of environmental diseases are:

- Nutrient deficiencies, such as not enough potassium
- Damage to plant parts, such as from improper plowing
- Chemical injuries, such as from incorrect use of chemicals
- Pollution injuries, such as from factory smoke

15–9. This corn has storm damage—an environmental disease.

- Weather-related conditions, such as damage from storms or effects of too much rain
- Naturally occurring genetic abnormalities

15–10. Yellowing of the leaves on these greenhouse-grown strawberries is due to a nutrient deficiency. (Courtesy, Education Images)

Tissue analysis and soil testing may be needed to help identify some environmental diseases. Some of these diseases may be caused by nutrient deficiencies in the soil. A frequent indicator of nutrient deficiency is leaf color. Yellow color on green leaves is a sure sign of a nutrient deficiency. Laboratory analysis is often needed to identify accurately the nutrient that is deficient. Adding the deficient nutrient to the soil or growing medium will usually correct the deficiency. However, by the time the discoloring of the leaves is evident, permanent damage may have already been done so that the plant will not be productive.

Diseases are caused by several organisms:

- **Fungi**—Fungi cause more plant diseases than any other type of parasite. They are small, one-celled organisms that grow on and in plants. Fungi cause plant mildew, rusts, smuts, and often fruit rot. Wind, water, insects, and other means can spread fungi. Unclean tools can spread disease. For example, trimming a healthy shrub with shears used on a diseased shrub may pass a disease. Not all fungi are plant pests; some are beneficial.

15–11. Tomato blight is caused by the fungus *Sclerotium rolfsii*. The disease is most damaging in hot, humid weather. (Courtesy, Jackie Mullen, Auburn University)

- **Bacteria**—Bacteria are small, one-celled organisms. Some bacteria are beneficial, and others cause plant disease. Bacteria often enter plants through cuts or breaks in the bark or epidermis. Some enter through flowers and natural openings in the stem and leaves. Apples

15–12. The discolored spots on these photinia leaves are caused by a viral disease.

and pears may get a bacterial disease known as fireblight. Rot in fruit and vegetables may be due to bacteria.

- **Viruses**—Viruses are tiny forms that do not have organized nuclei. They are not considered living organisms. They are very small and are visible only with powerful electron microscopes—regular microscopes do not have enough power. Viruses cause many diseases. Cucumber mosaic, citrus tristeza, and tomato ring-spot are examples. Viruses are spread by insects, equipment used to plow fields, and vegetative propagation.

WEEDS

A ***weed*** is a plant that is growing where it is not wanted. Weeds compete with crop plants for nutrients and space. The presence of many weeds in a field can reduce crop yields and lower the quality of the crop.

Plants can be useful in one place and weeds in another. In effect, a weed is a plant that is out of place.

15–13. Dandelions are pesky weeds in lawns. The seeds in the round heads are readily scattered about by wind.

Kinds of Weeds

Weeds are classified in much the same way as crop and ornamental plants. Knowing the growth characteristics of weeds helps in managing them.

Weeds may be classified based on life cycle, as follows:

- **Annuals**—Annual weeds go through a complete life cycle in one season. They can be classified as summer annuals and winter annuals.

15–14. The predominant weed growing in this corn is crabgrass, a summer annual.

Examples of summer annual weeds are morning glory, cocklebur, pigweed, and crabgrass. These weeds cause problems in crop plants that grow in the summer, such as corn and soybeans, vegetables, ornamental plants, and lawn grasses.

Examples of winter annual weeds are chickweed and henbit. These weeds cause problems in wheat, oats, rye grass, and winter turf grass.

- **Biennials**—Biennial weeds go through a complete life cycle in two years. Examples include thistle and wild carrot.

AgriScience Connection

GOOD CRITTERS VERSUS BAD CRITTERS

Not all insects are bad! Many rid the environment of damaging insects. For example, some species of wasps attack and destroy other forms of insects. Parasitic wasps usually attack only the targeted pests, such as the corn ear worm (also known as the tomato fruitworm and cotton bollworm).

This shows a wasp (*Microplitis croceipes*) laying an egg inside a caterpillar. The egg hatches in a few days, and the larva feeds on the caterpillar. The caterpillar is unable to grow to the pupa stage and dies. The wasp larva eats its way out of the caterpillar and becomes an adult in two weeks.

Using pesticides in a field kills both the beneficial insects and the harmful insects. Having 300 to 600 tiny wasps per acre of crop will usually provide adequate control of targeted pests. (Courtesy, Agricultural Research Service, USDA)

15–15. The poison ivy growing on this tree is a perennial weed. Caution: Some people are highly allergic to poison ivy. Avoid touching the plant and being in the vicinity where it is being cut.

- **Perennials**—Perennial weeds live more than two years. They are often difficult to control because of their root systems. Some, such as dandelion and plantain, have fleshy roots. Others have rhizomes that grow below the ground or stolons that grow above the ground. Examples are Bermuda grass and Johnson grass. A tree growing along the edge of a field is a weed if it damages by shading or taking water away from crop plants.

Table 15–1. How Weeds Cause Losses

Weeds:
• Compete with desired plants for nutrients, water, sunlight, and space. A few weeds may not cause much of a problem in a crop. A lot of weeds can cause crop plants not to grow.
• Waste fertilizer applied for crop plants when it is used by weeds.
• Lower the quality of harvested crops.
• Make harvesting harder and make it more expensive through increased fuel use and machinery wear.
• Harbor (hold) insects and disease that attack crop plants.
• Detract from the appearance of a field, lawn, or flower bed.

RODENTS AND OTHER ANIMALS

Rodents and other animals can attack and destroy plants. A ***rodent*** is a small mammal with two large front teeth. Rats and rabbits are examples. Other animal pests include birds and deer. These pests eat leaves, stems, fruit, and roots of plants. Nationwide, the loss caused by these animals is not as great as by other pests. Individuals, however, may have large losses.

15–16. A snail pest among plants in a greenhouse. (Courtesy, Education Images)

Losses may result from:

- Deer that eat soybean plants growing in a field
- Raccoons that climb cornstalks and get roasting ears

- Rabbits that bite off the tender leaves of new bean plants
- Birds that eat the grain from sorghum
- Rats that eat fruit growing near the ground
- Slugs and snails that eat fruit and leaves

Managing animal pests involves destroying their habitat and getting rid of the animals. Cutting weeds and bushes around fields destroys habitat. Since laws protect some species, killing the animals may be illegal.

ENTOMOLOGY AND PLANT HEALTH

Entomology is the branch of zoology that deals with insects and often with related small animals, such as spiders, mites, and centipedes. It is frequently studied in conjunction with agronomy, horticulture, and forestry.

Entomology includes identifying both beneficial and harmful insects. Beneficial insects, such as honeybees, do good things. Harmful insects, such as corn borers, white flies, and termites, cause damage to plants, animals, people, and structures. Ways to encourage the growth of beneficial insects are studied. Methods to control harmful insects are developed.

15–17. A beneficial wasp is laying eggs in the larva of an insect pest. The eggs will hatch and destroy the larva. (Courtesy, Agricultural Research Service, USDA)

INSECT CLASSIFICATION AND NAMING

Insects are the most diverse animals on Earth. More than a million species of insects have been described. Scientists estimate that there are several million more that have not been described. All species fit within the accepted hierarchy of classification. An example of classification of the European corn borer is:

Domain—Eukarya
Kingdom—Animalia
Phylum—Arthropoda
Class—Hexapoda (formerly Insecta)

Subclass—Pterygota (winged)
Division—Endopterygota (major body change in growth)
Order—Lepidoptera (butterflies and moths)
Family—Crambidae
Genus—*Ostrinia*
Species—*nubilalis*

"European corn borer" is the common name. The scientific name is *Ostrinia nubilalis*. Common names often vary; scientific names are consistent among scientists. The two-word scientific name is a binomen (binomial name). Insects are sometimes placed in subclasses based on the presence or absence of wings. The subclasses may be further divided on the basis of body changes during growth, with those insects undergoing metamorphosis being placed in the division Endopterygota. Most insects can be fairly easily classified insofar as their order is concerned. Beyond order, the finer classification requires the use of an identification key.

Sometimes the immature forms are most visible and recognizable, such as the tomato horn worm (also known as the tobacco horn worm). It is also the immature forms that often inflict the most damage to plants.

More than 86,000 species of insects have been identified in North America. Managing insect pests requires recognizing them and understanding some of their features. Detailed keys are available to help identify insects.

Technology Connection

QUADRAT SAMPLING

A quadrat is a device for intensely studying a small area of ground. Most quadrats are squares that are 1 meter on each side and made of PVC or wood. They can be used in conjunction with transect lines to ensure that an area is appropriately sampled.

Lawns, athletic fields, pastures, crops, and other areas can be divided into 1-meter squares for counting the population of weed plants, insects, diseased leaves, flowers, or other conditions. When counting, data should be kept on a pad or with a handheld computer. Quadrats can be used in conjunction with GPS and maps to accurately record locations of the sampling and population counts.

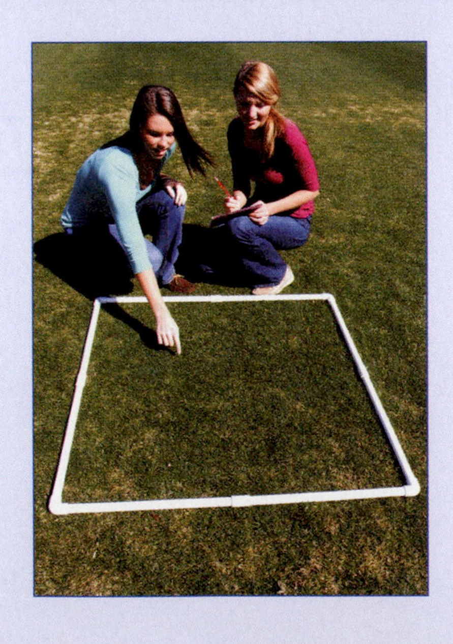

The newly emerged turf on an athletic field is being sampled to study signs of plant health. Counts are made of emerging weeds, the presence of insect forms, and other conditions. (Courtesy, Education Images)

INSECT BIOLOGY

Insects are best known as small animals with six legs, segmented bodies, and exoskeletons (invertebrates). The body segments of an insect are head, thorax, and abdomen. The head has a pair of antennae, eyes, and mouth parts. Antennae are sensory organs. The legs and wings, if any, are attached to the thorax. The abdomen usually has no attachments and may be divided into subsections.

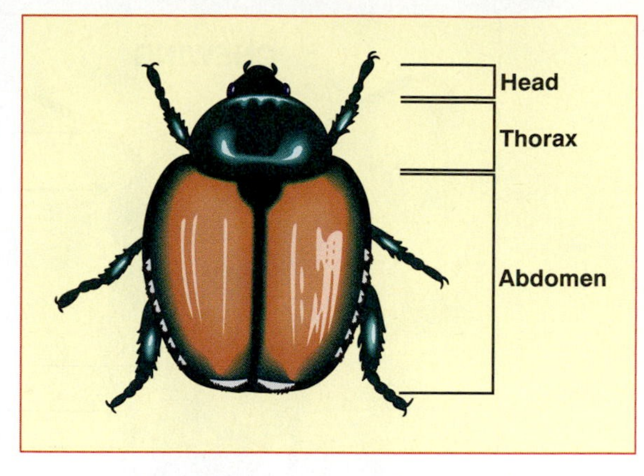

15–18. Body sections of an insect.

Sizes of insects range from very small to quite large. The fairy fly (*Dicopomorpha echmepterygis*) is cited as the smallest, with a length of less than 0.006 inch. The long stick (*Phobaeticus chani*) is cited as among the largest, at more than 22 inches long when mature.

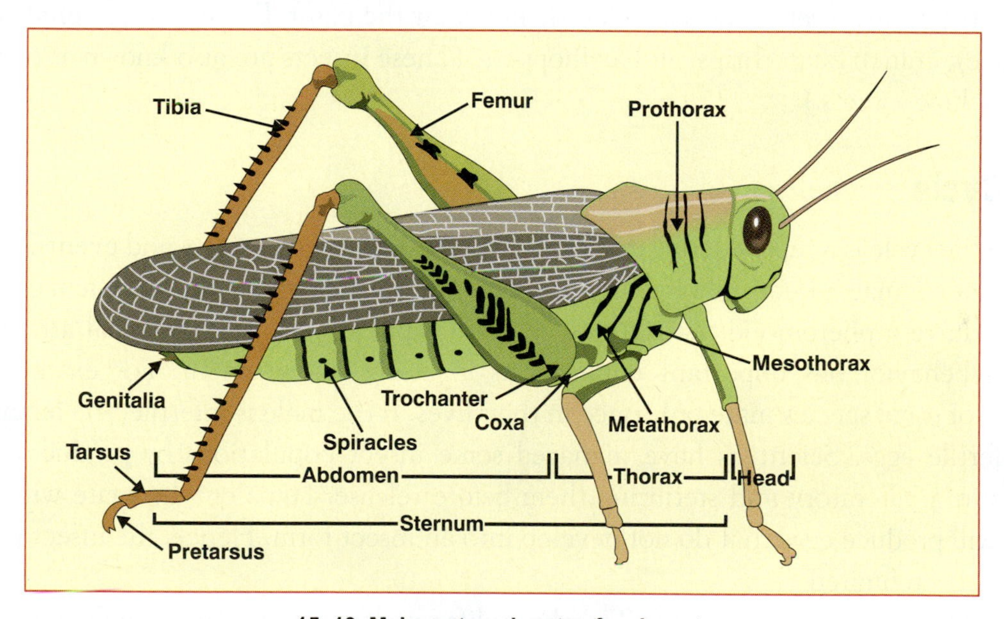

15–19. Major external parts of an insect.

Mouth Parts

How an insect's mouth is made determines the way it gets food. Plant damage is related to mouth structure, which is important in trying to limit insect populations.

Insects are classified as:

- ***Chewing insects***—Insects that bite off, chew, and swallow parts of plants. They eat holes in plant tissue. Examples are cutworms, bean beetles, and armyworms (cutworms and armyworms are the larva stage of moths).

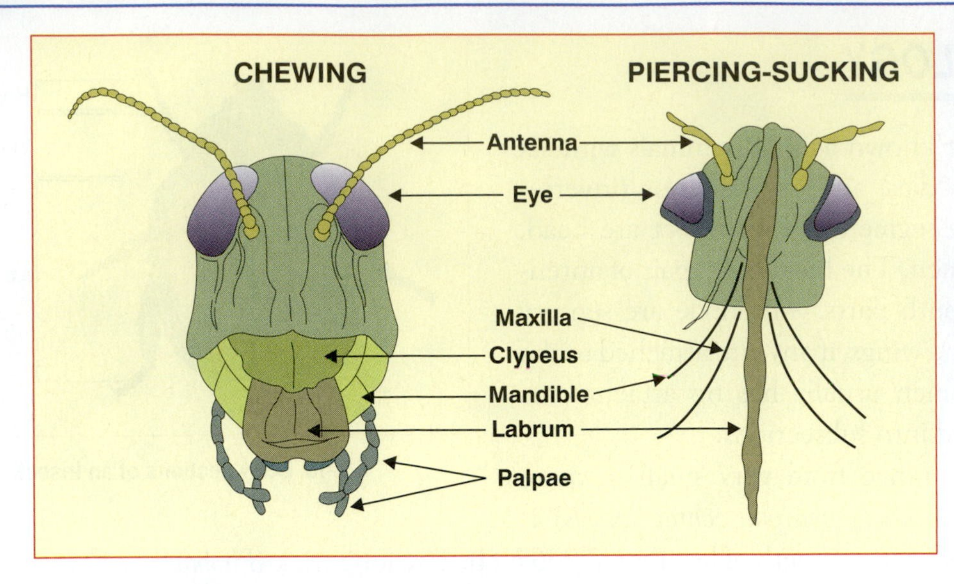

15–20. Insect mouth parts.

- ***Sucking insects***—Insects that suck the sap from a plant. They pierce the outer layer of the plant to get to the sap. They do not chew the plant. Examples are aphids (plant lice), chinch bugs, thrips, and leafhoppers. (These insects are also known as piercing-sucking insects.)

Life Cycle

The life cycle is a reproductive and maturation sequence of changes and events. Mating of male and female insects involves the transfer of sperm from the male to the female. Some females have a pheromone that attracts males. (A pheromone is a chemical attractant.) Mating behavior has important implications in insect management. For example, the females of some species mate only once in their lives. If the male is infertile, the female will lay infertile eggs. Scientists have managed some insect populations by producing male insects in a laboratory and sterilizing them before release. Females who mate with these males will produce eggs that do not develop into an insect form. Hence, the insect population has been limited.

Insects have definite life cycles. They begin life in one form and go through one or more other forms before maturity and death. The series of changes from one form to another that an insect goes through is a process known as ***metamorphosis***. Some insects have a complete metamorphosis; others have an incomplete (simple) metamorphosis.

Complete metamorphosis has four stages of development. The stages are so different that the adult form has little in common with the other forms. The stages of complete metamorphosis are:

1. Egg
2. Larva

3. Pupa (resting stage)
4. Adult

Examples of insects with complete metamorphosis are moths, beetles, and boll weevils.

With incomplete metamorphosis, the young and adult forms resemble each other. The stages of incomplete metamorphosis are:

1. Egg
2. Nymph
3. Adult

Nymphs appear similar to the adults. Nymphs are smaller, do not have well-developed wings, and have no reproductive organs. The grasshopper is a good example.

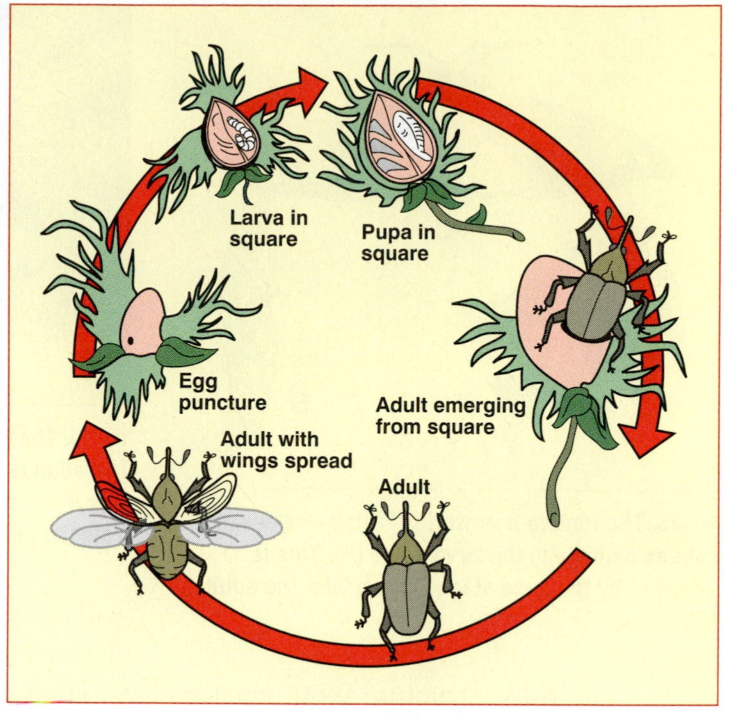

15–21. A complete metamorphosis is illustrated by the life cycle of a boll weevil.

Life Stages and Damage

Insects cause damage at different life stages. Often, it is at the larva stage that insects cause the greatest damage.

The tomato hornworm (*Protoparce sexta*) is an example of an insect that does damage in its larva stage. The adult moth does not cause damage. The adult lays eggs on a plant that

15–22. Tent caterpillars (larvae of the *Malacosoma americanum* moth) construct silk tents and can eat all the leaves off apple, cherry, and pecan trees.

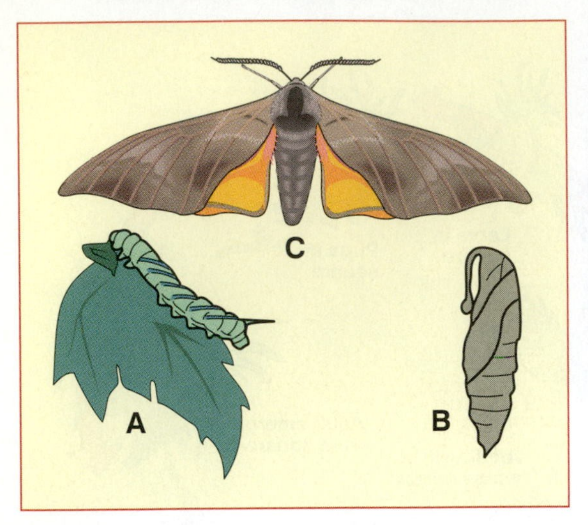

15–23. The tomato hornworm (*Protoparce sexta*) causes damage in the larva stage (A). This is followed by the pupa stage (B) and later the adult (C).

15–24. The Hessian fly (*Mayetiola destructor*) is a pest on wheat. The adult is only about 1/8 inch (3.2 mm) long. Eggs laid on wheat leaves hatch in five days. The developing maggots suck juice from the plants. (Courtesy, Mississippi State University)

will provide food for the larva. As an egg hatches a few days after laying, a worm 2 to 3 inches long with a hornlike structure at one end develops. This worm has a big appetite and can strip the leaves from a tomato or tobacco plant in a few days. In the pupa stage, it "rests" in the soil and may be dug up when the soil is plowed.

Knowing the life stages of an insect helps in controlling the plant pest.

CONDITIONS THAT PROMOTE PEST DAMAGE TO PLANTS

A ***plant pest*** is anything that causes injury or loss to a plant. Pests damage plants and make them less productive. Some pests are living organisms. Other pests are nonliving substances. Sometimes, more than one pest will attack the same plant. Combinations are even more damaging to plant health.

Pests of one kind or another are usually present anywhere plants are grown. Their presence does not mean that they will always damage plants. Certain conditions must exist. Since most crops are grown crowded together, pests may attack more easily and cause greater problems.

Three things are needed for a pest problem to develop:

- **A pest**—A pest must be present. If one does not exist where plants are grown, it will not cause a problem. Cultural practices can be used to deter pests. For example, crop rotation, or growing different crops on the same land in different years, helps to keep

15–25. These pomegranate trees show good health and productivity. They appear free of pests and diseases. (Courtesy, Education Images)

pests away. Tomatoes are often rotated with other crops because of blight and other pests that may get into the soil. By not growing tomatoes every year, the pests lose a host that helps them survive. A host is a plant that provides a pest with food.

- **A susceptible plant**—Some plants are more likely to be attacked by pests than others. Pests that feed on one species of plant can be present and not cause problems for other species. Once the susceptible plant is grown, the pest has a host to attack. Producers who know they have certain pest problems grow crops that are not affected by the pest. Again, the tomato is a good example. People who know that the cause of tomato blight is present in the soil will not grow tomatoes. They will grow other plants that the blight does not attack.
- **The right environment**—A pest and a susceptible plant can grow together without problems, particularly if the environment is not right for the pest to increase population and become a threat to the crop. For example, if a plant that prefers warm, dry

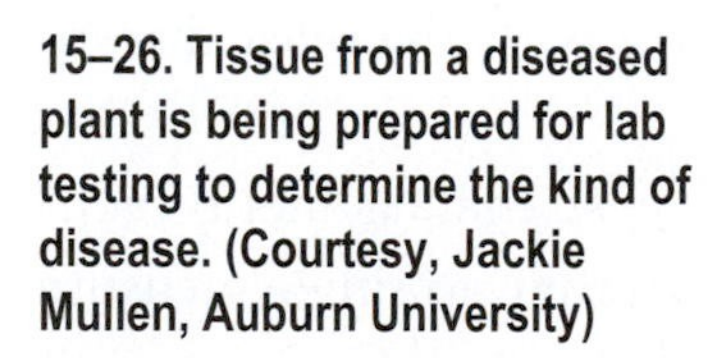

15–26. Tissue from a diseased plant is being prepared for lab testing to determine the kind of disease. (Courtesy, Jackie Mullen, Auburn University)

15–27. Damage to this tree was caused by a wire placed around it. The tree continued to grow and covered the wire. The strangling effect (often called girdling) of the wire will kill the tree because the xylem and the phloem will be unable to transport raw materials and food between the roots and the leaves. (Courtesy, Education Images)

weather is growing in cool, damp weather and a pest is present that does well in that type of weather, the pest may attack the plant. Growers can use practices that create a hostile environment for the pest. In some cases, varieties of crops that are resistant to disease can be grown. Pesticides (pest killers) are used to make the environment deadly to pests. Always plant crop varieties that are well suited to your climate.

INTEGRATED PEST MANAGEMENT

Many pest problems can be prevented. Good practices in producing plants will help reduce problems. Most practices are relatively easy to implement.

Integrated pest management (IPM) is a planned process for managing pests. It involves using methods that have the best outcomes for the well-being of society. A combination of measures is used to reduce pest damage with the least disruption of the environment.

IPM uses a blend of pest management techniques. Research has shown that no single measure works consistently over a long time. One reason is that pests develop resistance to pesticides. IPM is a combination of physical, chemical, and biological methods.

Internet Topics | Search

Selected topics for Internet discovery and reporting are listed here. Begin by searching for each topic. Next, read and learn about it. Conclude by preparing a brief report on your findings.

1. Insect scouting
2. Integrated pest management
3. Nematode

Emphasis is on reducing pest numbers to a level where they cause minor damage. IPM does not usually result in the elimination of all pests. It brings their populations down so they do not destroy too many leaves, buds, fruits, or other parts of plants. Fields must be checked regularly (scouted) for insects.

A field of crops is managed as an ecosystem. This means that all organisms live together and share the space and its resources. IPM attempts to keep the pests at a level where damage to plants is minimized.

IPM has fewer adverse (bad) effects on the environment and society. It is used to keep the environment clean and free of harmful materials. In years past, insect control was a major goal. This involved using methods to destroy all insects without regard to the impact of such methods on the overall environment.

15–28. Using adapted, genetically improved varieties and treated seed promotes good plant health. (Courtesy, American Soybean Association, St. Louis, Missouri)

PREVENTING PEST PROBLEMS

Preventing pest problems is far better than treating them. If there are no problems, there is no cost to manage them.

Some suggestions for preventing pest problems:

- **Use good seed**—Seed can carry pests. A pest of this type is known as a ***seed-borne pest***. Diseases are more likely to be in seed than other pests. Use seed from sources that are disease free. Use seed and plants that have been appropriately inspected and certified as disease free.

15–29. Scouting is a scientific method of determining the kinds and numbers of pests in a field before applying a pesticide. (Courtesy, Agricultural Research Service, USDA)

- **Scout and trap to know the situation**—Scouting is observing plants in the field or greenhouse for signs of pests when they may first occur. Traps capture insect pests and confirm their presence. Scouting and trapping allow management measures to be used before the pest population increases to cause major losses.

- **Destroy diseased plants**—Destroying diseased plants often destroys the pests that are in them. Cutting and moving trees with borers protects uninfected trees from the borers. Bury, burn, or otherwise destroy diseased shrubs.
- **Use the right fertilizer**—Plants can suffer from nutritional diseases. Provide the needed fertilizer to overcome these problems. Plants that don't have the needed nutrients are weaker and subject to attack by other pests.
- **Disinfect equipment**—Equipment can carry pests from one field to another. Wash equipment that was used in a field where pests were present before using it in another field. A solution of water and bleach may be used to disinfect hand tools and other equipment, such as lawn mowers.
- **Use good water**—Irrigation water can bring pests to a farm or field. Use water that is free of pests. Water that is polluted with chemicals can bring problems to crops. Water that is high in certain minerals can poison plants and the soil. For example, water treated with chlorine should not be used on tomato plants. Chlorine damages the plants.
- **Manage animal movement**—Animals can carry pests in their bodies, drop uneaten food, or pick up pests on their feet and bodies. Managing wild animals isn't easy, but good fences can keep out larger animals.

CAREER PROFILE

PESTICIDE APPLICATOR

Pesticide applicators operate equipment to apply pesticides. They must know the target pest and how best to control it. Pesticide applicators need skills in using equipment, formulating pesticide solutions, and following safe practices.

Pesticide applicators need education in such areas of agriculture as entomology, weed science, and agronomy. Some have college degrees; many have two-year or associate degrees. All take certification training and inservice to learn and stay up-to-date on regulations and pesticides.

Jobs are found where crops are grown. Pesticide applicators may work for large farms or applicator companies, or they may be in business for themselves. This shows apple trees being sprayed with gentle air flow to prevent drift. (Courtesy, Agricultural Research Service, USDA)

- **Use chemicals properly**—Plants can be damaged by the improper use of chemicals. They can be stunted or killed or may not produce fruit if harmful chemicals get on them. Always use chemicals in approved ways.
- **Use tests to check for pests**—If a pest is suspected, have the plant tested. Take samples of leaves or other tissue to a lab for analysis by trained people. Managing pests is easier if they are caught before a large area is infected.

15–30. An entomologist and a cotton producer check a boll weevil trap baited with boll weevil pheromone. (Courtesy, Agricultural Research Service, USDA)

REDUCING PEST POPULATIONS

If pests get into crops and other plants, they must be managed, or the crops may be lost. Several methods of pest management are used. The method chosen must be right for the crop and the pest.

Laws regulate practices that can be used. Some pesticides have been declared unsafe. They are no longer approved for use on crops, trees, and ornamental plants.

Mechanical Methods

Mechanical pest management is the use of tools or equipment to remove or destroy a pest. Examples of mechanical methods:

- **Plowing**—Plowing destroys some pests, particularly weeds. The pests are cut off or dug up. Conventional cultivation with hoes, harrows, and other equipment destroys weeds. This works best in hot, dry weather. Plants can be "set out" and grow again, if a rain immediately follows plowing.
- **Mowing**—Mowing cuts off weeds. Not only does it partially destroy the weeds, but it also destroys places where other pests can hide. Most weeds can sprout after mowing. Mowing may be needed several times in a growing season. Mowing is not used with some crops because it would destroy the crops.
- **Mulching**—Covering the ground with a layer of plastic, sawdust, or other material prevents weed growth. Mulching is used with ornamental plants and certain vegetable crops.

Cultural Practices

Cultural pest management is the use of techniques to keep pests out of a crop. Some practices are easy to do; others are more involved. Some examples:

- **Rotating crops**—Grow different crops on the land. Growing the same crop year after year allows pests that harm a particular crop to build up in the soil. Crop rotation involves planting one crop, such as soybeans, one year and another crop, such as corn, the next year. Without a host, the pest may die before a particular crop is planted again.
- **Roguing**—***Roguing*** is the removal of infected plants from a field, forest, or orchard. For example, a diseased tree can be cut down and moved. This prevents the infection of other plants.
- **Trap cropping**—***Trap cropping*** is planting a small plot of the crop near the field about two weeks before the main crop is planted. Pests will attack the trap crop. Those pests that reproduce in the earliest plants of the year go to the trap plots, where they can be killed before they damage the main crop.
- **Burning**—Setting dry fields on fire after they have been harvested destroys some pests. It also does away with plant remains that could serve as hiding places for pests. Of course, burning produces smoke, which damages the air.
- **Using resistant varieties**—Some varieties of plants have been developed that resist pests. Growers should plant these varieties to help reduce pest problems.
- **Cleaning around fields**—Weeds and bushes around fields can be hiding places for pests. Mowing and spraying can be used to keep pests down.
- **Cutting stalks**—Insect pests can hide in plant stalks. Cutting the stalks after harvest removes hiding places.

15–31. Cutting stalks after harvest is a good cultural practice to curtail insect populations. It reduces hiding places for over-wintering.

Pesticides

15–32. Many kinds of pesticides are sold in garden centers.

A ***pesticide*** is a chemical material used to control pests. Most pesticides are deadly—they are made to kill pests. As "killers," they can cause problems for other life. However, the advantages of pesticides far outweigh the disadvantages. They make it possible to produce more food!

Most pesticides are complex compounds. They are developed in laboratories and tested on trial plots. Various government agencies, such as the Environmental Protection Agency (EPA), must approve their use.

Chemicals are often mixed with other materials when they are used. Many chemicals are mixed with water. In some cases, a substance, such as a surfactant, is used to help a solution "work." A ***surfactant*** is a surface-active material that helps a solution spread over the waxy surfaces of leaves.

Insecticides

An ***insecticide*** is a chemical used to kill insects. Some insecticides are very poisonous. The material in an insecticide that kills the pests is known as the active ingredient. The label on the container describes the amount of active ingredient. Some are more powerful (concentrated) than others. Insecticides are mixed with other materials, such as water, for use.

Insecticides come in several forms: dusts, granules (small solid pieces), insecticide-fertilizer mixtures, various kinds of powders, and solutions. Each works best under certain conditions. Most of the powders and solutions are mixed with water for spraying on the affected plants. Applicator equipment must be properly adjusted and operated.

15–33. This airplane is applying insecticide to cotton. (Courtesy, Mississippi State University)

Since insecticides are designed "to kill," use them with great care. Follow all directions. Use only as approved by the federal government. Protect yourself and the environment. Also, remember that insecticides will usually kill beneficial insects as well as those that are pests.

Insecticides are classified by how they enter an insect's body.

- **Stomach poisons**—A ***stomach poison*** is a poison that must be eaten by the insect to be effective. Layers of the insecticide must be on the plant leaves or other materials the pest eats. Stomach poisons work best on chewing insects, such as leafworms, armyworms, and bagworms.
- **Contact poisons**—A ***contact poison*** is absorbed through an insect's skin. To work, a contact poison must come into contact with the insect. This means that the poison must be sprayed on the insect or on places where the insect will go. Contact poisons are best for controlling sucking insects, such as aphids.
- **Systemic poisons**—A ***systemic poison*** is a poison that is absorbed by a plant. It may be applied to the soil and taken up by the roots. Some systemics are sprayed on the leaves and stems and absorbed into the plant. The vascular system moves the poison all over the plant. When an insect bites the plant or sucks its juice, it gets poison.
- **Fumigants**—A ***fumigant*** is an insecticide in gas form. The poison enters the insect's body through the respiratory system. Fumigants can be used only in enclosed places. People and other animals must stay out during the treatment. Fumigants are good for managing insects in seed, such as wheat in a bin, or insects on plants in greenhouses. Sometimes they are used to treat the soil. When this is done, the ground is covered with a layer of plastic to hold the gas in contact with the soil. Some fumigants are quite hazardous. Their use is being increasingly restricted.

15–34. The soil under the plastic has been fumigated. Notice how the edges have been sealed to prevent loss of chemical.

Nematocides

A ***nematocide*** is a chemical used to control nematodes. Since nematodes are in the soil, treatments must be made that reach into the soil.

Fumigants are sometimes used. The soil is usually plowed and covered with a plastic sheet. The nematocide fumigant is sprayed under the plastic. In some cases, the nematocide is injected into the soil.

Tear gas (chloropicrin) may be used for fumigation. It kills nematodes, fungi, and weed seed. (Caution: Only use approved nematocides and in approved ways.)

ACADEMIC CONNECTION

BIOLOGY

Relate the content of this chapter to your study of biology. Include insects, viruses, fungi, and other organisms that are pests in plant production.

Protectants and Therapeutants

Many kinds of chemicals are used to manage plant diseases. The type used depends on the problem. An accurate diagnosis of the disease is essential. Without the diagnosis, the wrong chemical may be used, and the disease will not be controlled. When used on plants that are not diseased, a chemical is known as a ***protectant***. When used on plants that are diseased, it is known as a therapeutic chemical or a ***therapeutant***.

A ***fungicide*** is a chemical used to manage disease caused by fungi. The best fungicides are systemic in action. They are absorbed into the vascular system of a plant and reach all parts. Fungi are killed all over the plant. Fungicides that kill by contact must be sprayed directly on the growing fungi.

15–35. A fungicide is being applied to young poinsettia plants. (Courtesy, Agricultural Research Service, USDA)

A ***bactericide*** is a chemical used to manage bacteria. Such a chemical is also known as a germicide. Bactericides are frequently sprayed on plants. Caution should be used to avoid damaging the plants.

15–36. A small sprayer is labeled for use with a herbicide. Residues in the sprayer could damage plants if the sprayer were used to apply an insecticide, fungicide, or protectant.

Herbicides

A ***herbicide*** is a chemical used on weeds. Many kinds of herbicides are available. The right kind must be used for the intended results.

A selective herbicide is a herbicide that kills only certain kinds of plants. It will usually kill either broadleaf weeds or grasses, but not both. For example, cocklebMurs (broadleaf weeds) can be controlled in corn (crop with grass-type leaves) with herbicides that cannot be used on soybeans or other broadleaf plant crops. When used properly, selective herbicides will kill weeds without damaging the crop plants.

A nonselective herbicide is a herbicide that will kill upon contact. It is sometimes known as a contact herbicide. All living cells, whether weed or crop plant, are killed. Nonselective herbicides kill all the vegetation where they are applied. For example, a nonselective herbicide can be used along the edge of a walk in a lawn to keep grass from growing over the walk.

15–37. A herbicide was used on henbit (a winter weed) in a dormant Bermuda grass lawn (top—before use; bottom—after spraying).

A ***translocated herbicide*** is a herbicide that is absorbed and moved through the vascular system to all plant parts. This herbicide, also known as a growth regulator, upsets growth processes by interrupting the use of food. Action of the herbicide may take time. The plant has to move the herbicide throughout before it is killed. This herbicide often causes a plant to grow in twisted, unnatural ways the first few days after use. The plant usually turns yellow and brown as it dies in a couple of weeks. Translocated herbicides work best in warm weather when plants are growing rapidly.

Spraying is a common way to apply herbicides. They are also applied as granules mixed with fertilizer and applied to the soil. Herbicide action takes place slowly, as granules are dissolved. Herbicides may be applied before, during, or after planting.

Applying a herbicide before planting is preplant application or pre-emergence. This keeps weeds down as crop seeds germinate. Weed competition is reduced. The herbicide is sometimes incorporated (mixed) into the soil. Care is needed so the herbicide does not damage seed or tender plants as they grow.

15–38. Greenhouse tomatoes infested with white flies.

Some herbicides are applied at the time of planting. Sprayers are attached to the planter. A band 8- to 12-inches wide is sprayed on the drill after the seeds are covered with soil.

Post-emergence herbicides are used after crop plants are up. They are sprayed on the leaves of weeds. The herbicide used must be selective in action; otherwise, it will kill the crop plants.

Biological Methods

Biological pest management is the use of living organisms to manage pests. This control method might include insects that eat other insects, bacteria and fungi that attack pests, and alterations to the reproductive processes of the pests.

Not all insects are bad. Some do much good. They help maintain a good environment for plants. A ***beneficial insect*** is an insect that helps crops grow or provides other important benefits.

Many insects are beneficial in controlling other insects. Pests have natural enemies in the environment. Lady beetles are notorious for their roles in attacking a range of insect pests. Predatory mites have been used to control other kinds of mites on apples. Of course, many gardeners like to have a few toad frogs around to eat insects! Cau-

15–39. The honeybee is a beneficial insect that pollinates flowers and provides honey. Some people raise bees on a bee farm (called an apiary).

15–40. The brachonid wasp (*Cotesia congregata*) is an example of biological control. The tiny wasp lays eggs in other insect forms. The larvae emerge from the worm's body and form attached cocoons. Feeding on the worm causes gradual

tion: Insecticides should not be used on plants to kill insects, if predatory insects are present. The insecticide will kill the beneficial insects.

Forms of bacteria and fungi have been developed for release into the environment. These organisms attack pests and destroy them. An example is the bacterium *Bacillus thuringiensis*. When released in fields, the bacteria attack and kill various species of worms.

Some insect pests have been controlled by altering their reproduction. Since some insects mate only once, a female that mates with a sterile (no sperm) male lays eggs that will not hatch. The male insects are sterilized and released in areas with insect problems. The sterile males mate with the wild females. The eggs produced are infertile (will not hatch). This method has been used to control some pests, such as the pink bollworm, which attacks cotton.

15–41. Selected examples of insect pests are: 1—alfalfa plant bug, 2—Mexican fruit flies, 3—Colorado potato beetle, 4—sugarcane borer (adult), 5—tarnish plant bug, and 6—medfly. (Courtesy, Agricultural Research Service, USDA)

Genetic Methods

Genetic pest management is the development of crops that are resistant to pests. This means that the crop plants have some trait that repels the pests.

Science has had great success with plants that are resistant to disease. Wheat, tomatoes, potatoes, corn, soybeans, strawberries, and many other plants have been developed that resist at least one disease.

Some plants have been developed that are resistant to nematodes. Crops that resist nematode pests are more productive and economical to produce.

Pests can sometimes overcome the resistance of certain varieties. Consequently, new resistant plants must be continually developed.

More emphasis is being given to genetic methods of pest management. These are thought to be less damaging to the environment than other methods. However, some people fear that the new genetic forms may pose problems themselves.

SAFETY PRACTICES IN PEST MANAGEMENT

Many pest management methods pose hazards. They can injure people and animals, pollute the environment, and contaminate water and food. Laws have been passed to regulate the use of pesticides. Most pesticides are approved only for certain uses.

15–42. A Minnesota producer is checking navy beans for signs of damage. (Courtesy, Education Images)

GENERAL SAFETY PRACTICES

Always read the labels and use pesticides accordingly. Many pesticides can have damaging effects on the environment if used improperly. The well-being of the person applying a pesticide, as well as the well-being of other people and living organisms, must be respected. A few safety rules to follow in pest control:

- **Use only approved pesticides**—Government regulations allow only certain pesticides to be used. Follow the laws!
- **Get the training to be an applicator**—Training is needed to apply many pesticides. Arrange to have the training and gain certification as an applicator.
- **Know the pesticide**—People must be informed about the pesticides they use. Container labels offer much information. Use a pesticide according to directions on the label. Do not use it where it is not approved.
- **Use the pesticide with lowest toxicity**—Toxicity refers to how poisonous the pesticide is. Use the pesticide that will do what needs to be done but that is no stronger than needed.
- **Use pesticides only when needed**—Pesticides should be used when pests are a problem. Using a pesticide without need damages the environment and wastes money.
- **Do not contaminate resources**—Pesticides can pollute the environment. They should never be dumped into streams or on the ground. Leftover pesticides should be properly disposed of when no longer needed. Keep pesticides away from food and water.
- **Wear protective clothing and a respirator**—Protective clothing and a respirator (mask) should be worn. Clothing should be washed after it is worn. The Worker Protection Standard (WPS) law specifies how pesticides are to be used and labeled.
- **Wash the skin after contact**—Wash with soap and water immediately if any pesticide gets on the skin. Wear rubber gloves to reduce pesticide contact with skin.
- **Dispose of empty containers properly**—Empty containers should never be thrown into creeks or gullies. Some manufacturers recycle

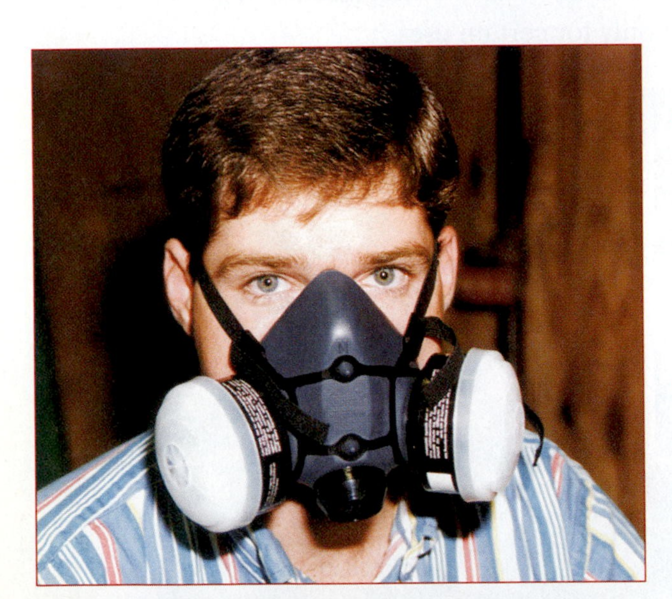

15–43. A respirator may be needed when applying pesticides.

empty containers. Generally, empty containers should be rinsed out three times and returned for recycling or sent to an approved waste facility.

- **Apply in good weather**—Pesticides should be used when they will have the most effect on pests. Wind blows sprays and dusts away. Sometimes, drifting poisons can damage other crops, water, or livestock.
- **Use the right equipment**—This includes funnels to help in pouring and measuring and proper mixing containers. Spraying equipment should be adjusted correctly so it applies no more than is needed.
- **Know the emergency measures**—People who use pesticides should know what to do in case of an accident. Local physicians know whom to contact when people are poisoned. It is a good idea to have the emergency telephone number handy.

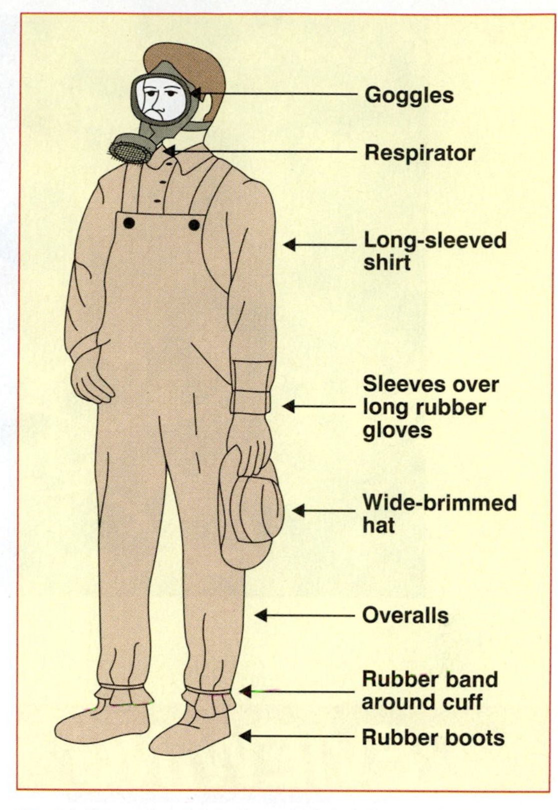

15–44. Proper protective clothing is needed when applying chemicals.

PROPER APPLICATION

Proper application is a responsibility of anyone who uses a pesticide. Here are a few principles to follow:

- Know the hazards of the pesticide and how to transport and store it properly.
- Know the approved uses of the pesticide.
- Use the pesticide only for the approved purposes with regard to both the crop to which it is applied and the pest being managed.
- Apply only the approved amount based on the formulation being used.
- Use equipment that is in good condition and properly calibrated.
- Consider nearby crops, animals, creeks and ponds, and other features that could be damaged by drifting pesticide.
- Wear the proper personal protection equipment.
- Properly clean equipment after an application.

15–45. Ultra-low volumes of pesticides can be applied using special application equipment. This reduces likely environmental damage. (Courtesy, Agricultural Research Service, USDA)

REVIEWING

MAIN IDEAS

Plant health is the condition of a plant as related to growth and productivity. It affects both the plant itself and the products it produces.

Plant pests cause injury or damage to plants. They may be living organisms or nonliving substances. The five major kinds of pests are insects, nematodes, diseases, weeds, and rodents and other animals.

Entomology is the study of insects and related small animals, such as spiders, mites, and centipedes. It is frequently studied in conjunction with agronomy, horticulture, and forestry. Both beneficial and harmful insects are included.

Three things are needed for a pest problem to develop: a pest, a host plant, and the right environment. If one is missing, there will be no problem. Pests damage plants in many ways. They cause reduced crop yields even if the plants are not destroyed.

Preventing pests is far better than trying to get rid of them after they get into a crop. Managing pests can be costly and cause other problems, such as damage to the environment. Methods of pest management should be selected that are most effective.

Integrated pest management (IPM) is a planned process for managing pests. It allows some pests to exist, with control measures used when the pests reach a certain level. Regular scouting of the crop by trained people is needed.

Always follow safety procedures when using pesticides. Protect people and the environment.

QUESTIONS

Answer the following questions, using complete sentences and correct spelling.

1. What is a plant pest?
2. What are insects? How are they identified?
3. What are the life stages of insects? How do the stages relate to plant damage?
4. What is a beneficial insect? List three ways insects are beneficial.
5. What is a nematode? How do nematodes damage plants?
6. What is a plant disease? What are the signs of plant disease?
7. What are the types of plant disease?
8. What is a weed? What are the kinds of weeds?
9. How do weeds cause problems for plants?
10. What conditions must exist for pests to damage plants? Explain each.
11. How do pests damage plants? How does the damage affect the producer?
12. How are plant pest problems prevented?
13. What methods are used to manage pests? Briefly describe each.
14. What is the relationship between the way an insect gets its food and the kind of pesticide that may be used?
15. Distinguish between selective and nonselective herbicides.
16. What is integrated pest management?
17. What are the important safety practices when using pesticides? Select three that you feel are most important.
18. What is entomology? What does it include?

EVALUATING

Match each term with its correct definition.

a. plant pest
b. environmental disease
c. weed
d. pesticide
e. insecticide
f. herbicide
g. fungicide
h. translocated herbicide
i. beneficial insect
j. biological pest control

______1. An insect that helps crops grow or provides other important benefits.

______2. A plant that is growing where it is not wanted.

______3. Anything that causes injury or loss to a plant.

______4. A chemical used to manage weeds.

______5. A chemical used to control disease caused by fungi.

______6. A chemical used to kill insects.

______7. A plant disease caused by factors in the environment that are not "right" for the plant.

______8. A herbicide that is absorbed and moved through the vascular system to all plant parts.

______9. A chemical material used to manage pests.

_____10. The use of living organisms to manage pests.

EXPLORING

1. At your home, in the school lab, or elsewhere, find a plant that has been damaged by a pest. Determine the nature of the damage and what caused it. Prescribe ways of eliminating future damage. Get the assistance of a local entomologist or other agricultural specialist. Prepare an electronic report on your findings, including images you have made of the damage.

2. List major plant pests in your community. For each, describe how the pest can be prevented and controlled. (Interview producers, use reference materials, and get the help of agriscientists with this activity.)

3. Scout a field crop, vegetable crop, or lawn area for pests. Identify the pests found. Prepare a recommendation on what should be done about the pests. Give an oral presentation in class on your findings.

4. Make a collection of brochures and pamphlets that describe plant pests found in your community. This should include information about the pests and how they are managed. Information is available from the Cooperative Extension Service, a university, or a supplier of pesticides.

Part Five

Animal Science

CHAPTER 16

Animal Care and Well-Being

OBJECTIVES

This chapter presents important principles of animal science. It has the following objectives:

1. Discuss major factors in animal production.
2. Explain animal well-being.
3. Explain the meaning and importance of quality-assurance programs.
4. Describe the role of veterinary care.
5. Identify practices in animal restraint and management.

TERMS

animal husbandry
animal well-being
diagnostic ultrasound
euthanasia
head squeeze
laboratory analysis
quality assurance
quality-assurance program
restraint
stethoscope
veterinary medicine
veterinary technology

16–1. A newly-hatched chick is being weighed. (Courtesy, Education Images)

MANY SPECIES of animals are found on Earth. So far, scientists have identified and named 1½ million species of animals. New species are identified each year. Animals provide a large number of useful food and fiber products. People also gain numerous benefits from animals, including companionship and service.

Animals vary in many ways. For example, some animals are very large; others are very small. Some live in water; others fly in the air and walk on the ground. Some have been domesticated—they live under the control of humans; others are wild—they may be hunted or admired for their beauty. Many wild animals are dangerous.

Over time, the knowledge gathered by producers, scientists, and others is most useful. Today, research continues to provide additional information about these animals.

ANIMAL PRODUCTION FACTORS

Animals are complex organisms. Knowing the needs of animals and how their bodies function helps us do a better job of caring for them. Some types of animals have been domesticated and raised for hundreds of years. Domesticated animals are removed from their native environments and raised in situations created by humans.

Caring for animals is often known as husbandry. ***Animal husbandry*** is the scientific management and control of animals. The needs of the animals are met in an environment that provides for their well-being. It is essential to understand animal needs to properly provide for their care.

16–2. Confined environments must provide for the needs of animals and promote their well-being. (This shows a modern hog facility. Note the feeder in the background and the suspended waterer. The floor is slated so wastes fall below into a pit keeping the area where the hogs are raised clean.) (Courtesy, Education Images)

NATIVE NEEDS

Animals are naturally found in a wide range of environments. They can be classified by habitat and by climate. Animals raised on farms and kept as companions should have their needs met. To do so, animal producers need to understand habitat and climate requirements and strive to provide these for their animals.

Habitat

Most animals are terrestrial or aquatic. Terrestrial animals live on land. Aquatic animals live in water or in a water environment. A few animals are amphibious, so they can live in water or on land. However, some of these animals are amphibious only during certain life stages.

Terrestrial animals need an environment that meets their needs. Food must be available. Appropriate protection from weather conditions may be required. The needs of animals vary. For example, cattle and hogs differ in food and environment needs.

16–3. Dolphins are adapted to aquatic environments.

Aquatic animals must have water appropriate to their needs. For instance, some animals need freshwater; others need saltwater. It is important to know the water environment for a species. For example, rainbow trout need cold, rapidly flowing water. Putting them in warm, nonflowing water does not meet their needs.

Climate

Animals vary in the climates they prefer. Temperature and rainfall are two major climate factors. Some animals, such as the polar bear, live well in very cold climates. Others, such as hogs and cattle, prefer temperate climates. Monkeys, gibbons, and jaguars prefer the warmer climates.

Animals have adapted to wet or dry climates. The animals that live in areas with a lot of rain live in water or have adaptations to help them escape from the water. Some monkeys have a long tail, which allows them to retreat into trees when water covers the ground. Ducks have webbed feet so they can swim in water. In contrast, camels, scorpions, owls, and

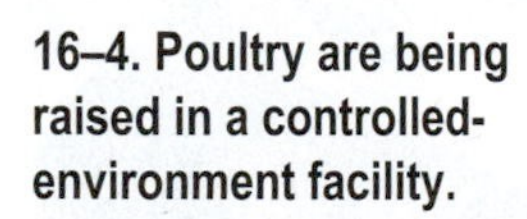

16–4. Poultry are being raised in a controlled-environment facility.

snakes are adapted to desert climates. Most cattle and other farm animals are adapted to temperate climates.

Cattle with humps are adapted to warmer, wetter climates than those without humps. For example, the Brahman breed of cattle is better suited to warm, wet climates than the Angus breed. Ranchers may cross the breeds to combine the best traits of both.

Some species need shelter; others are raised in enclosed houses. Housing varies by species. Poultry are raised in totally enclosed, climate-controlled facilities. Cattle may be provided sheds to escape inclement weather. Young animals, in particular, may need housing to protect them from cold, rain, or heat.

HUSBANDRY PRACTICES

Producers strive to meet the needs of animals in a productive environment that may vary considerably from the native environments of a species. Domesticated species no longer run wild and search for food. Producers keep the animals captive, control elements of their environments, and provide for their needs. If producers fail, animals will not thrive or grow and be productive.

Several husbandry practices apply in animal production, as follows: (Most of these are covered in more detail elsewhere in this book.)

- **Making good choices**—This involves choosing the species to produce and providing husbandry practices that meet the needs of animals. The species chosen should be one

Technology Connection

PROTECTING FROM THEMSELVES

Animal health care is sometimes challenging because animals do not understand. The native responses of animals to health issues must be considered. How do you keep a dog from tearing a bandage off and licking a wound? This is a behavior that must be controlled when providing care.

The Elizabethan collar is used on some animals, particularly dogs and cats. The device is attached to the regular collar worn by the animal. The purpose is to keep the animal from biting or licking at its body or scratching its head. Collars are typically sized so the animal can eat and drink. In some cases, the collars are removed when watering and feeding and put back on immediately afterward.

The Elizabethan collar is also known as a space collar or cone. The collars are named after the ruffs worn by women in the Elizabethan times. They can be obtained from a pet store or a veterinary supplies business.

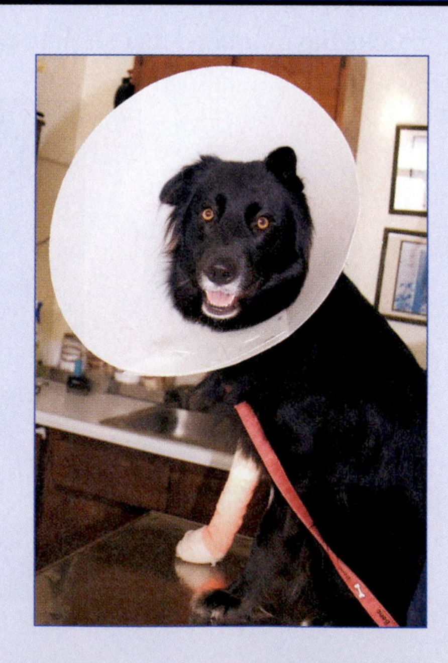

that is productive, meets consumer demands, and will thrive in the environment where it will be raised.

- **Providing nutrition and feeding**—Animals kept by humans do not have the freedom to seek and gain food as they would before domestication. They are often confined in pastures, pens, or buildings. Proper feed and water must be provided to meet the needs of the species. In some cases, animals will thrive on good pasture. In other cases, careful attention is needed to provide the appropriate feed. Regardless, the producer needs to have efficient growth so the animal enterprise is profitable.
- **Promoting good health**—Animal health is always a major concern of those in animal husbandry. It includes caring for animals in a manner that reduces or eliminates stress and injury. It also includes using practices that prevent disease, such as vaccination and sanitation. Providing the proper food and water is a major step in promoting good health. Some animals are confined in close quarters with many other animals, and this creates additional stress and disease issues.
- **Managing reproduction**—The success of animal producers often depends on the ability to gain efficient reproduction. New animals are essential. Some animals are kept for breeding; others are harvested for human products. Of course, some animals are kept as companions and are used for other purposes. With these animals, limiting reproduction is often a wise practice.
- **Protecting from hazards**—Most animals are subject to harm by predators, environmental conditions, and unique species endangerments. Housing or protection from the weather may be necessary. Producers may need to use practices to protect animals from harm. For example, a guard dog may be used to keep wolves away from sheep, or the needle teeth of baby pigs may be clipped to prevent injury to the teats of sows when the pigs nurse.

16–5. A muzzle is being placed over the goat's mouth to keep it from eating wood shavings on the floor of its pen. (Courtesy, Education Images)

- **Dealing with wastes**—Animals raised in confinement produce large amounts of wastes that require proper disposal. Removing the wastes promotes animal health and well-being. Also, proper disposal minimizes damage to the environment.

16–6. An aerial view of a North Carolina hog farm with barns and wastewater holding and treatment facilities. (Courtesy, Agricultural Research Service, USDA)

- **Being an ethical producer**—Animal ethics is caring for animals so their needs are met. They are never abused. Animals are fed and medicated properly. Practices are followed to ensure quality and wholesome products.

ANIMAL WELL-BEING

16–7. A hog that is kept in a way that promotes its well-being. (This young woman is grooming and training her pig for showing.) (Courtesy, Education Images)

Animal well-being is the state of an animal's health and comfort. We provide for an animal's well-being in ways that best meet its needs. Differences are considered when providing care. Producers of animals, as well as people who have companion animals, must practice animal well-being. Terms such as animal welfare and animal rights are sometimes used when referring to animal well-being. It is the opinion of some authorities that animal welfare is an acceptable term but animal rights may not be appropriate.

PROMOTING ANIMAL WELL-BEING

Producers, owners of pets, and others with animals can follow practices that promote the well-being of their animals. Promoting animal well-being includes:

- Providing food, water, and other nutrients that meet the needs of an animal

- Providing housing or protection from weather appropriate to the species and condition of an animal
- Respecting animals as living organisms and never abusing them
- Providing health care appropriate to an animal's needs
- Using humane methods in animal slaughter and other practices
- Holding reasonable expectations of an animal's ability, such as the speed of a horse
- Preventing unwanted reproduction so excess populations do not develop

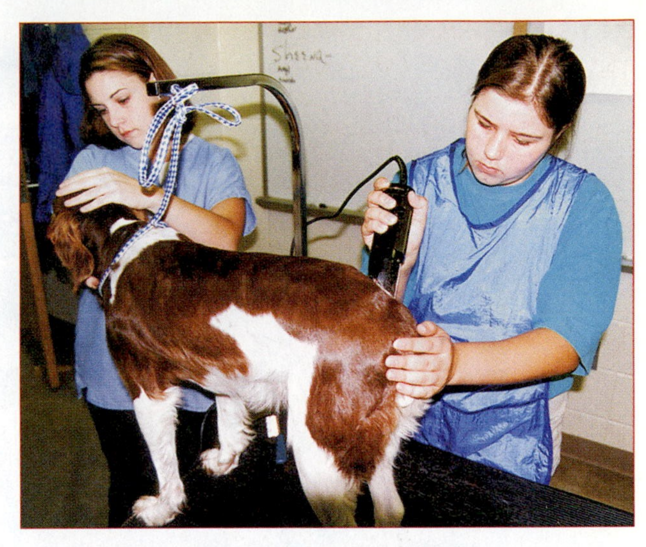

16–8. Grooming a dog helps provide for its well-being.

- Constructing barns, fences, and other confinements that protect animals from hazards and prevent injuries or exposure to polluted environments
- Transporting animals carefully to prevent injury
- Restraining an animal properly to prevent injury to yourself and the animal

CAREER PROFILE

SMALL-ANIMAL GROOMER

A small-animal groomer cares for the hair, skin, claws, and other exterior parts of an animal. The groomer may bathe the animal, clip and style its hair, trim its claws, clean its teeth, and otherwise improve the animal's appearance and cleanliness. Most small-animal groomers work with small animals and must be able to restrain and provide for their well-being.

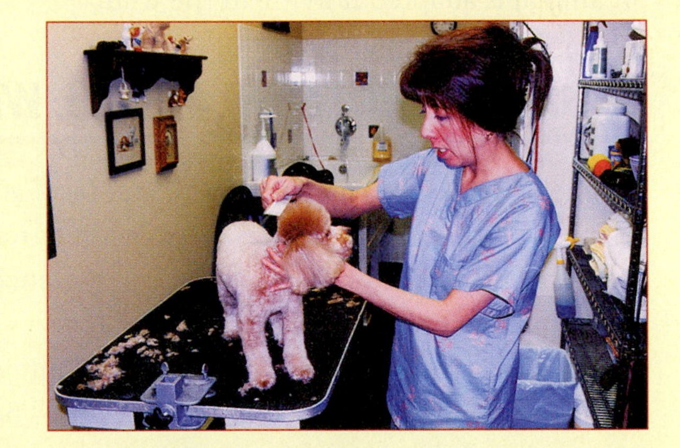

Small-animal groomers require on-the-job training and other preparation in how to handle and care for animals. A high school education is needed, though many have technical education in animal care or a related area. Preparation begins by taking science, business, and agriculture courses in high school.

Jobs are found in veterinary medical clinics, pet stores, and grooming clinics. A small-animal groomer may work for someone else and be paid a salary or may operate his or her own small-animal grooming business. This shows a poodle being groomed according to breed requirements. (Courtesy, Education Images)

16–9. Using a horse to work cattle on a ranch. (Courtesy, Texas Department of Agriculture)

- Respecting other people by confining an animal that might wander away or attack people

Some animals develop disease and suffer. These animals need veterinary medical care. In some cases, euthanasia is a practical solution. ***Euthanasia*** is the act of killing an animal to relieve it of suffering. Painless methods should be used. Only qualified people should practice euthanasia on animals.

Most animal keepers provide a far better environment than the animal would have if it were not domesticated. The overall living conditions should meet or exceed those to which an animal is adapted if living in the wild.

LEGAL ASPECTS OF ANIMAL WELL-BEING

The Federal government, states, and local governments have regulations that relate to animal well-being. These rules are sometimes referred to as animal welfare laws. They promote the treatment of animals in ways that encourage their well-being and do not subject them to unnecessary pain, injury, or death.

The U.S. Code (Title 7, Chapter 54) contains legal regulations on the transportation, sale, and handling of certain animal species. The species include those used as companion animals (pets), lab animals, performing animals, and fighting animals. The Animal Fighting Prohibition and Enforcement Act of 2005 prohibits interstate commerce of animals for fighting. It includes selling, buying, transporting, delivering, or receiving animals for the purpose of a fighting venture. Dogs and roosters have been used in fighting activities. Such uses are considered cruel and illegal in the United States. Agricultural animals are typically exempt from such laws.

State governments may have a wide range of laws to promote animal well-being. Performing animals and animals kept in parks or displays are often covered. Some regulations may apply to how animals are used and how they interact with people. States may prohibit allowing untrained riders on performing animals, such as having a citizen ride an elephant at a carnival.

16–10. Local regulations often include companion snakes. (Courtesy, Education Images)

Local governments often have ordinances about the keeping and treatment of animals as companions and for other uses. These ordinances include conditions in which the animals are kept. Improper confinement, failure to provide care, and abusive practices are typically prohibited. Local authorities may sometimes confiscate abused animals and charge their owners with neglect. Local governments may operate animal shelters for housing confiscated or unwanted animals.

Local regulations may require the payment of a tax or the purchase of a tag for some species of animals. Animal control officers may confiscate animals or cite owners with tickets and fines for violating the regulations.

QUALITY ASSURANCE

Quality assurance is an effort to promote quality products through good management practices. The goal is to provide consumers with high-quality products that are safe to eat or use. Quality assurance is sometimes abbreviated as QA.

QA may be set up as a program. A ***quality-assurance program*** is an organized effort with specific requirements promoted for compliance to occur. QA programs are offered through producer organizations with support from processors and government agencies. In the U.S. Department of Agriculture, the Animal and Plant Health and Inspection Service (APHIS) is involved.

QA is a highly valuable practice that promotes on-farm quality of products. Producers follow specified practices in producing animal products. If the practices are followed, the herd is conferred a status as based on the use of good production practices or standards. QA is voluntary, but producers should know that it may hold considerable benefit. Some processors will only use products from farms or ranches that have met QA standards. Animal species and products typically included are beef, dairy, swine, poultry and eggs, some fish species (e.g., trout), and others.

16–11. Assuring that animals are not the subject of contamination involves having regulations for individuals to enter a farm. (Courtesy, Education Images)

The implementation of QA is grounded in HACCP (Hazard Analysis Critical Control Point). HACCP involves the identification of points in the path of producing, processing, and marketing a product where contamination of food products may occur. (More is presented on HACCP in Chapter 27.)

Consumers are increasingly wanting information about the origin and path of the products they buy. The phrases "gate to plate" and "farm to fork" are sometimes used as related to the ability to follow or track a product. In the unfortunate case of a food-borne illness, tracking a product to identify the source of contamination is important.

TYPICAL QA GUIDELINES

Guidelines vary somewhat by species or by product produced. The guidelines are intended to promote the use of best management practices. In general, the guidelines are in five major areas:

- **Food safety control points**—The practices in this area focus on assuring consumers a quality product that is safe to eat or to use. With large animals, major areas are animal medicine residues, injection sites, and foreign object avoidance. Animal exposure to pathogens is also an important matter in terms of food product safety. Chemical residues may be from deworming, external parasite control, and application of pesticides to pastures that are ingested while grazing or through harvested feed materials. Needles that break off in an animal also pose problems.

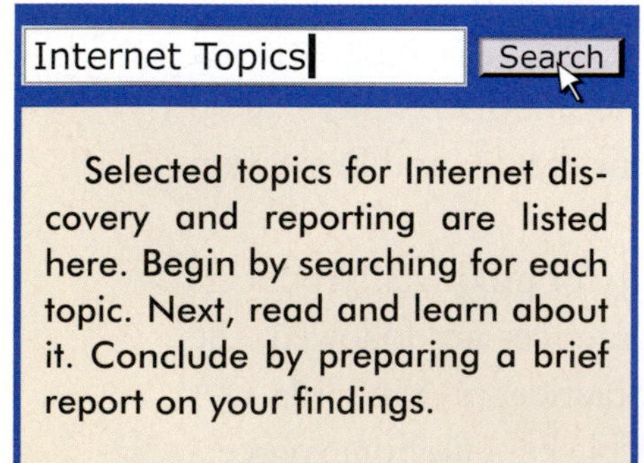

Selected topics for Internet discovery and reporting are listed here. Begin by searching for each topic. Next, read and learn about it. Conclude by preparing a brief report on your findings.

1. Feline bag
2. Cattle squeeze chute
3. Stethoscope

- **Product quality control points**—The practices in this area focus on the breeding and selection of animals and the use of animal health products. Careful selection of herd replacements and breed combinations may be related to product quality. The overall goal is to make sure consumers have a quality product.

16–12. Producers who use enclosed facilities and who follow QA guidelines can more readily ensure the quality of their animals.

- **Handling and processing control points**—This area addresses proper animal nutrition and removal of nonproductive or unsafe animals. For instance, animals that show certain diseases or other characteristics are culled. Some are not allowed to enter the human or companion animal food chain. Animals with certain diseased conditions may be used for human food with the part of the carcass that is diseased being trimmed away. An example is a cow with cancer eye. Corals, chutes, and head squeezes should be designed so animals are not injured. Such injuries may cause sores or bruises that damage quality.
- **Environmental control points**—This area deals with the disposal of dead animals, animal waste disposal, water quality, pesticide use, and soil and forage management. A goal is to use soil, water, and forage so resources are sustainable. Pastures are not overgrazed. The soil is not damaged. Water is protected from pollution. Good grazing management practices are followed.
- **Record keeping**—Accurate and up-to-date records of animal husbandry practices are needed. The records may be kept on paper or with computers. Such records should document important production factors, such as herd health and nutrition. With herd health, the records indicate what treatments have been used and when they were used. Exact data should be kept on what medicines were used, how much was given, how it was given (including the location of an injection on large animals), and when it was given. Records are helpful

Table 16–1. Treatment Record Information

Records of Treatments Administered to Animals Should Contain the Following:
• Date of treatment
• Identification of animal or group
• Animal weight or group average
• Product administered
• Serial number or lot of the product
• Earliest date the animal could clear withdrawal time
• Size or amount of dose given
• Route of administration (how given)
• Location of injection
• Name of person administering medicine(s)

if a marketed animal is found by a state or federal agency to violate residues present in the animal. Large animals are individually identified by number. Small animals are identified by groups or flocks. All records should be kept for a while after the sale or harvest of an animal, such as two years with beef animals.

Quality-assurance guidelines vary by species. Cattle are somewhat different from swine. Yet cattle and swine have major differences from poultry, poultry products, and fish. Of course, QA guidelines for cattle kept for milk are different from those for beef.

VETERINARY CARE

16–13. A veterinarian is using a stethoscope to listen to the internal sounds of a small dog.

Veterinary medicine is the branch of medicine that deals with animals. Individuals who provide veterinary medical care are known as veterinarians. These individuals have completed a doctor of veterinary medicine program at a university. Many veterinarians specialize in a particular area, such as large animals or small animals. Some veterinarians specialize by species, such as equine (horses) or swine.

Veterinary technology is the science and art of providing professional support service to veterinarians in their practice. People in this line of work must have basic skills in areas of animal health care. Jobs may involve veterinary assisting or veterinary nursing. Some veterinary clinics may also have grooming services.

Animals cannot tell people when they are feeling bad. Therefore, owners need to know the normal behavior of their animals and look for signs that the animals are not feeling well. It is a good idea to schedule annual or regular vaccinations, deworming, and other care for some animals. This will help prevent problems and ensure that the animals will live long, beneficial lives.

COMMON PRACTICES

Some of the practices used in veterinary medicine are included here. The practices listed are not complete but should help you become familiar with animal care. You will be in a better position to use veterinary medical services and to discuss animal care with your veterinarian.

- **Checking vital signs**—The vital signs are temperature, external respiration rate and effort, heart rate and rhythm, blood pressure and flow (sometimes known as perfusion), level of consciousness, and urine production. Temperature is taken with a thermometer or a similar device. Respiration and heart rate are taken by counting. Frequency of inhaling is used for the respiration count. Pulse is the beat of the heart that can be felt through the walls of the arteries. (The vital signs of selected species are presented in Chapter 20.)
- **Listening to internal sounds**—Internal sounds are important in assessing animal health and in identifying disease problems. A ***stethoscope*** is an instrument used to hear and amplify the sounds produced by the heart, lungs, and other internal organs. The modern stethoscope has two earpieces and a flexible rubber cord leading to them from the two-branched opening of the bell. Through this, sound travels simultaneously through both of the branches to the earpieces. Trained individuals know normal internal sounds and can detect if something is not "right" with an animal.
- **Laboratory analysis**—***Laboratory analysis*** is observing samples of fluids or tissues for abnormalities. Such testing often involves blood or urine. Lab tests may also be

AgriScience Connection

A LONG WAY FROM HOME

Transporting animals requires attention to their well-being. The needs of animals must be met and movement handled with minimal stress and no injury. Sometimes animals are transported long distances, requiring special care. Such is the case of Yang Yang.

Yang Yang is a Giant Panda moved from China to Zoo Atlanta—a distance of 7,526 miles! A special crate to provide enough space was constructed. Its dimensions were 63 inches long, 35½ inches wide, and 40 inches tall. The crate was made of steel and was designed so plenty of air could enter and leave. Most of the distance was covered in a large cargo airplane, with a rest stop in Anchorage, Alaska.

Yang Yang is now in an American home and resting at ease. (Courtesy, Education Images)

16–14. Laboratory analysis of blood from an animal may involve centrifuging the sample.

used to assess parasites, such as small mites in the ears or internal worms found in feces samples. Tests may also be made on tissues (e.g., skin, muscle, bone, and others), cells, milk, and semen. Samples of the fluids or tissues must be properly collected from an animal for accurate lab analysis.

- **Giving medications**—Animals are sometimes given medications to prevent or to treat disease. Some medicines are given by mouth; others are applied to the skin or hair. A few medicines work best if administered by injection because medicine given by injection usually acts rapidly. An injection involves using a syringe and a hypodermic needle to force fluid into the skin, muscle, or bloodstream. According to the location on the body and the type of medication, different types of injections are used. Common types of injections include intramuscular, intravenous, and subcutaneous. (More information on injections is in Chapter 20.)
- **Using diagnostic ultrasound**—New technologies allow a trained individual to view tissues inside the body without surgically opening the body. ***Diagnostic ultrasound***, or ultrasonography, is a noninvasive way of imaging soft tissues in the body. It records the reflections (echos) of the ultrasonic waves into the tissues. Low-intensity, high-frequency sound waves are sent into the soft tissues, where they interact with tissue interfaces. Sound waves are reflected back and are analyzed by a computer to make an image.

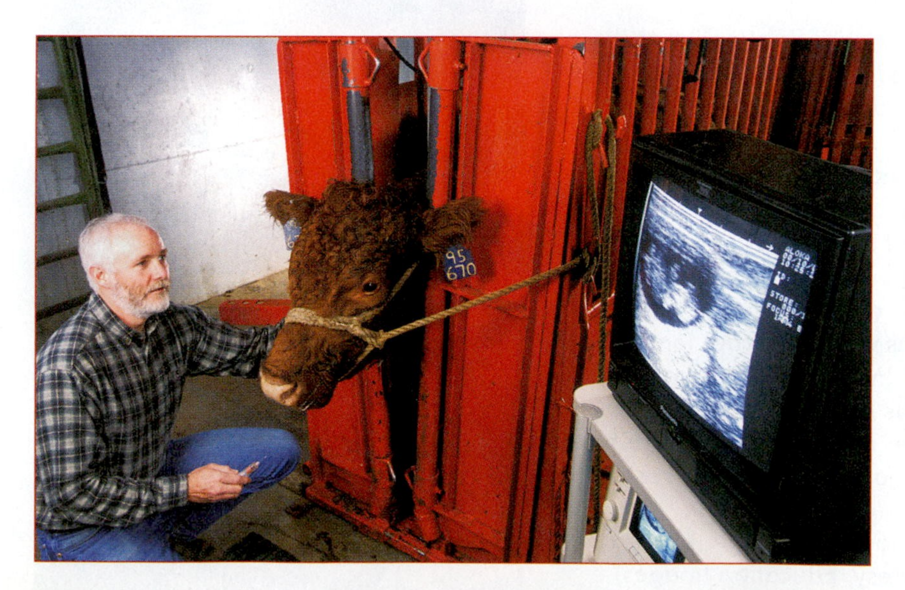

16–15. A reproductive physiologist is using ultrasound to assess the pregnancy of a cow. (A blood test for level of hormone production is also being made). (Courtesy, Agricultural Research Service, USDA)

WOUND TREATMENT AND SURGICAL PRACTICES

Veterinary practices include treating wounds and performing surgery. Procedures are followed to minimize pain and infection and to promote healing.

Wounds include abrasions to the skin and broken bones. These injuries may occur when an animal is attacked by another, runs into a fence, is hit by a car, falls, or in other ways. Each wound must be assessed, and the appropriate action must be taken for the greatest benefit. For example, should the leg of a dog be amputated, or should steps be taken to promote healing of the broken leg? Is euthanasia appropriate? The action depends on the severity of the situation. Only individuals trained in veterinary medicine should handle most of these situations.

Surgical practices may vary from routine herd management to major surgery to correct a health impairment. A veterinarian should perform all surgery other than a few exceptions of routine herd management. Some methods used are bloodless; others involve cutting and opening the body and skin. Asepsis is always needed with surgery to prevent infection and to promote healing. Producers often develop skills in performing routine management surgery, such as docking, castrating, and dehorning on young animals. Doing the work properly and using disinfectant promotes fast healing and minimizes the risk of infection. Of course, minimal invasion of the body also lowers the risk of infection.

16–16. A dog with a severe injury has been given mobility with a wheeled device.

With routine management surgery, areas where cuts are made should be disinfected. Common disinfectants used are iodine surgery spray, a 7 percent iodine solution, isopropyl alcohol, and a Lysol® solution. These products are available from veterinary supply houses or farm and ranch stores. Weak solutions may be poured on the area of the surgery; slightly stronger solutions may be wiped or brushed on with a cloth. It is sometimes recommended that the scrotum area be washed with soapy water and dried before castration.

ACADEMIC CONNECTION

ZOOLOGY

Relate your study of animal needs, well-being, and restraint in agriscience to your zoology class. Compare the requirements for growth and development with your agriscience studies. Prepare a report on your observations.

ANIMAL RESTRAINT AND MANAGEMENT

Restraint is the control of an animal so that it can be transported, examined, treated, groomed, or otherwise managed. Without proper restraint, these activities are inefficient and may be impossible and hazardous to the animal and to people.

Restraint may be used with animals of all sizes and species. Improper restraint can injure an animal, injure the people who are trying to administer to it, damage facilities, or result in the animal escaping.

NEED FOR RESTRAINT

The need for restraint depends on the temperament, size, and species of animal, as well as the nature of the work to be done. Large animals are strong. They require appropriate restraint measures. Animals with teeth and claws may require restraint measures that prevent biting and scratching. Some animals have poisonous bites and stings; they must also be managed carefully to prevent injury.

Animals may become frightened when placed in different environments or when they are being moved. Fear develops, which may be expressed as aggression. Biting, attacking, acting wild, and other forms of fear expression may occur. Animals can become aggressive around others of the same species. Males may readily fight if not kept separate. Boars can be vicious fighters with each other. Bulls, stallions, and others may also be quite aggressive. Mother animals with babies nearby may become aggressive to protect their young.

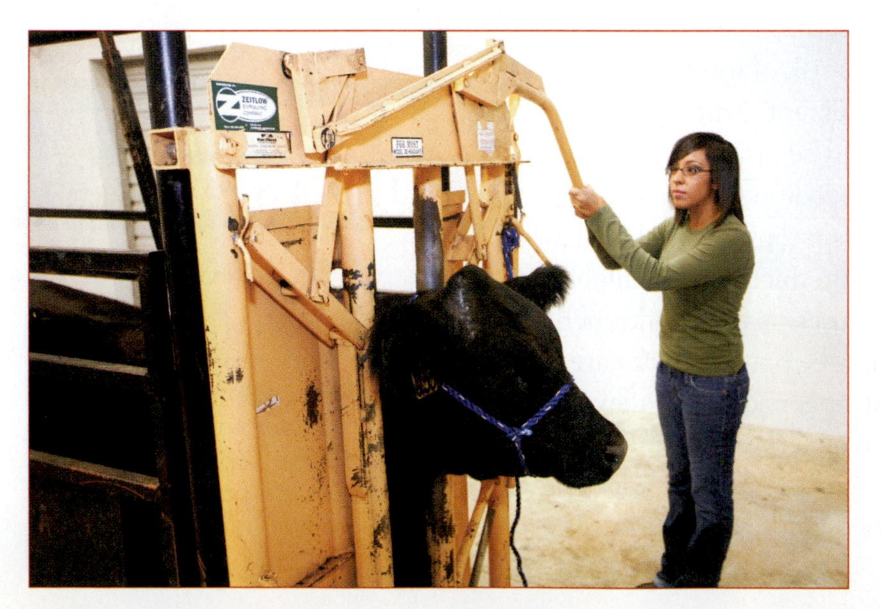

16–17. A large animal is being restrained with a head squeeze. (Courtesy, Education Images)

HOW TO RESTRAIN

Animals should be restrained in ways that prevent damage to themselves, other animals, people, and property. In most cases, animals are restrained because they need to be examined, treated, moved, or transported. How an animal is restrained depends on its species, size, and aggressiveness. Some common ways of restraining animals are:

- **Feline bags**—A feline restraint bag is used with a cat to reduce the likelihood of scratches. Restraint bags have openings and fasteners that allow access to certain parts of the body. In some cases, a cat can be wrapped in a heavy bath towel.
- **Head squeeze**—A ***head squeeze*** is a mechanical device at the end of a chute that closes around an animal's neck. It is used with large animals. Cattle, wild-mannered horses, sheep, and a few other species may be restrained with a head squeeze. A chute allows animals to move through one at a time for tagging or tattooing, vaccinating, and performing other activities. Cattle may need to be herded from a pasture into a corral equipped with a chute (restraint cage).
- **Chemical restraint**—Chemical restraint is the use of tranquilizers, sedatives, and other anesthetics. Such methods are most appropriate with wild animals or in the capture of horses or cattle that have escaped. A blow dart is fired from a gun some distance away into the skin of an animal. The impact of the needle moving into the skin releases the chemical.
- **Physical restraint**—This involves physically holding an animal in place. Halters, hobbles, nose rings, and other instruments may be used. With a small animal, such as a baby pig, human strength is adequate to hold the animal for most procedures. This method of restraint is often used to vaccinate, castrate, dock, or dehorn baby animals.
- **Diversionary restraint**—This is used with large animals to divert their attention away from the procedure. Horses are most commonly restrained using diversionary tactics, such as holding an ear tightly or pressing on a fold of skin. Sensation is such that the animal may not appreciably feel the insertion of a hypodermic needle for an injection or a blood sample.

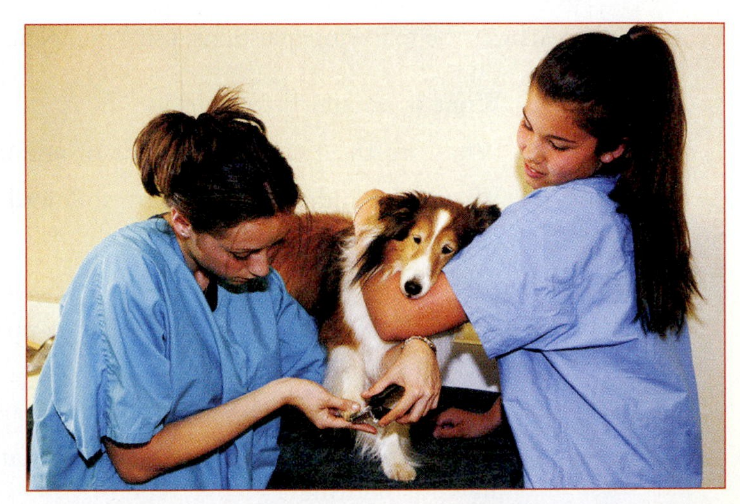

16–18. Physical restraint is used when clipping this dog's claws. (Courtesy, Education Images)

REVIEWING

MAIN IDEAS

Animal producers follow practices that ensure the efficient production of animals and products. The well-being of animals is considered throughout the growth and marketing process. Producers need to be educated in the husbandry of the species they are raising.

Proven husbandry practices should be followed. These practices include wisely choosing the animals to raise, feeding them properly, promoting good health, managing reproduction (breeding), protecting from hazards, properly disposing of wastes, and being ethical in all matters. A part of husbandry is the well-being of the animal. Well-being refers to the state of an animal's health and comfort. In addition, certain legal aspects may apply to the animals being produced.

Quality-assurance programs have been developed to provide consumers with the products they desire. The guidelines in quality assurance vary somewhat among the species, but all guidelines point to producing a product that is safe for the consumer. These programs are offered through producer associations and are voluntary. Producers who follow the guidelines may receive additional marketing incentives.

Veterinary care is sometimes needed by animals. Veterinary medicine is the branch of medicine that deals with animals. Some of the common practices are checking vital signs, listening to internal sounds, making analyses, giving medications, and using ultrasound. Sometimes injuries and surgical procedures are needed. These should always be done by qualified people who place animal well-being as a high priority. Proper animal restraint is needed with most veterinary medical practices.

QUESTIONS

Answer the following questions, using complete sentences and correct spelling.

1. What is animal husbandry?
2. What are the two native needs of animals? Briefly explain each.
3. What husbandry practices are followed with animal production? Name and briefly explain any three.
4. What is animal well-being?
5. What practices should be followed in promoting animal well-being? Name and briefly explain any three.
6. What is euthanasia? Why would it be used?
7. What is the name of the federal law that prohibits the interstate commerce with animals for fighting?
8. What is quality assurance from the perspective of an animal producer?
9. What is the focus of food safety control points in quality assurance?

10. What records are kept by animal producers in quality-assurance programs?
11. What is veterinary medicine?
12. What are five common practices in veterinary medicine used in assessing an animal?
13. What is restraint? Why is it important?
14. What are five methods of restraining animals? Briefly explain one of them.

EVALUATING

Match each term with its correct definition.

a. veterinary medicine	e. laboratory analysis
b. euthanasia	f. animal husbandry
c. stethoscope	g. animal well-being
d. diagnostic ultrasound	h. quality assurance

______1. The scientific management and control of animals.

______2. The state of an animal's health and comfort.

______3. The act of ending the life of an animal to relieve it of suffering.

______4. An effort to promote quality products through good management practices.

______5. The branch of medicine that deals with animals.

______6. An instrument used to hear and amplify internal sounds.

______7. Technology used to make images of soft tissues of the body.

______8. Using test procedures to observe fluids or tissues for abnormalities.

EXPLORING

1. Maintain a pair of small animals in the school laboratory or at your home. (They might be a part of your supervised experience.) Observe their behavior, including breeding, parturition, and the care received by the young. Be sure to consider their well-being. Keep records on the animals, including a log of events. Prepare a report on your observations.

2. Tour a dairy farm, and observe the milking of the cows. If you have never hand-milked a cow, ask the worker to show you how it is done. Try it for yourself.

3. Investigate a quality-assurance program for a particular species or product. You may gather information from organizations of producers, the Cooperative Extension Service, local producers, and others. Prepare a brief report that indicates the purpose of the program, producer requirements, and training needs of the producer.

CHAPTER 17

Animal Biology

OBJECTIVES

This chapter presents important principles of animal biology. It has the following objectives:

1. Identify and describe major features of animals.
2. Explain principles of animal anatomy, morphology, and physiology.
3. Identify organ systems, and describe the structure and functions of each.

TERMS

alveoli
anatomy
blood
bone
bony fish
breathing
cartilage
companion animal
conformation
crustacean
estrous cycle
exoskeleton
external respiration
fetus
gestation
internal respiration
invertebrate
involuntary muscle
lactation
mammal
mammary glands
mollusk
morphology
nonruminants
parturition
physiology
plasma
platelet
pregnant
puberty
receptor
red blood cells
ruminant
vertebrate
voluntary muscle
white blood cells

17–1. The study of animal structure often begins with a skeleton. (Courtesy, Education Images)

ANIMALS are popular and useful organisms that can be found in many shapes and sizes. They are very important to human beings. What would our lives be like without animals?

We use animals in many ways. Some animals are our companions (pets); others provide food, clothing materials, services, entertainment, and educational opportunities. In some cases, animals serve as the "eyes" of their human owners, such as a leader dog for the blind.

We need to understand animals so we can improve husbandry. No one knows exactly how many species of animals are on Earth. However, scientists have classified 1½ million species, but more than half of these are insects. The number of species in our daily lives is fairly small, with only a few used as sources of food and clothing products. A few other species are used for companionship, research, and service.

ANIMALS AND THEIR BODY FEATURES

All animals are members of the kingdom Animalia. Most animals have three common traits: they are made of many cells; they can move about; and they obtain food from other sources.

17–2. Important terrestrial agricultural animals. (Chickens and dairy cattle—Courtesy, Mississippi State University)

Animals have different body structures. The major groups are invertebrates and vertebrates. An ***invertebrate*** is an animal without a backbone. In contrast, a ***vertebrate*** is an animal with a backbone.

INVERTEBRATES

About 97 percent of all animals are invertebrates. They are a diverse group, including earthworms, spiders, butterflies, snails, and lobsters.

Most of the important agriculture invertebrate species are in the phylum Arthropoda. Other phyla, such as the Annelida and Mollusca, also have important animals.

Arthropods

An arthropod has a hard outer covering known as an ***exoskeleton***. It protects and supports the body, and it keeps the internal organs from being injured easily. The exoskeleton is made of chitin (a tough, carbohydrate material) and protein. Chitin is similar to human fingernails. Arthropods molt (shed the exoskeleton) as they grow.

The body and appendages (attachments, including legs and antennae) are jointed. These jointed segments allow the animal to move. Arthropods have simple body systems that carry out life processes.

The most important arthropods are the crustacea, uniramia, and chelicerata. (These are subphyla in the Arthropoda phylum.)

- Crustacea—A ***crustacean*** is an aquatic animal with an exoskeleton. There are 25,000 crustacean species, including shrimp, crawfish, lobsters, crabs, barnacles, and copepods. Crawfish and shrimp are often cultured. Lobster, crab, and others may be

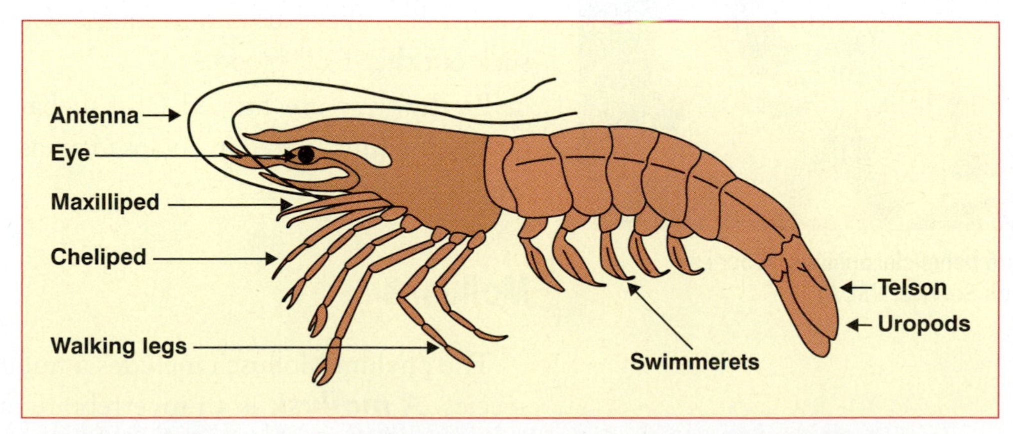

17–3. Major external parts of a shrimp—a crustacean. (Crawfish are similar but have much larger chelipeds, or pinchers.)

harvested from the sea as wild animals. Although some of these animals grow in saltwater, others grow in freshwater.

The crawfish (or crayfish, or crawdad) is an example of a crustacean. Their bodies, similar to shrimp, are in two sections: cephalothorax and abdomen. The cephalothorax is the head and thorax fused into one piece. On the other hand, the thorax is divided into 13 segments. Five pairs of legs are attached to the thorax. The first pair of legs is large and strong, known as the chelipeds or pincers. The other legs are used for walking. The abdomen is a muscular, seven-segment structure. Each segment—except the last—has swimmerets. It has a uropod (tail) that can be used to move backwards in water.

- **Uniramia**—This subphylum includes the insects, millipedes, and centipedes. Insects are the most important because approximately three-fourths of the animals on Earth are insects. Insects assume beneficial and harmful roles. Some insects are farmed. A good example is the honeybee. These insects help pollinate crops and produce honey.
- **Chelicerata**—This subphylum contains ticks, mites, spiders, scorpions, and other species. Many of these are pests that damage plants and animals. Most are in the class Arachnida, commonly called arachnids (spiders). There are more than 100,000 species of arachnids. They have four pairs of legs. Some spin webs for homes and to trap insect prey.

17–4. Earthworms are beneficial animals. (Courtesy, Agricultural Research Service, USDA)

Annelids

The phylum Annelida contains earthworms, blood-sucking leeches, and a few others. Some are beneficial, particularly earthworms in the soil. Leeches are often known as pests, especially when they attack fish. Some leeches have been used on human wounds to suck out the "bad" blood.

Earthworms are farmed for fish bait. The worms are often grown in enclosed areas where food is provided.

Mollusks

The phylum Mollusca includes a number of species. A ***mollusk*** is an invertebrate animal with a soft body and a hard shell for protection.

17–5. The scallop (*Argopecten irradians*) is a bivalve mollusk.

Slugs and octopuses have soft bodies and no shells. Oysters, snails, clams, mussels, and abalone have hard shells. There are about 100,000 species of mollusk. Most mollusks are aquatic animals.

A bivalve mollusk is a class of mollusk with two-sided shells hinged together. A strong, muscular foot opens and closes the shell. Oysters, clams, scallops, and mussels are examples. Most of these are "filter feeders," so they do not move around to obtain food. Instead, they receive food by filtering it from the water that flows through their shells.

Some mollusks are farmed; others are captured in the wild. Oysters must have good water. In feeding, they take in whatever is in the water. As a result, humans may become sick after eating oysters grown in "bad" water. Oysters harvested wild are more likely to carry disease. Some oyster producers use a purification process after harvest to remove potential dangers.

Others

Many other invertebrates affect life on Earth. They help plants and other animals grow. Some invertebrates cause disease and damage plants and animals. Others are parasites of larger animals. For example, roundworms and tapeworms are parasites of hogs, cattle, and dogs. Heavy infestations can result in poor growth of animals or death.

VERTEBRATES

The vertebrates are animals with backbones. They are in the subphylum Vertebrata of the phylum Chordata.

The vertebrates are named for the vertebrae (bones and cartilage) that surround the nerve cord. They have a hard skull that protects the brain. Internal skeletons provide a framework for their bodies. The part of the skeleton made up of the backbone and skull is the axial skeleton. Paired limbs (appendages) are attached to the axial skeleton. Muscles are attached to the skeleton and provide for movement of the animal.

Many important agricultural species are vertebrates. Examples of the major vertebrates follow.

Fish

Three classes of vertebrates are known as fish. The Osteichthyes is the class of ***bony fish***. The Chondrichthyes class includes sharks, rays, and a few others. The third class is the small Agnatha class of jawless fish, such as the lampreys.

17–6. Important aquatic animals. (Prawns—Courtesy, Mississippi State University)

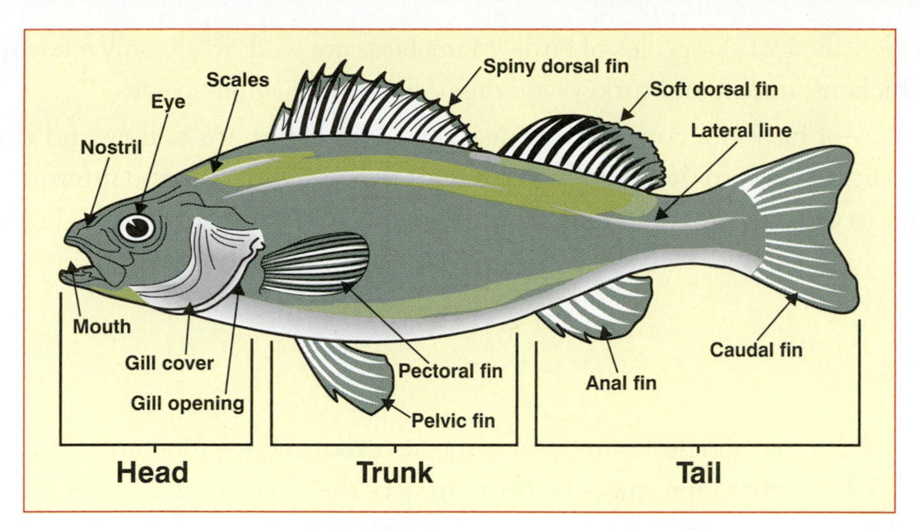

17–7. Major external body parts of a bony fish.

Approximately 25,000 species of bony fish have been identified. Some of these bony fish are important sources of human food. They are caught as wild fish from rivers, lakes, and oceans. Increasingly, fish are grown on aquafarms. The most popular cultured fish in North America is the catfish. Trout, tilapia, salmon, and a few others are aquafarmed. In addition, ornamental fish are kept as pets.

Fish are covered with scales or skin. Their bodies are divided into three primary parts: head, trunk, and tail. The main parts on the head are the mouth, eyes, and gills. The trunk is the part of the body that contains the major internal organs. The tail is a strong muscular structure used to help the fish move.

Fish have organ systems similar to other animals. One important difference is the way in which fish obtain oxygen. Fish have gills that "filter" oxygen from the water as it passes over them. Fish reproduce by laying eggs, which are fertilized after laying and then are incubated. Fish are ectothermic animals, so their body temperature is regulated by the temperature of the water in which they live.

17–8. The tilapia fish are often raised in tanks. The tanks provide a desired cultural environment by filtering and aerating the water. (Courtesy, Education Images)

Birds

Birds are in the class Aves. Birds are adapted to wider environments than most animals. Some species live in the air and on the land. Others live in and around water. Scientists have identified 9,000 species of birds. Many birds are wild, while only a few species are farmed. Chickens, ducks, and turkeys are raised more than other species.

The bodies of birds are covered with feathers. The bones are hollow and thin, which makes birds lightweight so they can fly. Birds have two legs with feet and internal organ systems similar to those of other animals. They reproduce by laying fertile eggs that hatch in a few weeks.

Mammals

The class Mammalia includes important species produced for food and fiber. A ***mammal*** is a vertebrate that reproduces by the mating of the male and female of the same species. An additional characteristic that sets mammals apart from others is the presence of mammary glands on females. ***Mammary glands*** are milk-producing glands by which newborns are fed. The female carries a developing embryo in her uterus for a time known as ***gestation***; birth follows.

Mammals have hair, a heart with four chambers, a lower jawbone with teeth, and a well-developed brain. Mammals are endothermic—their bodies maintain a certain temperature

Technology Connection

WRINKLED HOGS

Many breeds of hogs are found worldwide. Some breeds are quite different from those in the United States. Differences are not always bad, even when the hogs do not look like the Duroc, the Hampshire, or other common breeds. In fact, different breeds may have genetic traits that will be useful in improving breeds that we know about.

Perhaps no nation has more breeds of hogs than China. Some of the Chinese breeds have interesting and useful characteristics. Wrinkled skin may not be a particularly useful trait, but reproductive potential is definitely useful. The more pigs a female produces, the more productive the animal is considered to be. Further, some females of the Chinese breeds reach puberty at three months of age. Their first two litters average 14 baby pigs (piglets). Subsequent litters average 17 piglets—far more than typical breeds in the United States!

This photo shows a Chinese Fenjing boar imported for research and potential mating with females of U.S. breeds. Hope is that the matings will result in an increased litter size. (Courtesy, Agricultural Research Service, USDA)

regardless of the temperature of their environment. Examples include cattle, goats, sheep, hogs, horses, and other common farm species.

Companion Animals

A ***companion animal*** is an animal kept as a pet. It is a source of joy and entertainment to its owner. Its well-being is a major consideration. Many companion animals are expected to live in a human-dominated environment.

A wide range of species are used for companionship. Their scientific classes vary considerably. Some of these animals are terrestrial; others are aquatic. Some are mammals; others are birds, reptiles, and crustaceans.

ACADEMIC CONNECTION

ZOOLOGY

Relate your study of anatomy, morphology, and physiology in agriscience to your zoology class. Compare the anatomy and physiology of animals to that of the human body. Investigate similarities and differences. Prepare a report based on your observations.

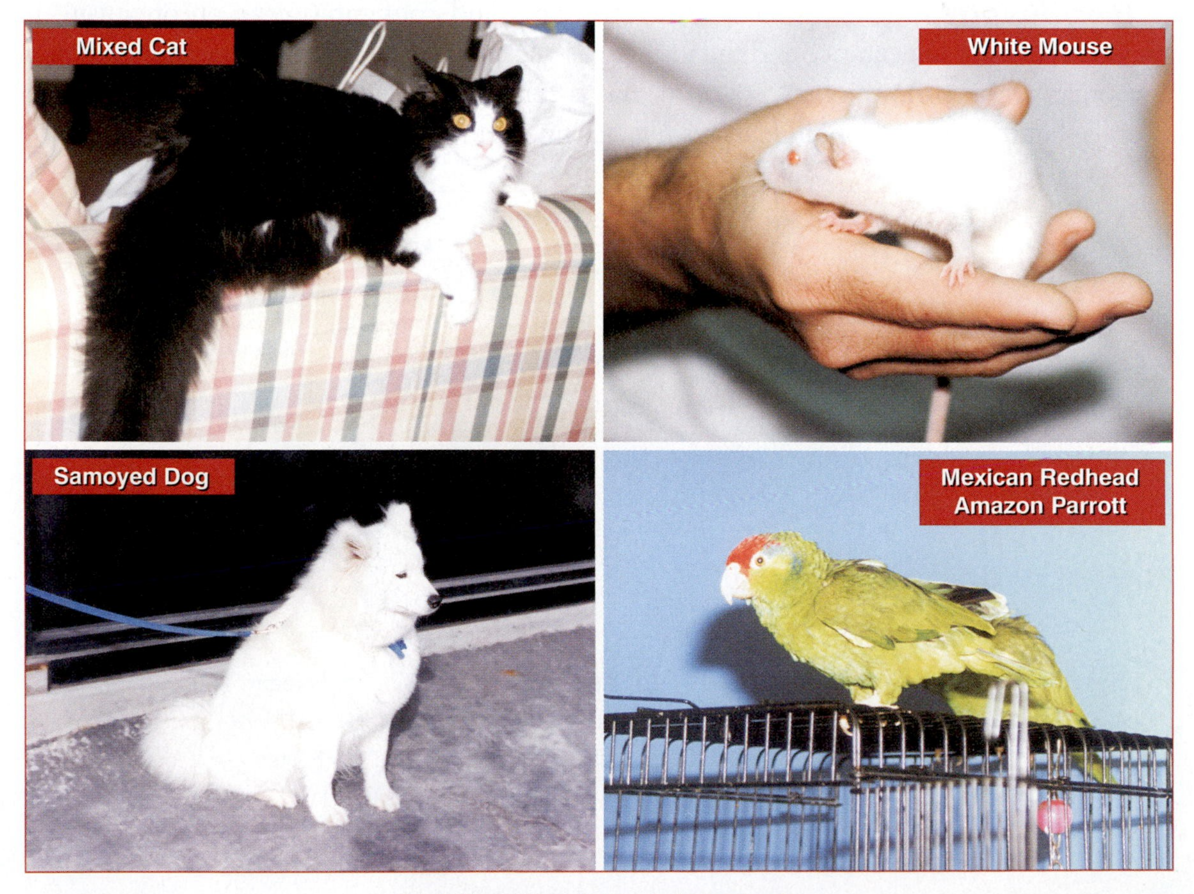

17–9. Examples of companion animals.

ANATOMY, MORPHOLOGY, AND PHYSIOLOGY

Animals are complex organisms. They take in, transform, store, and release energy from food.

As with all living organisms, cells are the basic building blocks of the animal. Cells divide and specialize to form masses known as tissue. The tissues form organs, and the organs form systems that work together for the animal to live.

Knowing about animals helps producers raise them properly. Anatomy, morphology, and physiology are basic areas of knowledge.

Anatomy is the study of the form, shape, and appearance of an animal. Gross anatomy deals with structures that can be seen with the human eye. Microscopic anatomy deals with structures that require magnification to be seen, such as cells and sperm.

External gross anatomy is useful in identifying animals. These features are also useful in determining the quality of the animal.

Morphology is a study of the form, structure, and configuration of an organism. It includes outward appearance and the form and structure of the internal parts, such as bones and organs. Gross morphology is a term that is sometimes used to describe an organism's overall shape, color, and main features but not fine details.

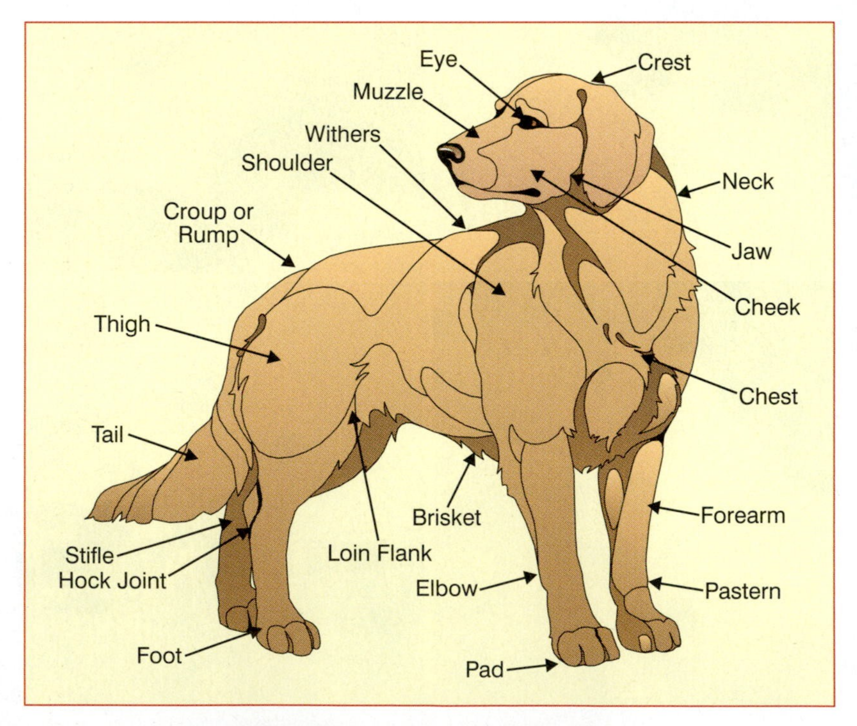

17–10. Major external parts of a dog.

Physiology is the study of the functions of the cells, tissues, organs, and systems of the body. It includes the study of the mechanical, physical, and biochemical functions of living organisms. Physiology also deals with relationships of the systems to each other. For instance, if the lungs of a pig do not take in oxygen, the blood cannot absorb oxygen in the lungs. As a result, the pig will die!

In animal husbandry, the term conformation is often used. ***Conformation*** is the proportion of an animal's body features in relation to one another or to the whole body of an ideal of the breed or species. Judges often compare conformation of animals to the ideal for a particular species or breed.

The major external parts of common animals are shown in this chapter.

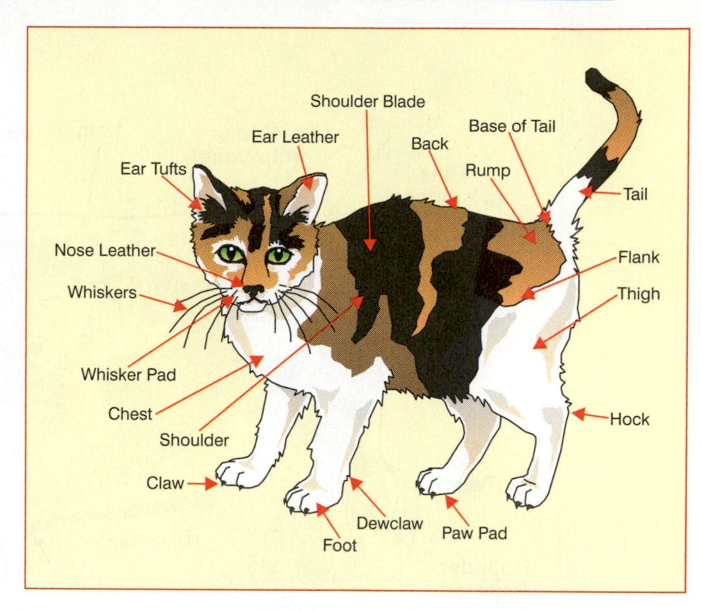

17–11. Major external parts of a cat.

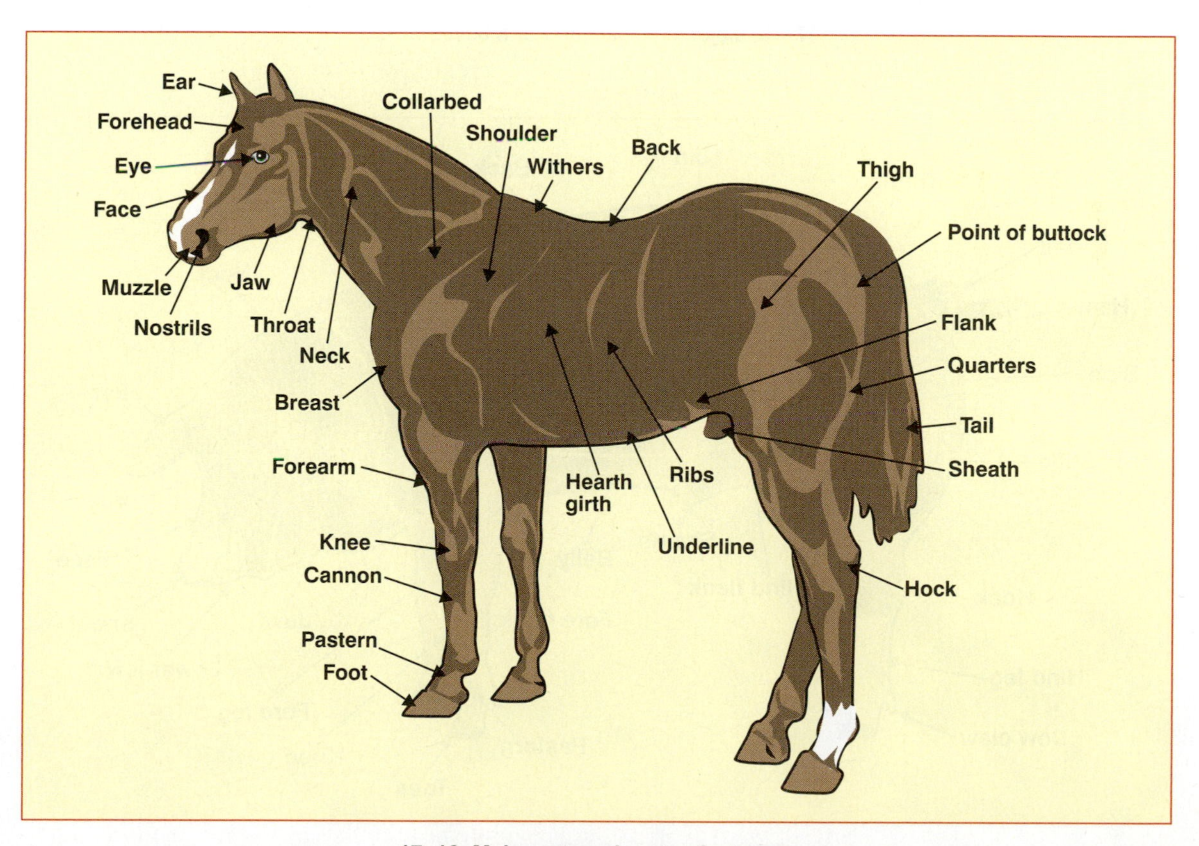

17–12. Major external parts of a horse.

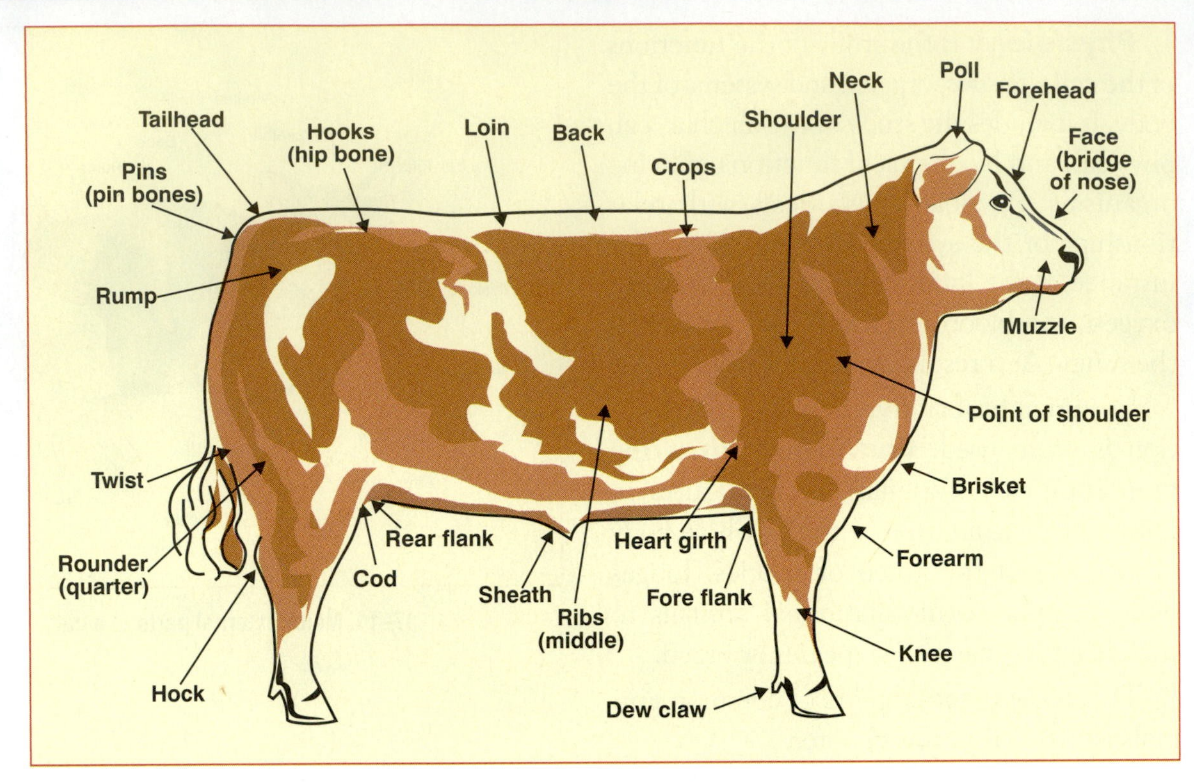

17–13. Major external parts of a beef steer.

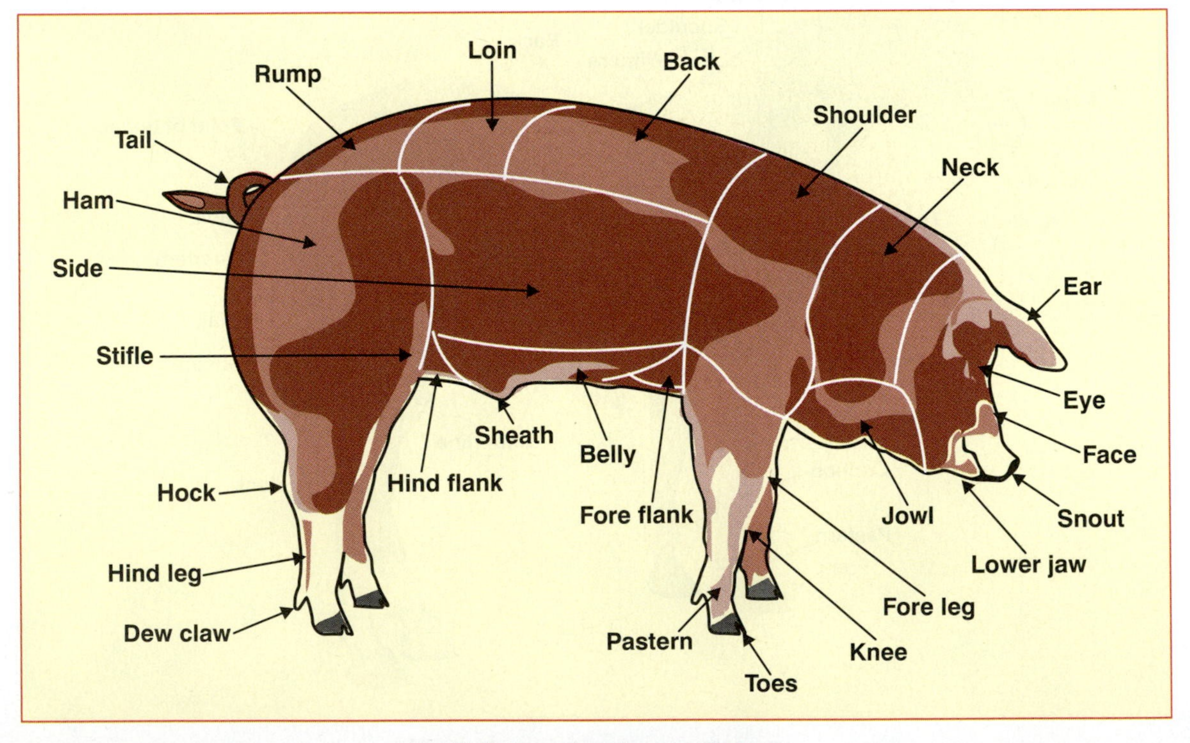

17–14. Major external parts of a hog.

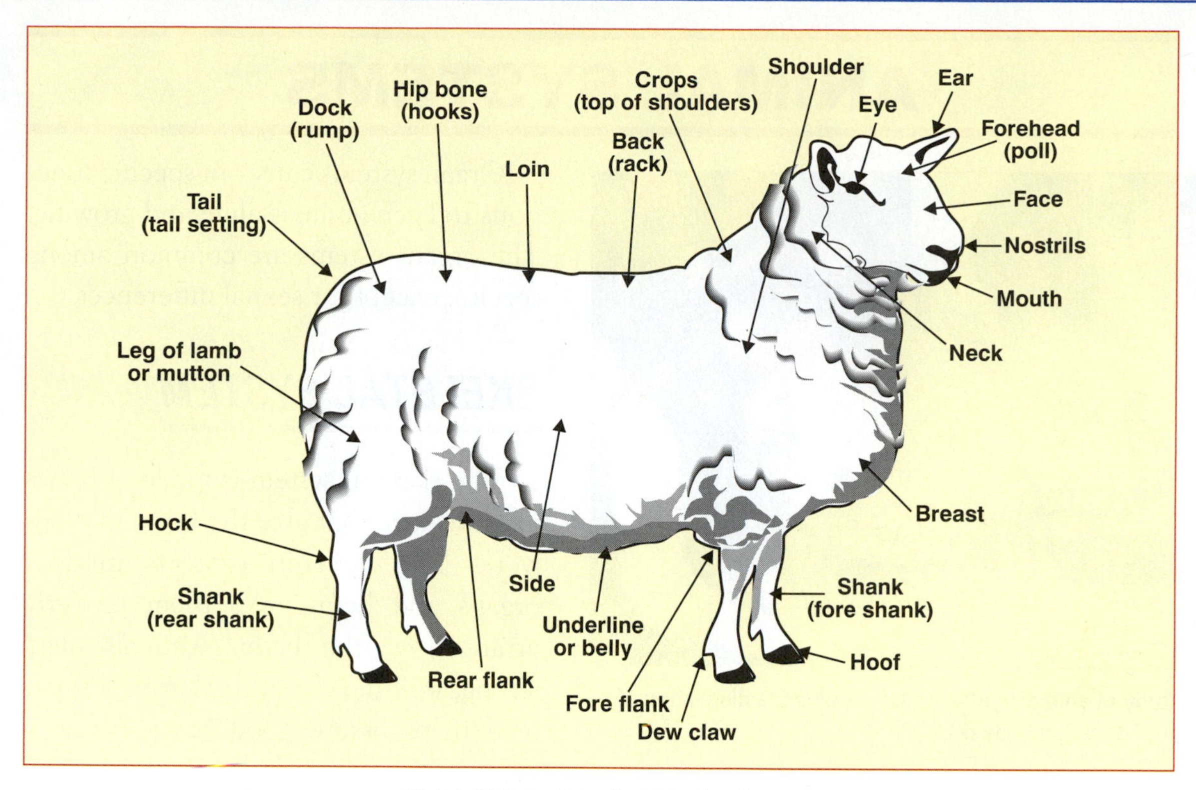

17–15. Major external parts of a sheep.

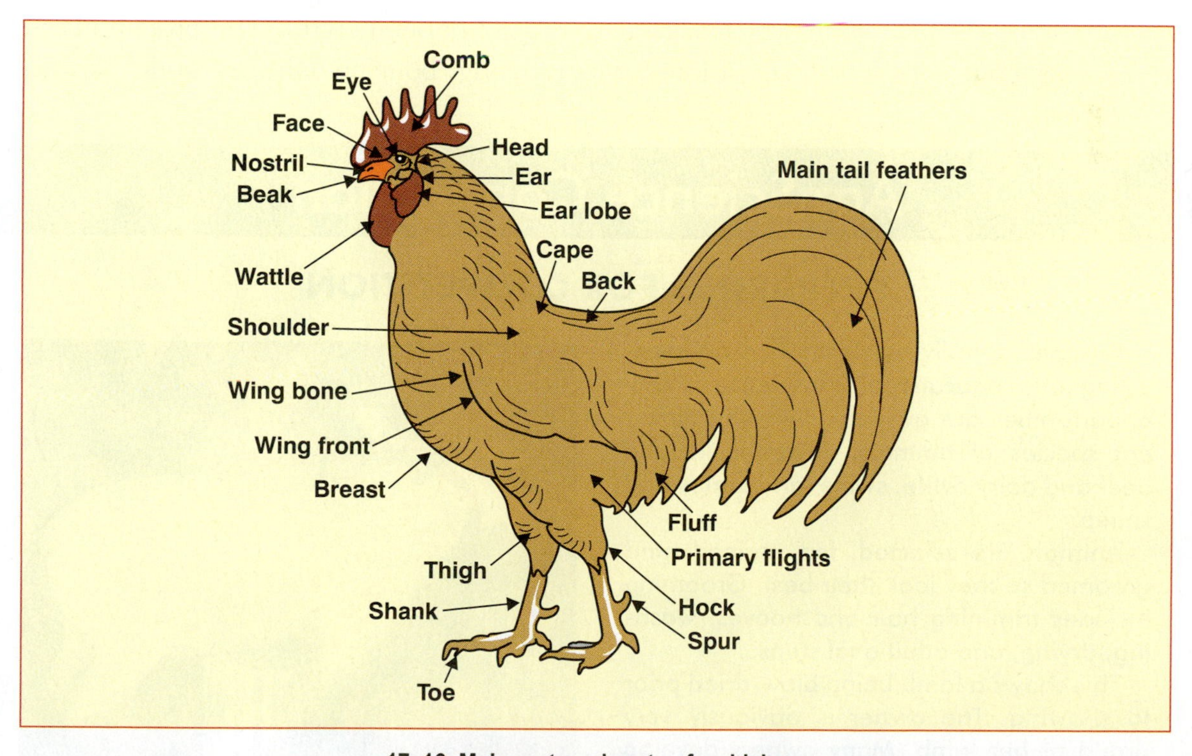

17–16. Major external parts of a chicken.

ANIMAL SYSTEMS

17–17. The study of animal systems may involve the dissection of a rat. (Courtesy, Education Images)

Organ systems carry out specific functions to keep animals alive and growing. The organ systems are common among species, except for sexual differences.

SKELETAL SYSTEM

The skeletal system is made of bones and cartilage that give the body a framework. The skeleton protects internal organs and helps keep them properly arranged in the body. Animals need strong, well-developed skeletons. Proper feed helps assure good bone development.

Bone is the hard part of a skeleton. Bones are made of calcium, phosphorus, and other substances. They have several layers: outer membrane (periosteum), which contains small blood vessels; compact bone, which is made of rings of mineral crystals and protein fibers; and spongy bone, which is the most interior part. Spongy bone contains red bone marrow,

AgriScience Connection

PREPARING FOR COMPETITION

Competing in livestock shows is an interesting and educational experience. Many opportunities are available with the different species of animals. Favorites include beef and dairy cattle, swine, and lambs and sheep.

Animals are selected, fed, trained, and groomed so they look their best. Grooming includes trimming hair and hooves, washing, drying, and additional steps.

This shows a lamb being blow-dried prior to showing. The owner is obviously very proud of her lamb. Many owners develop strong bonds with their animals.

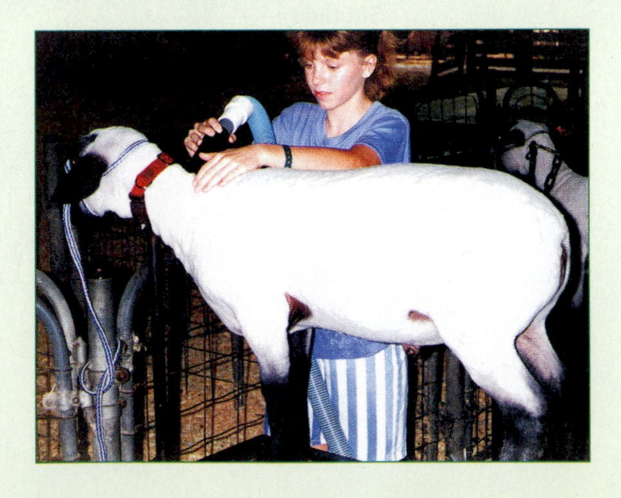

which produces blood cells, and yellow marrow, which serves as an energy reserve for the body.

Cartilage is a flexible material at the ends of bones. It lubricates the joints, cushions shocks, and protects bones from damage when they move against each other in a joint.

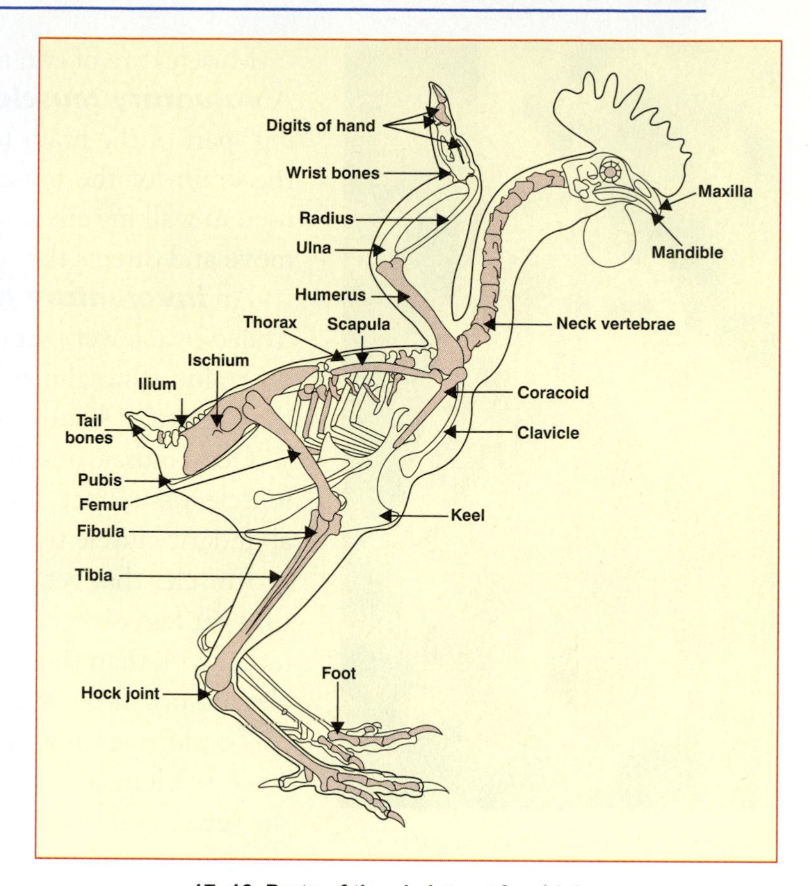

17–18. Parts of the skeleton of a chicken.

MUSCULAR SYSTEM

The muscular system is the largest body system. It makes up about 45 percent of the body weight of animals. Muscles are important in internal movement, locomotion, circulation, digestion, and breathing. The muscular system is made of muscles and connective tissue. Tendons connect muscles to the bones of the skeleton.

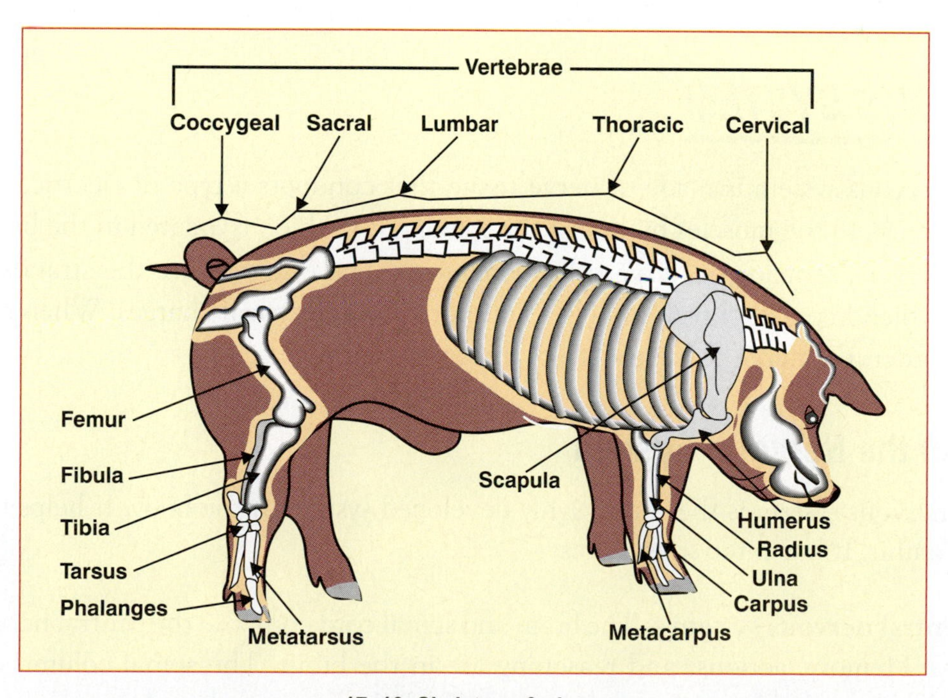

17–19. Skeleton of a hog.

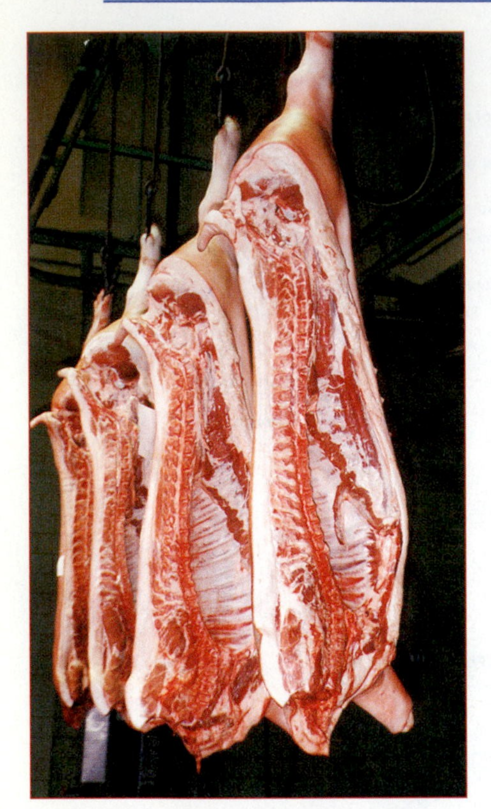

17–20. Hog carcasses showing good muscling with little fat.

Muscles are of two major types: voluntary and involuntary. A ***voluntary muscle*** is a muscle controlled by the "thinking" part of the brain (cerebrum). A message must be sent by the brain for the muscle to move. For example, the muscles used in walking are voluntary muscles. The brain tells them to move and directs the type of movement.

An ***involuntary muscle*** is a muscle automatically controlled by a lower part of the brain. These muscles operate the heart, intestines, lungs, and other organs. Animals cannot stop these muscles from moving.

Good muscle development is essential in meat animals. The muscle tissues are used for food. People want animals with abundant muscle tissue, small bones, and minimal fat.

Muscles that get a lot of exercise are not as tender as those that get less exercise. The muscles in the necks of animals do more work than those in the loin (back); therefore, neck muscles are not as tender.

Tendons are not as tender as muscles. Older animals have more tendons and less muscle tissue. Also, loin muscles and the "eye" muscles of the rib have less tendon and make the best meat (e.g., beef ribeye steak, filet mignon, pork loin, and pork chops).

Muscles respond to commands from the nervous system.

NERVOUS SYSTEM

The nervous system is made of nerve tissue that conducts a type of electrical impulse from the brain to the muscles by way of the spinal cord, which is located in the backbone.

The nerves are made of nerve cells or neurons. These nerves are the strands of cells through which impulses travel. Impulses are short-lived electrical charges. When the brain gives an order, the nerve tissue transmits the order to the muscles.

Parts of the Nervous System

The nervous system is the most highly developed system in the body. It helps the body work as a unit. It has three subsystems:

- **Central nervous system**—The brain and spinal cord compose the central nervous system. Memory, actions, and reasoning are in the brain. The spinal column is in the backbone and is the main tissue through which the brain sends and receives messages.

- **Autonomic nervous system**—This system contains the nerves connected to the involuntary muscles and organs. It provides for near automatic operation of the organs.
- **Peripheral nervous system**—This system deals with the voluntary muscles. It includes all of the nerves outside of the brain and spinal cord. The peripheral nervous system carries impulses to and from the brain. Muscle action is sent through these nerves by the brain.

Receptors

Parts of the nervous system collect information from an animal's environment. A receiver of information is a ***receptor***. Information perceived by receptors is processed by the brain. Neurons sense and respond to a stimulus and carry impulses from one end of a cell to the other. These actions include excitability and conductivity. Neurons send impulses to each other through synapses.

The five types of sensory receptors are sight, hearing, touch, taste, and smell. These senses are more highly developed in some animals than others. All have a good sense of touch. If something attacks an animal, the sense of touch becomes active and, on orders from the brain, responds. All information is quickly processed through the nervous system to the brain, where a response is issued.

17–21. Animals, such as this beagle, can be trained to do useful work. (The beagle is used to sniff luggage for illegal food brought into the United States on international air flights.) (Courtesy, USDA)

CIRCULATORY SYSTEM

The circulatory system moves digested food, oxygen, wastes, and other materials around the body. Individual cells need food and oxygen to live and divide. The circulatory system

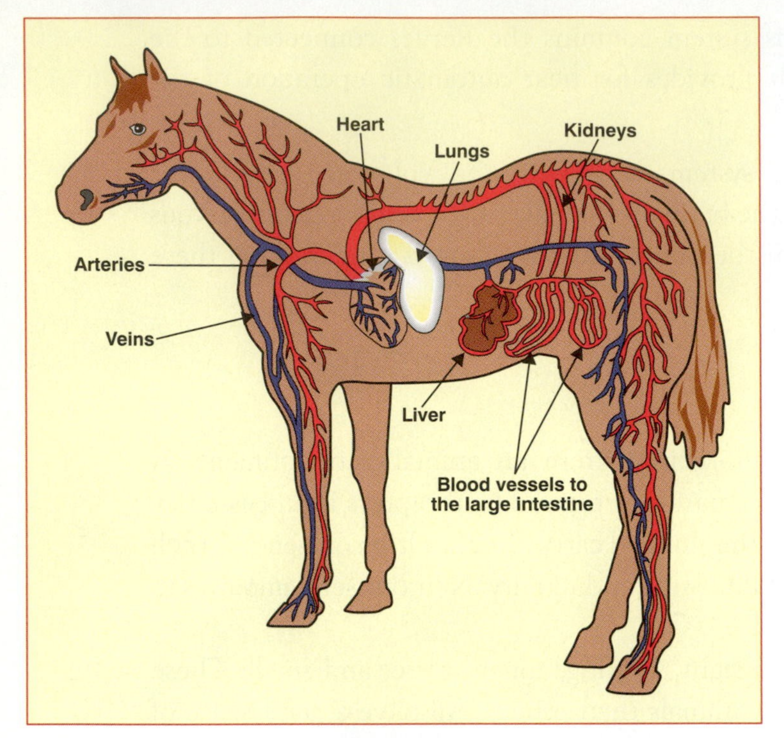

17–22. Circulatory system of a horse. (Most vertebrate mammals have a similar system.)

removes wastes from the cells and carries them to be released by the body. The movement of blood in the body is known as systemic circulation.

Blood

The "liquid" in the circulatory system is ***blood***, which is made of ***plasma*** (the liquid part) and various solid materials. Plasma is 90 percent water. The dissolved substances in the plasma include glucose, vitamins, minerals, and amino acids (proteins). Animals have different types of blood, but all blood tends to be similar in composition.

The solids in the blood include red blood cells (erythrocytes), white blood cells (leukocytes), and platelets (thrombocytes).

Red blood cells (corpuscles) are the blood cells made in the red marrow of bones. They contain a protein called hemoglobin, which is the molecule that carries oxygen in the blood. Red blood cells are much more numerous than white blood cells. Red cells cannot repair themselves and die after a time. The liver and spleen remove the dead blood cells. Yet the bone marrow makes more. The blood in one animal may contain several trillion red cells.

White blood cells are the blood cells that help fight off disease. An animal that is diseased will produce many more white cells than usual. The pus that forms at wounds is a collection of white blood cells. These cells are sometimes called soldiers or scavengers because of their role in limiting disease.

A ***platelet*** is a tiny cell fragment that is essential for blood to clot. If blood does not clot, a wounded animal may bleed to death.

System Parts

The circulatory system has four major parts.

- **Heart**—The heart is the "pump" that sends blood throughout the system. It is a strong, involuntary muscular organ that contracts (beats) continuously throughout the life-

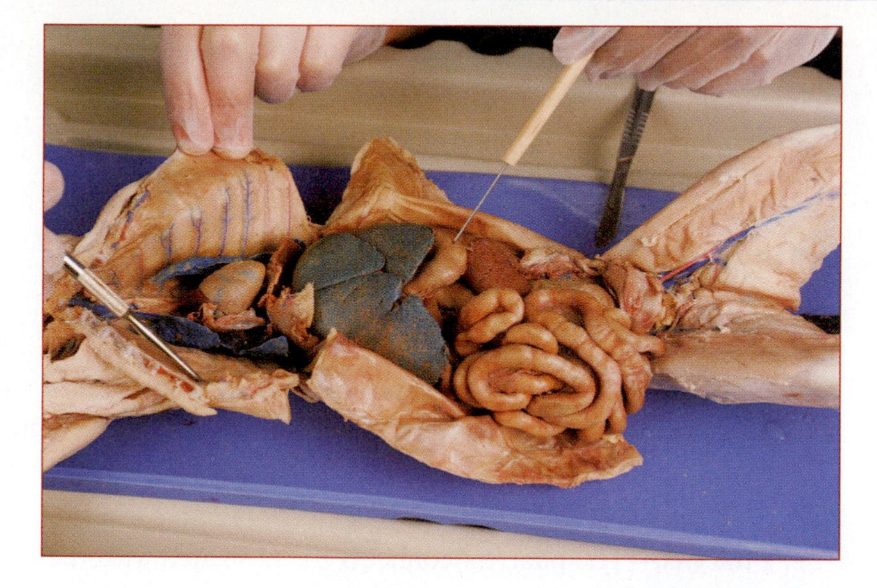

17–23. Several systems are observed in this opened and preserved cat specimen. (Courtesy, Education Images)

time of an animal. The beat slows when the animal rests. A heart may beat as many as a billion times during a lifespan.

- **Arteries**—The arteries are the vessels that carry blood from the heart.
- **Capillaries**—The capillaries are small branches from the arteries that carry blood to the cells. Capillary walls are only one cell thick. This allows for the easy exchange between the cells and the blood, which carries oxygen and energy to the cells and picks up wastes from the cells. After contact with the cells, the blood continues through the capillaries to the veins.
- **Veins**—The veins carry blood back to the heart. This blood has given its oxygen to the cells and picked up wastes.

The heart sends the blood to the lungs where the carbon dioxide is exchanged for oxygen. Lungs are a part of the respiratory system. After picking up oxygen in the lungs, the blood returns to the heart for circulation throughout the body. Sending blood to the lungs is known as pulmonary circulation.

RESPIRATORY SYSTEM

The respiratory system moves gases to and from the circulatory system. Respiration provides blood with oxygen and removes carbon dioxide.

Kinds of Respiration

Two types of respiration occur in animals: internal and external.

Internal respiration is the exchange of gases between the cells and the blood within the body. Cells break down glucose for energy.

External respiration is the exchange of gases in the lungs between the blood and the atmosphere. Everyone is familiar with the process of breathing.

Parts of the Respiratory System

The respiratory system of terrestrial animals has five parts.

- **Nostrils**—The nostrils are the openings near the mouth through which gases enter and leave the body. Incoming air is warmed, filtered, and moistened by the hairs and mucous membrane in the nostril area. Some things in the air cannot be filtered out, such as harmful carbon monoxide.
- **Pharynx**—The pharynx connects the nose area with the mouth area. It is located at the back of the throat.
- **Larynx**—The larynx is often called the "voice box." Animals produce sounds when air is passed over two ligaments, known as vocal cords. The pitch of a sound varies with the amount of air sent over the cords and the nature of the cords.

CAREER PROFILE

LIVESTOCK MANAGEMENT TEACHER

A livestock management teacher is an agriculture instructor who specializes in livestock. The work involves instructing high school students and adults in the fundamentals of livestock care, production, and management. Basic principles of zoology are included in the instruction.

Most livestock management teachers have strong bonds with animals. They enjoy animals and teaching about them. These teachers have college degrees in agricultural education, animal science, or a closely related area. Many of these teachers have master's degrees and some have doctorates. They are employed by local school districts, community colleges, technical schools, or colleges and universities. Livestock management teachers need to have had practical experience with livestock production.

You can begin preparing now for this career by taking animal science courses, biology, and anatomy and physiology in high school. Plan to enroll at a college or a university to earn a bachelor's degree. Be sure to complete the appropriate credentials to be a teacher.

This shows a livestock management teacher demonstrating how to use a specially-made tube to relieve bloat in a bovine. (Courtesy, Education Images)

- **Trachea**—Known as the windpipe, the trachea connects the pharynx with the lungs. The trachea branches into bronchi, which further divide into smaller tubes inside the lungs.
- **Lungs**—Gas is exchanged between the atmosphere and the blood in the lungs. The lungs are located in the chest cavity and are supported by the rib cage and diaphragm. The lungs are surrounded by mucus membranes that secrete liquids to make breathing easier.

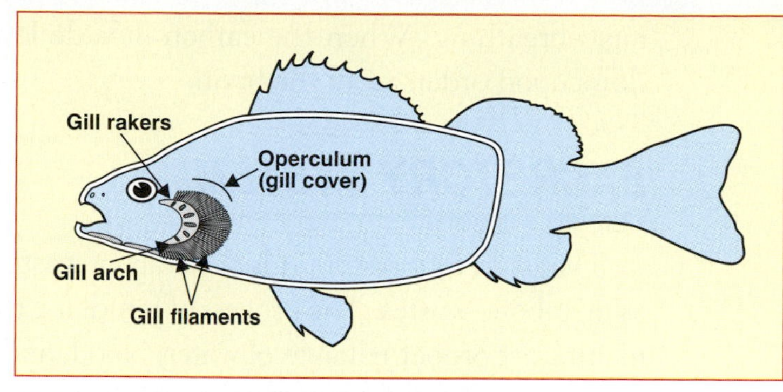

17–24. Gill structure of a fish. (Water flows through the mouth and over the gill filaments where oxygen is removed.)

Aquatic organisms have different types of respiratory systems. Most fish have a gill system. Oysters, shrimp, and other species vary. The respiratory systems of most aquatic organisms remove oxygen from the water in which the organisms live. Low levels of oxygen can cause stress or death. Producers of aquatic organisms use practices that maintain and increase the level of dissolved oxygen (DO) in water. DO is measured as parts per million (ppm) parts of water. Most aquatic organisms require at least 3.0 ppm of DO to survive. Best growth occurs at a slightly higher level. DO is measured with meters or other devices.

17–25. A DO meter is being used to measure the dissolved oxygen content of water in this tank used for culturing fish. (Courtesy, Education Images)

Breathing Processes

Breathing is the process of air entering and leaving the lungs. Taking air into the lungs is inspiration. Air moving out of the lungs is expiration. The rib cage, diaphragm, and abdominal muscles are used in breathing. Gases are exchanged in the lungs by osmosis.

The rate of breathing is regulated by the amount of oxygen required by an animal's body. An active animal breathes faster. As cells burn oxygen in converting glucose to energy, carbon dioxide is released in the blood. Nerves send a message to the brain that more carbon dioxide needs to be removed and more oxygen is needed. In response, the brain orders more rapid breathing. When the carbon dioxide level has been lowered, the rate of breathing slows upon orders from the brain.

EXCRETORY SYSTEM

The excretory system rids the body of wastes from cell activity. These wastes are known as metabolic wastes. The process of removing them is excretion. The excretory system helps maintain a proper balance of water, blood, and other substances.

The major products excreted are carbon dioxide, water, and nitrogen compounds from the breakdown of nutrients. Carbon dioxide and some water are given off by the lungs, which are a part of the respiratory system.

Major parts and functions of the excretory system are:

- **Skin**—The skin helps rid the body of wastes through perspiration. Water and minerals are given off through the skin. In animals that perspire, such as the horse, the skin helps regulate the body temperature. When the body is too warm, the skin perspires and is cooled by evaporation of the water. The water passes from the skin through pores. Some animals, such as chickens and dogs, do not perspire. They pant (breathe rapidly through the mouth) when they need to cool their bodies.
- **Kidneys**—The kidneys filter the blood. They remove wastes created by the cells in the body as well as excess water and minerals. The liquid produced by the kidneys is known as urine.
- **Bladder**—Urine is stored in the bladder (a muscular sac connected to the urethra) until excreted.
- **Urethra**—The urethra is the tube that carries urine out of the body.

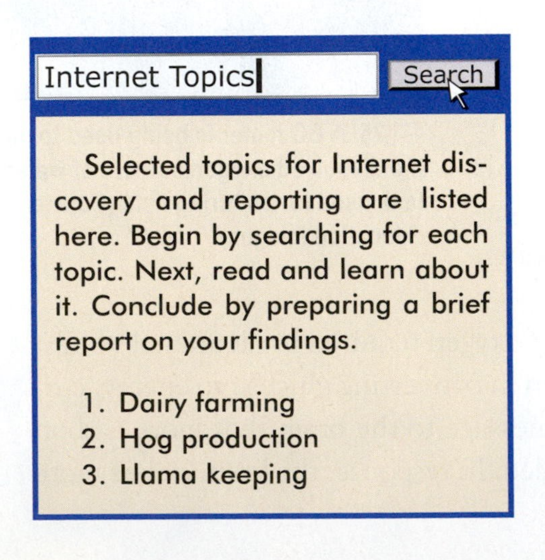

Internet Topics
Search
Selected topics for Internet discovery and reporting are listed here. Begin by searching for each topic. Next, read and learn about it. Conclude by preparing a brief report on your findings.
1. Dairy farming
2. Hog production
3. Llama keeping

DIGESTIVE SYSTEM

The digestive system prepares food for use by the body. Food is broken into molecules that the body can absorb. Digestion occurs in the alimentary canal (digestive tract).

Kinds of Digestive Systems

Animals are divided into two broad groups based on their digestive systems: ruminant and non-ruminant. A ***ruminant*** is an animal that chews a cud (small mass of food brought from the stomach for chewing); examples are cattle and goats. Its stomach has three or four compartments. ***Nonruminants***, such as hogs and chickens, do not chew a cud. Their stomachs have one compartment. Horses and a few other species are pseudoruminants, meaning their systems are "between" ruminants and nonruminants.

Ruminants use large amounts of roughage, which is forage, such as grass, clover, and other vegetation. Ruminant animals graze on land that cannot be used for other crops. They convert forage into higher quality food. Ruminants do not normally need large quantities of grain and other concentrated feeds. They are typically fed grain when in a feedlot for growth and tenderness in the meat. Nonruminants make poor use of roughage. They need grain and other concentrated feed for rapid growth.

Parts of Digestive Systems

Major parts of a digestive system are:

- **Mouth**—Food enters the body through the mouth by ingestion (eating). Teeth may chew the food to break it into smaller pieces. Saliva is mixed with food in the mouth to begin the digestive process and to make swallowing easier. Saliva is produced by salivary glands in the mouth.
- **Esophagus**—The esophagus connects the mouth and the stomach. Swallowing moves food and water from the mouth. It involves actions of the tongue and pharynx. Muscu-

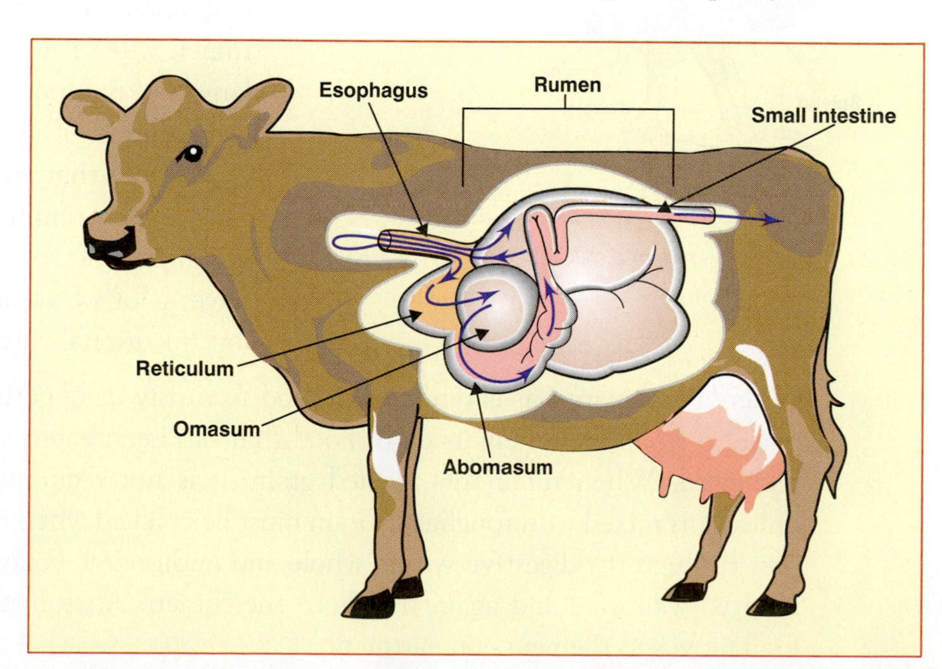

17–26. Digestive system of a cow.

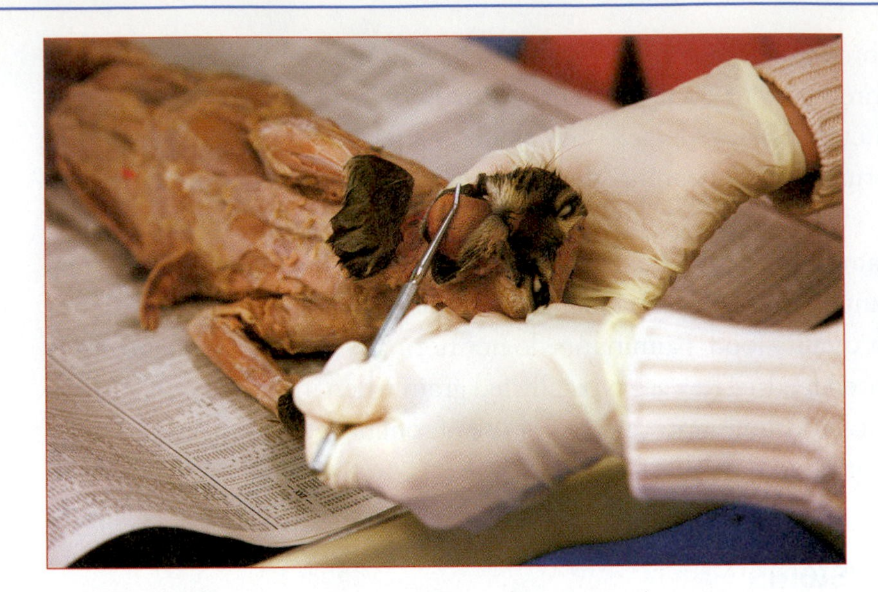

17–27. The open mouth of a partially dissected cat reveals internal structures that promote food intake. (Courtesy, Education Images)

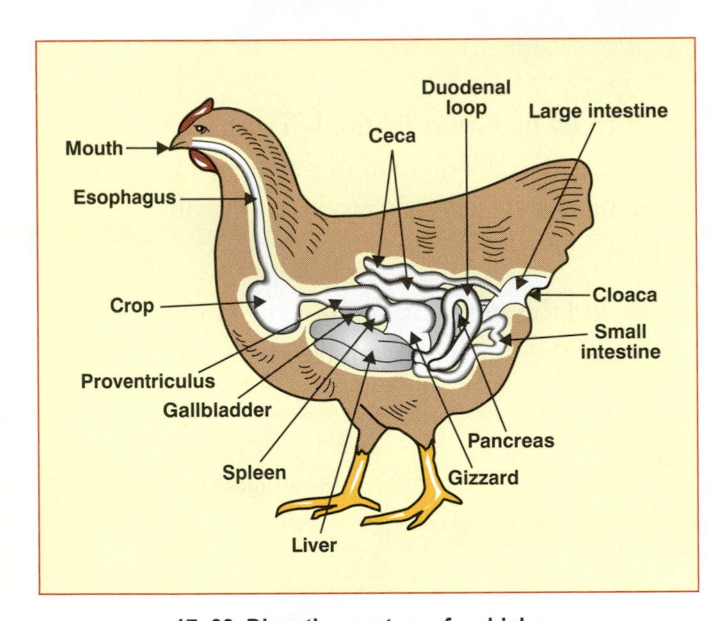

17–28. Digestive system of a chicken.

lar contractions in the esophagus move the food along. Food without sufficient moisture (saliva) may be difficult to swallow.

- **Stomach**—Stomachs may have one to four compartments. Ruminants have a compound stomach; it is divided into three or four parts. The compartments are:
    1. Rumen—The rumen or paunch is the first compartment. Food enters the rumen when it is swallowed. It is the largest of the four compartments and has billions of bacteria that break down foods that are high in cellulose. This is why ruminants can eat large quantities of roughage—feeds that have a lot of volume and relatively low nutritional value.

When a ruminant has eaten enough food to satisfy its appetite, it finds a quiet place to ruminate (chew its cud). Food that has been eaten is regurgitated and rechewed. When ruminants are fed grain, it is not regurgitated and chewed, unless it is mixed with roughage. Grain must be cracked when first eaten or it will pass through the digestive system whole and undigested. Following chewing, the food is swallowed and again travels to the rumen. After further digestion, the food moves to the next compartment.

2. Reticulum—The reticulum stores food and sorts out foreign materials. Sometimes cattle will eat nails, pieces of wire, or other harmful objects. They are retained in the reticulum.
3. Omasum—Also known as the manyplies, the omasum has strong, muscular walls that help break the food apart. After this is done, the food passes to the next compartment.
4. Abomasum—This is the fourth compartment in the stomach of a ruminant. Gastric juices are mixed with the food to help the digestive process. Next, the food moves to the small intestine.

   Nonruminants are known as monogastric animals. They have stomachs with one compartment. Digestive processes are similar to those in ruminants except that no cud is chewed. Nonruminants must have food that is of higher nutritional value. Horses have a digestive system part known as the caecum that allows the animals to use grass in their diets.

- **Small intestine**—The small intestine follows the stomach. Most absorption of nutrients and water occurs in the small intestine. Absorption is the process by which the end products of digestion are transferred into the blood. The liver and pancreas secrete substances into the small intestine that aid in digestion. The liver secretes bile, which breaks down certain foods, such as fat.
- **Large intestine**—The large intestine follows the small intestine and is the final organ of the digestive system. Slow contractions occur in the large intestine, moving the solid materials along. Mucus is secreted by the intestines and binds the solid materials together. The solid material is called feces.
- **Anus**—The anus is the opening in the body through which the large intestine expels solid wastes.

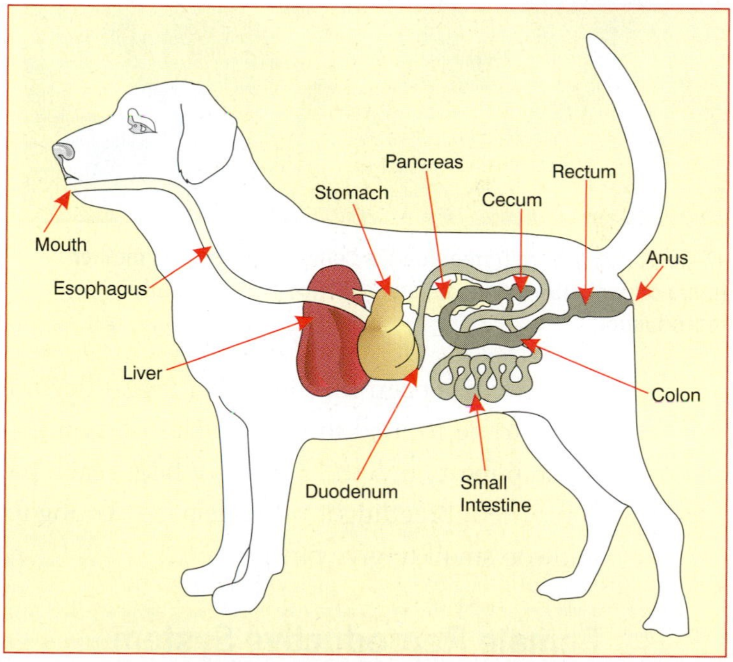

17–29. Digestive system of a dog—a monogastric animal.

Special Adaptations in Poultry

The digestive systems of chickens, ducks, turkeys, and other poultry vary from those of ruminants and nonruminants. Fowl have a mouth, crop (where recently eaten food is stored), glandular stomach, muscular stomach (gizzard), and intestines similar to those of other animals. The food is taken in the mouth and forced down the throat to the crop. The crop is a pouch-type structure that holds and softens food. From the crop, the food moves to the glandular stomach where it is mixed with gastric juices. The food travels from the glandular stomach to the gizzard, a strong and muscular organ that crushes and grinds the food. Afterward, it moves to the small intestine where much of the absorption takes place.

REPRODUCTIVE SYSTEM

Reproduction is carried out by the reproductive system. It begins with a male and a female animal mating. A sperm (male sex cell) unites with an egg (female sex cell) in a process called fertilization. With mammals, union of the sperm and egg occurs inside the reproductive tract of the female. The new animal is nourished and grows in the female's reproductive tract.

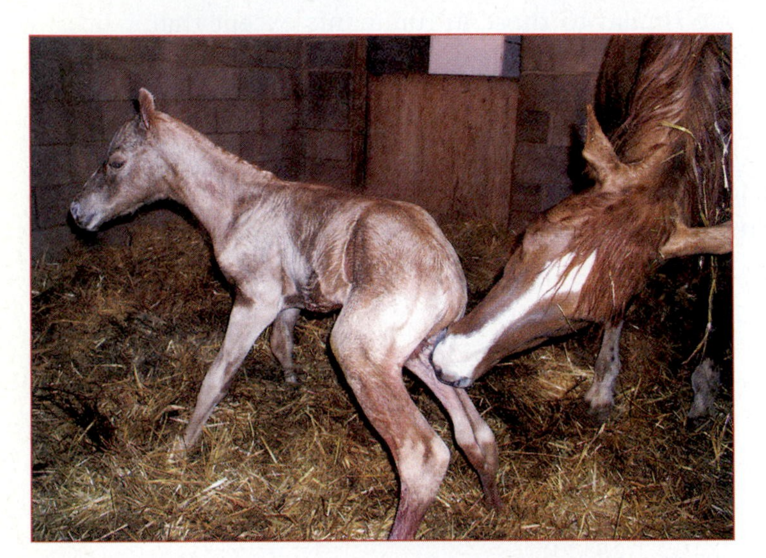

17–30. A newly born horse (foal) is being cleaned by its mother (mare or dam). (New animals are evidence of successful reproduction.)

The reproductive system is the only organ system that varies between male and female animals of the same species. These differences determine the sex of the animal. The male and female systems are designed to allow for mating and fertilization.

Puberty

Puberty is the age at which an animal is capable of reproduction. Sexual maturity has been reached. Ova (eggs) begin to develop in the ovaries of a female. Sperm begin to be formed in the testicles of a male. Animals vary in the ages of puberty. Cattle reach puberty in 8 to 12 months; hogs reach puberty in 4 to 7 months. Even though puberty is reached, breeding is often delayed. Young female animals may be too small for gestation and too small to give birth.

Female Reproductive System

The female reproductive system makes fertilization possible and provides a good environment for growth of the new animal. Major parts of the female system are:

- **Vulva**—The vulva is the external part of the female reproductive tract.

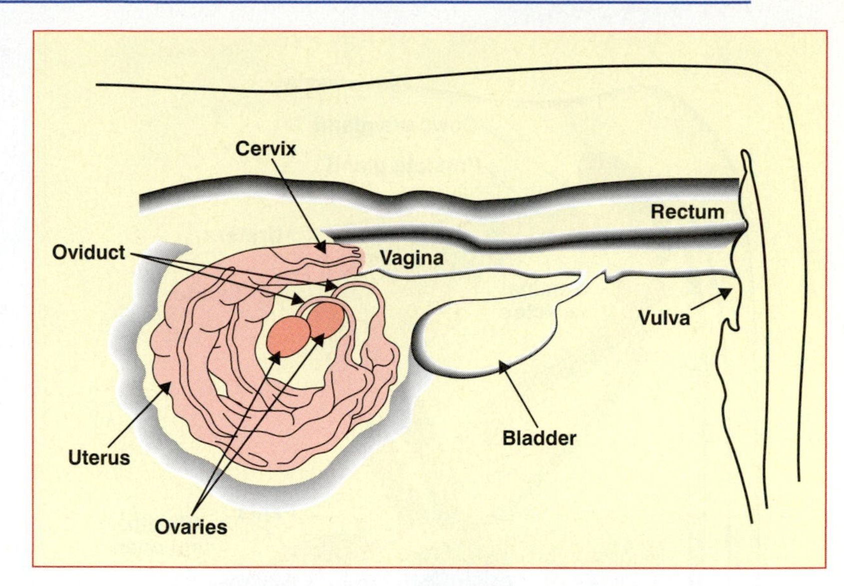

17–31. Parts of the reproductive system of a cow.

- **Vagina**—The vagina is the mating organ of the female. It is a muscular, tube-like organ that connects the vulva and cervix. The tube conducts sperm deposited in it by the male to the cervix and is the canal through which the fetus passes at the time of birth.
- **Cervix**—The cervix is the entrance to the uterus. It separates the vagina and uterus.
- **Uterus**—The uterus is a hollow organ with two parts known as uterine horns. This is where the fetus develops after fertilization.
- **Oviducts**—The oviducts are tubes that run from the uterus to the ovaries. They are also known as fallopian tubes. Fertilization (union of sperm and egg) usually takes place near the upper end of the oviduct.
- **Ovaries**—The ovaries produce the eggs. Females have two ovaries, each attached to an oviduct. Ovaries have many immature eggs (ova), which individually mature throughout the life of a female after puberty. The ovaries also produce hormones that regulate the reproductive system. Females usually ovulate (release eggs) on a regular cycle. Mating must occur at the right time for an egg to be fertilized.

Male Reproductive System

The male reproductive system is designed to produce, store, and deposit sperm in the reproductive tract of a female. Sperm have genetic information that joins with that of the egg. A sperm has a head filled with DNA and a flagellum (tail). DNA in sperm contains genetic traits of the male. The flagellum moves the sperm and receives energy from mitochondria located in the flagellum.

Parts of the male reproductive system are:

- **Penis**—The penis deposits sperm in the vagina of the female. Sperm are in a fluid known as semen. During copulation, semen is expelled by ejaculation. Millions of tiny sperm are released in one ejaculation.

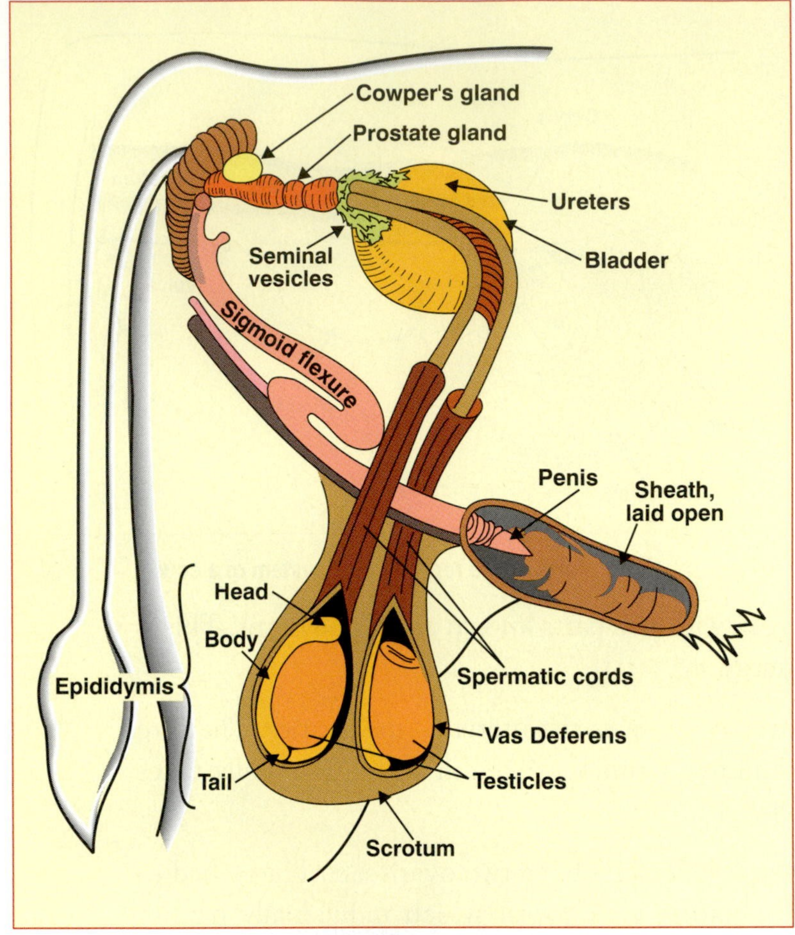

17–32. Parts of the reproductive system of a bull.

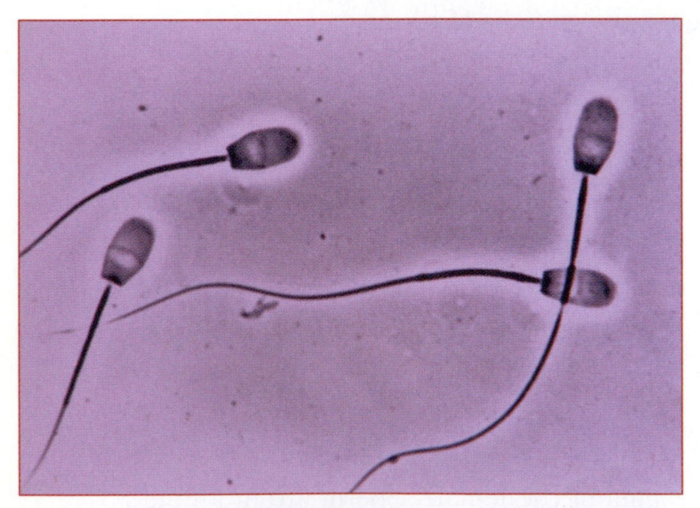

17–33. Image of sperm made through a microscope. (Courtesy, Elite Genetics, Waukon, Iowa)

- **Urethra**—The urethra is the tube that extends from the urinary bladder to the end of the penis. Semen passes through the urethra. Urine is expelled by the bladder through the urethra.
- **Seminal glands**—Two glands produce fluids that help males produce viable sperm. The cowper's gland is located on the urethra and secretes a fluid that precedes the passage of semen through the urethra. The fluid prepares the urethra for the semen and neutralizes any urine that could damage sperm. The prostate gland produces a fluid that helps carry sperm. The fluid, when mixed with sperm, is semen. The glands are located on the urethra near the bladder.
- **Seminal vesicles**—The seminal vesicles are near the bladder and are attached to the urethra. They produce a fluid that nourishes the sperm. The fluid is mixed with that from the prostate gland.
- **Sperm ducts**—The sperm ducts are tubes that connect the urethra with the testicles. The sperm ducts store sperm and move them to the urethra.
- **Testicles**—Testicles, also known as testes, produce sperm. Males normally have two testicles in the scrotum.
- **Scrotum**—The scrotum is a pouch of skin that holds the testicles outside the body cavity so they have a slightly

lower temperature because the lower temperature increases sperm production. Sperm production begins at puberty and continues throughout life.

Estrous Cycle

The ***estrous cycle*** is the cycle in the female reproductive system that prepares it for reproduction. During the "heat" part (estrus period) of the cycle, the female is receptive to breeding. The cycle begins with the ripening of an egg or eggs in the ovaries. When mature, the eggs rupture the ovary walls and pass into the oviduct. Mating must occur when a ripe egg is present; otherwise, no sperm will be present for fertilization. Eggs are fertilized in the oviducts. The length of the estrous cycle varies by species. Cattle and hogs have 21-day estrous cycles. The estrous cycle stops when the female becomes pregnant until after parturition.

The number of eggs released during the estrus period varies by animal species. Cattle and horses normally release only one egg. Sheep and goats may release two to four eggs. Hogs may release 15 to 20 eggs.

17–34. A female llama with her recently born baby. (A baby llama is known as a cria.)

After Fertilization

Following union of the egg and sperm, the fertilized egg moves through the oviduct and attaches to the wall of the uterus. Here, the fertilized egg receives nourishment and grows by cell division.

While a female has a developing animal in her reproductive tract, she is said to be ***pregnant***. In the first half of the pregnancy, the unborn animal is known as an embryo; it is called a ***fetus*** in the second half of pregnancy.

Gestation is the period between fertilization and full development of the fetus. It ends when the animal is born. In cattle, gestation lasts an average of 281 days. The process of giving birth is known as ***parturition***.

Male and Female Differences

Males and females vary in the frequency that they can fulfill their roles in reproduction. Males can contribute to reproduction more frequently than females because the roles of the sexes differ. Animal producers recognize these differences in animal reproduction.

17–35. A calf nurses its mother.

Females vary in the frequency of breeding and number of offspring produced at each breeding. Cows, for example, usually have one calf each year. Estrus cycles resume a few weeks after giving birth. Cows can be bred again at this time. Sows are usually bred to produce two litters of pigs per year. Litters are multiple births of pigs, with as many as 20 in a litter. Most hog producers prefer about 12 pigs in a litter. Female mammals are also responsible for providing milk for the newborn animals.

Males can be bred to a number of females each year. Sperm mature in a period of 12 to 24 hours. The male has no role in carrying the developing embryo and fetus to birth. One bull in good condition can mate with about 30 cows a year.

MAMMARY SYSTEM

Only female mammals have developed mammary systems. Male animals have rudimentary or undeveloped mammary tissues.

The mammary system is made of glands that produce milk. All female mammals produce milk as food for their young. In dairying, cows are milked to obtain milk for human food.

The mammary system consists of mammary glands containing millions of ***alveoli***. These structures use nutrients from the blood to make milk. A cow has four mammary glands and each has a teat through which the milk passes when feeding the young or during milking. The four mammary glands are bound together to form the udder. Hogs may have 10 to 18 mammary glands along the underside (belly) of the female. A mare has four mammary glands, but only two teats. Females must receive proper feed if they are to produce plenty of milk.

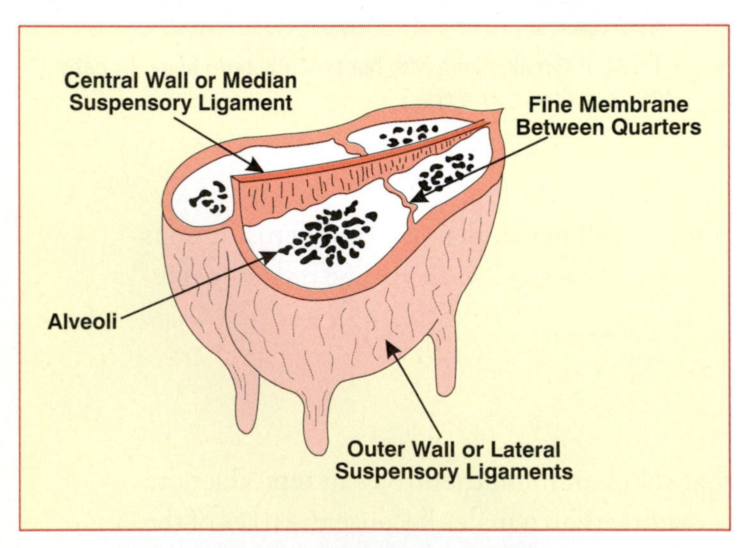

17–36. Structure of the udder.

The production of milk is known as ***lactation***. Females lactate only after giving birth to young. Lactation may last for several weeks or months. It stops when the offspring are weaned. When a young animal can eat other foods, it has reached the weaning age. Dairy farms use practices to ensure maximum lactation by the cows.

HOW POULTRY ARE DIFFERENT

Chickens, turkeys, and other poultry do not give birth to live young and do not have mammary glands. They lay eggs. If the eggs are fertile and properly incubated, they will hatch a new animal. Incubation is providing the proper temperature and humidity for the egg to develop. Chicken eggs have an incubation time of 21 days; other poultry eggs vary.

In chickens, fertilization requires the sperm of a rooster (male) uniting with the egg as it is being formed in the hen. The male deposits sperm in the hen's reproductive tract. Sperm from one mating may live for several days and fertilize several eggs.

17–37. Fertile chicken eggs in a commercial incubator are regularly rotated to assure embryo development.

17–38. A turkey hen oversees her babies in a natural environment.

A sperm fertilizes the female germ cell, which is located on the outer edge of the yolk. Fertilization must occur before the egg shell is fully formed. The female lays the egg and, with incubation, it will hatch. Eggs that are not fertilized will not hatch.

Young poultry start eating small particles of solid food shortly after hatching.

REVIEWING

MAIN IDEAS

Animals can be grouped based on several criteria: body structure (vertebrate or invertebrate), where they live, and the products they provide.

The most common farm animals are vertebrates. They have backbones and extensive body systems that perform important functions. Anatomy, morphology, and physiology are important areas of study in animal biology. Additionally, animal producers are concerned with conformation, which is the relationship of an animal's body features to that of the ideal animal of that species.

The major body systems are: (1) the skeletal system, which supports the body and protects delicate organs; (2) the muscular system, which provides animals with locomotion and aids in internal movement, circulation, digestion, and breathing; (3) the nervous system, which sends messages (impulses) and helps the body work as a unit; (4) the circulatory system, which moves digested food, oxygen, wastes, and other materials around the body; (5) the respiratory system, which moves gases to and from the circulatory system; (6) the excretory system, which rids the body of wastes; (7) the digestive system, which prepares food for use by the body; (8) the reproductive system, which creates new animals; and (8) the mammary system, which develops only in females and provides milk for young animals.

QUESTIONS

Answer the following questions, using complete sentences and correct spelling.

1. What are vertebrates and invertebrates? Distinguish between the two.
2. What are the distinguishing characteristics of the phylum Arthropoda? List examples of species in this phylum.
3. What important animals are in the phyla Annelida and Mollusca?
4. What important animals are in the phylum Chordata?
5. What important animals are in the following classes: Osteichthyes, Aves, and Mammalia?
6. Define anatomy, morphology, and physiology.
7. List the major body systems of animals. Briefly describe two important functions of each.
8. Distinguish between voluntary and involuntary muscles.
9. What are the major parts of the nervous system?
10. What is blood? What liquids and solids does it contain?
11. Distinguish between internal and external respiration.
12. What is breathing? What determines the rate of breathing?
13. What wastes are given off by the excretory system?
14. Distinguish between ruminants and nonruminants. List a major advantage and disadvantage of each.
15. What is puberty?

16. What is the major purpose of the female reproductive system and the male reproductive system?
17. What occurs after an egg is fertilized?
18. How does the frequency of breeding vary between males and females? Why is this important to producers of animals?
19. Why do females have mammary systems? What is lactation?
20. Since chickens and turkeys do not have mammary systems, how do their babies receive needed nutrition?

EVALUATING

Match each term with its correct definition.

a. lactation
b. gestation
c. puberty
d. estrous cycle
e. fetus
f. blood
g. vertebrate
h. exoskeleton
i. mammary glands
j. anatomy

______1. Milk-producing glands by which the newborn of mammals are fed.
______2. The secretion of milk by mammary glands.
______3. The study of the form, shape, and appearance of an animal.
______4. The time in which a female mammal carries a developing embryo in her uterus.
______5. An animal with a backbone.
______6. The cycle in the female reproductive system that prepares it for reproduction.
______7. The outer, hard covering that protects and supports the body of an arthropod.
______8. The age at which an animal is capable of reproduction.
______9. An unborn animal in the second half of gestation.
_____10. Liquid in a circulatory system.

EXPLORING

1. Dissect a fetal pig, fish, or other small animal to observe its organ systems. (Be sure to follow the appropriate procedures in this activity. Your teacher can demonstrate how it is done.)

2. Make a list of the animals raised in your community. Indicate the kinds of products they provide. Prepare a report on your findings.

3. Maintain a pair of small animals in the laboratory. Observe their behavior, including breeding, parturition, and the care received by the young. Be sure to consider the well-being of the animals.

4. Tour a dairy farm, and observe the milking of the cows. If you have never hand-milked a cow, ask the worker to show you how it is done. Try it for yourself.

CHAPTER 18

Animal Nutrition and Feeding

OBJECTIVES

This chapter presents the principles of nutrition and feeding of animals. It has the following objectives:

1. Describe the feed needs of animals.
2. List and explain the functions of nutrients.
3. Identify and select feedstuffs that provide nutrients.
4. Explain the characteristics of good feed.
5. Discuss the role and quality of pasture forages.
6. Describe how animals are fed.

TERMS

additive
animal unit
balanced ration
carbohydrates
carrying capacity
concentrate
cracking and rolling
diet
digestible nutrient
energy
fats
feed
feedstuffs
forage
free access
grinding
hay
implant
maintenance
meal
mineral
mixed feed
palatability
pasture
pellet
permanent pasture
protein
protein supplement
rangeland
ration
roughage
scheduled feeding
silage
stocking rate
temporary pasture
vitamin
working animal

18–1. Healthy, productive animals require proper nutrition. (This cow and calf are utilizing this pasture as a major source of nutrients.)

ANIMALS must have the right feed to live, grow rapidly, and be productive. Animals that do not get what they need may be stunted, get sick, or die. Producers and animal keepers should always strive to meet the nutrient needs of their animals.

Feed needs vary with the species, age, and condition of an animal. An animal's biological makeup is important. Ruminants, such as cattle, can eat large amounts of roughage because they have stomachs that convert low-quality feedstuffs into higher-quality nutrients. They are fed differently than other animals. Nonruminants, such as dogs, usually have stomachs with one compartment and this affects what they are fed. Of course, poultry are different from hogs, horses, sheep, cattle, and fish.

Animals have preferences about feed just as people have their favorite foods. These need to be considered. Cattle will eat about anything if molasses is added to it!

FEED NEEDS OF ANIMALS

Animals should have feed that meets their needs. What an animal needs depends on its age and other conditions. Young animals need feed that promotes growth. Other animals need feed that helps them perform efficiently and have good health. Even the weather has something to do with feeding animals!

18–2. A horse needs sufficient energy for competitive and strenuous events.

Animals use feed to maintain their bodies and for growing, reproducing, lactating, working, and, in some cases, for providing other products.

- **Maintenance**—Animals use nutrients to keep the body warm, replace old cells, run the internal organs, and have locomotion. Even an animal that is lying still is using energy—its heart is beating! Animals use energy 24 hours a day.

  Maintenance is the level of nutrition needed to keep an animal from losing weight when not producing milk or other products. Maintenance does not involve any gain or loss in weight.

  The feed needed for maintenance varies with the species of animal. Cattle and sheep need nearly 10 percent more feed when standing than when lying down. This is why animals that are being fed-out (finished) are kept in small pens and encouraged to lie down.

18–3. A ewe and her lamb need nutrients for maintenance, growth of the lamb, and lactation by the ewe.

- **Growth**—Feed for growth varies with the stage of life. Young animals need feed that helps them develop. Growth is an increase in the size of bones, muscles, internal organs, and other body parts. Young calves, pigs, and chickens are expected to grow rapidly. Their diets are modified as they grow to meet their needs better. The diet of a very young, growing animal needs careful attention. The nearer maturity an animal gets, the more the diet can be like that of an adult. Adult diets have less protein and provide nutrients for maintenance, reproduction, and lactation.

18–4. Automated free-access systems provide feed to rapidly growing broilers.

- **Reproduction**—An animal production operation depends on animals reproducing. The goal is to produce many well-developed animals. Species vary in how they reproduce. Nevertheless, their needs must be met for reproduction to occur.

  With mammals, the females give birth to live young. The pregnant female must get the feed needed to support her body as well as her developing baby or babies. An example is the dog. A pregnant female dog may have several puppies developing in her reproductive tract. Nutrients are essential for the health of the female as well as proper development of the unborn puppies. Male mammals need a proper diet. If they are too fat, they may be incapable of reproducing. Not enough feed lowers sperm production, resulting in females not being bred.

  With chicken operations, the emphasis is on egg production. The eggs may be fertile and used to hatch chicks. Most eggs, however, are used as human food. Fish also require proper nutrients to reproduce.

- **Lactation**—A female mammal must receive a proper diet to produce milk. Dairy cows have special diets to maximize milk output. A cow cannot produce milk without the right feed. Females nursing young animals need to have feed that helps them provide milk for their young. Sows may have 10 or more offspring to feed. Horses and cows usually have one foal or one calf, respectively, whereas sheep may have two or three lambs.

18–5. To produce sufficient milk for her litter, a lactating sow must receive appropriate feed. (Courtesy, Agricultural Research Service, USDA)

- **Working**—A ***working animal*** is an animal used for power, pleasure, or other purposes, such as a herd dog that watches sheep. Animals need feed for maintenance and energy for activity. Animals used for power include horses, oxen, and water buffalo. These animals need feed that provides enough energy. Pleasure animals, such as those used for riding and racing, need feed for activity and to maintain the body.

- **Other products and uses**—Animals that produce wool and similar products need nutrients associated with production. If the animal does not receive the right feed, not only is the rate of growth reduced, but the quality of the wool may be low. Animals used as pets or for showing must have diets that result in the desired animal. Calves, lambs, and pigs raised for showing receive special feeds to help them grow.

NUTRIENTS AND FUNCTIONS

A ***feed*** is a product containing nutrients. Animals ingest (eat) feed so that it can be digested. The digestive process breaks food down into nutrients that are used by the body. The ingredients used to make the feed contain various nutrients.

18–6. Raising lambs for showing requires careful attention to their nutrient needs. (Courtesy, Education Images)

A nutrient is a chemical substance in feed that supports the life processes of an animal. Nutrients nourish animals so they maintain themselves, grow, and reproduce. Feeds must contain adequate nutrients for an animal.

A ***ration*** is the total amount of feed an animal gets in a 24-hour period. The feed can all be given at one time or at several smaller feedings throughout the day. A ***balanced ration*** is a feed that provides all the nutrients needed in the right amount and proportion for a 24-hour period. Providing too much of some nutrients results in waste. A deficiency of one nutrient can limit animal growth and production.

A ***diet*** is the type and amount of feed and water an animal eats. Diets vary according to the nutrient content of the feed ingredients. Diets should be based on the needs of the animal being fed.

The major nutrient needs in the diets of animals are energy nutrients, protein, minerals, vitamins, and water.

ENERGY NUTRIENTS

Energy is a force that supports all the life processes of animals. A deficiency of energy (not having enough) stunts growth, lowers production, and causes loss of body tissue. Animals without enough energy cannot do as much work as they could if fed a diet high in energy nutrients.

Two main nutrients provide energy: carbohydrates and fats. In addition, some energy comes from protein.

18–7. Information on energy, protein, and other nutrients is on the label of most feed, including pet food. (Courtesy, Education Images)

Carbohydrates

Carbohydrates are sugar, starch, and related substances. They are formed in plants by photosynthesis. Animals that eat leaves, stems, roots, and seeds of plants get the carbohydrates. Animal bodies convert carbohydrates to glucose, which is found in the blood and provides much of the energy animals need. The remaining carbohydrates are stored as body fat.

Carbohydrates are made of carbon, hydrogen, and oxygen. The major sources of carbohydrates are corn, oats, hay, soybean oil meal, and grain sorghum. Most adult cattle can get adequate carbohydrates from the grass and other roughages in a good pasture.

Fats

Fats are concentrated sources of energy. Fats have 2.25 times as much energy as carbohydrates. Excess fat in the diet of animals is stored as body fat. Animals being fattened for harvest (slaughter) have diets high in fats and carbohydrates.

The major sources of fats are meat scraps, tankage (processed meat and bones), cottonseed, and fish meal.

PROTEIN

Protein is the substance needed for maintenance, growth, reproduction, and other functions. Protein is used to build new tissue and repair old tissue. Young, growing animals especially need protein. When not needed for building cells and tissue, it can be used for energy.

18–8. Animals may receive foods prepared especially for their species, such as the ferret. (Courtesy, Education Images)

Table 18–1. Essential Amino Acids

• Arginine	• Methionine
• Histidine	• Phenylalanine
• Isoleucine	• Threonine
• Leucine	• Tryptophan
• Lysine	• Valine

Animals need protein to form muscle tissue (meat). Bones, hair, hooves, skin, organs, and other parts of the body require protein. Producing milk, eggs, and wool also requires protein.

The major sources of protein are tankage, soybean oil meal, legume hay, blood meal, feather meal, fish meal, and skim milk.

Protein quality varies according to composition. Protein is made of amino acids. Twenty-three amino acids have been found in protein. Ten amino acids are essential; the others are nonessential. Some feeds have small amounts of certain amino acids; others have higher amounts. Proteins that provide the right balance of amino acids must be included in the diet.

Different animals make use of different protein sources. Ruminants form some amino acids in their digestive processes. Roughages low in protein are converted to higher-quality protein in the rumen. Animals with simple stomachs cannot make protein, so it must be in their diets.

When animals get more protein than they need, the liver converts the excess to energy. The kidneys excrete the nitrogen in the protein (this makes the urine yellow). Since protein costs more than fats and carbohydrates, balanced diets should be fed to keep costs down.

Technology Connection

VEGETARIAN FEEDS

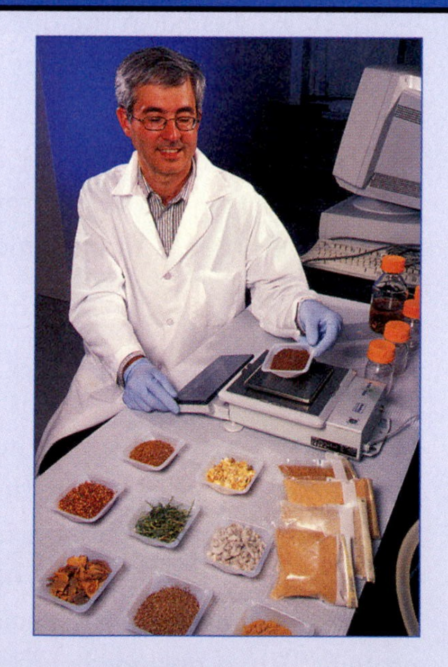

Cattle are herbivores—plant eaters. Grain, pasture forage, hay, silage, and other plant-origin feeds are the natural foods of cattle. As cattle feeding was domesticated, cooked and ground meat wastes were sometimes used in cattle feed as a source of protein. They were known as animal products.

Scientists now know that diseases can be transferred in the meat, though it has been processed. A good example is BSE (bovine spongiform encephalophy)—or mad cow, as it is commonly known. Ingredients in feed for livestock are increasingly being checked. Processed meat scraps are no longer acceptable in many feeds.

Cattle in feedlots now receive vegetarian feeds—feedstuffs of plant origin. This prevents the spread of disease through feed containing meat scraps. Scientists have developed new approaches in manufacturing and testing feed to assure it is safe for cattle. This photo shows a chemist conducting a test on feed to determine if any meat is present in the feed ingredients. (Courtesy, Agricultural Research Service, USDA)

MINERALS

18–9. Feed for ornamental fish is specially formulated to provide nutrients and appeal to fish. (This shows feed manufactured in a flake form for feeding to tropical fish.) (Courtesy, Education Images)

A ***mineral*** is a chemical element or compound needed for maintenance, growth, reproduction, and other body functions. It is used for teeth and bones, in body fluids and secretions, and in soft tissues of the body.

Without enough minerals, the body will not grow properly. Deficiencies result in weak and crooked bones, less milk production, fewer eggs, breeding problems, and disease. Mineral disease problems may be indirect or direct. Indirect disease problems occur when animals are weak, allowing causes of disease (bacteria, etc.) to invade the body. Diseases caused directly by a lack of minerals are known as nutritional diseases.

Animals need eighteen minerals. They need some in larger quantities, called major or macrominerals. Six minerals are macrominerals and are needed in amounts of one or more grams in a ration (daily amount). Calcium and phosphorus are needed in the largest amounts. Animal bodies contain slightly more than 1 percent calcium and slightly less than 1 percent phosphorus. Calcium and phosphorus are primarily found in teeth and bones.

Minerals needed in smaller amounts are trace or microminerals. Most feedstuffs grown in soil that contains these minerals will provide enough to meet the needs of animals.

Table 18–2. Essential Minerals Needed by Animals

Macrominerals (Major)	Symbol	Microminerals (Trace)	Symbol
Calcium	Ca	Chromium	Cr
Salt (sodium chloride)	NaCl	Cobalt	Co
Phosphorus	P	Copper	Cu
Magnesium	Mg	Fluorine	F
Potassium	K	Iodine	I
Sulfur	S	Manganese	Mn
		Molybdenum	Mo
		Selenium	Se
		Silicon	Si
		Zinc	Zn

18–10. Quality pasture grasses surround a livestock watering pond on a cattle farm.

Feedstuffs high in minerals include alfalfa and soybean hay, cereal grains, bone meal, molasses, and salt. Mineral supplements may be provided, such as the block of mineral salt placed in a pasture for cattle to lick. Minerals are often added to manufactured feeds as finely ground limestone, oyster shell, and ground rock phosphate.

VITAMINS

A ***vitamin*** is an organic substance that performs an important function in an organism. Vitamins regulate body processes, keep the body healthy, and help it develop resistance to

AgriScience Connection

AMOUNTS AND RECORDS

Animals often need specific amounts of feed. You would not want to overfeed or underfeed your show animal. Either might have an undesirable effect on your animal. Based on the needs of your animal, you may need to carefully weigh feed amounts. By doing so, you can provide exactly what your animal needs. Slightly increasing the feed amount might result in more fat or faster growth. Slightly decreasing the amount might have the opposite effect.

Feed given an animal may be weighed before it is fed. Weighing feed also allows you to keep accurate records. The weight of each day's feed can be recorded. These students are weighing feed for an animal in their school lab. (Courtesy, Education Images)

disease. Most rations contain enough vitamins. Rapid growth may require vitamin supplements. The absence of a vitamin can cause problems, such as disease or the inability to breed.

Plants naturally make some vitamins, such as vitamin C. Animals that eat the plants ingest the vitamins.

Vitamins are divided in two groups based on solubility (how dissolved): those soluble in water and those soluble in fat. Fat-soluble vitamins can be stored in the fat of animals' bodies. Water-soluble vitamins are not stored in the bodies of animals. Those not stored must be in a daily ration.

Table 18–3. Vitamins Classified by Solubility

Fat-Soluble Vitamins	Water-Soluble Vitamins
• Vitamin A	• Biotin
• Vitamin D	• Cholin
• Vitamin E	• Folic acid (folacin)
• Vitamin K	• Inositol
	• Niacin
	• Vitamin B-1 (thiamin)
	• Vitamin B-2 (riboflavin)
	• Vitamin B-3 (pantothenic acid)
	• Vitamin B-6
	• Vitamin B-12
	• Vitamin C (ascorbic acid and dehydroascorbic acid)

18–11. High-quality hay is a source of provitamin A for cattle. (Courtesy, Education Images)

Synthetic vitamins, known as crystalline vitamins, are pure forms that may be sprayed on or mixed with feeds. Vitamins A and D and the B vitamins are most important in animal health.

- **Vitamin A**—All animals require vitamin A. They make the vitamin in their bodies from carotene. Carotene is sometimes called provitamin A. Plant leaves and yellow foods are high in carotene. Yellow corn is a common source of carotene. Cattle grazing on green pastures can make all of the vitamin A they need. High-quality grass or legume hay and silage also provide provitamin A.
- **Vitamin D**—Most animals require vitamin D. Some feed contains little vitamin D; however, this does not usually pose a problem. The skin of animals contains provitamins, which act in the presence of ultraviolet (sun) light to form vitamin D. Hay and roughages cured in the sun are sources of vitamin D for ruminants.

- **B Vitamins**—Eleven different substances are known as B vitamins. Numbers are used after the "B" to distinguish between the forms of vitamin B. Some B vitamins are known by chemical names, such as thiamin. Animals vary in their need for B vitamins. Ruminants make them in their compound stomachs. Animals with simple stomachs cannot make B vitamins. They must get B vitamins in their feed. Since B vitamins are water soluble, they are not stored in the body fat of animals. This means that B vitamins are needed every day in a ration.

18–12. Most animals are provided free access to water. (This Congo African Grey Parrot is provided free access to containers of clean water and feed.) (Courtesy, Education Images)

WATER

Water is the most important nutrient. Most animals can live longer without feed than water. Animal bodies contain high amounts of water. A market-ready hog is 40 percent water. A newborn calf is 70 percent water.

Water is needed for the body to function. It is in all cells and body fluids and acts as a carrier to move materials around the body and eliminate wastes. Water helps regulate body temperature and is a part of all chemical processes. It softens food and aids in digestion. Excess water leaves the body as urine, in perspiration, in feces (solid body wastes), and as water vapor from the lungs.

Water should be free of dangerous chemicals, microorganisms, and excess minerals. Water sources should be dependable because animals need a steady supply. Animals usually have free access to water in tanks, waterers, or ponds.

ACADEMIC CONNECTION

HEALTH

Relate your study of animal nutrition and feeding in agriscience to your health class. Compare animal nutrition with that of the human body. Investigate similarities and differences. Prepare a report on your observations.

NUTRIENT SOURCES

Many different materials are used as feed. Some animals, particularly cattle and sheep, get sufficient nutrients by grazing pasture. Other animals, especially those confined (kept in pens), are fed prepared rations.

Balanced rations vary according to the kind of animal, its age, and its activity. Various ingredients are used in feeds to provide a balanced ration. Some ingredients are high in nutritional value; others are low.

Feed ingredients, also known as ***feedstuffs***, are classified by fiber content. Fiber is what remains of feed after the sugar, protein, and other substances have been removed. Fiber is a plant cell product, such as cellulose and lignin. Two general classes of feedstuffs are roughages and concentrates. Table 18–4 shows the nutritional value of a few feedstuffs.

18–13. An African hedge hog is being fed its favorite food—a worm!

Table 18–4. Nutritional Value of Selected Feedstuffs[a]

Feedstuff	Digestible Protein (%)[b]	Total Digestible Nutrients (%)[c]	Fiber (%)
ANIMAL PROTEIN SOURCES			
Concentrates:			
Blood meal	58.4	60.4	0.9
Fish meal	53.6	70.8	0.9
Tankage	50.5	65.8	1.9
PLANT PROTEIN SOURCES			
By-products:			
Cottonseed meal	35.9	72.6	11
Soybean meal	42	78.1	5.9
Corn gluten meal	36.7	79.1	3.8
ENERGY			
Grains:			
Corn	11.2	91	2.6
Grain sorghum (milo)	11.4	86	2.5
Oats	13.3	77	11.9
Wheat	11.1	80	2.6

(Continued)

Table 18–4 (Continued)

ROUGHAGES			
Hay, Grasses, Legumes, Silage:			
Alfalfa hay (mature)	15.22	55	29.3
Bahiagrass pasture (green)	8.9	54	30.4
Sugar beet pulp	8.8	67	26.3
Bermuda grass hay	9.22	43	28.4
Sorghum (milo) silage	7.4	18	17.9
Soybean hay	14.12	49	31
Wheat straw	3.22	40	41.7

Source: *Applied Animal Nutrition*, 2nd ed., Upper Saddle River, New Jersey: Prentice Hall, Inc.

[a]Some information has been rounded.

[b]Reported as crude protein for a few feedstuffs.

[c]Primarily reported for ruminants.

18–14. Reading information on a feed bag provides details about ingredients and nutrients. (Courtesy, Education Images)

ROUGHAGES

A ***roughage*** is a feedstuff that is high in fiber and low in energy. Roughages have low digestibility. A ***digestible nutrient*** is the part of a feedstuff that can be digested, or broken down. The fiber in roughage is not easily digested. Ruminants and horses are efficient users of roughage. This makes roughage an appropriate feed for these animals. These animals are fed grain to gain weight rapidly or for energy.

Examples of roughages are hay, green pasture grasses and legumes, and silage. Some by-products are used as roughage, such as dried sugar beet pulp and the leftovers from vegetable processing. (More information is given on roughages later in the chapter.)

CONCENTRATES

A ***concentrate*** is a feed low in fiber (less than 18 percent) and high in energy. Concentrates have high digestibility, known as total digestible nutrients (TDN). Feedstuffs used for concentrates are grains and protein supplements.

Grains

Grains are high in TDN, but they do not provide a balanced ration. Grains provide energy, but they are low in protein and must be supplemented. Grains are used to fatten animals. Cattle, hogs, sheep, and poultry that will be harvested are fed diets high in grain. Fat makes meat tender and better tasting.

18–15. Six common feedstuffs.

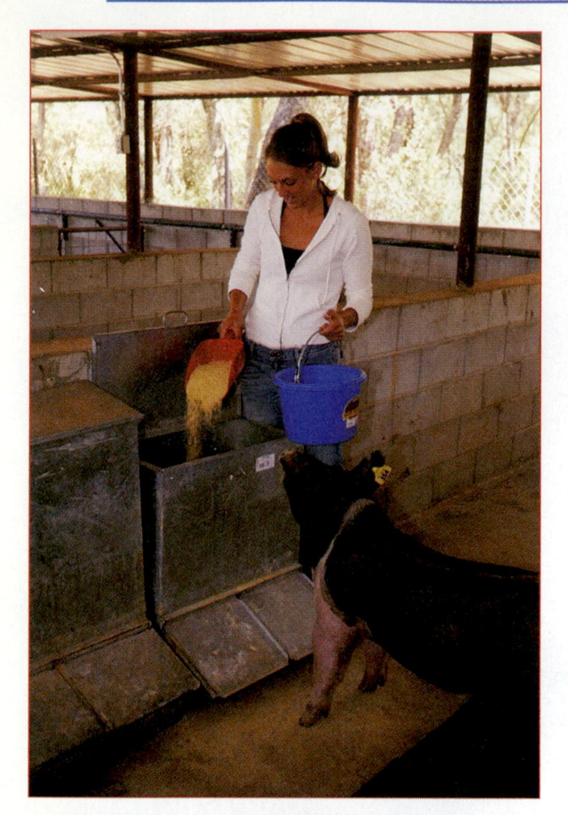

18–16. Corn is a widely used feedstuff. (Courtesy, Education Images)

Grain concentrates include corn, oats, wheat, and grain sorghum. Corn is the most widely used grain, especially for beef cattle and hogs. Corn is low in fiber and high in TDN. Use of grain sorghum has increased as a feed, particularly for poultry. Dairy cattle, horses, and other animals may receive corn. Feed for catfish, tilapia, and trout is often high in corn and soybean content.

Protein Supplements

A ***protein supplement*** is a feedstuff high in protein and TDN. These supplements may be mixed with grain or fed separately. Young, rapidly growing animals are usually fed rations high in protein supplements. Three types of protein supplements are used: animal, plant, and synthetic.

Animal sources of protein supplements are fed to hogs, chickens, fish, and cattle. Protein content ranges from 40 to 60 percent. These supplements correct protein deficiencies in rations high in grain. Examples of animal proteins are meat scraps and tankage (from meat packing plants), blood meal (dried blood collected from packing plants), fish meal, and skim milk. Caution is needed in feeding animal proteins. The meat products must be cooked to prevent the spread of disease, such as mad cow.

Plant sources of protein supplements include soybean oil meal, cottonseed oil meal, and various grain by-products from manufacturing, such as corn gluten meal. Digestible protein content ranges from about 20 to 45 percent for these. Beef cattle, dairy cattle, horses, and sheep do well on plant protein supplements. Soybean meal can be used for hogs.

Synthetic supplements may be used to feed ruminants. Examples include urea, molasses, rice hulls, and citrus pulp that have been treated with ammonia. Urea is a by-product of nitrogen fertilizer plants and widely used in supplements for ruminants. Exposing hay to a gas form of ammonia enhances nutrient content and digestibility.

SUPPLEMENTS

Mineral and vitamin supplements are often added to feed to ensure that animals get essential nutrients. The method used depends on how the animals are fed. When cattle are pastured, mineral blocks or supplement feeders may be used. Cattle have free access to the

18–17. Pastured cows may be provided free access to supplements using a well-constructed box that keeps rain out.

supplements. In mixed feeds, supplements are blended with the concentrates as the feed is being manufactured.

18–18. Feedstuffs may be stored outside in dry Southwestern U.S. climates.

FEED ADDITIVES AND IMPLANTS

Animal growth and production are important to producers. Various products are available to help animals grow better or produce more. Using these products may increase profits for animal producers. In many cases, they are added to feeds or are given to the animals.

Additives

An ***additive*** is a substance placed in feed during manufacture to preserve it and enhance animal growth. Many kinds of additives are used. All must be approved by the Food and Drug Administration (FDA). They should only be used for the approved purposes.

Additives are used for various reasons:

- **Medications**—Antibiotics and other medications may be added to feed to keep disease under control among the animals.

Internet Topics | Search

Selected topics for Internet discovery and reporting are listed here. Begin by searching for each topic. Next, read and learn about it. Conclude by preparing a brief report on your findings.

1. Amino acids
2. Pasture carrying capacity
3. Hay

- **Wormers**—Wormers are added to some feeds during manufacture to control internal parasites of animals.
- **Marketing enhancement**—Consumers often want products with certain characteristics. Using chicken feeds high in carotene and xanthophyll makes the leg shanks and egg yolks have a darker yellow color.
- **Antioxidants**—Antioxidants are put in feed to keep the feedstuffs from spoiling. These are widely used in fish meal, tankage, and other animal concentrates.

Implants and Injections

A few products may be placed directly into the body of an animal to increase production or protect it from dangers in the environment. An injection is the act of using a hypodermic needle and syringe to get a substance into the body system of an animal. An ***implant*** is a solid material placed under the skin to release substances over time. Food and Drug Administration rules cover the use of implants and injections.

Animals may be injected with hormones to increase growth or productivity. The use of bovine somatotropin (bST) increases milk production in dairy cows. bST naturally occurs in cows but, when they are given more, their bodies undergo change. The cows eat more feed and give more milk. A similar product has been used with hogs to increase milk production in nursing sows and growth in pigs. Female fish may be injected with human gonadotropin to encourage spawning (egg laying).

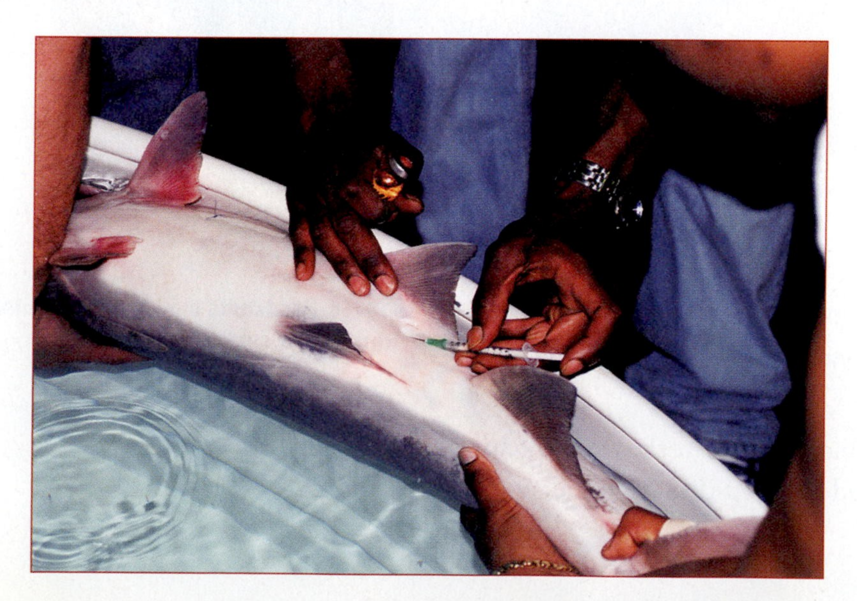

18–19. A fish being injected with a hormone. (Courtesy, Education Images)

GOOD FEEDSTUFFS

Some feedstuffs are better than others. Animals require different kinds of feeds. Select quality feedstuffs for the specific animals to be fed.

FEED REQUIREMENTS

Feed should be selected based on the needs of an animal and the ingredients in the feed.

Nutrient Content

Feed must provide the nutrients needed by an animal and be nutritionally complete. Protein must be in a form animals can use in their digestive systems. Energy must be provided. Since energy is measured in calories, lab procedures can be used to assess calories in feed. (A calorie is the amount of heat required to raise the temperature of 1 gram of water 1°C.) Minerals, vitamins, and water must be provided.

Some feedstuffs have many nutrients; others do not. Grain and other concentrates are highly nutritious. Wheat straw, corn stalks (fodder), and similar plant materials have little nutrition.

Animals should receive balanced rations. Feedstuffs have been chemically analyzed to determine nutrient content. This information is used in selecting and mixing feeds. It should result in a balanced ration for the animals being fed. (Books on animal science and feeding give information on how to balance a ration.)

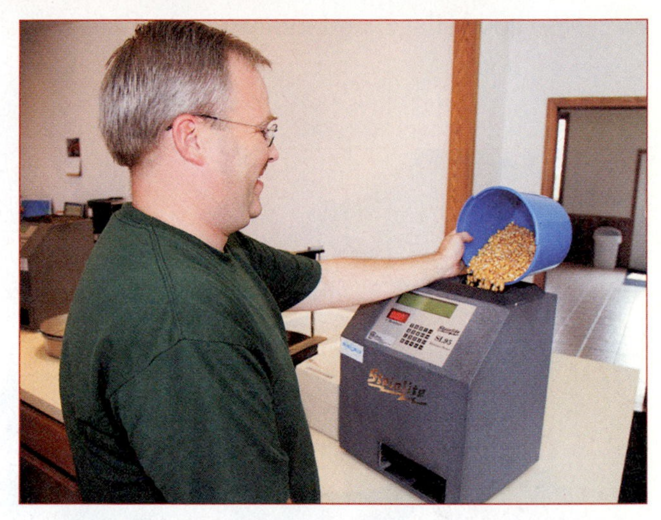

18–20. Corn quality is being assessed, with one step being passing a sample through a moisture meter. (Courtesy, Education Images)

18–21. Assessing feed quality assures that animals get good feed. On the left, whitish and dry wheat silage has molded. Good silage on the right is moist and fermented well.

Palatability

Palatability is the way a feed feels in the mouth and tastes. Animals like feed that has good palatability. Taste, odor, temperature, and texture are important. A good ration is of no benefit if the animal refuses to eat it!

In general, animals like feeds they are accustomed to eating. If feed is to be changed, do so gradually. Introduce some of the new feed along with the old.

Some of the feeds that cattle like best are green pasture grasses and legumes, corn, oats, good silage, and molasses. Animals will choose feeds that are palatable, over those that are not.

18–22. A water trough that should be cleaned of growing algae.

Free of Hazardous Materials

Poisons, spoiled feed, and contamination with foreign materials can make a feed dangerous to animals. Feed ingredients should be free of poison residues. Never store feed near pesticides.

Feed should not be contaminated with metal fragments or other materials. For example, a small nail in feed can injure a cow's stomach and cause a condition known as "hardware disease."

Water should be clear—not muddy. It should be free of pesticides and other hazards.

Variety

Some variety of feedstuffs is needed. A ration should contain several ingredients. Variety helps to ensure that all essential amino acids are present. Animals with simple stomachs need variety in their feeds more than ruminants do. Variety makes feed more palatable, which means that animals will eat more. When they eat more, they grow faster! Variety also helps provide needed vitamins and minerals.

Bulkiness

Bulkiness is the proportions of concentrates and roughages. Roughages are bulky, whereas concentrates are not. Ruminants especially need feeds that are bulky. Chickens and hogs need less bulk.

Cost

Feed cost is important. It determines if the producer will make a profit. The cheapest ration may not always be best because the quality of the feed may be low. The best way to assess cost is to look at the net returns (profit) when animals are sold.

18–23. Concentrates stored on-farm are moved from the bins to a wagon for hauling to the cattle.

Feed Storage

Feed should be protected from moisture, rodents and insects, and from contamination by fertilizer and pesticides. Some feed should be used shortly after manufacture because the vitamins and other ingredients lose their quality in storage. Feed is often hauled from mills to feed bins or troughs. This saves loading and unloading the feed and retains nutrients until eaten. Feed may be stored in bags or in specially designed bins in bulk form.

18–24. Feed may be put into paper bags or other containers during manufacturing.

FEED MANUFACTURING

Feed may be manufactured on the farm or ranch or by a commercial feed mill. Commercial feed mills are in the business of making feed that they sell to animal producers.

Manufacturing Processes

Manufacturing is preparing feed in the form that is needed and includes: (1) acquiring the feedstuffs (ingredients), (2) grinding or otherwise preparing the feedstuffs, (3) mixing the ingredients in the proper amounts based on nutrients needed in the feed product, and (4) preparing the feed in the form that it will be fed.

18–25. Large amounts of feed are delivered to farms in bulk and unloaded directly into a feed storage bin.

Feed should be formulated to provide the needed nutrients. Feed is bagged or stored in bulk. Proper labeling of bags is essential.

On-Farm Manufacturing

Larger ranches and feedlots may prepare their own feed. This requires buying the manufacturing equipment and hiring people to operate it. The ingredients may be grown or purchased. At least some of the ingredients, such as mineral and vitamin mixes, must be purchased. The feeds must be formulated to provide the needed nutrients.

It is usually impractical for small farms to have feed processing equipment. Farmers may form cooperatives to make feed for members, but they do not sell to the public.

FORMS OF FEED

Feeds are manufactured in various forms. A ***mixed feed*** is a feed made from a variety of ingredients. The ingredients provide the nutrients needed by animals. Feed may be prepared

CAREER PROFILE

FEED MILL OPERATOR

A feed mill operator is in charge of all processes in manufacturing feed. Modern mills are computer controlled and involve automated processes. The work involves monitoring equipment to assure proper operation and production of the desired feed.

Feed mill operators need backgrounds in computers, mechanical operations, and feedstuffs. A college degree in animal science, agricultural engineering, or related area is usually needed. Begin in high school by taking courses in agriculture, computers, physics, and chemistry. Practical experience in a feed mill is essential.

Jobs are typically in large feed mills near poultry, cattle, or hog feeding operations. This photo shows an operator using a computer system to control and operate equipment.

in different forms or particle sizes. The form selected should meet the needs of animals and provide necessary nutrients. Convenience of feeding is also a consideration.

Roughages

Roughages may be fed to cattle, sheep, and horses without harvesting. The animals graze in the pasture and do the harvesting themselves! The most common harvested roughages are hay and silage. Forms of roughage include:

18–26. Large round bales of hay are protected from rain by storing under a shelter or wrapping in plastic.

- **Bales—*Hay*** is the dried leaves and stems of plants. It is cut, dried, and usually made into bales. The bales can be moved for storage and feeding. Increasingly, hay is stored in the field and covered with plastic or a tarpaulin. Hay may not be covered in dry climates. When stored outside in climates with rain, hay quality deteriorates. The hay gets wet and begins to decay. Its color and consistency change. Weather-damaged hay is neither palatable nor nutritious.
- **Loose-chop**—Silage is usually cut and chopped while green into small 1- to 2-inch pieces. It is not bound together in any way. The chopped silage is stored in a silo where water is added and it ferments. This process is known as ensiling. Silos are upright, round structures or trenches in the earth, or they are built above the earth out of concrete.
- **Pellets and wafers**—Roughages are sometimes made into pellets or wafers. A ***pellet*** is a feed piece shaped by forcing ground material through holes in a heavy steel plate. A binder is added to hold the pellet together. Pellets for cattle are about 1 inch in diameter and 1 to 2 inches long. Several feedstuffs may be mixed together into pellets to assure a more nearly complete diet. For example, some corn and soybean meal may be added during the grinding process. Pelleting adds to the cost of feed because the ingredients must be finely ground.

18–27. Three forms of feed are shown here (from the left): meal, small pellets, and range cubes. The meal is a range supplement of cottonseed meal and salt for grown cattle. The pellets are high-protein feed for young heifers. The range cubes are large pellets intended for feeding to cows on pasture to supplement winter grazing.

Wafers are similar to pellets, except slightly larger, and they do not require the ingredients to be ground as finely. This method is increasing in use with good-quality hay crops.

Concentrates

Concentrates have high nutritional quality, are easy to use, and provide animals with needed nutrients. They are prepared in several ways.

- **Cracking and rolling**—***Cracking and rolling*** is a process that breaks the hard outer coatings on grain. Animals more easily digest grain when the outer covering is broken. Uncracked grains can pass through the digestive tract whole and not be digested. Corn and oats are examples of grains that are cracked.
- **Grinding**—***Grinding*** is a process that makes the particle size of feed smaller. Small animals require small feed sizes. For example, day-old chicks must have feed in particle sizes that they can eat. Grinding also cracks grains open so they are more digestible. It permits feed ingredients to be mixed together and made into other forms, such as pellets. Ground feed is sometimes known as ***meal***.
- **Extruding**—Extruding is a preparation method that makes feed in particles similar to pellets. Ingredients are ground, mixed, heated with steam to soften the particles, and forced through a steel plate to form pieces of feed. Extruded pellets may range in diameter from ⅛ to ½ inch. The length may range from less than ½ inch to 1 inch or more.

  Extruding allows feed ingredients to be mixed together to provide a proper diet. These are known as "all-in-one rations." They are often formulated for animals at different stages of life, such as puppy or calf chow.

  Fish feed is a good example. The extruded pellets contain all the nutrients a fish needs. The amount of protein and other nutrients is varied depending on the age and size of the fish. Most fish need feed that has 28 to 40 percent protein; small, young fish need higher protein.

Supplements

Supplements come in forms easy for the producer to use in meeting the nutritional needs of animals. Most supplements are used for cattle on pasture. The goal is to provide nutrients not adequately available in the plant materials eaten by the animal. Supplements may be provided in several forms.

- **Blocks**—Minerals and other materials may be compressed into blocks. Some blocks weigh 50 pounds and are about 8 inches wide, 8 inches thick, and 12 inches long; others may weigh as much as 500 pounds. Many cattle producers use 50-pound salt and mineral blocks for cattle to lick to get the needed nutrients.
- **Liquids**—Liquid supplements often contain water, molasses, and other ingredients that carry minerals and nutrients. Molasses makes the liquid sweet and palatable to cattle. Liquid supplements are 50 to 75 percent molasses. Minerals and vitamins are mixed with the liquid material.
- **Mixes**—Various mixes of loose, ground ingredients may be fed in troughs or other containers. Cattle have free access to them. Salt, protein sources (such as cottonseed meal), and other feed materials may be in the mixes.

18–28. Bags of vitamin premix await mixing with feedstuffs for poultry in a feed mill.

FEED LABELING

Commercial feeds must meet certain standards. On bagged feeds, a label is printed information that provides details about the feed. Labels tell the ingredients, nutritional content, and the weight of the feed product. Other information about the feed may be included.

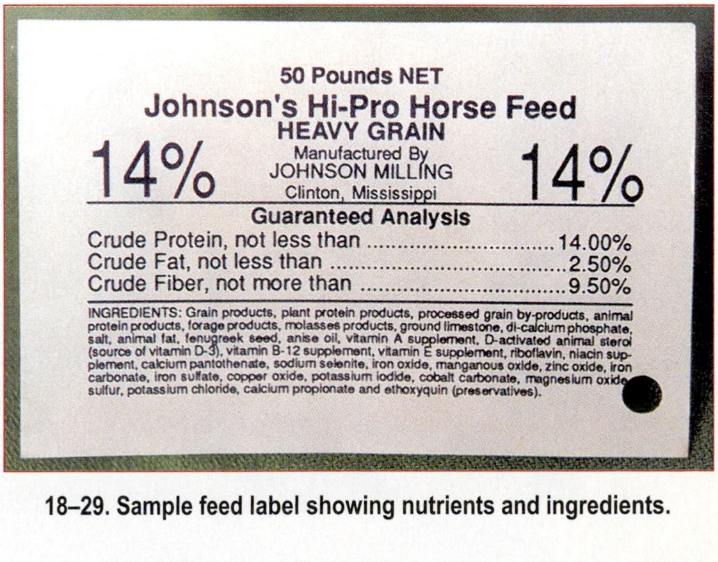

50 Pounds NET
Johnson's Hi-Pro Horse Feed
HEAVY GRAIN
14% Manufactured By JOHNSON MILLING Clinton, Mississippi 14%

Guaranteed Analysis
Crude Protein, not less than 14.00%
Crude Fat, not less than 2.50%
Crude Fiber, not more than 9.50%

INGREDIENTS: Grain products, plant protein products, processed grain by-products, animal protein products, forage products, molasses products, ground limestone, di-calcium phosphate, salt, animal fat, fenugreek seed, anise oil, vitamin A supplement, D-activated animal sterol (source of vitamin D-3), vitamin B-12 supplement, vitamin E supplement, riboflavin, niacin supplement, calcium pantothenate, sodium selenite, iron oxide, manganous oxide, zinc oxide, iron carbonate, iron sulfate, copper oxide, potassium iodide, cobalt carbonate, magnesium oxide, sulfur, potassium chloride, calcium propionate and ethoxyquin (preservatives).

18–29. Sample feed label showing nutrients and ingredients.

PASTURES AND FORAGE CROPS

Pastures and forage crops have major roles in the nutrition of ruminant animals. Plants that grow on land of little use for cultivated crops can be productive in providing food for animals. Animals, such as beef cattle, convert the plants into high-value food products.

PASTURE

Land where grasses and other plants grow for animals to graze is ***pasture***. It is a popular way of providing feed for cattle, sheep, and horses. Nearly 30 percent of the land area in the United States is in pasture. Pasture is of little value to chickens and turkeys. Adult breeding hogs are sometimes pastured.

Rangeland is prominent in some areas of the United States, particularly in the Western states. All states east of the Mississippi River except Florida have no land classified as rangeland. ***Rangeland*** is land that grows native forage plants. The kind and quality vary with the moisture, soil fertility, altitude, and other conditions. Rangeland is primarily used for pasturing animals. Most rangeland requires little maintenance, which fosters increased forage production. The climate is often very dry, and harsh winters may occur. Nearly half of the rangeland in the United States is owned by the federal government. The Bureau of Land Management and the U.S. Forest Service oversee the land, which is typically rented to ranchers.

The nutritional value of pasture depends on the kinds of plants and stages of growth. Pastures in areas with high rainfall are often planted with grasses and legumes that grow rapidly. Tender plants are high in water (sap). Dry areas (less than 20 inches of rainfall a year) may

18–30. Sheep benefit from good pasture.

have native plants that grow more slowly and contain less water. Pasture plants must be adapted to the environment.

18–31. This pasture is short and dead leaving nothing for the cow to eat. She will likely soon die of starvation.

Kinds of Pasture

Two kinds of pastures are used: permanent and temporary. Livestock producers may have both types.

- **Permanent pasture**—A ***permanent pasture*** is land with grasses and legumes that live and grow for years. Many pastures grow native plants. Pastures in areas with more rainfall may be seeded, fertilized, and otherwise improved. Permanent pastures are often seeded with warm season grasses and legumes. Bermuda grass, fescue, Bahia grass, Kentucky bluegrass, and other native grasses may be used. Common legumes are red clover, alsike clover, and sericea lespedeza.

  Pastures vary by region of the U.S. In the South, pastures may be seeded with grasses and legumes, such as bermuda grass and white clover. Pastures with warm season grasses provide little or no grazing in the winter and, therefore, the animals must receive hay, concentrate, or other feed. Some pastures are overseeded in the fall with cool season grasses that will provide some winter and early spring grazing. In the Northeast, Plains states, and Western states, pastures may have shorter grazing seasons. Winter may require greater supplemental feeding. Since most pastures are localized to the climate, authorities on pasture production should be consulted.

- **Temporary pasture**—A ***temporary pasture*** is land planted for winter, summer, or semipermanent grazing.

  Winter temporary pastures are planted with crops that grow in the winter. Rye grass, oats, wheat, and other cold-tolerant plants may be used. Winter pastures are often used in the southern part of the United States.

  Summer temporary pastures are planted in the spring for grazing in the summer and early fall. Millet and grain sorghum are widely used in summer pastures.

  Semipermanent pastures are planted with grasses and legumes and grazed for several years before being plowed. This type of pasture is not used as widely as permanent and temporary pastures.

18–32. Cows at this mixed-herd organic dairy farm are grazing short grass. (They will soon be rotated to another pasture.) (Courtesy, Education Images)

Carrying Capacity

Carrying capacity is the maximum stocking rate on pasture land that is consistent with maintaining or improving vegetation on the land. Producers may use stocking rate as a measure of carrying capacity. ***Stocking rate*** is the number of animal units that graze or otherwise use an acre of land for a specific time. An ***animal unit*** is one animal at the weight of 1,000 pounds. For example, a beef cow weighing 1,000 pounds is one animal unit. (Other units are: beef cow/calf, 1.3; mature bull, 1.5; horse, 1.23; sheep/goat (nonlactating), 0.2; and weaned lamb/kid, 0.15.) The stocking rate on permanent pastures may vary from about 3 acres per cow and calf in the Southeastern United States to 50 or more acres in the Western states.

Several factors influence stocking rate. These include:

- **Quality of pasture**—Pastures can often be improved by seeding with productive plants (such as grasses and clovers), fertilization and liming, irrigation, weed management, and other practices. Shade and water should be available.
- **Species and condition of animals**—Species and condition of the animal grazing the pasture is a part of the animal unit calculation. Obviously, a large beef cow needs more forage than a small goat. Further, a pregnant and/or lactating cow needs more pasture than a dry cow. Competition from wild animals that may invade a pasture should be considered. In some areas, deer, elk, antelope, and bison may compete with the cattle, horses, or other animals in the pasture.
- **Degree of nutrition from pasture**—Some animals are on pasture and receive supplemental feed; others rely solely on pasture for their nutrients. Obviously, supplemental feeding allows more animals per acre of land, but the cost may make it unwise in some cases.

- **Grazing management practices**—Grazing management refers to whether animals continuously graze a pasture or are rotated to other pasture areas. Rotation allows a pasture area to be more heavily grazed for a short period. The animals are moved to another pasture and the forage regrows.
- **Season**—Some grasses and legumes grow only in the warm season; therefore, the land has pasture benefits of consequence only during this time. Select the species and mix of grasses and legumes when establishing a pasture that will achieve your goals.
- **Climate**—Climates vary widely throughout the United States. Dry areas produce less forage from pastures than areas with greater precipitation. The same is true with temperatures, as areas with warmer climates have the potential of greater forage production.

Maintaining the correct stocking rate on pastures is an important consideration in animal production. Overstocking stunts animal growth and leads to health issues, such as short pastures increase parasite infestation. Overstocking also damages the land, plant population, and creates soil erosion that may damage stream quality. It also leads to loss of income for the livestock producer.

Pasture Management

Pasture management is the activity of establishing and maintaining quality forage. Establishing a pasture calls for preparing the land and selecting the plant varieties. Building fences is also a part of pasture establishment.

Pastures are maintained by keeping weeds down and applying fertilizer. Plant diseases may need to be controlled. Permanent pastures are clipped once or twice a year to cut weeds and remove old growth from pasture plants. Herbicides are sometimes applied to help manage weed pests. These chemicals should be used properly so that livestock are not exposed to them. Livestock may be rotated to prevent overgrazing, which results when the plants are eaten too short.

18–33. A windrower is cutting and conditioning a hay crop. (The crop is left in a windrow for curing.) (Courtesy, AGCO, Duluth, Georgia)

HAY

Hay is the leaves and stems of plants that have been cut and dried for feed. Hay is cut with a mowing machine or conditioner and dried in sunlight. Special hay crops may be planted or hay can be cut from pastures that grow more vegetation than cattle eat. Hay is best fed to ruminants.

Leaves and tender stems of legume plants make the best hay. Legume hay is higher in minerals and protein than non-legume hay. Examples include alfalfa and soybeans. Hay of lower nutritional value is made from rye grass, oats plants with immature grain heads, Bermuda grass, and Johnson grass. Older plants with many stems are not as good as younger plants with few stems.

The best quality hay is free of weeds. Some weeds are poisonous and can kill livestock. To retain quality and prevent nutrient loss, hay should be protected from rain after it is cut and stored. It needs to be dry before baling and storing. Hay that is not dry enough may rot or catch fire by spontaneous combustion. Hay may be baled in rectangular or large round bales. It is usually cut in the spring and summer and stored for feeding in the winter. Some farms feed hay all year, such as those raising dairy cattle.

18–34. Special crops may be used as hay for animals that need particular nutrients. (This is a bale of perennial peanut hay—high in protein and good for show goats.) (Courtesy, Education Images)

SILAGE

Silage is a crop of green plants that are chopped into small pieces and placed in a silo to ferment. Water is usually added and fermentation of the chopped plants preserves the nutrients. Silage is cut in the spring and summer when the plants are green and tender. Fermentation may result in an unpleasant odor. Silage is often used to feed dairy cattle and in feedlots for beef cattle.

Corn, grain sorghum, and wheat are frequently used for silage. Corn is cut for silage after the ears have formed but before they begin to dry. All silage has high water content.

18–35. Corn silage is chopped stalks, leaves, and ears, with some corn kernels.

OTHER ROUGHAGES

Four examples of other roughages are:

- **Crop residues**—Crop residues are the parts of plants left in a field after harvest. For example, only the seeds of soybeans are harvested, leaving pods, stems, and leaves in the field. Cattle may graze on the fields, or the residue may be gathered and brought to the cattle.
- **Haylage**—Haylage is a form of silage with low moisture. The grass or legumes are wilted before they are put in the silo. Haylage use is increasing for dairy cattle.
- **Soilage**—Soilage is green plants that are cut and chopped in the field and hauled to feed animals that are confined, such as in feedlots. This is also known as green chop. Soilage is fed without storage or fermentation.
- **Poultry litter**—Chickens are raised in houses where the floor is covered with wood shavings and sawdust. After the chickens are grown and sold, the house is cleaned out. Molasses may be added to the mix of wood shavings, chicken manure, and uneaten feed and fed to cattle.

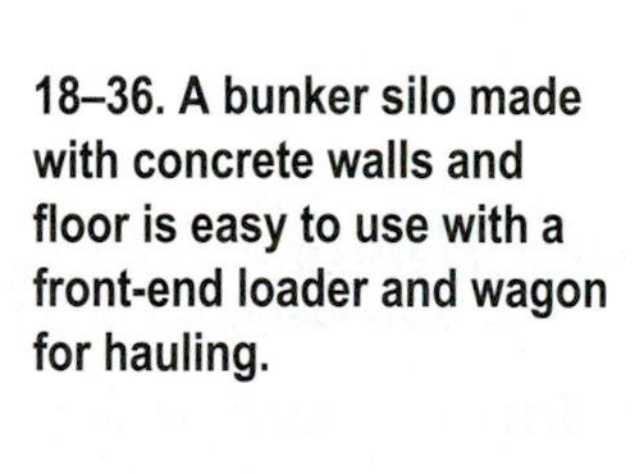

18–36. A bunker silo made with concrete walls and floor is easy to use with a front-end loader and wagon for hauling.

HOW ANIMALS ARE FED

Animals are fed in several ways. The goal is to provide feed in an efficient manner to meet the needs of animals. Producers follow different production practices.

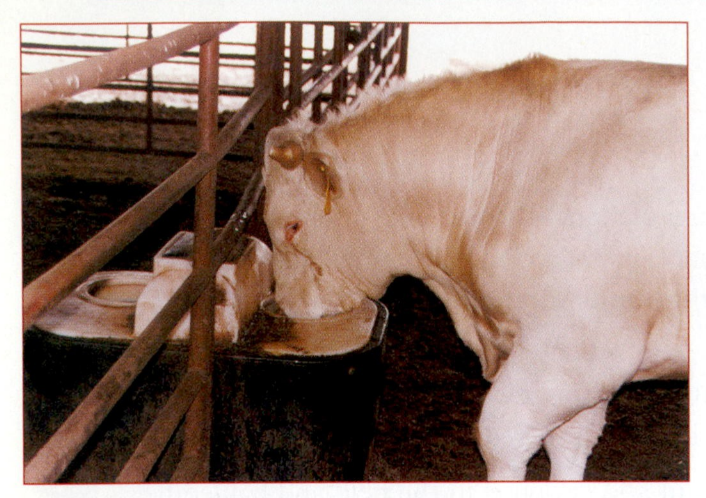

18–37. An automatic waterer provides free access for a bull in a test facility.

18–38. Caged laying hens have free access to feed. Note the automated feed trough just above the eggs.

18–39. A timed feeder schedules feeding fish in a tank.

FREE ACCESS

Free access is a feeding system that provides animals with access to the feed when they want it. Feed in feeders or troughs is always available. Hay may be placed in racks where animals can get to it. Of course, cattle, sheep, and horses on pasture can graze when they want!

Livestock on pasture may have free access to mineral supplements. With free access, the feed and supplement must be protected from the weather and damage. Special feeders may be used.

Automated systems may regularly add feed to the containers. Computers control some; others require hand feeding.

Animals in confinement (pens) must be provided the feed they need. Hogs being fed for market in pens, cattle being fed in feedlots, and chickens being grown in houses for meat are provided access to all the feed they will eat. Rapid growth occurs when they get all they want with the right nutrients.

SCHEDULED FEEDING

Scheduled feeding is a feeding system that provides feed at certain times of the day. It is often used with cattle and fish. Dairy cows are often fed at the time of milking. This allows the cow to eat and "letdown" the milk. Dairy cows that are on pasture may only be given a concentrate supplement at milking. Those in confinement have hay or silage throughout the day.

Beef cattle producers may provide supplemental feeds at certain times of the day, such as each afternoon in the winter.

Fish are fed based on their needs or on a demand basis. Tiny, newly hatched fish (fry) are fed once each hour throughout the day. The feed is in very small particles. Their digestive systems are small and will not hold much. Little energy is stored in their bodies; therefore, they must be fed frequently. As they grow, the intervals between feedings are lengthened. Fish several inches long are fed once or twice a day and no more than they will eat in a few minutes. Uneaten feed wastes money and causes water problems.

18–40. With a demand feeder, fish have feed any time they wiggle the "trigger" suspended into the water.

SPECIAL ASSISTANCE

Animals that are going to be exposed to stressful situations may be given special nutrient supplements to build up their resistance to diseases associated with stress. Shipping is often a stressful time for animals. Some animals have been said to develop shipping fever. Being penned, loaded, hauled, and unloaded in a strange environment is stressful. Special conditioning can be used to minimize the likelihood of an animal contracting disease during shipping.

Another stressful activity is being exhibited in an animal show. Special supplements of vitamins as well as antibiotics can be administered. These are usually given orally to individual animals.

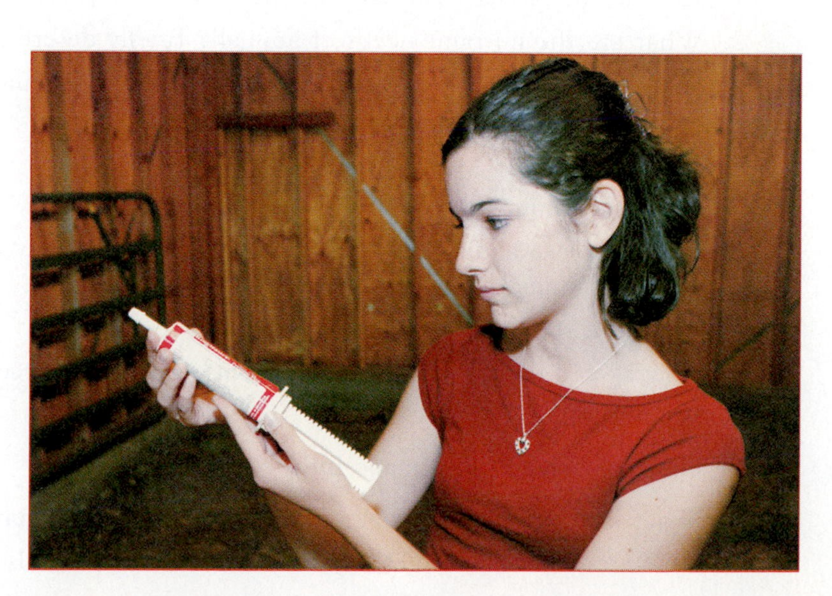

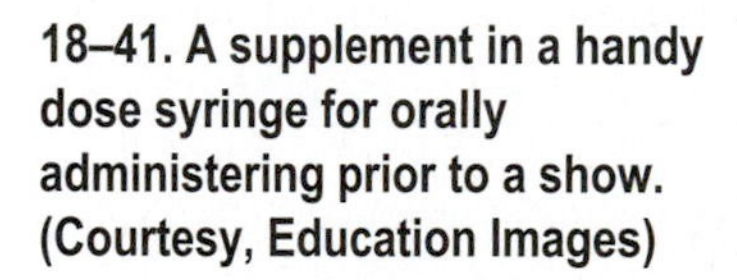

18–41. A supplement in a handy dose syringe for orally administering prior to a show. (Courtesy, Education Images)

REVIEWING

MAIN IDEAS

Animals need feed for several purposes: maintaining their bodies, growth, reproduction, lactation, working (race horses, draft horses, and herd dogs), and to produce other products, such as wool.

Feed provides the nutrients animals need. Nutrients are chemical substances that support life processes. Feeds make up a ration, which is the amount of feed an animal gets in a 24-hour period. A diet is the amount and type of feed the animal eats. The major nutrients needed by animals are: energy, which is primarily provided by fats and carbohydrates; protein, which is used for growth; minerals; vitamins; and water.

Major feed ingredients are roughages and concentrates. Roughages are high in fiber and make good feed for ruminants. Roughages are provided by pastures and other forage sources. Concentrates are low in fiber and are needed by hogs, chickens, and fish. Supplements and feed additives may be used to help animals grow.

Animals need a balanced ration. The ingredients must be palatable to the animal. A balanced ration should be free of pesticide residues and weeds, provide a variety of feedstuffs, have bulkiness appropriate for the animal, and be economical.

Feed may be manufactured on the farm or by commercial feed mills. The feed form should be appropriate for the animal. Animals may be fed on a free-access basis or on a schedule. Producers must select the best system for their situations.

QUESTIONS

Answer the following questions, using complete sentences and correct spelling.

1. How do animals use feed? How do these uses vary with the age and size of the animal?
2. What are the nutrient needs of animals? Briefly describe each.
3. What feedstuffs provide the nutrients needed by animals? Distinguish between the feedstuffs and give examples.
4. What kinds of pasture may be used?
5. What determines the quality of hay?
6. What are feed additives and implants? Distinguish between how the two get into the bodies of animals.
7. What are the basic requirements of feed?
8. How is feed obtained? What is involved with the sources?
9. What is a mixed feed?
10. How are concentrates made into feed?
11. What is a feed label? What information does a label have on it?
12. How are animals fed? Distinguish between the methods.

EVALUATING

Match each term with its correct definition.

a. feed
b. digestible nutrient
c. ration
d. protein
e. carbohydrates
f. roughage
g. permanent pasture
h. concentrate
i. palatability
j. stocking rate

_____1. A feedstuff high in energy and low in fiber.

_____2. The total amount of feed an animal gets in 24 hours.

_____3. A product eaten by animals that contains nutrients.

_____4. The number of animal units that graze an acre of land.

_____5. The part of a feedstuff that can be digested, or broken down.

_____6. A feedstuff low in energy and high in fiber.

_____7. A quality of feed that describes how well an animal likes to eat it.

_____8. Grasses and legumes that live and grow for years and are used for grazing.

_____9. Sugar, starch, and related substances formed in plants during photosynthesis; a source of energy.

_____10. The substance needed for maintenance, growth, reproduction, and other functions.

EXPLORING

1. Visit a feed store. Determine the kinds of feed sold. Look at how the feed is packaged and the ingredients in the feed. Determine the protein, energy, mineral, and vitamin content of a feed. Prepare a written report of your findings. (Remember, most supermarkets have pet food sections.)

2. Tour a feed mill. Determine the feedstuffs used, how the feed is made, and the forms of the final product. Write a report on your findings.

3. Visit a farm or ranch that uses pasture for animal feed. Interview the operator about the kind of pasture used and the practices followed in growing it. Prepare a report on your findings.

4. List the plants that can be grown in your area for animal feed. Name the kinds of animals that eat the plants. Briefly describe how the parts of the plants are prepared as feed.

CHAPTER 19

Breeds and Breeding

OBJECTIVES

This chapter focuses on breeds and the reproduction of animals. It has the following objectives:

1. Explain breed and breeding systems.
2. List examples of common animal breeds.
3. Identify the sexual classification of animals.
4. Discuss production systems used with selected species.
5. Describe management in breeding animals.

TERMS

artificial insemination
beef-type cattle
bloodline
breeding
breeding system
broiler
castration
colostrum
cow-and-calf production
crossbreeding
dairy-type cattle
debeaking
draft horse
estrus synchronization
feeder pig
feedlot
inbreeding
lamb
light horse
market hog
mutant
mutton
natural insemination
needle teeth
outcrossing
pedigree
placenta
polled
pony
pregnancy testing
presentation
production system
purebred
purebred breeding
spay
spaying
steer
stocker calf
straightbreeding
upgrading

19–1. Scientists are using ultrasound in assessing muscle area and backfat thickness of a lamb. (Such information can be used in selecting animals for breeding and showing.) (Courtesy, Agricultural Research Service, USDA)

PRODUCING new animals is an important part of animal science. We need new animals to maintain a constant supply of meat, wool, milk, eggs, and other products as well as services. This requires that animal producers be able to promote desired animal reproduction.

Efficient production is important. Each year the average person in the United States consumes about 160 pounds of red meat (beef, pork, and lamb and mutton). Multiply the pounds by 306 million (the number of people), and the total is more than 490 billion pounds of meat. Many animals are needed to provide that much meat!

Reproducing animals is the process of creating new life—more animals. Animals need to be properly cared for to reproduce efficiently. Animal owners must be able to select animals for mating and to manage reproduction.

BREEDS AND BREEDING SYSTEMS

Animals are produced for different purposes. Most farm animals are produced for the benefit of humans. Breeds and bloodlines have been developed to have animals that yield more of what people want. Various crosses and hybrids have evolved to improve productivity. In addition, some animals are being improved by artificial genetic manipulation.

Most animals have similar sexual reproductive processes. Horses, cattle, sheep, and hogs are mammals, with males and females of the same species mating and the females giving birth to live young. Chickens and fish are not mammals and reproduce by hatching fertile eggs. Some mate for internal fertilization, and others fertilize eggs outside the female's body. A producer's goal is to achieve efficient reproduction by using appropriate breeding management. Breeding is helping animals reproduce to gain desired products in a timely manner. (Chapter 17 covered the fundamentals of the sexual reproduction process.)

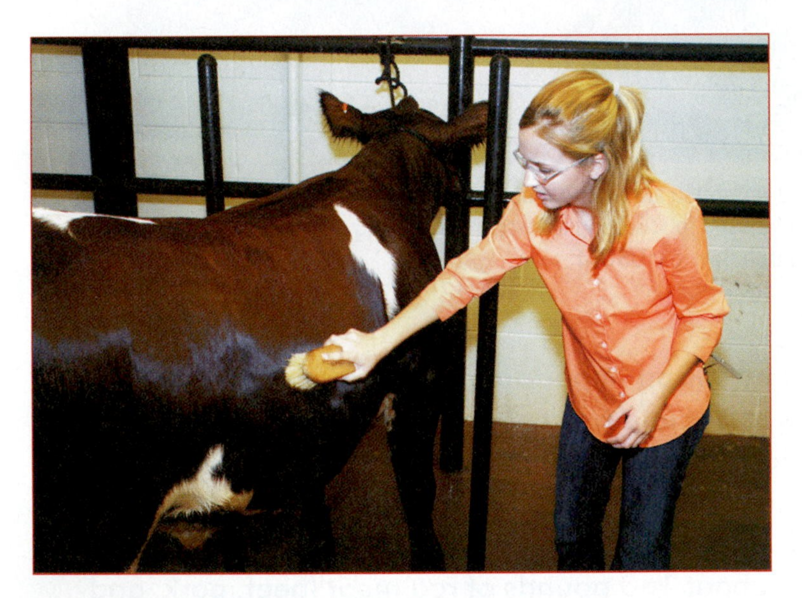

19–2. A young woman is preparing her calf for showing. (The calf was carefully selected as a much smaller animal based on heredity and conformation.) (Courtesy, Education Images)

INHERITED TRAITS AND MUTATIONS

Offspring inherit traits from both parents. The traits are those of the species and breed. Inherited traits include:

- **Color**—Animals of the same breed usually have the same color. Red parents produce red offspring! Spots and other configurations in color are also a part of coloring, as seen in a Dalmatian dog or a Hampshire hog.

- **Milk capacity**—Animals that produce a lot of milk will generally have offspring that produce a lot of milk.
- **Horns**—Cattle with horns usually produce offspring with horns. Likewise, cattle that are polled (naturally without horns) usually produce offspring that will not grow horns. (An animal that had its horns removed [undergone de-horning] would be considered horned.)
- **Size**—Large parents produce offspring that will grow to be large animals. The reverse is true for small animals.
- **Type**—Animals may have characteristics that make them useful. Dairy cattle are useful because of the large amount of milk they give. Beef cattle are good for meat because they have a high proportion of muscle tissue. Dogs and cats also conform to specific body types for their breed, such as a Greyhound dog with long legs for running.
- **Other traits**—Depending on species, other traits are important. One example is the egg-laying ability of chickens, such as the White Leghorn breed. An example of a possible defect in breeding Boer goats is the occasional presence of defective teats.

19–3. Longhorn cattle are known for at least one distinctive characteristic: long horns. (Courtesy, Education Images)

CAREER PROFILE

DAIRY SCIENTIST

A dairy scientist studies dairy cattle and milk production. The work may involve research in the field or in dairying facilities. Dairy scientists assure that quality milk is produced.

A dairy scientist needs a college degree in dairying, animal science, or a related area. Many have master's degrees and doctorates. Practical experience on a dairy farm and/or in a milk processing plant is beneficial. Begin preparation in high school by taking agriculture and science classes.

Jobs for dairy scientists are typically with colleges, research stations, government agencies, and agribusinesses involved with dairying. (Courtesy, Agricultural Research Service, USDA)

Exceptions occur in inherited traits. Offspring may have traits that are not obvious in their parents. These traits were likely present two or more generations earlier and have occurred again. Sometimes a trait might show up for the first time. An offspring that has a trait different from its parents is a ***mutant***. Mutations can occur naturally, or they can be created by radiation or chemicals. A mutation is typically a fairly major characteristic that results in an individual that is noticeably different from its parents.

19–4. Donkeys have distinct traits: big ears, small frames, brownish to grayish colors, and stubborn dispositions. This animal is also known as a burro.

Purebreds

A ***purebred*** is an animal registered with a breed association or eligible for registry. The breed must be recognized and criteria established for registration. Purebred offspring are eligible for registration if both parents were registered. Registration is not automatic. Papers about the offspring must be filled out and submitted to the respective breed association. Most associations charge fees to register animals. The American Kennel Club (AKC) and the American Cat Fanciers Association (ACFA) are two examples of organizations that register companion animals.

Pedigree

The genetic traits of an animal come from its ancestors. The names of the ancestors of an animal form its pedigree. Most pedigrees go back no more than five generations. A ***pedigree*** is proof of the genetic ancestry of an individual and is important in some breeding systems. Producers of purebred animals must keep careful pedigree records of their animals.

BREEDING SYSTEMS

Breeding is the process of helping animals reproduce. Systems of breeding are used to maintain and improve animal quality, with the goal of more effectively meeting consumer demand. Breeding is sometimes used to create new breeds, such as Santa Gertrudis cattle.

A ***breeding system*** is the way animals are selected for mating to get certain results. Breeding systems help producers control the heredity of their animals. One system is not necessarily better than the other; producers should use the system that helps them meet their goals.

Two breeding systems are used: straightbreeding and crossbreeding. Each has variations.

Straightbreeding

Straightbreeding is the mating of animals of the same breed. The bloodlines of different breeds do not become mixed. Three approaches are used in straightbreeding: purebred breeding, outcrossing, and inbreeding.

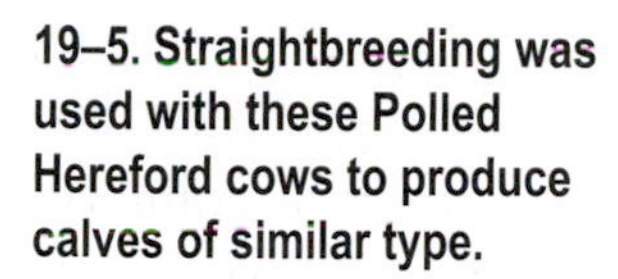

19–5. Straightbreeding was used with these Polled Hereford cows to produce calves of similar type.

Purebred Breeding. ***Purebred breeding*** is the mating of a purebred animal with another purebred of the same breed. Purebred animals are raised to provide breeding stock of known quality and ancestry. Purebred animals may be eligible for registration with a breed association.

The offspring of purebred parents of the same breed should be eligible for registration as a member of the breed. The exact parents of the offspring must be known. Controlling the access of males to females is essential. Only one male should be with a female to be certain of the offspring's parents. If two bulls are in the pasture with a herd of cows, the exact sire (father) of the calves will not be easily known. The offspring cannot be registered.

Occasionally offspring are not eligible for registration because of disqualifications. These may be color markings, ear shapes, or other undesirable traits. Such traits occur because of recessive genes. A breed association sets registration standards.

19–6. Pug dog breeders are careful to mate animals that are purebred to assure the unique features of the Pug breed in the offspring.

A goal in purebred breeding is to emphasize the desired traits of animals. It is used to produce high-quality animals. Just as the desired traits are emphasized in purebred animals, the undesired traits may also be made more prominent. Some animals are homozygous, which means that they have identical genes for a trait.

If the goal is to furnish products that people want, purebred animals may not be the best choice. Crossing purebreds with other breeds or with grade (not purebred) animals may result in the best animals for meat. This has been widely done with hogs and chickens.

Outcrossing. ***Outcrossing*** is the mating of animals of the same breed but of different families within the breed. Families are animals that share the same bloodline and have similar pedigrees. With outcrossing, the mated animals do not have a close relationship in their pedigrees.

Outcrossing is a good way to expand the genetic background of purebred animals. This approach is used to reduce or limit a weakness or undesirable trait. Outcrossing may conceal undesirable genes.

A variation of outcrossing is linecrossing, which is mating animals of the same breed that are as different as possible. Pedigrees of animals must be studied to determine common ancestry. With linecrossing, as much difference as possible is wanted in the ancestry.

Inbreeding. ***Inbreeding*** is the mating of closely related animals of the same breed. For example, sires may be mated with their daughters, or sons may be mated with their mothers. Some producers of foundation (original) purebred stock use inbreeding. Inbreeding emphasizes both the desired and undesired qualities of animals.

19–7. Crossbreeding often produces hogs with more desirable traits, such as this Yorkshire-Hampshire–cross gilt.

Crossbreeding

Crossbreeding is the mating of animals of different breeds of the same species. It may involve mating two purebred animals or a purebred male with a grade (not purebred or a mixed breed) female. The goal is to improve the yield or quality of offspring. Crossbreeding may be used to improve the rate of growth, use of feed, or resistance to pests.

Crossbreeding is often used for ***upgrading***, or grading up to improve the quality of offspring. For example, a grade cow mated with a purebred bull will produce a calf with half the inheritance of the bull. This approach is used to have calves with more meat or no horns or for other specific genetic purposes. A bull is important—a bull is sometimes said to be half the herd!

EXAMPLES OF BREEDS

A breed is a group of animals of the same species that share common traits. Members of a breed have a common origin and similar characteristics. The similarities, passed from generation to generation, are known as inherited traits.

A ***bloodline*** is a group of animals within a breed with similar pedigrees. All members of a bloodline tend to have one common ancestor. It might be an outstanding animal, such as a horse that won the Kentucky Derby.

BEEF CATTLE

Beef cattle are raised to produce beef. Consumers want a lot of lean muscle with only enough fat to provide a desired flavor. ***Beef-type cattle*** are cattle with good muscling in the loin and hind quarters. This is where the most valuable meat cuts are located.

Most breeds of beef cattle originated in Europe. A few breeds have been developed in North America by crossing European breeds with cattle imported from India. Many beef cattle breeds are raised in North America. A few are more widely known than others; these are the major breeds.

- **Angus**—Originating in Scotland, the Angus breed is popular in North America. The cattle are black, ***polled*** (naturally without horns), and have a smooth hair coat. They grow well and produce high-quality beef. Since Angus cattle have a dominant poll gene, nearly all offspring of crossbreeding an Angus with another breed will be polled. One variation in the Angus is the Red Angus, which has a red color.

19–8. Angus bull. (Courtesy, American Angus Association)

- **Belted Galloway**—The Belted Galloway has become more popular in

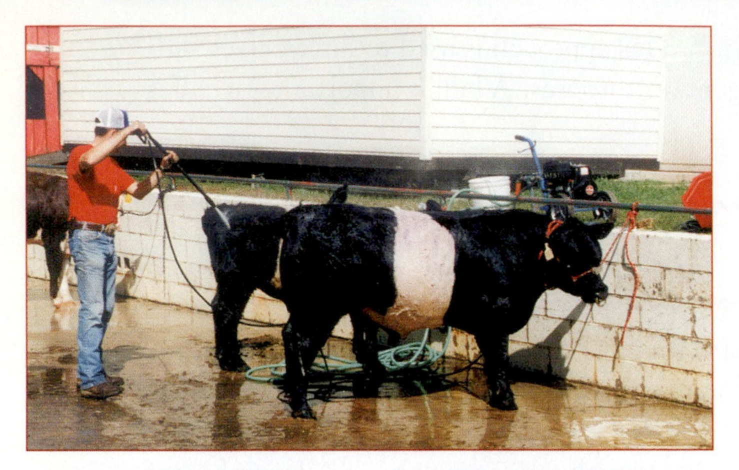

19–9. A Belted Galloway heifer is being washed for showing.

19–10. Brahman cattle have humps and loose skin.

19–11. Brangus cows readily show the traits of their breed. (Courtesy, Education Images)

the last decade with people who have small herds. Its distinct color—black, with a tinge of dun (grayish brown to yellow)—along with a white belt encircling the body between the shoulders and hooks, is appealing. This breed originated in the Galloway District of Scotland. The first Belted Galloways were brought to the United States in 1950. They are polled, and their heavy coats of hair make them suitable for cooler climates.

- **Brahman**—Brahman cattle are known for loose skin and large humps over the shoulders. Color varies from light gray to nearly black, but some have red coloring. They are tolerant of heat and insects. Brahman cattle are often crossed with other cattle. The breed was developed in the United States from cattle imported from India. Mature bulls weigh up to 2,200 pounds and cows up to 1,400 pounds. They are sometimes known for a mean disposition—ready to fight!
- **Brangus**—The Brangus breed was developed by crossing Brahman and Angus cattle. Brangus cattle are solid black and polled. They tend to have the skin and hump traits of the Brahman but not to the same extent.
- **Charolais**—Charolais cattle originated in France. They are white to a light straw color, with pink skin. Charolais cattle are among the larger beef cattle, with bull weights reaching 2,500 pounds and cow weights reaching 1,800 pounds. Most are naturally horned. Charolais cattle make good use of feed and have a lot of muscle.

- **Chianina**—Originating in Italy, Chianina cattle were first brought to North America in 1971. The Chianina is white except for its black switch (end of the tail). Its skin has a black pigment. The breed is gentle and tolerates both hot and cold weather. Chianina cattle are the largest beef breed, with bulls reaching as much as 4,000 pounds and cows reaching 2,400 pounds. Chianinas are often used in crossbreeding because they make good use of pasture.
- **Hereford**—The Hereford breed originated in England and was brought to the United States about 1817 by Henry Clay. The Hereford has a white face and red body. It also has white on the legs, switch, and belly. A horned breed, Herefords have been popular in North America. They are hardy and vigorous. Mature males weigh about 1,800 pounds, while females weigh about 1,200 pounds. They are sometimes called "white faces."
- **Polled Hereford**—The Polled Hereford has the same traits as the Hereford, except members of the polled breed do not have horns. Polled Herefords were developed in the United States by selective breeding of Herefords that did not have horns.
- **Limousin**—The Limousin breed originated in France and was first brought to North America in 1969. Colors vary. In France, most Limousins are red, but in North America, they may be light yellow or black. Variations are

19–12. Charolais cow and calf. (Courtesy, Gary and Pam Naylor, Missouri)

19–13. Horned Hereford bull in a holding pen. (Courtesy, USDA)

19–14. A Polled Hereford obviously has no horns. (Courtesy, USDA)

19–15. Limousin cow with calf. (Courtesy, North American Limousin Foundation, Englewood, Colorado)

19–16. Grade Santa Gertrudis cattle.

due to selection of the breeding stock. The Limousin is lean and has a large loin area (muscle along the back). Weights range up to 2,400 pounds for bulls and 1,350 pounds for cows. They usually have horns.

- **Santa Gertrudis**—Developed by the King Ranch in Texas, the Santa Gertrudis breed is a cross of the Shorthorn and Brahman breeds. Santa Gertrudis cattle are cherry red, usually horned, and have loose hides, similar to the Brahman. The bulls have small humps over their shoulders. They are similar in size to the Brahman.
- **Shorthorn**—Shorthorn cattle originated in England and were brought to Virginia in 1783. The cattle are red and white, with a red-white mix known as roan. Shorthorn cattle have horns, except for the Polled Shorthorn breed. The cattle are adapted to a range of climates. The cows make good mothers. Mature weights

19–17. Grooming a steer of Shorthorn breeding. (Courtesy, Education Images)

are up to 2,400 pounds for bulls and 1,500 pounds for cows.

- **Simmental**—Originating in Switzerland, Simmentals were brought to Canada in 1967 and the United States in 1971. Simmental semen was introduced in the United States in 1968. Their faces of Simmentals are white or light straw in color, and their bodies are red to dark red and may appear spotted. Bulls weigh up to 2,600 pounds, and cows weigh as much as 1,800 pounds. Simmentals are known for rapid growth on roughage. They develop good muscle with little fat.

19–18. A Simmental heifer being shown at a major livestock exposition. (Courtesy, USDA)

- **Other beef breeds**—Beefmaster, Braford, Red Brangus, Maine-Anjou, Gelbviah, Charbray, Devon, Pinzgauer, Ankole-Watusi, and Texas Longhorn.

DAIRY CATTLE

While dairy cattle are raised for milk production, some are harvested for beef, including cows that are no longer milked. ***Dairy-type cattle*** are cattle with the capacity to produce large amounts of milk. There are five major breeds of dairy cattle in North America.

- **Ayrshire**—The Ayrshire breed was introduced into North America in 1822 from Scotland. The colors of Ayrshires vary. They may be shades of cherry red, mahogany, brown, or white. Black and brindle (blackish streaks) are not acceptable. Ayrshires have horns of medium length that turn up and out. The breed ranks third among the dairy breeds in the amount of milk given at 11,700 pounds a year with 4.0 percent milk fat.

- **Brown Swiss**—Brown Swiss cattle originated in the Alp Mountains in Switzerland and were brought to North America in 1869. Their color is brown, though it ranges from light to dark. They have black noses and tongues. Medium horns are turned forward and upward. Brown Swiss cattle rank second in average milk production per cow at 12,100 pounds a year with 4.1 percent milk fat.

- **Guernsey**—Guernsey cattle were brought to the United States in 1831 from the Isle of Guernsey off the coast of France. Guernseys may be any shade of fawn with white markings. Their skin is yellow. Horns turn outward and toward the front. Guernseys rank fourth in volume of milk produced at 10,600 pounds a year with 5.0 percent milk fat. Guernsey milk has a slight golden color, which lends the name "Golden Guernsey."

- **Holstein-Friesian**—This old breed of cattle from the Netherlands was brought to North America in the mid-1600s. Distinctive black and white color patterns make it easy to identify. Holsteins are good grazers and can eat a lot of feed. They are the pre-

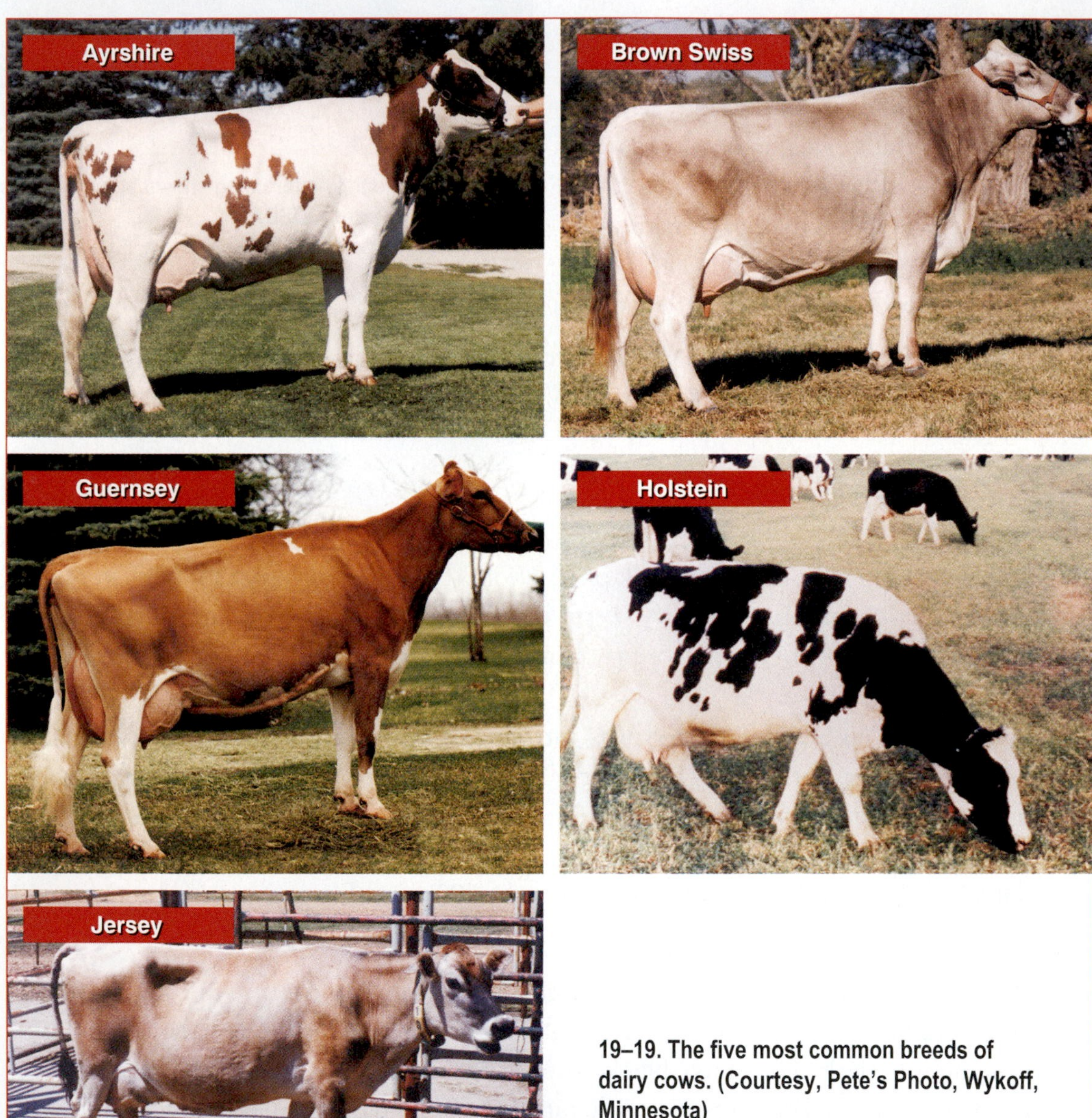

19–19. The five most common breeds of dairy cows. (Courtesy, Pete's Photo, Wykoff, Minnesota)

19–20. These Holstein calves are being carefully raised to become dairy cows.

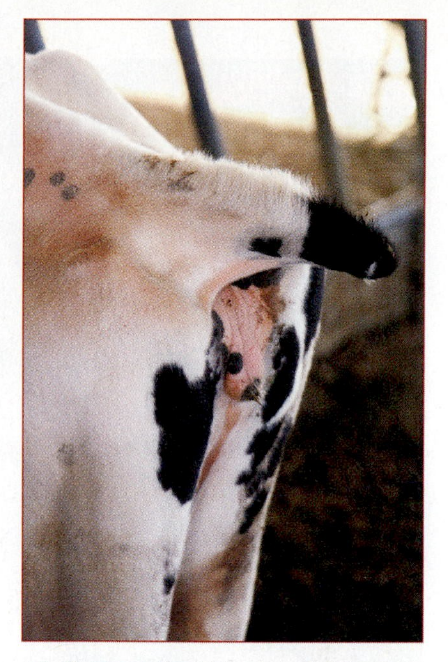

19–21. The tail of this dairy heifer was docked when she was a young calf. (Docking is done on dairy farms that have milkers that attach from the rear, where a tail would be in the way!) (Courtesy, Education Images)

dominant breed of dairy cattle in the United States, with an annual milk production of 14,500 pounds per cow with 3.5 percent milk fat.

- **Jersey**—The Jersey breed ranks fifth in pounds of milk produced, but it is first in the percent of milk fat. Jerseys average 10,000 pounds of milk a year with 5.4 percent milk fat. The breed originated on the Isle of Jersey in the English Channel and was first brought to North America about 1815. Jersey colors range from cream to almost black. The cows are the smallest of any of the dairy breeds. They have horns that curve in toward the head and forward.

HOGS

While there are about 400 breeds of hogs (swine), only a few breeds are produced in the United States. Except for producers of registered stock, the emphasis on breeds has declined. Crosses are used because of better growth and type (more muscle and less fat). The preferred hog is one with a lot of lean (known as meat type) and little surplus fat. A hog with a large amount of fat is known as a lard-type hog. Another is the bacon-type hog—a long, lean hog.

- **Berkshire**—A Berkshire's ears stand erect over its eyes. Its color is black with six white points: each foot, some white on the face, and a white tail switch. Berkshires do not grow as large as other breeds. They have long, deep sides and are moderately wide

19–22. Boars of several common hog breeds. (Courtesy, Prairie State Semen, Inc., and www.showpigs.com)

across their backs. Berkshires sometimes have small litters, but this is being overcome by selective breeding.

- **Chester White**—This is a popular breed in the northern United States. Chester Whites are white, but bluish freckles on the skin are acceptable. The females are exceptionally good mothers and produce large litters.
- **Duroc**—The Duroc is the most popular breed in North America. Durocs are various shades of red. The ears of the Duroc droop over the eyes. Durocs have several traits that make them popular: large size, good appetite, and hardy and large litters. Though body type varies within the breed, Durocs are good meat-type hogs.
- **Hampshire**—This breed is easy to identify. The Hampshire is black with a distinctive white belt around the shoulders. It is a meat-type hog. The breed is trim, with little extra fat. The ears stand erect above the eyes. Hampshire sows raise a high proportion of their pigs and are well suited to confinement. They are known as active hogs.
- **Poland China**—The Poland China breed is black with six white points: the feet, the tip of the tail, and the nose. Years ago, the Poland China was a short, fat-type hog. Through selective breeding, it is now a good meat-type hog. Poland Chinas are often crossed with other breeds of hogs.
- **Yorkshire**—Yorkshires are white, and those animals without freckled skin are preferred. Erect ears are important in the identification of Yorkshires. Yorkshire sows farrow (give birth to) large litters and raise them well, because they are good milkers (give a lot of milk). The Yorkshire body is long and has long, deep sides. A hog of this breed yields a quality carcass (yield good meat).
- **Other breeds and hybrids**—Spotted Poland China, Tamworth, and Hereford. Improved hybrid-type hogs have been developed by crossbreeding. Several hog companies have developed

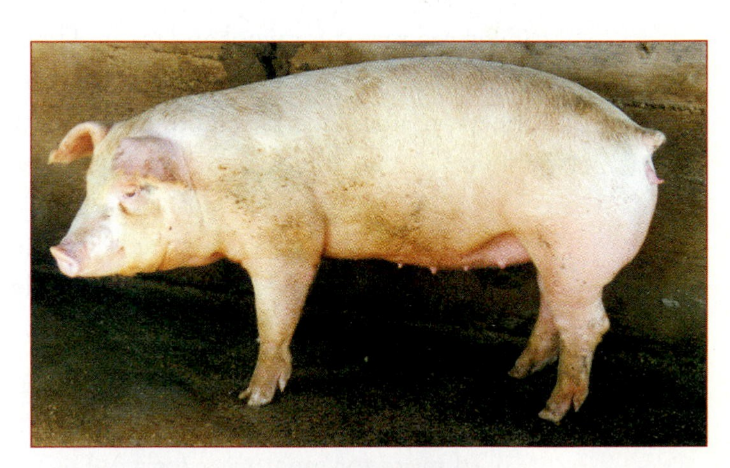

19–23. Yorkshire gilt.

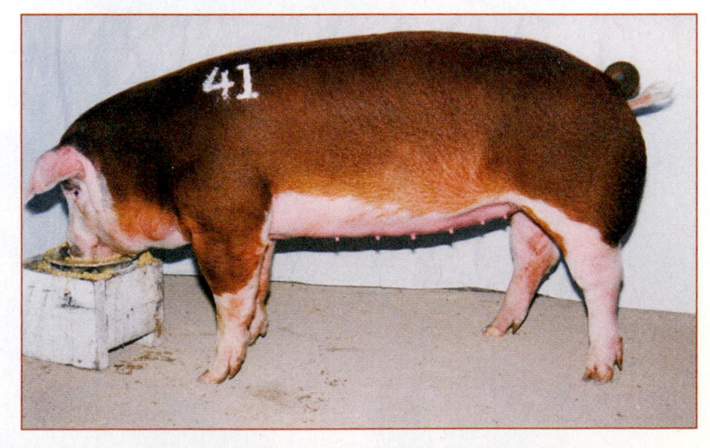

19–24. A Hereford gilt. (Courtesy, National Hereford Hog Record Association)

19–25. A genetically improved boar for producing high-performance meat hogs.

hogs with exceptional ability to produce high-quality meat. The hogs have genetic material from two or more breeds. The boars are sold to pig producers for natural breeding or artificial insemination. A hybrid hog often goes by the name of the company that produced it, such as a "DeKalb" for a boar from DeKalb Swine Breeders. Improved hogs predominate in large factory-type pork production.

SHEEP

Sheep are raised for wool and meat. The meat of a young sheep is known as ***lamb***, while the meat of an older sheep is called ***mutton***.

Sheep are classified based on the quality of their wool. Examples of breeds in each wool type follow.

- **Fine-wool sheep**—The breeds of fine-wool sheep include Rambouillet, American Merino, Delaine Merino, and Debouillet. These sheep produce wool with a heavy yolk (oil content) and are also used for meat.
- **Medium-wool sheep**—Breeds of medium-wool sheep include Cheviot, Dorset, Finnish Landrace, Hampshire, Shropshire, Southdown, and Suffolk. This group of sheep is used more for meat than for wool.

19–26. Dorset sheep are medium-wool sheep.

- **Long-wool sheep**—Examples of long-wool sheep are the Cotswold, Leicester, Lincoln, and Romney. These breeds produce a long wool with coarse fiber.
- **Crossbred wool sheep**—Breeds of crossbred wool sheep include Columbia, Panama, and Southdale. These sheep have excellent banding (flocking together) instincts and are adapted to the range in the western United States.
- **Carpet-wool sheep**—Only one breed of carpet wool sheep is found in the United States: Black-faced Highland. The fleece (wool) is coarse, wiry, and tough.
- **Fur sheep**—The Karakul breed is the only sheep used for fur in the United States. Young lambs are killed, and the fur pelts are used to make coats.
- **Hair sheep**—Katahdin is the only breed of sheep with hair of any importance in the United States. It is not necessary to dock (to remove the tail) Katahdin lambs, because hair does not get dirty as wool does.

19–27. Suffolk sheep have distinct black faces and legs.

19–28. A Katahdin ewe with lamb.

GOATS

Goats are raised for milk, meat, and mohair. Like sheep, goats are ruminants. Goats are often raised on land where brush and weeds grow. Goats eat shoots, leaves, twigs, briars, and other things avoided by cattle. Some goats have even been accused of eating cans!

Goat breeds are in three classes or groups.

- **Mohair and cashmere goats**—Mohair is the long hair of Angora goats. Cashmere is a fine-quality hair produced by the Cashmere goat, mostly in China and Iran. Mohair is used for clothing, furniture upholstery, and other products. The mohair breed in North America is the Angora. Most Angora goats are grown in Texas and other southwestern states.

19–29. The Alpine goat is kept for milk.

19–30. Fitting a Boer goat for showing. (Courtesy, Education Images)

- **Dairy goats**—Dairy goats are raised for milk, usually on small farms. The breeds of dairy goats include LaMancha, Alpine, Nubian, Saanen, and Toggenburg. Dairy goats are often white with gray, cream, tan, and black coloring.
- **Meat goats**—Two breeds or groups of goats used for meat are Boer and Spanish.

  The Boer goat was developed in the early 1990s in South Africa and brought to the United States shortly afterward by way of New Zealand. The Boer goat is popular for showing in some areas. It has appealing conformation and colors—a white body and brown head. Boer goats are hardy and adapt well to a wide range of climates. Mature Boer bucks weigh 240 to 300 pounds, while mature does weigh 200 to 220 pounds. Relative to the weight of cattle, Boer goats need about twice the amount of feed. Pasture and good-quality hay should be supplemented with a concentrate. An issue with some does is that their teats do not develop so that their babies can nurse without assistance.

  Spanish goats are kept to produce meat and control unwanted vegetation. They need little care and are prolific (produce many offspring). Spanish goats have many colors, ranging from solid white to black and with spots.

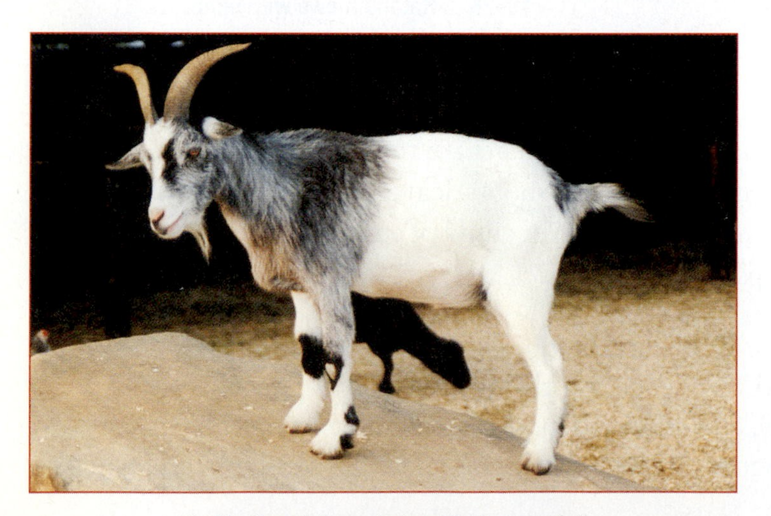

19–31. Spanish goats tend to be playful.

HORSES

Many breeds of horses are found in North America. Formerly horses were needed to do work (provide power). They are now kept as pets, for riding, to race or work rodeos, or to herd cattle. Mules are crosses of horses and donkeys.

19–32. Paint horses have distinctive color markings.

19–33. A Quarter Horse mare with its colt.

Horses are classified as light horses, ponies, and draft horses. The distinction is size. Horse size is measured in hands and weight. A hand is the width of a human hand, or 4 inches (10.37 cm). Measurements in hands are taken from the ground to the withers (top of shoulders). Inches are used with hands—for example, 14-2 is 14 hands 2 inches.

- **Light horses**—A ***light horse*** is 14-2 to 17 hands and weighs 900 to 1,400 pounds. Light horses are used for riding, driving, racing, and other purposes. They are typically long-legged and able to move fast.

19–34. A blue-ribbon-winning American Saddlebred. (Courtesy, American Saddlebred Horse Association, Inc., and photographer Jamie Donaldson)

The Quarter Horse is a popular breed for pleasure and to work cattle. Other breeds include the Arabian, Paint Horse, American Saddlebred, Appaloosa, Tennessee Walking Horse, and Palomino.

- **Ponies**—A ***pony*** is a small horse of 14-2 hands or less and typically weighs less than 900 pounds. Miniature horses, such as the Shetland Pony, are included in this group.
- **Draft horses**—A ***draft horse*** is a horse developed to pull heavy loads, such as wagons, plows, or logs. The use of animal power declined with the increased use of tractors and other vehicles with engines. Draft horses are typically 15-2 to 17-2 hands and weigh upwards of 1,400 pounds, with some being much heavier. Draft horses are also kept for recreation, such as competition in pulling contests. Breeds include the Clydesdale, Belgian, Percheron, and Shire.

19–35. A team of three Belgian draft horses is harnessed to a plow.

19–36. Chickens have been bred for meat production. (Courtesy, Mississippi State University)

POULTRY

Poultry is a broad group of domestic birds (fowl) that includes chickens, ducks, turkeys, and others. Of these, chickens are the most important. Turkeys and ducks are raised for meat; chickens are raised for meat and eggs. Other poultry animals include quail, guinea, ostrich, and emu.

Breeds of poultry are often divided into varieties. More than 200 varieties of chickens are found in the United States. Of these,

19–37. Large-breasted white turkeys are widely raised. (Courtesy, Minnesota Turkey Growers Association)

19–38. The White Pekin duck is the most widely grown breed in North America. It weighs about 8 pounds when mature.

five breeds are of major importance: White Leghorn, White Rock, Rhode Island Red, Barred Rock, and New Hampshire. Chickens have been selected and bred to produce eggs and meat. A chicken used to produce eggs is not a meat type, though it may be used for meat.

Crosses and hybrid versions of the White Leghorn are used for egg production. Likewise, the White Rock has served as the major breed for developing meat chickens.

AgriScience Connection

ANIMAL IDENTIFICATION

Owners of animals need to be able to identify each animal. This is especially true when pedigree or production information is important.

Several methods are used. Ear tags with numbers are common. Each animal in a herd has a tag with a different number. Other methods of identification include tattoos, nose prints, and DNA tests. With hogs, ears may be notched to help with identification. A brand is used to identify the owner of an animal rather than a specific animal.

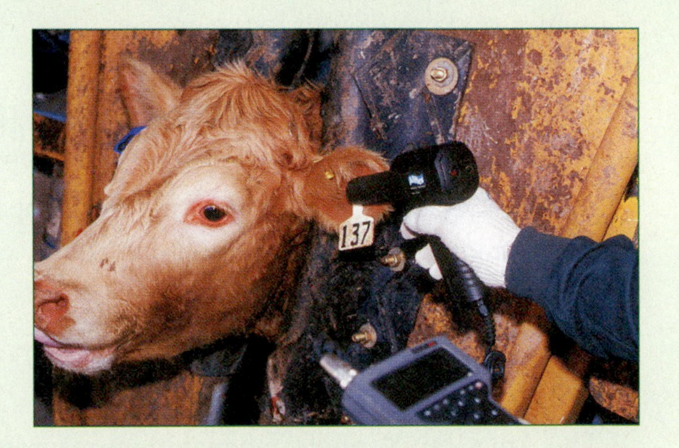

New ways of individual animal identification have been developed by Allflex USA. Electronic methods are used to identify and track animals. A handheld reader detects a number in an electronic ear tag. The information instantly goes to a computer where records are stored. (Courtesy, Allflex USA)

19–39. More channel catfish are produced than any other species of fish in North America. (Courtesy, *Progressive Farmer Magazine*)

19–40. The production of tilapia is increasing in the United States.

AQUACROP SPECIES

Aquacrops are aquatic species that are farmed. Catfish, trout, tilapia, salmon, oysters, shrimp, frogs, crabs, alligators, crawfish, and a few other species are used as aquacrops. Research in the last few years has improved fish stocks. Hybrid striped bass have been developed by crossing white bass and striped bass.

In 2001, a genetically altered catfish known as "NWAC 103" was released by the National Warm Water Aquaculture Center in Stoneville, Mississippi. This fish grows faster than other catfish. This means greater efficiency in fish production, better quality for the consumer, and more profit to the grower.

COMPANION AND SERVICE ANIMALS

Several species of animals are kept for companionship, service, and other uses. Most common among these are dogs and cats. Birds, exotic fish, reptiles, and other species are also kept for similar purposes.

Dogs

The American Kennel Club maintains records on about 160 breeds of dogs. Dog breeds are classified into eight groups, as follows:

- **Sporting**—The Sporting Group includes breeds of dogs kept for sporting purposes. Pointers, setters, spaniels, and retrievers are within this group.

- **Herding**—The Herding Group includes breeds of dogs that are easily trained to herd larger animals, such as cattle and sheep, in pastures. The common breeds include the collies, the German Shepherd Dog, and the Australian Cattle Dog.

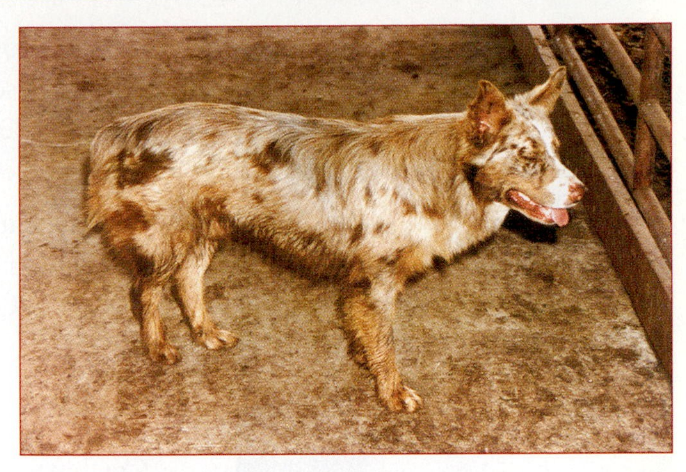

19–41. The Australian Cattle Dog is popular on ranches to help manage livestock.

- **Hound**—The Hound Group includes breeds of dogs that have the ability to follow the scent left by another animal. The breeds in this class may be used in hunting or in finding a missing person who is lost in a forest. The Beagle, the Dachshund, and the Greyhound are examples.
- **Working**—The Working Group includes breeds of dogs that perform various useful activities, such as pulling sleds, guarding property, and sniffing for drugs or explosives. Examples of breeds in this group are the Alaskan Malamute, the Rottweiler, and the Saint Bernard.

19–42. A purebred Rottweiler is nursing her pups.

- **Terrier**—The Terrier Group includes breeds that are known for chasing small animals that go in the ground. Breed examples are the Scottish Terrier and the Airedale Terrier.
- **Toy**—The Toy Group includes breeds that are small and weigh between 4 and 16 pounds. Breeds include the Chihuahua and the Poodle.

19–43. A long-haired cat.

- **Non-sporting**—Dogs in the Non-sporting Group are primarily used as pets or companions. They typically have no other specific purposes. The Dalmatian and the Bulldog are two examples.
- **Miscellaneous**—The Miscellaneous Class has breeds not included in the other groups established by the American Kennel Club.

Cats

Cat breeds are classified by hair length. Examples of the short-haired breeds are the Rex, Manx, and Siamese. Examples of the long-haired breeds are the Maine Coon, Persian, and Balinese. Many people who have cats are not especially fond of particular breeds. They own what are known as house cats, which are often of mixed breeding.

SEXUAL CLASSIFICATION OF ANIMALS

Most animals are classified by age and sexual condition. Sexually mature animals are used for breeding. Only those capable of regular breeding should be kept. Feeding animals that do not reproduce is expensive!

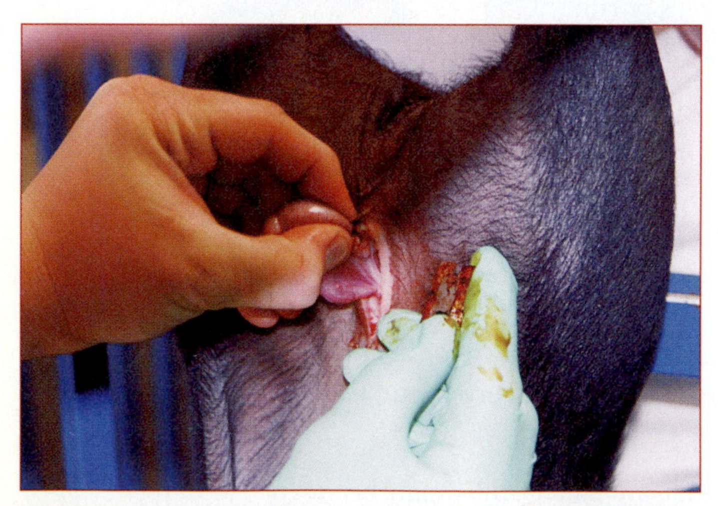

19–44. Surgically castrating a young pig. (Courtesy, Casa Robles High School and Education Images)

Animals are sometimes sexually altered (neutered) by removing essential reproductive organs. Males are neutered by ***castration***, which is removing the testicles. Females are neutered by ***spaying***, which is removing the ovaries. A female animal that has been neutered is known as a ***spay***. Pets, such as dogs and cats, are spayed. Male livestock animals are more likely to be neutered than female livestock animals. Castration is done fairly easily, whereas spaying is more invasive because surgery into the body cavity is needed. With a male, the testicles are normally in a scrotum outside the body cavity. The surgical process is less invasive and less likely to result in complications.

Animals are neutered to prevent unwanted reproduction. With meat animals, castrated males produce better meat and do not try to breed the females. With companion animals, females are neutered to prevent the birth of unwanted babies. This lowers the population of unwanted dogs and cats.

New technology is allowing for gender pre-selection and the ability to alter many animals sexually. Gender pre-selection is a process to control the gender of animal offspring at the time of conception or as newly hatched babies. An example in fish is with tilapia. Growers have found that same-sex populations of tilapia grow faster and do not reproduce. To assure that all individuals in a tilapia population are of the same gender, sex reversal methods are used on newly hatched fry.

Another example is with dairy cattle, in which demand is greater for females. Gender pre-selection involves sorting sperm based on the desired sex of offspring to be produced.

Internet Topics | Search

Selected topics for Internet discovery and reporting are listed here. Begin by searching for each topic. Next, read and learn about it. Conclude by preparing a brief report on your findings.

1. Colostrum
2. Light horse
3. Stocker calf

Technology Connection

PREVENTING UNWANTED CATS

Cats may produce far more new cats than needed. A queen may have three or more kittens each time she gives birth. What happens to the kittens? The owner might take care of them. Often, however, the extra cats are unwanted, run wild, and do not receive care. A cat owner has the problem of how to get homes for the extra cats.

Gestation of a queen is 60 to 65 days. She will nurse the kittens for about 25 days and cycle into heat. A queen that runs free or is kept where males are present may have three or more litters a year. Since cats reach puberty fairly quickly, the female cats that were born will be having kittens in about a year. Soon there are many, many cats.

Steps must be taken to prevent unwanted cats. Confining a female cat away from male cats is one approach. Castrating the males that a female may come into contact with is another. Still another is to spay the female cat. This removes her ovaries and makes her incapable of reproduction. Spaying is done by a veterinarian, using sanitary practices in the surgical facility of a veterinary clinic or hospital. This photo shows a veterinarian preparing to spay a sedated queen. (Courtesy, Education Images)

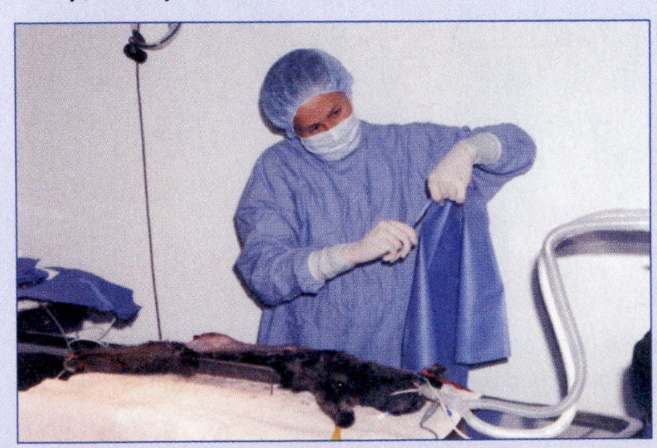

This is very effective in shaping the gender of an animal since gender is determined by the sex cell contributed by the male (the sperm) in the reproduction process.

Table 19–1. Sexual Classification of Selected Animals

Species	Young Animal[a]	Mature		Castrated Male	
		Female	Male	Young	Mature[b]
Cattle	calf[c]	cow	bull	steer	stag
Hog	pig	sow	boar	barrow	stag
Sheep	lamb	ewe	ram	wether	
Goat	kid	doe	buck	wether	
Chicken	chick	hen	rooster	capon	
Horse male female	foal colt filly	mare	stallion	gelding	stag
Dog	puppy	bitch	stud	neuter	
Cat	kitten	queen	tom	gib	
Rabbit	kit	doe	buck		

[a]Young animal of either sex, except for horses.
[b]Castrated after secondary sexual characteristics have developed.
[c]A female calf is known as a heifer.

PRODUCTION SYSTEMS

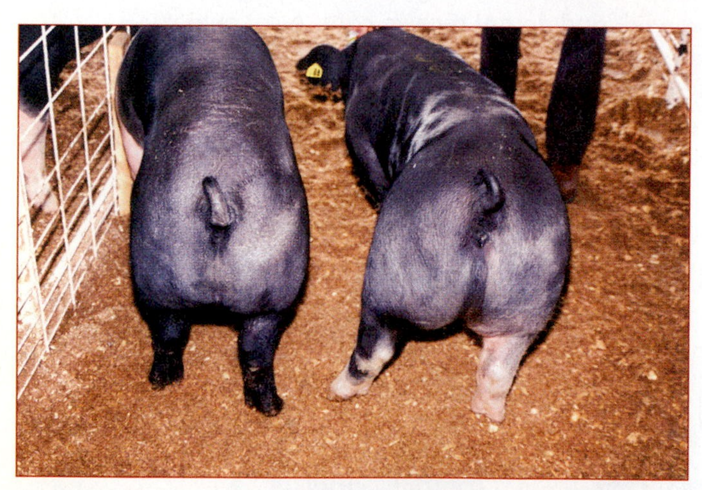

19–45. A barrow (left) and gilt of Poland China breeding. How would you rate muscle development and fat in the hams?

A ***production system*** is the approach used in producing animals. The outcome is a desired product. Production systems are used with most animals. There are similarities, but each system is unique to the animal species.

The production systems for beef cattle are purebred, cow-and-calf, and finishing or fattening. Hogs and sheep have similar systems. For each species, there are producers of (1) purebred animals, (2) young animals for growing into meat animals, and (3) meat animals ready for harvest.

PUREBRED PRODUCTION SYSTEMS

Purebred production systems produce purebred animals. Purebred producers must keep careful records. Many purebred producers have one breed. They may compete with other producers at shows to see who has the "best of show."

Purebred producers raise both male and female animals. Careful screening of the animals is used to select the best ones. With cattle and hogs, those that are not the best purebreds may be grown for meat. Top-quality male purebred animals are in greater demand because of their use in breeding. Males usually bring more money when sold.

MEAT-ANIMAL PRODUCTION SYSTEMS

Meat-animal systems produce offspring that are raised for meat, milk, or other products. Cows are kept to produce calves, sows to produce pigs, and ewes to produce lambs.

Cow-and-Calf Production

Cow-and-calf production is a production system in which cows are kept to produce calves. Calves are born weighing 75 pounds or more and weaned at about 500 pounds (six, or so, months of age). After weaning, they may be harvested as baby beef or used as stockers. A ***stocker calf*** is a calf that is put in a feedlot or on pasture for additional growth and fattening. A ***feedlot*** is a space where cattle are confined and fed for maximum gain. Stocker calves develop muscle (lean meat) and fat to assure quality beef. The cattle are said to be "commercial" or "grade," because they are not purebred. Often, a purebred bull is used with grade cows in producing stockers.

19–46. A Limousin cow-calf herd on quality pasture.

A number of management practices are used to help cows reproduce and to help calves grow well. In general, the cows graze pasture much of the year. In some climates, they are fed hay and supplements in the winter. Cows are bred so calves are born in late winter or early spring. This calving time allows cows to use the tender springtime vegetation for milk production so that their calves grow rapidly. Calves should weigh about 500 pounds by late fall of the year. The cows are bred in the late spring to calve again in winter or early spring.

Some producers select the best female heifer calves as herd replacements. A cow will produce about 10 calves in her lifetime. Producers continually replace cows that do not calve. Producers who do not raise herd replacements must buy them from other producers.

Male calves are usually castrated when a few days old. Testicles are removed so the animals do not develop the characteristics of bulls. A calf castrated before sexual maturity is a ***steer***. Steers cannot breed.

Calves that weigh 400 to 700 pounds at weaning may be harvested as baby beef. The meat is lean and tender. Other calves are sent to feedlots for more growth and fattening or put on pasture.

Feeder Pig Production

A ***feeder pig*** is a pig being grown into a meat-size hog. A feeder pig weighs 30 to 60 pounds. A meat-size hog weighs 240 to 260 pounds—somewhat heavier than in the late 1900s. Feeder pig production is keeping sows to farrow (have pigs). Sows are selected to produce large litters of quality pigs. Producers like to have about 12 pigs in a litter. Boars of the desired type are mated with the sows. Producers may use genetically improved boars. Most sows produce two litters a year.

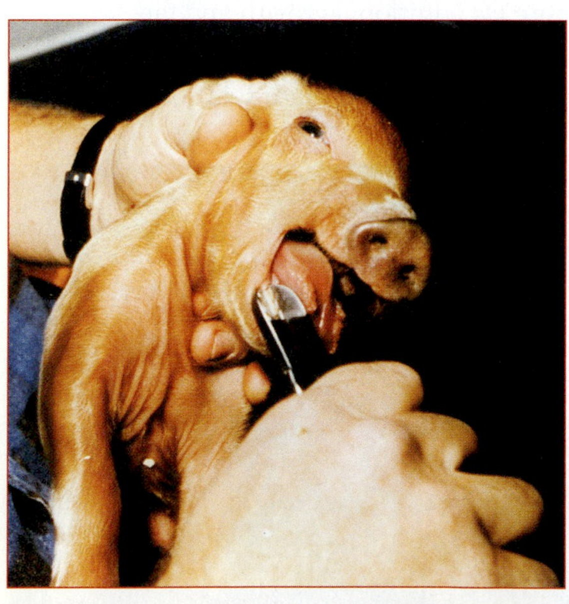

19–47. Clipping a baby pig's needle teeth. (Courtesy, Jim Floyd, Extension Veterinarian, Auburn University)

Replacement females may be selected from the female piglets. Selection is guided by looking at type, strength of body, and number of nipples. A female pig should have at least 12 well-developed nipples (6 on each side) so large litters can be fed.

Male pigs are typically castrated when a few days old. Barrows (male pigs castrated before sexual maturity) can be fed in pens with females.

Good management is needed to raise feeder pigs. Baby pigs are fragile and must be protected from harsh weather and from the sow. A sow may accidentally mash and kill baby pigs when she lies down. Newborn pigs have eight sharp teeth, known as ***needle teeth***. These teeth are usually clipped off after birth. If not clipped, the teeth will puncture the teats of the sow when the pigs nurse. Baby pigs are also given iron and other injections to keep them healthy.

FINISHING SYSTEMS

Finishing systems feed young animals to the desired size and weight for harvesting. Finishing involves feeding diets high in concentrates so animals will put on body fat. Fat improves the flavor and palatability of meat. Of course, too much fat is sometimes objectionable. People want meat that has just enough fat to provide marbling (small fat streaks in the lean). Too much fat in the human diet may be unhealthy.

Beef Cattle

Beef cattle are finished on pasture and in feedlots. The aim is for animals to grow body muscle and fat. High-energy rations are fed. Most feedlots can handle more than 1,000 head at a time.

Cattle finishing takes place mostly in Texas, Nebraska, Iowa, Kansas, California, and Colorado. The grain (corn and grain sorghum) used as feed is grown in these states. For feeding, stocker calves are trucked to feedlots from the cow-and-calf systems of the South and West.

Stockers go into finishing at weights of 400 to 700 pounds. High-energy rations are fed for 90 to 120 days or longer until the animals weigh 1,000 pounds or more. Feedlots expect to get 2.5 to 4 pounds of gain per day. Feedlots need large amounts of feed and water. Waste disposal requires lagoons and other measures. Feedlot wastes cannot be put directly into streams. Finished cattle are sold to packing plants.

19–48. Cattle in a California feedlot.

Market Hogs

A pig goes into finishing weighing 30 to 60 pounds and leaves as a 240- to 260-pound ***market hog***. Pigs are placed in confinement areas for feeding out. A confinement building may be on a concrete slab and have a partial roof or be completely enclosed. Free-access feed and water are provided. Buildings should be built for easy cleaning and moving of hogs.

Hogs are fed high-energy rations of corn and related feedstuffs. Finishing hogs must grow as well as put on fat. Most hogs are on feed for 70 to 100 days, depending on their size and the daily rate of gain. Packing plants reduce the price paid for hogs that are below or above the desired weight of 240 to 260 pounds.

Disease control is essential. A disease outbreak can quickly kill a large number of hogs in confinement. Hogs are usually grown in pens of 20 to 50 animals.

MANAGING ANIMAL BREEDING

Animals need careful management to produce new animals efficiently. Knowing when to breed an animal and how to manage the pregnant female is important. Knowing the procedures at the time of birth and how to care for the newborn animals and the mother after the birth assures good survival.

INSEMINATION

Management includes assuring conception (union of sperm and egg). Insemination is an important procedure in gaining conception. Insemination is placing semen in the reproductive tract of the female. Two methods are used: natural and artificial.

Natural Insemination

Natural insemination is the ejaculation of semen (fluid containing sperm) by the male in the vagina of the female during copulation. The semen must be present during the fertile time of the female for conception to occur. Males may be in the same pasture or pen, or they can be kept separately and the females brought for breeding during heat. Heat is the time in the reproductive cycle of females in many species when the females are receptive to copulation. Ova are released near the end of heat.

ACADEMIC CONNECTION

ADVANCED BIOLOGY

Relate your study of animal breeding to the study of reproduction in advanced biology class. Use your learning in agriscience to support your mastery in advanced biology.

Artificial Insemination

Artificial insemination (AI) is the collection of semen from a male and the depositing of it in the reproductive tract of the female. The technique is used with beef and dairy cattle, horses, hogs, turkeys, and other animals. The first major use of AI was with dairy cattle. Now it is used with many species.

The use of AI has increased in recent years. Semen producers collect, process, and market semen from superior males. The males are better quality than most producers could afford to own. In some cases, producers own the males and collect semen. One male can breed more females with AI than naturally. As obtained, a volume of semen contains millions of tiny sperm. Since only one sperm is needed for an egg, semen can be diluted so more females are bred with AI.

With fish, eggs may be artificially fertilized in a lab or hatchery. The process typically involves stripping eggs from the females and sperm from the males. The sperm are placed in a container with the eggs. After union occurs, the fertilized eggs are moved to a hatching facility for incubation.

Semen Quality. The semen used in artificial insemination must be of good quality. It should be from a male of proven ability to produce desired offspring. Semen is collected and stored to keep the sperm alive. Sperm that are defective or dead will not unite with an egg. Semen can be stored for a week at 41°F (5°C) or for several months frozen at –320°F (–196°C) in liquid nitrogen.

Semen collection methods should assure good sperm. The general procedure is to use an artificial vagina warmed with water. With cattle, a bull mounts a steer, and the penis is placed in an artificial vagina.

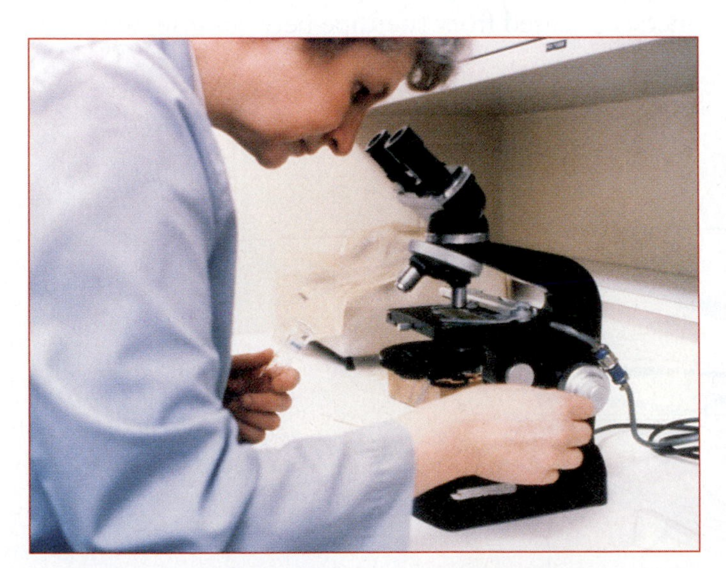

19–49. Sperm motility (movement) is examined with a microscope. (Courtesy, American Breeders Service, DeForest, Wisconsin)

19–50. Cross-section view of a semen tank where liquid nitrogen is –320°F (–196°C). (Courtesy, Jim Floyd, Extension Veterinarian, Auburn University)

Electrical stimulation may be used to promote ejaculation. After collection, semen is checked with a microscope for live, motile (active) sperm.

Portable semen tanks and other equipment are used to maintain quality. The semen for one breeding is placed in a "straw" at a semen lab. Most semen is provided through cooperatives or associations, such as the American Breeders Service, of DeForest, Wisconsin, and through private semen companies, such as Prairie State Semen, of Illinois.

Estrous Manipulation. The estrous cycle prepares the reproductive system of the female for breeding. The female is receptive to breeding during heat, or an estrus period. The four periods in estrous are estrus, metestrus, diestrus, and proestrus.

For natural breeding, the female in heat will allow the male to mount. In artificial insemination, breeding is done at the time eggs are released by the ovaries for fertilization. Best conception is inseminating the cow about 12 hours after heat is first detected. The ability to detect heat is essential with artificial insemination.

Hormone injections are used to change the estrous cycle. ***Estrus synchronization*** is the use of hormones to get several females to come into heat at the same time. It is useful with artificial insemination, advanced breeding procedures such as superovulation (getting the female to produce a number of eggs at one time), and embryo transfer.

Insemination Procedures. Artificially inseminating a cow is placing semen into the cervix. The rectovaginal method is most common. The procedure involves inserting one glove-covered hand into the rectum of the cow to guide the insemination tube, or straw, through the vagina into the cervix. Semen is released from the tube between the middle of the cervix and the uterus.

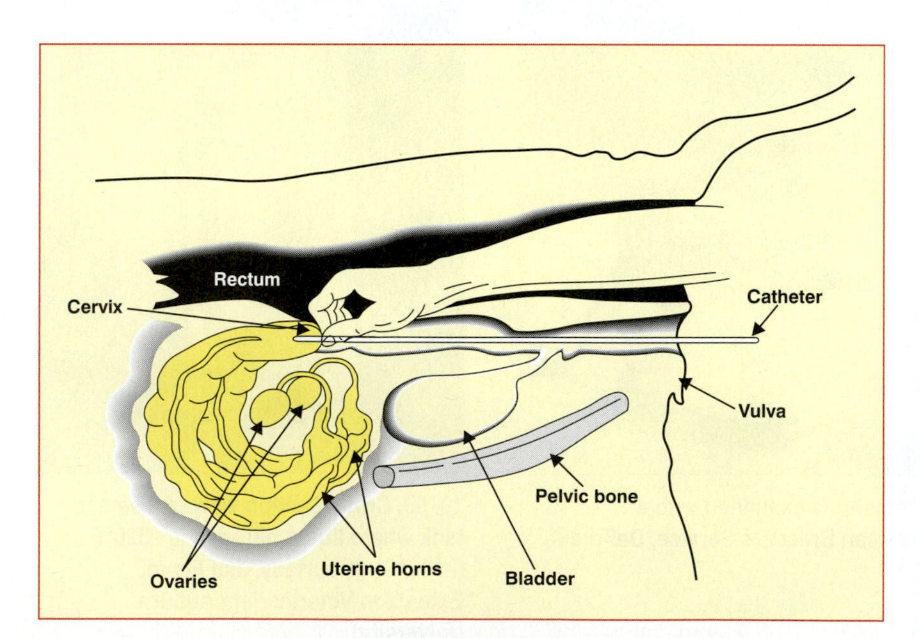

19–51. Artificial insemination in a cow requires that the semen be placed properly.

All processes in artificial insemination require training. Safety practices should be followed. Animals can be dangerous. Training programs in artificial insemination are held by breeding companies and agricultural colleges.

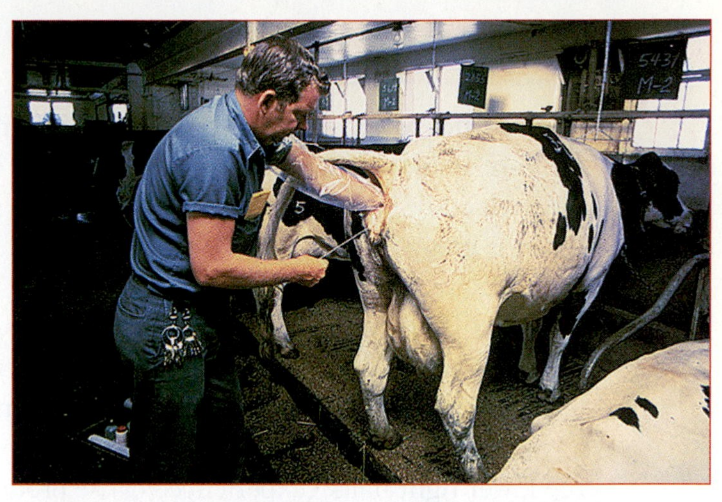

19–52. A dairy cow being artificially inseminated. (Courtesy, USDA)

WHEN TO BREED

Females should be bred at the time when they are most capable of carrying the embryo and fetus to birth. This varies by species.

Table 19–2. Breeding Ages and Schedules

Species	Recommended Age for First Breeding	Estrous Cycle (days)	Gestation (days)
Cow/heifer	14 months	21	283
Sow/gilt	12 months	21	114
Ewe	17 months	17	148
Doe	18 months	20	151
Mare	2–3 years	21	336

With cattle, a heifer is first bred when mature enough to carry a fetus and give birth. Heifers should weigh 600 to 750 pounds when bred. They should calve when they are about two years old. A heifer that is not mature and is small will have difficulty calving. The bone structure of a small heifer may not be large enough to allow a calf to be born. Some producers use bulls that sire smaller calves on first-calf heifers.

Cows are bred the first heat after calving or so calves will be born at the desired time. Most cattle producers want calves born in the winter or early spring. This means that cows are bred about 283 days (9½ months) before the calves are to be born. (Hogs have a gestation period of 114 days—3 months, 3 weeks, and 3 days.)

Following conception, the female is said to be pregnant. During pregnancy, the fertilized ovum (egg) forms cells that divide and specialize to create a new individual. The time of pregnancy is known as gestation. Gestation is the period from conception until birth. Gestation length varies by species. The period for a horse is longer than for a dog. With fish and birds, the time between fertilization and hatching is known as incubation.

PREGNANCY TESTING

Pregnancy testing is a process to determine if females are pregnant. Several kinds of tests are available. Some require lab testing of blood or urine samples. Milk can be tested to see if a lactating cow is pregnant. Levels of the hormone progesterone in milk increase when a cow is pregnant.

For cows, the most common and reliable way of pregnancy testing is using the rectal method. The hand is covered with a rubber glove and inserted into the rectum several inches. The hand feels the uterus through the rectum wall. Changes in the size and location of the uterus indicate pregnancy. Training is needed to use this test accurately.

Several other methods are used, such as "bumping"—pushing inward with the hand on the lower right flank (experienced cattle producers can detect a fetus)—and checking for a fetal heartbeat by using a stethoscope.

Pregnancy testing determines the status of the female. She might need to be bred again. Some producers sell non-pregnant females to packing plants. There is no need to feed an animal that is not producing!

MANAGING THE PREGNANT FEMALE

A pregnant (gestating) female must have proper nutrition to develop one or more baby animals in her body. Good pasture may provide adequate nutrition for cows and sheep. Other animals may require concentrates. The female usually does not need additional special care.

As the time of parturition (giving birth) gets close, the female should be regularly observed. In some cases, the female may need to be moved to an appropriate place. Most cows can remain in the pasture. Hogs may need to be placed in farrowing crates. Regardless, the goal is to help females give birth to healthy babies that will live and grow rapidly.

19–53. Good pasture usually provides adequate nutrition for gestating cows. Note that this herd of purebred Limousins has a black cow (1 out of 100 Limousins is typically black). (Courtesy, Education Images)

19–54. Individually penned sows are carefully fed to assure maximum development of embryos and maximum litter size. (Courtesy, USDA)

Signs of approaching parturition are enlarged udder, swelling of the vulva, and hollowness in front of the pin bones. Nervousness, voluntarily leaving the herd, and muscular exertion and distress are signs that the birth is near.

BIRTH

Most pregnant mammals give birth without assistance. Sometimes, assistance may be needed, particularly for pigs and lambs. With a cow, the calf should be born in about an hour after abdominal straining begins. If the time extends more than an hour, the cow may need help. Seek the services of a veterinarian or qualified animal technician or breeder to assist with the birth process.

A calf is delivered with the head between the two front legs. This is the normal ***presentation***—the position of the fetus during birth. If other parts are visible, the birth is not normal. A cow that has a calf in an abnormal fetal position may need help. If she does not get help, the calf and/or the cow may be lost during parturition.

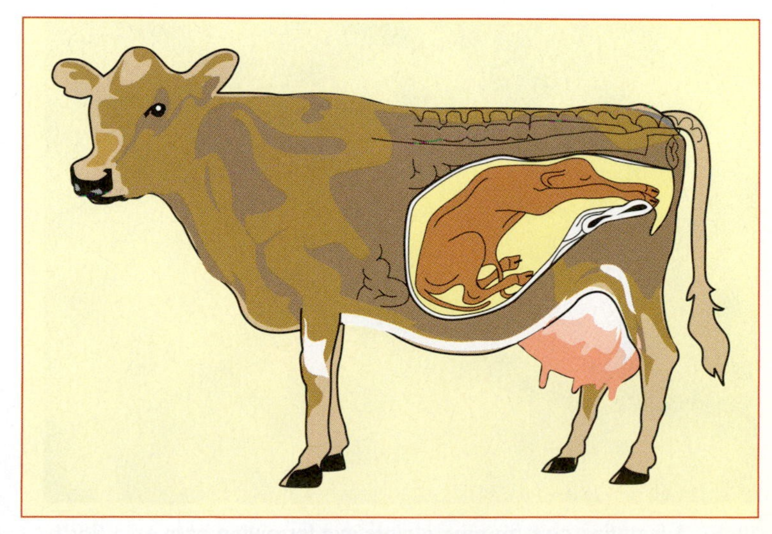

19–55. Normal presentation of calf in uterus.

Many cattle producers are skilled in assisting with the birth process. If a producer lacks skill, a veterinarian can do what is needed. Pulling on the new animal may help the female in giving birth. This should be done carefully and only in an approved manner. Sanitation is very important. People can contract diseases from animal fluids and should protect themselves.

AFTER THE BIRTH

Most baby animals can take care of themselves after they are born. Pigs may need special attention. All newborn animals should be protected from harsh weather. Sometimes, a newborn may need help to start nursing its mother.

The first milk after giving birth is ***colostrum***. It is high in antibodies and other substances that help the new animal survive. Early nursing is essential to be sure that the new animal gets the colostrum.

19–56. A ewe has given birth to four lambs. Note that part of the placenta is attached inside the ewe. (Courtesy, University of Minnesota–Crookston)

19–57. A lactating sow nursing piglets in a farrowing barn on a North Carolina swine farm. (Courtesy, USDA)

In the uterus, the fetus develops inside the ***placenta***—a baglike structure that nourishes and protects the developing animal. The female expels the placenta during or shortly after giving birth—usually within three to six hours. Sometimes, the placenta (commonly called the afterbirth) is retained by the female. Only an animal technician or a veterinarian should remove a retained afterbirth.

Some animals require management practices shortly after their birth. These practices include clipping the needle teeth of pigs, identifying the animals, and castrating the males.

Cows, sows, and other mammal mothers need proper feed to help them produce milk for their young. Feeding a female during lactation is more critical than during gestation. Cows on pasture usually get enough nutrients. Dairy cows need feed supplements high in energy, protein, minerals, and vitamins. Sows need feed with grain, protein, minerals, and vitamins. All nursing females need free access to water.

19–58. Some animals, such as kittens, need special protection because they are born with their eyes closed.

MANAGING THE BREEDING MALE

More attention is given to managing the breeding female than the male. The male, however, needs to be fed and managed to assure reproductive capacity.

Bulls may be kept in separate pens from cows or pastured with them. If kept separate, the bulls should be in well-fenced pastures or pens. They should get plenty of exercise. When pastured with cows, bulls may be with them only during the breeding season. The remainder of the year, the bulls are kept in separate pens.

Bulls need feed with energy, protein, minerals, and vitamins. They should be healthy and in vigorous condition. Without adequate nutrition, bulls may fail to breed.

Boars are usually kept separate from sows except at breeding time. They should get plenty of exercise and a diet with energy, protein, minerals, and vitamins. Some boars are allowed access to green pasture and provided a concentrate feed supplement. Boars should neither get too fat nor be neglected.

CONSIDERATIONS WITH POULTRY

Producing poultry is different from producing beef cattle, hogs, and other animals. There are also differences between producing chickens for meat and for eggs. Most meat is from broilers. A **broiler** is a chicken 6 to 12 weeks old weighing 2½ pounds or more.

Production begins when fertile eggs are produced on breeder farms. The eggs are taken to a hatchery for incubation. Eggs hatch into chicks in 21 days. The chicks may be subject to several management practices, such as debeaking and injections. **Debeaking** is the

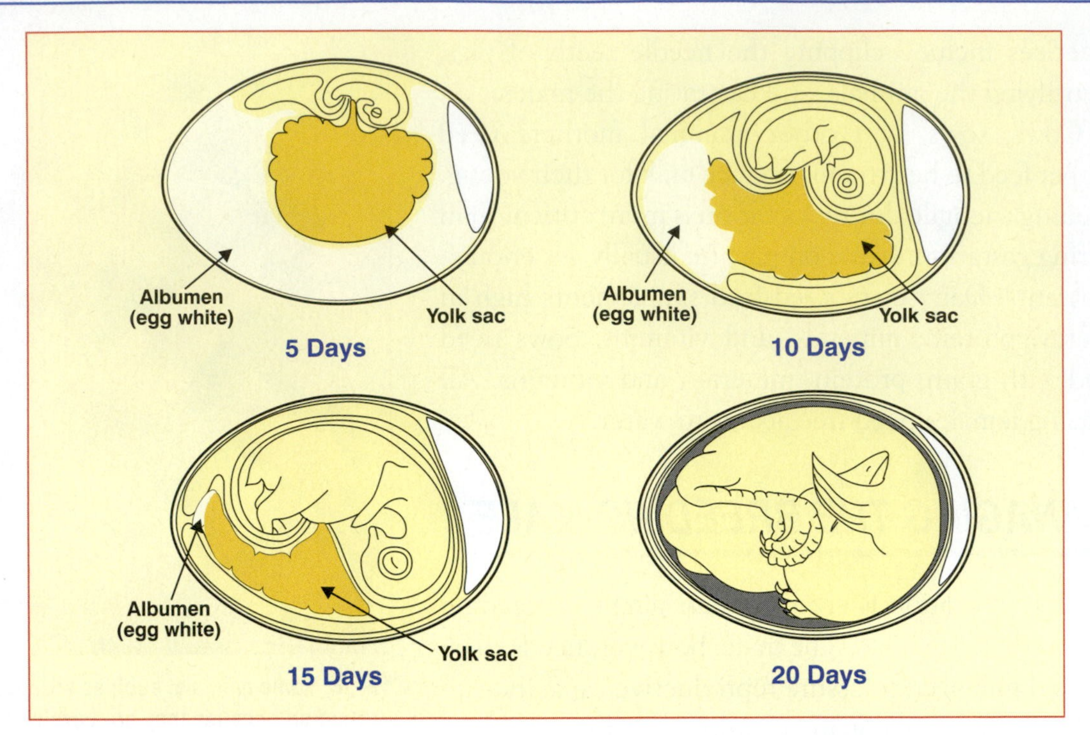

19–59. Development of a chicken embryo in an egg. A chick hatches 21 days after incubation begins.

removal of the tip of the beak to keep birds from attacking and pulling feathers out of each other.

A day or two after hatching, the chicks are placed in a broiler house (building where broilers are raised). Several thousand chicks may be in one house. Feed and water are pro-

19–60. Preparing fertile eggs for placing in an incubator.

19–61. Day-old chicks. (Courtesy, Education Images)

vided free access. Heating or cooling may be needed, depending on the weather. In five to six weeks, the birds will weigh 3 to 5 pounds. Those that are larger are roasters.

REVIEWING

MAIN IDEAS

The goal of breeding animals is to provide food and other products for people. Knowing what is involved helps producers do a better job of raising animals.

Several breeds exist among all species of commonly produced animals. Some producers raise purebred animals; others raise grade animals. Various breeding systems are used to improve animal quality. These include variations of straightbreeding and crossbreeding.

Production systems are used to produce young animals and raise them to the right size. Beef cattle typically begin with cow-and-calf systems. After weaning, the calves go to a feedlot for finishing. In finishing, the calves are fed high-energy feed for rapid growth and the development of body fat. Hogs go through similar procedures, with feeder pigs being produced for finishing in confinement. Poultry is different from beef cattle and hogs, though it also involves producing and raising young.

Efficient and timely animal reproduction is essential for the producer. With some animals, the breeding process is managed to ensure that the desired animals are produced. Some animals are neutered (made incapable of reproduction) to prevent the birth of unwanted animals or to gain a desired growth rate. Pregnant females need proper nutrition. An appropriate place for giving birth should be provided. Care may be needed depending on the species and on the nature of the birth process.

QUESTIONS

Answer the following questions, using complete sentences and correct spelling.

1. What is a breed? Distinguish between bloodline and breed.
2. What are the common breeds of beef cattle? Name one major trait that would help you identify each breed.
3. What are the common breeds of dairy cattle? What are the advantages of each breed?
4. What are the common breeds of hogs? What major traits help identify them?
5. What are the classes of sheep on the basis of wool quality? Give an example of one breed in each class.
6. What are the three classes of goats? How do the classes differ?
7. What are the three groups of horses? Briefly distinguish between the groups. Which is most widely found today?
8. What is a breeding system? Name and distinguish between the common breeding systems.
9. What is a production system? How do production systems differ from breeding systems?
10. What production systems are used to produce beef? Briefly describe each.
11. What production systems are used to produce pork? Briefly describe each.
12. What two methods are used to inseminate animals? Distinguish between the methods.
13. What are the major management practices in breeding animals? Briefly describe each.
14. What is the general production system for broilers?

EVALUATING

Match each term with its correct definition.

a. feeder pig
b. inbreeding
c. pedigree
d. bloodline
e. needle teeth
f. pregnancy testing
g. colostrum
h. placenta
i. presentation
j. castration

_____1. The position of the fetus during parturition.

_____2. The mating of closely related animals of the same breed.

_____3. The eight sharp teeth that can damage the nipples of a lactating sow if not removed.

_____4. The names of the ancestors of an animal.

______5. The first milk after giving birth.

______6. A pig under 60 pounds used for growing into a meat-size hog.

______7. Removal of the testicles of a male animal.

______8. A process to determine if a female is carrying a developing embryo or fetus.

______9. The baglike structure that nourishes and protects the developing animal.

_____10. A group of animals within a breed.

EXPLORING

1. Review a copy of the book entitled *Introduction to Livestock & Companion Animals,* available from Pearson. Note approaches in animal breeding. Compare several species of animals. Prepare a report that summarizes your findings.

2. Survey your community, and identify the animals that are produced. List the breeds and/or crosses involved and the production systems used.

3. Visit a local farm or ranch, and observe the artificial insemination process. Of course, many farms do not use artificial insemination. Determine why artificial insemination is used on the farm you visit.

4. Observe the management practices used on a farm or ranch in breeding animals. Practices may include clipping needle teeth, castrating, and giving injections, among many others.

5. Collect labels from feed for newborn, young, mature, and lactating animals. Study the amounts and kinds of nutrients in the feed. Compare the feedstuffs that are used.

6. Plan an animal/livestock production enterprise. State an overall goal or purpose of your enterprise, such as to produce market animals. Identify the species and/or breeds you would select. List and explain the factors you would consider in selecting the animals. Give an oral report in class on what you propose and how you would select the livestock or other animals.

CHAPTER 20

Animal Health

OBJECTIVES

This chapter presents general information on helping animals have good health so that they grow and produce as they should. It has the following objectives:

1. Explain health and biosecurity, and list health signs.
2. Describe environmental influences on health.
3. Identify losses caused by poor animal health.
4. Explain how good health is promoted.
5. Name the classes of diseases, and discuss examples in each class.
6. Describe how animals defend against disease.
7. Select methods of disease control.

TERMS

acquired immunity
acute disease
antibody
antibiotic
biosecurity
body system disease
chronic disease
contagious disease
disease
ectotherm
endotherm
external parasites
health
immunity
infectious disease
injection
internal parasites
isolation
leukocyte
lymphocyte
natural immunity
noncontagious disease
parasite
pathogen
phagocyte
preconditioning
primary defenders
prion
pulse rate
respiration rate
sanitation
secondary defenders
symptom
systemic medicine
topical medicine
vital sign
zoonosis

20–1. Healthy animals are fun and enjoyable to own.

HEALTHY ANIMALS grow and go about life in a productive manner. You have probably seen a diseased animal, such as a dog, pig, or bird. You suspected a health problem because of a change in behavior, slow growth, or lack of production. These changes can all be signs that an animal is not healthy.

The role of a producer is to provide a good environment for an animal. Not keeping animals healthy costs animal producers billions of dollars a year in the United States. It also results in lower-quality meat and other items. Products from diseased animals are not good and might pose dangers to people if consumed. Meat scientists estimate that the steaks and roasts of one beef carcass will potentially have 542 consumers. If an animal is bad, 542 people have bad meat!

Producers can carry out many useful practices in keeping animals healthy. They begin by knowing the signs of health, following practices to promote health, and properly handling animals with disease. Sometimes, they may need the help of animal technicians or veterinarians.

ANIMAL HEALTH AND BIOSECURITY

Health is the condition of the body and a measure of how well the functions of life are being performed. Good health means that all life processes are being performed normally.

The health condition of animals can vary widely. Some animals are healthy—free of disease and pain. Others may be unhealthy and, to some extent, suffer disease and pain. Some ill health can be prevented and easily corrected.

A ***disease*** is a condition of pain, an injury, or the inability to function normally. A diseased animal deviates from normal health. Producers use basic knowledge of disease in effectively producing animals.

20–2. Healthy, growing turkeys reflect a diet enriched with vitamin E. Researchers have found that increased levels of vitamin E boost the immune response and clear bacteria known as *Listeria monocytogenes* from digestive systems. (*L. monocytogenes* cause a food-borne infection that also affects humans.) (Courtesy, Agricultural Research Service, USDA)

In recent years, national concern over the maintenance of a safe and sustaining food supply has led to creation of the term and practices known as biosecurity. ***Biosecurity*** is the use of approaches to manage risk and to ensure the production of disease-free animals and other products. The focus is to prevent the introduction of pathogens (infectious causes of disease) into a herd or flock. Disease prevention and control practices used in good production operations are usually sufficient. Every producer has a role in biosecurity. Of course, the major advent of the term and practices occurred following outbreaks of terrorism in the United States and other countries. The content of this chapter addresses most of the major practices in biosecurity.

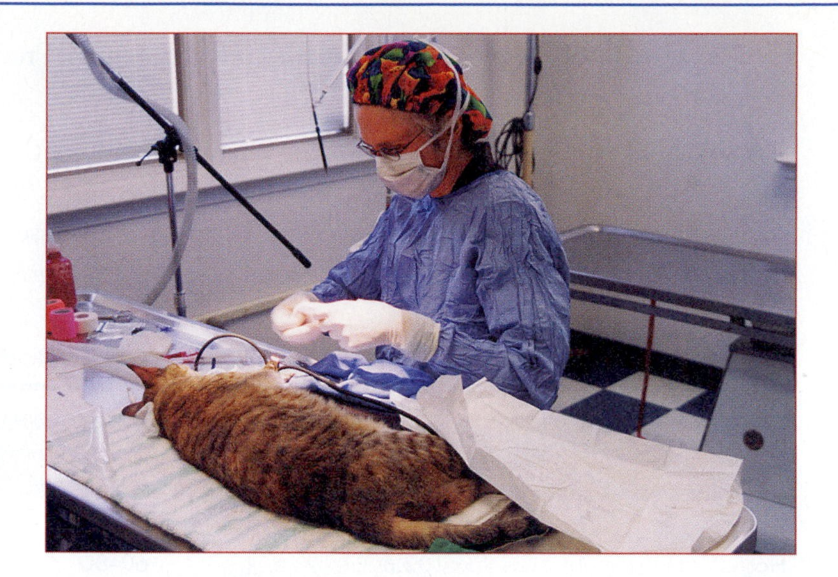

20–3. A veterinarian prepares for surgery on a cat. (Courtesy, Education Images)

GOOD HEALTH

Good health is the absence of disease. The animal has normal body conditions. Producers of animals need to know the signs of good health.

Signs of good health include:

- **Good appetite**—A healthy animal has a good appetite and will eat when feed is offered. If it does not eat, disease should be suspected. In ruminants, chewing the cud (cudding) is a sign of good health. Healthy fish will quickly eat floating feed.
- **Responsive and content**—Healthy animals are responsive to what is going on around them. They are content and appear satisfied. Healthy hogs have curled tails, and healthy cattle will stretch when they get up. Nervous animals may have health problems.
- **Bright eyes and a shiny coat**—Eyes reveal much information about health. They should be clear and alert. Membranes around the eyes should be pink and moist. There are no discharges from healthy eyes. The hair is bright and has good color. Old hair sheds when a new coat is grown.
- **Normal feces and urine**—Blood, pus, or mucus in the feces is a sign of problems. The consistency of feces should

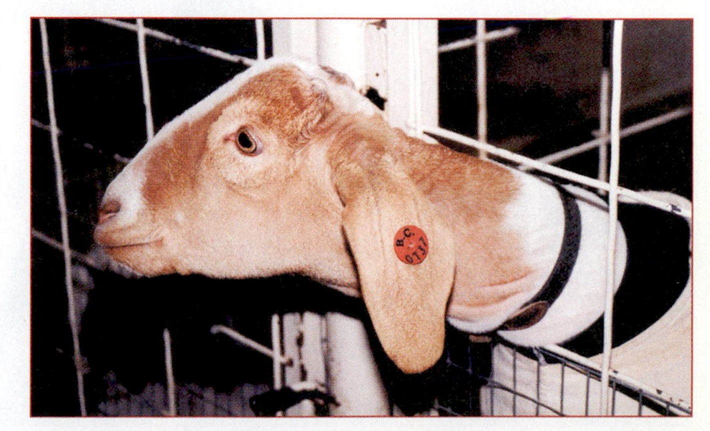

20–4. Bright eyes, shiny coat, clear nostrils, and other facial features are signs that this goat has good health.

be normal for the animal and the feed it is consuming. Producers usually know the normal characteristics of feces and urine.

- **Normal vital signs**—A ***vital sign*** is an indication of the living condition. Important vital signs are breathing (respiration rate), pulse rate, and body temperature. Changes in the vital signs indicate a problem. Vital signs are assessed according to the species and the activity of the animal. Several normal vital signs are in Table 20–1.

Table 20–1. Normal Signs of Selected Animals

Species	Average Normal Temperature (rectal °F)	Normal Pulse Rate (rate/min.)	Normal Respiration Rate (rate/min.)
Cattle	101.5	60–70	10–30
Hog	102.6	60–80	8–15
Sheep	102.3	70–80	12–20
Horse	100.5	32–44	8–16
Chicken	106	200–400	15–30
Dog	101.3	60–120	10–30
Cat	103	140–240	20–40

Pulse rate is the movement of the arteries as a result of the heartbeat. All animals have a range in which the normal pulse rate should fall. Rates that are too slow or too fast are signs of possible problems. A pulse rate is measured as the number of heartbeats per minute. Pulse rates are normally taken on the soft part of the leg, just above the foot. The artery should be prominent and easily felt with the fingers to count the heartbeats.

Respiration rate is the number of breaths taken in a minute. Increased breathing indicates that the animal is stressed or that the animal has been exercising. Respiration rate is determined by observing the rise and fall of the animal's sides (flanks).

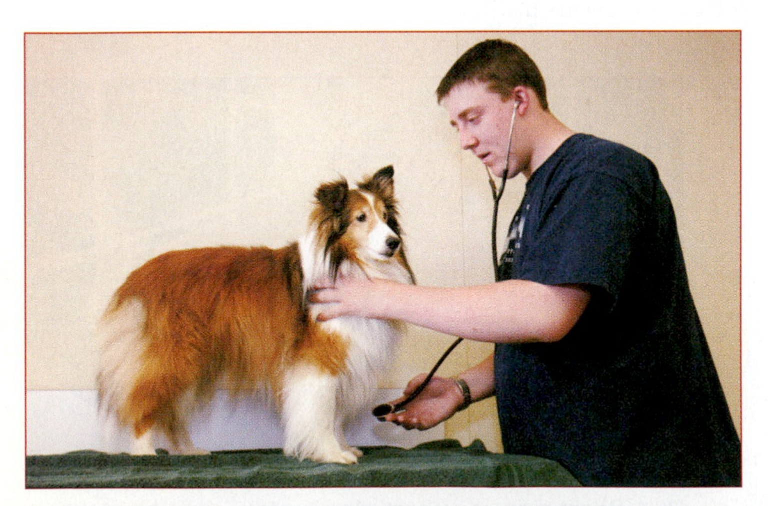

20–5. Using a stethoscope to listen to the heart, the lungs, and other internal sounds of a dog. (Courtesy, Education Images)

Use a clock or stopwatch to measure an exact minute to count pulse and respiration rates.

All animals have normal body temperature ranges. Temperature is usually taken using a rectal thermometer. Temperatures are reported in degrees F or C. Increases in temperature indicate a health problem, such as an infectious disease.

- **Reproduces**—Normal sexually mature animals reproduce if allowed to mate. Reproductive diseases often result in failure to reproduce. Females may not cycle and, therefore, will not become pregnant after mating. Males may not produce sufficient sperm. Several diseases are associated with reproductive failure.

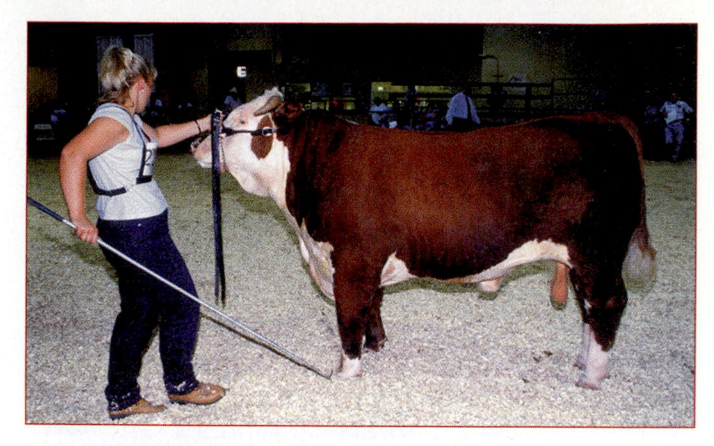

20–6. A Hereford bull being shown at the Eastern National Livestock Exposition has signs of good health. (Courtesy, USDA)

ILL HEALTH

If the signs of good health are not present, the animal is said to be ill. Producers observe animals each day for health signs, also known as disease symptoms. A ***symptom*** is the way a disease shows itself. Some diseases may have several symptoms.

AgriScience Connection

DISPOSING OF DEAD ANIMALS

Even with good care, animals will occasionally die. Properly disposing of dead animals is important. Proper disposal minimizes odor and prevents the spread of disease. Fish and poultry producers often dispose of dead animals.

Poultry producers commonly dispose of dead birds by incineration and composting. Incineration is burning the remains of an animal to reduce it to ash and to destroy all disease-causing organisms. A fuel-powered incinerator is used (shown lower right).

Composting is using natural decay processes with dead animals. The dead animals are placed in a compost pit or bin along with wood shavings and soil high in bacteria. The remains of dead animals will decay in a few days. A compost bin is shown in the top photograph.

20–7. Bloody areas on the skin and fin of a fish are signs of disease. (A diagnostic lab may need to examine the fish to determine the exact disease.)

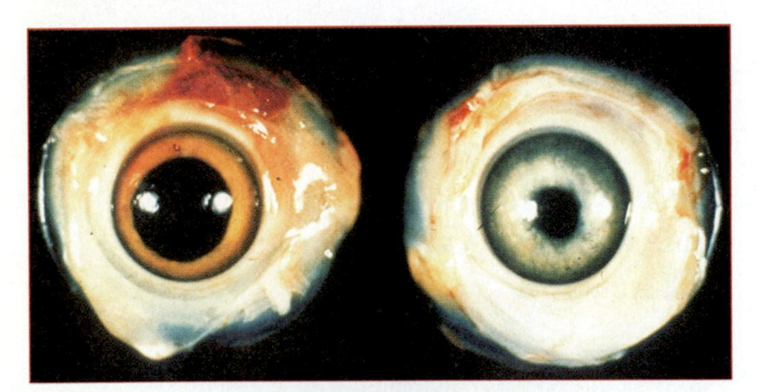

20–8. The eyes of an animal often reveal its health status. Sometimes eyes are diseased; other times they reflect the condition of the animal. This shows two eyes from a chicken. The one on the left is normal; the one on the right has lesions and an irregular pupil caused by Marek's disease. (Courtesy, Agricultural Research Service, USDA)

Signs (symptoms) of ill health include:

- not eating normally or not drinking water
- sunken eyes and possible discharge from the eyes
- discharge from the nostrils or mouth
- inactive—stands in one place and drawn-up
- rapid breathing
- rapid pulse rate
- higher than normal body temperature (fever)
- dull hair
- lumps or protrusions on the body
- open sores
- moving away from the rest of the animals; seclusion
- unusual feces or urine—may contain blood
- loss of production, including a drop in milk production and not gaining weight

ENVIRONMENTAL INFLUENCES ON HEALTH

The environment influences an animal's health. Stress results when environments change or an animal is in an environment for which it is not suited. Stress occurs because of an animal's inability to adjust to its environment. While stressed, the animal is susceptible to disease.

Environmental health factors include temperature, light, moisture, transport, and pollution.

20–9. Animals may be isolated as they recover from disease or to prevent them from exposing healthy animals to disease. (Courtesy, Education Images)

TEMPERATURE

Animals live best within certain temperature ranges. Temperatures that are too high or too low may cause health problems. Some animals, such as newly hatched chicks, require protection from the weather.

20–10. These chickens, with open beaks, are panting because they are too warm. (Courtesy, Mississippi State University)

Most farm animals are endotherms. An ***endotherm*** is an animal that maintains a certain body temperature. Fish and most other aquatic animals are ectotherms. An ***ectotherm*** is an animal with a body temperature that adjusts to its environment.

Endothermic animals adjust as best they can to hot and cold weather. They may sweat, pant, or wade into water to cool their bodies. In cold weather, endotherms burn more energy to keep warm. They must be fed better and must be allowed to exercise. When an animal shivers, the muscles are contracting involuntarily to burn energy and keep warm.

Young animals cannot tolerate cold weather as well as mature animals. In cold weather, young animals often need to be protected from low temperatures. For example, young pigs may be kept in heated barns. Calves take cold better than pigs, but they should be sheltered from extremely low temperatures.

During warm weather, confined animals need care to keep their body temperatures down. Chicken houses may have ventilation fans. Hogs may have a fine mist of water sprayed on them. Cattle in pastures will generally move under a shade tree in hot weather.

20–11. Temperature and light are regulated in a broiler house with electronic controls.

Extreme temperatures stress animals. However, animals that are properly fed can better withstand the extremes.

LIGHT

Sunlight and length of day influence animals. Some animals, particularly those with light skin pigments, may develop skin problems from exposure to the sun. Cattle with light hair and light skin around their eyes are more likely to have eye diseases.

Lengthening the day can increase the growth and production of some animals. Chickens are sensitive to light. Both broiler and layer houses may be lighted for a few hours after dark. In laying hens, the light causes more hormones to be dumped into the bloodstream by the pituitary gland. This results in chickens growing faster and laying more eggs. Of course, chickens do have limits in their production.

MOISTURE

Humidity and precipitation (rainfall, snowfall, etc.) are moisture forms most likely to cause animal health problems. Air with high humidity does not cool an animal as well as air with low humidity because perspiration does not evaporate as rapidly in high humidity.

Confined animals may be more affected by humidity. Chicken houses and barns can build up a lot of moisture on the inside. One cow breathes nearly 1½ gallons of water into the air in a day. Ventilation is needed to move the excess moisture out and to bring in drier air.

Disease agents live better in moist environments. Most causes of disease do not survive as well in low humidity.

Animals may need to be protected from rain and rising water. Cattle can drown when water overflows a creek into their pasture.

MOVING

Animals may be stressed when they are corralled, hauled, or otherwise moved. Chasing animals to move them into pens or barns raises their body temperature. Loading and hauling in trucks can expose them to wind, heat, cold, rain, and other conditions.

Animals in sale barns or feedlots are also exposed to health hazards. The concentration of animals from many different places promotes disease. While stressed, an animal has lower resistance and is more likely to contract a disease. Some producers of stocker cattle precondition them so the stress of moving does not make them sick. Animals in good condition can respond better to stress.

POLLUTION

Pollution and other hazards in the environment may create disease. Animals are sometimes exposed to poisons in water and feed. Pastures sprayed with pesticide can cause poisoning. Industrial wastes in the air and water can damage animals. Storing feed in places where poisons have been kept can contaminate the feed. In addition, hauling feedstuffs in boxcars or trucks with chemical residues can contaminate feed.

OTHER CONDITIONS

A wide range of environmental conditions can threaten animals. These conditions vary with the species and the environment needed. Terrestrial and aquatic animals vary considerably. Fish, for example, need appropriate water. Among other qualities, the water must have adequate dissolved oxygen (DO). The oxygen level can be measured with a DO meter.

Technology Connection

PIGS HAVE ALLERGIES

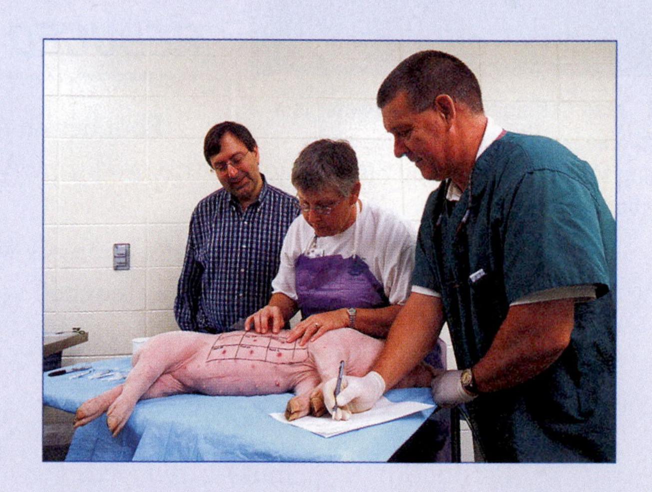

Pigs and people have some similar attributes, such as some of the same allergies. A few of the similarities are so great that research for the benefit of humans can be done on pigs.

A few people are allergic to soybeans and soybean products. The number is small, with no more than 1 to 2 percent of adults having the allergy. Scientists set about to identify the gene in soybeans that is responsible for the allergy. Once the gene was identified, a technique known as gene silencing was used to produce a soybean variety that should not result in an allergy.

Testing of the new soybean was on pigs. The pigs that had allergies to soybeans were identified and used in the study. The accompanying photo shows scientists performing an allergy test on an anesthetized pig. A marker was used to draw a grid on the pig's skin to aid in making measurements. (Courtesy, Agricultural Research Service, USDA)

LOSSES CAUSED BY POOR ANIMAL HEALTH

Poor animal health results in losses. Some losses are easy to see, such as the death of animals. Other losses may go unnoticed, such as those caused by internal parasites. The greatest losses may be those that are not easily seen.

20–12. Dead fish floating on the water is a sign of disease. Such a presence gives meaning to the term "belly up." (Courtesy, Education Images)

DEATH

The most obvious loss is death. Dead animals cannot be sold. Their disposal creates additional costs. Chicken producers frequently have deaths among birds. Chicken farms have compost pits, incinerators, or other means of disposing of the dead birds.

The proper disposal of dead animals is essential. Burying and burning are the best methods. Producers should obtain information on the approved ways of disposing of dead animals in their local area. Dead animals that are not properly disposed of can lead to more disease. Decaying animal bodies create bad odors and attract unwanted animals, such as buzzards and wild dogs.

REDUCED PRODUCTION

Poor health results in reduced production and less profit to the producer. A few examples are:

- Animals with disease can be quarantined, resulting in reduced value or no value.
- Animals fail to breed and bear young.
- Animals grow slowly.
- Production (e.g., milk) is reduced.
- Products, such as meat, are of lower quality and may be condemned if diseased.
- Costs of production increase, such as drugs and labor to treat animals.

Healthy animals provide their owners with greater returns—more money!

20–13. Signs of eye disease include a swollen eyelid and discharge from the eye.

HUMAN DISEASE

Some animal diseases can be transmitted to humans. Any disease that can be transmitted to humans, as well as to other animals, is known as a **zoonosis**. Some zoonosises cause severe problems in humans.

Producers of cattle and consumers of cattle products are aware of brucellosis, also known as Bang's disease. Humans can contract undulant fever by drinking milk from cows with the disease or by letting fluids from the animal enter open wounds on the human body. Dairy cattle are regularly checked for Bang's disease. Milk from cows with the disease cannot be used. Pasteurization (heating milk to 145° to 150°F for 30 minutes) helps make milk safe.

Leptospirosis is a disease that affects humans and other animals. Though not often thought of as a disease of humans, each year 100 to 200 human cases of leptospirosis are reported in the United States. Leptospirosis is spread through contact with any medium that has been infected by an animal with the disease. Contact with urine or infected tissue is the most common way in which humans contract the disease. The leptospirosis organism can enter the human body through skin cuts or scratches and through mucous membranes of the mouth, eyes, nose, or genitals.

Another example is trichinosis, which is caused by a small worm in the flesh of some hogs. It can be transferred to humans if pork meat is not cooked thoroughly. Today, few hogs going to market have trichinosis. Thus, this disease has been virtually eliminated as a threat in the United States.

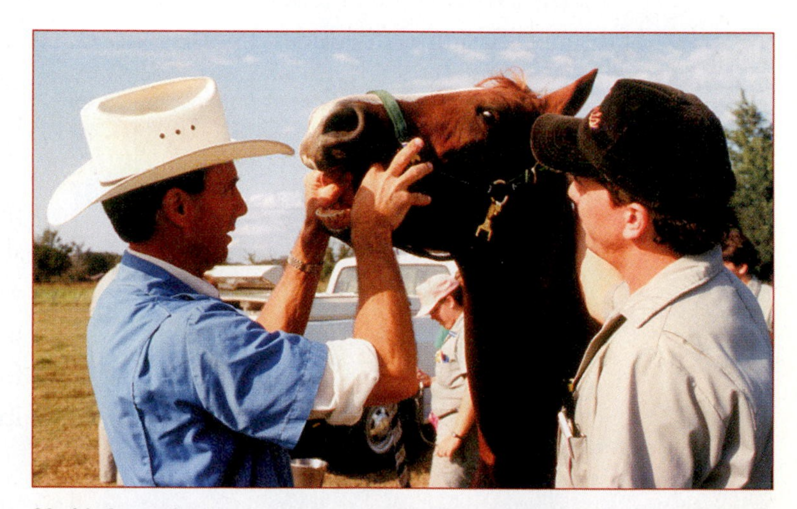

20–14. An oral exam is used to check a horse for disease. (Courtesy, College of Veterinary Medicine, Texas A&M University)

Other examples of animal diseases that can be transmitted to humans are rabies, anthrax, and sleeping sickness. Every animal disease should be viewed as a possible zoonosis.

PROMOTING GOOD HEALTH

Animal health can be promoted by producers. Practices to minimize disease should be part of the production system. The goal is to help animals have good health.

20–15. Computers and electronic systems have made it easier to keep records and prepare reports on animals. (Courtesy, Education Images)

MEDICAL RECORDS

Medical records on new animals may be used to promote good health in a herd. If a new animal has a history of immunization and is disease-free, it can usually be introduced into a herd after a period of isolation. Herd or flock history of the originating producer is beneficial. If other animals have had a disease, the new animal is suspect, even though it appears disease-free. The animal is kept in isolation long enough for any infectious diseases to manifest themselves.

SANITATION

Sanitation is the practice of keeping areas clean. Filth harbors disease and the carriers of disease. Sanitation reduces disease sources. Animals that do not come into contact with the causes of disease are less likely to get the disease.

20–16. These hogs are kept in an unsanitary wallow that promotes disease. (Courtesy, USDA)

Dead animals should be removed promptly and disposed of properly. Barns should be cleaned. Manure and other wastes should be disposed of in a safe manner. Many farms use fenced lagoons for waste disposal from pig parlors, feeding pens, and dairy barns.

Rodents should be controlled. Rats, mice, and similar animals are carriers of disease. Keeping their numbers down will keep the diseases away.

Some producers use disinfectants as part of their sanitation programs. A disinfectant is a substance or process that destroys the causes of disease. Animal supply stores have disinfectants for sale.

Sanitation is also an important practice in veterinary medical clinics. Animal cages, examining areas, surgical equipment, and other areas and articles must be sanitized. Proper cleaning is essential. Some equipment, clothing, and other items that come into contact with animals during particularly vulnerable times must be sterilized. This is done by placing these items in an autoclave for exposure to heat that will kill harmful organisms.

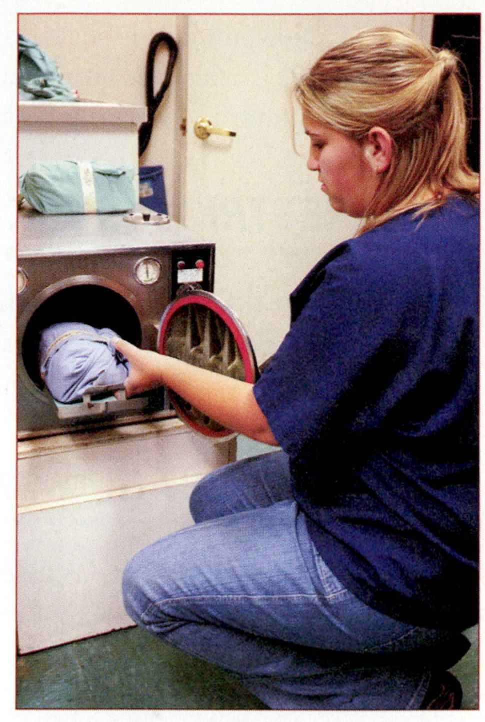

20–17. A folded veterinary surgical gown is being placed in an autoclave for sterilizing. (Courtesy, Education Images)

PROPER NUTRITION

Animals that receive proper feed and water get the nutrients needed for good health. Good nutrition helps an animal resist disease. Poorly fed, weak animals do not have the resistance they need.

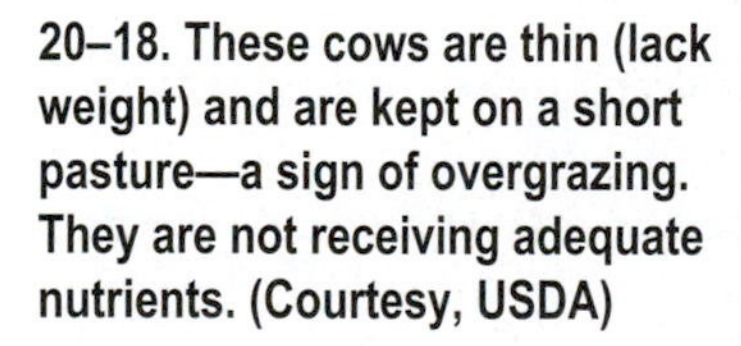

20–18. These cows are thin (lack weight) and are kept on a short pasture—a sign of overgrazing. They are not receiving adequate nutrients. (Courtesy, USDA)

ISOLATION

Isolation is the act of separating diseased and non-diseased animals. Sick animals are removed from the herd because animal contact often spreads diseases. Isolation protects healthy animals from contact with diseased ones.

New animals brought to a farm or a ranch should be isolated for at least three weeks. While in isolation, an animal will likely show signs of disease, if it has one. If no disease is evident after the isolation period, the animal can be introduced to the herd or flock.

RESTRICT TRUCK AND EQUIPMENT TRAFFIC

Trucks, tractors, and other vehicles can bring disease to a farm or ranch. Trucks should be disinfected between trips to different farms, feedlots, or other animal producers. Mud on a truck as well as manure and sawdust in the back of a truck can transport disease.

RESTRICT HUMAN ACCESS

People may carry diseases on their shoes or boots. Every place they go, they pick up and leave small particles that stick to their boots. Some places have boot tubs at entrances to animal facilities. A solution of disinfectant is placed in the tub for dipping the feet. A few places use disposable plastic boots. Some farms do not allow visitors.

Internet Topics | Search

Selected topics for Internet discovery and reporting are listed here. Begin by searching for each topic. Next, read and learn about it. Conclude by preparing a brief report on your findings.

1. Undulant fever
2. Prion
3. Zoonosis

Table 20–2. Levels of Biosecurity*

Level	Practice
1	Closed herd or flock, known as SPF (specific pathogen free)
2	No entry or reentry of animals
3	No entry of new animals, but reentry allowed
4	Entry of new animals based on known medical records and following isolation
5	Entry of new animals based on known medical records but without isolation
6	Entry of new animals without medical records or isolation

* The levels are based on the amount of risk an animal owner/producer is willing to take. Level 1 offers the greatest security and reduces risk to its lowest; Level 6 offers the least security and incurs the highest risk. The levels are courtesy of Oklahoma Cooperative Extension Service, Oklahoma State University.

PRECONDITIONING

Preconditioning is the process of preparing an animal for stress. It is used with stocker calves and other animals that are to be transported or otherwise placed under stress.

Preconditioning is doing routine management well before stress to reduce the likelihood of stress-related problems. For example, castration, dehorning, and other practices should be performed by two months of age. Calves should be weaned and started on feed 30 days before shipping. The length of time in hauling and moving animals should be kept to a minimum. Other health-care practices to prevent disease can be used.

IMMUNIZATION

Immunity is the condition of an animal being resistant to a disease. It can be developed in some animals for certain diseases. The animal can withstand exposure to the disease if it comes into contact with it.

Animals may naturally develop immunity or may be immunized with vaccines. Natural immunity often develops if an animal has a disease and recovers from it. In most cases, animals must be artificially immunized against certain diseases.

20–19. Colostrum, the milk given right after parturition, is high in antibodies and substances that help young animals develop immunity. (Colostrum is sometimes referred to as "first milk." This calf is only 13 minutes old and is already nursing.)

Immunizing helps animals develop immunity. Animals may be injected with vaccines, consume the vaccine in their feed, or be given the vaccine in another way. Most vaccines contain dead or live organisms, or their products, that cause the disease. The animal's body uses these in developing immunity.

CLASSES AND EXAMPLES OF DISEASES

Diseases are classified as contagious or noncontagious. Knowing how a disease is classified helps in preventing and controlling it.

CONTAGIOUS DISEASES

A ***contagious disease*** is a disease spread by direct or indirect contact. It is caused by microscopic organisms, such as bacteria and fungi, or by larger organisms, such as worms. Controlling the organisms helps prevent contagious diseases from attacking animals.

A disease caused by organisms that get inside an animal's body is known as an ***infectious disease***. These organisms enter, grow, and are active in the animal's body. Some symptoms of infectious diseases are fever, abnormal breathing, and sores.

Some infectious diseases are caused by microorganisms. A microorganism that enters the body and causes infectious disease is known as a ***pathogen***. Infectious diseases are not always contagious. Sometimes they begin without being caught from another animal. Examples of contagious diseases in cattle are blackleg, Bang's, and anthrax.

The major causes of contagious diseases are:

- **Bacteria**—Bacteria that cause disease are known as pathogens or germs. Some bacteria are beneficial and do not cause disease. Pathogenic bacteria produce poisons within the body of an animal. Some poisons may be isolated to one place on the body, such as a sore. Other poisons may be moved throughout the body by the blood in the circulatory system. Tuberculosis and tetanus are two examples of bacterial diseases.
- **Protozoa**—Protozoa cause diseases that are different from those caused by bacteria. Anaplasmosis in cattle and coccidiosis in poultry are two examples.
- **Viruses**—Viruses are very small, incomplete cells. They require a host (another cell or organism) to survive. Animals infected with bacterial diseases are susceptible to viral diseases. Examples of viral diseases are rabies, hog cholera, foot and mouth disease, and distemper.

CAREER PROFILE

POULTRY SCIENTIST

A poultry scientist studies the production of poultry. The work may involve studying feeds, nutrition, facilities, and equipment needed in poultry production. It may include developing improved practices in breeding, hatching, and marketing poultry.

A poultry scientist needs a college degree in poultry science or a related area. Many, especially those in research, have master's degrees and doctorates. Take biology, agriculture, and related courses in high school.

Jobs for poultry scientists are found in areas where poultry are produced. Colleges, research stations, poultry integrators, pharmaceutical companies, and others may have jobs. (Courtesy, Animal and Plant Health Inspection Service, USDA)

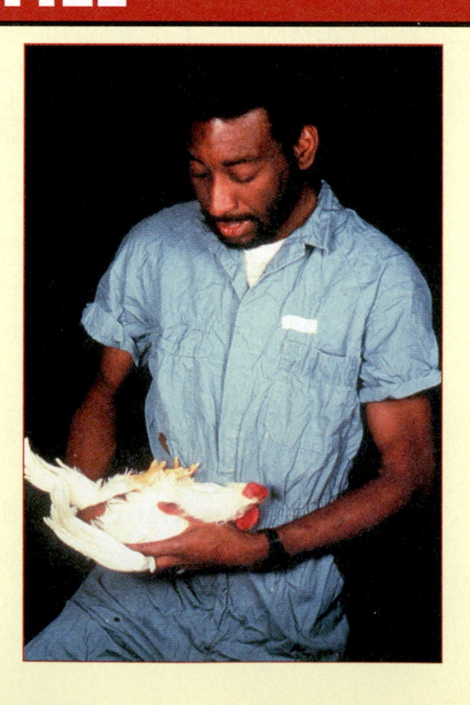

- **Prions**—A ***prion*** is a tiny particle that causes brain diseases. Scientists do not fully understand prions. They consist mostly of a kind of protein. Prions have no genetic material. Mad cow disease is likely caused by prions.

- **Fungi**—Fungi are one-celled organisms that primarily cause diseases on the external parts of the body. Fungal diseases are seldom fatal. Examples include ringworm and certain ear and eye infections.

- **Parasites**—A ***parasite*** is usually a multi-celled animal organism that lives in or on other animals. A few are single-celled, such as protozoa. Parasites receive their nutrition from the animals they live on, known as their hosts.

  Internal parasites are parasites that live inside an animal and that may be in the digestive system, muscles, or other tissues of an animal. Some internal parasites are taken in the mouth by ingesting contaminated feed or water. Most internal parasite

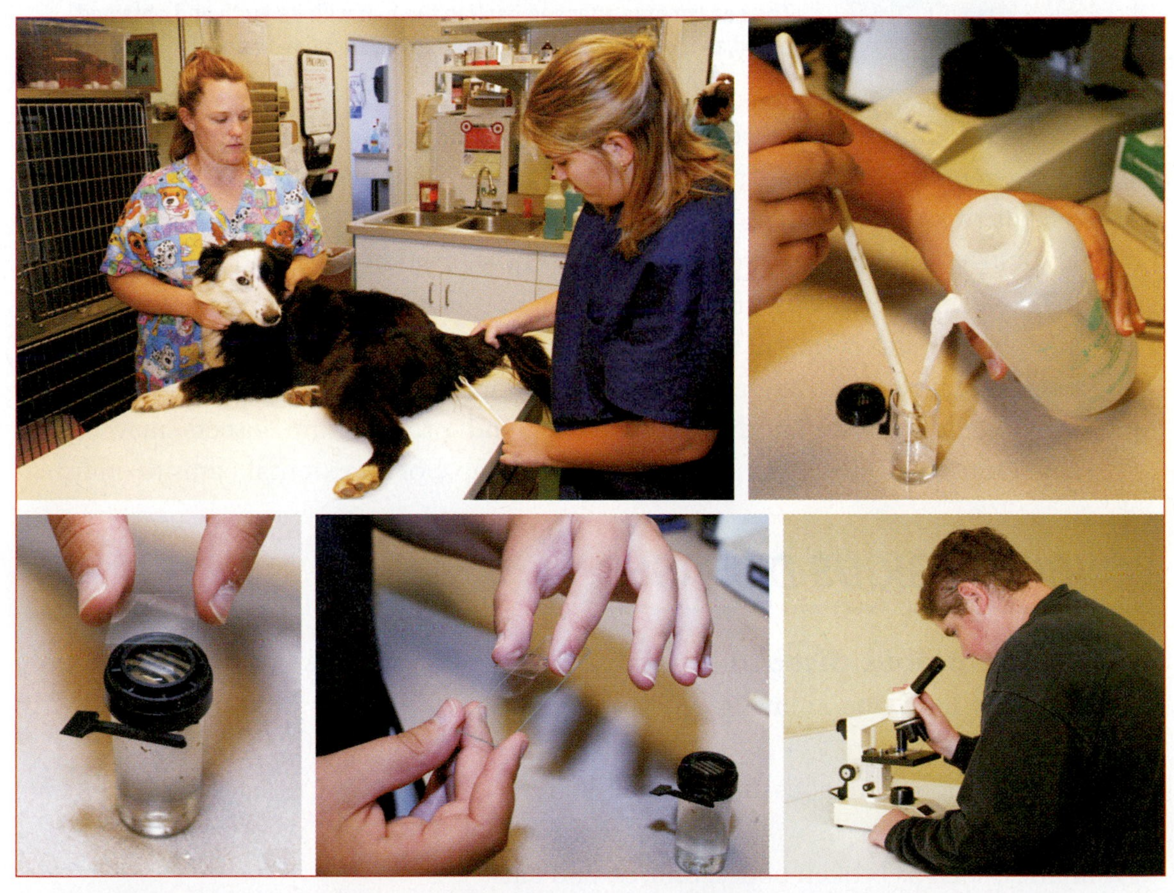

20–20. A fecal loop is used to gather a sample of feces that will indicate the presence of certain internal parasites. The loop is carefully inserted into the dog's rectum and removed. The specimen is washed in a small vial filled with water. A slide cover is placed on top of the filled vial so that it touches the water. Parasite eggs and other life forms become attached to the slide cover after a few hours. The cover is placed on a slide for microscopic viewing to determine the presence of parasites. (Courtesy, Education Images)

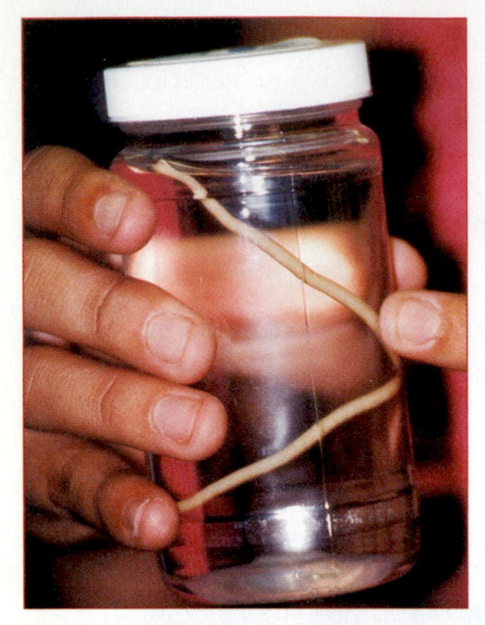

20–21. A preserved roundworm specimen.

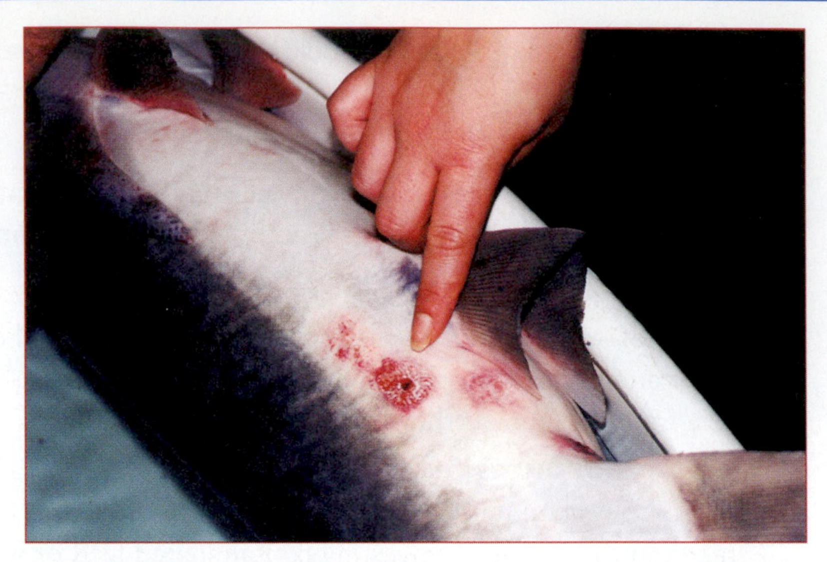

20–22. The lesion on this fish is due to a sea lamprey, an external parasite. The sea lamprey attached itself to the fish with its teeth and obtained blood as food from its host. (Courtesy, Education Images)

contamination is due to feces in the feed. An example is cattle coming into contact with manure while grazing. Short pastures force an animal to eat closer to the ground, which increases the likelihood of the animal getting some manure in its mouth. Examples of ingested parasites are tapeworms, trichina, roundworms, and hookworms.

External parasites are parasites that live on the external parts of an animal. They obtain their food from the blood and tissue of the animal. Examples include ticks, fleas, lice, and mites. External parasites can transfer contagious diseases or parasites from one animal to another. Mosquitoes and other biting insects may spread anaplasmosis among cattle. The leech is an example of an external parasite on fish.

NONCONTAGIOUS DISEASES

A ***noncontagious disease*** is a disease caused by conditions or substances that are not transferred from one animal to another. Animals do not give these diseases to each other.

Examples of noncontagious diseases are:

- injuries, such as broken bones, bruises, and cuts
- poisons, such as pesticides, fertilizers, and substances in plants (e.g., prussic acid, which forms in sorghum and Johnson grass grown under certain conditions)
- faulty body processes, such as milk fever, heart failure, and acetonemia

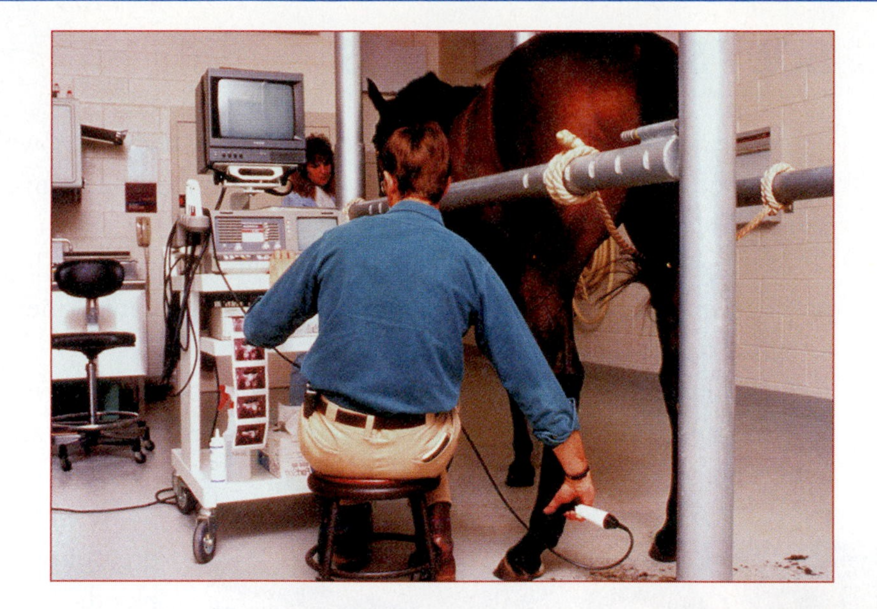

20–23. Ultrasound is used to assess a horse's leg. (Courtesy, College of Veterinary Medicine, Texas A&M University)

- congenital diseases or those present at birth, such as defects in parts of the body
- inadequate nutrition, such as mineral and vitamin deficiencies

Hardware disease is a form of injury resulting from feed that contains metal pieces, such as nails or parts of machines. These objects can lodge in the digestive system or puncture holes through the stomach or intestine walls.

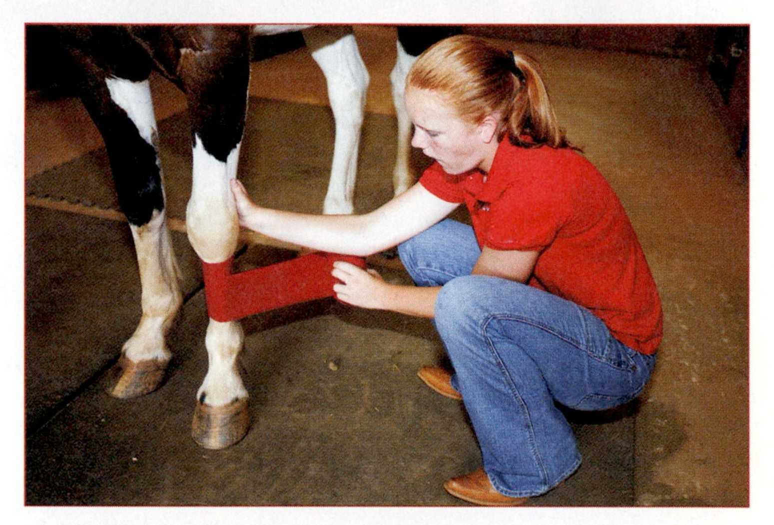

20–24. Placing a wrap on a horse's leg to promote healing. (Courtesy, Education Images)

TIME OF AFFLICTION

Some diseases last for a short time; others last longer. A **chronic disease** is a disease that afflicts an animal for a long time. An **acute disease** is a disease that afflicts an animal for a short time and may be quite severe.

Chronic diseases may gradually begin and cause few or no symptoms for days, weeks, or months. For example, a few worms in a hog may cause no real problem. However, chronic diseases can weaken animals so they are susceptible to acute diseases.

Acute diseases begin quickly and show symptoms almost immediately. They are often the diseases that are likely to cause loss of production and death of an animal in a short time.

BODY SYSTEM DISEASES

A disease that attacks only certain systems of the body is known as a **body system disease**. The disease may be confined to the organs in these systems. All body systems are subject to disease.

Diseases of the digestive system occur when the system fails to function properly. The problems may be associated with the mouth, stomach, intestines, and the glands or organs attached to the digestive system.

Genitourinary diseases are in the reproductive or urinary systems of animals. Reproductive diseases result in the animal not reproducing properly, if at all. Urinary problems may be infections or blocked urine flow.

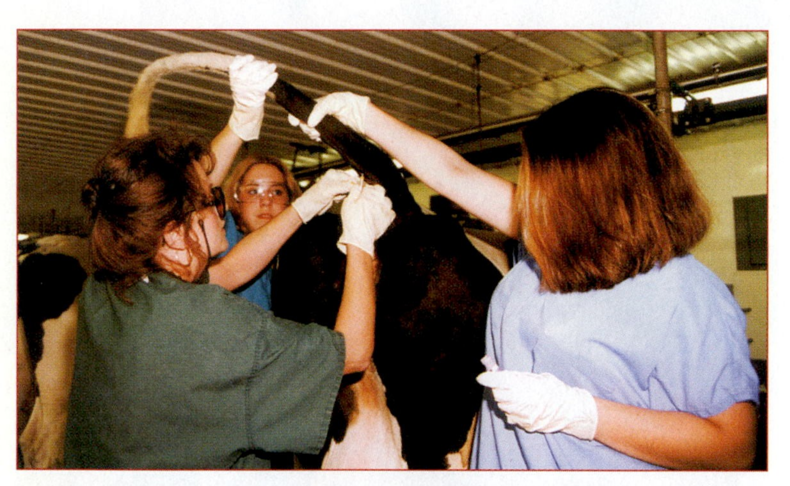

20–25. A blood sample is being drawn under a cow's tail for lab analysis.

Respiratory diseases interfere with breathing and the oxygenation of blood.

Circulatory system diseases interfere with blood circulation or qualities of the blood. Restrictions in the flow of blood damage all parts of the body. A common circulatory system disease in cattle is anaplasmosis.

Nervous system diseases attack the brain and spinal cord or the nerves that spread throughout the body. Movements of the body may be impaired.

COMMON DISEASES

Some diseases attack all species; other diseases attack only certain species of animals. For example, roundworms are found in hogs, cattle, sheep, horses, poultry, and fish. Blackleg is primarily a disease of cattle, but sheep and goats are sometimes affected.

Anaplasmosis

Anaplasmosis is a parasitic disease caused by protozoa that attack red blood corpuscles. Cattle are primarily affected. In chronic anaplasmosis, animals are anemic and may overcome the disease on their own. In acute forms, animals have rapid heartbeat, muscular tremors, and loss of appetite. Sick animals may become aggressive and want to fight. Death may be in a few days. The disease can be prevented by immunizing cattle. Some treatments are available if the disease is caught early.

Anthrax

Anthrax is an acute infectious disease that affects most endothermic animals. It most frequently affects cattle during the summer when they are on pasture. Affected animals have fever, rapid respiration, and swellings on the neck. Animals usually die suddenly. Prevention involves vaccinating animals, controlling flies, and sanitizing the area. Treatments are effective if large doses of penicillin are administered early.

Blackleg

Blackleg is an acute, highly infectious disease that usually results in death. It primarily affects cattle, though sheep and goats can get it. Symptoms include high fever; swelling in the neck and shoulder area; muscles in the neck, shoulder, and thighs crackling when mashed; loss of appetite; and death. Few animals recover from blackleg, but it can be prevented by vaccinating calves.

Bang's Disease (Brucellosis)

Bang's disease causes abortion in pregnant cows, ewes, and sows. It can be a major problem in cows because cows may have to be bred several times to conceive, and the pregnancy may end in a few weeks due to the disease. Bang's can be prevented by vaccinating young heifers. Sanitizing, testing cattle, and bringing only Bang's-free cattle into a herd help control the disease. Bang's has been nearly eradicated in the United States but is reoccurring near wild herds of bison and other species that carry the disease. Humans can contract undulant fever by consuming raw milk and in other ways from infected cows.

Coccidiosis

Coccidiosis is a parasitic disease of poultry, cattle, and other species. It is caused by protozoa. With poultry, the disease affects chickens, turkeys, ducks, geese, and game birds and can cause severe losses. Symptoms in poultry include bloody droppings (feces), ruffled feathers, unthrifty (sick)

ACADEMIC CONNECTION

HEALTH

Relate your study of animal health to the study of human health in a health class or an advanced biology class. Use your learning in agriscience to support your mastery in health or advanced biology.

appearance, and pale coloring. This disease is said to cost the cattle industry more than $100 million a year in the United States. In cattle, coccidiosis presents itself as acute diarrhea with or without blood. The disease is primarily a problem with young cattle and is transferred from one animal to another by infected fecal material entering feed, water, and soil. Medications known as anticoccidials are put into feed and/or drinking water to prevent and treat coccidiosis. Since infected birds transmit it, wild birds should be kept away from poultry flocks, and infected areas should be sanitized. Sick livestock should be isolated from the herd and treated by feeding medicated feed. Several therapeutic and preventive drugs are available.

Equine Sleeping Sickness

Sleeping sickness in horses is caused by a virus transmitted by biting insects. Affected animals walk aimlessly, crashing into things; later they appear sleepy, cannot swallow, grind their teeth, and possibly go blind. Some horses recover, but many horses die within two to four days. No effective treatment has been developed, but horses can be vaccinated against the disease.

Foot and Mouth Disease

Animals that have cloven (divided) feet get foot and mouth disease (FMD). It is a highly contagious disease caused by a virus. There is no known treatment or vaccine, so quarantine is currently the best control. No animals or meat can be imported to the United States from countries with the disease. Animals with the disease get watery blisters in the mouth and on the skin around the feet. Teats of females may have blisters. Animals will have high fever. Foot and mouth disease created great problems in Great Britain in 2001; thousands of animals were killed and burned.

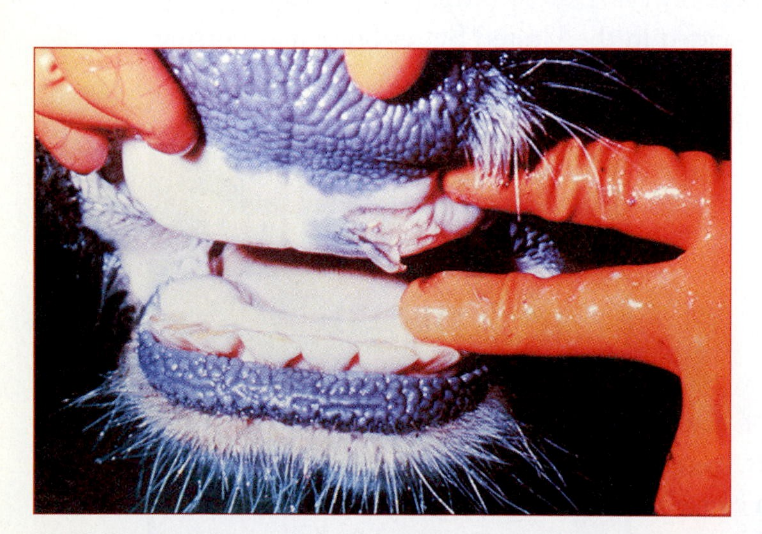

20–26. Diseased tissue inside the mouth of a cow is a sign of foot and mouth disease. (Courtesy, USDA)

Grubs

Cattle grubs (warbles) are internal parasites caused by heel flies. The fly lays eggs in the summer around the heels of cattle. When the eggs hatch as larva, they enter the skin and move through the body until they reach the back in a few months. In the back, the grubs (larva) cause bumps to swell in late winter and spring. The larva (about 3/4 of an inch long) will come out of the bumps and grow into an adult fly and repeat the cycle. Fly control will prevent grubs, which damage the hide by making holes in it and in the flesh around it.

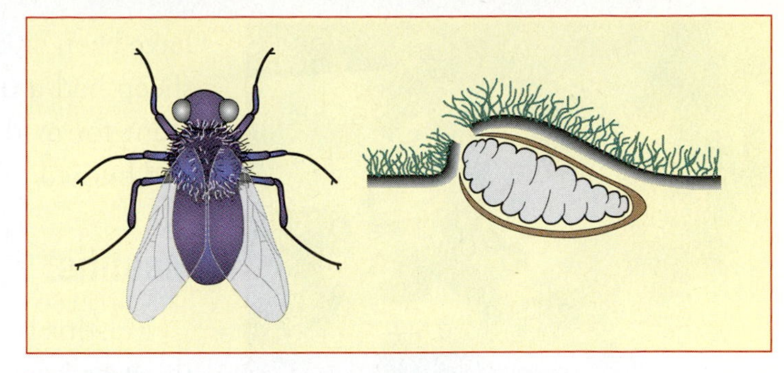

20–27. The cattle grub is the larva of a fly that lays eggs on the heels of cattle. After hatching and entering the body, the larva later shows up under the skin of the animal.

Leptospirosis

Leptospirosis is a bacterial disease that affects cattle, sheep, and most other farm animals, including dogs. Among the symptoms are high fever, poor appetite, bloody urine, and pregnant females that abort their fetuses. Without prompt treatment of the disease, liver and kidney damage may occur. Antibiotics are sometimes used to treat leptospirosis. Animals can be vaccinated against the disease. Infected animals, particularly dogs, should be isolated because leptospirosis can be spread through contact with the urine from infected animals. As stated earlier, humans can also get the disease. Signs in people are sudden onset of fever, headache, chills, red eyes, and severe muscle aches or cramps. Diagnosis by a physician is essential in gaining treatment.

Lice

Lice are external parasites of cattle, hogs, and other species. Lice are small insect-like pests that suck the blood of their host. As a result, animals (hosts) may become anemic. Animals with lice often scratch or rub against trees or posts because the bites itch. Lice usually appear in the winter. Back rubbers or other means can be used to administer pesticides on cattle to control lice.

Mad Cow Disease

Mad cow disease (bovine spongiform encephalopathy, or BSE) became a major threat in cattle production in Europe and Great Britain in the late 1990s. Isolated instances occurred in the United States in 2003 and 2004. Careful testing is needed to confirm the disease. Infected animals should never be used for food or as feed ingredients for other animals. Mad cow is a neurological disease evidenced by muscle twitching, aggression, nervousness, inability to stand, and other signs. It eventually causes death. Examination of dead animals infected with mad cow disease shows that the brain contains prions. The exact cause of the disease is unknown, but it appears to be related to prions. Outbreaks of mad cow disease

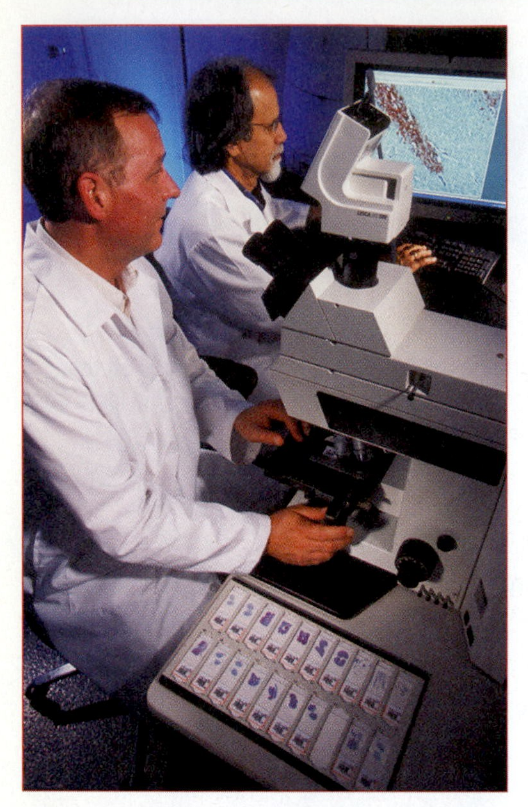

20–28. Veterinary medical scientists are using high-powered microscopic equipment to examine brain tissues of cattle infected with prions related to BSE. (Courtesy, Agricultural Research Service, USDA)

have been linked to feeding scraps of sheep meat when the sheep had a disease known as scrapie. There is no treatment for mad cow disease, so infected animals are killed and burned.

Mastitis

Mastitis is a bacterial disease of the udder. Milk may be thick or lumpy in cattle with chronic mastitis. A sign of acute mastitis is fever and a hard, swollen, and very warm udder that is sore. Milk from an infected udder is abnormal or watery in appearance and may contain flakes and clots. It is caused by bacteria entering through the canal and infecting the tissues inside the udder. Bacteria may produce a toxin that damages milk production cells. Diagnosis involves a stripcup test, California mastitis test, Wisconsin mastitis test, and electronic somatic cell counts. Acute mastitis that is untreated can result in death. Treatment is intramammary—by

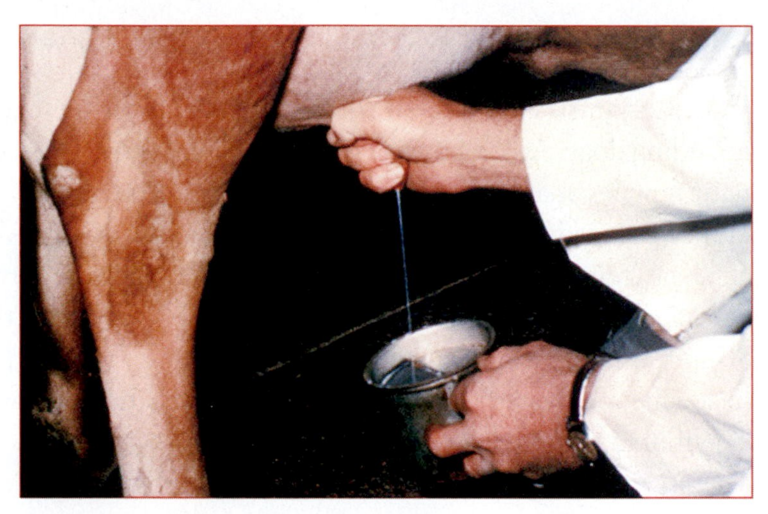

20–29. Strip cup tests are used to identify cows with mastitis. (Courtesy, Mississippi State University)

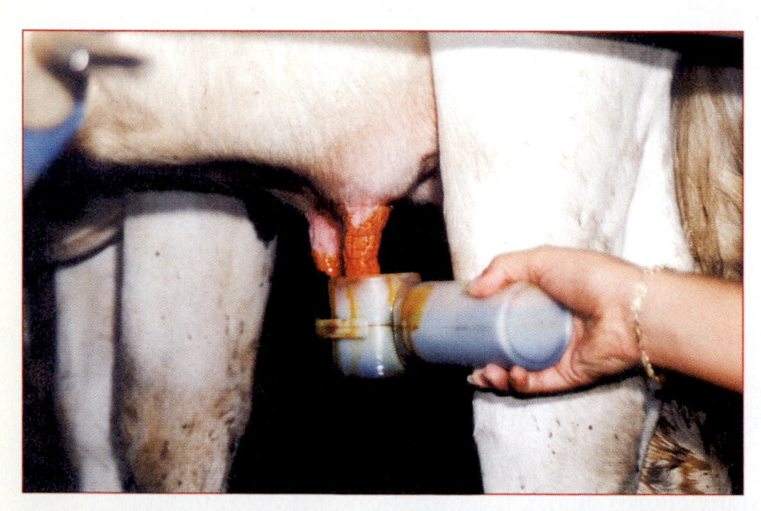

20–30. Dipping teats in an iodine solution after milking helps prevent mastitis and other diseases.

injecting medicines into the udder through the openings in the teats. Milk from infected cows cannot be used. This makes mastitis a serious concern for dairy producers. It affects female cattle, sheep, goats, and swine.

Newcastle Disease

Newcastle disease is an infectious disease in birds. Some forms of Newcastle are more severe than others. A form known as END (Exotic Newcastle Disease) caused major losses in the poultry industry in 2002, with some 3.5 million chickens and other poultry being euthanized. END is a fatal disease in birds caused by a virus that has attracted considerable attention. A vaccine has been developed that appears promising in preventing END. However, research on END and other forms of Newcastle is continuing.

20–31. A microbiology researcher is experimenting with a virosome vaccine to intranasally ("through the nostrils") vaccinate a chick against Newcastle disease. (Courtesy, Agricultural Research Service, USDA)

Porcine Reproductive and Respiratory Syndrome (PRRS)

PRRS is a disease of hogs. It primarily manifests itself as reproductive failure in sows (farrowing early with piglets that are stillborn, weak, or mummified) and as the development of pneumonia-like symptoms in unweaned piglets, resulting in high mortality. PRRS is caused by a virus that is spread by pig-to-pig contact of mucous, manure, urine, and semen. PRRS appears to have entered the United States in the late 1980s, because there are no known cases before 1986. It had been in Europe and other countries prior to the United States. The virus was first isolated in the Netherlands in 1991. Before its isolation, the disease was referred to as Mystery Swine Disease. The National Pork Board is leading the research effort to understand the disease better. A PRRS MLV (modified live virus) vaccine is available for pigs, gilts, and sows. It should not be given to boars. As with all medicated animals, pigs should not be harvested for human consumption before 21 days have passed following vaccination.

Rabies

Rabies is a highly infectious viral disease of the nervous system. All endothermic (warm-blooded) animals may get it. Rabies is transmitted in saliva, with animal bites being the major method of transmission. Dogs, cats, skunks, and other animals with rabies pose haz-

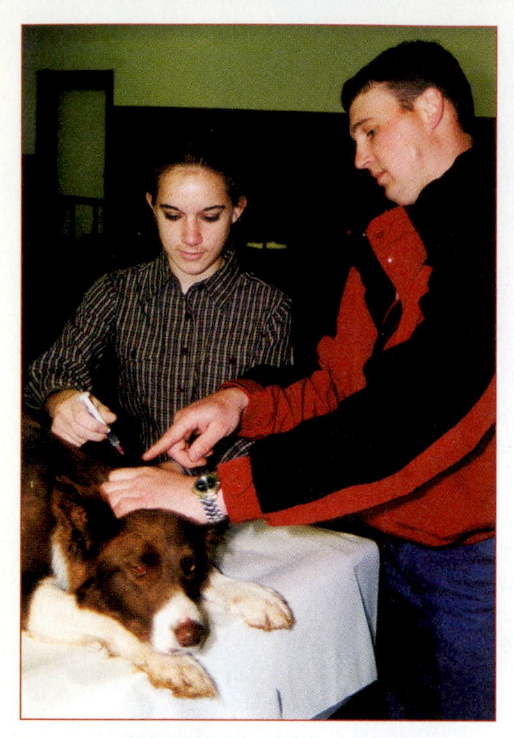

20–32. A student is learning to vaccinate a dog under the direction of an agriculture teacher. (Courtesy, Education Images)

ards to humans through biting. In humans, rabies can be fatal if not treated properly. Human treatments involve a fairly severe series of vaccinations. Because infected animals have a fear of water, rabies is sometimes known as hydrophobia. Infected animals wander about, attack without reason, frequently bark or call out, and develop paralysis before death. All dogs and cats should be vaccinated against rabies. Note: Capture and cage an animal suspected of having rabies. If necessary, kill the animal and quickly transport it to a veterinary laboratory for diagnosis. Humans who have been bitten or scratched by an animal that may have rabies should immediately seek medical attention and, if needed, begin treatment.

Roundworms and Tapeworms

Roundworms, tapeworms, and other worms are parasites of the stomach and intestines of animals. Tapeworms can grow several feet long. A worm this large uses a lot of feed that the animal has eaten. Yet animals with low infestations of small-size worms may not show symptoms. As the number and size of worms increase, the animal loses weight, becomes anemic, and may have diarrhea. Sanitation is a good control procedure. Treatments are available, but only approved products should be used.

Scours

Scours can be a problem with most animals and is especially a threat to young calves. The signs include diarrhea, weakness, and dehydration (loss of body fluid). Forms of scours vary depending on their cause, such as by bacteria or by a virus. Providing clean environments, feeding colostrum to newborn calves, and, in some cases, vaccinating can prevent scours. Infected calves can be given antibiotics and should have milk to overcome dehydration.

Shipping Fever

Cattle and sheep of all ages can get shipping fever, though it is a particular problem for younger animals. It is caused by conditions animals encounter when hauled or sold through a sale barn. Signs include high temperature, discharge from the eyes and nostrils, and cough. Most infected animals have difficulty breathing. It is more likely to afflict thin, underfed animals that are hauled long distances by truck. Cattle should be vaccinated three to four weeks before being moved as a precaution. Some cow and calf producers use a preconditioning program to help animals stay healthy while being shipped.

Swine Erysipelas

Swine erysipelas is an infectious bacterial disease. Its forms range from mild to quite harmful. Animals affected with the mild or chronic form may have arthritis and diamond-shaped lesions on the skin. The valves and lining of the heart may show inflammation. Severe forms of swine erysipelas appear as rapid death of 50- to 200-pound pigs. Affected pigs may have high temperatures of 104° to 108°F, have poor appetites, appear depressed, and be reluctant to move because of stiffness. Vaccination with killed forms (bacterins) and modified live forms of the bacteria can be used to prevent the disease by helping pigs develop immunity. Vaccination, however, is not 100 percent effective in preventing the disease.

20–33. Cows are sick from eating toxic alkaloids in tall fescue grass. (Courtesy, Agricultural Research Service, USDA)

Others

Many other diseases may affect animals. Companion animals, such as dogs and cats, have their own diseases. A preventive health care approach should be used. All diseases reduce productivity of animals.

HOW ANIMALS RESPOND TO DISEASE

The bodies of animals defend themselves against infectious diseases. Strong animals with immunity can usually avoid or overcome most diseases.

PRIMARY DEFENDERS

The ***primary defenders*** against disease are the skin and mucous membranes of the body. Skin forms a barrier against disease. If it is cut, disease can enter the wound.

20–34. Newly hatched chicks are automatically vaccinated at the hatchery before delivery to a broiler grower.

Mucous membranes that line the soft tissues at openings to the body secrete fluids that defend against disease. The stomach secretes gastric juices, the eyes have tears, and the urinary tract is washed out by the urine.

SECONDARY DEFENDERS

If a pathogen gets past the primary defenders, the body has ***secondary defenders*** against disease. These defenders may be natural responses or responses triggered by immunizations.

One form of secondary defender is the ***antibody***. Antibodies are proteins produced by organisms in response to an invasion of pathogens. The body may produce up to six different kinds of antibodies, depending on the disease.

Another kind of secondary defender is the leukocyte. A ***leukocyte*** is a white blood cell. There are two types of leukocytes: phagocyte and lymphocyte. A ***phagocyte*** is a cell that surrounds and destroys disease-causing microorganisms. Some phagocytes move around the body in the blood and are known as white blood cells. Wounds on the body often collect pus, which is white blood cells. Other phagocytes do not move about in the body. They attack pathogens that come past them in certain organs, such as the liver, spleen, and bone marrow. A ***lymphocyte*** is in the vertebrae immune system and is important in the immune system of an organism. Some are known as natural killer (NK) cells; others are known as T and B cells. NK cells distinguish between infected cells and uninfected (normal) cells. Upon activation, NK cells release cytotoxic granules that destroy altered cells. T and B cells are the major components of the adaptive immune response. Once T and B cells identify an invader, they unleash responses to eliminate it and pathogen-infected cells.

DISEASE CONTROL

Preventing a disease is far better than trying to treat it. Producers should focus on prevention, though they may need to treat infected animals.

IMMUNITY

Immunity is the ability of an animal to resist a disease. It is important in keeping animals healthy.

Natural immunity is an immunity that an animal develops without vaccination. Some animals will not get certain diseases; other animals get the disease and overcome it.

Acquired immunity is an immunity an animal develops by having a disease and developing a resistance to it. Acquired immunity may be permanent or temporary. Permanent immunity means that the animal will never get the disease again. Temporary immunity lasts for a short period; afterward, the animal may again be susceptible to the disease.

Producers can help animals develop immunity by using immunizing agents, also known as biologicals. Common kinds of immunizing agents are:

- **Vaccines**—Vaccines contain some form of the organism that causes the disease. The organism in the vaccine may be live, killed, or modified. The animal's body uses the vaccine to develop immunity. Animals develop immunity 7 to 21 days after vaccination. Some vaccines create permanent immunity; others protect for only a few weeks or months. Animals must be vaccinated again if the immunity lasts only a short time.

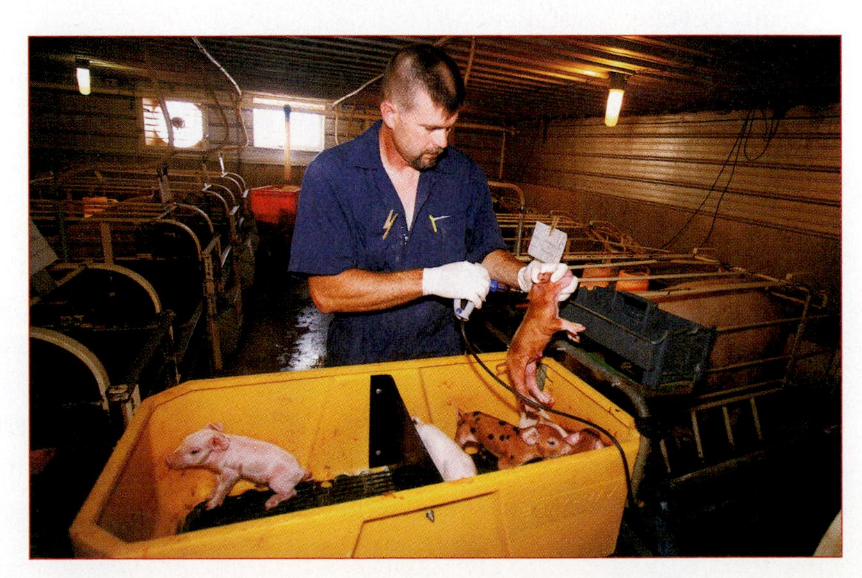

20–35. All day-old pigs in this hog farrowing facility are vaccinated to promote immunity. (Courtesy, Education Images)

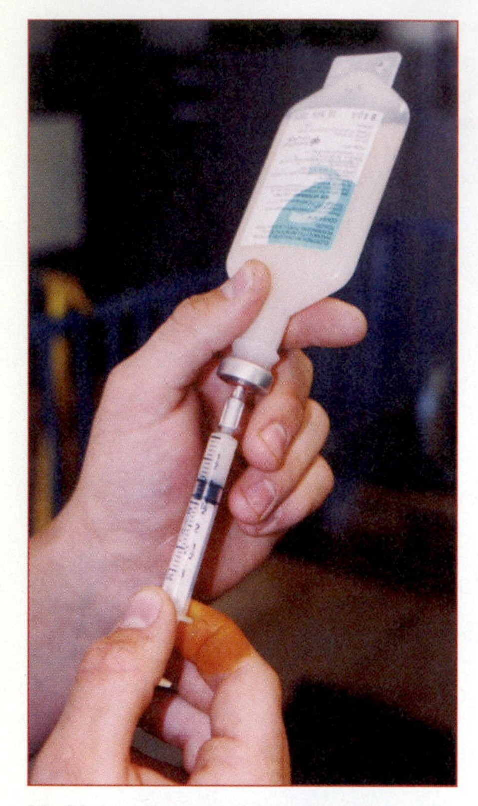

20–36. Loading a syringe with a dose of vaccine for blackleg, malignant edema, tetanus, enterotoxemia, and other diseases in sheep.

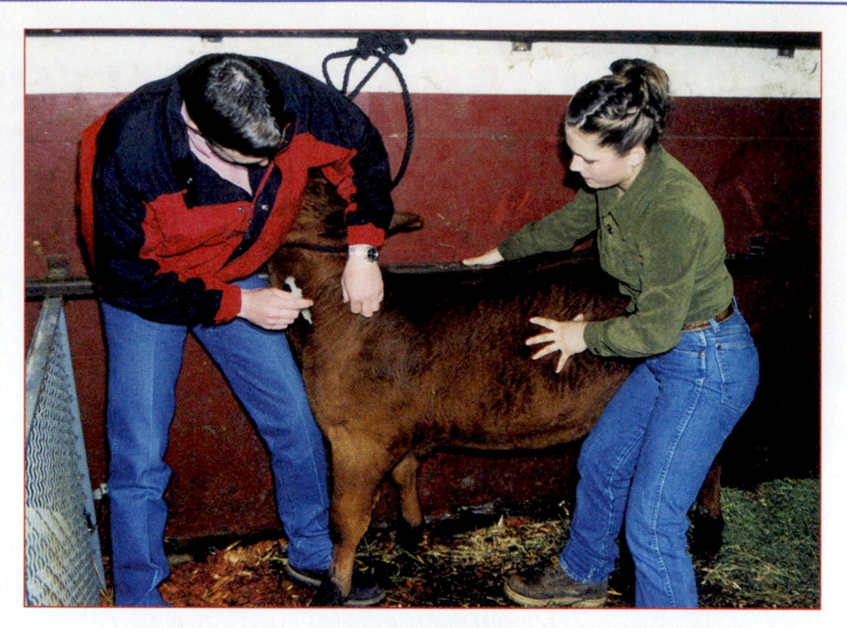

20–37. Vaccinating a calf. (Courtesy, Education Images)

- **Serums**—Serums are made from the fluid part of the blood that contains antibodies. Serums do not contain pathogenic organisms, as do vaccines. Serums are injected into the body and help the body's secondary defenders in their work.
- **Bacterins**—Bacterins are fluids that contain dead pathogenic organisms. They are sometimes known as suspensions. Animals are injected with the fluid and may gradually develop resistance.
- **Toxoids**—Toxoids are poisons that are no longer poisonous. They are produced by pathogens and altered to eliminate the poison effect.

DESTROYING PATHOGENS

Once the cause of a disease is present, various procedures and substances can be used to try to eliminate it. Sanitation, isolation, and other practices are effective in preventing diseases.

After an animal gets a disease, it may need help to overcome it. Consider the location in the body and the cause of the disease. For example, most internal parasites are not controlled with materials for use on external parasites.

Several ways of destroying pathogens are used.

Antibiotics

An ***antibiotic*** is a germ-killing substance made from bacteria and fungi. The best-known antibiotic is penicillin, made from a mold known as *Penicillium notatum*. Other well-known antibiotics are Terramycin and streptomycin. These medicines are useful in treating some diseases that do not respond to penicillin.

Most antibiotics are administered by injection. Some forms are added to feeds, especially when the animals are small and hard to inject. For example, fish and poultry may have antibiotics mixed with their feed.

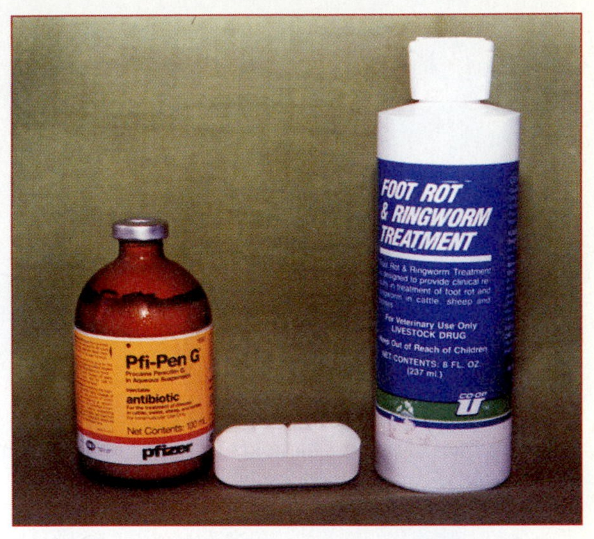

20–38. Forms of animal medicine are an antibiotic (left), which is injected with a hypodermic needle and a syringe into the bloodstream of an animal; a bolus (large pill) containing wormer (center), which is placed in the digestive tract of an animal; and a topical liquid (right), which is applied to diseased areas on the skin.

Pesticides

Pesticides include insecticides, which kill insects; miticides, which kill mites; and other poisons. Pesticides are used to kill external and internal parasites.

Dietary Supplements

Animals may be given vitamins, minerals, and sugar solutions, such as dextrose. The supplements help animals overcome anemia and provide energy. They are used with weak animals and treat diseases, such as acetonemia in cattle.

Immunizing Agents

Immunizing agents may also be used to treat disease. These agents include vaccines, serums, bacterins, and toxoids.

ADMINISTERING MEDICATIONS

Medications must get to where they can help the animal overcome the disease. The nature of the disease is a major factor in determining the appropriate medicine to use. Only use medications approved by the Food and Drug Administration.

- Systemics—A ***systemic medicine*** is a medicine that is absorbed into the bloodstream. The medicine is transported by the blood to all parts of an animal's body.

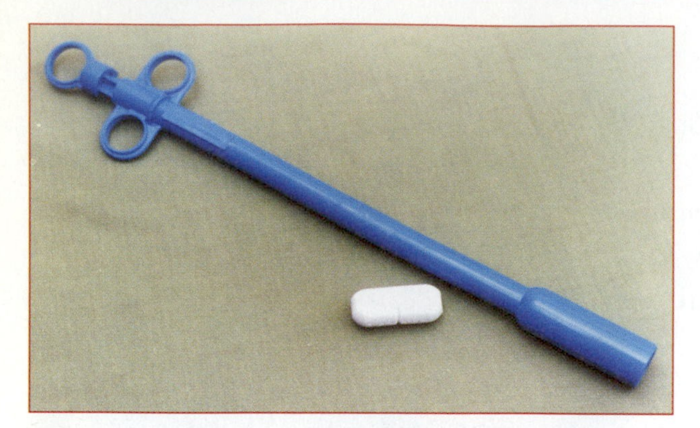

20–39. A balling gun is used to get the bolus into the digestive tract of cattle. The bolus is placed in the end of the gun and released when the gun is in the animal's throat.

Systemics may be poured on an animal, given by mouth or rectum, given by injection through the skin, or applied in other ways.

- **Topical applications**—These therapeutants are applied to the skin or outer surface of animals. Absorption into the body is very slow. In most cases, topicals are applied to control diseases on the skin.

A ***topical medicine*** is a medicine that is sprayed, dusted, or poured on an animal to control diseases or to treat wounds. Animals are sometimes dipped into big tubs or vats of pesticide solution. Backrubbers may be put in pastures and near places where cattle gather. The fabric in the backrubber may be saturated with an approved pesticide. Animals rub against the material and receive the pesticide on their bodies. Dipping and spraying are thought to be the most effective ways of applying topical medicines. Some topical medicines are in easy-to-apply aerosol cans.

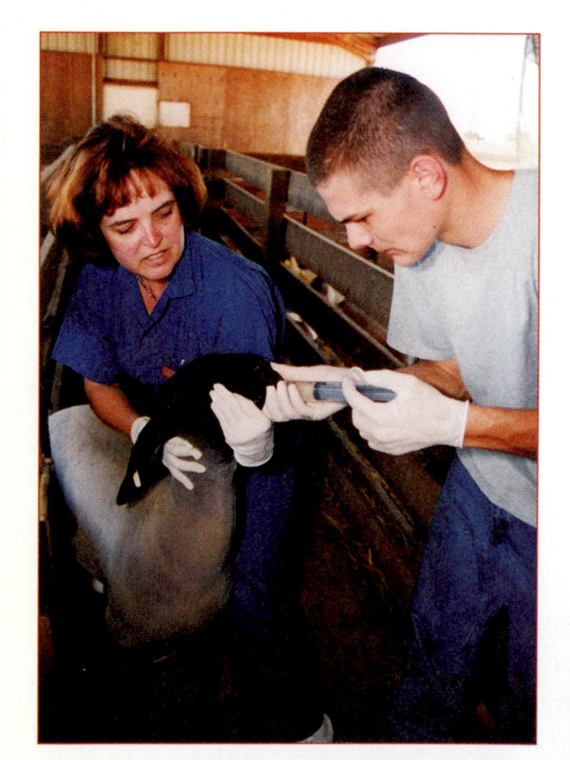

20–40. A dose syringe is being used to administer liquid medicine to a sheep.

- **Internal medicines**—Internal medicines are substances that are placed inside the animal's body. They may be added to feed or injected into the body. In some cases, they are systemic medicines. Sometimes they are given through the mouth (orally) as liquids or solids, such as a big tablet or bolus. Dose syringes or balling guns may be

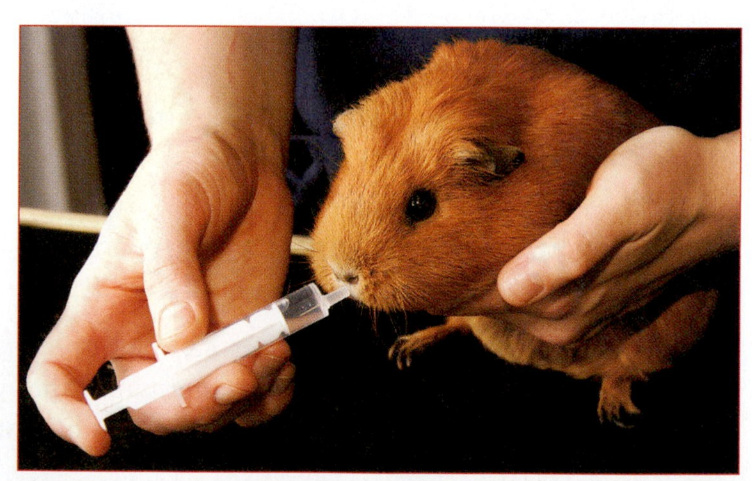

20–41. Oral medications can be given to a small animal, such as this guinea pig, with a syringe with the needle removed.

needed to give the medicine. These procedures may be used to treat parasitic or pathogenic diseases. For example, dose syringes are often used to administer liquids to control worms.

Kinds of Injections

A fast way of getting antibiotics and other substances into the body systems of an animal is with an injection. An ***injection*** is a method of administering medicines directly into the bloodstream by using a syringe with a hypodermic needle, or other device, to puncture the skin and dispense the medication. How injections are made depends on the animal and the kind of medication. Sterile techniques should be used with injections to prevent the spread of disease. Needles should be properly positioned when giving injections.

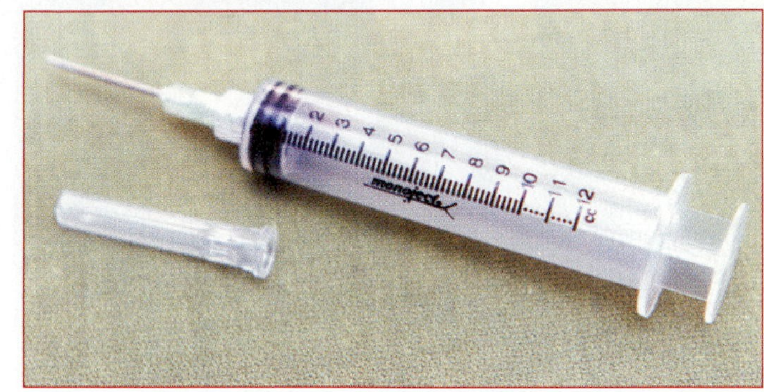

20–42. Injections are given with a syringe and a hypodermic needle. Note that the barrel of the syringe has been calibrated for ease in measuring the dosage.

Common kinds of injections are:

- **Intradermal injections**—These injections are administered just below the outer layer of skin or epidermis.
- **Subcutaneous injections**—These injections are administered just beneath the skin. They are given where the skin is somewhat loose and in easy-to-get-to places, such as the neck or behind the shoulder of beef cattle.

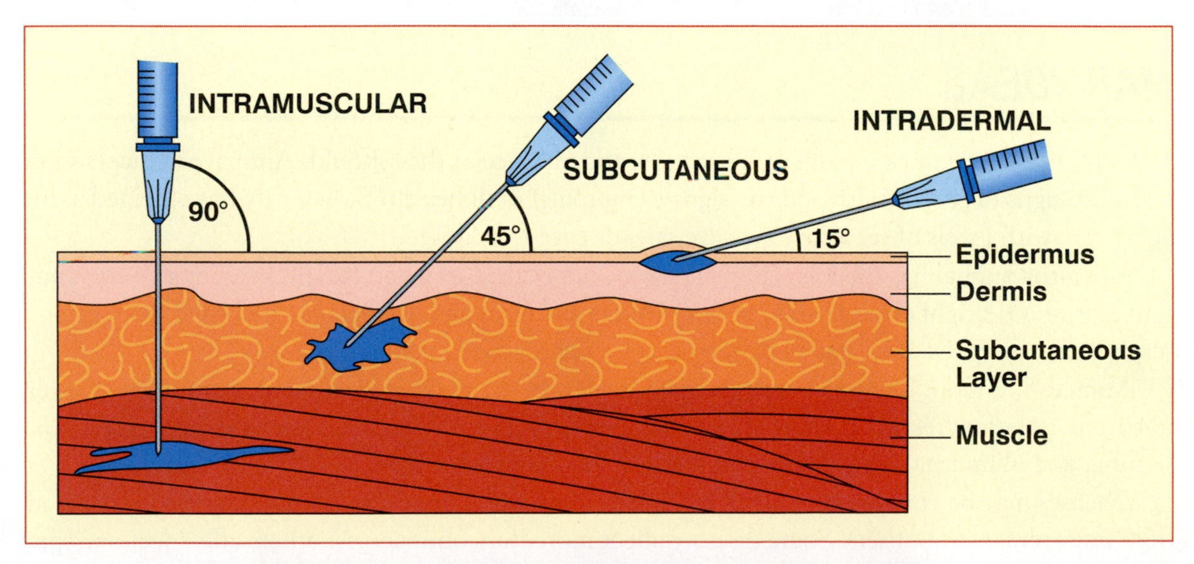

20–43. Sites of injections as related to skin and muscle structure.

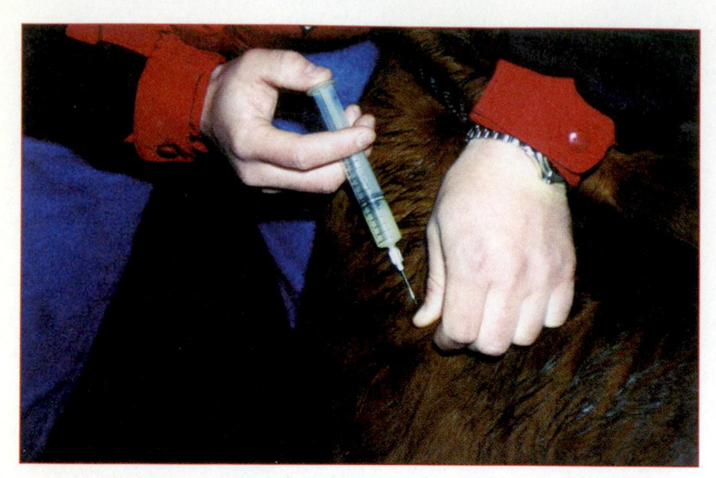

20–44. Holding a syringe to inject a calf in a fold of skin on its neck. (Courtesy, Education Images)

- **Intramuscular injections**—These injections are administered through the skin into the muscle. Penicillin is injected into the muscle of the hindquarters and shoulders of most animals.
- **Intravenous injections**—These injections are administered into the veins of an animal. Intravenous injections are used when a quick response is needed to save the life of an animal.
- **Intraperitoneal injections**—These injections are administered through the belly wall of cattle and a few other animals. Intraperitoneal injections are given in the hollow of the flank. They are used when a quick response is needed.
- **Intranasal injections**—These injections are administered through the nose of cattle and other species.
- **Intramammary injections**—These injections are administered to treat diseases in the udder, such as mastitis in cattle.

REVIEWING

MAIN IDEAS

Good health is a must if animals are to grow and produce as they should. Animal producers soon learn the signs of good health and the signs (symptoms) of ill health. Biosecurity has emerged as an approach, with levels of security to ensure safe, disease-free animals.

A healthy animal has the following: (1) a good appetite (eats when fed); (2) responsive and content nature; (3) bright eyes and a shiny coat; (4) normal feces and urine; and (5) normal vital signs—respiration rate, pulse rate, and body temperature appropriate for the species.

Animals are more likely to maintain good health in the right environment. Sanitation, proper nutrition, isolation, restriction of truck and equipment traffic, restriction of human access, preconditioning, and immunization are important in keeping animals healthy.

Diseases may be contagious (spread by direct or indirect contact) or noncontagious (not spread from one animal to another). Some diseases affect more than one species; others affect just one species.

Contagious diseases are caused by organisms that infect the animal—known as infectious disease—and usually produce poisons, or they live in (internal) or on (external) the animal.

Some diseases attack a certain body system. Of course, when one system does not work well, the entire animal is sick. Diseases may be chronic or acute. Chronic diseases often last a long time and are less severe. In contrast, acute diseases develop quickly and are more severe.

Animals respond to disease. Their primary defense is their skin and mucous membranes, which keep out pathogens. Their secondary defense is the ability to produce antibodies and phagocytes, especially white blood cells.

Prevention is far better than trying to cure the disease after animals become infected. Animals develop natural immunity and can be immunized.

Medications help animals fight off disease. Medicines need to get to where they can help the animal. Some medicines are systemics and are carried by the blood throughout the body. Others kill the causes of disease on contact. Various methods are used to administer medications including oral, rectal, topical, and injection.

QUESTIONS

Answer the following questions, using complete sentences and correct spelling.

1. Define health, disease, and biosecurity.
2. What are the general signs of a healthy animal?
3. How does the environment influence animal health?
4. What losses occur due to poor animal health?
5. What practices can be followed to help assure good animal health?
6. Distinguish between contagious and noncontagious disease.
7. What is an infectious disease? Pathogen?
8. How are diseases related to body systems?
9. What are some common diseases of animals? What animals can have each of these diseases? What are the symptoms? How are the diseases prevented and treated?
10. How do animals defend themselves against disease?
11. What is immunity? What are the common kinds of immunizing agents?
12. What may be used to help an animal overcome disease?
13. What kinds of medications may be used?
14. What are the kinds of injections? Describe how they are used.

EVALUATING

Match each term with its correct definition.

a. acquired immunity
b. antibiotic
c. infectious disease
d. immunity
e. sanitation
f. isolation
g. vital sign
h. pulse rate
i. health
j. disease

_____1. An indication of the living condition; breathing, pulse rate, body temperature.

_____2. An immunity an animal develops by having a disease and developing immunity to the disease.

_____3. A condition of pain, an injury, or the inability to function normally.

_____4. The movement of the arteries as a result of the heartbeat.

_____5. The condition of the body and a measure of how well the functions of life are being performed.

_____6. The condition of an animal being resistant to a disease.

_____7. A disease caused by organisms that get inside an animal's body.

_____8. The act of separating diseased and non-diseased animals.

_____9. The practice of keeping areas where animals are being raised clean.

_____10. A germ-killing substance made from bacteria and fungi.

EXPLORING

1. Investigate the meaning and importance of biosecurity. An international Web site for beginning your investigation is the Food and Agriculture Organization of the United Nations at **www.fao.org/biosecurity**. Prepare a report on your findings.

2. Tour a veterinary medical clinic. Observe the equipment used, procedures followed, and the treatment of an animal. Write a report on what you saw.

3. Interview a local animal producer. Determine the health problems that have existed and how they were controlled. Determine if the producer participates in a quality-assurance program and, if so, how following the guidelines relates to improved animal health. Prepare a written report on your findings.

4. Check the vital signs of an animal. Record its pulse, respiration rates, and rectal temperature. Compare what you found with what is normal for the species.

5. Observe an animal that is in good health. List the signs of good health. Compare your list with those in a book or with the opinion of a person experienced with the species.

Part Six

Earth Science and Technology

CHAPTER 21

Earth Science

OBJECTIVES

This chapter presents basic information about earth science as it relates to sustainable practices in producing plants and animals. It has the following objectives:

1. Describe the major features of the earth.
2. Explain changes that occur in the earth.
3. Describe atmosphere and its importance in agriscience.
4. Identify and describe the major factors in weather.
5. Explain the importance of climate in agriscience.
6. Explain succession and its relationship to the earth.

TERMS

air pressure
atmosphere
barometer
bathymetry
biotic community
chemical weathering
climate
climate zone
cloud
continental crust
core
crust
dew
dew point
equinox
front
galaxy
humidity
mantle
map
mechanical weathering
polar climate zone
precipitation
primary succession
relative humidity
revolution
rotation
secondary succession
solar system
stratosphere
succession
temperate climate zone
thermometer
topography
tropical climate zone
troposphere
universe
weather
weathering
wind

21–1. Scientists are drawing soil water samples in studying the movement of bacteria in soil. (Courtesy, Agricultural Research Service, USDA)

ALL LIVING THINGS gain support for life from Earth's resources. Though Earth may seem large, space and resources are limited. Living and nonliving things on Earth must share and exist within these limits. All living things must exist together and in harmony with human life.

Plants, animals, and other organisms depend on the earth for life-supporting nutrients, with most being relatively scarce. The available resources are used to support life processes. Plants and animals respond differently in obtaining and in using resources. For example, plants use nutrients and photosynthesis to make sugar and other substances. Animals eat plants for their food. When their life cycles have been completed, both return nutrients to the vast nutrient storehouse on Earth. New plants and animals grow to complete the life cycle.

As mentioned, space and resources on the earth are limited. Yes, the earth is big—25,000 miles (40,074 km) around at the equator! However, many people need food, fiber, and shelter. Using resources wisely to sustain life indefinitely is essential.

EARTH'S FEATURES

Earth is part of the universe. The **universe** is all that exists in space—both known and unknown. Of course, recent exploration has lead scientists to learn more about the universe.

Within the universe are galaxies. A **galaxy** is a system of stars, dust, gases, and other matter held together by gravitational forces. Our solar system is in the Milky Way galaxy, which contains more than 100 billion stars. A **solar system** is a star and a group of large bodies, called planets, that revolve around it. The star at the center of our solar system of nine planets is the Sun. A moon is a body that revolves around a planet. There are more than 40 moons in our solar system. Although Earth has only one moon, some planets have several.

21–2. Natural resources include features of Earth's surface.

Earth is unique among planets in its solar system. It is the only planet known to support life. On Earth, life as we know it requires water and oxygen. Fortunately, the earth has liquid water and an abundance of oxygen. All the earth's energy is from the Sun. The energy can be stored, as in oil, or can be used daily for life processes, such as photosynthesis.

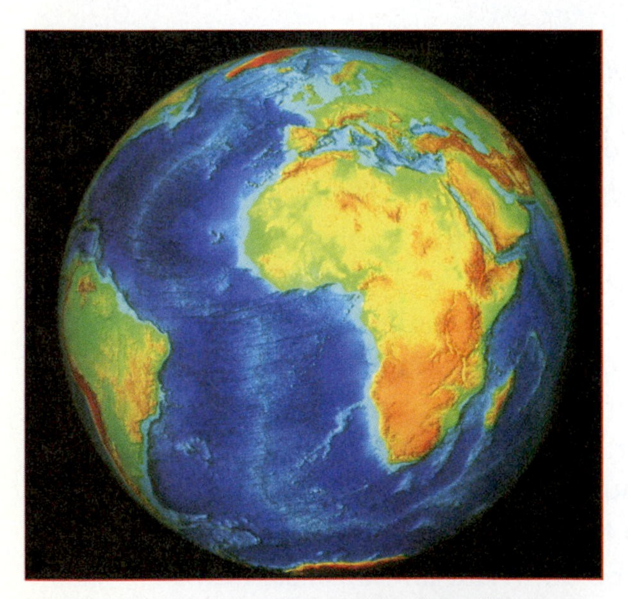

21–3. A computer-generated satellite view of Earth, showing water amounts in the Atlantic Ocean and the African Continent. (Courtesy, National Geophysical Data Center)

EARTH'S STRUCTURE

Earth can be viewed as a large, almost-round object with three major parts: a surface, an interior, and an atmosphere. Life on Earth uses resources from all three.

The Surface

The surface of the earth has mountains, valleys, plains, streams, lakes, and oceans. Combined, these are known as the earth's **crust**, which is the outer layer of Earth's surface.

Some of the crust is above water; oceans cover other parts of the crust. The part of the earth's crust that is land is known as ***continental crust***. This is the area where crops are grown, animals are raised, and humans live. Some continental crust may be frozen year round, known as permafrost.

Drivers, hikers, and others often use maps to find their way to locations. A ***map*** is a drawing, photograph, or other representation of the features of the crust. Maps may show elevations, the presence of water, and other natural features. ***Bathymetry*** is the measurement of the crust beneath large bodies of water, such as oceans. It includes the depth of the water. ***Topography*** is the surface configuration of land areas. It includes elevations, streams, and other features.

21–4. Contour strip cropping has been used to sustain productivity of Earth's crust. (Courtesy, Natural Resources Conservation Service, USDA)

The crust beneath the ocean is the oceanic crust. Of course, water supports aquatic animal and plant life, such as fish and seaweed. As a result, some aquatic plants grow attached to the oceanic crust.

The continental crust consists of inorganic and organic materials. Organic materials include remains of plants, animals, and other organisms. Carbon is an element present in all organic materials, such as coal. Many organic materials are found in the soil.

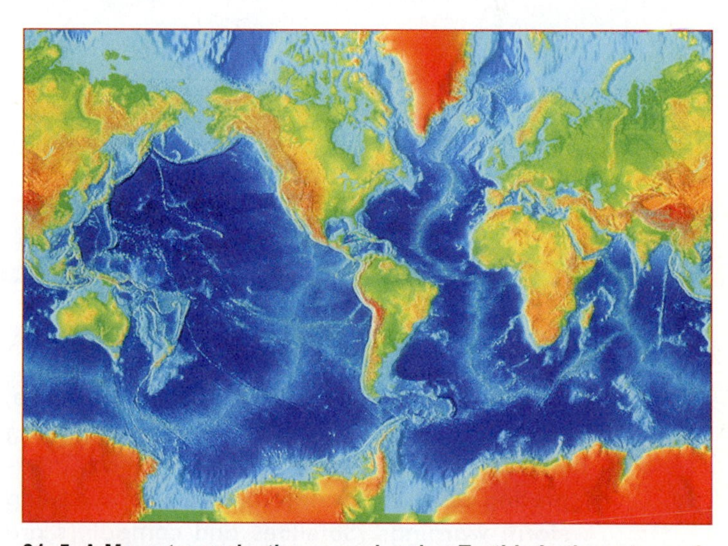

21–5. A Mercator projection map showing Earth's bathymetry and topography. Colors show water amounts, with dark blue areas having much water and red very little or no water. (Courtesy, National Geophysical Data Center)

More than 2,000 kinds of minerals are in the crust. Fewer than 20 minerals compose most of the crust. Minerals are typically thought of as rocks or rocklike substances. The continental crust varies from 12.4 to 43.5 miles (20 to 70 km) thick. (The crust is thinner under the deep oceans.)

The surface has water that may be in liquid, solid (frozen water or ice), or gas (vapor or steam) form. Liquid water is found in lakes and streams on the surface, as molecules in the

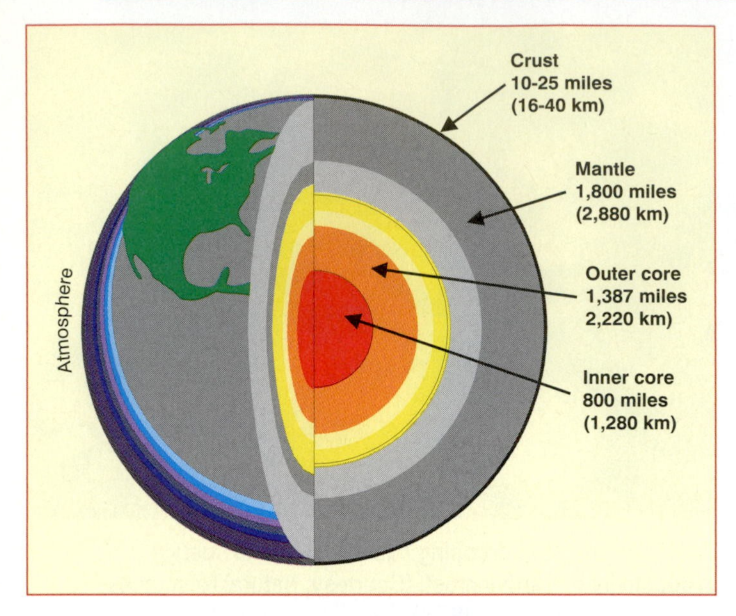

21–6. Structure of Earth's interior.

soil, or as aquifers deep in the earth. Frozen water is found in higher elevations and in colder climates, and gaseous water is found in the atmosphere.

The Interior

As a large, nearly round ball, the earth averages 7,908 miles (12,735 km) in diameter. The earth's center has a hot ***core***—7,772°F (4,300°C)—made of dense materials, such as iron. The materials become less dense near the crust.

The layer between the earth's core and crust is known as the ***mantle***. This also is a hot area, with some molten (melted) rock. The temperatures are as high as 2,030°F (1,110°C).

CAREER PROFILE

METEOROLOGIST

A meteorologist studies the atmosphere, weather, and climate. Work may involve using weather stations, collecting information, interpreting data, preparing forecasts, and issuing storm warnings. Satellites and computers are also used in gathering and reporting information.

Meteorologists need a college degree in meteorology, climatology, or a related area. Many meteorologists have bachelor's degrees. However, a master's degree or doctorate is typically needed for research positions. Some complete government training in weather forecasting.

Jobs for meteorologists are with the National Weather Service, colleges, research stations, and the media, including television stations. This photo shows a weather station being installed. (Courtesy, Agricultural Research Service, USDA)

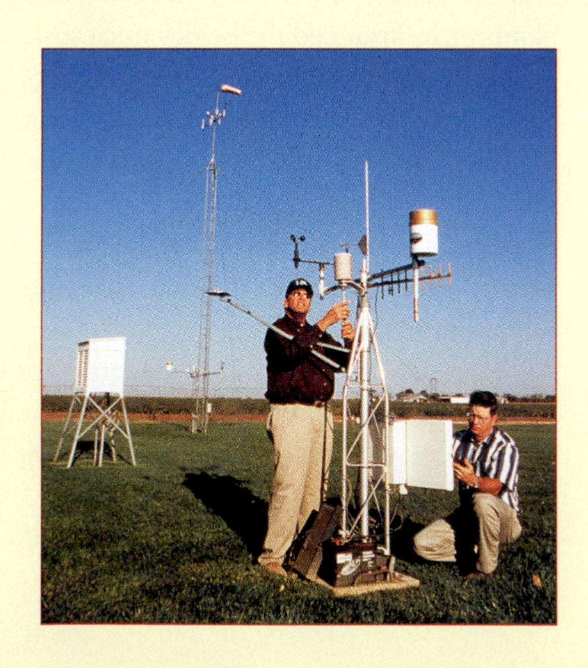

The Atmosphere

Atmosphere is the air that surrounds the earth. It is a collection of invisible gases, water vapor, and solid particles (particulate). Oxygen and nitrogen are most abundant near the earth's surface. Nitrogen composes slightly more than 78 percent of the air, with oxygen making up nearly 21 percent. The remaining gases are in small amounts, such as carbon dioxide at 0.03 percent.

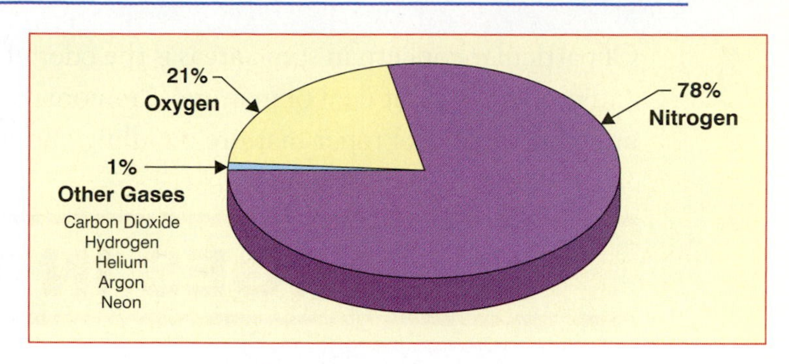

21–7. Gas composition of the air.

Gravity holds gases near Earth. Few gases are in the atmosphere above 30 miles from the earth's surface.

Amounts of water and solids in the air vary with the surface composition beneath the atmosphere and with the emissions from factories, homes, and vehicles. Water vapor is greater over a tropical forest than over a desert. The amount of solid particles, such as dust and smoke, is greater above cities and dry cropland. These particles pollute the air and make it less desirable.

People have become increasingly concerned about air pollution because it causes the contents of the air to be altered from the natural proportions of gases. Polluted air may be unhealthy for living things. In recent years, concern has increased about the "greenhouse" gases that are released by the use of fossil fuels, industrial activities, and daily living. The greenhouse gases alter the atmosphere. They absorb and emit radiation, which creates the greenhouse effect (global warming).

Air pollution is from many sources, such as engines, heating systems, electrical generating facilities, and factories that release steam, smoke, and other substances into the air. Agricultural activities also create air pollution. Examples of agricultural pollutants are dust from plowing fields and harvesting crops, pesticides applied to crops, and wastes from mills that process products.

21–8. Wastes from a hog production facility go into a lagoon waste management system. (Courtesy, Natural Resources Conservation Service, USDA)

Of particular concern in some areas is the odor of manure. Livestock and poultry operations can produce a great deal of manure. Ammonia and hydrogen sulfide are the two odors that are most obvious. Proper manure handling and disposal help minimize the odors.

CHANGES IN THE EARTH

21–9. Movement and features of Earth can be studied using a globe.

Earth moves and changes. These movements cause day and night and the seasons of the year—both important to living things.

MOVEMENT IN THE SOLAR SYSTEM

Earth moves in several ways. It changes its position in the solar system and undergoes changes in its composition.

Day and Night—Rotations

Rotation is the turning or spinning movement of Earth. One complete rotation (turn) is made every 24 hours—one day.

Almost everyone is familiar with a globe or model of the earth. Locations on the globe include the North and South Poles. An invisible axis extends from each of these poles, and the earth rotates on this axis. A good way to see how this works is to stick two pencils into an orange at opposite sides and turn the pencils between your fingers. The orange rotates!

Seasons—Revolutions

Revolution is the movement of Earth in space around the Sun. A revolution requires one year (365.24 days). Seasons are based on the position of the earth in a revolution.

The equator (an imaginary circle around the middle point of the earth) divides the earth into two parts: Northern Hemisphere and Southern Hemisphere. When a hemisphere is tilted toward the sun, the season in that region is summer. Seasons are reversed in the hemispheres, meaning that summer in the Northern Hemisphere is winter in the Southern

Hemisphere and vice versa. When the sun is directly over the equator, day and night are of equal length; this time is known as **equinox**.

Spring begins in the Northern Hemisphere on the day the North Pole begins to slant toward the sun. This is the vernal equinox, usually March 20 or 21.

Fall (autumn) begins in the Northern Hemisphere when the earth tilts away from the sun and the Southern Hemisphere tilts toward the sun. This is the autumnal equinox and is usually September 21.

21–10. Seasons influence plant and animal growth. In the fall, the leaves of some trees turn appealing shades of red, orange, and yellow.

Importance of Revolutions and Rotations

Earth's movement regulates life processes. Light from the sun affects plants and animals because they use the sun's energy.

Day Length. Hours of daylight vary. In the Northern Hemisphere, days are longer in the summer and shorter in the winter. Day length affects living organisms.

Plants must have light for photosynthesis, and longer days provide more light for the process to occur. This allows them to grow and store food. As the days become shorter, plants carry out less photosynthesis. Day length also affects flowering, fruiting, leaf color, and other plant conditions.

Animal life processes occur more rapidly in light. Darkness is associated with resting. Poultry producers are well aware of the importance of light in helping birds grow rapidly and lay more eggs.

Seasons. Plant and animal life cycles are related to seasons of the year. Some prefer warm weather; others prefer cool weather.

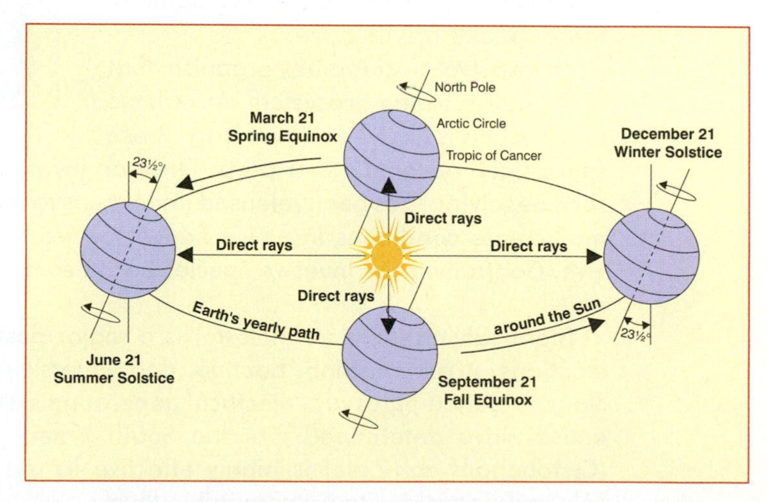

21–11. Seasons result from how the earth revolves around the sun.

The major growth of warm-season plants, such as corn, occurs in the spring and summer. Warm-season plants stop growing in late fall and winter.

Cool-season crops, such as winter wheat, prefer the shorter days of winter. Cool-season plants may stop growing and die when the days become longer and the weather becomes warmer. Winter wheat produces a seed head (grain), dies, and is harvested in late spring.

Animal production is based on seasons of the year. Cows are bred to give birth in the late winter or in the spring. The reproductive cycles of some animals are related to seasons and hours of light. Baby animals are not stressed by cold weather when born at warmer times of the year.

INTERNAL MOVEMENT

The earth's exterior features are a result of changes in its interior, which create movement, and weathering. Mountains, valleys, and plains were caused by movement and weathering.

Internal layers of rock in the earth heat and cool. As they warm, they shift positions and sometimes become molten. The molten rock pushes upward through cracks and rock layers to form volcanoes. These forces cause the surface to change.

Technology Connection

INVASIVE SPECIES

Sometimes organisms that are not native to the United States get here. They are known as nonnative species. Some of these species are invasive.

An invasive species is any organism that is not native to the ecosystem and whose introduction is likely to cause harm. These species can become major pests. Once an invasive species, such as salvina, has been released and begins to take over, it may create conditions in which native species cannot survive. Controlling the invasive species becomes a high priority.

Giant salvina (*Salvinia molesta*) is a major pest in some locations. It ruins fishing, boating, and waterskiing. It also clogs irrigation pipes and electrical generating systems. Scientists have determined that the South American weevil (*Cyrtobagous salviniae*) is highly effective in reducing the salvina infestations to an acceptable level.

This image shows the infestation of the salvina pest in a tank used for research. The inset is of the South American weevil, whose introduction is a potential approach to controlling pesky salvina. (Courtesy, Agricultural Research Service, USDA)

Rocks sometimes break under great pressure. This produces fractures (breaks) and faults. A fault is the continuing movement of the surface after being broken. Movement may occur only occasionally and may not be noticed. Major movements result in earthquakes that shake the earth's surface.

Earthquakes are caused by the sudden shifting and breaking of sections of Earth's interior. Some earthquakes are powerful and are felt for many miles. Others are less powerful. The power of an earthquake is related to the amount of shifting that occurs. Most earthquakes occur along natural fractures or breaks deep under the surface. An earthquake produces seismic waves that move out from the focus or site where the earthquake begins. Usually the focus of an earthquake is less than 45 miles deep.

Earthquakes themselves rarely take human life. Death and injury result from the side effects of the earthquakes, such as falling buildings and other structures. An earthquake under the ocean creates a tsunami, which is a large wave created by the movement associated with the earthquake. As the wave approaches land, it creates a large wall of water that can travel inland far beyond usual tide and wave levels. This wall results in tremendous property loss and may claim many human lives. (Wildlife animals typically escape these hazards because of their ability to sense a change in conditions.) One example is the tsunami that occurred in December 2004 in the Indian Ocean. Likely the deadliest natural disaster in modern times, it claimed an estimated 310,000 human lives in several countries bordering the ocean. The exact number will likely never be known, as many bodies were swept out into the ocean as the tsunami subsided. The strongest U.S. tsunami in the past 100 years was in 1964 in Alaska.

21–12. Drought may cause lakes to dry up and streams to have little water flow. (This lake contains very little water. The bottom soil has dried and cracked.) (Courtesy, Natural Resources Conservation Service, USDA)

WEATHERING

Weathering is the effect of weather on the rocks and minerals on the crust of the earth. Rocks break into smaller pieces when exposed to water, air, and temperature changes. Weathering is a major factor in the formation of soil.

Chemical weathering is the alteration of rocks on the crust of the earth as a result of chemical reactions. Water, oxygen, and carbon dioxide are often involved. Oxygen may join with minerals to form oxides, like iron oxide (rust). Water acts as a solvent to dissolve

21–13. Water flow in the Soque River gradually weathers rock. (Courtesy, Education Images)

minerals from the rocks. Carbonic acid can be formed by carbon dioxide dissolving in water. This acid causes more rapid weathering. Other substances in the water may speed decomposition.

Mechanical weathering is the breaking of larger rocks into smaller ones. The rocks become smaller and smaller until they are the size of soil particles. Rocks break apart when water around them freezes. They also break apart after falling or bumping around in flowing water.

Weathering is a slow process. Thousands of years may be required for rocks to become particles of soil. However, acid rain, sudden temperature changes, and other conditions may cause faster weathering. Heavy rainfall speeds weathering by increasing water flow over rocks.

Rocks of different minerals weather at different rates. Limestone is relatively soft and weathers more quickly. Granite is a hard rock that weathers slowly.

THE ATMOSPHERE

The atmosphere has a definite structure. Events in the atmosphere affect living and nonliving things on Earth.

The atmosphere has five layers. The two most important on Earth are:

- **Troposphere**—The atmospheric layer closest to the earth is the ***troposphere***. It extends 4.4 to 10 miles (7 to 16 km) out from the earth (generally thought of as up).
- **Stratosphere**—The ***stratosphere*** is the layer above the troposphere that extends out about 30 miles (50 km) from the earth. As the distance from the earth increases, the temperature cools. For example, the temperature in the stratosphere averages about –67°F (–55°C).

Atmospheric layers help life on Earth by protecting it from radiation. For example, ozone is the outer part of the stratosphere. It protects Earth from the ultraviolet rays of the sun. Life could not exist without ozone to absorb ultraviolet rays. Unfortunately, the ozone layer is being destroyed by pollution from life on Earth.

WEATHER

Weather is the general condition of the atmosphere as it pertains to temperature, air pressure, wind, moisture, clouds, and precipitation. These factors affect life on Earth. Weather information is very useful in agriculture. Forecasts help make decisions about when to plant and when to harvest, particularly hay, which cures best in low-humidity sunlight.

21–14. Newspapers carry detailed information about the weather.

TEMPERATURE

Temperature is a measure of the heat energy of molecules as they move. Gas molecules in the air are continually moving. The faster they move, the higher the temperature. Molecules move faster in sunlight because energy is being transferred to them.

Temperature is measured with a ***thermometer***. Different kinds of thermometers are used. Most have a small sealed bulb at the bottom of a tube that contains an alcohol solution. In the past, thermometers contained mercury, but its use has been phased out because of possible hazards if the thermometer were to be broken. Temperature is measured using several scales with degrees Celsius (°C) and Fahrenheit (°F) being most common.

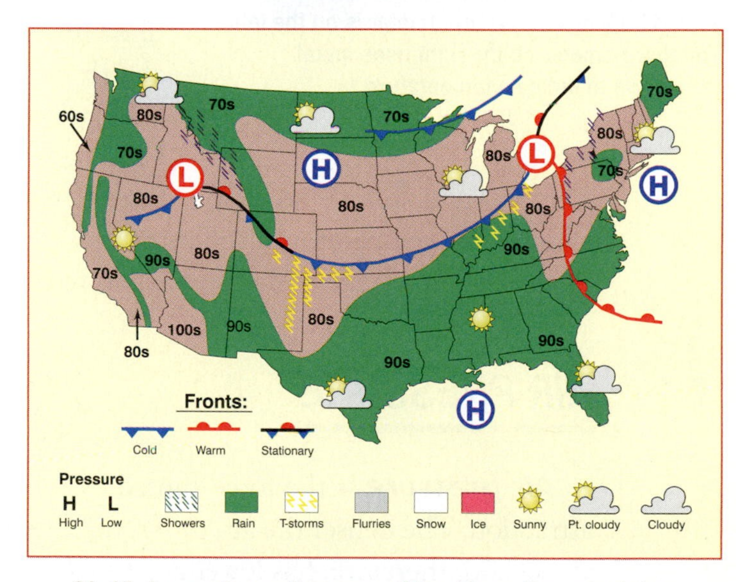

21–15. A sample weather map of the 48 contiguous states.

Radiation heats the earth and its atmosphere. It is the energy of the sun that travels as waves. The earth can also absorb and transfer energy to other objects, including the air. Three methods are used to transfer heat: radiation, conduction, and convection. Conduction is the transfer of heat from one object to another when the objects are touching. Con-

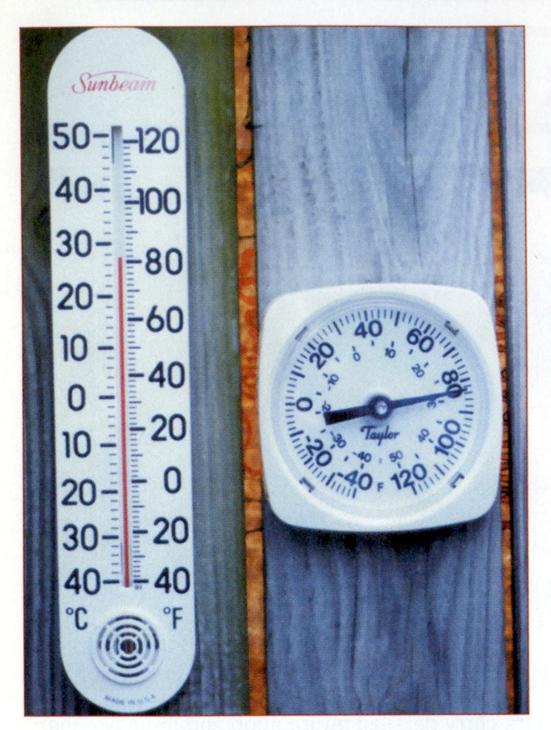

21–16. Thermometers measure temperature. A sealed bulb mercury thermometer is on the left. The thermometer on the right uses metal expansion to indicate temperature.

21–17. A soil thermometer with a shaft 4 inches (10 cm) long that is pushed into the soil to measure the soil temperature.

vection is the uneven transfer of heat by a carrier, such as when air moves about the earth creating wind currents.

Temperature varies on Earth. The portion of the earth tilted toward the sun is warmer. Also, different materials absorb heat differently. Dark colors absorb heat, while light colors reflect heat. In warm climates, the roofs of buildings are often painted white or silver to reflect heat.

Plants and animals are adapted to different temperatures. Some plants and animals can tolerate low temperatures; others cannot. Selecting which plants and animals to produce requires a knowledge of their temperature requirements.

AIR PRESSURE

Air pressure is the force caused by the weight of air due to the earth's gravitational attraction. The denser the air, the higher the air pressure. The top of a mountain has less air above and, therefore, has lower air pressure.

Air pressure is measured with a **barometer**. Four types are used: water-based, mercury, aneroid, and barograph (which plot barometric information on paper). A typical mercury-based barometer has a glass tube sealed at one end; the other end rests in mercury (or a similar liquid). As the air pressure goes up, the mercury rises higher in the tube because the air pressure outside the tube is greater than inside. Air pressure is measured in pascals (Pa), which are units of the metric system, or pounds per square inch (psi), which are units of the English system. Average air pressure at sea level is 101,300 Pa. Equivalent units of air are

760 mm Hg (mercury) = 101,300 Pa = 1 atm (atmosphere). Measures on a typical barometer range from 28 to 31 inches. (Metric measures are also used.)

21–18. A barometer with its needle pointing to a barometric pressure of 30, indicating fair weather.

Aneroid barometers are most commonly found today. They operate on the basis of a flexible metal box called an aneroid cell. Changes in barometric pressure result in the cell expanding or contracting. Tiny movements are amplified and displayed on the face of an aneroid barometer. The measures are typically reported in inches, just as with water-based and mercury systems.

Barometric pressure is used to determine elevations (distances above sea level) and changes in the weather. As elevation increases, such as when climbing a mountain or flying an airplane, barometric pressure decreases.

Variations in air pressure help in forecasting weather. Lower pressure means that the weather is humid, clouds are forming, and rain is likely. Increasing barometric pressure indicates clearing weather. Low pressure usually comes before a storm that could damage crops and livestock.

Areas of high and low air pressure are used to forecast the weather. Maps on television and in newspapers show the locations of highs and lows. As fronts move across North America, weather forecasts are made for areas ahead of and behind the pressure systems.

21–19. Farm land has been flooded along a small river due to recent heavy rains. (Courtesy, Natural Resources Conservation Service, USDA)

Information about barometric pressure is used to make decisions about plants and animals. For example, hay should not be cut if the barometric pressure is going down because rain is likely. Rain will damage hay before it has cured.

Air pressure influences the boiling point of water. At sea level, water boils

ACADEMIC CONNECTION

ENVIRONMENTAL SCIENCE

Relate your study of earth science in agriculture to the study of environmental science. Explore the important role of protecting natural resources as a part of environmental science. Use your learning in agriscience to support your mastery in environmental science.

at 212°F (100°C). It boils at lower temperatures as the elevation (altitude) increases. Water on high mountains may boil below 200°F! Cooking and keeping food safe requires longer boiling to kill germs.

WIND

Wind is the movement of air. Differences in air pressure and heat cause wind. The wind is measured in two ways: direction and speed. Wind direction is measured with a wind vane; wind speed is measured with an anemometer.

There are three types of winds: local, global, and jet stream.

Local winds are caused by differences in temperatures between land and water and elevations in the land. Land warms faster than water, so breezes often blow from bodies of water toward the land. This is more likely to happen during the day. At night, land cools faster than water, so the wind may blow from the land to the ocean on the coastline.

21–20. Wind speed is measured with an anemometer (left) and direction with a wind vane (right).

Global winds affect large areas of the earth. These winds result from rotations of the earth and differences in temperature on the earth. Global winds cause conditions that lead to major storms, such as hurricanes.

Jet streams are high-velocity winds at the top of the troposphere. Jet streams are often shown on weather maps so forecasts can be made for several days. In the Northern Hemisphere, when the jet stream moves down toward the equator, colder weather can be expected.

HUMIDITY

Humidity is the amount of moisture in the air. The moisture is in vapor form, known as water vapor. Measuring humidity requires knowing the temperature. Warmer air will hold more water vapor.

The amount of water in the air compared with the air's ability to hold water vapor is known as ***relative humidity***, and it is stated as a percentage. With higher humidity, the water vapor in the air is nearing 100 percent. Rain or other precipitation occurs at 100 percent humidity. As air cools, its capacity to hold water decreases, and rain is likely. This is the reason rain or snow comes before cooler weather.

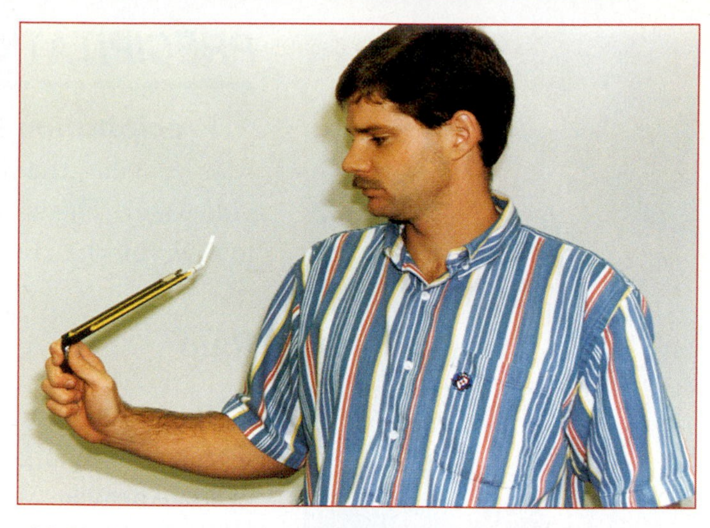

21–21. Using a sling psychrometer to assess relative humidity.

A ***cloud*** is a mass of condensed water vapor, droplets, and pieces of ice floating in the air. Some clouds have more moisture than others. Stratus clouds create overcast weather, block out sunlight, and bring rain. Cirrus and cumulus clouds generally indicate fair weather. Cirrus clouds occur at high altitudes and appear wispy. Cumulus clouds are fluffy clouds that are found at all altitudes. Cumulonimbus clouds are often called thunder heads. They reach great heights and extend 10 to 12 miles from base to top. Cumulonimbus clouds produce heavy rain, lightning, thunder, and, sometimes, a tornado.

21–22. Examples of clouds.

21–23. Rain gauges, such as this simple post-mounted model, are emptied before cold weather freezes the water and cracks the gauge.

PRECIPITATION

Precipitation is any form of water that falls to the ground. Rain provides the most water, but snow, sleet, and hail also provide water. The water from melted snow and ice is often collected in reservoirs for crop irrigation, city water systems, and other uses.

Rain

Raindrops form as clouds cool, and the tiny cloud droplets merge to form larger droplets or ice particles. As they merge, they become heavier. When they get too heavy to stay afloat, they form a teardrop shape and fall toward the ground. Some will evaporate before reaching the ground; others fall as rain.

Precipitation ranges from a few inches in desert areas of Nevada to several hundred inches on mountains in Hawaii. The eastern half of North America receives an average of 30 to 60 inches of precipitation per year. Rainfall amounts decrease west of the Mississippi River to 10 inches or less in the drier areas. Some Pacific Coast areas have high rainfall, such as Puget Sound in the Pacific Northwest.

Precipitation is measured with rain gauges that give the depth of the rainfall in inches or centimeters. Rain gauges should be located carefully for accurate measurement. Buildings, trees, and other structures interfere with rain or cause runoff that enters the gauge. When this happens, the rainfall is not measured accurately.

Simple rain gauges can be put up at home. Weather stations use more sophisticated gauges.

21–24. Mountain snow accumulations melt in the spring and create water for irrigating crops and other uses.

Frozen Precipitation

Snow, sleet, and hail are frozen or solid forms of water. Snow and sleet fall in cold weather. Sleet is usually the size of raindrops. It forms when rain leaves the cloud and freezes before reaching the ground.

Hail falls in warm weather during severe storms. Hail pieces may be an inch or more in diameter and may cause damage to crops and buildings. Strong upward wind lifts the falling particles back into the cloud where they become larger. When too large for the wind to support, they fall to the ground.

Snow is an important source of water in many areas. For example, snowfall in the mountains of California is needed to supply water for crop irrigation in the valleys. Some mountains in California may have 10 feet or more of snow during the winter.

21–25. A weather station rain gauge collects precipitation in the top and funnels it to the gauge.

Precipitation Contents

Precipitation brings the contents of the air with it. Rain sometimes brings nitrogen; other times it brings smoke particles, acids, and other solids. In stormy weather, rain may appear to bring large objects. For example, a tornado can lift water from a pond or a stream. Fish and other objects are lifted with it. When the water falls, people think it has "rained fish." Actually, storm debris has fallen to the earth.

Acid rain is viewed as a problem. This rain contains pollution that forms acid in water. The acid damages crops, soil, and structures on which it falls.

Dew Point

Dew is a form of liquid precipitation. It is water that forms on the leaves of plants and other materials when the air temperature cools so that it cannot hold as much moisture. Plant leaves become wet with dew drops. Dew forms on clear nights when there is no wind. It is dried the next day by the sun.

Dew point is the temperature at which dew will begin to form. It is lower than the temperature of the air, or it is the same as the air if relative humidity is 100 percent.

Frost forms as tiny ice crystals when the weather is cold and the dew freezes. Tender crops are damaged by frost.

21–26. Mustard green plants with frost damage.

In humid climates, as much as 5 inches of precipitation may be formed annually by dew, which is an important source of moisture in desert areas.

Agricultural operations may be halted or delayed by dew. In the early morning, dew may cover cut hay in the field. However, the hay cannot be baled until the sun dries the dew. Harvesting grain and cotton also may be delayed until the dew has dried. Harvesting at night may have to be stopped when dew forms.

21–27. Soil with abundant moisture has frozen and spewed ice crystals.

WEATHER CHANGES

Weather changes result from atmospheric conditions. Changes in temperature and air pressure are likely to cause other weather conditions.

Large air masses can develop. Some have warm air, and others have cold air. When two masses touch, fronts result. A ***front*** is the boundary between the two masses. Fronts may be warm or cold. Warm fronts follow cold fronts, so warm air replaces cold air. Cold fronts result as cold air replaces warm air.

Changes in the air occur all the time and vary with the seasons and the atmospheric conditions. Thunderstorms can develop when cold air bumps into warm air. The cold air pushes the warm air upward, where it forms heavy clouds and creates wind. Lightning may occur when electrical charges develop in clouds.

Tornadoes are violent, funnel-shaped storms. They have high winds that can damage trees, crops, buildings, and other structures. Hurricanes are large tropical storms that can do extensive damage; they are known as typhoons in the Pacific Ocean area.

Scientists continually study the weather. There is much to learn, such as why hurricanes form. People are also interested in ways to create rain. In dry weather, crops often need rain. Lacing clouds with chemicals (salting) has been used as a way to make rain. In addition, silver iodide and sodium chloride have been used with positive results.

CLIMATE

Climate is the average of all the weather conditions for an area over time, usually a year or more. Climates vary in temperature and amount of rainfall. Various regions of the earth are known for certain climates because of latitude (distance from the equator); altitude (height above sea level); and surrounding geography (e.g., mountains, valleys, and nearby lakes or oceans).

21–28. Good crop harvests, such as with soybeans, require adapted varieties and good weather in the growing season.

CLIMATE ZONES

The earth is divided into climate zones. A **climate zone** is a large area of the earth's surface with similar temperatures. The zones are tropical, temperate, and polar.

A **tropical climate zone** is an area where the average temperature is above 65°F (18°C). The tropical zone is a wide band around the earth on both sides of the equator.

The **temperate climate zone** is a band around the earth between the tropical and polar zones. The average temperature in temperate zones is based on winter and summer average temperatures. Temperate zone temperatures average above 65°F (18°C) in the summer and below 50°F (10°C) in the winter.

The **polar climate zone** is an area where the average temperature for a year is below 50°F (10°C). The polar zones include the North Pole and the South Pole.

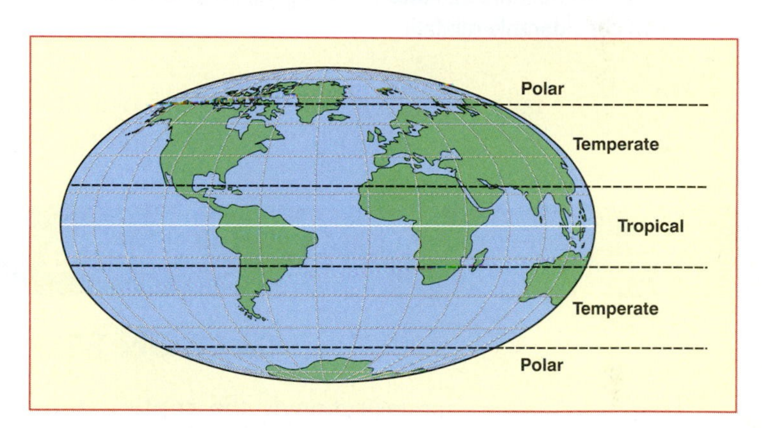

21–29. The locations of climate zones on a map of Earth.

CLIMATE FACTORS

Weather conditions in an area are described by measures of the temperature and precipitation. These conditions are the result of several factors.

- **Sun rays**—Rays that strike the earth directly, as they do at the equator, are warmer. Sun rays that strike the earth at an angle are cooler. The round shape of the earth and how it is tilted in relation to the sun cause differences in the temperature and amount

21–30. Hot, low-moisture climates have vegetation different from areas with considerable rainfall.

of light. The angle of the sunlight rays is related to latitude, which is the distance north or south of the equator. It is measured in degrees.

- **Altitude**—Altitude is the elevation or distance above sea level. The higher the elevation, the cooler the weather. The tops of tall mountains may stay so cool that snow is on them all year. For example, mountains in the Rocky Mountain National Park in Colorado have snow year round. The peaks are above the tree line, so only tundra plants grow there.
- **Water areas**—Large areas of water influence climate. Oceans and large lakes tend to make the climate near them less extreme. Water is slower to cool than the land. On warm days, the water cools the air as it moves to the land. On cold days,

AgriScience Connection

CARBON SEQUESTRATION

Carbon sequestration is the capture and storage of carbon that would otherwise be released into the air as carbon dioxide (CO_2). CO_2 is one of the "greenhouse" gases that is considered a threat to the environment. It is released by burning fossil fuels and by other activities.

Carbon sequestration is storing the captured CO_2 in deep, underground natural structures such as salt formations. A carbon credit program has been established by the Chicago Climate Exchange (CCX). Industries can purchase carbon credits to offset the carbon they release into the air.

Tilling the soil exposes surface areas that allow the release of CO_2 into the air. Using practices of minimum tillage or no-till reduces the release of CO_2. Some states have enrollment plans whereby farmers can sign up. In addition, some plans are available through the Conservation Reserve Program of the USDA.

This shows a scientist studying the influence of woody plants on soil carbon storage. He is taking a 10-foot deep sample of soil for analysis. (Courtesy, Agricultural Research Service, USDA)

the water warms the air as it moves toward land. Large areas of water tend to keep the nearby land area nearly the same temperature year round.

Sustaining life requires the use of resources. Plants and animals live together using these resources. The physical features of the earth promote growth and, at the same time, resist change. The earth has ways of dealing with changes. For instance, new ecosystems (communities of organisms) may grow when old ones are destroyed.

Internet Topics

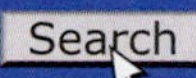

Selected topics for Internet discovery and reporting are listed here. Begin by searching for each topic. Next, read and learn about it. Conclude by preparing a brief report on your findings.

1. NOAA
2. Carbon sequestration
3. Weather

SUCCESSION

Plants and animals follow patterns of growth. The pattern of growing and changing is ***succession***. It is the natural and continual replacement of organisms.

Disasters can destroy an ecosystem. Fires burn range land. Hurricanes destroy forests. Volcanic eruptions destroy everything in the path of the lava. After a disaster, the resources of the earth begin to build a new ecosystem.

Replacement is always going on in an ecosystem. New plants and animals are produced and grow. Old ones mature and die. The decomposition of the old plants and animals provides organic matter. This material reacts with the earth's minerals to cause change in the composition of the crust of the earth.

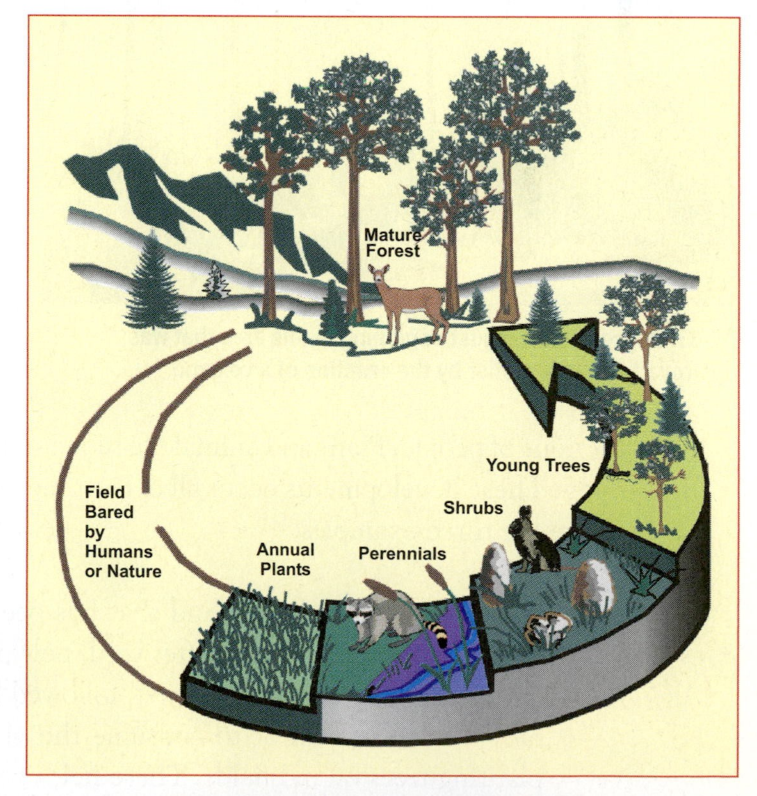

21–31. Succession of plant and animal life.

PRIMARY SUCCESSION

Primary succession is the development of a biotic community where none existed. A ***biotic community*** is

21–32. Wild fires can cause major damage to land. Many years are needed for land in some areas to overcome the destruction of plants and other organisms. (Courtesy, Agricultural Research Service, USDA)

21–33. Succession was just beginning in this area that was destroyed 20 years earlier by the eruption of a volcano.

a group of plants and animals that live together. In most places, established communities have been around for thousands of years.

This is how it might have happened in North America: Glaciers that covered much of the region began to disappear about 15,000 years ago. Large areas of land and rock were exposed. Low forms of life began to grow first, such as lichens. These forms of life were gradually replaced by mosses, which were followed by ferns, grasses, shrubs, and trees. As the trees grew, the grasses and other small plants were shaded out. First insects, and then other animals, began to grow as the plants provided food. Mature plants and animals died, and new ones grew in their places. Meanwhile, rocks and minerals were weathering. Soil was being formed that could support additional life. The decomposition of the living organisms also added humus to the soil.

SECONDARY SUCCESSION

Secondary succession is the regrowth of living organisms in an area destroyed by a natural disaster or by the actions of people. Plant and animal life sequentially develop much as during primary succession. These developments occur all of the time, but it is easy to overlook what is happening. Here are three examples.

- **Abandoned field**—Cropland that has been tilled for years may be abandoned. The land is no longer farmed. Afterward, new plants gradually grow and develop. Grasses and other small plants grow first, followed by small trees and larger trees. Animal life moves in to live on or to consume the plants. Some people promote succession by planting trees on old fields. These fields become “tree farms” and remain productive land.

- **Clear-cut forest land**—Some methods of harvesting timber involve cutting all of the trees on the land. No trees are left. Tree farmers will further clear the land and plant young trees, known as seedlings. The seedlings will grow in a few years to dominate the land. Until the trees are large enough, grasses and weeds will grow and compete with them for space. Undesirable trees may also grow and should be cut out to help the new forest grow properly. If the land is not planted with trees, natural succession will take place. Grasses and small plants will grow. Different species of trees will also grow and overtake the grasses and small plants. Some tree species will be dominant; others will be suppressed.
- **Burned land**—Lightning may set fire to a forest or range area. The fire may be intense and may destroy all of the trees and other plants. The land may have nothing left growing on it. In one growing season, grasses and small plants will begin to sprout. In another season, trees will begin to grow. Over time, the forest or range land will regrow. Those species that are dominant will survive until they die and new ones replace them. (Note: Fires are particular problems in dry areas. In addition to destroying plants and animals, fires cost money. People who own timber lose what it was worth. Always be careful with fire!)

21–34. Clear-cut forest areas expose soil to erosion and allow increased runoff from precipitation.

21–35. Lightning during thunderstorms may set fires in forests or range areas.

21–36. Humans can help protect damaged land by reseeding it with native plants. (Courtesy, Agricultural Research Service, USDA)

REVIEWING

MAIN IDEAS

Earth is in the Milky Way galaxy. It is in a group of other planets, moons, and stars with one sun that forms a solar system.

The earth has three major parts: the surface or crust, the interior, and the atmosphere. The crust is important because it supports plant and animal life. Through weathering, rocks and other minerals have formed soil. The interior of the earth has a hot core of heavy materials. Between the core and the earth's crust is a layer known as the mantle. The atmosphere is made up of the air that surrounds the earth.

Many natural changes occur in and on the earth. First, the earth is constantly moving. It rotates and revolves. One rotation is made every 24 hours, and one revolution is made every 365.24 days. Rotations and revolutions are measured in days and years. These movements are important in producing plants and animals.

Events in the atmosphere affect the environment on the earth. Weather is the general condition in the atmosphere. The weather consists of temperature, air pressure, wind, humidity, and precipitation.

The climate is the average of all the weather conditions over a year or several years. The earth has three climate zones: tropical, temperate, and polar. The tropical climate zone is a band on both sides of the equator that extends around the earth. The temperate climate zones are found on both sides of the tropical zone. The polar climate zones are located between the temperate zones and the poles.

The earth's resources are used to support a succession of plant and animal life. Succession is the pattern of growth and change in plant and animal life, known as a biotic community. Primary succession is the development of a biotic community where none existed. Secondary succession follows primary succession and occurs after an area is disturbed by a natural disaster or by people.

Understanding earth science helps in using the earth's resources to produce plants and animals. It provides a background for learning about soil—the part of the earth's crust that supports terrestrial plant and animal life.

QUESTIONS

Answer the following questions, using complete sentences and correct spelling.

1. What is the universe? How is the earth a part of the universe?
2. What are the parts of the earth's solar system?
3. What are the three major parts of the earth? How do these parts relate to each other?
4. Distinguish between continental and oceanic crust.
5. How does Earth move in the solar system? Name the kinds of movements, and explain each. Why are these movements important in producing plants and animals?
6. What is weathering? What are the two major kinds of weathering? Why is weathering important?
7. What are the major elements of the weather? Describe how each is measured.
8. What is climate? List and briefly explain three climate zones.

9. What is succession?
10. What is primary succession? When does it happen?
11. What is secondary succession? When does it happen?
12. How can agriscientists get involved in succession?

EVALUATING

Match each term with its correct definition.

a. climate
b. weather
c. front
d. precipitation
e. humidity
f. cloud
g. wind
h. weathering
i. rotation
j. revolution

_____1. The turning or spinning of Earth on its axis; one day.
_____2. Moving air.
_____3. The effect of weather on rocks and minerals on the crust of the earth.
_____4. The general condition of the atmosphere as it pertains to temperature, air pressure, wind, moisture, clouds, and precipitation.
_____5. The average of all the weather conditions for an area over time, usually a year or more.
_____6. The movement of Earth in space around the Sun.
_____7. The amount of moisture in the air.
_____8. Any form of water that falls from the atmosphere to Earth.
_____9. A mass of condensed water vapor, droplets, and pieces of ice floating in the air.
_____10. The boundary between two air masses.

EXPLORING

1. Collect the weather maps from the local newspaper for a week. Review the maps, and note trends in the weather. Prepare a report on your findings.

2. Develop a log of the weather at your home for a week or longer. Use a thermometer to measure and record the temperature at the same time each day. If possible, use a rain gauge to measure precipitation. Other measurements can be taken, if you have the proper equipment. (Alternative: Rather than taking and recording the readings, compile the information from the newspaper. For each day, chart the high and low temperatures, as well as precipitation, barometric pressure, and relative humidity.)

3. Using reference materials, determine if your home is in the tropical, temperate, or polar climate zone. Assess the features of the climate where you live when making the determination. Write a report on your findings.

CHAPTER 22

Soil and Land Science

OBJECTIVES

Knowing how soil is formed and its nature will help manage it when producing field crops, ornamental plants, and forestry. This chapter has the following objectives:

1. Describe soil and its components.
2. Discuss the nature of soil.
3. Explain how soil is formed.
4. Explain the meaning and importance of soil profiles.
5. Identify and distinguish between the kinds of soil water.
6. Describe soil pH and salinity as related to plant growth.
7. Describe important soil management practices.
8. Explain the meaning of land and how it is classified.
9. Discuss how land is measured and described.

TERMS

acre
agricultural drainage
artificial erosion
base lines
boundary
capillary water
contour plowing
crop rotation
gravitational water
hardpan
horizon
humus
hygroscopic water
infiltration
internal drainage
irrigation-induced salinity
land capability classes
land description
land forming
land surveying
loam
loess
meridian
metes and bounds
mineral soil
minimum tillage
mulching
organic soil
parent material
percolation
rectangular survey system
soil pH
soil profile
soil salinity
soil structure
soil texture
soil tilth
strip cropping
terracing
water table

JUST ABOUT ALL living things depend—in one way or another—on the soil. The products of soil help meet their needs. The well-being of humans is certainly related to soil productivity because having plenty of food and other essentials requires productive soil.

Using soil poses a threat to its productivity. For example, bulldozing and plowing loosen the surface so it can be washed away by rain. In addition, crop plants take nutrients out of the soil. Practices should be used to return nutrients. If nutrients are not returned, the soil loses its fertility.

22–1. The dark green color of this corn indicates that the soil is providing plenty of nitrogen. (Courtesy, Agricultural Research Service, USDA)

Today, we often hear about sustaining the soil—using practices to ensure that the soil will be productive for many years to come. In reality, we mean an indefinite period. You would not want to use the soil in ways that would cause it to fail in the future!

Soil varies from one place to another. Different practices are needed for different soil types. The irrigated soil of western North America must be managed differently from the soil in eastern North America where producers do not rely as much on irrigation.

SOIL COMPONENTS

22–2. Properly planting a shrub in good soil will promote its growth. (Courtesy, Education Images)

Soil is the top few inches of the earth's crust and the medium in which plant roots grow. Artificial growing media are sometimes used. These media are made of materials that meet the needs of growing plants, similar to soil.

Four major soil ingredients are minerals, organic matter, water, and air. Minerals and organic matter are the solid particles in the soil. Water and air fill the spaces between the particles. Soil also contains living organisms; some are very tiny, and others are large.

MINERAL MATERIALS

Most of the soil in North America is mineral soil. ***Mineral soil*** is soil high in mineral content. It has more minerals than other materials. Some soil is high in organic matter and is known as ***organic soil***, which has more than 20 percent organic matter. Mineral soils usually have 6 to 12 percent organic matter.

Minerals are inorganic elements or compounds that naturally occur. They sometimes form crystals, such as quartz and feldspar. Mineral materials in soil come from rocks and other materials that have weathered. These materials form particles known as sand, silt, and clay. Other minerals are also found in soil.

- **Sand**—Sand is the largest-size mineral particle in soil. The particles are 0.05 to 2.0 millimeters (mm) in diameter. Most sand is made of quartz, but it can be made of many different materials. Its chemical name is silicon dioxide (SiO_2). Sand is an important ingredient in soil. The amount varies; some soil is mostly sand. Soil with more than 85 percent sand is known as sand. Soils that are high in sand dry out quickly because the sand parti-

Soil Contents

25% Air

45% Minerals (clay, sand, or silt particles)

25% Water

5% Organic Material (living and dead plants and animals)

22–3. Typical soil contents.

cles do not hold water well. Soils high in sand are not fertile because water moves through them quickly and washes the nutrients away.

22–4. Three soil components (from left to right): organic matter, clay, and sand.

- **Silt**—Silt particles are smaller than sand, but they are larger than clay. Silt particles range from 0.002 to 0.05 mm in diameter. The particles fill spaces between sand particles. Silt particles are sometimes deposited by water when land is flooded or washed into the oceans where streams empty. The large, fertile Mississippi River Delta is a product of silt settling from the flood water of the Mississippi River. Hundreds of years may be required for large land areas to develop by silting.

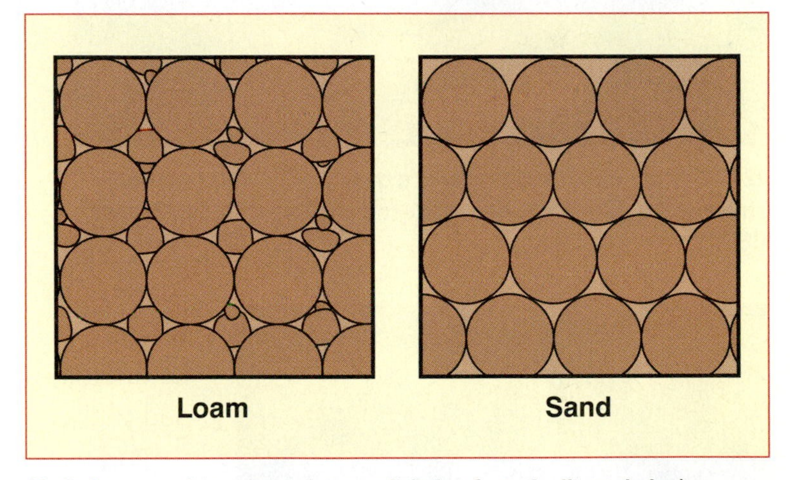

22–5. Compared to sand, a loam soil (mix of sand, silt, and clay) provides more particle surface area for moisture, air, and nutrients.

- **Clay**—Clay is the smallest particle in soil. It is less than 0.002 mm in diameter. Various minerals are found in clay; kaolin and vermiculite are two examples. Clay fills spaces between silt and sand particles. It gives soil the ability to hold water. Soils that are high in clay are slow to dry out after a rainfall. Clay also holds minerals that plants need for growth. Since clay particles are quite small, they have more total surface area and space between them than the larger sand particles. This space helps clay hold water.
- **Other minerals**—Other minerals in soil are important in plant growth. In fact, these are the minerals that become plant nutrients. They are often added in fertilizer to give plants more nutrients. The major minerals, in element form, include calcium, phosphorus, potassium, and nitrogen. Some of these are easily lost from the soil, such as nitrogen; others tend to remain in the soil longer, such as calcium.

ORGANIC MATTER

Organic matter is plant and animal remains in various stages of decay. The decaying organic matter releases nutrients. Well-decomposed organic matter is ***humus***. Soils with a high humus content are usually high in nutrients; peat and muck are high in humus. These soils are dark brown or black and hold a lot of water. Practices should be used that encourage the development of organic matter. Leaving crop residue and plowing it into the soil incorporates organic matter.

22–6. Fertile soil is used to produce quality products, such as strawberries. (Courtesy, Natural Resources Conservation Service, USDA)

Soils with high organic matter are more productive. The soil is more easily tilled and made into a seedbed for planting. High organic matter increases a chemical capacity of soil, known as ionic exchange, which increases nutrients available for plant growth.

Manure from cattle feedlots or hog feeding facilities may be applied to fields. Litter from chicken houses also may be spread over the land. Homeowners often buy bags of dried manure to add to the soil in flower gardens!

Forests naturally provide organic matter. Leaves shed at the end of summer fall to the ground and begin decay. Depending on the kind of tree, a few years may be required for the leaves to become humus.

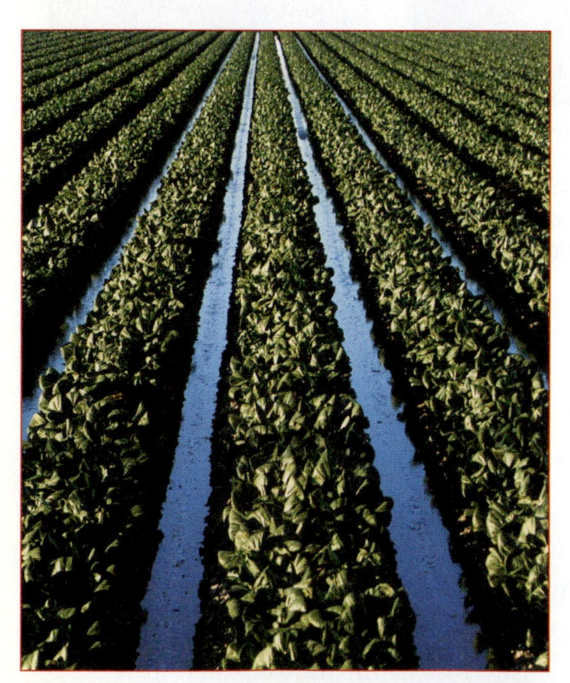

22–7. Irrigation water is being added in the middle between rows of lettuce in southern Arizona. (Courtesy, Natural Resources Conservation Service, USDA)

WATER

Water is an important nutrient for plant growth. It is held between soil particles and adheres (sticks) to their surfaces. Soils high in clay hold more water. Soil that is too wet or too dry will not support the growth of most crops and ornamental plants. The amount of moisture needed varies with the kind of plant. Some plants, such as watercress, can grow in an aquatic environment.

Terrestrial plants, such as lettuce, are sometimes grown with their roots in a water solution. This is known as hydroponics.

In dry areas, soil moisture is supplemented with irrigation. Some locations would not be able to produce crops without added water. Many of the vegetable crops are produced in climates where irrigation is essential.

AIR

Just as water fills spaces between soil particles, air also fills spaces. Adding water to soil drives air out. As soil dries, air returns to it. Plant roots need air to live and grow. Some plants adapt to different environments. For example, cypress trees grow in wet places. For these trees to have air, primarily oxygen, the roots grow structures known as "knees." The knees rise a few feet above the water level.

22–8. Mulch-till rotary hoe equipment may be used to prepare a seedbed and to add air to soil. (Courtesy, Case Corporation)

Most plants need a balance of air and moisture in the soil to grow well. Crops in soil that is too wet may develop a yellow color rather than a healthy green. For example, if water stands around corn plants for a few days, the leaves turn yellow. Once the water is gone and air returns to the soil, the corn plants will regain their green color. Meanwhile, however, yellowing of the leaves slows growth of the plant and reduces its yield.

22–9. Some potted plants grow better in "soil" specifically designed to meet their needs, known as a growing medium. (Courtesy, Education Images)

THE NATURE OF SOIL

The proportions of components in soil give it distinct characteristics. Soil has physical, chemical, biological, and drainage characteristics.

PHYSICAL NATURE OF SOIL

Soil has important physical qualities that determine how the soil is used and the additives needed for it to be more productive. They give it the workability often required in culturing crops.

Soil Texture

Soil texture is the proportion of sand, silt, and clay in the soil. These materials give soil its physical appearance. All soils are mixtures of sand, silt, and clay. Some soils may be

AgriScience Connection

DOES COLOR MATTER?

Plant producers have wondered how the color of the soil or mulch influences plant productivity. They knew that color influenced light reflectivity. They also knew that wave length of the light that reaches plants was related to colors surrounding the plants.

Research has focused on how soils and mulches of various colors influence plant development, including roots, stems, leaves, fruits, and seeds. The research included the role of color on flavor, aroma, and nutrient content of products. Scientists have found that tomato plants grown over red mulch had yields 20 percent higher than plants in black mulch. In addition, carrots grown with yellow mulch had higher concentrations of vitamin C and beta carotene. Cotton grew longer fibers when exposed to light reflected from yellow mulch. Of course, the color of the mulch alters light waves to a color that is not that of the mulch, such as yellow mulch reflects a red photosynthetic light.

This shows scientists studying mulch and soil color. The plants will be assessed as the growing season progresses. Question: Since soil color varies throughout the United States, do producers with soils of a certain color have advantages over others? (Courtesy, Agricultural Research Service, USDA)

almost exclusively one kind of material, such as sand or clay. One approach in determining texture is by feeling a soil sample, known as the ribbon test. Soil high in sand is gritty. Soil high in clay is smooth and soupy when moist.

Soils are named on the basis of content. **Loam** is soil that is nearly equal parts of sand, silt, and clay particles. Different names are used to indicate the percentages of the various materials. The following are a few examples:

- **Sandy loam**—Sandy loam is made mostly of sand with a little silt and clay. When dry, this soil forms clods that are easily broken.
- **Silty clay loam**—Silty clay loam is made mostly of silt with some clay and a little sand.
- **Silt loam**—Silt loam is made mostly of silt, but it has some clay and sand. Dry clods of silt loam are difficult to break. When wet, the soil is "buttery" or "velvety."
- **Sandy clay loam**—Sandy clay loam is made mostly of sand with some clay and a little silt.
- **Clay loam**—Clay loam is made mostly of clay with some silt and sand. Dry clods are very difficult to break.
- **Sand**—Sand is soil that is 85 percent or more sand. It may contain less than 15 percent silt or less than 10 percent clay.
- **Loam**—Soil materials are present in approximately equal amounts.
- **Clay**—Clay soil is made of 40 to 100 percent clay. Some sand and silt may be present. Clay soils with sand feel gritty when touched. Clay soils with silt feel smooth when wet.

The soil triangle is used to explain the makeup of soil. The triangle illustrates how soil textures are related to the different materials in the soil.

22–10. Soil high in sand. The sample holds little water and has a gritty feel.

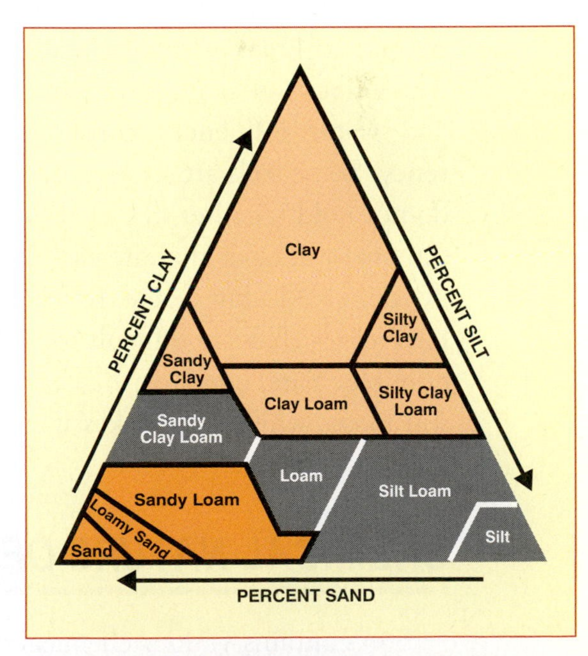

22–11. The soil triangle is used to describe the makeup of soil.

22–12. A cut along a roadway shows structures forming in this high-clay soil.

Soil Structure

Soil is more than its parts of sand, silt, and clay. It also has structure. ***Soil structure*** is the arrangement of the soil particles into shapes or pieces in undisturbed soil. Soil structure may be platy (in layers), columnar, prismatic, blocky, and granular.

Soil Tilth

Soil tilth is the physical condition of the soil. It is related to soil content and is often made artificially by plowing. Various disks, harrows, and other implements are used to provide good tilth, which is important in preparing a seedbed.

Soil with good tilth is loose and not too wet or dry. There are no large clods. It is easily planted and otherwise used in production.

22–13. Soil in this flower bed is being worked to develop good tilth.

Soil Consistency

Soils vary according to their feel and the amount of effort needed to break a small clod. The amount of sand, silt, clay, and water influences consistency. Some soils are loose and do not hold together in a clod. Others form clods that are easily crumbled, but some soils have clods that are difficult to crumble. Sand tends to make soil loose, while clay makes it hard to crumble.

CHEMICAL NATURE OF SOIL

Soil contains various chemical elements. Some elements, such as nitrogen, potassium, and phosphorus, are essential for plant growth. Others are merely present in the soil, though a few elements may damage plants. Too much of any element can pose a problem.

In addition to the elements used for plant growth, soil also has an overall chemical quality known as pH. The pH of soil refers to its acidity or basicity. Most plants have a preferred range of pH in which they grow best. (Soil pH is covered in more detail later in this chapter.)

BIOLOGICAL NATURE OF SOIL

22–14. Earthworms help develop good soil.

Most soil contains living organisms, such as bacteria, fungi, plants, and animals. These organisms vary from very small to large, with the larger being earthworms, ants, and rodents. Organisms do valuable things in the soil.

- **Break down organic matter**—Bacteria, fungi, and algae act on leaves, stems, and plant and animal remains. They cause decay! Organic matter in the soil is a result of the decay process. Years may be required for materials to decay and become soil. Without action by microorganisms, plant and animal remains would accumulate and not decay. Other animals help break down organic matter. Beetles and termites break wood apart. Birds and squirrels peck and dig into the ground. The activities of these animals cause change in soil.

Technology Connection

SALT SNIFFER

Much farm land is gaining salinity. The salt may occur naturally or may be applied in irrigation water. Land with high salinity lacks productivity and creates a serious problem for agricultural producers.

Scientists are studying salinity in soil in attempts to deal with the problems that are caused. New salinity-assessment field equipment has been developed. Known as the "salt sniffer," the mobile rig travels over fields using an electromagnetic inductance meter to find the salty spots. The meter is connected to GPS and relays information to a data logger. The information is used to produce maps of salt locations in fields.

This photo shows the mobile rig equipped with GPS, a conductivity meter, a drilling rig for collecting soil samples, and other features. (Courtesy, Agricultural Research Service, USDA)

22–15. Sprinkling a mild formaldehyde solution on soil helps in studying its biological nature. The solution causes living organisms to come to the surface. (Courtesy, Agricultural Research Service, USDA)

- **Aeration of the soil**—Good soil contains air. Tiny cracks and other breaks in the soil allow air to enter. Tillage is an artificial method of soil aeration. Some areas are treated with aeration equipment. Air is needed by soil organisms and plant roots. Small animals live and burrow in the soil. The holes and tunnels they make let air and water into the soil. Earthworms, ants, crawfish, and other animals move the soil around and improve tilth. Particles from several inches or feet below the surface may be moved.
- **Add fertility**—Some microorganisms add fertility to the soil. For example, legume plants, in the presence of rhizobium bacteria, fix nitrogen from the air in the soil. Legumes include clover, alfalfa, soybeans, and all beans and peas.

SOIL FORMATION

Soil develops gradually over many years. It is a process that goes on constantly. Several activities make soil: changing parent material, weathering, and varying climate in addition to plant and animal life, slope, and drainage.

PARENT MATERIAL

Parent material is the material from which soil develops. These materials include rocks, specific minerals, and, in some places, peat. (Peat is decaying plant matter found in wet places, such as bogs.) Most parent material is a few feet below the top of the ground. In some places, outcroppings of parent material show up at the surface.

Common parent material includes shale, limestone, and sandstone. Shale and slate form soils that are high in clay. Limestone forms shallow soil that may not be good crop land. Sandstone forms sandy soils.

Parent materials may be moved about before soil is formed. Areas of North America were once under water and may have marine deposits as parent material. Shells and other fossils may be found in the material. Some parent material may be washed from one place to another, such as in the delta areas along rivers. Other parent material may be blown from one place to another by huge, longtime winds. Windblown material is known as **loess**. Large areas of loess (a deep, fertile soil) are found in areas just east of the Mississippi River. Glaciers, gravity, and lake deposits also help make parent material.

22–16. Through weathering, large rock formations in the background have broken apart to create smaller rocks and mineral soil in the foreground.

WEATHERING

Weathering is the process of changing materials into soil. It is important to remember that soil does not suddenly appear; it goes through stages of gradual development. Rocks break apart and change their forms. Water movement through the soil leaches (carries) materials with it.

Older soils have often lost nutrients to leaching and plant growth. The rate of plant growth slows down. Nutrients need to be added to keep the soil productive.

CLIMATE

Long-term weather conditions in an area influence how fast soil is formed and lost. Heavy rains, freezing temperatures, and other events cause processes to go fast or slow.

Heavy rains can leach nutrients out of the soil. This is especially true of sandy soils. Producers often use fertilizer to return nutrients to the soil.

PLANTS AND ANIMALS

The organisms that grow on and in soil influence its development. Plants make leaves, stems, roots, and other structures that decay to form organic matter. Dead animals also decay to help form soil.

Some animals have a big influence on soil formation. The pounding hooves of walking cattle alter the soil. Rooting hogs move soil. Beavers build dams in creeks, causing flooding of surrounding land and thereby altering habitats. Overall, plants and animals have big impacts on soil.

ACADEMIC CONNECTION

ENVIRONMENTAL SCIENCE

Relate your study of soil to the study of environmental science. Explore the important role of sustaining soil resources, and investigate actions that can be taken in sustaining the soil. Use your learning in agriscience to support your mastery in environmental science.

22–17. A good vegetative cover protects the soil and ensures clear water in the stream.

SLOPE AND DRAINAGE

The lay of the land has much to do with how soil is formed. Land that slopes will have faster water runoff and will dry out more quickly. In contrast, land in creek and river bottoms will stay wet longer and will form a different kind of soil.

Hilly land often loses fertile topsoil to erosion. Water running down the hills picks up soil particles and carries them off. Level to gently rolling land is usually more fertile.

Soils that hold water have poor internal drainage. They are slow to form and low in fertility. Air cannot enter the soil because of the water. A lack of air slows the decay of the organic matter, and the decay must occur to release nutrients into the soil.

SOIL PROFILE

A ***soil profile*** is a view of a vertical section of ground at a particular place. The profile shows layers of materials from the surface down several feet. Profiles are divided into horizons.

HORIZONS

Soils have four horizons. A ***horizon*** is a distinct layer of soil materials. Horizons are easy to see in most soil profiles. The horizons are:

- **A horizon**—The A horizon is the topsoil or upper few inches that is high in nutrients. The roots of most grasses and other small plants grow in topsoil. Thickness of topsoil varies. In areas where the soil has eroded, there may be no topsoil and fertility may be lost. In other places, the topsoil may be several inches thick. Topsoil is usually gray to brown or black in color. The A horizon joins the B horizon. The exact location is sometimes difficult to determine because the two horizons may blend together.

- **B horizon**—The B horizon is known as subsoil. It is below the topsoil and contains more clay and less organic matter. The subsoil usually has a brighter color because it is higher in minerals and lower in organic matter. Colors may be red, yellow, brown, or various combinations. The B horizon influences the percolation (movement) of water in the soil. Sometimes this horizon becomes hard and compacted, forming a layer known as a ***hardpan***. Hardpans restrict water percolation and plant root growth. Subsoiling is used to break up hardpans with chisel plows at a depth of about 18 inches. Some hardpans are caused by heavy machinery compacting the ground.

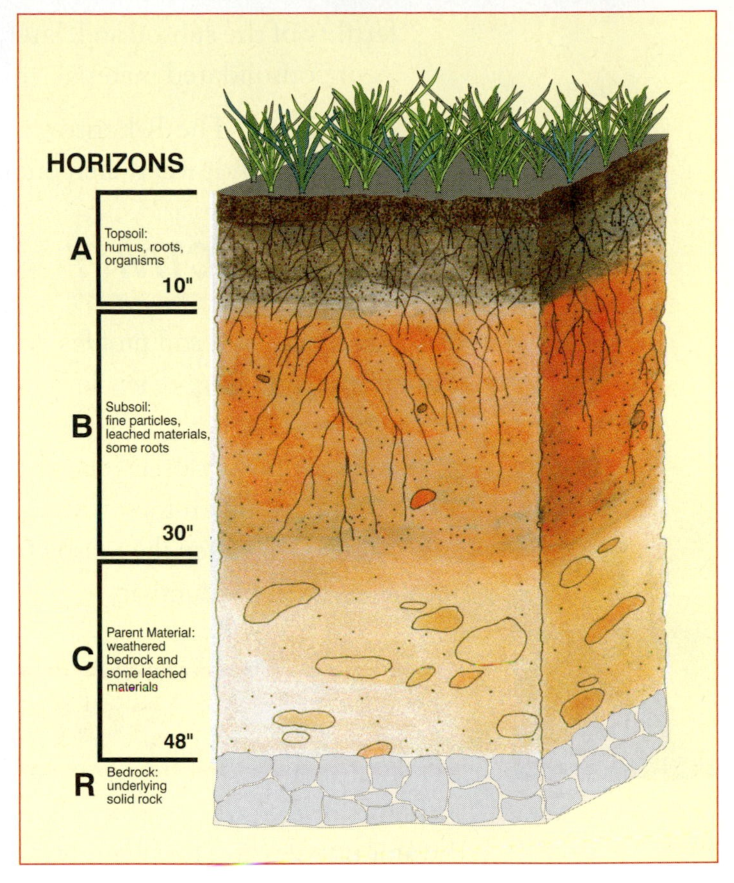

22–18. Locations of soil horizons.

- **C horizon**—The C horizon is the parent material. It lies between the B horizon and the earth bedrock. The C horizon is important because it will later become soil. It influences the

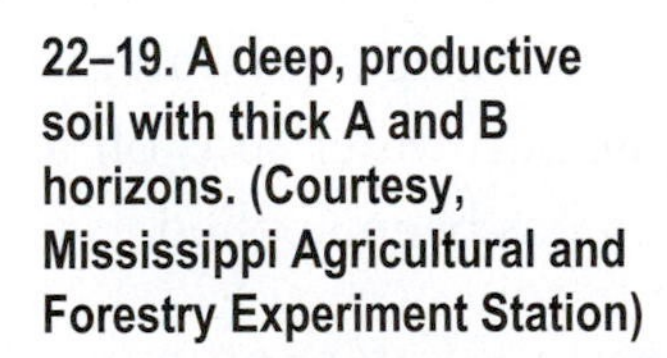

22–19. A deep, productive soil with thick A and B horizons. (Courtesy, Mississippi Agricultural and Forestry Experiment Station)

22–20. A cut-away profile shows the relationship of above- and below-ground parts of a daffodil to the soil.

fertility of the subsoil and, later, the soil. The C horizon is sometimes known as unconsolidated material.

- **R horizon**—The R horizon is the bedrock that is below the C horizon. It usually has not weathered enough to be parent material.

SOIL PRODUCTIVITY

Productivity and soil profiles are related. A soil with a thick "A" horizon has more topsoil and is more fertile. The "B" horizon influences soil fertility because it becomes topsoil.

A soil's profile is closely related to how it can be used. Soils with a thin A horizon do not have much topsoil and will not produce large crops. Producers need to use practices that keep the topsoil from being lost. Internal drainage of the soil has a lot to do with productivity.

SOIL WATER

Plants need water to grow. They receive it from the soil. However, soils vary in the amount of water they have and can provide for plants. Water enters the soil in two ways.

- **Soaking downward**—Water soaks into the soil after precipitation or irrigation. This soaking action is known as ***infiltration***. The downward pull of gravity on water after infiltration is ***percolation***. When soil receives all the water it will hold (no more infiltration or percolation), the balance runs off. The usual goal is to maximize infiltration and percolation. Applying too much irrigation water is a waste. If irrigation water runs off a field, too much has been applied.

20–21. A small earthen dam holds water to promote infiltration (soaking into the soil).

- **Soaking upward**—Water can move upward in the soil from higher concentrations deeper in the ground. However, this is not the way most crops receive water.

WATER TABLE

The ***water table*** is the depth of the natural level of free water below the surface of the earth. In areas with high rainfall, the water table may be only a few feet down. In dry areas, such as deserts, the water table might be hundreds of feet below the surface.

Land near large bodies of water, such as lakes, oceans, and rivers, is likely to have a higher water table than land without water nearby. Sandy soils near bodies of water may have a water table that is near the surface.

The water table is determined by digging a hole in the ground. The distance down to the point where water will collect and stand in the hole is the water table.

FORMS OF SOIL WATER

Soil with good tilth usually contains about 25 percent water. Plants need plenty of water to grow and produce large crops. Three forms of water are found in soil.

- **Capillary water**—Soil particles are coated with thin layers of water, known as ***capillary water***. The water moves from one soil particle to another by capillary action, which results because water sticks to the surface of soil particles. Smaller particles have more surface and have greater capillary pull. As a result, soils high in clay (the smallest soil particles) hold more water.
- **Hygroscopic water**—***Hygroscopic water*** is a thin layer of water that sticks to soil particles and does not move about. Even the driest soils have some hygroscopic water. Plants can use some of this water, depending on the amount in the soil. Clay and silt soils are better in dry weather because they hold more hygroscopic water.
- **Gravitational water**—***Gravitational water*** is the water that fills cracks and air spaces between soil particles. In low places after rains, water may stand on the land. This water causes problems when trying to grow crops because the plants may become yellow and die of a lack of air in the soil.

Internet Topics | Search

Selected topics for Internet discovery and reporting are listed here. Begin by searching for each topic. Next, read and learn about it. Conclude by preparing a brief report on your findings.

1. Land forming
2. Nutrient management planning
3. Land surveying

WATER LOSS

Soil loses water in several ways. Lost water is not available for plant growth.

- **Runoff**—Runoff is the flow of excess water from the land. Runoff does not soak into the ground. Heavy rain and melting snow create runoff. Streams may overflow

22–22. A carefully formed field holds winter rainwater for infiltration and to prevent runoff.

22–23. Cutting and hauling corn silage from a field removes considerable water. Question: What is the water content of corn silage? (Courtesy, Natural Resources Conservation Service, USDA)

their banks if runoff is too great. Earthen structures, such as terraces and ponds, can be built to slow runoff rates and to allow time for water to soak into the soil.

- **Evaporation**—Water that turns to vapor (gas form) and enters the atmosphere is lost. Large amounts of water can be lost from the soil on hot days through evaporation. Mulch helps prevent evaporation.
- **Transpiration**—Transpiration occurs as plants release water into the air through their leaves. The tiny stomata on the leaves open and close to help the plant adjust to changing temperatures.
- **Percolation**—Percolation is the downward pull of gravity on water. The water soaks into the soil at the surface and goes down into the lower soil profiles by percolation. Water that percolates deeply may be lost to shallow-rooted crops.
- **Harvested crops**—Many plants contain a high percentage of moisture. When harvested, the moisture is carried with the crop. Tomatoes and watermelons are much higher in moisture content than wheat and soybeans.

INTERNAL DRAINAGE

Internal drainage is the movement of water through soil. Some soils allow little water movement. Hardpans and shallow bedrock may hold water. Internal drainage is also influenced by soil texture.

Land is classed by how it drains internally. Some land has too much drainage, which means the soil does not hold enough water for crops to grow. Soils with a lot of sand tend to have too much internal drainage. Other soils have very poor internal drainage because they have a lot of clay and hold water on the surface for a long time. Layers of material under the surface may restrict the percolation of water. Most crops grow better on well-drained land that retains adequate moisture.

Ditches and drain tiles are sometimes used on land that has poor drainage to make it better.

22–24. The hardpan shown is 9 inches below the surface.

SOIL pH AND SALINITY

Chemical conditions in soil determine how it is managed. The kinds of crops and other uses are related to a careful balance of chemical elements.

pH

Soil pH is a way of expressing the acidity and alkalinity of soil. pH is the hydrogen ion concentration in the soil solution. A 14-point scale is used to express pH. Zero is extremely acidic, 7.0 is neutral, and 14.0 is extremely basic. All soil has a pH of between 0 and 14. Most soil is in the pH range of 4.0 to 8.0 (strongly acid to slightly alkaline).

Important chemical reactions take place between soil and plants. The pH of soil affects this reaction. Some nutrients are not available at the "wrong" pH. The best pH range for a good supply of nitrogen (N) is 6.0 to 8.0. Phosphorus is tied up when the pH is below 5.5 and above 7.5. Even when the soil is in the best pH range, the nutrients

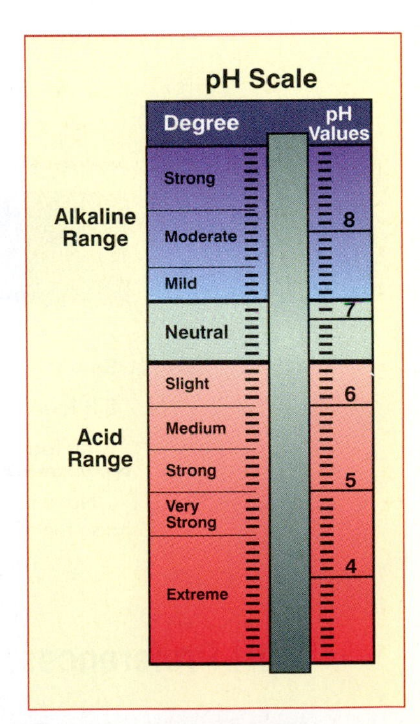

22–25. Soil pH as related to alkalinity and acidity.

are not available if they are not present in the soil. More plant nutrients are available at a pH level of 6.5 than any other pH.

- **Determining pH**—Soil pH is determined by testing a sample. This may require using a pH testing kit in the field or sending a sample to a soil laboratory for analysis. Careful interpretation of test results is essential. Trained specialists should be consulted to assess what to do to solve soil pH problems.
- **Modifying pH**—Soil pH can be changed by applying materials that alter soil acidity or basicity. Select materials and amounts for the change desired. Soils that are too acidic are made less acidic by adding alkaline materials. For example, ground limestone ($CaCO_3$), which is often called agricultural lime, can be added to acidic soil to raise the pH and make it less acidic. Lime is high in calcium and neutralizes acid.

  The amount of lime to use depends on the kind of soil and the change to be made in pH. Sandy soils need less lime than clay soils. The size of the lime particles also affects how well they work. Finer lime particles act faster, but they are used more quickly. Examples of lime amounts are shown in Table 22–1.

  Soils that are alkaline are made more acidic by adding sulfur, which forms sulfuric acid in the soil. Sulfur may be added in granular or powder forms. Aluminum sulfate [$Al_2(SO_4)_3$] and iron III sulfate [$Fe_2(SO_4)_3$] are sometimes used. The amount to apply depends on the results of soil testing and how much the pH is to be lowered.

Table 22–1. Amount of Lime Required to Raise pH

Kind of Soil	Amount to Raise pH From 4.5 to 5.5 for 1,000 sq. ft.	Amount to Raise pH From 4.5 to 5.5 for One Acre
	(pounds to use)	*(tons to use)*
Sandy loam	45	1
Silt loam	80	1.75
Clay loam	100	2.2

Note: The information represents approximate amounts of agricultural lime to use to raise the top 7 inches of soil from 4.5 to 5.5 pH.

pH Preferences

Crops vary in pH preferences. Some grow better in acid soils; others prefer slightly alkaline soils. The best plant growth occurs when crop species and pH are matched.

Some plants will grow in acidic or alkaline soil, but their flowers or fruit may vary. For example, the flowers on an ornamental shrub like a hydrangea will be pink if grown in basic soil and blue if grown in acidic soil. Therefore, sulfur can be added to get pink hydrangea flowers, or lime can be added to get blue flowers.

Table 22–2. Suitable pH Ranges for Selected Plants

Strongly Acidic (pH 5.0–5.5)	Medium Acidic (pH 5.5–6.0)	Moderately to Slightly Acidic (pH 6.0–6.5)	Slightly Acidic to Slightly Alkaline (pH 6.5–7.5)
azaleas	Bermuda grass	apples	alfalfa
Irish potatoes	oats	cantaloupes	cabbage
watermelons	peaches	corn	red clover
	pecans	cotton	spinach
	rye	peanuts	
	strawberries	soybeans	
		tomatoes	
		wheat	

Effects of Fertilizer on pH

Some kinds of fertilizers cause changes in soil pH. Fertilizers with nitrogen and sulfur tend to cause acids to form in the soil and lower the pH. Continuous use of acid-forming fertilizer will require the addition of lime to neutralize the acid.

Soil testing is used to determine the kind and amount of fertilizer to use. The same testing can also provide pH information. Recommendations are made by soil specialists who will need to know the kind of crop to be grown.

Processes in the soil are complex. As a result, it is necessary to use the best practices available when applying materials to modify soil.

22–26. Granular dolomitic limestone is often used on soil to reduce acidity.

SOIL SALINITY

Soil salinity is the amount of salt in the soil. If the soil has too much salinity, it can be difficult or impossible to grow crops. Salinity may be naturally present in soil or may be added through contamination. Salinity is more likely in areas that have low rainfall. In North America, approximately 43.2 million acres (16.0 million ha) are affected by salt.

Most salts in soil are water soluble (dissolved in water). Kinds of salts that may cause problems include sodium chloride (NaCl), calcium chloride ($CaCl_2$), magnesium chloride ($MgCl_2$), and potassium chloride (KCl). Always test irrigation water for salts before using it.

How Salinity Affects Plants

Plant roots absorb water by osmosis. The process uses osmotic pressure. Water enters roots when the osmotic pressure in the roots is greater than in the soil. Salt in the soil raises the osmotic pressure of the soil. Even a little salt means that less water can flow into the roots. Plants cannot carry out their functions as well as they could if they received more water.

Excessive salt can kill plants. Some plants are more tolerant of salt than others. The stage of growth of a crop is a factor in salt tolerance. Some plants will grow in salty soil, but their seed will not germinate and grow.

Sources of Salinity

Salts enter soil in several ways. Soil near areas of natural salt outcroppings has salts. Industrial accidents can pollute the soil with salts. The major source of salts on irrigated land is salts in the irrigation water, which is referred to as ***irrigation-induced salinity***.

Selecting irrigation water is an important decision. Even a small amount of salt in water can eventually add up to large amounts of salts on an acre of land. One acre-foot of irrigation water containing 750 ppm salt will add a ton of salt on an acre of land!

22–27. An aerial view of an area in the San Joaquin Valley of California shows fields with high salinity that are not supporting much plant growth. (Courtesy, Agricultural Research Service, USDA)

SOIL MANAGEMENT PRACTICES

Practices are used to conserve the soil and to ensure it will be productive in the future. These practices promote crop yields. Important practices include managing nutrients, preventing erosion, and draining wet land.

22–28. A soil scientist is collecting a water sample to measure nutrient content, particularly in runoff from surrounding crop land. (Courtesy, Agricultural Research Service, USDA)

NUTRIENT MANAGEMENT

Nutrient management is used to ensure future soil and land productivity and to protect the environment. It is most widely practiced in areas where animals are produced in intensive situations, such as cattle on a feedlot or in a dairy barn and chickens in a broiler house. The goal is to dispose of manure in a way that provides important soil nutrients without damaging the environment. Applying more manure and other animal wastes to land results in pollution of the soil and of the water that runs from it after rain falls or snow melts. Excess nutrients and chemical materials can also impair soil productivity.

Producers of animals who dispose of wastes by applying them to the land must have a comprehensive nutrient management plan (CNMP). Assistance in developing such a plan is available through the Natural Resources Conservation Service (NRCS) of the U.S. Department of Agriculture. Local NRCS offices are found throughout the nation.

Overall, a CNMP is a grouping of practices and management activities that ensure production and protect the soil and other natural resources. Each farm's plan must meet established standards. Three areas must be specifically addressed in a plan: manure management, crop nutrient needs, and water runoff. The resulting CNMP is a fairly detailed document that may be 15 pages or more. A plan must include an over-

22–29. Liquid wastes from a hog farm are being applied to cropland. (Overall, the amount of nutrients applied to land can be no greater than the amount removed by crops.) (Courtesy, Natural Resources Conservation Service, USDA)

22–30. Rainfall-simulation is being used to study bacteria levels in runoff water from a grass buffer strip. (Grass strips filter up to 90 percent of the bacteria from runoff water, according to the USDA's Beltsville Agricultural Research Center in Maryland.) (Courtesy, Agricultural Research Service, USDA)

all statement of the farm operation; a map showing various farm features; and information on production activity, manure collection and storage, application practices, and practices to assure that details of the plan are being met. Fairly detailed records must be kept.

The amount of nutrients applied to land in manure usually should not be greater than the amount removed by the plants that grow on the land. This requires knowing the nutrients used by crops and the nutrient content of the manure. Nutrients in commercial fertilizer that may be applied and nutrients in manure must be considered. For example, manure contains phosphorus. If phosphorus washes from the land into streams, the ecosystem in the streams may be destroyed. Testing of runoff water is a desired feature.

Runoff from land where manure has been applied may contain pathogens (causes of disease). Manure-borne pathogens include *Escherichia coli* and *Salmonella*. These organisms cause food poisoning in humans. Not all bacteria in manure are harmful. Scientists have determined that 1 pound of cow manure contains about 1 million bacteria, known as fecal bacteria. No one would want runoff to carry these bacteria into stream water!

EROSION

Erosion is the loss of soil. It may be worn or carried away by water, wind, or other means. Soil is made vulnerable to erosion by clearing land for factories, businesses, homes, and roadways. Plowing soil also makes it more easily lost. Crop producers are careful to use practices that minimize losses. Soil covered with grass or other plants is less likely to erode.

The kind of erosion varies with the climate. Large, dry fields not protected by a crop cover are most likely to be eroded by the wind. Excessive rain or other precipitation can wash soil away, especially on sloping land.

22–31. Gully erosion in this field could be stopped with good management. (Courtesy, Natural Resources Conservation Service, USDA)

Kinds of Erosion

Erosion is a natural process in all land—even that which has not been disturbed. The rate of natural erosion is usually very slow. Some erosion is part of the weathering process.

Artificial erosion is erosion caused by people disturbing the natural cover on land. Cutting trees, bulldozing for a road or building, and other activities promote soil erosion.

In agricultural land, erosion may be in sheets or gullies. Sheet erosion is the loss of layers of soil, with wind and water being the culprits. Sheet erosion may go unnoticed. Gully erosion is more easily seen because of the crevices or small ditches in the land. These may become larger and larger until any use of the land is impossible. Gully erosion is most common on hilly land.

Controlling Erosion

Erosion is controlled by keeping the soil in place. Protection from wind and water runoff will help slow erosion. Practices are available to control erosion.

22–32. A silt fence has been constructed to reduce erosion and to prevent silt from entering runoff. (Courtesy, Education Images)

- **Minimum and no tillage cropping**—Soil that is not plowed for crops is less likely to be lost. ***Minimum tillage*** is the cropping practice in which seeds are planted with very little plowing. With no tillage cropping, the land is not

22–33. A no-till drill is used in seeding. (Courtesy, Case Corporation)

22–34. Leaving crop stubble on a field in the winter helps prevent soil loss.

22–35. Strip cropping is used on this farm to prevent soil loss. (Courtesy, USDA)

plowed at all. Seeds are planted in crop stubble from the previous year. Chemicals are used to control weeds. The crop is not cultivated.

- Mulching—***Mulching*** is the practice of covering the soil with a layer of protective material. Crop stubble may be left on a field as mulch. Small gardens and flower beds may be covered with sheets of paper, plastic, ground tree bark, or sawdust. In addition to protecting the soil, mulch prevents the evaporation of water.
- Terracing—***Terracing*** is the practice of building earthen ridges or embankments that slow the rate of water runoff. Terraces are made with gradual slopes. They divert flowing water to grassy areas that do not wash like plowed fields. Terraces are covered with grass to prevent washing. Terraces should be built so equipment can cross them easily.
- Strip cropping—***Strip cropping*** is the practice of planting crops in strips. Plowed crops may be planted in alternating strips with crops that do not require plowing. The strips slow the speed of the wind and the rate of water runoff.
- Contour plowing—***Contour plowing*** is the practice of plowing around or across a slope rather than "with" the slope. The plowing is done so the rows follow a gradual decline and water flows slowly. Rows of crops should not go up and down hills; instead, the rows

should be in a direction that slows runoff. As a result, the rows are at angles (contours) that allow water runoff but slow its speed and retain the soil.

- **Crop rotation**—***Crop rotation*** is the practice of alternately planting different crops on the same land. Plant species use different nutrient amounts, so crop rotation conserves soil fertility. Soil organic matter is increased because different plants provide stems, leaves, and roots that hold the soil in place. For example, planting corn one year and wheat the next involves two different kinds of vegetation being left on the soil.
- **Diversion ditches and levees**—Water can sometimes be diverted around a field or its flow rate can be reduced. Small ditches and levees can be built to aid in this diversion. Levees may form ponds that hold water for animals to drink, for irrigation, and for restoring groundwater.
- **Grass waterways**—Grassed waterways may be established in and around fields. These waterways are low areas where runoff drains naturally. Grass is planted in the strips and protects the soil from erosion during runoff.

22–36. Grassed waterways help protect soil and retain water. (Courtesy, USDA)

DRAINAGE

Agricultural drainage is the process of removing excess water from land. Open ditches and underground drain tiles or tubes may be used. Land with poor surface runoff and internal drainage can be made more productive.

Drainage systems need to be carefully designed and installed. Ditches and tiles that are in the wrong place do not work properly. Consequently, care must be taken so the natural flow of area streams is not changed by the addition of excess water.

LAND FORMING

Land forming is the group of activities in which the surface of the land is shaped so it is more productive. Low places in fields may be filled and high places planed off.

22–37. Land forming with laser-guided equipment is used to obtain good drainage and to gain the desired surface qualities.

Land forming is necessary for some crops and irrigation systems. For example, rice requires land that will allow water to cover the surface of the field evenly. Fields where flood irrigation is used need to be land formed.

Land forming is also used to help remove excess water from land. Low places are filled so puddles of water do not stand after the rest of the land has dried. Water puddles can slow field preparation and planting.

Large land planes and laser systems are used in leveling or forming land.

LAND

Land is the term for all the characteristics of a site that make it suitable for use. Land is more than soil, but the soil is a major part. Land is also climate, slope, water supply, and location.

All the factors that compose land contribute to its productivity. Agricultural land is classified on the basis of productive capability into ***land capability classes***. Factors considered include the soil as well as other features.

LAND CLASSIFICATION FACTORS

Agencies of the U.S. Department of Agriculture have established classes of land based on capability. Major factors in capability are:

- **Surface texture**—This is the texture of the soil on the surface. It is the percentage of sand, silt, and clay. The soil triangle is used in texture classification.
- **Internal drainage**—This is the permeability of the soil. It is the ability of the soil to take up water and air and make it available for use by growing plants. The internal drainage of soil is classed as very slow, slow, moderate, and rapid. Of course, clay soils are slow and sandy soils are rapid. Parent material influences internal drainage. In addition, layers of rock can block the internal movement of water.

- **Depth of topsoil and subsoil**—Depth refers to the part of the soil that plant roots can use readily. Deep soils usually have topsoil and subsoil more than 30 inches deep. Moderately deep soils are 20 to 30 inches, shallow soils are 10 to 20 inches, and very shallow soils are less than 10 inches deep. Hardpans can restrict plant growth and limit soil depth.

- **Erosion**—Erosion refers to the amount of soil that has been lost. Land that has lost topsoil has eroded. It is not as productive as land that has plenty of topsoil. Plants grow best in topsoil that has the essential nutrients. No erosion or slight erosion exists when less than 25 percent of the topsoil has been lost and there are no gullies. Moderate erosion is a loss of 25 to 75 percent of the surface soil with the presence of a few small gullies. Severe erosion exists when 75 percent of the surface soil has been lost and small gullies are present. Very severe erosion exists when more than 75 percent of the topsoil has been lost and large gullies are present.

22–38. The loss of topsoil by erosion has reduced the capability of this land.

- **Slope**—Slope refers to unlevel land. The land may have gradual sloping, be hilly, or have severe slopes, as on the sides of mountains. In classing land, slope is based on the number of feet land falls over a distance of 100 feet. How land is classed varies a little from one state or region to another. In general, land that falls less than 2 feet in 100 feet is nearly level. Land that falls more than 12 feet in 100 feet is steep or very steep. Other slopes are gently sloping, which is a 2 to 5 feet fall in more than 100 feet of distance, and moderately sloping, which is a 5 to 8 feet fall in more than 100 feet of distance.

- **Surface runoff**—This is the rate at which water soaks into and runs off the surface of the soil. It deals mostly with excess water—that which does not soak in. Surface runoff can be good (no water problems), fair (water is removed slowly), poor (water is removed so slowly that soil is wet much of the time), and excessive (water is removed too rapidly, causing erosion). Level or near level land near creeks or rivers may have poor surface runoff. Water may stand on the land and interfere with its use.

LAND CLASSES

The eight land classes that are used are based on the factors in land capability. The eight classes are divided into those that can be cultivated and those that are unsuited for cultivation.

Land Suited for Cultivation

Four of the eight classes are suited for cultivation. Some of the classes require conservation practices to retain fertility. The classes are:

- **Class I: Very Good Land**—Class I land has few limitations. It is nearly level, has deep soil and good internal drainage, and the surface drains well. It is suitable land for most crops if the climate is satisfactory.
- **Class II: Good Land**—This class is not quite as good as Class I land, but it is very good. The slope may be up to 8 percent, so some practices may be needed to conserve the soil. The land can be used to grow many crops.
- **Class III: Moderately Good Land**—This land has slopes of up to 10 percent and may have slow internal drainage. Special conservation practices should be used. The land can be used for certain crops and for pasture.

CAREER PROFILE

SOIL PHYSICIST

A soil physicist studies the structure and processes in soil. The work may include identifying soil formations, developing management practices, and investigating water movement. This photo shows soil physicists using a vacuum pump to study soil water content. (Courtesy, Agricultural Research Service, USDA)

Soil physicists need college degrees in soil science or a closely related area that emphasizes soils. Many soil physicists have a master's degree and a doctorate in soil science. Begin in high school by taking physical science, agriscience, and related classes. Practical experience in growing crops is also beneficial.

Colleges, research stations, government agencies, and fertilizer and irrigation companies employ soil physicists.

22–39. Land being prepared for a grain crop—a use based on its suitability. (Courtesy, Case Corporation)

- **Class IV: Fairly Good Land**—This land has severe limitations. Slope and erosion, as well as internal drainage, may restrict the use of this land. It is best suited for permanent pasture, tree farming, hay, and wildlife habitat.

Land Unsuited for Cultivation

Four classes of land are unsuited for cultivation. This land tends to have a high percentage of slope or to be very low and wet.

22–40. Land limited by slope is sometimes used for a vineyard. (Courtesy, Natural Resources Conservation Service, USDA)

- **Class V: Unsuited for Cultivation**—This land can be used for pasture, hay crops, and tree farming. It may be wet and eroded. In addition, it may have a slope that causes problems.
- **Class VI: Not Suited for Row Crops**—This land is too steep to use in growing row crops. It may easily erode and have a rugged, hilly surface. Class VI land is best suited for tree farming and pasture.
- **Class VII: Highly Unsuited for Cultivation**—This land should not be cultivated. Erosion and other problems make this land best suited for tree farming and wildlife habitat. Sometimes, it can be used for pasture.

22–41. Fish ponds are a good use of this land. (The land is not flooded by a nearby stream and has soil high in clay that will hold water. It is not good for row crops because of poor internal drainage and slope.)

22–42. Land in swamps is unsuited for crops but may provide waterfowl habitat.

- **Class VIII: Unsuited for Plant Production**—This land may be very wet and in marshes or bayous. It is used for wildlife and recreation. Some of it may have a use in aquaculture. Areas known as "wetlands" may be in this class.

MEASURING AND DESCRIBING LAND

Land is a valuable resource. It needs to be measured accurately and described properly for legal purposes. Land measurement is sometimes referred to as surveying.

Land surveying is the process of measuring and marking real property. Maps and field notes may be prepared along with the written description. The survey serves as the basis for preparing a legal description of land.

Accuracy is essential in all aspects of land surveying. Measurements must be properly made, and boundary lines must be carefully marked. Plats and computer-based information must be accurate because many dollars of investment hinge on land surveys.

MEASURING LAND

Measurements with land include distances, angles, elevations, and areas. Distance measurements are often known as linear measurements.

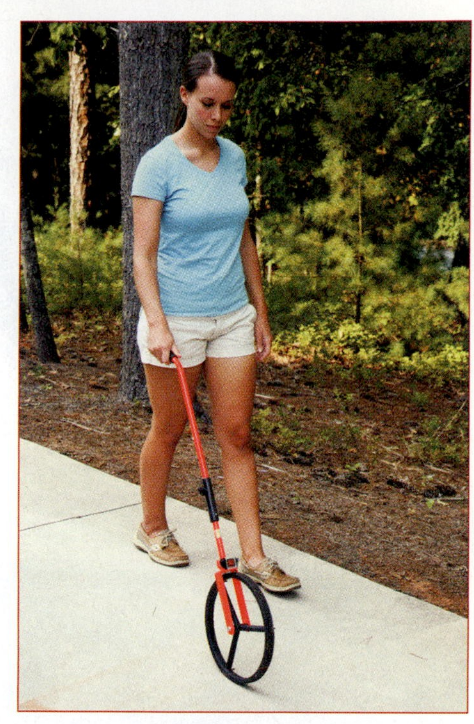

22–43. Using a measuring wheel (odometer wheel) to determine linear distance. (Courtesy, Education Images)

Linear Measurement

Linear measurement is determination of the distance between two points. Several points may be added together to obtain the overall linear measurement of a large or irregular area. The chain has been the common unit of land linear measurement. It is 66 feet long. A chain is divided into 100 links, and each link is 7.92 inches long. A surveyor's steel tape is typically 2 chains long—132 feet. Ten square chains equal 1 acre. A distance of 80 chains is 1 mile.

Linear measurement is accomplished in several ways:

- **Pacing**—Pacing is walking or stepping-off a distance. The length of a person's stride varies. Measure off 100 feet and count the number of paces needed to pace the distance. Afterward, always be consistent in the length of your step.
- **Chaining**—Chaining involves using a 66-foot chain divided into 100 links. Two people are needed so that pins are put down and taken up. The number of chain lengths is counted and multiplied by 66 to get the distance in feet. (A tape measure could be used similarly to a chain. Always consider its length and how it is used to ensure accuracy.)
- **Odometer wheel**—Also called a measuring wheel, the odometer wheel is rolled along the distance measured, and a counter records the distance. Always set the odometer wheel at zero before you begin. Go straight, and be consistent in wheel movement.
- **Instruments**—Some types of survey instruments provide distance information. An example is the total station surveying instrument that electronically senses horizontal distances and elevations.

Direction Measurement

Direction measurement is referencing a line in terms of true north. A **compass** is an instrument used to make direction measurements. Accuracy is very important. A compass mounted on a tripod is more accurate than a hand-held compass.

22–44. A compass provides direction measurement.

Compasses point to the magnetic north, which is not the same as true north. True north is where the North Pole is located. True north is about 1,300 miles from magnetic north. Also, the distance varies a small amount from one year to the next. The angle of difference between true north and magnetic north is known as declination and is determined using observations of the Sun. Compasses used in surveying land should correct for declination. If they do not, calculations will be needed to make the adjustment. The assistance of a certified surveyor or civil engineer may be required.

22–45. A clinometer can be used to determine slope (elevation). (Courtesy, Natural Resources Conservation Service, USDA)

Elevation Measurement

Elevation measurement is determination of the altitude of a point on land above (or below) sea level. Most land is above sea level. Yet some land is below sea level, such as areas in the Death Valley region of California.

Elevation measurement is typically done using differential surveying, which is establishing the difference in elevation between two points. Often, several steps are needed to make determination between the points. Total station surveying instruments can be used for elevation measurement.

Area Measurement

Area measurement is determination of the amount of land within set boundaries. It is based on horizontal surface area. Slope is often not considered. Land area in the United States is typically measured in square feet, acres, sections, and townships. One ***acre*** is 43,560 square feet or 10 square chains, a section is 640 acres, and a township is 36 sections.

Area measurement reveals the square units within boundaries. One approach is to make linear measurements of land boundaries and to use mathematical formulas to calculate the area. For example, if a rectangular residential lot is 100 feet by 300 feet, the lot contains 30,000 square feet. This is equal to 0.69 acre (30,000 divided by 43,560—the number of square feet in an acre).

Angle Measurement

Angle measurement is the determination of the degrees of an angle. Various approaches are used in measuring angles, such as using a surveying instrument, a chord, a tape-sine, or the 3-4-5 method. The 3-4-5 method is common in squaring the foundations of structures as they are being staked out, with most of the angles being 90° (square).

22–46. Using a survey instrument to measure land for a structure. (Courtesy, Natural Resources Conservation Service, USDA)

DESCRIBING LAND

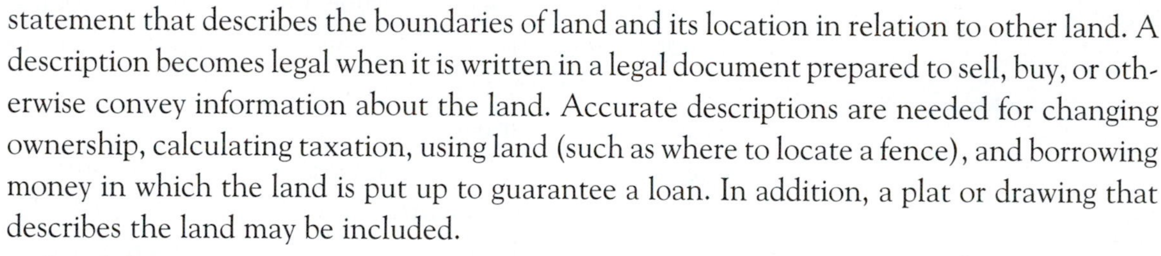

Land must often be described in writing. A ***land description*** is a written statement that describes the boundaries of land and its location in relation to other land. A description becomes legal when it is written in a legal document prepared to sell, buy, or otherwise convey information about the land. Accurate descriptions are needed for changing ownership, calculating taxation, using land (such as where to locate a fence), and borrowing money in which the land is put up to guarantee a loan. In addition, a plat or drawing that describes the land may be included.

Land descriptions are important in surveying and establishing land boundaries. A ***boundary*** is the limit or line of land. Boundaries help landowners know the exact location of the property they own.

In general, two survey systems are used in describing land: the metes and bounds system and the rectangular survey system.

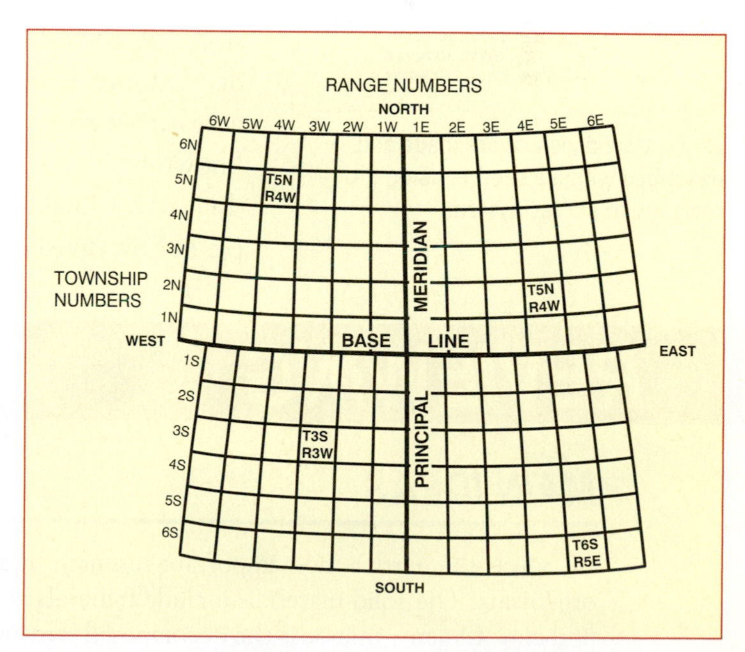

22–47. How township and range lines are drawn.

Metes and Bounds

Metes and bounds is a system of describing land in which a known starting point is used to establish and describe lines forming the property boundaries. The property is not referenced to a map or a lot on a map. The known starting point is the point of reference. Lines run a certain direction for a number of feet and then another direction for a number of feet. This is repeated until the property is described in its entirety.

Rectangular Survey System

The ***rectangular survey system*** is a method of describing land based on two fixed lines that are at right angles to each other. One line goes in a north-south direction; the other line goes in an east-west direction. Accurate use of a compass is essential with the rectangular survey system, which evolved through engineering and government laws as settlers moved into the Midwest and west. Today, 30 states use the rectangular survey system, including Alaska and Hawaii.

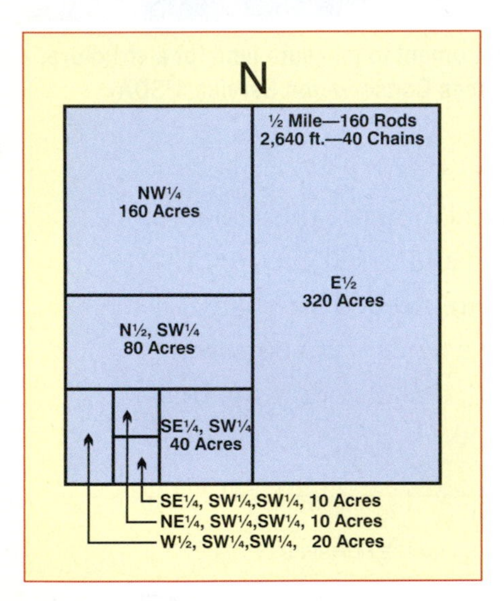

22–48. How divisions are made and described within a section using the rectangular survey system.

The two kinds of lines are the meridian and base line. A ***meridian*** is a true north-south line. Meridians were set by surveyors from identifiable points and are not in agreement with geographic longitude lines.

The east-west or horizontal lines are known as ***base lines***. They are sometimes known as latitude lines, but they do not follow latitude lines on the globe. Because Earth is round, correction lines are used every 24 miles to account for the curvature.

This system makes it easy to have a network of squares and rectangles set at intervals of 24 miles apart on Earth's surface. Since Earth is round, some correction will be needed. The 24-mile tracts are divided into 16 smaller tracts known as townships. A township is 6-miles square. Townships are further divided into 1-mile squares, which are known as sections (640 acres). Sections are further divided into quarters, which are known by location within the section. Quarters can be further divided for smaller tracts of land. Figure 22–48 shows an example of how divisions within a section are made and used.

REVIEWING

MAIN IDEAS

Soil is the earth's most important resource. It is made of solid materials, water, air, and living organisms. The solid materials include minerals and organic matter. Major minerals are sand, silt, and clay. Organic matter is the decomposed remains of plants and animals, often known as humus. Soil contains chemical elements, such as nitrogen, phosphorus, and potassium.

Soil has physical, chemical, biological, and drainage characteristics. Physical characteristics refer to the soil texture and structure. Chemical characteristics refer to the presence of chemical elements and general conditions that they cause. Biological characteristics refer to all of the living plants and animals found in soil.

Soil composition varies. The proportions of minerals and organic matter can be quite different. Soil texture is based on the amounts of the major ingredients that are present. Texture also has much to do with how the soil is managed. Clay soils tend to hold water better, while sandy soils do not. Most of the nutrients in soil are in the top few inches, known as topsoil. Conservation practices need to be used to keep the topsoil from eroding.

Soil management can help retain its fertility for future generations. Managing nutrients, controlling erosion, and properly draining and forming land can help maintain productivity.

Land is classified on the basis of how it is best used. Classification includes soil as well as other features, such as slope and water supply. Land measurement involves determining linear, direction, elevation, angle, and area measurements. Land description involves metes and bounds or the rectangular survey system.

QUESTIONS

Answer the following questions, using complete sentences and correct spelling.

1. What is soil? Where is it found?
2. What are the ingredients in soil? How do these ingredients determine the nature of the soil?
3. What is organic matter? How is it formed?
4. What is the soil triangle? Why is it useful?
5. Why is biological activity important in the soil?
6. How is soil formed? What processes are involved?
7. What is a soil profile? What horizons are found?
8. Explain water table.
9. What forms of water are found in soil? Distinguish between the forms.
10. How is water lost from the soil? What can be done to reduce the loss?
11. What is internal drainage? Why is it important?
12. What is pH? What pH ranges are found? Why is pH important?
13. How is pH changed?
14. What is land classification? What factors are used in determining the class of land?
15. Distinguish between the two major classes of land.
16. What is erosion? How is it prevented?
17. What management practices may be used to make land more productive?
18. What measurements are made of land, and how are these commonly made?
19. What two systems of describing land are used? Distinguish between the two.

EVALUATING

Match each term with its correct definition.

a. horizon
b. humus
c. land forming
d. land capability classes
e. terracing
f. water table
g. soil pH
h. acre
i. irrigation-induced salinity
j. soil texture

_____1. Soil salinity resulting from irrigation water with salts.

_____2. The proportion of sand, silt, and clay in soil.

_____3. A distinct layer of soil materials.

_____4. Depth of the natural level of free water in soil.

_____5. A measurement of the acidity or alkalinity of soil.

_____6. Shaping the surface of land to improve productivity.

_____7. Well-decomposed organic matter.

_____8. Building earthen ridges or embankments that slow the rate of water runoff.

_____9. The classification of agricultural land on the basis of its productive capability.

_____10. 43,560 square feet or 10 square chains.

EXPLORING

1. Contact the NRCS office in your area to obtain information on nutrient management. Interview a technician about the preparation of a Comprehensive Nutrient Management Plan (CNMP). Ask to review a sample plan. Prepare a plan for the school farm or a farm in your local community.

2. Determine if your state or school participates in land judging. If so, request a copy of the regulations. Then form a land judging team.

3. Collect a sample of soil, and then test it for pH. Run the test in your school lab, or send it to a soil testing laboratory. Determine the change needed in the pH, if any, for selected plants.

4. Using a transit, lay out a contour line on the school grounds or on nearby land. Your teacher or a soil conservation technician can help with this activity.

5. Study the profile of the soil in your community. Determine the depth of the horizons, general texture of the materials, and the parent material for the soil.

Part Seven

Physical Science and Technology

CHAPTER 23

Chemistry in AgriScience

OBJECTIVES

This chapter introduces important principles of chemistry as related to agriscience. It has the following objectives:

1. Explain the concept and properties of matter.
2. Describe the major kinds of matter.
3. Explain the meaning of compounds and how they are formed.
4. Discuss the importance of carbon and organic chemistry.
5. Distinguish between solutions and suspensions.
6. Explain acids, bases, and salts.
7. Identify common chemical processes in agriscience.
8. Describe the importance of measurements in chemistry.

TERMS

acid
alkane
area
atom
atomic mass
atomic number
base
bonding
Celsius scale
change in state
chemical property
chemical reaction
colloids
compound
corrosion
customary measurement system
diffusion
element
emulsion
energy levels
Fahrenheit scale
hydrocarbon
ion
isotope
Law of Definite Composition
linear measurement
mass
matter
measurement
metal
metric system
mixture
molecule
nonmetal
organic chemistry
oxidation
parts per unit
Periodic Law
Periodic Table of the Elements
pH
physical property
plasma
pressure
salt
soluble
solute
solution
solvent
suspension
temperature
volume
weighing scale
weight

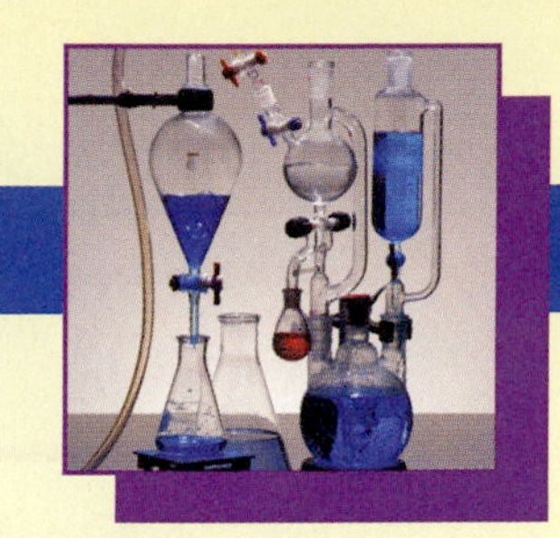

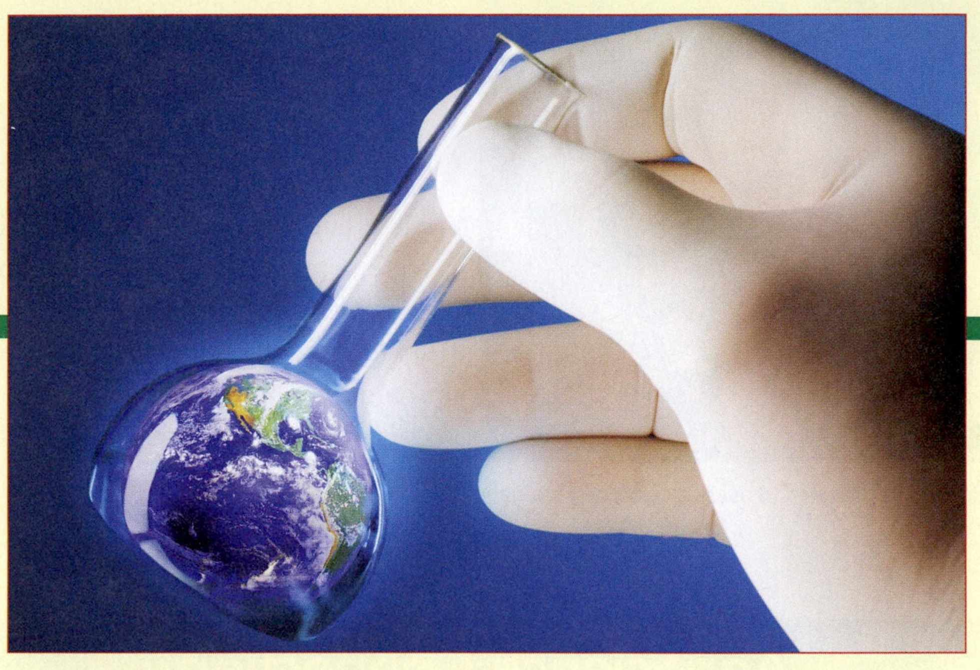

23–1. Chemistry is used to determine the makeup of the structures and processes of all things that compose the earth.

ALL LIVING and nonliving things are made of chemical structures. Such importance emphasizes chemistry as a central part of all sciences, including agriscience. Chemical elements are a part of all the structures and processes found in living things. ***Chemistry*** is the study of the composition, structure, properties and processes of matter, and the energy involved. Much of chemistry relates to biological science.

What is a material's composition? What happens when we heat, cool, or mix a material with other substances? Why? How are chemicals involved in everyday processes in living things? What chemicals are the building blocks that make living structures? The study of chemistry in agriscience helps us answer these questions.

MATTER

23–2. Laboratory analysis can be used to assess common materials found in a home.

Matter is anything that has volume and mass. Knowing the meaning of volume and mass helps understand matter. It could be a piece of coal, a diamond in a ring, or a tank of LP gas.

Volume is an amount of space measured in cubic units, such as cubic centimeters (cc) or cubic inches (in^3). For example, amounts of injections are measured as cc in a syringe. In metric units, volume can also be measured in common units, such as liters and milliliters. Volume is usually obvious: Milk in a jar takes up space, and the capacity of a jar to hold milk is its volume. Most of the things that

AgriScience Connection

COAGULATION: IRREVERSIBLE TRANSFORMATION

Heat changes materials in several ways. One material that can be changed by heat is the chicken egg. A raw egg is very watery. By exposing the egg to heat for a few minutes, a chemical process occurs inside the egg so it is no longer watery. Coagulation occurs.

A common way of heating an egg is by placing it in water and gradually raising the temperature of the water to near the boiling point. (Although it is called a boiled egg, it should not actually be boiled because undesirable chemical changes may occur in its color.) After a few minutes, the egg is said to be "done." The process of coagulation has been completed. The proteins that were in a liquid state have been changed to a drier, more solid state. Since the egg is enclosed in a shell, little water is lost through evaporation. A chemical process has occurred, and the process cannot be reversed.

This shows a carton of eggs and an inset of a boiled egg that has been sliced in half. Coagulation has worked well. (Courtesy, Education Images)

take up space can be seen. However, some things, such as gases in the air, are not readily visible. Regardless, matter has volume.

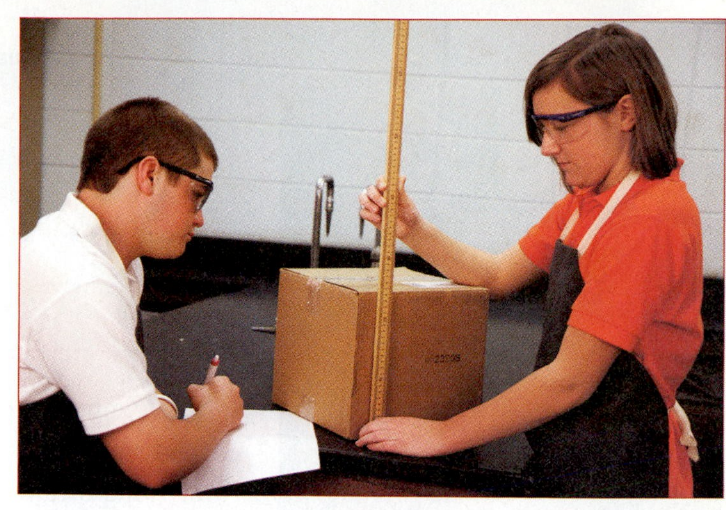

23–3. A meter stick can be used to measure the length, width, and height of a box to calculate volume. (Courtesy, Education Images)

Mass is the amount of matter that an object contains. Most often, mass is referred to as weight, but it is different from weight. ***Weight*** is a measure of the gravitational pull on matter. The amount of matter in an object remains constant, but the weight can change depending on the object's location. An object on the moon weighs less than an object on Earth because the gravity on the moon is less. Yet the amount of matter remains the same. In the metric system, mass is usually measured in kilograms (kg), but grams (g) and milligrams (mg) are also used for amounts less than a kilogram. English units of mass are pounds (lb) and ounces (oz).

PROPERTIES OF MATTER

Matter has two types of properties: physical and chemical. Each type of matter has distinct identifying properties.

23–4. Studying the physical properties of matter includes the study of crystalline structures of rocks, such as quartz.

Physical Properties

A ***physical property*** can be determined without changing the identity or composition of matter. Physical properties include color, odor, solubility (how much dissolves), melting and boiling points, hardness, density, and crystal formation. Differences in matter can be determined by observing the physical properties. Matter does not always have all of the types of physical properties. For example, rocks may be odorless.

Changes in State or Phase Changes. Melting point and boiling point are part of an important class of physical changes called changes in state. A ***change in state*** or

23–5. Many substances can be separated from each other in the laboratory by their melting or boiling points. (Courtesy, Education Images)

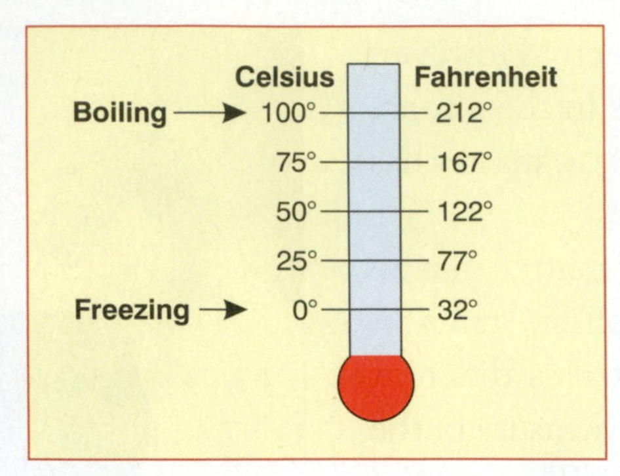

23–6. A comparison of temperature using Celsius and Fahrenheit scales.

phase change is a physical change from one form to another. The three common states of matter are solid, liquid, and gas. A fourth important state of matter is ***plasma***, which is a high-temperature gaseous state in which atoms lose most of their electrons—particles that are part of atoms. Most plasma is found in the gases of the universe. A common location in our environment is in fluorescent bulbs.

Two conditions are involved when a phase change occurs: temperature and pressure.

23–7. Antifreeze is mixed with water in the cooling systems of engines to prevent freezing in cold weather. (Courtesy, Education Images)

- **Temperature**—***Temperature*** is the presence of heat and is evident in the activity of the molecules in matter. As heat increases, molecules move more rapidly. At a certain temperature, they are so active that they no longer stay together. As a result, they become a gas. The liquid state occurs at a temperature range between a gas and a solid. When molecules move so slowly that they stop being a liquid, they become a solid.

  Temperature is measured on two types of scales. The scale used in the English system is the ***Fahrenheit scale***. On this scale, the freezing point of water is 32°F, and the boiling point of water is 212°F. Two scales are used in the metric system. The Kelvin scale is used only when working with gases that require mathematic calculations. The commonly used metric scale is the ***Celsius scale***. The boiling point of water on this scale is 100°C, and the freezing point is 0°C.

Matter varies in the heat required for different phases. Water becomes a solid at 32°F (0°C) and a gas at 212°F (100°C) at sea level. When water becomes a solid, it expands (increases in volume or becomes bigger). The force of expansion can crack rocks, burst water pipes, or break the iron parts of engines. Most other liquids contract when they solidify.

Substances mixed with water change the temperature at which water freezes and boils. This is why antifreeze is added to water in engine cooling systems in very cold or very hot weather. Most antifreeze is ethylene glycol. It lowers the freezing point of water and raises the boiling point. The higher the concentration of antifreeze, the lower the freezing point and the higher the boiling point.

Other kinds of matter change phases at different temperatures. Iron becomes a liquid at about 2,800°F (1,500°C). It returns to a solid below this temperature. Making steel involves heating iron to a liquid form, adding other matter, and cooling it in various shapes and forms.

- **Pressure—*Pressure*** is the force on a solid or a liquid. This includes atmospheric pressure as well as pressure in a fuel tank, a can of food, or a pipeline. Atmospheric pressure affects the temperature at which substances change from solids to liquids and gases. A good example is water. As the atmospheric pressure decreases, the temperature needed for it to boil and become a vapor also decreases.

  Gases kept in storage tanks are often compressed. Examples include fuels, such as butane and refrigerants that help make coolers and refrigerators cool. Solids and liquids can be compressed very little. Some gases will become liquids when compressed, such as liquefied petroleum.

  Gases tend to occupy all of the space available to them. A tiny can of fumigant can be used in a large greenhouse to control pests. The gas spreads throughout the greenhouse by ***diffusion***, which is the movement of a substance from an area of greater concentration to an area of lesser concentration.

  Solids will turn into liquid forms at lower temperatures when under pressure. Ice will become a liquid below the normal freezing point if pressure is applied. An example is pressure under the blades of ice skates.

23–8. A press creates pressure on substances, flattening them into thin, filmy powders for infrared spectroscopy analysis.

23–9. A robot is removing and assessing hazardous chemicals in soil samples. (This provides information about the soil in the field from which the sample was taken. It also protects humans from the hazardous chemicals.) (Courtesy, Agricultural Research Service, USDA)

Chemical Properties

A ***chemical property*** involves a substance's ability to go through changes that transform it into a different substance. Chemical properties are easiest to see when substances react to form new substances with a different appearance. When charcoal (a solid) burns in the air, it combines with the oxygen to become a new substance: carbon dioxide (a gas). The properties of the original substances are gone. When substances experience these changes, it is called a ***chemical reaction***. Some matter reacts violently as when fuel burns. Other matter is relatively chemically inactive, such as gold or the noble gases. They combine with few other elements.

Information about how materials react is helpful when using them. Most materials have labels that describe their dangers. For instance, ammonium nitrate fertilizer can be an explosive. Material safety data sheets (MSDS) often accompany chemicals with detailed information. (Refer to Chapter 6 for more information on MSDS.)

KINDS OF MATTER

Matter is classed into three general groups: elements, compounds, and mixtures.

ELEMENTS

An ***element*** is a substance that cannot be broken down into simpler materials by ordinary means. Elements are pure substances that always have exactly the same characteristic properties. They serve as the building blocks of matter. Currently, there are 115 elements. Of these, 92 are natural elements, and the others are artificially made. Some artificial elements may not have permanent names and are assigned letters until the name is definite. Most artificial elements are radioactive and were isolated after the creation of the atomic bomb.

Each element has a name as well as a symbol. The major elements in the crust of the earth are oxygen, silicon, iron, aluminum, calcium, sodium, potassium, and magnesium. Oxygen is present in the largest amount; it makes up about 46 percent of the earth's crust.

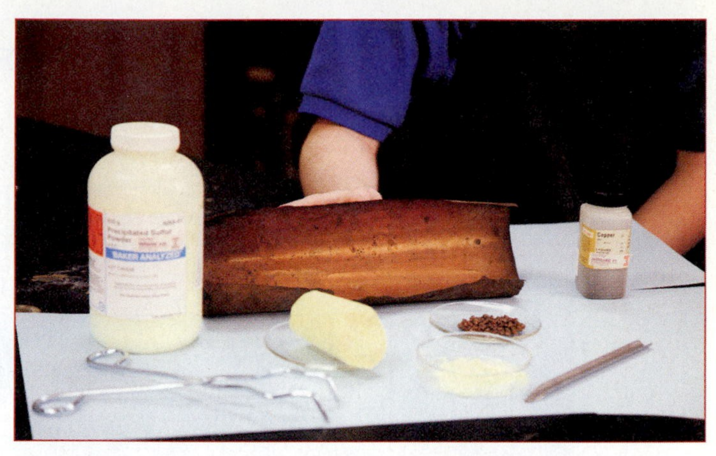

23–10. Metal elements, like copper, are identified by their color and metallic shine. Nonmetals, like sulphur, are brittle and can be crushed into powder. This shows copper and sulfur. (Courtesy, Education Images)

Table 23–1. Common Elements and Their Symbols

Element	Symbol	Element	Symbol
Aluminum	Al	Magnesium	Mg
Arsenic	As	Manganese	Mn
Calcium	Ca	Mercury	Hg
Carbon	C	Nickel	Ni
Chlorine	Cl	Nitrogen	N
Chromium	Cr	Oxygen	O
Cobalt	Co	Phosphorus	P
Copper	Cu	Potassium	K
Fluorine	F	Silicon	Si
Gold	Au	Silver	Ag
Hydrogen	H	Sodium	Na
Iodine	I	Sulfur	S
Iron	Fe	Tin	Sn
Lead	Pb	Zinc	Zn

The Periodic Table

The ***Periodic Table of the Elements*** is a table used to group and organize the elements for study. Elements are organized into groups based on similar chemical properties. Properties of elements can be predicted from the table. The table includes the symbol for

PERIODIC TABLE OF THE ELEMENTS

Metals — Transition Elements — Nonmetals

Period	1	2	3	4	5	6	7	8	9	10	11	12	13	14	15	16	17	18
1	1 H Hydrogen 1																	2 He Helium 4
2	3 Li Lithium 7	4 Be Beryllium 9											5 B Boron 11	6 C Carbon 12	7 N Nitrogen 14	8 O Oxygen 16	9 F Fluorine 19	10 Ne Neon 20
3	11 Na Sodium 23	12 Mg Magnesium 24											13 Al Aluminum 27	14 Si Silicon 28	15 P Phosphorus 31	16 S Sulfur 32	17 Cl Chlorine 35	18 Ar Argon 40
4	19 K Potassium 39	20 Ca Calcium 40	21 Sc Scandium 45	22 Ti Titanium 48	23 V Vanadium 51	24 Cr Chromium 52	25 Mn Manganese 55	26 Fe Iron 56	27 Co Cobalt 59	28 Ni Nickel 59	29 Cu Copper 64	30 Zn Zinc 65	31 Ga Gallium 70	32 Ge Germanium 73	33 As Arsenic 75	34 Se Selenium 79	35 Br Bromine 80	36 Kr Krypton 84
5	37 Rb Rubidium 85	38 Sr Strontium 88	39 Y Yttrium 89	40 Zr Zirconium 91	41 Nb Niobium 93	42 Mo Molybdenum 96	43 Tc Technetium 98	44 Ru Ruthenium 101	45 Rh Rhodium 103	46 Pd Palladium 106	47 Ag Silver 108	48 Cd Cadmium 112	49 In Indium 115	50 Sn Tin 119	51 Sb Antimony 122	52 Te Tellurium 128	53 I Iodine 127	54 Xe Xenon 131
6	55 Cs Cesium 133	56 Ba Barium 137	57-71 **	72 Hf Hafnium 178	73 Ta Tantalum 181	74 W Tungsten 184	75 Re Rhenium 186	76 Os Osmium 190	77 Ir Iridium 192	78 Pt Platimun 195	79 Au Gold 197	80 Hg Mercury 201	81 Tl Thallium 204	82 Pb Lead 207	83 Bi Bismuth 209	84 Po Polonium 209	85 At Astatine 210	86 Rn Radon 222
7	87 Fr Francium 223	88 Ra Radium 226	89-103 ##	104 Rf 261	105 Db 262	106 Sg 263	107 Bh 264	108 Hs 265	109 Mt 268	110 Uun 269	111 Uuu 272	112 Uub 269		114 Uuq XXX		116 Uuh XXX		118 Uuc XXX

Elements marked with XXX have recently been reported

Lanthanide Series **	57 La Lanthanum 139	58 Ce Cerium 140	59 Pr Praseodymium 141	60 Nd Neodymium 144	61 Pm Promethium 145	62 Sm Samarium 150	63 Eu Europium 152	64 Gd Gadolinium 157	65 Tb Terbium 159	66 Dy Dysprosium 163	67 Ho Holmium 165	68 Er Erbium 167	69 Tm Thulium 169	70 Yb Ytterbium 173	71 Lu Lutetium 175
Actinide Series ##	89 Ac Actinium 227	90 Th Thorium 232	91 Pa Protactinium 231	92 U Uranium 238	93 Np Neptunium 237	94 Pu Plutonium 244	95 Am Americium 243	96 Cmn Curium 247	97 Bk Berkelium 247	98 Cy Californium 251	99 Es Einsteinium 252	100 Fm Fermium 257	101 Md Mendelevium 258	102 No Nobelium 259	103 Lr Lawrencium 262

23–11. A Periodic Table of the Elements. (Elements 113, 115, and 117 have not been discovered, but they will be artificial elements.)

each element, atomic numbers, and atomic masses. The modern Periodic Table of the Elements is arranged in columns and rows. The horizontal rows are known as periods, and the vertical columns are called groups or families.

- **Atomic numbers**—All elements have atomic numbers by which they are identified. An ***atomic number*** is the number of protons in the nucleus of an atom. An ***atom*** is the smallest unit in which an element can exist. Some atoms can exist alone, such as gold (Au); others exist in pairs or different combinations, such as oxygen (O_2). The nucleus is the central part of an atom. It has a positive charge and contains most of the mass of the atom. The positive charge is due to the presence of protons.

  The properties of elements are dictated by their atomic numbers. This is stated as the ***Periodic Law***: *The chemical and physical properties of the elements are periodic functions of their atomic numbers.*

  The ***atomic mass*** or mass number of an element is the total of the number of protons and neutrons. The neutrons are in the nucleus along with the protons. Neutrons are particles that have a neutral electrical charge and contain mass. Protons have a positive charge. Most elements have one or more forms called isotopes. An ***isotope*** is an atom of the same element that has a different mass. Isotopes all have the same number of protons and electrons. However, they have different numbers of neutrons and, therefore, different mass numbers. An example is the three isotopes of hydrogen: protium, deuterium, and tritium.

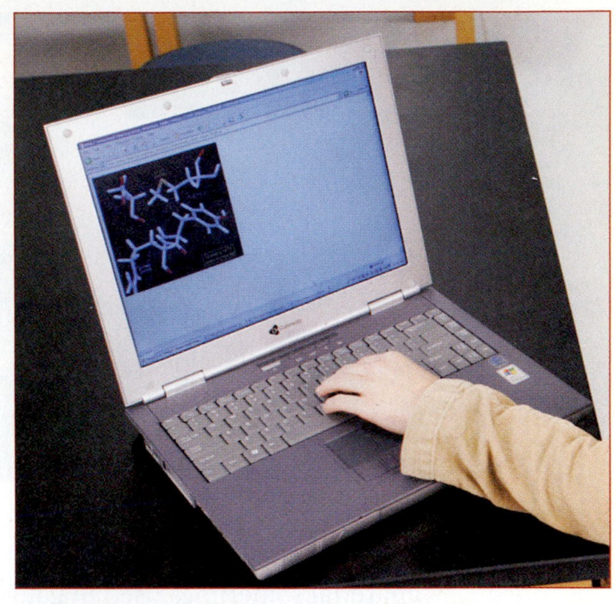

23–12. Computers are useful tools in the laboratory for studying structures of chemical compounds and for running virtual labs. They also allow for practice in writing equations and in studying electron configurations. (Courtesy, Education Images)

- **Electron cloud theory**—Atoms have electrons that move in electron clouds like swarms of bees around the nucleus. These areas are the most probable locations in which an electron can be found at any given moment. Electrons have a negative charge and are important in the combination of elements to form compounds. Based on the amount of energy the electrons have, they move at different distances from the nucleus in areas that we call ***energy levels***. In earlier years, these areas were called orbitals. The outer energy level of an atom may have an incomplete number of electrons. Trying to obtain a stable number of electrons causes matter to give electrons to or take electrons from other elements when forming chemical bonds. The electrons involved in bonding are called valence electrons. The valence of an element determines how it will react chemically—with which elements and with how many other elements it will join to form a compound.

Categories of Elements

Each element has a unique set of characteristics. However, some similarities exist among the characteristics of elements. Based on their similarities and differences, the elements can be grouped into two categories: metals and nonmetals.

- **Metals**—A ***metal*** is an element with a shiny appearance, known as metallic luster. Most metals have a silvery or grayish white appearance, but a few metals have distinct colors, such as gold (Au) and copper (Cu). Metals are good conductors of heat and electricity. They also can be formed into many shapes, such as iron into plows, and can be drawn into thin wires. Metals transfer electrons when they react. Most metals are mined from nature as ore.

23–13. Malleability of copper (Cu) allows it to be hammered into sheets, pulled into a thin wire, or made into pellets. (Courtesy, Education Images)

Most metals are solids at normal temperatures. One exception is mercury (Hg)—a material sometimes used in gauges to measure temperature and pressure. Mercury is a liquid at room temperature.

- **Nonmetals**—A ***nonmetal*** is an element that is a poor conductor of heat; it is brittle and is not easily shaped. Nonmetals gain electrons in chemical reactions. Most nonmetals are solids at room temperature, such as iodine (I) and phosphorus (P). All the Noble gases [neon (Ne), argon (Ar), krypton (Kr), xenon (Xe), radon Ra)] plus common gases like Chlorine (Cl), hydrogen (H), nitrogen (N), fluorine (F), and oxygen (O) are found in gaseous form at room temperature. Only one nonmetal, bromine (Br), is a liquid at room temperature.

Technology Connection

SOIL NUTRIENT BUILDUP

Collecting a soil sample and doing a laboratory analysis is not always essential. Equipment that makes use of an electromagnetic geoconductivity meter to measure soil conductivity is shown in operation here. The process determines electromagnetic induction (EI) as related to soil nutrient movement.

When the equipment is connected to a GPS unit and a computer, field maps and other useful data can be prepared. Year-to-year studies allow the investigation of nutrient movement within soil. Without such equipment, many soil samples and laboratory analyses would be needed each year. (Courtesy, Agricultural Research Service, USDA)

The metalloids are a section of the nonmetals that have characteristics of metals and nonmetals. They tend to be semiconductors of electricity and are found along the metalloid line that separates the Periodic Table of the Elements into metals and nonmetals. An important metalloid is silicon, which is used in circuitry in computers.

COMPOUNDS

A ***compound*** is a substance made of two or more elements that have chemically combined. Compounds can be broken into two or more simpler substances by ordinary chemical means. While combined, the two elements lose their separate chemical identities. An example is water. It is made of two elements—hydrogen and oxygen—that are gases. When chemically combined, they form water, which is a liquid.

Pure elements rarely occur in nature. They are usually found in compounds. For example, pure sodium does not exist in the earth's crust. It is found combined with other elements as compounds, like table salt (sodium chloride—NaCl).

A compound is always made of the same elements in a definite relationship. Thus, the ***Law of Definite Composition***: *All compounds have a definite composition by mass*. Because the same proportions of raw materials are always used to produce a product, the amounts of ingredients necessary for making a compound do not change.

23–14. Table salt is a compound with definite composition. (Courtesy, Education Images)

MIXTURES

A ***mixture*** is a blend of two or more kinds of matter, each of which keeps its own identity and properties. Air is a mixture because it is made of several gases. Each gas in the air is different and could be separated from the others, usually according to their different boiling points.

Complete fertilizers are mixtures. Compounds that provide nitrogen, phosphorous, and potassium are blended. The amounts of each substance in a blend are based on the intended analysis of the fertilizer. None of the fertilizer ingredients are used in elemental form; they are used as compounds. The ele-

23–15. Roundup® herbicide is a mixture of ingredients.

ment separates and becomes available to plants in the soil, depending on pH and other soil characteristics.

FORMING NEW COMPOUNDS

Atoms of elements combine to form compounds. The same compound is always composed of the same atoms in the same ratio. Chemical formulas are a shorter method of writing compounds.

Table 23–2. Common Compounds Used in AgriScience

Compound	Formula	State as Found in Earth Environment
Acetic acid (vinegar)	$HC_2H_30_2$	liquid
Ammonium nitrate (fertilizer)	NH_4NO_3	solid
Calcium carbonate (limestone)	$CaCO_3$	solid
Carbon dioxide	CO_2	gas
Citric acid (lemon juice)	$C_3H_4(OH)(COOH)_3$	liquid
Hydrochloric acid (muriatic acid)	HCl	liquid
Sodium bicarbonate (baking soda)	$NaHCO_3$	solid
Sodium chloride (table salt)	$NaCl$	solid
Sodium hydroxide (lye)	$NaOH$	liquid
Sodium nitrate (fertilizer)	$NaNO_3$	solid
Sulfuric acid	H_2SO_4	liquid
Water	H_20	liquid, gas, and solid

ACADEMIC CONNECTION

CHEMISTRY

Relate your study of chemical applications in agriscience to your studies in chemistry class. Explore the important role of chemical processes in plant and animal production. Use your learning in agriscience to support your mastery in chemistry.

Compounds can be formed as molecules. A ***molecule*** is the smallest group of atoms that acts together to form a stable, independent substance. The simplest molecules contain only two atoms. An example is O_2 (free oxygen). In nature, oxygen is stable only when bonded with another oxygen atom.

Many molecules contain more than two atoms. An example of a molecule with three atoms is H_2O (water). One molecule of water is made of two atoms of hydrogen and one atom of oxygen. A molecule of glucose sugar, as made by plants, has 6 atoms of carbon, 12 atoms of hydrogen, and 6 atoms of oxygen ($C_6H_{12}O_6$).

IONS

An atom has energy levels in which electrons move around a nucleus. The outer energy level can be incomplete. It can gain or lose one or more valence electrons. When an atom gains or loses a valence electron, it develops a charge. This charged particle is known as an ***ion***. For example, a chlorine atom can gain one electron to become the chloride ion and have a negative charge. A sodium atom can lose one electron to become a sodium ion.

BONDING

Bonding is the action of joining two or more elements to form a compound. Valence electrons are shared or transferred by the atoms forming chemical bonds in the compound. Ionic bonds are formed when electrons are transferred. Sodium chloride (NaCl) is an example of ionic bonding. The compound is held together by the attraction of the oppositely charged ions. Covalent bonding occurs when electrons are shared. The overlapping energy levels of the shared electrons hold the molecule together. Natural gas (methane, or CH_4) is an example of covalent bonding.

Bonding usually results because two or more atoms with an unstable number of outer energy level electrons form a stable compound. For example, water is more stable as a compound than oxygen and hydrogen are as elements.

23–16. Some of the compounds used in agriscience must be handled carefully, such as sulfuric acid (H_2SO_4). (Courtesy, Education Images)

CHEMICAL REACTIONS

Chemical reactions involve bonding and the formation of compounds. Four types of chemical reactions are common:

- **Composition—**Composition occurs when two or more substances join to form a more complex substance. For example, $2Na + Cl_2 \rightarrow 2NaCl$.
- **Decomposition—**Decomposition occurs when one substance breaks down to form two or more other substances. For example, $2NaCl \rightarrow 2Na + Cl_2$.
- **Single replacement—**Single replacement results when one element replaces another element in a compound. For example, $Zn + 2HCl \rightarrow ZnCl_2 + H_2$.

23–17. Setting up an apparatus to run a reaction in a fume hood.

23–18. Using a rotary evaporator to remove solvents under pressure.

- **Double replacement (ionic)**—Double replacement reactions occur when two compounds exchange elements. For example, $NaOH_{aq}$ + HCl_{aq} → $NaCl_{aq}$ + H_2O. (Note: The symbol $_{aq}$ means aqueous or "in water.")

Some chemical reactions involve oxidation and reduction. This causes them to depart from these exact patterns. (Oxidation and reduction are covered later in this chapter.)

ORGANIC CHEMISTRY

Carbon is an element found in all living things and their foods. Carbon is different from other elements in that it can bond in almost endless chains, ring designs, and complex shapes.

The study of the compounds of carbon is known as ***organic chemistry***. It includes the use of carbon in living organisms and their structures. All organic compounds contain carbon and involve covalent bonding. Carbon ranks 17th among all elements in terms of volume found on the earth.

Carbon is important in other compounds that are usually formed by formerly living materials. Coal, oil, natural gas, diamonds, graphite, and other products contain carbon. Of course, wood and paper also contain carbon.

Important gases contain carbon. Carbon dioxide is used by plants and is given off by animals. Burning wood and fuel produces gases that contain carbon, such as carbon dioxide and carbon monoxide.

23–19. An aspirin (acetylsalicylic acid) molecule is depicted by a model using colors for elements: red is O, black is C, and white is H.

HYDROCARBONS

Many materials in our environment are hydrocarbons. A **hydrocarbon** is a simple organic compound made of two elements: hydrogen and carbon. The common hydrocarbons are found in crude oil and natural gas and their products, including gasoline, lubricating oil, plastic, artificial fibers, and rubber. The difference between these products is the number of atoms, how the carbon bonds are arranged, and the type of covalent bonding.

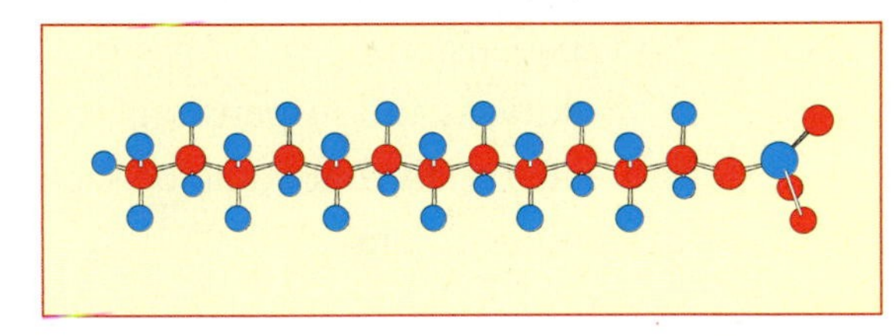

23–20. A hydrocarbon molecule generated by a computer for lauryl sulfate ($CH_3(CH_2)_{10}CH_2OSO_3-$), a substance used in detergents and toothpaste.

A good example of a hydrocarbon is natural gas. It is mostly methane (CH_4). Natural gas and petroleum were formed millions of years ago by plants and animals that decayed under pressure below the earth's surface.

Petroleum is refined to make usable forms, such as gasoline, diesel fuel, and engine oil. These items are produced at different stages in the refining process.

SUBSTITUTION COMPOUNDS

An **alkane** is a hydrocarbon that has single covalent bonds only. One of the hydrogen atoms can be replaced with an atom of another element. The halogen elements (chlorine, fluorine, bromine, and iodine) can combine easily with hydrocarbons to form halide compounds. Minor changes in the carbon bonds result in different products that are similar in some respects.

The compounds known as alcohols are made from organic halides. Methanol and ethanol are used as fuel or as additives to fuel. Some of these are made from grains and may be combined with other fuels. An example is gasohol, which is a combination of 10 percent ethanol and 90 percent nonleaded gasoline. Fuels made from grain tend to be more expensive than those made from petroleum.

Other forms of alcohol include ethylene glycol (a form used as antifreeze) and glycerol or glycerin (a form used as an ingredient in products such as medicines, cosmetics, and soap).

More complex bonding in hydrocarbons forms double and triple covalent bonds. A double covalent bond is the sharing of two sets of electrons. A triple covalent bond shares three sets of electrons. All these variations of bond types and substitutions of elements results in a vast number of organic compounds.

SOLUTIONS AND SUSPENSIONS

Solid materials are sometimes dissolved in water or other fluids. If the solid will dissolve in the liquid, it is ***soluble***. Dissolving breaks apart a solid into particles that are loosely attached to the solvent. The solid materials are diffused in the liquid in two forms: solution and suspension.

Solutions and suspensions are not new chemical compounds. They are mixtures of two or more materials that could be separated into their original forms. Yet separating them may be difficult or impractical.

23–21. When table sugar dissolves in water, it forms a colorless solution that is uniformly sweet. Agitating the solution with a spoon makes the sugar dissolve faster. (Courtesy, Education Images)

SOLUTIONS

A ***solution*** is a uniform mixture of two or more substances that have the same composition and properties throughout. The amounts that can be mixed together vary. The amounts that will go into a solution are usually limited. Solutions can exist as gases, liquids, or solids. An example of a gaseous solution is air. Each type of gas in air is evenly dispersed throughout the air. An example of a solid solution is copper in nickel, which is used to form an alloy. Sugar in water is an example of a solid in a liquid.

The material that does the dissolving is a ***solvent***, and the substance that is dissolved is a ***solute***. Molecules of the solute are distributed evenly throughout the solvent.

Use a few drops of food coloring in a glass of water for an easy-to-see example of a solution. The food coloring spreads evenly throughout the water by diffusion. After a time, all of the water will have an equal amount of color.

In a liquid solution, the substance dissolved and the solvent pass through various filters. A salt or sugar solution will pass through ordinary filters with the water. For the same reason, special water-conditioning equipment is needed with water that contains iron and other minerals.

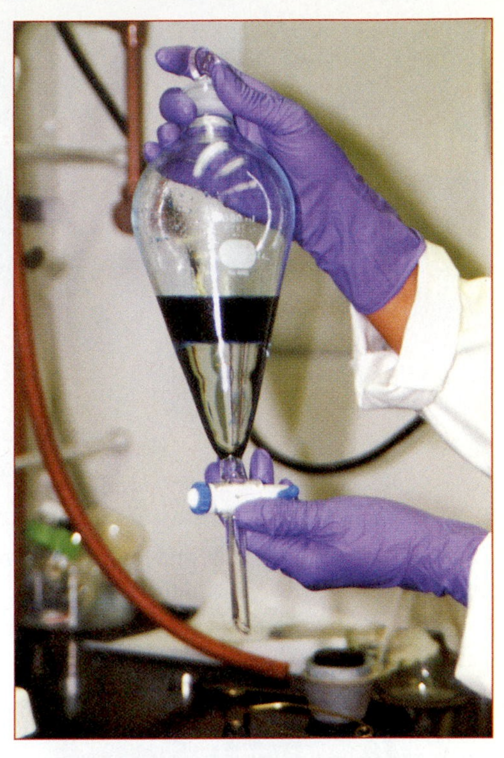

23–22. A separatory funnel is used to separate some kinds of fluids. Here, chloroform (clear fluid at bottom) is being separated from an aqueous dye solution (dark blue floating on clear fluid).

SUSPENSIONS

Solutes and solvents do not always form good solutions. Solutes may be made of large particles that settle out unless the mixture is constantly stirred or agitated. This type of mixture is called a ***suspension***. The solution will have a cloudy appearance, and the materials may settle to the bottom of the liquid. The particles of a suspension can be separated from the solvent by using a filter.

Colloids

Particles that are intermediate in size between those in solutions and suspensions form mixtures known as colloidal dispersions or ***colloids***. After the large particles settle out of muddy water, the water is still cloudy because colloidal particles remain dispersed in the water. If the cloudy mixture is poured through a filter, the colloidal particles will pass through, and the mixture remains cloudy. The colloidal particles are small enough to remain suspended in the solvent by the constant movement of the surrounding molecules.

23–23. A bottle of bacterin shows it is a suspension. (The level of the liquid is near the top of the bottle. The solid materials have settled to the bottom half of the bottle. The bacterin must be shaken before use to keep the particles in suspension.)

Emulsions

An ***emulsion*** is a colloidal suspension made of two solutions. One liquid is suspended as tiny droplets in the other liquid. For example, mayonnaise is an emulsion of oil droplets in water; the egg yolk in the mayonnaise acts as an emulsifying agent, which helps keep the oil droplets dispersed.

23–24. Salad dressing, milk, and mayonnaise are examples of emulsions. (Courtesy, Education Images)

Milk is another example of an emulsion. The butter fat (which is oil) rises to the top as cream. Milk processing homogenizes milk by breaking the fat particles into smaller pieces that stay suspended in the liquid milk.

APPLICATIONS

Many solutions and suspensions are used in agriculture. Knowing the characteristics of each will help in their use.

Filtration Removes Suspensions

Filters are used to remove substances from suspensions. Some are used to protect human health; others are used to remove contamination. A few examples follow.

23–25. A glass bowl fuel filter separates water from fuel by collecting water at the bottom. Any water is easy to see and is removed by a small valve underneath.

- **Fuel filters**—Fuel should pass through a filter before it reaches an engine. If it does not, delicate parts of the engine could be damaged. For example, diesel injector nozzles could be clogged with small particles of rust, soil, bacteria, or other solids. Filtering removes materials suspended in the fuel.
- **Dust**—Some people work in dusty environments. The dust is made of tiny particles of substances (soil, pollen, fertilizer, grain, etc.). Dust is a suspension, so particle masks over the nose and mouth can filter it from the air. Respirators are needed in particularly hazardous situations.
- **Chemicals**—Protection is needed when applying chemicals in fields, greenhouses, or other places. Use a mask to protect the respiratory system from pesticide particles suspended in the air. Tiny droplets of liquid will be filtered out of the air; as a result, they

will not reach the lungs. However, solutions of gas, such as a gaseous fumigant, may not be removed by filtration. Masks with charcoal filters are more likely to filter out gaseous particles because the particles stick to the charcoal.

Achieving Uniformity

Materials sprayed on crops or livestock or injected into animals often use water as a carrier or a solvent. However, some materials are oil-based and do not mix well with water. Oil and water do not mix because of polarity, which is the condition of substances having different negative and positive charges. Water molecules are strongly attracted to each other. In contrast, oil molecules have a weak attraction for each other. Even though oil and water are liquids, they do not mix because water has a higher polarity. Water is denser than oil; when the two are mixed, the water rapidly separates from the oil and goes to the bottom.

- **Surfactants**—Surfactants are known as wetting agents that help oil and water mix evenly. Plant leaves have waxy coatings that resist the flow of a water-based pesticide over them. A surfactant helps overcome the failure of oil and water to mix. Small amounts of surfactants are often added to pesticides as they are being mixed. Some people use soap as a surfactant.
- **Tank agitation**—Pesticide application equipment should keep the solution well mixed as it is applied. Materials that do not mix well should be used in tanks with agitators. The agitators move the liquid around and keep it mixed well.
- **Biologicals**—Liquid medicines given to animals must provide a uniform dosage. For example, an injectable antibiotic may be an aqueous (water) suspension. If the bottle sits in the refrigerator for a few days, the antibiotic material settles to the bottom. The top of the bottle's contents will be clear, and the bottom will be white. The bottle must be shaken to get all of the antibiotic in suspension.
- **Orally administered liquids**—Animal medicines, such as wormer, may be given by mouth using a dose syringe. A large number of animals may be given medicine from the same

23–26. The active ingredient must be uniform throughout a pesticide solution. (Here, the herbicide is being added to a tank of water. Agitation in the tank will ensure a uniform mixture of the material.) (Courtesy, USDA)

23–27. An elemental sulfur burner is used in a greenhouse to help manage insects. (The system is timed to release the sulfur vapor when no workers are present. This shows a sulfur burner being examined for proper operation.) (Courtesy, Education Images)

container. The medicine should be thoroughly mixed before the first dose is given and should be kept mixed thereafter. If it is not mixed properly, some animals may not receive enough wormer. As a result, it may not rid them of worms.

- **External therapeutants**—Medicines are sometimes mixed with water or other liquids before being applied to animals. An example is using formalin to treat fish in a tank. The formalin must be mixed throughout the water in all parts of the tank. If it is not, some of the fish will not be treated properly.

23–28. The blooms on hydrangea plants can be blue or pink based on the acidity of the soil. (Acid soils with a pH below 5.5 produce blue flowers; above 5.5 pH, the flowers are pink. Hydrangeas with white flowers are not affected by pH.)

ACIDS, BASES, AND SALTS

Groups of substances that form compounds and solutions with electrical charges are known as acids, bases, and salts. The solutions they form contain ions and are sometimes known as electrolytes. An important distinction between acids and bases is the ***pH***, which is the measurement of the acidity of a substance.

Acids, bases, and salts can be dangerous to people, animals, property, and the environment. Extreme care is essential to prevent burns and other injuries.

ACIDS

An ***acid*** is a compound that gives up protons to water molecules to form hydronium ions. Protons are designated as H^+ in water. It is a hydrogen atom that has transferred its negatively charged electron to another atom, leaving only the positively charged nucleus made of one proton. This makes the H^+ ion have a strong attraction to the polar water molecule forming the hydronium ion (H_3O^+). An acid is a proton donor. Acids have pH measurements below 7.0. Because acids contain ions with charges, they conduct electricity.

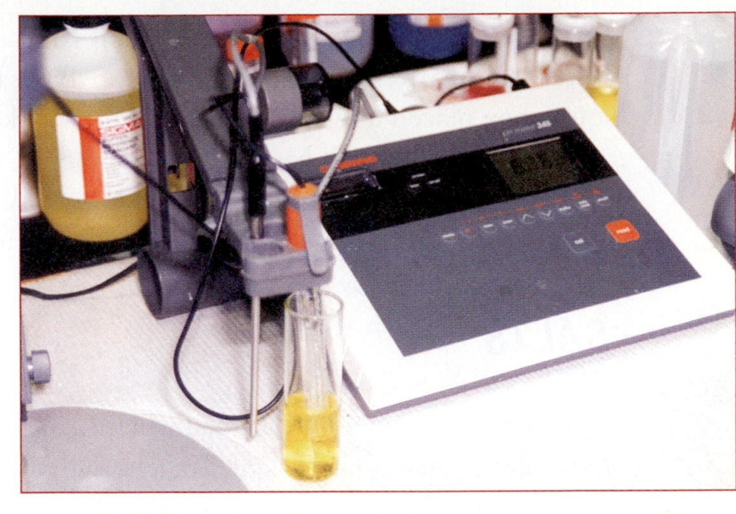

23–29. The pH of a liquid is being precisely determined.

Acids vary in strength. Vinegar (acetic acid) is a weak acid used in food preservation. Lemon juice (citric acid) is often used in foods and beverages for flavor. All acids have a sour taste. Other acids have agricultural uses, such as hydrochloric acid (HCl). In addition, some fertilizers will form acids in the soil, such as sulfur, which is used to increase the acidity of soil for certain crops. Sulfur forms sulfuric acid (H_2SO_4), which lowers the pH of the soil.

BASES

A ***base*** is a compound that produces hydroxide ions (OH^-) in water. Household ammonia (NH_4OH), lye (NaOH), and

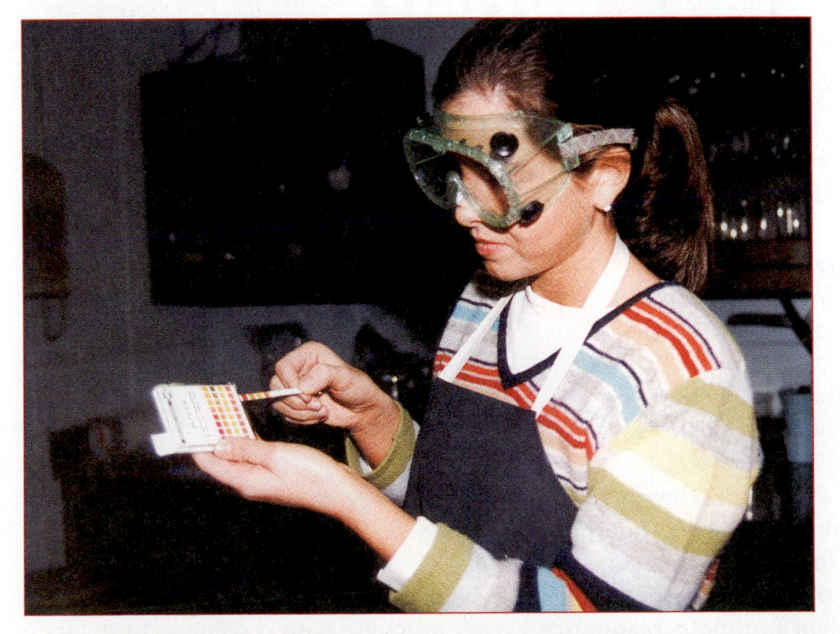

23–30. Easy-to-use kits may be helpful in testing substances.

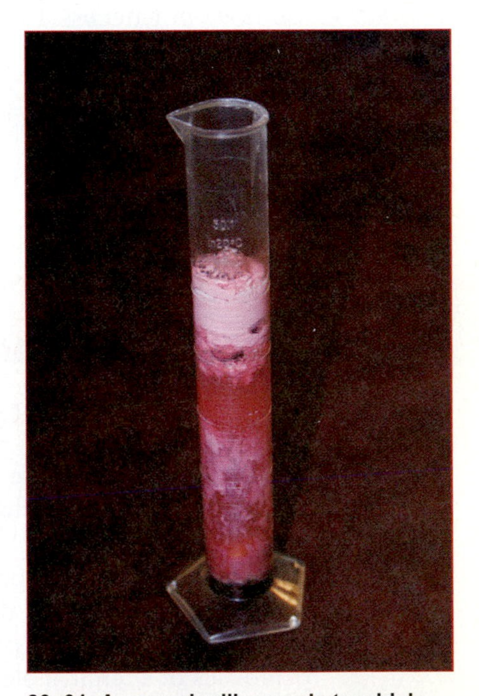

23–31. A soured milk sample to which litmus powder has been added indicates that the soured milk is acidic. (If the sample were blue, it would have been basic.) The pH of soured milk depends on the substances formed by the organisms that cause spoilage, with some acidic and others basic. (Courtesy, Education Images)

slaked lime [$Ca(OH)_2$] are three examples of bases. A base is a proton acceptor. Bases taste bitter and have a slippery feeling like soap. Bases have pH measurements above 7.0 and form ions when placed in water solutions, which makes them conduct electricity.

A water-containing base, such as ammonia, is used to clean and control the growth of microorganisms. Compounds that are basic (alkaline) are added to the soil to raise the pH (make it less acid). For example, slaked lime or crushed limestone is often used on soil.

SALTS

A **salt** is a compound formed when strong acids and bases are combined. This occurs most often when an acid is neutralized with a base. The combined pH totals 7.0, which is considered neutral. For example, when sodium hydroxide (lye) and hydrochloric acid are combined, sodium chloride is formed along with water. Sodium chloride has neither an acid nor a basic taste; it tastes salty! However, there are many other kinds of salt.

Salts tend to have a neutral pH, but the pH of some salts varies. Mixing an acid and a base does not necessarily produce a neutral (7.0) pH. The pH of the salt depends on the strength of the acid and base that are combined. A strong acid and a weak base will form a salt that has an acid pH. In contrast, a strong base and a weak acid will form a salt that has a basic pH.

CAREER PROFILE

CHEMIST

A chemist studies the characteristics of substances in our world. Many chemists specialize in a particular area, such as organic chemistry or physical chemistry. Chemists test materials, develop new products, and determine the presence and characteristics of chemical elements.

Chemists need college degrees in chemistry. Many chemists have a master's degree and a doctorate. Some chemists engage in study beyond a doctorate, known as post-doctoral study. Begin preparing in high school by taking mathematics, chemistry, and physics classes.

Chemists may be employed by chemical companies, government agencies, colleges, research laboratories, and other places where chemistry is used. This shows an Emory University chemist loading a nuclear magnetic resonance (NMR) machine with a spinner to determine the contents of a compound.

The operation of some kinds of batteries (known as dry cells) uses salt-forming processes to make electrical energy. A paste of ammonium chloride (NH_4Cl) is placed inside a zinc casing. Turning on the switch (of a flashlight or radio, for example) causes contact to be made so the electrical charges move and conduct an electrical current.

CHEMICAL REACTIONS

A chemical reaction occurs when a substance changes to become another substance with different characteristics. A compound may result that is different from the elements that combined to form it. Sometimes a compound reacts with another compound to form a substance. Some compounds and elements do not react with each other. Four easily observed changes usually indicate that a chemical reaction has occurred: the emission of heat or light, the production of a gas, the formation of a precipitate (a solid), or a color change.

Chemical changes are a natural part of life processes. Other chemical changes are created by people. In many cases, chemical changes can be sped up or slowed down by changing the temperature, amount of materials involved, and other conditions.

PLANT AND ANIMAL REACTIONS

Important chemical changes are carried out by plants and animals. Our food supply depends on creating conditions for them to complete these chemical processes.

Photosynthesis

In photosynthesis, plants use carbon dioxide from the air and water from the soil to produce sugar and elemental oxygen. Chlorophyll and sunlight must be present for the process to occur. The sugar is stored in various forms in the plant, including starch.

23–32. Growing cotton plants in sand tubes allows for the testing of needed elements and the effects of deficiencies.

Photosynthesis is the most important chemical change on the earth! Agriscience uses methods to encourage the process. For example, plants can be given fertilizer and water. Good conditions enable plants to be more productive in sugar making.

23–33. Horses breathe in oxygen from the air and exhale carbon dioxide on a cold winter day.

Respiration

Animals reverse the process of photosynthesis. They eat the sugars (and carbohydrates) made by plants and use the oxygen in respiration. Digestion breaks sugars into products the body can use. Animals inhale (breathe in) oxygen and exhale (give off) carbon dioxide and other substances. In photosynthesis, plants use the carbon dioxide given off by animals.

Agriscience involves helping animals have the right nutrients and keeping them healthy so the processes can occur.

CORROSION: LOSS OF PROPERTY

Corrosion (rusting) is a chemical reaction that destroys and weakens property. It typically involves changes in a metal of some type, such as iron or silver. In most cases, two processes are involved: oxidation and reduction.

Oxidation is the process of atoms of a metal giving up electrons. Most of the time oxygen is involved. The metal becomes more positive in its oxidation state.

Reduction is the process by which electrons given up in oxidation are captured by atoms of the same or another metal. The metal attains a more negative oxidation state. Oxidation and reduction are illustrated by rusting.

23–34. The brown color of this water tank is a sign that the tank is beginning to rust.

Rusting

Rusting is primarily associated with iron and some of its alloys (mixtures). As iron rusts, it weakens and loses its strength. Rusting occurs when iron combines with oxygen in the presence of moisture. Unprotected structures made out of iron will rust rapidly if left out in the rain.

Rust is a brownish material formed when iron or steel is exposed to damp air. The chemical reaction that occurs produces hydrated iron oxide ($Fe_2O_3 \cdot 3H_2O$). Oxygen from the air combines with the iron in the presence of water.

Machinery should be painted and stored inside. Shiny plow points can be coated with oil to protect them from rusting.

Some metal products are galvanized to prevent rusting. Galvanizing is applying a coat of a material that will not rust, such as zinc. For example, zinc-coated metal roofing and zinc-coated barbed fencing wire last longer. Zinc does not oxidize, and it protects the iron underneath from oxygen and water.

Acid Reactions

Acids react with metals, weakening and destroying them in the process. Some contact may be unavoidable; however, steps can be taken to minimize damage.

Fertilizer materials, particularly those containing nitrogen, often form acids. As with rusting, moisture speeds the process. Improperly stored fertilizer can cause metal buildings and equipment to corrode. Equipment used to apply fertilizer should be cleaned after each use to remove any residues that cause corrosion. In addition, plastic-lined storage areas and equipment can help prevent corrosion.

Batteries are used in many machines to provide electrical energy. A battery operates on the basis of chemical reactions. Sometimes, this reaction gets to places where it is damaging. A small amount of the sulfuric acid in a battery may reach its outside surface. The acid reacts with the lead and copper, forming corrosion. Corroded material should be washed away with water containing baking soda because the baking soda neutralizes the acid. A thin layer of oil should be applied to the battery post and cable when it is dry. (Care should be taken to avoid contact with the acid or corrosion. They can burn skin and make holes in clothes.)

23–35. Chemical reactions inside a battery generate electrical power to turn the starter motor when cranking the engine on this tractor.

PREVENTING SPOILAGE: FOOD PRESERVATION

Various chemicals are used to keep food safe and to improve its taste. Acids, bases, and salts are often added to food. Mild forms of certain chemical compounds improve the flavor of food. Some compounds prevent the growth of microorganisms that cause spoilage.

Pickling, salting, and using bases are examples of ways chemicals are used to preserve food. More information on these is presented in Chapter 27.

MEASUREMENTS IN CHEMISTRY

Chemistry often involves making measurements of very small quantities. These measurements must be carefully made and accurate. Errors lead to inaccurate results.

23–36. A graduated cylinder can be used to measure liquids.

MEASUREMENT

Measurement is a method of determining the number of units or quantity of something. A unit is a single quantity that represents magnitude, size, or amount. Units are the same size in all situations. For example, a measurement of 1 foot is always 12 inches long.

Different units of measurement are used. Units must often be converted from one to another. Conversions are usually easily done because measurement involves numbers. Units can be counted and divided into parts. Measurement has more meaning when a number is used with the unit, such as 4 cc water.

Uniform instruments are used in making measurements. They must provide the same value for the same quantity every time they are used. All of us are familiar with scales that fail to give an accurate weight every time. They are no good!

MEASUREMENT SYSTEMS

Two measurement systems are used in agriscience: the customary or English system and the metric system. The customary system is widely used in the United States and Canada in nontechnical areas of agriscience. The metric system is universally used by scientists.

Customary System

The ***customary measurement system*** is the measurement system used in everyday life. Agricultural measurements are often made in the customary system. However, measurements involving more scientific areas are in the metric system. Although both customary and metric systems are used in agriscience, more emphasis is being given to metric measurements.

Common units in the customary system are inch, foot, yard, mile, ounce, pound, quart, pint, gallon, and acre.

Metric System

The ***metric system*** is a decimal system; units increase or decrease by 10s. Its official name is the International System of Units, often known as SI because of its French origin. Common measurements in the metric system are grams, meters, liters, hectares (for land), and seconds.

The metric system uses prefixes to decrease or increase numbers. The prefix is placed before a unit, such as "centi" is added to "meter" to form centimeter (one-hundredth of a meter).

Internet Topics | Search

Selected topics for Internet discovery and reporting are listed here. Begin by searching for each topic. Next, read and learn about it. Conclude by preparing a brief report on your findings.

1. Atomic number
2. Electronic balance
3. Oxidation

Table 23–3. Prefixes Used in the Metric System

Prefix	Symbol	Factor by Which Multiplied
Giga	G	1,000,000,000.0 (one billion)
Mega	M	1,000,000.0 (one million)
Kilo	k	1,000.0 (one thousand)
Hecto	h	100.0 (one hundred)
Deka	da	10.0 (ten)
Deci	d	0.1 (one-tenth)
Centi	c	0.01 (one-hundredth)
Milli	m	0.001 (one-thousandth)
Micro	μ	0.000001 (one-millionth)
Nano	n	0.000000001 (one-billionth)

MAKING MEASUREMENTS

Accurate measurement is extremely important in chemistry. Several measurements are used. Two of the measurements are similar to those made with land: linear and area. ***Linear measurement*** is the distance between two points—measurement of length. The process is similar in chemistry and land measurement, but it is done with smaller measuring devices in chemistry, such as rulers, meter sticks, and tape measures. ***Area*** is the measurement of surfaces. It is determined by multiplying length times width. Area is reported in square units, such as square feet (ft^2) or square meters (m^2). (Refer to the section in Chapter 22 on land measurement.)

Volume

As indicated earlier in the chapter, volume is the total size of an object. Volume, as the amount of space something takes or holds, is calculated by multiplying length times width times height. It is reported in cubic units, such as cubic feet ($ft.^3$) or cubic meters (m^3). An example is the volume of a beaker. The volume of a beaker is measured in mL.

23–37. Volume measurement is easily done when a solution is poured into a beaker graduated in milliliters. (Courtesy, Education Images)

Weight or Mass

Weight is the heaviness of something. It is related to gravity, which is the pull of the earth. Mass is the amount of matter something contains. Weight may change at different elevations, while mass is constant. The weight measured on top of a mountain may be less than at sea level due to a greater distance from the center of gravity. People in space may float in air, as they have less weight than when on earth; however, they still contain the same amount of mass. In a laboratory setting, the terms weight and mass are sometimes used interchangeably since the location does not change. Weight or mass may be measured using the customary system (pounds [lb]) or metric system (kilograms [kg]).

The base mass unit in the metric system is the gram. As with linear units, the value can be increased or decreased using the prefixes shown in Table 23–3.

A gram is the mass or weight of 1 cubic centimeter of water, which is the same as 1 milliliter of water. This makes it easy to convert from volume to weight and vice versa. This is often expressed as: 1 g = 1 cc = 1 mL. (Note: cm^3 is common in science and equals 1 cc.)

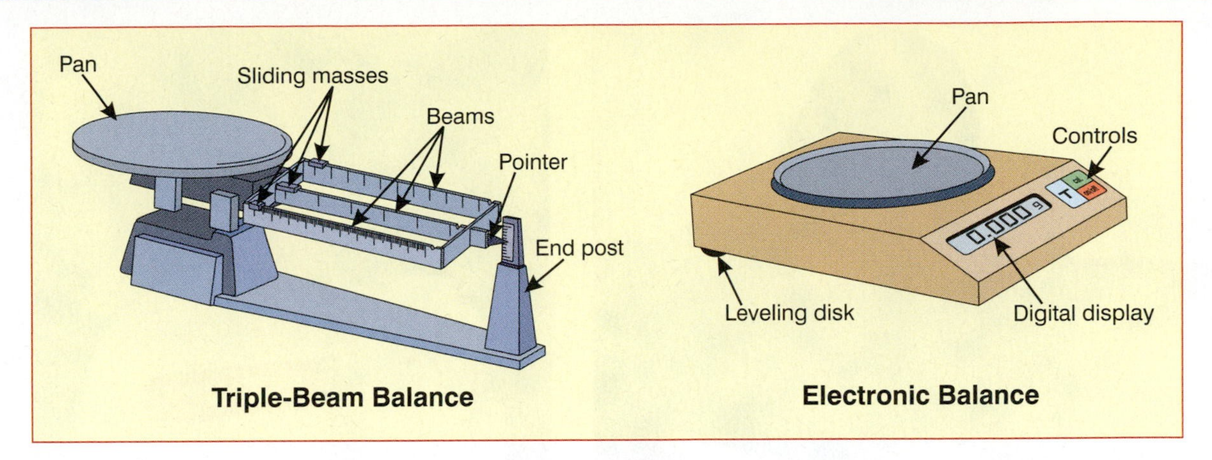

23–38. Major parts of triple-beam and electronic balances.

A machine that determines weight is a **weighing scale** (or scale). Accuracy is a must. Scales that make errors cost money and may cause harm to plants and animals. Scales must be used properly to provide accurate weights.

Scales need to be calibrated. This means that the weight they give is standardized for the unit being used. Objects of known mass or weight are used to adjust scales that need to be calibrated.

Many kinds of scales are used. Some are designed for very small amounts of substances, and others are for large items. Three major types of scales are balance, mechanical, and electronic.

- **Balance scales**—The balance scale is the oldest, dating back to Egypt in 2500 B.C. There are two kinds of balance scales: equal-arm balance and bar balance. Of these, the bar balance scale is most widely found. It involves using a bar that is anchored near

23–39. Using a triple-beam scale. (Courtesy, Education Images)

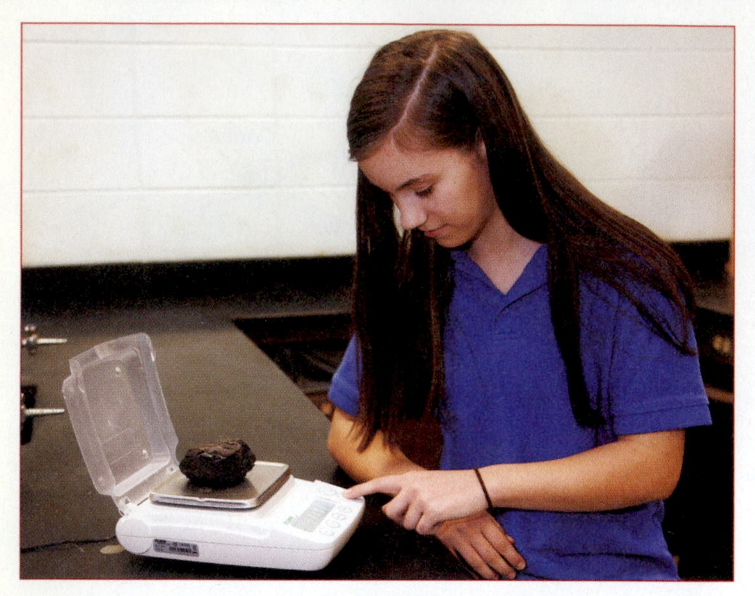

23–40. Using an electronic balance. (Courtesy, Education Images)

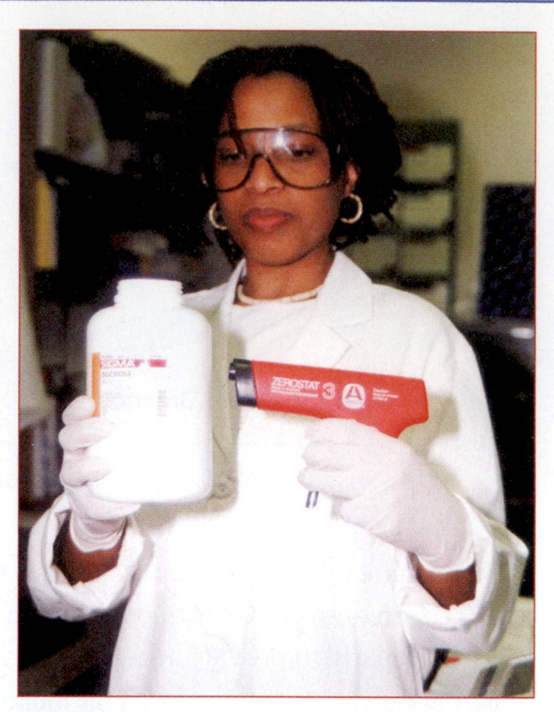

23–41. A Zerostat may be used to remove static electricity from a powder before weighing. This prevents the powder from getting in the air and collecting on the equipment.

one end so a small mass can be used to balance the bar. The bar balance allows weighing heavy objects with small, light masses on the bar. Common examples in agriscience are the triple-beam scale and the pan balance scale.

- **Mechanical scales**—Mechanical scales were developed in the 1700s. They use levers that reduce the load to a smaller force. This means that weights equal to the weight of what is being measured are not needed. Some of these scales involve beams, springs, and pendulum dials.
- **Electronic scales**—Developed in the 1950s, electronic scales are rapidly replacing other kinds. They can weigh objects quickly—far faster than the other scale types. Many electronic scales have built-in computer chips. These scales can do many things in addition to weighing, such as sorting and calculating price. Many meat markets, vegetable stands, processing plants, and other facilities use electronic scales.

MEASUREMENTS IN AGRISCIENCE

Many special kinds of measurements are used in agriscience. Several important areas are presented here.

Laboratory Work

Agriscience often involves working in laboratories where measurements are very important. Extremely small quantities may be measured. Accuracy and precision are essential.

Nearly all laboratory measurements are in the metric system. The metric system is easy to use because of the system of decimals and the relationships of the different units.

Conversions

Conversions from linear measures to volumes and weights are easy with the metric system. The original basis of measurement was with water, as follows:

1 cubic centimeter (cm^3 or cc) = 1 gram
1 cubic centimeter = 1 milliliter
1 gram = 1 milliliter

Table 23–4. Common Conversion Information in AgriScience

1 in.	=	2.54 cm
1 m	=	39.37 in.
1 kg	=	2.2 lb
1 L	=	1.06 qt

Measurements as Parts

Required measurements are sometimes stated as ***parts per unit***, which means that to a given number of units, another set of units needs to be added. Parts per unit are typically stated in three ways:

- **Parts per thousand (ppt)**—number of parts in a thousand

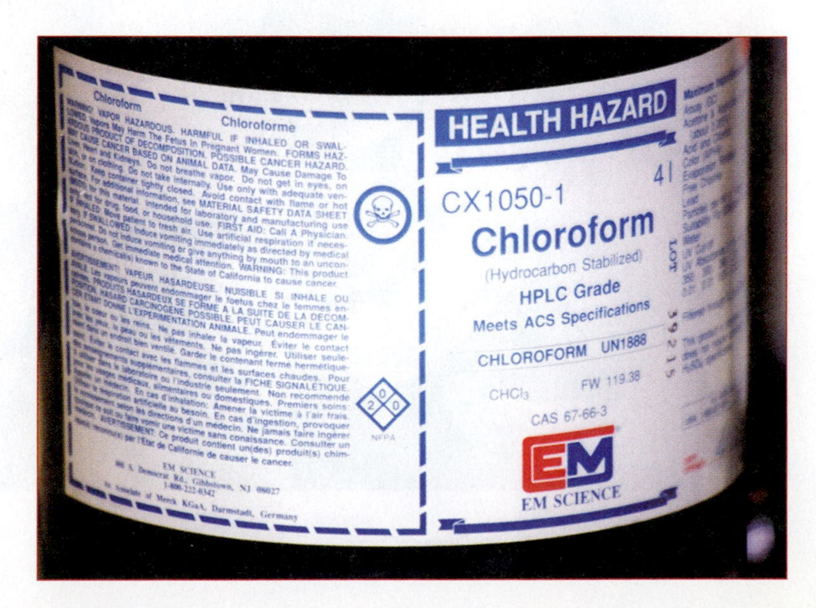

23–42. Always read and follow the instructions on chemical labels.

23–43. Acids should be stored in an approved safety cabinet that can be securely locked. (Courtesy, Education Images)

- **Parts per million (ppm)**—number of parts in a million
- **Parts per billion (ppb)**—number of parts in a billion

Instructions for mixing materials may specify that a given number of parts be mixed with a quantity of another substance. For example, fish farmers must often mix therapeutants (fish drugs) in the water. If the rate to be used in the treatment is 5 ppm, 5 parts of the therapeutant should be mixed with 999,995 parts of water (5 + 999,995 = 1 million).

Determining how much to use involves these steps:

1. Measuring the vessel (such as a fish tank) containing the water
2. Determining the volume of water in the vessel
3. Calculating the amount of material that is to be mixed with the water (For example, a dosage of 5 ppm would require 5.0 milligrams per liter of water.)

23–44. Always read and follow step-by-step instructions when carrying out an experiment. (Courtesy, Education Images)

Here are a few simple conversions that will help with measurements defined as parts:

1 ppm = 2.72 pounds per acre-foot of water (acre-foot equals 1 acre of water 1 foot deep)
= 0.283 grams per cubic foot of water
= 8.34 pounds per million gallons of water

1 ppt = 2,718 pounds per acre-foot
= 3.80 grams per gallon of water
= 28.3 grams per cubic foot of water

REVIEWING

MAIN IDEAS

Chemistry is the study of the substances that make up living and nonliving things on the earth. Scientists have identified 115 elements; 92 of these occur naturally.

Matter is anything that has volume and mass. This means that it occupies space. Mass is the amount of matter that something contains. Temperature and pressure can cause matter to change from solid to liquid, liquid to gas, etc.

Chemical elements can be placed in two categories: metals and nonmetals. Metalloids are a special group of nonmetals. Elements form compounds with different characteristics by chemically combining. Mixtures are different in that they do not chemically interact. Mixtures are only physically combined. Each ingredient keeps its own chemical properties.

Carbon is an important element in our world. Even though it is not one of the most abundant elements, it is found in many compounds. Compounds of hydrogen and carbon (hydrocarbons) are useful in many ways. Carbon is found in all organic compounds, which make up structures and processes in living things.

Acids, bases, and salts are substances with electrical charges. Salts are formed when bases and acids are mixed. Most salts have a near-neutral pH. Acids have pH ranges below 7.0, while bases (alkalines) have pH ranges above 7.0.

Solutions and suspensions are important. Solutions are made when something is dissolved in a different substance. Many times the solvent is a liquid. The substance dissolved is the solute. Water is the most common solvent. Suspensions involve solid materials (colloids) in a liquid. Emulsions involve mixing two solutions, such as water and oil. Sometimes they do not mix very well and need an emulsifying agent to help them.

Many applications of chemistry are found in agriscience, ranging from soil fertility to disease prevention, preservation of food, and protection of property from rust and corrosion.

Precise measurements are often needed in chemistry, such as volume, mass or weight, or other measures. Various equipment is available to assist in making measurements.

QUESTIONS

Answer the following questions, using complete sentences and correct spelling.

1. What two conditions are involved in changes in state?
2. Name and describe the two properties of matter?
3. What is an element? How many elements are known to exist naturally? Artificially? Name five common elements, and give their symbols.
4. Describe the differences between the Fahrenheit and Celsius temperature scales.
5. What is the Periodic Table of the Elements? How is it used?
6. What is the Periodic Law?
7. Name and distinguish between two categories of elements. What is a metalloid?
8. Identify the three main parts of an atom and where they are located in the atom.
9. What is a compound? How is a compound related to the Law of Definite Composition?
10. Identify the four main types of chemical reactions.
11. What is a mixture? Give examples.
12. What is organic chemistry? What two major areas are involved?
13. Explain the differences between solutions, solvents, and solutes.
14. How do suspensions, colloids, and emulsions differ?
15. Explain the following as related to agriscience: filtration, surfactants, and agitation.
16. What are acids, bases, and salts? How do they relate to each other?
17. What important agriscience chemical reactions occur in oxidation and food preservation?
18. Identify the two kinds of measurements used in chemistry in agriscience?

EVALUATING

Match each term with its correct definition.

a. measurement	f. emulsion	k. plasma
b. oxidation	g. compound	l. soluble
c. salt	h. mass	m. area
d. solvent	i. element	n. pH
e. solution	j. hydrocarbon	

_____1. A colloidal suspension made of two solutions.

_____2. A mixture of two or more substances.

_____3. The liquid material in a solution.

_____4. A method of determining the number of units in something.

_____5. A compound formed when strong acids and bases are combined.

_____6. The process of oxygen combining with a metal.

_____7. Organic compounds made only of hydrogen and carbon.

_____8. The amount of matter an object contains.

_____9. A substance that cannot be broken down further by ordinary methods.

_____10. A substance made of two or more elements that have chemically combined.

_____11. The fourth state of matter in a high-energy gaseous form found in fluorescent bulbs.

_____12. The measurement of surfaces calculated by length times width.

_____13. The measurement of the acidity of a substance.

_____14. When a substance such as a solid dissolves in a liquid.

EXPLORING

1. Test the effects of antifreeze on water. Add antifreeze to one of two containers with equal amounts of water. Place both containers in a freezer. Regularly check to see which freezes first. Use containers that do not easily crack upon freezing. Prepare an oral report on your findings.

2. Visit a supermarket and make a list of products that are solutions, colloidal suspensions, and emulsions.

3. Examine the label on a pesticide container. Write down the chemical formula for the material. List the elements found in the active ingredient.

4. With the assistance of your teacher, test a compound in the laboratory to determine the elements that are in it. Use chemistry books and other materials for references. Always follow safe practices in doing laboratory work.

CHAPTER 24

Physics in AgriScience

OBJECTIVES

This chapter introduces important principles of physics and how they are applied in agriscience. It has the following objectives:

1. Identify and explain areas of physics used in agriscience.
2. Explain work and power.
3. Name and describe simple machines.
4. Explain mechanical advantage.
5. Describe the use of thermal energy.
6. Explain the use of electrical power.
7. Describe the use of compression power.

TERMS

acceleration
alternating current
ampere
circuit
conductor
current electricity
deceleration
direct current
draft
electricity
energy
force
heat
heat engine
horsepower
hydraulic power
inclined plane
insulator
kinetic energy
Law of Conservation of Energy
lever
machine
mechanical advantage
mechanics
motion
optics
physics
pneumatic power
power
pulley
screw
sound
speed
thermal energy
volt
watt
wedge
wheel and axle
work

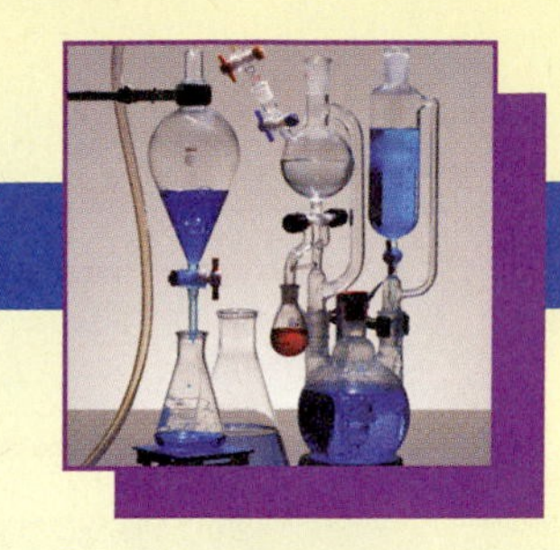

24–1. A grain elevator with dryer involves many parts, systems, and processes related to physics. (Courtesy, Education Images)

A GRAIN ELEVATOR is a complex of bins, dryers, graders, and handling equipment used in the storage of grain. Structures have gradually changed over the years from concrete or wooden facilities to corrugated metal. The structures, as well as the processes involved, make use of a number of physics principles.

Today we have the advantage of using the many machines that have been developed over the years. The agricultural industry depends on these new technologies in many ways. Plowing, planting, harvesting, marketing, and many other activities associated with plant and animal production involve the use of these technological advancements. Machines allow agriculture to mass produce products for use by countless people.

Agriscience applies many principles of physics. These include the basic principles of simple machines, power, and energy and ways they can be made to do work for us.

AREAS OF PHYSICS

24–2. Properly designing and using peanut drying facilities provides quality products for consumers. (Courtesy, USDA)

24–3. Transistors, integrated circuits, and other applications of physics have made the modern computer possible.

Physics is the study of matter and energy and how the two relate. The focus is on material things and how people can use them to their advantage. Physics involves studying and using energy in various forms. Of course, mathematics has many applications in physics. Skills in mathematics are very important in making calculations associated with physics principles.

Almost everything people do involves the use of physics in some way.

The application of physics in agriscience is often referred to as mechanical technology. Another important area is computer technology.

Physics is often divided into several areas of study. Some of these have considerable applications in agriscience. Six areas are discussed here.

- Mechanics—***Mechanics*** is a division of physics that deals with the forces exerted by objects when moving or at rest. Objects in motion are known as dynamic, while those at rest are known as static. Energy is required for an object to move. Falling and spinning objects, friction, and weight are all subjects studied in mechanics. Many mechanical devices are used in agriscience. These apply various principles of motion and of energy use in their operation. For example, a tractor uses diesel fuel in its engine to supply the energy necessary to till the soil.
- Heat—***Heat*** is a form of energy that results from the motion of atoms and molecules. ***Kinetic energy*** is the result of an object's mass and motion. Heat refers to the total kinetic energy of an object. This can be compared with temperature, which is the average kinetic energy. Heat is the flow of energy caused by temperature differences. Temperature measures change in heat. Temperature and heat are related but are not the

24–4. Ultrasound images are formed with reflected sound waves—an application of physics principles. (This shows ultrasound being used on a steer to determine size of loin, muscle area, and backfat.) (Courtesy, Agricultural Research Service, USDA)

same thing. Heat is important in agriscience—for example, in internal-combustion engines, crop drying equipment, and methods of keeping young animals warm in cold weather.

- **Sound**—***Sound*** is the result of vibration through a medium, such as air or water, creating waves. Human ears pick up these vibrations and translate them into sound. The use of sound waves gives us musical instruments, radios, iPods, car speakers, and many applications in medical areas. Sound that is too loud is managed in different ways. A muffler is put on an engine to reduce the sound given off as the fuel suddenly burns. People who work near loud noises should wear hearing protection devices.
- **Magnetism and electricity**—Together, magnetism and electricity are often known as electromagnetism. Electricity can produce magnetism. Magnetic forces can be used to produce electrical fields, such as the fields around the coils in engines or speakers in radios. Magnetism can also be used to produce electricity. Electrical power plants use the mechanical energy of falling water or steam to operate generators that produce electricity (electrical energy). Electric motors produce electromagnetic fields. Large windings of wire are used around magnets in building electric motors.

24–5. Electric motors are used in many areas of agriscience. This cutaway shows the coils of wire, magnets, and other parts of a motor.

24–6. A laser level is often used in construction and land forming.

- **Light**—Light makes vision possible. Light illuminates chicken houses, greenhouses, and our homes. It gives us the reflections we see in home and car mirrors. The use of laser equipment in agriculture is another practical application of light.

  The blending of rays of light produces the spectrum of colors we call a rainbow. We can control the direction of light through polarized lenses in sunglasses. The study of the use of light is known as ***optics***.

- **Electron physics and nuclear physics**—Electron physics and nuclear physics are often treated as two separate areas. Electron physics deals with understanding the structure of atoms and molecules. Nuclear physics is concerned with understanding the atomic nucleus, particle acceleration, and radioactivity. Among the uses of these processes are production of energy, preservation of food, and treatment of diseases.

WORK AND POWER

Work and power are often needed to accomplish a task in agriculture. In physics, work is a little different from the duties of a job.

WORK

24–7. Pushing a pickup truck produces work only if the truck is moved.

Work is the moving of an object through a distance when there is some resistance to its movement. Resistance (also known as inertia) is opposition to movement. The more mass an object has, the greater the inertia. A heavy bag of feed doesn't slide easily on the warehouse floor—it offers resistance.

A plow moving through the soil offers an example of resistance. The plow must turn the soil as well as break roots of plants that may be growing in it. The resistance of a plow to movement is often called ***draft***. Several things can happen if the plow gets too deep and the soil is hard to turn. The engine of the tractor will labor under the load, and more fuel will be used. The plow or some part might break if the load is too great. If the draft is excessive, the tractor could stall.

Two things are necessary for work to occur: Force must be applied to the object, and the object must be moved a distance. The object must move for there to be work. If an object is pushed very hard but doesn't move, no work has been accomplished.

24–8. A tractor with considerable power is needed to overcome the draft on each disk blade in the soil. (Courtesy, AGCO, Duluth, Georgia)

Force

Force is the push or pull on an object. The object is either moved or damaged by force. What happens when an automobile crashes into a tree? The force of the impact damages the automobile and may break or damage the tree, depending on the size and force of the car.

Formula

Just as with most principles of physics, work can be expressed as a mathematical formula. Work is the force applied, multiplied by the distance the object is moved. The formula is:

$$\text{work} = \text{force} \times \text{distance} \; (W = F \times D)$$

Both metric (SI) and English measures of work are used. In the SI system, force is measured in newtons (N), and distance is measured in meters (m). This means that work is designated in newton-meters. One newton-meter is equal to one joule (j). Work is calculated as joules.

In the English system, force is measured in pounds (lb), and distance is measured in feet (ft). Work is calculated as foot-pounds. The same formula is used for both systems.

24–9. The force exerted on the handle of a shopping cart causes it to move forward. (The more you place in a cart, the more force that is needed to move it.) (Courtesy, Education Images)

Here is an example of how work is calculated: Suppose a person pushes a small pickup truck 10 meters using 200 newtons of exerted force. How much work was done?

$$\text{work pushing the pickup} = 200 \text{ N} \times 10 \text{ m}$$
$$= 2{,}000 \text{ newton-meters (joules)}$$

Here are three useful conversions from SI to English:

$$1 \text{ meter} = 3.28 \text{ feet}$$
$$1 \text{ newton} = 0.225 \text{ pound}$$
$$1 \text{ foot-pound} = 1.356 \text{ joules}$$

(In this example of work, the distance the pickup is moved is 10 m × 3.28 ft/m, or 32.8 feet, and the force is 0.225 lb/N × 200 N = 45 pounds.)

$$\text{work pushing the pickup} = 45 \text{ lb} \times 32.8 \text{ ft}$$
$$= 1{,}476 \text{ foot-pounds (ft-lb)}$$

In scientific calculations, the factor-label method is used. The advantage of the method is that it allows thinking through the concepts of the labels before the calculations. This

CAREER PROFILE

AGRICULTURAL ENGINEER

An agricultural engineer studies and applies scientific knowledge with the goal of improving agricultural production. Agricultural engineers may specialize in machinery, implements, irrigation, soil and water, and other areas. This photo shows an agricultural engineer studying nozzle size for an aerial applicator. (Courtesy, Agricultural Research Service, USDA)

An agricultural engineer needs a college degree in agricultural engineering or a closely related area. Many have master's degrees and doctorates. Begin in high school by taking mathematics, physics, and agriculture classes. Practical experience with machinery can be useful.

Jobs can be found with research stations, colleges, machinery manufacturers, and government agencies.

helps to organize the problem logically. A problem is set up so labels cancel out in the denominator and numerator, leaving the label of the answer.

Here is how the problem in pushing the pickup truck would be worked using the factor-label method:

force in lb	200 ~~N~~	0.225 lb	= 45 lb
		1 ~~N~~ <	

$$W = F \times d$$
$$W = 45 \text{ lb} \times 32.8 \text{ ft}$$
$$W = 1{,}476 \text{ ft-lb}$$

Energy

The ability to do work is known as ***energy***. Fuels are frequently used to create energy so work can be done. For example, gasoline is burned in an internal-combustion engine to provide energy to push the piston down in the crankshaft. Heat is created as a byproduct when the fuel burns.

POWER

Power is the rate at which work is done. Time is a factor in doing the work. A force moves an object a distance in a certain amount of time. Work is the same if an object is moved slowly or quickly. Power varies with the speed of movement.

Power is stated as units of work divided by units of time. In SI, power is stated in watts. A ***watt*** is the power needed to do one joule of work in one second.

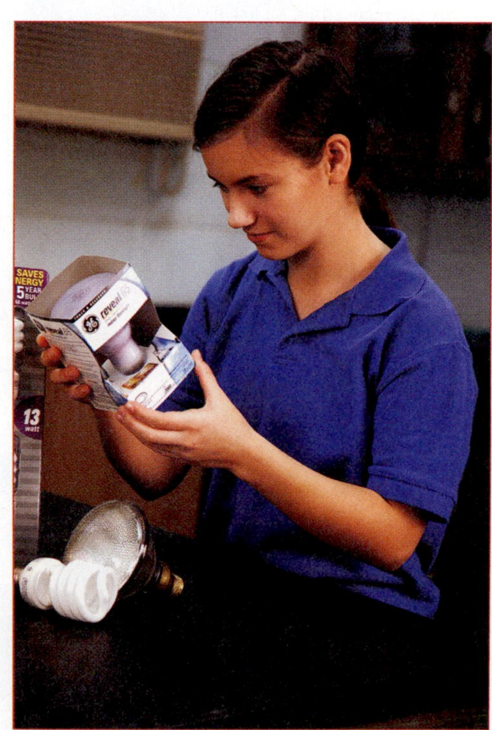

24–10. The wattage of a light bulb indicates the rate at which energy will be used (converted). (Courtesy, Education Images)

Formula

The formula for power is:

$$\text{power} = \text{work} \div \text{time, or } w/t = F \times d/t$$

In the above example of pushing the pickup truck, assume that 20 seconds (s) were needed to move the truck the distance of 10 meters. The amount of power in SI was:

$$\text{power to push the pickup} = 2{,}000\ \text{N} \cdot \text{m} \div 20\ \text{s} = 100\ \text{watts.}$$

(A watt is a joule per second.)

In the English system, power is measured in foot-pounds per second. The amount of power to move the pickup in the English system was:

$$\text{power to push the pickup} = 1{,}476\ \text{ft-lb} \div 20\ \text{s} = 73.8\ \text{ft-lb/s}$$

Horsepower

Power is often stated as horsepower. **Horsepower** is the amount of power used to do 550 foot-pounds of work per second. (It is 746 watts in SI.) The term *horsepower* originated with the steam engine. The power of a steam engine was compared with the power of a typical draft horse by James Watt, inventor of the steam engine. Engines used in tractors, pickup trucks, and other power equipment are given horsepower ratings.

The formula for horsepower is:

$$\text{horsepower} = \text{power} \div 550\ \text{ft-lb/s}$$

In the example of the pickup truck, the horsepower used to move it in the English system was (remember, the power calculated above was 73.8 foot-pounds per second):

$$\begin{aligned}\text{horsepower to move pickup} &= 73.8\ \text{ft-lb/s} \div 550\ \text{ft-lb/s/horsepower} \\ &= 0.134\ \text{horsepower}\end{aligned}$$

24–11. Considerable horsepower is needed to operate a hay cutter and conditioner. (Courtesy, KMN Modern Farm Equipment, Inc.)

The horsepower to move the pickup in the SI system was (remember, power was 100 watts):

$$\begin{aligned}\text{horsepower to move pickup} &= 100 \text{ watts} \div 746 \text{ watts/horsepower}\\ &= 0.134 \text{ horsepower}\end{aligned}$$

Both SI and English calculations result in the same horsepower required to move the pickup.

Since the pickup was on a smooth street and pushed by one person, the amount of horsepower was very small—only a little more than 10 percent of one horsepower. If the pickup were traveling 50 miles per hour down a highway, the horsepower would be much greater.

The factor-label method of working this problem in the English system is:

$$\begin{array}{c|c} 7.38 \ \not{\text{ft-lb/s}} & 1 \text{ horsepower} \\ \hline & 550 \ \not{\text{ft-lb/s}} \end{array} = 0.134 \text{ horsepower}$$

The factor-label method of working this problem in the metric system is:

$$\begin{array}{c|c} 100 \ \not{\text{watts}} & 1 \text{ horsepower} \\ \hline & 746 \ \not{\text{watts}} \end{array} = 0.134 \text{ horsepower}$$

MOTION

Motion is the process of moving or changing position and is a part of work and power. Motion is rectilinear (straight) or curvilinear (circular).

24–12. Motion lifts harvested potatoes into a wagon.

Examples of rectilinear motion are a tractor clipping pasture, a truck going down the highway, and a handsaw cutting a board into two pieces.

Examples of curvilinear motion are a wheel turning on an axle, a fan ventilating a chicken house, a lid being screwed onto a jar, and a crankshaft turning in an engine.

Everything on the earth is in motion all the time. The motion is often undetectable in everyday life, but it is ongoing.

The motion that people can detect is of one object compared with another. A tractor can move through a field cultivating the rows of crop plants. The plants appear to stay in the same place, and the tractor passes over them.

24–13. Using power equipment is a speedy way to harvest peas on a Wisconsin farm. (Courtesy, USDA)

Speed

Speed is the rate of motion. It is a calculation of how far an object travels in a set amount of time. Some things move quickly; others move slowly. A race horse runs fast and must run faster than the other horses to win the race. A horse race is usually timed to see how long it takes to run the race.

Tractors and other powered equipment are capable of certain speeds in a field. They can go faster when doing some jobs than when doing others. Tractors pulling heavy plows may go slower than those raking hay.

Speed is measured in miles per hour or kilometers per hour. Usually a motor vehicle has a speedometer that indicates the speed of the vehicle. Many jobs performed on crops are done with tractors moving at speeds of 5 to 10 miles per hour (mph). Speed must be adjusted to the job. Tractors cultivating tiny plants may need to travel slower than those cultivating larger plants. Faster speed tends to move the soil more, and the soil could be moved on top of tiny plants and destroy them. Larger plants stand taller and are stronger.

Speed is important in determining how long it takes to do a job. For example, if a 50-acre pasture is being clipped with a 6-foot cutter, how long will it take to clip the pasture at a speed of 8 mph? An acre is approximately 210 feet long and 210 feet wide. The cutter will have to make 35 passes over an acre. Each pass will be 210 rectilinear feet long. Using 6-foot swaths, there will be a total of 7,350 rectilinear feet in an acre. Fifty acres would contain 367,500 rectilinear feet (50 × 7,350). The total rectilinear distance for the pasture is 69.6 miles (367,500 ÷ 5,280 [feet in a mile]). At a speed of 8 mph, nearly 8.7 hours would be required to clip the pasture.

The factor-label method of calculating this problem would be:

50 ~~acres~~	35 ~~passes~~	210 ~~ft~~	1 ~~mi~~	1 hr	= 8.7 hr
	1 ~~acre~~	1 ~~pass~~	5,280 ~~ft~~	8 ~~mi~~	

Acceleration and Deceleration

Speed may change. ***Acceleration*** is an increase in the speed of an object over time. ***Deceleration*** is a decrease in the speed of an object over time.

Speed is an important factor in carrying out many agricultural jobs. Tractor speed can be varied to suit the situation. Fuel consumption and power are related to speed. Labor requirements are also related to speed.

Refer to the previous example of using a tractor to clip 50 acres of pasture. If the speed of the tractor was increased to 12 mph, how much time would be required to clip the pasture? At 12 mph, 5.8 hours would be needed to clip the pasture.

Using the factor-label method, the calculations would be as follows:

$$\frac{50\ \cancel{\text{acres}}}{} \times \frac{35\ \cancel{\text{passes}}}{1\ \cancel{\text{acre}}} \times \frac{210\ \cancel{\text{ft}}}{1\ \cancel{\text{pass}}} \times \frac{1\ \cancel{\text{mi}}}{5{,}280\ \cancel{\text{ft}}} \times \frac{1\ \text{hr}}{12\ \cancel{\text{mi}}} = 5.8\ \text{hr}$$

MACHINES, MACHINERY, AND SIMPLE MACHINES

A ***machine*** is a device that helps transmit power, force, and motion. Machines do work! They are simple things that we use in our daily lives. Machines increase power, make more force, or increase or decrease motion.

AgriScience Connection

DUST IN THE AIR

Many agricultural and industrial operations create dust that gets into the air. When the air is inhaled, so is some of the dust. The dust also coats plants, buildings, and other structures with a thin layer of "dirt." What can be done to reduce dust?

Begin by studying the nature and composition of dust. Use physics applications to develop devices to sample dust and study air quality. Develop equipment that reduces dust. Perform work so that dust is minimized.

In this image, scientists are examining filters in air samplers for the presence of dust. The dust will be closely analyzed to determine its content and particle size. You will note field operations underway in the background creating dust. (Courtesy, Agricultural Research Service, USDA)

Several simple machines can be joined to make a complex machine. Large, complex machines are sometimes known as machinery because they involve many different uses of simple machines. In agriculture, large hay balers, planters, and harvesters involve many simple machines.

All machinery is a combination of six simple machines. These six machines provide work with less human effort. A machine can increase (or decrease) the force acting on an object at the expense (or gain) of the distance moved, but the product of the two—the work done on the object—remains constant. An example is the amount of force necessary to move a refrigerator onto a flatbed truck. The amount of force would be much greater if the refrigerator were lifted straight up and then onto the truck, but the distance moved would be shorter. If a ramp were used, the force would be less, but the distance moved would be longer. The total amount of work done would be the same both ways. The simple machines are the lever, the wheel and axle, the pulley, the inclined plane, the wedge, and the screw.

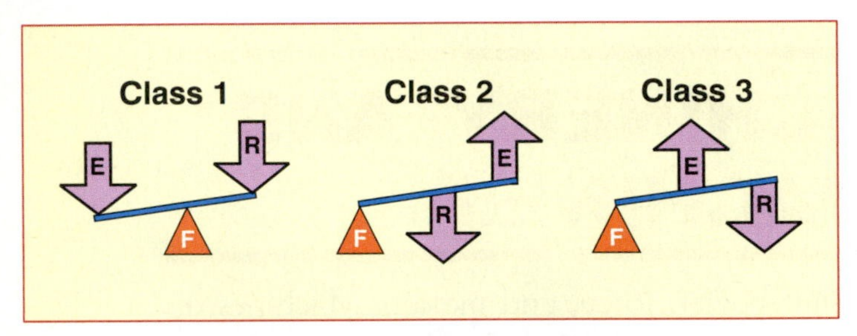

24–14. Three classes of levers. Note: In the figure, F is fulcrum, R is resistance, and E is effort.

24–15. The pry bar is an example of a first-class lever. Note the locations of the effort, fulcrum, and resistance.

LEVER

A ***lever*** is a rod or bar that rotates on a fulcrum, or pivot point. Levers help lift loads with less effort. A person weighing 150 pounds can lift 1,000 pounds with a lever!

A lever has two parts: the effort arm and the load arm. The effort arm is the distance between the force and the fulcrum. The load arm is the distance between the fulcrum and the load.

A lever is classified by the position of the fulcrum. With a first-class lever, the effort arm is the long part of the lever, and the load arm is the short part of the lever. A good example of a first-class lever is a crowbar. This is a simple machine used to pry boards apart and pull nails out of boards. The effort arm between the fulcrum and the force is the longest part of the lever. This makes it much easier to take boards apart and remove nails. A first-class lever changes the direction of the effort force by moving the load in the opposite direction. Another example of a first-class lever is a claw hammer used to pull a nail out of a board. When you push down on the arm of the hammer, the claw pushes up on the head of the nail. Where the head of the hammer rests on the board is the fulcrum.

A second-class lever has the fulcrum at one end and the effort arm at the other end. The load arm is between the fulcrum and the effort arm. In a second-class lever, the effort is exerted in the same direction as the load is moved. A heavy load can be moved with less effort if the distance to the fulcrum is increased. The wheelbarrow is an example.

A third-class lever has the fulcrum at one end and the resistance arm at the other end. The effort arm is between the two. A good example is the hammer when used in driving nails. The hand is the fulcrum, the arm supplies the effort force, the length of the hammer is the lever, and the driving of the nail is the resistance or load. Another example is the lifting of weights in one hand. The weights are the load, the forearm is the effort, and the elbow is the fulcrum.

24–16. A hammer driving nails is an example of a third-class lever. (Courtesy, Education Images)

WHEEL AND AXLE

The ***wheel and axle*** is a simple machine used in most moving vehicles. Wheels are usually thought of as round, such as the wheels on pickup trucks and tractors. Other uses of wheels and axles involve handles that make round motions. Turning a doorknob, turning a knob to tune a radio, and turning off a water faucet are common examples of using a wheel and axle. A handle attached to a shaft is a wheel-and-axle machine, such as the crank on a hand-operated grinder or the bit brace used in drilling holes. Another example is the pedal system on a bicycle. The pedals go around like wheels and are attached to an axle.

24–17. The wrench on the bolt nut provides power like a wheel and axle. (The nut on the bolt is a wheel.) (Courtesy, Education Images)

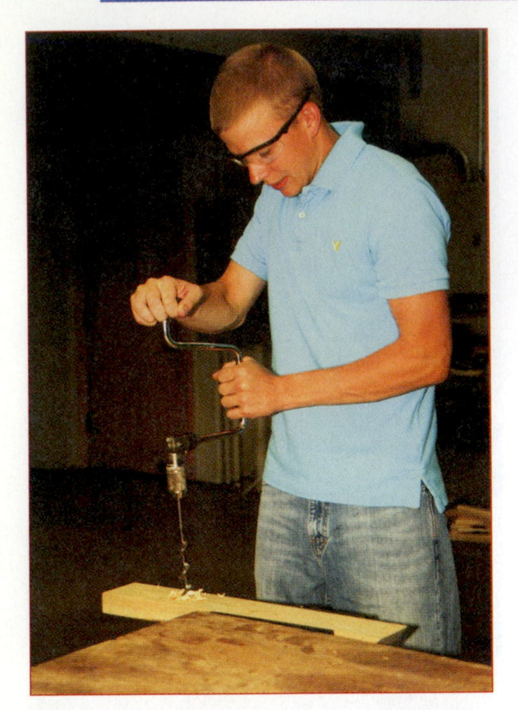

24–18. A brace with bit used to bore holes is an example of a wheel and axle. (Courtesy, Education Images)

The use of wheels and axles increases power or speed but not both. Turning a big wheel on a small axle increases power. Turning a smaller wheel on a larger axle increases speed.

PULLEY

A ***pulley*** is a special kind of wheel. It changes the direction of force by using a belt or rope. The wheel is free to spin on an axle. The wheel usually has a groove in the middle for holding the belt in position. A load is placed on one end of the rope, the rope is placed over the pulley, and effort is exerted on the other end. Pulling down on the rope through a greater distance is easier than picking the load straight up. Elevators are common examples of the use of pulleys. Belts are examples of pulleys that are widely used on agricultural equipment to transfer power.

Pulleys are used to provide a mechanical advantage in lifting. Pulleys with belts that provide the power to turn the parts of an engine change the direction of force of the crankshaft. Several pulleys used together result in much less effort being necessary to lift a heavy object.

24–19. The belt on a pulley of a gasoline engine is used to transfer power.

24–20. A fixed pulley on the left and a double pulley on the right are being used to lift weights. (Courtesy, Education Images)

Pulleys arranged as a block and tackle can be used to lift heavy loads with less effort. A good example is the block and tackle used to remove a heavy engine from a piece of machinery.

Pulleys may be fixed or movable. A fixed pulley is mounted in one place, while a movable pulley rises and falls with its load. Movable pulleys are used to increase force. Pulling on one end of a rope can lift an object much heavier in relation to the force that is applied.

INCLINED PLANE

An ***inclined plane*** is a straight, sloping surface. Moving an object up a sloping surface requires less effort than lifting it straight up, but you move the object over a greater distance.

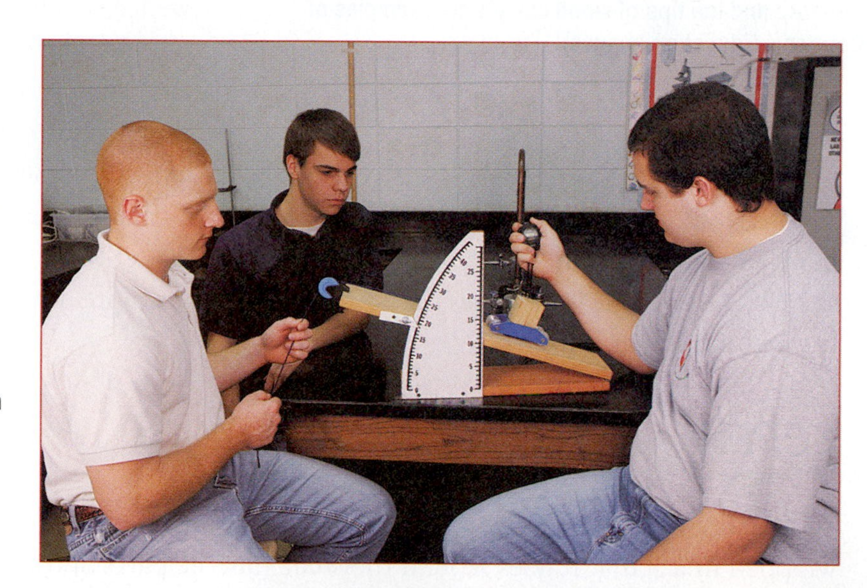

24–21. An inclined plane is being used to investigate the speed at which a wheeled vehicle travels the length of the incline at any of various angles. A stop watch is used to determine the time needed. (Courtesy, Education Images)

A ramp in a feed mill or an access ramp to a building is an inclined plane. People can move much heavier weights by pushing them up a ramp than they can by lifting them straight up. Inclined planes also help when loading products, such as bags of feed or crates of vegetables.

24–22. A loading chute is an example of an inclined plane.

WEDGE

A ***wedge*** is a solid piece of material, usually wood or metal, that is thick at

24–23. A doorstop and the tips of wood chisels are examples of wedges. (Courtesy, Education Images)

one end and thin at the other, much like an inclined plane. A wedge has sloping surfaces that taper to a point. It has been described as a double inclined plane. The greater distance you move a wedge, the greater the force it applies on the object. For example, the deeper the head of an axe moves into a block of wood, the more force it applies, causing the wood to split. Eventually, the two pieces are pushed apart. Other useful wedges include doorstops, plows, and chisels. The opener on a planter pushes the soil apart so the seed can be dropped into the ground. A covering wheel closes the drill opening.

Wedges vary in angle. Angle refers to the amount that a wedge thickens away from the point. Wedges that are long and gradually taper are easier to drive into wood, but they split the wood more slowly. Wedges that get thick faster are more difficult to drive, but they split the wood more quickly.

SCREW

Like the wedge, the **screw** is a modified inclined plane. The inclined plane on a screw is wrapped in a spiral. It goes around the shaft (middle part) of the screw. When the screw is rotated, a small force is applied over the long distance along the spiral inclined plane. The screw applies a large force through the short distance it is pushed. Screws are used in many ways: fasteners to hold materials together, augers to move grain through tubes to bins, and the lifting mechanisms of jacks for cars with flat tires.

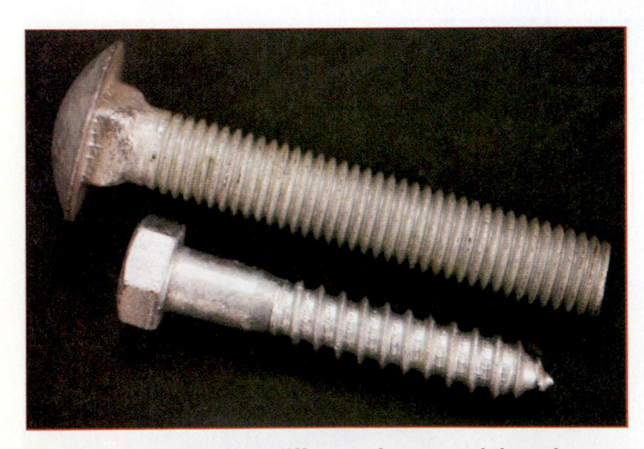

24–24. Screws showing different shapes and thread gauge. (The screw with the smaller threads has greater mechanical advantage.) (Courtesy, Education Images)

The pitch of a screw is the angle of the thread (incline) around the shaft. A screwdriver uses rotating force to move a screw into the wood. A wrench may be used to put a bolt and a nut together. With effort, the screw can be made very

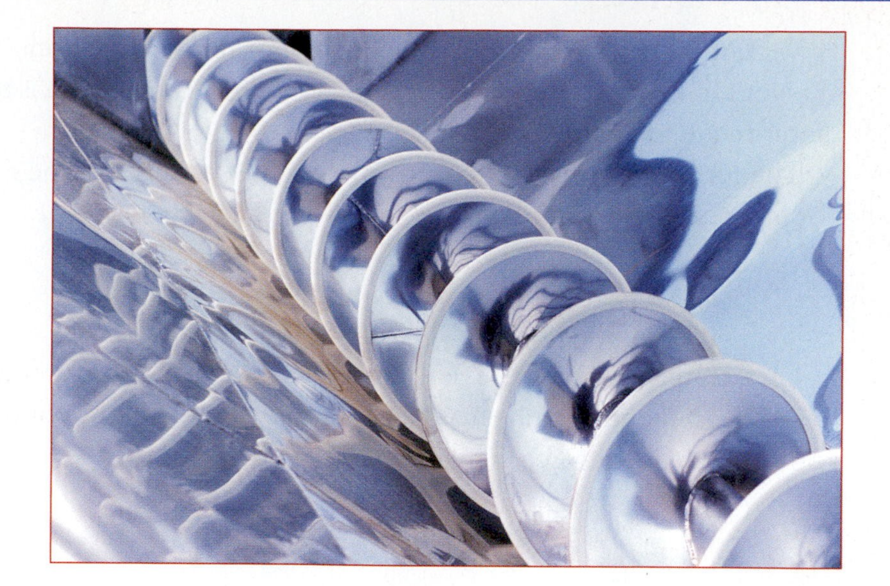

24–25. An auger is a kind of screw. (This stainless steel auger at a food processing plant is used to move raw products inside for processing.)

tight so it will hold heavy loads securely. Sometimes, lock washers are put on screws to keep them from working loose by turning backward.

ADVANTAGES OF MACHINES

Machines cannot create energy, but they can control and multiply the force. A machine never does more work than the energy put into it. Machines can convert energy from one form to another, such as diesel fuel into mechanical power. This makes them very useful in many agricultural jobs.

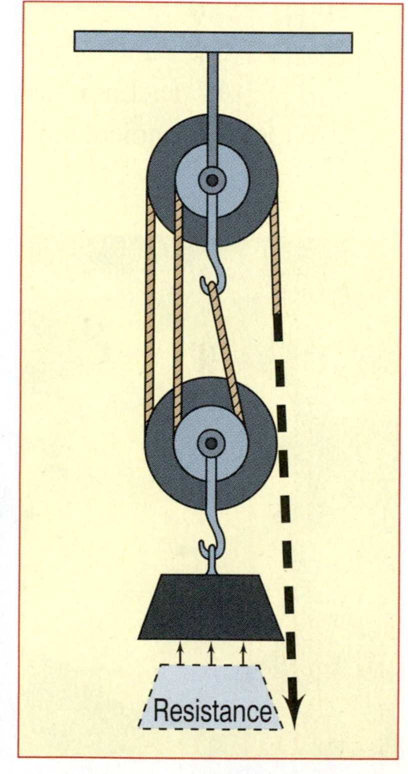

24–26. Simple machines, such as a pulley system, illustrate mechanical advantage in moving objects. Counting the number of supporting ropes indicates that in this system, MA = 4. The power exerted in pulling the rope is increased four times. MA makes jobs easier!

MECHANICAL ADVANTAGE

Mechanical advantage (MA) is the way in which a machine multiplies the force it receives. With mechanical advantage, a small force can become a larger force—often a much larger one. Mechanical advantage tells you how many times the machine multiplies the force. A good example is a long handle on a wrench. A little force on the end of a long lever can move a heavy weight on the other end.

With mechanical advantage, the lever must rest on a fulcrum. A fulcrum is the support or pivot for a lever that allows it to turn. The closer the fulcrum is to the load, the less effort that is required to move the load.

The formula for mechanical advantage (MA) calculates the ratio of input and output power. Here is the formula:

$$MA = \text{output force} \div \text{input force}$$

For example: Suppose a lever is used to lift one side of a farm tractor. The side to be lifted weighs 1,000 pounds, known as the output force. The effort put on the lever is 20 pounds, known as the input force. The mechanical advantage is:

$$\begin{aligned} MA &= 1{,}000 \text{ lb} \div 20 \text{ lb} \\ &= 50 \end{aligned}$$

The same formula can be used with the SI system by substituting metric weights.

All machines have mechanical advantage. If they didn't, who would use them? Formulas for mechanical advantage for all simple machines are found in introductory physics books.

24–27. When a light bulb is on, some energy is converted into thermal energy. This makes the bulb feel warm. Part of the electrical energy is converted into light. Other energy is converted into thermal energy because of friction in the wire. New fluorescent bulbs are more energy efficient and save money.

WORK PRODUCED

Work is a product of force and distance. A machine will not increase both the force and the distance moved. The lever is a good example. A person using a lever to lift an object must move the lever a much greater distance than the object will be lifted. The long part of the lever (known as the effort arm) gives an advantage in lifting a load a short distance with less force.

The ***Law of Conservation of Energy*** states that *energy can be neither created nor destroyed*. Energy can be changed from one form to another, but the different forms of energy in a system always add up to the total amount of energy. Some of the energy is changed to heat energy because of friction in the machine. Other energy is required to rotate the lever around its fulcrum! No machine is 100 percent efficient. Efficiency is sacrificed to make jobs easier.

The length of a lever and the location of the fulcrum are important in mechanical advantage. Distance is measured from the fulcrum to each end of the lever. Here is how it is done:

output force = F_2 (weight lifted)
output distance = D_2 (distance the object is moved)
input force = F_1 (weight exerted on the lever)
input distance = D_1 (distance the long end [effort arm] of the lever is moved)

The formula is:

$$F_1 \times D_1 = F_2 \times D_2$$

In the example of lifting one side of a tractor, how far would the effort arm (D_1) need to be moved if the tractor was lifted 0.2 foot (D_2)? To calculate: (Remember, 20 pounds of force was applied to the long end of the lever, and this is known as D_1. The weight lifted was 1,000 pounds, known as F_2.)

$$20 \text{ lb} \times D_1 = 1{,}000 \text{ lb} \times 0.2 \text{ ft}$$
$$20D_1 = 200 \text{ ft-lb}$$
$$D_1 = 200 \text{ ft-lb} \div 20 \text{ lb}$$
$$D_1 = 10 \text{ ft}$$

THERMAL ENERGY

Thermal energy is energy in the form of heat energy that is produced when energy is converted from one form to another. Thermal energy produced as a result of friction is not useful energy and is not used to do work. Sometimes thermal energy results when fuel burns, such as gasoline in an engine. A car engine converts most fuel into work. Some waste energy results that exits the car engine through the radiator and the exhaust pipe mostly in the form of heat.

24–28. Thermal energy (heat) is used to dry grain. (Courtesy, USDA)

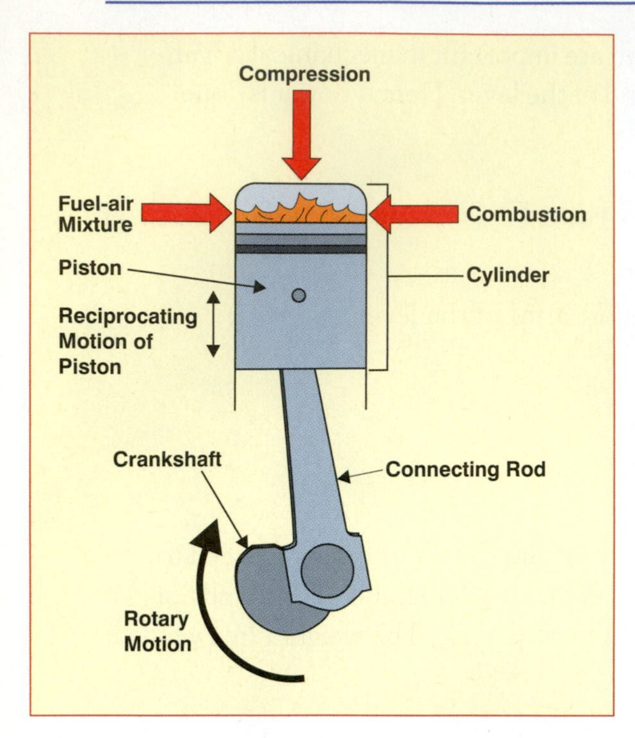

24–29. How an internal-combustion engine operates.

Heat causes gases to expand rapidly. When inside a strong container (such as an engine cylinder), rapid expansion is used to move mechanical parts. This process converts thermal energy to mechanical power in heat engines.

INTERNAL-COMBUSTION ENGINES

Engines on lawn mowers, tractors, and pickup trucks are examples of heat engines. A ***heat engine*** is a type of engine that burns fuel (gasoline, diesel, etc.) to produce hot gases. The gases create pressure, which is converted to mechanical power.

An engine works like this: Fuel is placed in one end of a strong cylinder and ignited (burned). Burning creates gases that rapidly expand inside the cylinder. One end of the cylinder is closed. A piston that can move back and forth is in the other end. The rapid expansion pushes the piston, creating motion. The piston is connected by a lever to a crankshaft that is rotated by the moving piston. The crankshaft is connected to axles and wheels, pulleys, and other machines that can do work.

24–30. Checking the oil level of an engine in golf course maintenance equipment. (Courtesy, Education Images)

FUELS

Three kinds of fuels are commonly used in internal-combustion engines in agriscience: gasoline, diesel, and liquefied petroleum gas.

Gasoline

Many engines use gasoline as fuel. Gasoline engines are in pickup trucks and small equipment. In a gasoline engine, a gasoline and air mixture is compressed in the cylinder and burned with a spark from a spark plug.

The ratio of air to gasoline in the fuel mixture in a cylinder is 9,000 to 1. This means that 9,000 gallons of air are needed for 1 gallon of gasoline. That is a lot of air!

Gasoline is made by refining petroleum (also known as crude oil). Most gasoline is a blend of many kinds of hydrocarbons. Each hydrocarbon burns differently in an engine.

ACADEMIC CONNECTION

PHYSICS

Relate your study of physics applications in agriscience to your studies in physics class. Explore the important role of physics principles in plant and animal production, particularly as related to hand tools and to power tools and equipment. Use your learning in agriscience to support your mastery in physics.

Diesel

Diesel fuel is used in engines with more power. The engines are heavier than gasoline engines and are designed to run long hours and pull heavy loads.

Diesel engines do not have spark ignition systems. In a diesel engine, a piston creates high compression in each cylinder. The high pressure causes heat, which ignites the fuel when it is injected into the cylinder. Heat causes fuel to burn rapidly. Burning creates gas pressure that forces the piston backward in the cylinder.

Liquefied Petroleum Gas

Liquefied petroleum gas (LP gas) is made of butane and propane. It is stored as a liquid under pressure in bottles or strong tanks. Tractors, pickup trucks, and other lighter-weight engines can be run on LP gas.

Gasoline engines are sometimes converted to LP gas. All processes of an engine are the same except the way LP gas is stored and gets into the engine. The fuel requires a spark to ignite. An engine fueled with LP gas usually costs less to operate than one that uses gasoline.

Other Fuels

Other fuels are less commonly used to power engines. Alcohol, biodiesel, methane, and hydrogen gas are examples. Electricity is used to power vehicles on a small scale, such as golf carts, small experimental cars, and hedge trimmers. New hybrid cars use a combination of electricity and fuels to power the engines.

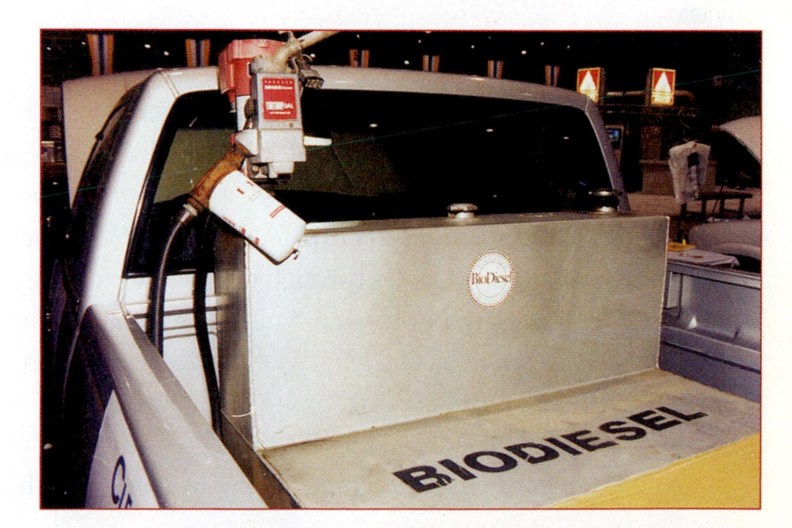

24–31. A pickup truck that uses biodiesel fuel.

ELECTRICAL POWER

Electricity is the flow of electrons, which are small units of negative electrical charge. All materials are made of protons and electrons that have electrical charge.

CURRENT ELECTRICITY

Current electricity is electricity caused by flowing electrons through a wire or other transport material. It is the most important kind of electricity. It is the electricity used by light bulbs, electric motors, and heating elements. Current electricity is produced by electrical generators, batteries, and solar cells.

24–32. Energy in reservoirs is converted to electricity at the Rocky Reach Dam generating plant in Washington as the water falls over turbines.

Another form of electricity is static electricity. It is the result of friction. You may have felt static electricity when combing your hair or when walking on a carpet on a cold, dry day. The sparks produced can be a source of danger. In a grain elevator with dust in the air, a single spark can result in a huge explosion.

Kinds of Current Electricity

The two kinds of current electricity are direct and alternating.

Direct current (DC) is electricity that always flows in the same direction. It is produced by batteries and some kinds of electrical generators. DC electricity is used in the ignition systems of tractors, trucks, and other engines.

Alternating current (AC) is electricity that regularly reverses, or alternates, its direction of flow. AC goes through cycles. Two changes in direction are one cycle. The AC cycle is measured in hertz. One cycle per second is 1 hertz (Hz). Most of the current electricity in North America is 60 Hz. AC electricity is used in homes, on farms, and in agribusinesses for light, heat, and cooling. Standby electrical generators may be needed in case the power goes off.

Current can be changed from AC to DC with a rectifier.

How Current Is Measured

Electrical current is measured to determine its rate of flow, its pressure, and the amount available. Most people are familiar with the electric meter used by a power company to measure the amount of electricity consumed. Customer charges are based on this measurement.

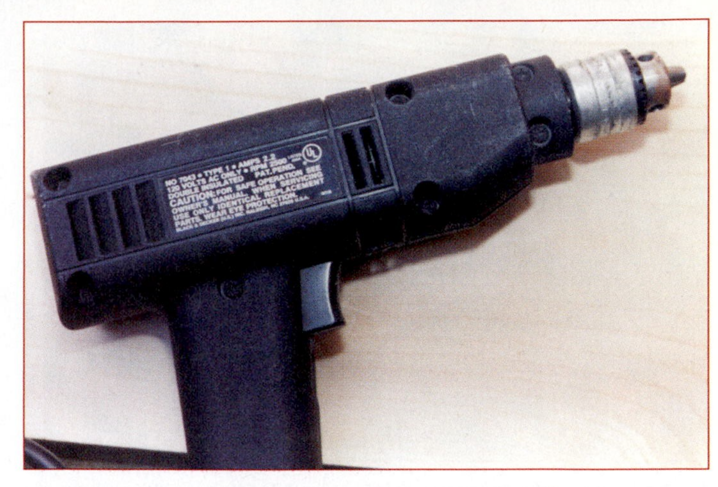

24–33. Electrical requirements for a small drill are listed on the name plate.

- **Ampere**—An ***ampere***, or amp, is a measure of the rate of flow of electrical current. One ampere is made of a huge number of electrons that flow every second. Amperes are measured with an ammeter. Most electrical devices require wires that can conduct a certain flow of amperes. Larger wires will conduct more amperes of electricity, much like larger pipes will allow more water to flow.
- **Voltage**—The measurement of the potential energy of electricity is the ***volt***. It is the difference in electrical potential between two points in an electrical circuit. The difference in potential that pushes electrical charges through a conductor is known as voltage. Most light-duty electrical equipment in North America runs on 110 volts. Heavier equipment, such as welders and electric heaters, may need 220 volts. Some electrical power lines carry very high voltage. The voltage is reduced with transformers to the amount needed. High voltages and large wires make it efficient for a power company to send electricity. It is "stepped down" when sent to the customer.
- **Wattage**—The amount of electrical energy supplied or used is measured in watts. Most electrical appliances, including light bulbs, have wattage ratings. For example, household light bulbs are usually 60-, 75- or 100-watt, although some manufacturers are lowering these standard wattages slightly to conserve energy.
- **Combinations**—Amperes, watts, and volts are used in various combinations to calculate electrical use and demand. The name plates of most electrical devices provide information about the amperes, watts, and volts needed for operation. Proper installation of a wiring system requires knowing the electrical demands on the system.

Conductors and Insulators

All electrical current must have a conductor. A ***conductor*** is a substance that allows electrons to flow freely. Copper, gold, and aluminum are good conductors of electricity.

24–34. Electrical wire made up of neutral (white), positive (black), and ground (bare) wires. Insulation keeps the wires from touching. All three are wrapped together with more insulation to keep them from coming into direct contact with other objects.

For safety purposes, all electrical wires and other devices must be properly insulated. This is done with an ***insulator***, a material that does not conduct electricity. Rubber, plastic, and glass are commonly used as insulation. Electrical wires are usually covered with a layer of plastic insulation. Without insulation, a wire is dangerous and could cause electrocution as well as fires and other damage.

Electrical Circuits

A ***circuit*** is the path electricity follows. Most circuits are made of wires and other materials that are good conductors.

A circuit must have a minimum of two wires. One wire carries the current from its source to where it is used. The other wire carries the current back. In addition, most electrical circuits have a third wire, or ground wire. A ground wire prevents shock in case the other wires fail, a short occurs, or an excess charge builds up. A ground wire is attached to a metal rod driven several feet into the ground. The ground wire serves as a pathway for the discharge of electricity when there is an interruption in the normal route.

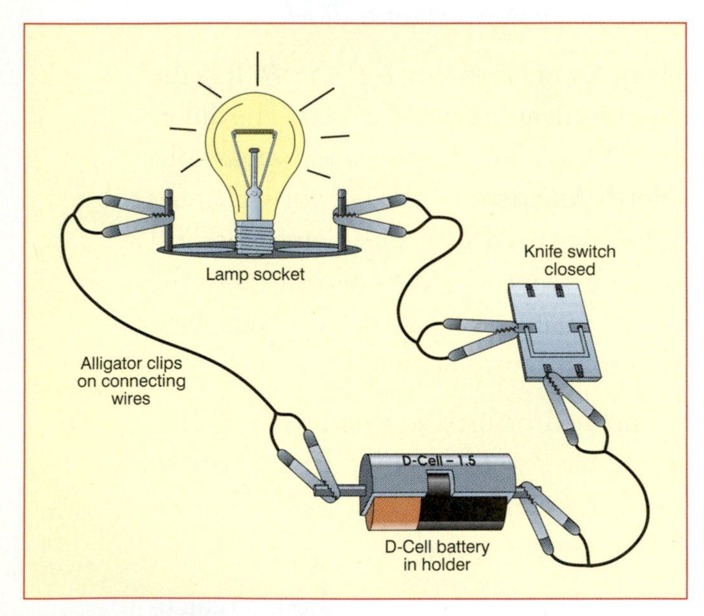

24–35. This diagram of an electrical circuit shows that the source of the electricity is a battery with wires leading to a bulb and a switch.

Two kinds of circuits are commonly used: parallel and series. A parallel circuit has more than one path the electricity can follow. The electricity can flow to several output devices (lights, appliances, computers, etc.) along the path of the circuit. Any device on a circuit connected in parallel can be on or off without affecting the others. Household and business wiring is connected in parallel.

A series circuit has only one path the electricity can follow. The electricity must flow through each output device located on the circuit. Any break in the circuit prevents all the output devices from working. This kind of circuit has limited uses. If used to connect a string of lights, for instance, when one bulb burns out or becomes loose, the entire string goes out.

Electrical Controls

Many kinds of devices are used to regulate and control electrical power. Switches turn electrical power on and off. These range from simple toggle switches that operate lights to complex switches that control other devices.

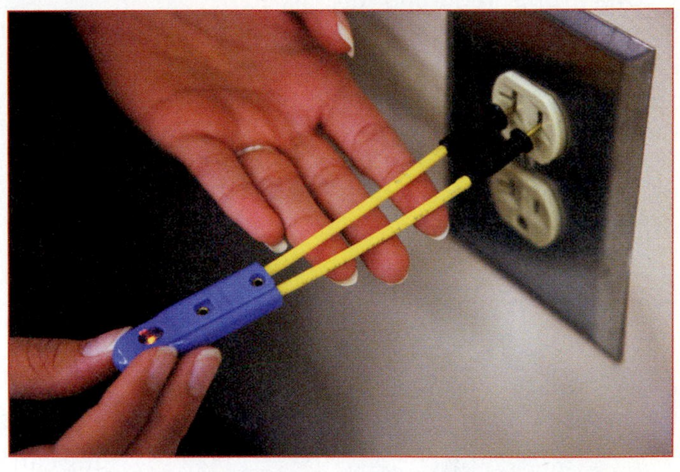

24–36. An electrical circuit tester can be used to determine if a circuit is energized. (Courtesy, Education Images)

Thermostats are used to regulate the amount of heat and cooling, depending on the result wanted. To keep something cool, a thermostat is set to the desired temperature. If it gets too warm, the thermostat trips a switch that provides power to operate the cooler. When it gets cool enough, the thermostat turns the power off.

Technology Connection

ANTI–FRUIT FLY ENERGY

Fruit and melon flies are destructive pests in some places. They attack a large number of valuable crops, particularly fruits and melons. Many food products are damaged by these flies each year. Managing the populations of these pests is a particular challenge. Several different approaches have been used.

Some scientists have turned to high-energy particles known as electromagnetic waves. These waves are produced by radioactive substances and particle generators. The pupae (immature forms) of male melon flies can be grown in a lab and sterilized using these waves. Once the adult stage is reached, the male flies are released. The females that have grown in nature mate with the sterile males and, therefore, produce infertile eggs. A female fruit fly mates only once. If that is with a sterile male, no offspring will result.

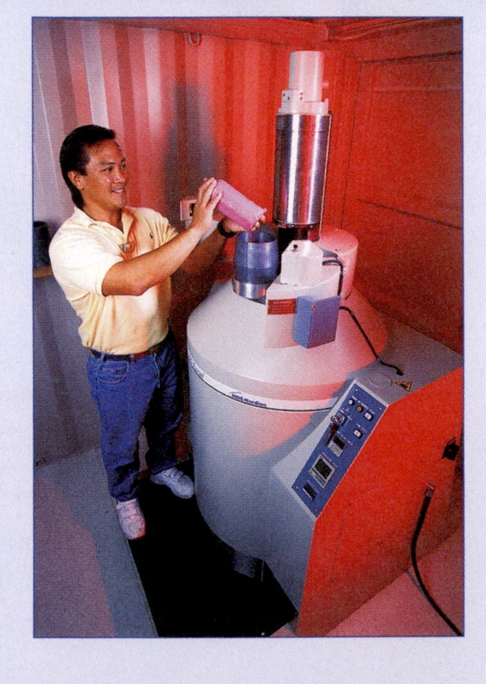

This shows a scientist preparing to place a container of male melon fly pupae into a particle generator for sterilization. (A similar approach has been used to virtually eradicate the boll weevil in the cotton-growing areas of the United States.) (Courtesy, Agricultural Research Service, USDA)

Internet Topics

Search

Selected topics for Internet discovery and reporting are listed here. Begin by searching for each topic. Next, read and learn about it. Conclude by preparing a brief report on your findings.

1. Alternating current
2. Pneumatic power
3. Energy-efficient light bulbs

Timers may be used to operate lights and other equipment. A timer may turn the lights on and off in a greenhouse so the plants get more light than the day is providing or in a poultry house so the birds have longer days. Timers are set to turn lights and other equipment on or off at a certain hour of the day or night.

Security lights may have sensors that turn the lights on at night when it gets dark and off in the morning in response to sunlight.

REGULATIONS AND SAFETY

24–37. A ground fault circuit interrupter (GFCI) on a power supply cord.

Electricity can be dangerous to humans and property. It can cause injury when used improperly. A building can burn if a short circuit creates heat that sets wood or other material on fire.

Various regulations, known as codes, have been established regarding the use of electricity. The National Electrical Code serves as the basis for electrical work. Local cities and power companies may also have codes. Always follow the codes and observe safety practices when using electricity.

All electrical devices should be grounded to protect the people who use them. Avoid using electrical appliances in wet conditions.

A small appliance may have a ground fault circuit interrupter, known as a GFCI. The GFCI is designed to protect people from shock when using electricity in proximity to water. The GFCI may be built into a breaker box, or it may be a part of the cord where the appliance is plugged into a receptacle.

COMPRESSION POWER

Liquids and gases may be compressed to create power. Machines that do the compression are known as compressors. They are usually pumps with tanks or reservoirs filled with liquids and gases (often air) used to create compression.

HYDRAULIC POWER

Hydraulic power is compression power that uses liquids to create pressure inside cylinders. In some cases, gases are a part of hydraulic power. Pressure can be used to transfer or increase a force. Liquids, such as water and oil, can be placed under pressure. This allows them to lift heavy weights, operate hydraulic motors, and do other work.

24–38. Wheels on a grain drill are raised and lowered by hydraulics. The cylinder is easy to see in this photo.

Pressure in a hydraulic system is often measured as pounds per square inch (psi). Leaks in the system reduce the pressure and may result in the loss of liquid. Also, leaks can be dangerous and injure people. Hoses and connections should be kept tight and in good condition.

Many farm and horticultural tractors have hydraulic systems that lift implements. Also, equipment may need power at places that cannot be easily reached with levers, belts, or axles. Hydraulic pressure can be transferred in oil through hoses to hydraulic motors that convert the pressure into movement.

A hydraulic system operates using a pump powered by a heat engine or electric

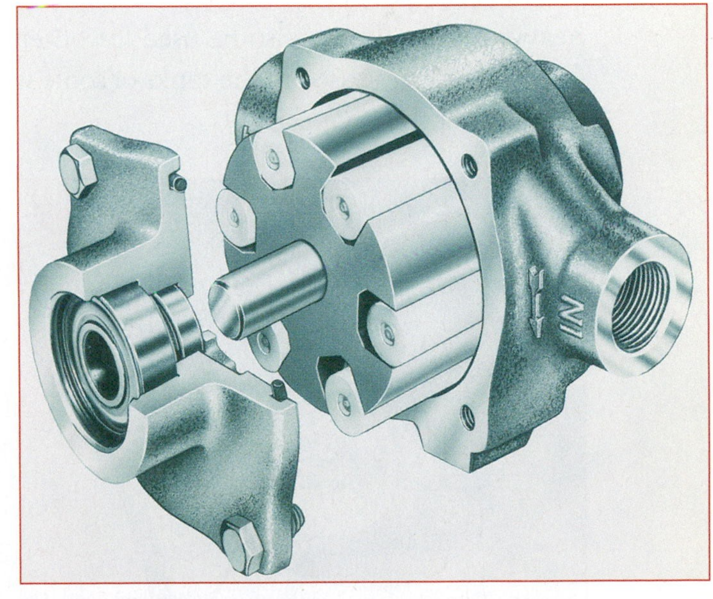

24–39. Cutaway view of a roller pump. (Courtesy, Hypro Corporation)

motor that creates pressure on the oil in the system. The oil may be moved to other locations, where pistons in cylinders are moved. Movement of the pistons causes a lever to move. The lever is connected to wheels, axles, or other levers. This series of actions causes lifting.

For example, a disc harrow may have wheels that can be raised or lowered. When the wheels are lowered, the blades of the disc harrow are raised and do not touch the soil. The weight of the harrow rests on the wheels. Wheels are lowered when moving the harrow from one field to another. When the wheels are raised, the blades go into the soil. The position of the wheels regulates depth of tilling.

COMPRESSED AIR

Air can be used to do work. Air is compressed when it is put under pressure. The use of compressed air to do work is known as **pneumatic power**.

Lowering air pressure creates a vacuum. A vacuum operates on the basis of the pressure inside a container being lower than the pressure of the air in the surrounding area. Suction is created.

Air pressure is measured in pounds per square inch (psi). A tank, pipe, or other device filled with air usually has a maximum pressure. It will burst like a balloon if the pressure is excessive. Air pressure is measured with an air gauge.

Air is compressed to create pressure. Air under pressure is used to fill tires that support heavy weights. It can also be used for other purposes, such as to operate a paint sprayer. Compressed air is used in the tanks of some water supply systems to create pressure that will cause water to flow.

24–40. A tire pump uses a piston inside a cylinder to create air pressure.

Pneumatic tools are powered by compressed air. Examples are wrenches that remove lugs from wheels, riveters, and air hammers that break concrete. Some pneumatic tools are used in woodworking to drive nails and other fasteners.

The suction of air (lowered pressure) is used to move some products. Air suction is used to load and unload cotton or move products in a processing plant.

Air pressure and suction can be created with piston or rotary pumps. The hand-operated tire pump is an example of a piston pump. When the piston in the cylinder is pushed down, the pressure in the cylinder increases, and air is forced out through a small tube connected to a tire. The pressure in the cylinder causes the air to move into the tire.

A rotary pump has rollers that turn inside a pump housing. Rollers create a steady pressure, whereas pistons may create pressure in spurts.

24–41. Using a pneumatic nailer in construction.

REVIEWING

MAIN IDEAS

Many areas of physics are applied in agriscience. The areas of mechanics, sound, magnetism and electricity, light, and electron and nuclear physics are all areas of physics that can be studied for applications in agriscience. These applications involve mathematical formulas and calculations.

Physics is used to increase the ability of people to do work. Mechanical power has replaced human and animal power. One person can do much more work with a tractor than could be done without one.

Work is the moving of an object through a distance when there is some resistance to its movement. Force is applied to make it move. If the object doesn't move, no work has been done.

Power is the rate at which work is done. Time is an important factor in computing the amount of work that can be done. Horsepower is a measure of power that involves doing 550 foot-pounds of work a second. Motion is an important part of work and power. Speed is the rate of motion. Acceleration is an increase in speed, and deceleration is a decrease in speed.

Simple machines are used to accomplish work. The six simple machines are lever, wheel and axle, pulley, inclined plane, wedge, and screw. These machines can be put together in complex combinations to form machinery that will do a lot of work. A machine provides mechanical advantage. This means that the machine requires less input force for the output force it will provide.

Thermal and electrical energy are used in agriscience. Thermal energy involves changing heat into mechanical energy. Internal-combustion engines use various forms of fuels to produce heat energy. They then use that energy to perform specific jobs. Electrical current can be used for many purposes, including light, power, and heat.

Compression power includes hydraulics and pneumatics. Hydraulics is the compression of liquids. Hydraulics is used for lifting and other operations on mobile and stationary equipment. Pneumatics is the compression of air. It is often used for nailing and fastening nuts and bolts on construction sites.

QUESTIONS

Answer the following questions, using complete sentences and correct spelling.

1. What is physics? What areas are included?
2. What is work? Power? Distinguish between the two.
3. Calculate the amount of work if 5 pounds of force are used to push a lawn mower 500 feet.
4. Calculate the amount of power if it takes six minutes to push the lawn mower in question 3.
5. What is horsepower?
6. What is motion? Speed?
7. How does speed relate to the time required to do a job?
8. What are the six simple machines? Name examples of the use of each.
9. What is mechanical advantage?
10. What is a heat engine? Why is heat important in the engine?
11. What is current electricity? What are the two kinds, and how do they differ?
12. How is electrical current measured?
13. Distinguish between an electrical current conductor and an electrical current insulator.
14. Why is a GFCI important?
15. What is hydraulic power? How is it used in agriscience?
16. What is pneumatic power? How is it used in agriscience?

EVALUATING

Match each term with its correct definition.

a. kinetic energy
b. work
c. force
d. power
e. horsepower
f. speed
g. machine
h. mechanical advantage
i. heat engine
j. circuit
k. Law of Conservation of Energy
l. acceleration

_____1. The rate of motion.

_____2. The path electricity follows.

_____3. The amount of power used to do 550 foot-pounds of work per second.

_____4. The result of an object's mass and motion.

_____5. The rate at which work is done.

_____6. The push or pull on an object.

_____7. The moving of an object through a distance when there is some resistance to its movement.

_____8. The way in which machines multiply the force they receive.

_____9. A device that helps transmit power, force, and motion.

_____10. A type of engine that burns fuel (gasoline, diesel, etc.) to produce hot gases.

_____11. An increase in the speed of an object over time.

_____12. The principle of physics stating that energy can be neither created nor destroyed.

EXPLORING

1. Visit a building construction site where electrical wiring and controls are being installed. Observe the procedures used in putting the wires, switches, and other devices in place. Prepare a report on your observations.

2. Construct a simple electric motor. Refer to the Activity Manual for information on how to build a motor.

3. Use a gauge to determine the air pressure in the tires on a vehicle. Determine the recommended pressure for the tires. Add or reduce the pressure so the tires are properly inflated.

CHAPTER 25

Mechanics in AgriScience

OBJECTIVES

This chapter introduces important areas of agricultural mechanics and initiates skill development in these areas. It has the following objectives:

1. Explain the meaning and areas of agricultural mechanics.
2. Identify materials used in agricultural mechanics.
3. Discuss the roles of a plan and bill of material.
4. Identify fasteners, and select the fastener for a job.
5. Identify common hand and power tools used in agricultural mechanics.
6. Use woodworking skills to construct a project.
7. Use welding and cutting skills to fabricate metals.

TERMS

acetylene
agricultural mechanics
bill of material
cast iron
drawbar power
drill
fabrication
fastener
green lumber
harvesting
lumber
mild steel
plan
planter
plywood
power take-off
scale
seasoning
shielded arc welding
tillage
tool
veneer
welding

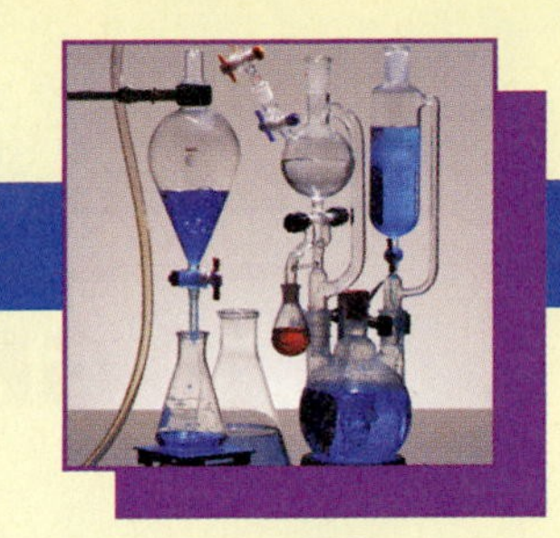

25–1. Welding skills are needed in a wide range of agricultural mechanics construction and repair jobs.

PEOPLE have developed ways of accomplishing more with less effort. Just think of the number of acres one person can plow with a tractor in a day—10, 20, 30, or more. Compare this to digging in the soil with a hand shovel.

In the agricultural industry, power, machinery, structures, and devices are used in many ways. Plowing, planting, harvesting, marketing, and other activities associated with plant and animal production use these technologies. Using them helps meet the food, clothing, and shelter needs of countless people. If we did not have these abilities, few people would receive the things they want in the way in which they have them today.

Construction projects help provide needed devices and structures. Some projects help manage animals, and others help with plant production and natural resource practices. Wood and metal are commonly used in these projects. All mechanical activities require attention to safe practices.

AGRICULTURAL MECHANICS AND MECHANIZATION

Agricultural mechanics is the use of machinery, tools, and associated electronic devices to perform agricultural jobs. It includes the design, construction, operation, and repair of agricultural engines and implements as well as areas of agricultural structures, land and water management, and electrical applications.

25–2. Equipment used on golf courses is highly sophisticated. This fairway mower is an example. (Courtesy, Education Images)

Considerable emphasis is on mechanization—the use of machinery to replace human labor. Today, machinery has "smart" devices (e.g., computers, sensors, and controllers) to operate, monitor, and record field conditions and applications. Global positioning systems are often integrated with these "smart" devices.

Much of the work in agriculture is carried out with powered implements. Some implements are complex. Other implements lack complexity, but they require careful attention during operation.

POWER

All implements require a source of power. Typically, an engine, electric motor, or other power source is used. The power source may be stationary (not movable) or mounted on a frame with wheels that allow it to move. Farm tractors are power units that move on wheels, on tracks, or in other ways.

Power on most tractors and implements is provided by an internal combustion engine. Such engines use gasoline, diesel fuel, LP gas, and other energy sources. Gasoline engines are used for smaller tractors and implements; diesel engines are used for larger tractors.

Tractors are equipped to pull implements and to provide rotating power. Tractors have two types of power: drawbar power and PTO power.

- **Drawbar power—*Drawbar power*** is the power of a tractor in pulling a load attached to a drawbar. The tractor itself must move. As it moves, the drawbar moves with it. A drawbar is a heavy, strong steel bar designed for attaching plows, wagons,

25–3. A tractor with a 3-point hitch and a hitch for wagons and other pull-behind implements.

and other implements. Drawbar power results from a combination of engine power and traction of the wheels or tracks. Some tractors have 3-point hitches. These do not have a drawbar, but implements are attached for ease in lifting and turning. Drawbar power applies to tractors with 3-point hitches.

- **PTO power**—Most tractors have a rotating shaft powered by the engine. It is known as a ***power take-off*** or PTO shaft and is typically located at the rear of the tractor. The shaft rotates and provides power to operate implements, such as mowers, pumps, and hay balers. The rotation of PTO shafts is standardized at 540 and 1,000 revolutions per minute. The rotating power is transferred to implements with a shaft and universal joints.

25–4. A lightweight 3-point hitch clipper showing the PTO connection with a shaft and universal joints.

EQUIPMENT

Specialized implements perform a variety of agricultural tasks. Equipment can be categorized by its unique use.

25–5. A reversible plow breaks soil in primary tillage. (Courtesy, Deere & Company)

25–6. A row crop cultivator in soybeans provides secondary tillage. (Courtesy, AGCO Corporation, Duluth, Georgia)

Tillage Equipment

Tillage is the slicing, breaking, or cutting of the soil. This activity prepares a good seedbed, controls weeds, or creates a soil mulch. Tillage equipment is grouped as primary or secondary.

Primary tillage is the cutting and shattering of the soil to prepare it for growing a crop. It requires powerful implements that go deeply into the soil. Plows, rippers, listers and bedders (form rows), and rotary tillers are common primary tillage implements.

Secondary tillage is the tillage that follows primary tillage. It further prepares a seedbed or cultivates crops. Equipment includes harrows (disk as well as toothed) and cultivators. Secondary tillage equipment is used to break apart clods, pulverize and loosen the soil, and control weeds.

There are many variations of tillage equipment. Design differences are based on the nature of the soil and the crop to be grown.

Seeding Equipment

Seeding equipment is used to disburse seed in a field, lawn, or other environment to gain desired seed germination and plant growth. Several types of equipment are used.

25–7. A large, multiple-row planter. (Courtesy, Case Corporation)

A ***planter*** is seeding equipment that places seed in the soil with accuracy of rate, depth, and spacing. Planters are most often used with crops grown in rows and may be known as row-crop planters. Some planters are quite complex and are computer-controlled to ensure proper seed placement. Planters are equipped to open the soil, place the seed, and cover the seed with soil. Some planting equipment also applies fertilizer and pesticides.

A ***drill*** is a piece of seeding equipment that provides solid planting—the seeds are not placed in rows. Drills open the soil, place the seeds, and cover the seeds to assure germination. Some solid planting is done with air-operated or cyclone seeders.

Application Equipment

Application equipment is used to apply fertilizer, pesticides, growth regulators, and other materials. The equipment may use dry or liquid materials. Proper calibration is essential. Without proper calibration, too little or too much may be applied. Sprayers apply liquid

25–8. A sprayer with field and roadway maneuverability. (Courtesy, Case Corporation)

materials, so they must have pumps and nozzles. Dusters apply dry powder materials and must be properly calibrated.

Harvesting Equipment

The unique features of different crops are an important consideration in harvesting. ***Harvesting*** is the reaping, picking, or otherwise gathering of a product that is ready for marketing or some other use. The differences in the varieties of harvest-ready cotton require specialized equipment as do the differences in grain crops as well as leaf, root, and nut crops.

The major categories of harvesting equipment are:

25–9. Hay bailing equipment promotes efficient forage harvesting. (Courtesy, AGCO, Duluth, Georgia)

- **Forage harvesting**—This equipment is used to harvest hay and silage. It may include cutters, dryers, balers, choppers, and other specialized equipment.
- **Grain harvesting**—Corn, wheat, barley, rice, and similar crops are typically harvested with combines. A combine cuts the stalk off and moves it through cleaners and separators. The grain is moved into a grain bin on the implement, and the stalk is discarded back on the land.
- **Root harvesting**—Sugar beets, potatoes, and other root crops must be removed from the soil without damage. This requires digging, separating, and lifting the crop in addition to discarding the stalks in the field.
- **Cotton harvesting**—Mature cotton bolls have fiber and seed, known as seed cotton. This is removed from stalks and transported on the picker to a wagon that hauls it to a gin. A gin separates the lint (fiber) and the seed.

25–10. A combine is used to harvest wheat. (Courtesy, AGCO Corporation)

- **Other harvesting**—Most crops can be harvested mechanically or with the assistance of mechanical devices. Vegetables, fruits, and nuts require specially designed harvesting equipment. In some cases, such as tomatoes, the plants have been bred to make mechanical harvesting satisfactory. Hand labor is still needed with some crops, such as lettuce, okra, and cucumbers.

25–11. Handheld computers and GPS units are increasingly used in crop production. (Courtesy, Communication Services, North Carolina State University)

Equipment With Other Uses

Many other kinds of equipment are used, including equipment in horticulture, forestry, and animal production. It also includes newer applications of variable rate technology, such as global positioning and computer applications.

MATERIALS USED IN AGRICULTURAL MECHANICS

The materials used in agricultural mechanics are primarily those in structures, tractors, and implements. Structures include sheds, barns, bins, growing houses, fences, corals, and warehouses. The kinds of materials include wood, metal, glass, and plastic.

WOOD

Several kinds of wood materials are widely used. Some of these are made directly by sawing or cutting harvested wood. Others are made by manufacturing processes that re-make the natural grain and other wood qualities.

Lumber

Lumber is pieces of wood made by sawing logs. Cuts are made to take advantage of the strength and attractiveness of wood. Logs are assessed before sawing to determine the direc-

25–12. Mixed dimension, rough sawn, green hardwood lumber.

tion of the grain and the presence of knots or other defects. Lumber may be made out of softwood or hardwood. Softwood lumber is often from pine or fir. This lumber is widely used in construction work. Hardwood lumber includes oak, gum, ash, and poplar as well as the more expensive walnut, maple, and cherry. Hardwood lumber is often used for furniture manufacture.

Rough lumber is the quality of the lumber when it is first sawed. It is not smooth, but it has straight sides and edges. After sawing, rough lumber is graded and seasoned. Grading involves sorting on the basis of knots and defects. Modern saw mills have scanners that automatically grade, sort, and mark lumber. Otherwise, it is done by an experienced lumber grader.

Rough lumber is usually sawed from freshly-cut logs that contain sap. The logs are said to be "green." Consequently, the lumber produced is **green lumber**. The high moisture content of green lumber makes it unusable for most purposes. Green lumber often twists and curls as it dries.

CAREER PROFILE

AGRICULTURAL WELDER

An agricultural welder uses metal fabrication to meet the needs of the agricultural industry for equipment or structures made out of metal materials. Agricultural welders must be able to prepare and use plans, select and use materials, and properly use a wide range of welders and other equipment. This photo shows a welder with all personal protective equipment in place who is ready to go to work. (Courtesy, American Welding Society)

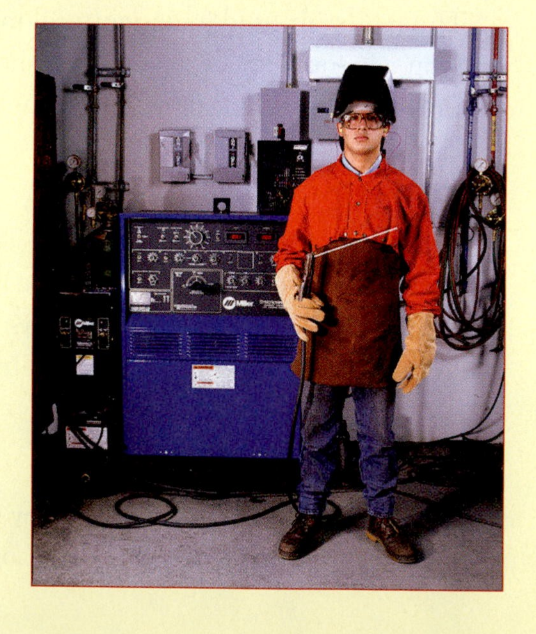

Agricultural welders usually have high school diplomas and specific training in welding and other areas of metal fabrication. On the job experience is very beneficial. They may work for other people or be entrepreneurs. Practical experience with machinery is useful.

Jobs can be found wherever agricultural equipment is manufactured or repaired.

Seasoning is the process of drying green lumber. The excess moisture is removed from it. Seasoning may be by air drying or by kiln drying. Air drying (also known as sun drying) is allowing stacks of lumber with spaces between each board to dry naturally. Kiln drying is using a heated building or chamber to speed the drying process. All lumber should be thoroughly dried before it is used. After seasoning, lumber is planed.

Planing is used to dress lumber—make it smooth and ready to use. Dressed lumber is smooth, is free of splinters, and has uniform dimensions of the pieces. It is graded and marked to assure uniform quality.

Most lumber is planed to have square edges. Some lumber is milled into special shapes during the dressing process. Planing helps ensure uniform dimensions of finished lumber materials.

Lumber dimensions in the United States are measured in inches of thickness and width and feet of length. Dimensions are always stated in the same order of thickness, width, and length. Width and thickness are nominal measurements, which means that the actual dimensions of lumber are slightly less. Rough lumber is cut in nominal or named sizes somewhat larger than the intended planed lumber. For example, a dressed board that is a 1 × 6 is only ¾" × 5½". Common lumber dimensions are 1 × 4, 1 × 6, 1 × 8, 1 × 10, 1 × 12, 2 × 4, 2 × 6, and 2 × 8.

Length is an actual measurement and, in some cases, may be slightly more than actual. This means that a 1 × 6 × 14 is a full 14 feet long. Softwood lumber is sold in lengths of 6 through 20 feet in increments of 2 feet, such as 8, 10, 12, 14, 16, and 18.

Softwood lumber of varying sizes is also known by other terms. Boards are 1-inch thick and 6 or more inches wide. Pieces less than 6 inches wide are known as strips. Planks are 1¼ or more inches thick and 8 or more inches wide. Lumber that is 2 inches or more thick and wide is known as dimensional lumber. Timbers, beams, and posts measure at least 5 inches in their smallest dimension.

25–13. Moving a sheet of plywood to a job.

Plywood

Plywood is a wood material made by gluing alternating layers of thin wood together. The layers are called plies, hence the name plywood. The glued layers are matched so the grains are in opposite directions to assure strength and quality.

Plywood may be made out of softwood or hardwood. Common softwoods used

ACADEMIC CONNECTION

MATHEMATICS

Apply skills in preparing a bill of material in mathematics classes. Relate this to calculating materials costs, taxes, and total sum due.

in plywood are southern pine and Douglas fir. A number of other softwoods are used in plywood manufacture. About 80 different species of hardwoods are used to make plywood, with oak, red gum, poplar, birch, and cherry being the most common.

Manufactured Wood Materials

Some wood materials are made by breaking natural wood apart and gluing it back together. These materials are often known as manufactured wood. They have varying uses and are increasing in popularity. Various species of pine are often used in making manufactured wood products.

Particle board is sheets of wood material similar in size to plywood. It is made by gluing wood chips, splinters, and sawdust together. Due to its hard and brittle nature, particle board is used underneath counter tops and in some furniture. It does not hold screws and nails well. Tungsten carbide-tipped tools should be used to cut particle board.

Wafer board (also known as flakeboard) is made by gluing wood chips into sheets similar to plywood. The natural appearance of the wood chips is evident in the surface. It is often used for sheathing on buildings and as paneling. Wafer board is not as hard as particle board and can be cut with regular tools. It has fairly good nail-holding ability.

Hardboard is made by gluing wood fibers into sheets similar in size to plywood. In some cases, it is made in long pieces similar to lumber. Hardboard is brittle and does not hold nails or screws well. It should be cut with tungsten carbide-tipped tools. Hardboard is used to make doors, walls, siding, tabletops, and other products. A common form of hardboard is pegboard, which is often used in displays.

25–14. Thickness of a layer of poplar veneer compared to a quarter. (This veneer will be used in making doors.)

Veneer

Veneers are made similar to plywood. Sometimes veneer is thought of as a kind of plywood. A ***veneer*** is made by gluing a thin layer of high-value wood, such as walnut, over other wood materials of lesser value. This allows the beauty of scarce wood materials to be more efficiently used.

METALS

A metal is a chemical element that reflects light, has a shiny appearance, and conducts electricity. Most metals are ductile, which means they can be made into wire. Metals are also malleable, which means they can be formed into thin sheets by hammering or rolling. These and other characteristics of metals make them useful. Metals used in agriculture are primarily made with iron. Copper, aluminum, and a few other metals are also used. This chapter will introduce metals made with iron.

Iron

Several kinds of iron products are useful. Pig iron is cured iron, which is a direct product from blast furnaces. ***Cast iron*** is a product of pig iron and contains considerable carbon (2.2 to about 4.3 percent) and some impurities. It is brittle and granular in structure and cannot be welded with the usual techniques. Cast iron is sometimes used in the manufacture of machinery, although there is a trend toward the use of steel and other tough materials. Hard blows from a hammer or other object will cause cast iron to crack.

25–15. A metal cutoff saw is being used to cut metal studs for a building.

Wrought iron is a product of pig iron that has had most of the carbon removed. Wrought iron is a two-component metal consisting of a high-purity iron and iron silicate, which is known as slag and is composed of several materials that decrease corrosion. The carbon content of high-quality wrought iron is usually around 0.02 or 0.03 percent, while some wrought iron may contain 0.08 to 0.10 percent carbon.

Wrought iron is used in machinery and equipment construction. It is easily bent when cold or hot, can be threaded and drilled, and is easy to weld. Wrought iron has good tensile strength, will withstand shock, and is not brittle. Wire, nails, bolts, nuts, chains, pipes, rods and bars, strap iron, and other products are made of wrought iron.

Mild steel is often called soft steel. It is a product of pig iron and contains about 0.10 to 0.50 percent carbon. This type of steel is often used in agricultural mechanics because it is strong, and it is cheaper than wrought iron. Mild steel is a good metal to use in making braces and trusses for machinery used in agriculture. Bolts are also frequently made of mild steel.

Tool steel is made of various carbon contents, with a common carbon content being 1.0 percent. It can be tempered to various degrees of hardness and is used for making chisels, punches, drills, wrenches, and other tools.

Sheet metal is made of steel and is often galvanized. The thickness of sheet steel varies from 1/64 inch to ¼ inch. Galvanized steel is coated with another metal, such as zinc, to resist rust.

Identifying Metals by Spark Testing

Spark testing is determining the kind of metal by creating tiny sparks from the metal. Careful observation of the sparks is necessary. Because sparks are important in testing, proper eye protection is essential.

High speed grinders work best with spark testing. Make spark tests by holding the specimen metal against a grinding wheel so the sparks will fly horizontally. Best results are obtained when the sparks are examined against a dark background, such as an unlighted corner of the shop.

The color, shape, average length, and activity of the sparks are characteristics to study. Spark testing can be an accurate method of identifying metals, but practice is needed to become an expert.

25–16. A portable grinder is being used to smooth welds on this project. (Note the color of the sparks, indicating the material is steel.) (Courtesy, Education Images)

Fabrication

Fabrication is the process of making an implement from metal, plastic, or another material. It also includes repairing implements when they break or modifying them so they are more efficient.

Fabrication may involve bending, cutting, making holes, joining, and otherwise creating useful characteristics in materials. Sometimes the materials can be fabricated while cold. At other times, fabrication requires heating the materials.

PLANS AND BILLS OF MATERIAL

A ***plan*** is an outline or drawing that describes the intended outcome of a project. Plans may have lines, descriptions, symbols, or other means to convey ideas and procedures. Plans range from simple sketches to detailed drawings.

25–17. Large structures such as barns, sheds, and houses require detailed plans.

GRAPHIC REPRESENTATIONS

A sketch is a simple or quickly prepared drawing that gives the major features but not the details of an intended project. Sketches may be used to gain the general idea of a project before detailed drawings are prepared. With simple projects, the detail of the drawing may be little more than a sketch.

Mechanical drawings are needed for many construction projects. A mechanical drawing is detailed and is carefully and accurately drawn to scale using various drawing instruments. Such drawings are the means of communication between those who prepare plans and the builders. Architects and engineers often refer to mechanical drawings as technical drawings, blueprints, or construction drawings. (Some differences exist in these kinds of drawings.)

If an object is drawn in proportion to its actual size, it is drawn to ***scale***. This helps in understanding how the sizes of various parts of a project relate to each other.

Scales vary depending on the measurement system used. Many scales for buildings are based on 1 inch on the drawing representing 1 foot on the building. This allows small drawings to represent larger structures.

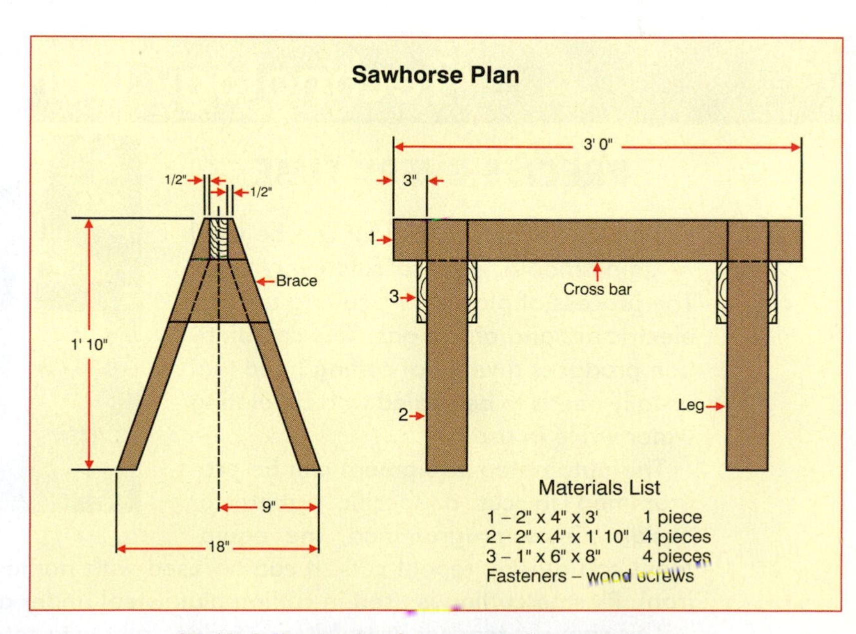

25–18. A simple plan for constructing a sawhorse.

BILLS OF MATERIAL

A ***bill of material*** is an itemized list of the materials needed to construct a project. It will include the number of pieces by dimension of lumber, plywood, and other wood materials as well as fasteners, hardware, and other materials. For example, a bill of material for a shed contains a list of all the materials necessary for building it—lumber, nails, roofing, etc.

Lumber dealers have paper pads or order books for writing orders for lumber. An order contains the name and address of the buyer, the date, the name of the project, the number of pieces of lumber, the kind of wood, the name of each piece, the dimensions of each piece, the number of board feet, and the cost. A bill of material should be prepared for each project before it is started. This indicates how much the project will cost.

To efficiently plan construction work, it is important to know the kinds of materials required, the amount of each material needed, and the total cost. If people can figure bills of material, they will not have to ask the lumber dealer for assistance. They will also be able to check the accuracy of the dealer's bill.

FASTENERS

Most types of construction work involve the use of fasteners. A ***fastener*** is a device that is used to hold two or more parts together. The primary fasteners are nails, screws, and bolts. Other methods include gluing (cementing) and welding. In addition, some projects require

Technology Connection

PRECISE EVERY TIME

Plasma cutting equipment can be used to gain smooth, precise cuts every time. The process of plasma arc cutting uses an electric arc and argon gas. This combination produces a very hot cutting head that usually needs to be cooled with circulating water while in use.

The automated equipment can be programmed to cut a specific pattern or design. Once programmed, the equipment can quickly repeat cuts. It can be used with nonferrous metals (those not containing iron). Plasma cutting is used in cutting plate steel under automated conditions.

This shows a teacher instructing a student in how to set up PlasmaCam equipment for cutting designs in plate steel. (Courtesy, Education Images)

the use of hardware such as hinges, hasps, and locks. This chapter deals with nails, screws, and bolts.

NAILS

Many kinds of nails are available, but only a few are widely used. Some nails are quite specialized, such as those used to install roofing or to attach electric fencing. Power nailers are changing how materials are fastened together, especially in wood construction.

Common nails and brads are sized by inches of length. In the past, nail size was designated by the penny system. Some people continue to use the penny system. The symbol for "penny" is the letter "d." A six-penny nail is designated as 6d. The length in inches of common nails through the ten-penny may be determined by dividing the "penny" by 4 and adding ½. The length of a 6d nail may be found as follows: 6 divided by 4 equals 1½ plus ½ equals 2 inches. The use of the penny sizing system is disappearing in favor of sizing nails by length, such as a nail is 2 inches long.

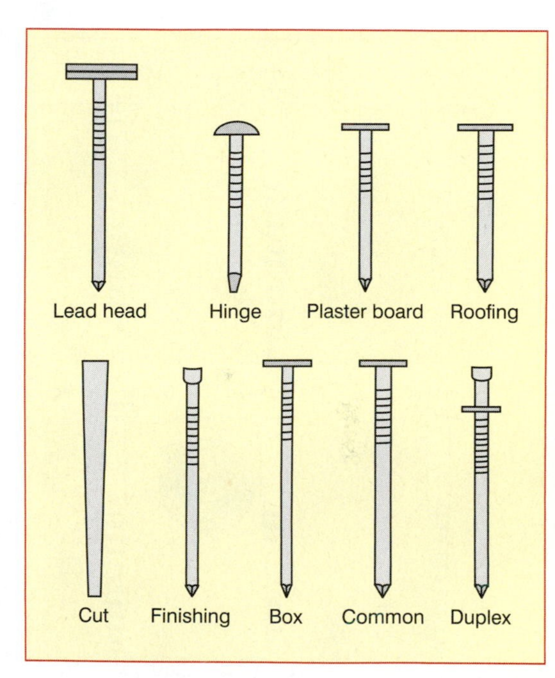

25–19. Examples of kinds of nails.

Many kinds of nails are available. Most of these are designed for specific uses. In recent years, nails with threaded shanks have increased in use. The grooves or notches on the shanks increase the holding power of nails, particularly shorter nails in softer wood.

- **Common nails**—Common nails are used more than any other type of nail. They are made of wire and range from two-penny (2d) to sixty-penny (60d). The eight-penny nail is the size most frequently used for light jobs with 1-inch lumber. Common nails have flat heads. They are generally used for nailing such materials as framing, sheathing, shiplap, fencing, and barn boards. Wood framing is often nailed with 16d common nails.
- **Box nails**—Box nails are slightly smaller in diameter than common nails and are better suited for use in lumber that splits easily. The size ranges from two-penny to forty-penny. The six- and the eight-penny are the sizes most frequently used. They are often used for nailing siding. Years ago, box nails were used for making wooden boxes for products, such as apples and ammunition.
- **Finishing nails**—Finishing nails have small heads and are used when the heads of the nails should not show. The heads are usually sunk or "set" in the wood and the holes

are filled with wood putty. They are used extensively for interior finishing, for ceilings, and for cabinet work. Sizes range from two-penny to twenty-penny.

- **Others**—Other kinds of nails that may be used in agriculture include brads, casing nails, shingle nails, roofing nails, plasterboard nails, hinge nails, and concrete nails.

SCREWS AND BOLTS

Screws and bolts are small metal shanks with threads. The shanks of screws are made with a point at one end and a flattened or rounded head with a slot on the other end. Bolts have a flat head with square or hexagonal shoulders at one end and are blunt at the other end of the shank. A nut is placed on the threaded shank and is turned to gain tightness and a secure fasten.

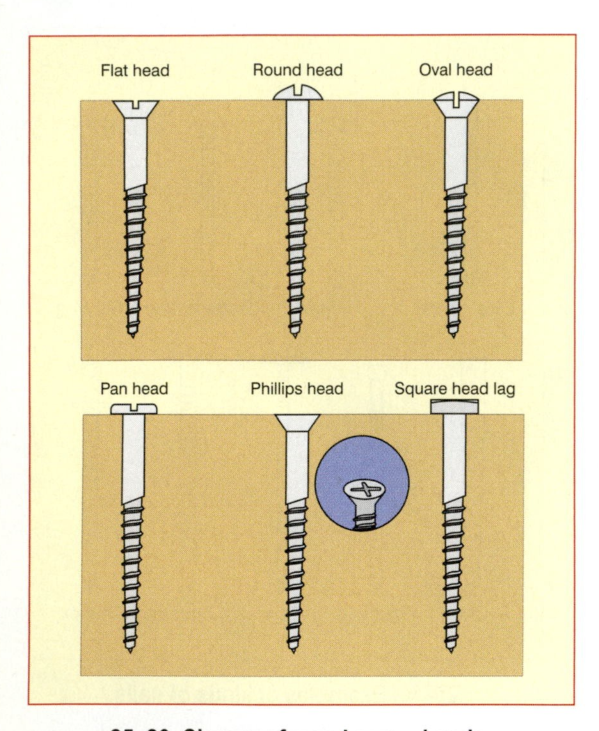

25–20. Shapes of wood screw heads.

Screws

Screws are easily classified by the shape of head, the kind of material they are designed to hold in place, and the size (length and gauge).

- **Head shape**—This includes the kind of slot in the head and the outer shape. Two kinds of slots are most common: slotted and Phillips. Other types are used, such as square, clutch, and tamper-proof. Some screws have combination heads, so they are both slotted and Phillips. The common head shapes are flat, round, pan, and oval. Some have square or hexagon heads and no slots. These screws are turned with a wrench or nutdriver.
- **Kind of material**—Screws are made for use in certain kinds of material. Typically, wood screws have fine threads that "bite" into the wood. Metal screws have wider threads that allow the thin metal to fit between the threads. Lag screws, sometimes known as lag bolts, have coarse threads that make them particularly useful in holding large timber, such as in wooden fence construction.
- **Size**—The size of a screw is based on two things: length and diameter. In the United States, length is measured in inches. Diameter is measured by gauge numbers ranging

from 2 to 24. Common screw diameters are numbers 6, 8, and 10. Usually, longer screws have a greater diameter or a higher number. In ordering screws, specify length, gauge number, type of head, and material used in making the screw. For example, a "1 in., no. 6, wood, steel screw" is a screw that is a gauge no. 6 screw, 1" long, made of steel, and designed for use in wood.

Bolts

Bolts may have fine or coarse threads. The latter are the more frequently used type in agriculture. The fine thread is usually used in motors. Never try to put a nut onto a bolt that has different threads. For example, a nut with coarse threads should not be forced onto a bolt with fine threads. The threads will be stripped out, and the fasten will not be secure.

Washers are frequently used on bolts. A washer may be at the nut or at the nut and head ends of the bolt. Some washers lock the nut to prevent loosening. Other washers are flat and provide additional surface for contact of the bolt nut and head with the material being fastened.

Many types of bolts are used, but the most common are machine, carriage, and stove bolts.

- **Machine bolts**—Machine bolts have hexagonal heads. Bolts of this type are ordinarily used in assembling machinery. Sizes range from ¼ inch by 11½ inches to 11¼ inches by 30 inches.
- **Carriage bolts**—Carriage bolts have rounded heads and square shanks to fit square-slotted holes in machinery. They are used frequently in heavy wood construction, such as livestock feeders, wagon boxes, and other equipment where nails would not hold.
- **Stove bolts**—Stove bolts have round or flat heads. They are available in a number of small sizes. They have slotted or Phillips heads, which permit the bolts to be tightened or loosened with a screwdriver. They are used for lightweight structures of metal or wood.
- **Special bolts**—Two special kinds of bolts included here are expansion shields and toggle (lag) bolts. Expansion shields and lag bolts are used in fastening materials to concrete, brick, or stone walls. A star drill is used to drill the holes. When a lag bolt is tightened, the

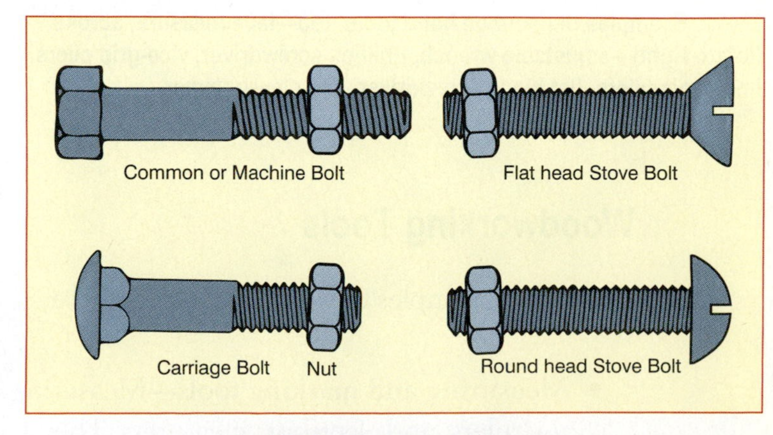

25–21. Examples of bolt types.

collar around it expands, making it secure. An expansion shield consists of a metal tube with inside threads and a lead collar around the tube. A small bolt screws into the shield. Toggle bolts are used to fasten objects to walls and ceilings. A hole is drilled in the wall or ceiling, and the toggle bolt is inserted with the head closed. When the head reaches the opposite side of the wall or ceiling, it opens. The bolt may then be tightened with a screwdriver.

HAND AND POWER TOOLS

A ***tool*** is an implement used to perform a mechanical job. A hand tool is a small, powerless type of tool used to do a task that would be impossible with bare hands, such as turning the nut on a bolt. A power tool is a somewhat larger tool operated by an electric motor or other power source. This chapter will have introductory content on hand and power tools.

HAND TOOLS

Hand tools are of two major types: woodworking and metal working. The differences are primarily the kind of material with which the tool was manufactured to be used. For example, woodworking tools are for use with wood, and metal working tools are for use with metal. Some tools may be used with wood or metal.

Always know how to use a tool safely before attempting to do a job. Be sure to use appropriate personal protective equipment. Never use a tool to do a job it was not intended to do.

25–22. Examples of common hand tools: top—tape measure; across (left to right)—adjustable wrench, Phillips screwdriver, vice-grip pliers, lineman's pliers, flat blade screwdriver, and claw hammer.

Woodworking Tools

Several examples of woodworking tools are:

- **Measuring and marking tools**—Measuring devices include rulers, tapes, zigzag folding rulers, and electronic measurers. They are used to determine the linear distance of

25–23. Using a tape measure to mark a board for cutting. (Courtesy, Education Images)

materials and spaces. Once a measurement has been made, a mark is often needed to indicate where a cut should be made or where an object should be placed. Pencils and other devices are used to make lines or other marks.

- **Squares**—Squares are devices used to get angles and to draw lines. The lines help to make proper cuts or to place materials in position. Examples are framing, try, bevel, combination, and speed squares.
- **Saws**—Saws are tools for cutting materials. Most saws have metal blades with teeth on one edge. Several kinds of handsaws are crosscut, ripsaw, compass,

25–25. Using a handsaw to cut a 2 × 4. (Courtesy, Education Images)

25–24. Using a try square to mark a board. (Courtesy, Education Images)

25–26. When driving a nail, always position the hammer to strike the nail head squarely. (Courtesy, Education Images)

and coping. The crosscut and ripsaws are used to make straight cuts. In contrast, the compass and coping saws are used to make cuts that involve curves.

- **Hammers**—Hammers are used to drive nails or other work in construction. A claw hammer has a face made of hardened steel for driving nails and a curved double claw for pulling nails or objects. Proper use is needed to prevent damage to the handle, face, or claws.
- **Screwdrivers**—Screwdrivers are tools for turning screws so the threads engage and tighten into wood or metal. Several kinds are available. Some kinds are based on the shape of the slot in the head of a screw: flat blade or Phillips head (cross). Special kinds of screwdrivers are sometimes used, such as the offset screwdriver.

AgriScience Connection

REGULAR CHECKS

Most power equipment has fluids that should be checked regularly. These include coolant, lubricant, and braking and hydraulic fluids. If the fluid levels are low, more of the proper kinds should be added to bring the levels up to the proper amount. The photo shows a dip stick being checked for the oil level in an engine.

If fluids are frequently low, further assessment is needed. There is a reason why the fluids are low. Is the system leaking? Is the system wearing and using up fluids in operation? For example, an engine uses oil if it is worn and the oil enters the combustion chamber of the engine.

Some fluids should be regularly changed. An example is oil lubricant. The oil should be drained out, a new filter installed, and the proper can of oil used to fill the engine to the desired level. Oil in engines operated in dusty conditions, such as fields, should be changed more frequently. (Be sure to properly dispose of used engine oil.) (Courtesy, Education Images)

- **Others**—Other kinds of hand woodworking tools include planes, wood chisels, wood rasps, drawing knives, bits, clamps, and levels. The tool used depends on the nature of the work to be done.

25–27. A handy, portable tool set with a variety of tools.

Metalworking Tools

A metalworking tool is one designed for the unique needs of working with metal. Properly selecting, using, and caring for tools is essential if tools are to be maintained.

- **Wrenches**—Wrenches are tools for gripping and turning bolts, nuts, and other fasteners. Some have fixed jaws; others are adjustable. The most common types of fixed-jaw wrenches are socket, straight, and box-end. Allen wrenches have special shapes to fit inside bolt heads. In addition, adjustable wrenches have jaws that can be moved so they will fit varying sizes of bolts or nuts.

25–28. A combination box-end and open-end wrench.

- **Pliers**—Designed with two levers that oppose each other, pliers can grip and turn nuts, bolts, and other devices. Some can be used to cut wire, sheets of metal, or other materials. In using pliers, never allow them to slip on what is being gripped. If slippage occurs, the bolt or nut will be damaged by teeth in the jaws of the pliers.
- **Drilling**—Hand drills are sometimes used to make holes in metal. Most drills are powered with electricity; therefore, drills are not hand tools.
- **Holding devices**—Several kinds of holding devices are used, but vises and clamps are most common.
- **Files**—Files are flat or round metal devices with teeth along the shank. The teeth are used to cut away other metals or materials.
- **Hacksaws**—Hacksaws are hand devices for cutting metals or other materials.

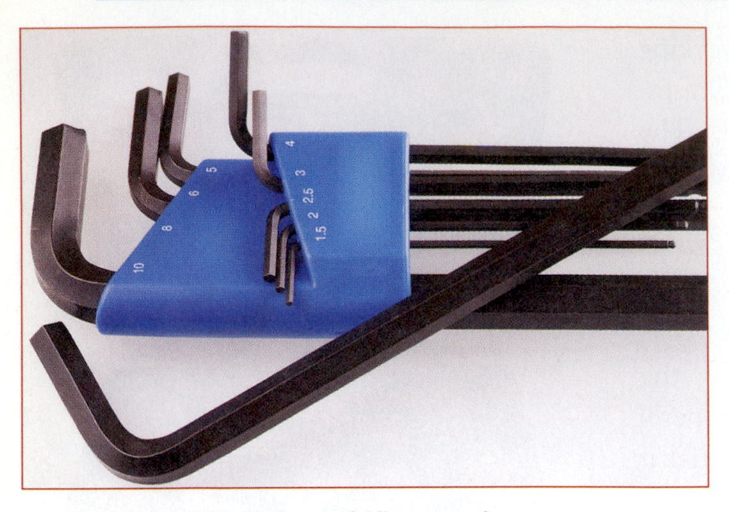

25–29. A set of Allen wrenches.

- **Punches and chisels**—Punches are short bars of steel shaped for a particular use in metal working, such as making a slight indention to guide a drill. Metal chisels are short bars of steel with a cutting edge on one end.
- **Others**—Several other hand metal working tools may be used, such as tape measures, scribes, dividers, and calipers.

POWER TOOLS

Power tools typically have electric motors. Other power sources can be used, such as pneumatics, hydraulics, and engines. Some power tools are portable; others are fixed. The portable power tools are small, so they can be moved easily. Of the small power tools, some are cordless, so a battery provides the power.

Safety with power tools is essential. Many of them move rapidly and are capable of inflicting wounds on users or causing property damage. Therefore, users should always know how to safely use a power tool before attempting to do a job. People should also wear personal protective equipment when using tools or when near others using tools.

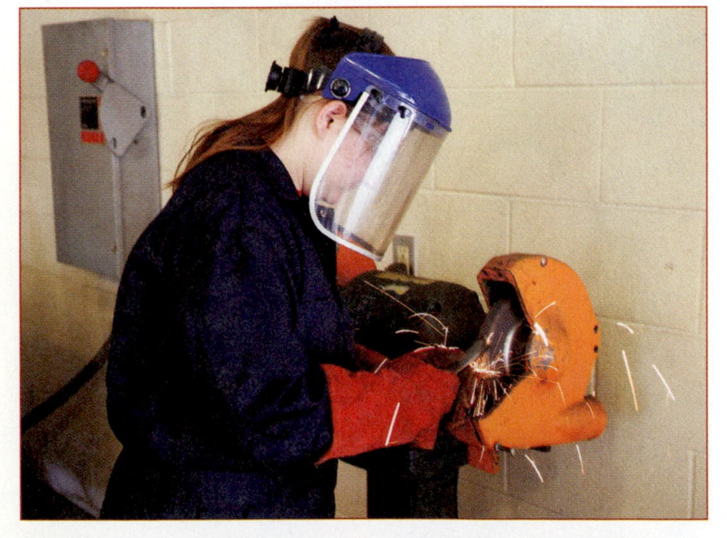

25–30. Using a grinder to sharpen a chisel. (Courtesy, Education Images)

- **Grinders, sanders, and buffers**—These power tools may be used for woodwork or metal work. The wheels or disks on such tools vary with wood and metal. Grinders are used to shape metal devices, such as the cutting blade of a lawn mower. Sanders are used to smooth wood or metal materials. Buffers are used for light sanding or for polishing materials.
- **Boring tools**—Power drills are used for boring holes in materials, such as wood and metal. Some power drills

25–31. Using a drill press to bore a hole in steel. (Courtesy, Education Images)

25–32. Using a lightweight portable drill to bore a hole in wood. (Courtesy, Education Images)

are portable; others are stationary. The drill press is a stationary drill designed for precise and heavy duty drilling jobs.

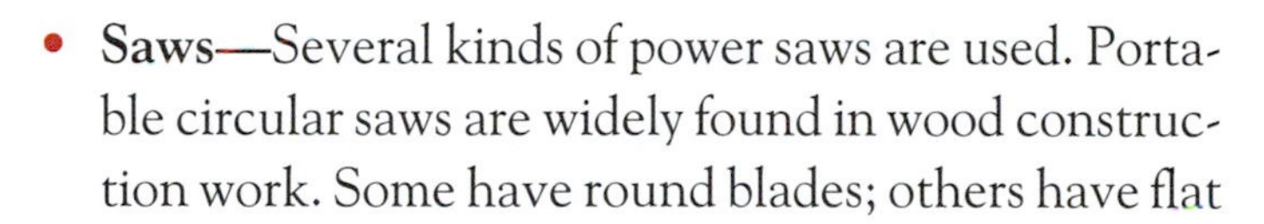

- **Saws**—Several kinds of power saws are used. Portable circular saws are widely found in wood construction work. Some have round blades; others have flat blades that move back and forth, such as the sabre saw and reciprocating saw. Chain saws are used for less-precise cutting of logs or poles and are powered by gasoline engines or by electric motors. Stationary saws include the radial arm saw, table saw, band saw, and cutoff saw. Saws may be designed for cutting wood, metal, plastics, or other materials. A saw (and its blade) should be used only for the material it was intended to cut.

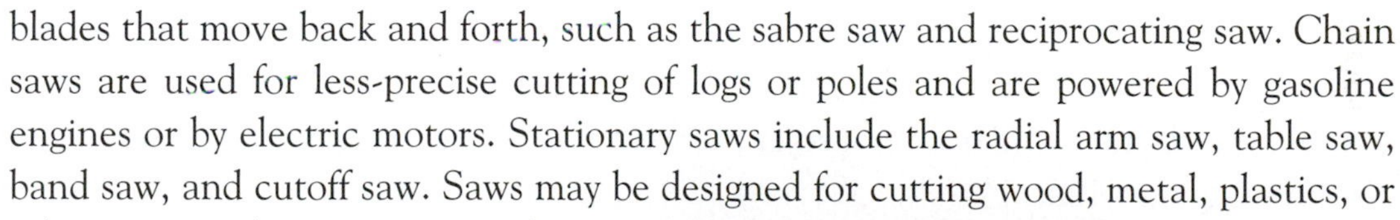

- **Planers and joiners**—These power tools are used for smoothing the surfaces and edges of wood. They are fairly large and have sharp blades that rotate at high speeds.
- **Others**—Lathes, milling machines, and other power tools are sometimes used. These items may be for woodworking or metal working.

25–33. Using a table saw to cut a piece of wood.

WOODWORKING

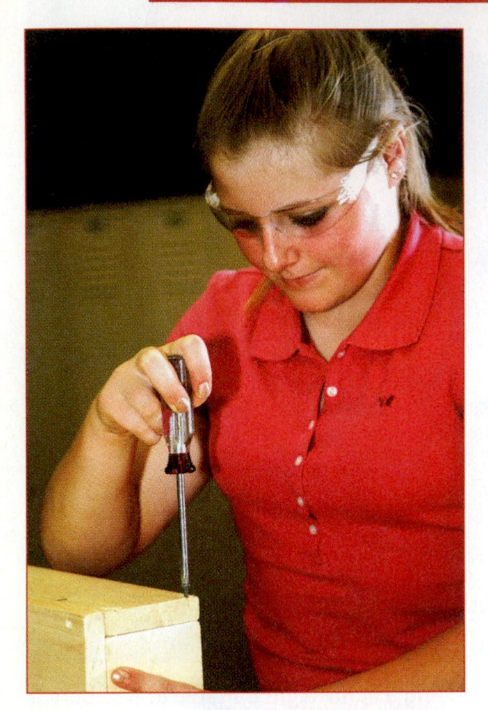

25–34. Using a woodscrew in fastening a project. (Courtesy, Education Images)

Wood is a popular material for constructing a wide range of devices and structures. In agriculture, wood is used in constructing barns, feed troughs, livestock handling facilities, and more. It may be used to frame a greenhouse or to construct a work bench.

Some general procedures to follow in woodworking are:

- **Begin with a plan**—It is good to have as many details as possible.
- **Prepare a bill of material**—Carefully assess the plan and determine the materials needed, including lumber, plywood, fasteners, paint, and other materials.
- **Carefully translate the plan to your materials**—Mark the wood.
- **Make cuts on the marked lines**—Be sure dimensions are measured and sawed accurately.
- **Assemble the project with care**—Be sure all pieces fit properly.
- **Appropriately use fasteners**—Use the right kind and size of fastener for your work.
- **Appropriately finish your project**—Sand the surfaces. Apply paint or some other coating material.

Before you begin a project, be sure you know how to properly use the tools involved with it. Follow all safety procedures.

WELDING

Heat is often used to manufacture and repair equipment. A wide range of metal and plastic materials are included in this equipment. Nearly all agricultural, horticultural, and forestry equipment, in some way, involves welding, casting, and bending in its manufacture. Breakdowns often occur because of failures of metals. These breakdowns may be repaired with welding.

Heat processes may be used to separate or cut a large piece of material into two or more smaller pieces and to fuse or bond the pieces together. The most popular heat process is welding. Other heat processes include soldering and brazing.

Welding is fusing two pieces of metal or plastic together. Because a heat process is used, extreme caution is needed to ensure safety when welding or when in the area where welding is underway. Safety goggles and helmets, leather gloves, leather aprons, and leather shoes must be worn. As mentioned previously, heat processes can also be used to cut metal.

Heat for welding and cutting metal is typically from two sources: electricity and gas.

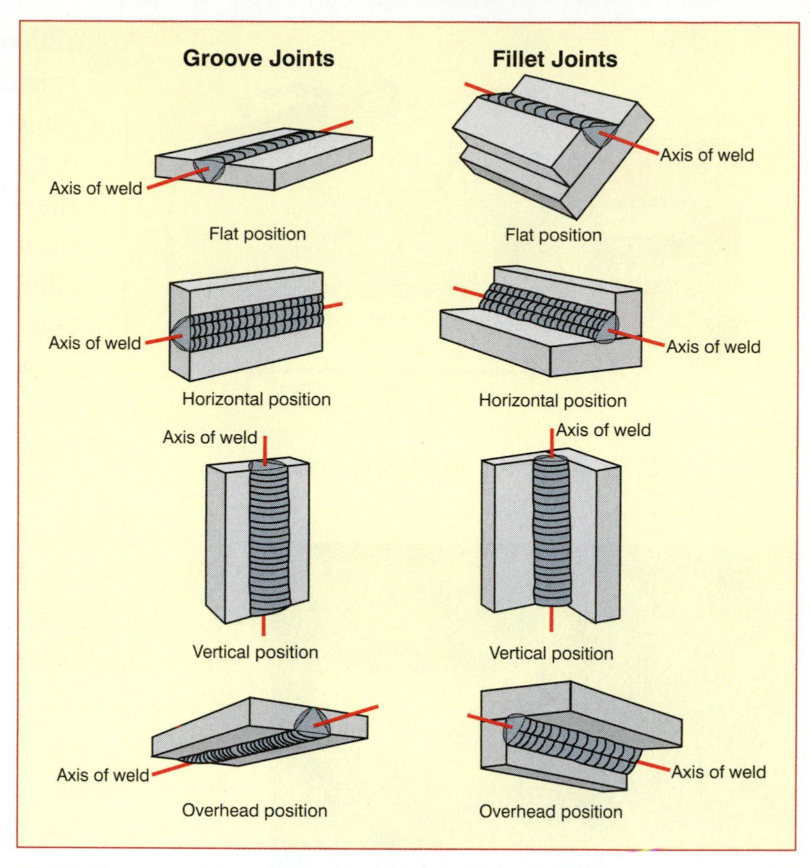

25–35. Shapes and locations of welds vary. This shows the common groove and fillet welding joints and positions.

ELECTRIC WELDING

Electric welding may be carried out using several processes. The most common process is shielded arc welding. Other processes include metal inert gas (MIG) welding, tungsten inert gas (TIG) welding, and submerged arc welding (SAW). Each of the processes involves generating an electric arc between a base metal and an electrode of some type to create high heat. The heat is so great that the two pieces being welded become molten (liquid-like) and fuse together. The focus in this section of the chapter is on shielded arc welding.

Shielded arc welding is the welding process in which, to create heat, an electric arc is used between a coated electrode and the metal being welded. A welding machine is used to manage the process. Such a machine is usually hooked to a

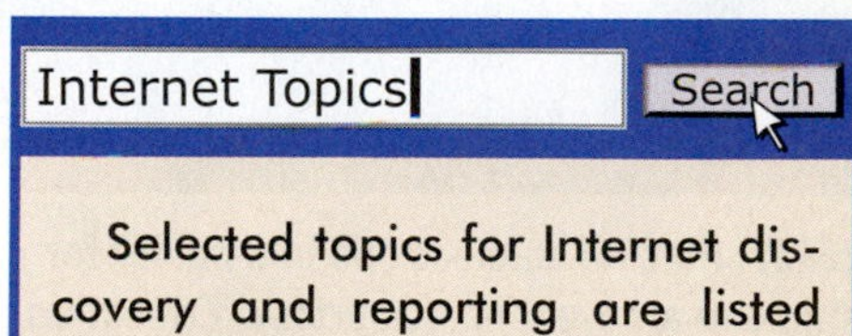

Selected topics for Internet discovery and reporting are listed here. Begin by searching for each topic. Next, read and learn about it. Conclude by preparing a brief report on your findings.

1. Board feet
2. Electrode coating
3. Shielded arc welding

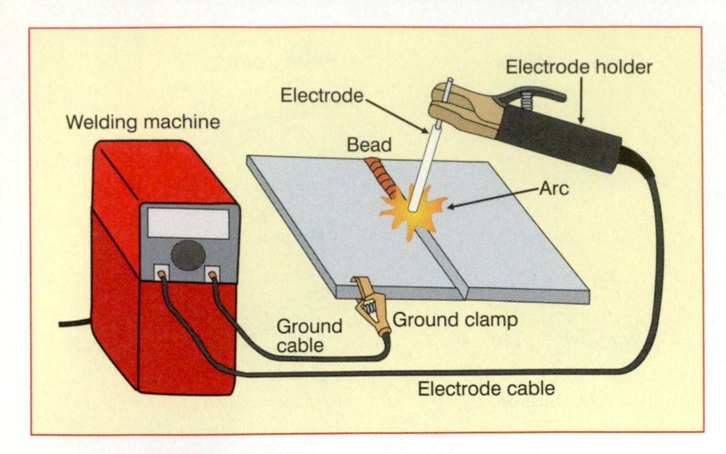

25–36. A shielded arc welding system.

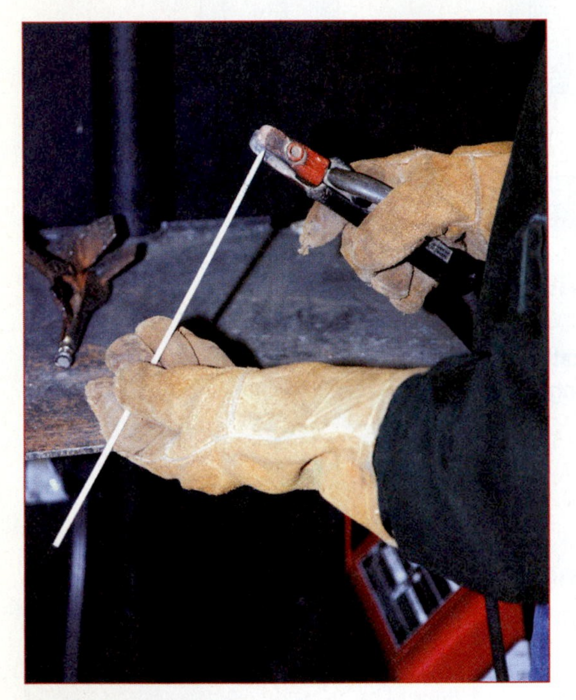

25–37. Placing an electrode in a holder for shielded arc welding. (Courtesy, Education Images)

220-volt electric power source. Most machines use alternating current (AC); some use direct current (DC); and others use a combination (AC/DC). Cables from the welding machine connect to an electrode holder and a ground clamp. These allow the electric circuit to be completed and the arcing to occur.

An electrode is placed in the electrode holder. Several kinds and diameter sizes of electrodes are available. The kinds are based on the coatings as well as the metals in the electrodes. Kinds of electrodes are designated by color codes and numbers. Common kinds are E-6011, E-6013, and E-7018. The E-6011 is the most widely used for general purpose steel welding. Common sizes of electrodes are 3/32 inch and 3/16 inch in diameter. Most are 14 inches long. A short section of an electrode has no coating because that is the part to be attached inside the electrode holder.

The ground clamp is attached to the metal being welded. To create an arc, a good connection must be made. Creating an arc is called striking. It involves lightly touching or tapping the metal with the electrode and then slightly raising the tip from the metal being welded. This will produce an intensely hot electric arc, causing the metal to melt and then fuse together as it cools. The metal “laid down” (deposited by the electrode) is known as a bead. Some of the base metal is included in a bead. Slag covers a bead and should be chipped away. Shielded arc welding processes can also be used to cut metal.

Good, strong welds require several practices:

- **Preparing the base metal**—The base metal must be cleaned of rust, paint, and other dirt that would interfere with its ability to secure a good bond. The pieces must be fitted together properly to result in a product that is within the desired dimensions. Pieces often need to be clamped in place for welding.

25–38. Properly adjust the welding machine for the thickness of metal and the kind of electrode. (Courtesy, Education Images)

- **Selecting the proper electrode**—The size (diameter) and other electrode characteristics must be considered. The diameter of the electrode depends on the thickness of the metal being welded and the amp setting of the machine. Choose an electrode made for use in welding the kind of steel or other metal with which you will be working.
- **Setting the machine at the appropriate power**—Machine setting refers to the amps involved in the welding process. Thicker metals and larger diameter electrodes require higher settings.
- **Positioning and moving the arcing electrode to lay down a good bead and create strong fusion between the pieces being welded**—Place the electrode in the holder correctly (usually at a 90-degree angle). Grasp the holder properly in your hand. (Refer to Figure 25–39.)
- **Cleaning after welding**—Use a chip hammer and a wire brush to clean flux from the weld. Always wear safety goggles when chipping and brushing.

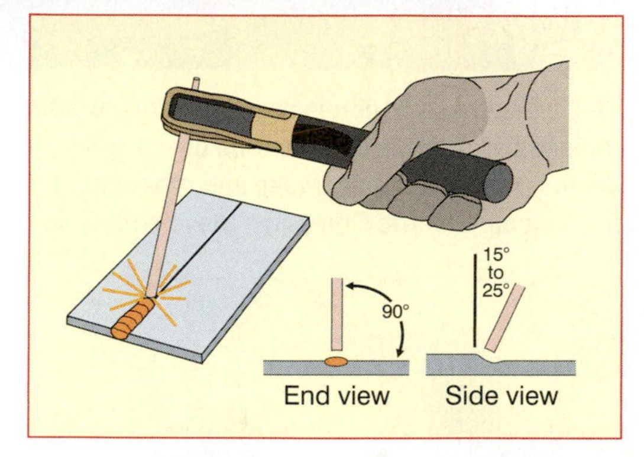

25–39. How to hold an electrode holder for the proper welding angle.

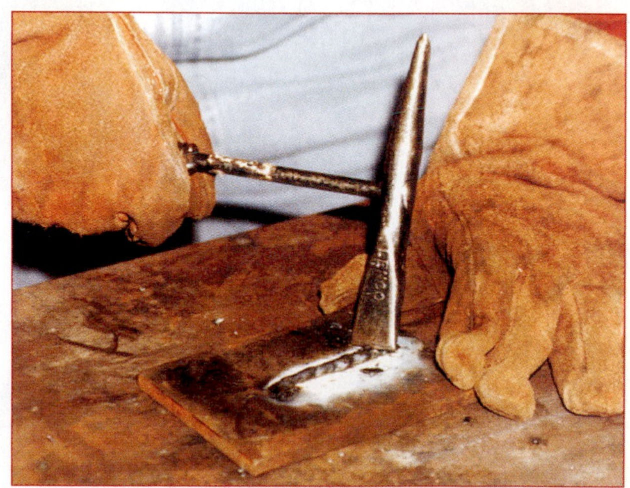

25–40. Using a chip hammer to remove slag from an arc weld. Caution: Never hold hot metal with leather gloves. Use tongs.

25–41. Good ventilation is essential when welding in an enclosed area. Note that the tube opening is very near the welding site to remove smoke and other gases. (Courtesy, The Lincoln Electric Company, Cleveland, Ohio)

- **Being safe**—Follow all safety procedures. Refer to a book with more details on welding for a complete list of safety precautions. Always use the proper helmet, gloves, and other protective devices. Never weld in damp areas. Be sure the welding machine is in good condition. Never take risks. Follow all safety practices.

GAS WELDING

Gas welding is using combustible gases that provide a hot flame to cut and fuse metals. Several different gases can be used. Oxyacetylene welding is the most popular type of gas welding. The combination of acetylene and oxygen gases produces a very hot flame that may reach 6,300°F. This is hot enough to melt most metals. **Acetylene** is a colorless gas that combusts with oxygen, producing a high temperature. Oxygen is necessary to support the combustion process.

25–42. Preparing to do oxyacetylene welding involves setting up the equipment, checking regulators, and ensuring that the blowpipe or cutting tip is clean. (Courtesy, Education Images)

25–43. Adjusting the oxygen regulator in an oxyacetylene system.

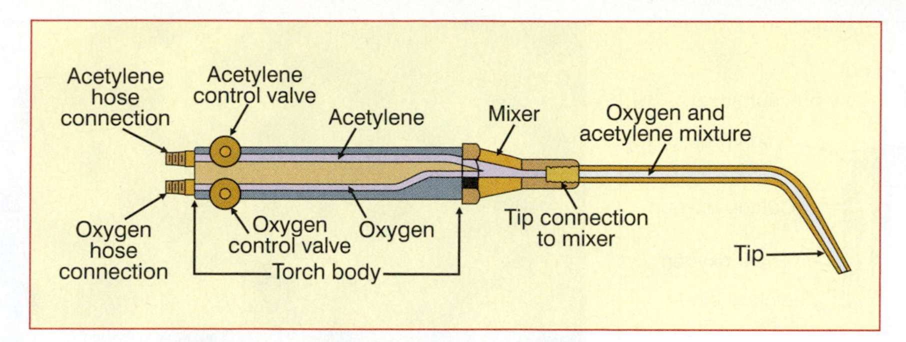

25–44. Parts of an oxyacetylene blowpipe for welding.

A proper oxygen-acetylene mixture is essential to assure that an appropriate temperature is reached.

Oxyacetylene welding may use a rod, or it may involve working to fuse two pieces of metal without a rod. Various tip sizes are used to produce the desired flame. Acetylene is also widely used to cut metal, particularly steel. Special tips are needed to cut. As with all welding, extreme caution is critical. Acetylene adds the extra dimension of a highly combustible gas.

The gases are stored in large cylinders. Regulators are used to ensure the proper flow and pressure of each in the welding blowpipe or cutting torch. When the cylinders are empty, they must be returned to a source of the gases, which is typically a welding supply business. These businesses usually have routine routes for delivering filled cylinders and other supplies and for picking up empty cylinders.

Oxyacetylene cutting requires a cutting torch that is attached to the hoses after the blowpipe has been removed. Additional pressure, particularly of oxygen, is needed in cutting. Flame adjustment is critical, and the proper positioning of the tip is essential. Backfires and flashbacks are hazards that can be prevented by properly using equipment that is in good condition.

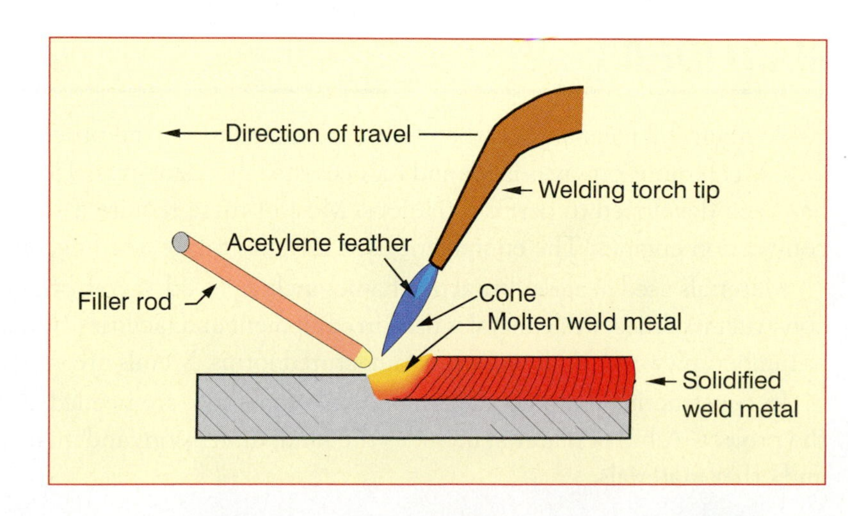

25–45. The oxyacetylene process using a filler rod.

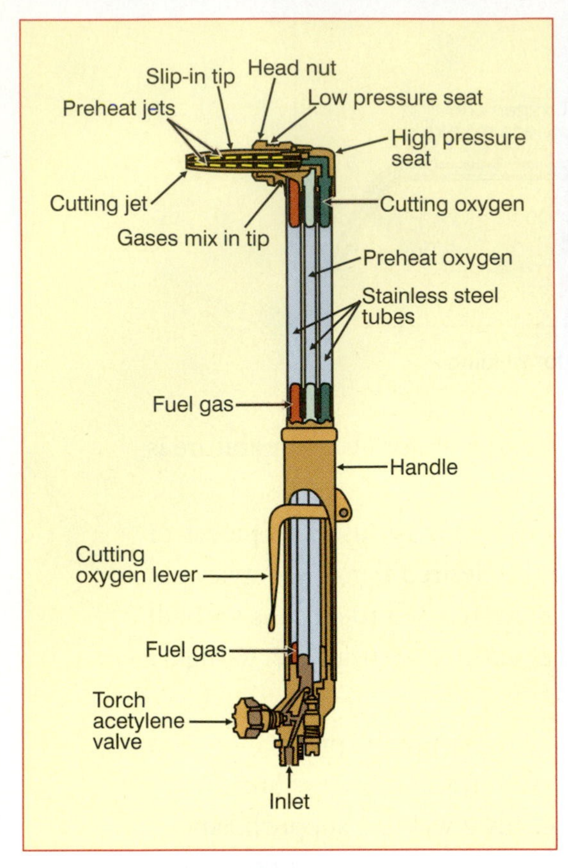

25–46. Parts of an oxyacetylene cutting torch.

25–47. Using an oxyacetylene torch to cut steel.

Note: For more details on using heat to fuse metal, refer to *Introduction to Agricultural Mechanics* and *Modern Agricultural Mechanics* (both available from Pearson Prentice Hall Interstate). Get instruction and demonstrations from your teacher. Always follow safety rules. Never abuse equipment or use it to try to perform work it was not intended to do.

REVIEWING

MAIN IDEAS

Agricultural mechanics is the use of machinery, tools, and other devices to perform agricultural jobs. Mechanization saves labor and reduces the labor demands of jobs. A wide range of implements has been developed to perform the jobs. Most of these require a source of power, such as internal combustion engines. The equipment used includes tillage, seeding, application, and harvesting.

Materials used in agricultural mechanics include wood, metal, plastics, and glass. Wood is used to construct facilities, while metal is used in equipment and facilities. Wood materials are typically used as lumber, plywood, and various manufactured forms. Metals are typically forms of steel or iron.

In constructing a project, a plan and bill of material are needed. A plan shows how to assemble the project. A bill of material includes the kind, dimension, and quantity of wood, metal, fasteners, and other materials.

Hand and power tools are used in wood and metal work. Always select and use a tool properly. Follow safe practices, including the use of personal protective equipment. After use, clean and properly store tools.

Welding is used to fuse metals and plastics. Two forms are widely used based on the source of heat: electric and gas. Shielded arc welding is the most widely used type of electric welding. Oxyacetylene is the most widely used form of gas welding. Both require careful attention to safety as well as quality of work.

QUESTIONS

Answer the following questions, using complete sentences and correct spelling.

1. What is agricultural mechanics?
2. What are the two kinds of power provided by tractors? Distinguish between the two.
3. What are the four major kinds of agricultural equipment? Distinguish between the four kinds.
4. What is lumber?
5. What is the distinction between rough and dressed lumber? Green and seasoned lumber?
6. What is plywood?
7. What is mild steel? Why is it widely used in agriculture?
8. What is fabrication?
9. What is meant when a plan is prepared to scale?
10. What kinds of fasteners are used?
11. What is a tool? Distinguish between hand and power tools. List examples of each.
12. What general procedures should be followed with woodworking projects?
13. What is welding? What are the sources of heat for welding?
14. What is shielded arc welding? How is it done?
15. What two gases are used in common gas welding and cutting? Why are these used?

EVALUATING

Match each term with its correct definition.

a. acetylene
b. drawbar power
c. mild steel
d. welding
e. veneer
f. bill of material
g. lumber
h. fastener
i. plywood
j. shielded arc welding

_____1. Pieces of wood made by sawing a log.

_____2. Wood products made by gluing thin layers of wood together.

_____3. Using an electric arc to create heat.

_____4. An itemized list of materials needed to construct a project.

_____5. The power of a tractor in pulling a load.

_____6. A colorless gas that combusts with oxygen producing a high temperature.

_____7. A wood product made by gluing a thin layer of valuable wood over less valuable wood.

_____8. The process of fusing two pieces of metal together.

_____9. A device that holds two or more parts together.

_____10. A material often used in manufacturing agricultural equipment.

EXPLORING

1. Observe the demonstration of welding equipment. Be sure to follow all rules related to eye and personal safety. Apply the physics principles of heat and fusion to the welding process. Before you try welding, refer to other sources of information, such as *Introduction to Agricultural Mechanics* and *Modern Agricultural Mechanics* (both available from Pearson Prentice Hall Interstate), and become very familiar with the safety practices that should be carried out.

2. Visit a building construction site where electrical wiring and controls are being installed. Observe the procedures used in putting the wires, switches, and other devices in place. Prepare a report on your observations.

3. Observe machinery that is used in some area of the agricultural industry. Make a list of the different simple machines that you see. Write a brief overall description of what the machinery does.

4. Construct a simple woodworking project, such as a sawhorse or birdhouse. Locate a sketch, or prepare a sketch. Make a bill of material. Carry out the construction work properly to assure a quality product. Display your product, and discuss the procedures you followed.

Part Eight

Food and Fiber Technology

CHAPTER 26

Management and Marketing in Free Enterprise

OBJECTIVES

This chapter introduces important areas of marketing technology as related to agriculture, horticulture, and forestry. It has the following objectives:

1. Discuss free enterprise and ways of doing business.
2. Describe the meaning and importance of agricultural marketing technology.
3. Explain the ways agricultural products are marketed.
4. List and explain the major functions in agricultural marketing.
5. Describe the role of marketing infrastructure.

TERMS

advertising
agricultural marketing technology
assembling
blemish
crating
custom harvesting
custom order marketing
distribution
e-marketing
economic system
free enterprise
grading
grain elevator
harvesting
infrastructure
inspection
labeling
management
marketing function
marketing infrastructure
niche market
packaging
packing shed
portioning
principle of supply and demand
promotion
retail market
roadside market
selling
transporting
vertical integration

26–1. The exchange process in marketing is depicted by a customer using money to buy clothing.

HAVE YOU EVER gone to a store to buy something, and the store didn't have it? What did you do? You had little choice. You might have gone to another store, bought a substitute item, or bought nothing at all.

Although the reason for a shortage may not be obvious, several things could have happened. The product could have been out of season or not available because of a crop failure. The absence of an item in a store is usually due to problems other than production. Chances are good that it was produced and could have been available. But, it wasn't!

The culprit might have been marketing. It is more common for problems to arise in transporting an item to the right place, in the right form, and at the right time. Solving marketing problems helps assure that consumers have the desired products when they want them.

FREE ENTERPRISE: THE SETTING FOR AGRICULTURAL MARKETING

Governments provide the setting for agricultural marketing. This setting is the economic foundation for all of the agricultural and non-agricultural industry. Within this setting, the exchange of goods and services occurs. (The study of this area is sometimes referred to as agricultural economics.)

ECONOMIC SYSTEM

An **economic system** is the way goods are created, owned, and exchanged. Countries vary in the economic system used, with examples being capitalism, communism, and socialism. None of these exist in pure form, as there is considerable variation. Trade is made possible by cooperation of nations. As nations trade among themselves, a single economic system tends to evolve.

A major distinction between economic systems is the ownership and control of property. The roles of governments vary. In all systems, governments establish regulations under which agricultural production and marketing occur.

The United States uses a modification of capitalism known as free enterprise. **Free enterprise** is the system that allows individuals to organize and go about business activity with a minimum of government regulations. It also allows individuals to own the property used in production, as well as what is produced. Individuals try to go about business ventures so that a profit results. In recent years, unscrupulous business activities have caused the need for greater oversight of business activities within free enterprise.

26–2. A role of government is to promote efficiency in the economic system.

WAYS OF DOING BUSINESS

Free enterprise in the United States is carried out to promote the creation and profitability of business ventures. Among the many businesses are farms and ranches, manufacturing companies, and marketing enterprises. Of course, some ventures are organized without the intent of making a profit and are known as nonprofit organizations. The nonprofit ventures are publicly owned and are often organizations formed for particular functions.

The major ways of doing business are:

- **Sole proprietorship**—The sole proprietorship is a business owned and managed by one individual. The individual who owns it is often referred to as the proprietor. All profits and losses belong to the owner. The owner is fully responsible. Examples of sole proprietorships are lawn-care businesses, cattle production operations, and greenhouse operations. The proprietor must have the money needed to establish and operate the business. Sole proprietors often borrow needed money from banks or other lending agencies.
- **Partnership**—In a partnership, two or more people join together to own a business. An agreement is established among the partners. Profit, if any, is divided among the partners. Losses are also the responsibility of all the partners. The roles of partners may vary. Some partners may provide finances and not be involved in the daily operation of the business.
- **Corporation**—A corporation is a way of doing business that involves a number of people owning shares of the business venture. A charter is obtained from a state government office (often the secretary of state's office) that creates a corporation as a legal entity. A corporation is owned by the individuals who buy stock in it. The employees,

AgriScience Connection

PRODUCT HARVESTING

A major step for the agricultural producer is to harvest the product. The nature of products varies widely. Some products are plants, and others are animals. In some cases, the harvested product is something produced by animals, such as milk or eggs.

Nearly all cow's milk today is taken from the cow with a milking machine. The advent of the milking machine resulted in saving a large amount of hand labor. Hand milking requires several minutes per cow. Good hand milkers moved rapidly, but time was still required. The grasping and squeezing action of each hand developed strong muscles in the wrist and arm.

Today's modern milkers are fast and efficient. They are also clean and sanitary. This results in a quality product for processing and sale to the consumer.

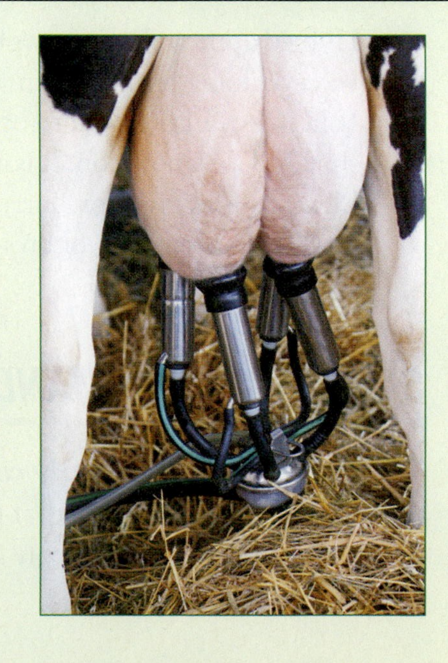

26–3. Producing a quality product that meets the needs of consumers is essential in successful marketing. (Courtesy, Education Images)

including the manager, are hired. Profit, if any, is divided among the shareholders as dividends or other means based on the number of shares owned. Individuals who own shares of stock can lose no more than the amount of money they have invested in the stock. All phases of the agricultural industry may be incorporated.

- **Cooperative**—A cooperative is an association formed by interested individuals to meet a particular need. The individuals pool their buying or selling into larger amounts and gain a price advantage. Agricultural cooperatives may focus on marketing or purchasing. A marketing cooperative helps producers gain greater returns from the sale of their products. A purchasing cooperative combines purchases from a number of smaller farms or other agricultural enterprises into larger-scale purchases at lower prices.
- **Hybrid approaches**—Laws have been passed that allow various modifications to these ways of doing business. An approach now widely used is the limited liability company (LLC). An LLC is said to be a hybrid because it has characteristics of both a partnership and a corporation. It is suited for one-owner businesses. As with a corporation, an LLC has articles of organization on file with the appropriate government office. The professional services of an accountant may be needed in meeting financial requirements.

SUPPLY AND DEMAND

The ***principle of supply and demand*** influences the potential for a profit from a business enterprise. It states that prices vary with the interaction of the supply of a product and the demand for it. As supply goes up, demand tends to go down, and vice versa. Supply is the amount of a product that is available for sale in the market. Demand is the amount that will be bought at a given price at a particular time. Agricultural marketing specialists are well aware of supply and demand. For example, a drought can reduce crop yields, resulting in higher prices for the crops that are produced.

The relationship between supply and demand is sometimes shown as a graph or curve. A supply-and demand curve illustrates the quantity of a product that will be bought at a given price and time. Exchange occurs where the demand curve and the supply curve cross in the graph. Figure 26–4 is an example.

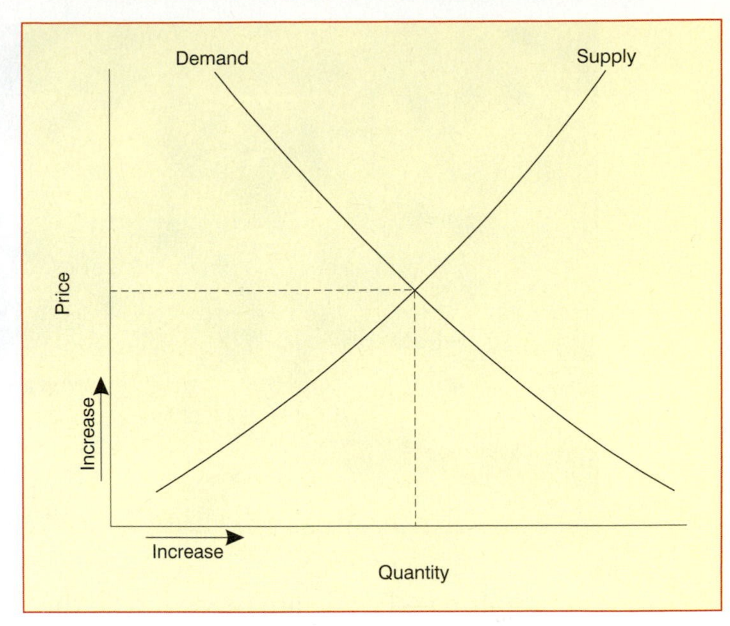

26–4. A supply-and-demand curve illustrates the quantity that will be bought at a given price.

MANAGEMENT

All business enterprises, from the smallest to the largest, require management, regardless of their nature. The success and profitability of an enterprise is closely related to the quality of the management that is provided.

Management is the use of resources to achieve the objectives of an enterprise. Levels of management vary. Sometimes, a small, one-animal enterprise is involved. Other times, it is a very large corporation. Regardless of the size of an enterprise, the goal is to achieve efficient operation or production so that a profit or other desired result is obtained.

Management is sometimes said to consist of five functions:

- **Planning**—This involves choosing a course of action. Planning is devising means for carrying out the mission and accomplishing the objectives of a business venture. It applies to both small and large enterprises.

- **Organizing**—This deals with establishing a framework to implement what is planned. It is concerned with structure, employee functions, and relationships within the business venture. Overall, approaches that yield efficient and profitable production should be used.

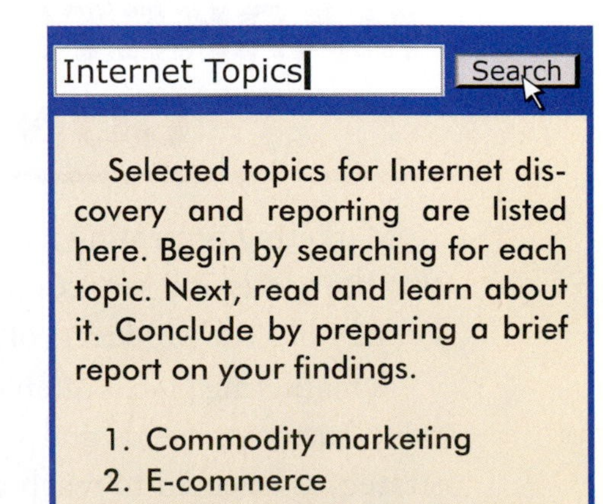

Internet Topics
Search
Selected topics for Internet discovery and reporting are listed here. Begin by searching for each topic. Next, read and learn about it. Conclude by preparing a brief report on your findings.
1. Commodity marketing
2. E-commerce
3. Niche marketing

26–5. Management involves decisions about what is produced and about how and when it is produced. (This shows a windrower harvesting almonds.) (Courtesy, Natural Resources Conservation Service, USDA)

- **Staffing**—People are needed to do the work of a business venture. One person might do all that is needed with a very small business, and that would likely be the owner. In other cases, workers will need to be hired to do the job. The manager wants to be sure that the best, most efficient workers for the particular job are hired.
- **Leading**—Leading involves using the personal qualities of the manager to inspire and motivate others to do what needs to be done. Good communication is essential. Good knowledge of human behavior and skills in relating to other people are needed.
- **Evaluating (controlling)**—This management function deals with assessing whether the desired outcomes have been realized. Was a quality product produced? Was production carried out efficiently? Was a profit realized? Based on answers to these and other questions, decisions are made that may result in redirection of the enterprise.

MARKETING TECHNOLOGY AND CONSUMER DEMAND

Marketing is creating and delivering desired goods and services. It is the process of getting the goods and services that consumers want to them in the forms that they want. Marketing links the producers of goods and services with people who use them.

A marketing plan is often prepared to guide marketing. Market research is used to determine customer needs and wants. The specific market group is pinpointed. A marketing strategy is selected to reach consumers with product information. Successful marketing is related to the 4Ps: product (having the right product), place (having the product in the right place for consumers), price (pricing the product to appeal to consumers), and promotion (providing information and motivating purchasing).

AGRICULTURAL MARKETING TECHNOLOGY

Agricultural marketing technology is all the processes in providing products of an agricultural origin to consumers in the forms they want when they want them. It is the movement of food, fiber, and shelter materials from the producer to the consumer—the person who uses them.

Agricultural marketing technology varies with the kinds and forms of products. The focus is on meeting the demands of consumers. Many things have to be done to get food ready for consumers. Few people do their own processing—not many people buy live chickens and take them home to be plucked! People want to buy their food ready to cook, their clothes ready to wear, and their home needs ready to use.

26–6. Agricultural marketing helps provide an abundance of food. (Courtesy, Aramark Services)

In addition, inputs used in agricultural production are marketed to producers. Manufacturers carefully assess what producers need. Factories produce accordingly. Inputs include feed, seed, fertilizer, tractors, implements, and other supplies and services.

When Marketing Begins

Marketing begins with a decision to produce something. The producer must study the situation. Only products that can be sold at a price high enough to make a profit should be produced.

Decision making is not easy. Producers get the best information they can from experts on agricultural marketing. They read news articles on market trends. They assess what exists in their community that is needed to make marketing possible. They use the Internet and computer models to help make assessments. They may also sign agreements with buyers to assure that their products can be sold after they are produced.

A demand for a product must exist. This means that someone must want what is produced.

26–7. Shoppers in supermarkets make decisions based on the selection and quality of products as well as their meal plans. (Courtesy, USDA)

Consumers Are Important

Millions of people consume huge amounts of food, fiber, and shelter. They have preferences about what they want. They will not buy just any product; they want specific things. Agricultural marketing requires studying consumer preferences and producing what consumers want.

Demand is the amount of a product or service that will be bought at a given price. Consumers must want it enough to spend their money to get it. People change their wants from time to time. Styles and notions about foods change. Consumers will reflect these changes in what they demand.

Producers Are Commercial

Farmers and ranchers are commercial producers. They produce plants and animals, as well as the products of plants and animals, that can be sold. If a product cannot be sold, it should not be produced!

If something that cannot be sold is produced, the producer is stuck with it. Why would a producer grow something that he or she could not sell? Sometimes producers make poor decisions about production.

26–8. Marketing information helps in making decisions.

COMMUNICATION IN AGRICULTURAL MARKETING

Communication technology is all the ways people exchange information—broadcast, print, Web-based means, and others. Information is needed and used in the marketing process. Regular reports of market information are provided on many plant and animal crops. Information about supplies, services, and other areas is also useful. Several media are used. One-to-one information is also important in relationships between sellers and buyers.

Marketing may occur great distances from producers. Some products are traded at commodity exchanges. These are located in central cities where buyers and sellers get together. The prices in these markets have big influences on decisions about production and marketing in agriculture.

Communication in agricultural education may involve:

- **Media reports**—Information on radio and television and in newspapers and specialized publications aids in the marketing process. Daily market information is reported and found useful in making decisions about the marketing of a product.
- **Computer networks and the Internet**—Many aspects of agricultural marketing are now handled with computers and the Internet. Information can be immediate and wide spread. Status information about crops, weather, and other details is readily available through links to various Web sites.
- **Brokers and consultants**—These individuals keep up with the market and help producers and others in making decisions. Such individuals need wisdom and honesty.
- **Inspectors and graders**—These individuals help assure quality, uniform products in the marketing process. They may be employed by government agencies, associations, or the enterprises themselves.

26–9. Garden supply stores maintain tools for the homeowner.

HOW AGRICULTURAL PRODUCTS ARE MARKETED

Marketing technology provides a channel or link between the producer and the consumer. Different marketing channels are used. Some producers only want to produce a product and get it to a processor. Other producers like direct marketing and dealing with customers. The quantity produced is also important in selecting a marketing channel.

ROADSIDE AND RETAIL MARKETS

In a roadside or retail market, the producer sells directly to the consumer. This adds to the tasks involved in producing plants and animals. The producer harvests, grades, packages, and weighs the product(s), and runs the market.

26–10. A roadside coffee market operated by Bay View Farm in Hawaii.

26–11. A roadside stand on a California strawberry farm.

Some kinds of products are better suited to roadside and retail markets. Examples of produce sold through these markets are fresh vegetables and fruit, eggs, melons, and nuts. Except for fish, meat products are not usually marketed to the consumer by the producer. This is because of the amount of processing required and the regulations that must be followed. Permits are needed in most locations to operate retail stores.

A ***roadside market*** is an on-the-farm retail facility. This market works best for a farm that is on a major road or state highway. The producer operates the retail market in addition to producing what is sold.

A ***retail market*** is an off-the-farm retail facility. It may be in a nearby town or city. Sometimes, retail markets are part of a planned farmers' market, where many different growers have small stalls or booths for selling their produce. Retail markets can also be separate stores operated by the producer.

CENTRAL LOCATIONS—PACKING SHEDS AND ELEVATORS

Harvested produce may be delivered to packing sheds and grain elevators, where it is bought from many growers. All the produce is assembled for shipment to processing plants. Before shipping, the produce is graded and put into uniform lots.

The packing sheds and elevators negotiate with processing plants to get the best possible prices. Sometimes, several growers may cooperatively own these facilities.

A ***packing shed*** is a market outlet that receives, sorts, and combines produce from a number of different farms. This provides an advantage to growers—combining what they have produced separately into a larger volume. Larger volumes are more appealing to processing plants and may command higher prices.

A ***grain elevator*** is a marketing facility that receives, grades, drys, stores, and sells grain to mills. Elevators have long been used in marketing corn, wheat, and other grains. They are usually located on railroads or major highways.

Grain elevators receive the grain and prepare it for further marketing. They may dry grain that has too much moisture. Elevators carefully store grain until it is sold to mills. The mills manufacture grain into the desired products for food or animal feed.

26–12. Trucks are being unloaded at a grain elevator.

PROCESSING PLANTS

Some producers market directly to processing plants. These facilities convert the raw products into forms that consumers will buy. Sometimes, a processing plant will convert a product into a form that may require additional processing at a retail store, such as fresh meat that will be cut into portions for customers.

Usually, the processing plants want large volumes of produce. They also want produce that meets their needs and standards. Some processors may contract with growers to produce certain kinds of crops. These growers know they can sell what they produce before they plant it.

CUSTOM ORDERS AND VERTICAL INTEGRATION

Custom orders and vertical integration use similar marketing technology but are different. Both are based on knowing ahead of time that an outlet for the product exists.

Custom order marketing is a process in which buyers (processing plants and others) contract with producers to grow a certain plant or animal. For example, a pickle company may sign growers to grow a certain kind of cucumber. In addition, the pickle plant may specify growing methods and the exact size the cucumbers must be when harvested.

Vertical integration is a process in which a nonfarm agribusiness owns the processing plant and all the sources of supplies for producing a particular plant or animal. The farm producer merely provides land, including facilities, and labor.

Vertical integration is the major way chickens are produced. A broiler processor will sign up growers and specify exactly how the birds are to be raised. A grower is paid a set amount for each bird. The processor provides the baby chicks and feed, sends a field representative to supervise the operation, and hauls the chickens to the plant for processing. People who grow chickens on their own, without a contract, cannot sell them to a processor.

Hog and catfish production have rapidly moved into forms of vertical integration. With hogs, the approach is sometimes referred to as "factory farming." The benefit is a quality food product produced in a carefully controlled environment—far different from a hog rooting in a muddy water hole! With catfish, processors are assured of a quality product raised with good water to drink and fed a carefully designed diet.

NICHE MARKETS

26–13. Organic vegetables are niche marketed at this street-side stand.

Some growers develop special markets to fit their situations. A ***niche market*** is one that meets a unique need for a particular product. Here are a few examples:

- **Lambs**—In some areas, lambs can be grown and marketed live for religious and holiday celebrations. This is becoming more difficult because local ordinances forbid the harvesting of animals in residential areas.
- **Goats**—A few goat growers produce for the barbecue season of the year. Young goats are sold live, and the buyer does the harvesting or uses a custom harvesting business.
- **Recreational fish**—A few aquafarmers sell fish to fee lake operators. The fish may be sold for slightly higher prices than a processing plant or local fish market would pay.

E-MARKETING

Marketing using the Internet is called ***e-marketing***. A Web page may be designed using a domain name obtained specifically for a business. Web pages may allow order taking via the Internet or provide telephone or facsimile numbers and mailing addresses for placing orders.

The use of the Internet in marketing has rapidly emerged since the early 1990s and gained market share in the early 2000s. Some producers use the Internet to market their products. Other e-marketing businesses connect buyers and sellers. This applies in agriculture, horticulture, and other areas. Often, an e-marketer never takes physical delivery of a product. A Web page provides worldwide access. This may create special shipping circumstances in some cases. Ways of receiving payment and assuring accurate shipping are essential.

26–14. The Internet is increasingly used in agricultural marketing. (Courtesy, Education Images)

Several sites are available for selling and buying farm products and input supplies. Grocery stores may offer Internet shopping for a restricted delivery area.

MARKETING FUNCTIONS

Many activities are involved in getting products to consumers in desired forms. These are sometimes known as marketing functions. A ***marketing function*** is an activity or process that converts a product or performs a role in making a product readily available for the consumer. The functions are the links between the producer and the consumer.

Marketing is more than selling what has been produced. Many people view marketing only as changing ownership of products or services. It is far more than that. The ownership of a product may change several times between the producer and the consumer.

Nine marketing functions are briefly described.

> **ACADEMIC CONNECTION**
>
> **ECONOMICS**
>
> Relate your study of marketing in agricultural areas to your studies in economics class. Explore the important role of marketing in economics, especially as related to a free marketing economy. Use your learning in agriscience class to support your mastery in economics.

HARVESTING AND ASSEMBLING

The harvesting and assembling functions begin at the point of production. They are the first steps in getting what has been produced ready for the consumer.

Different crops require different methods of harvesting. ***Harvesting*** is digging, cutting, picking, reaping, or other-

26–15. Cauliflower is being harvested in California.

wise removing products. Some products are harvested with machines; others are harvested by hand. Examples of harvesting are picking apples, digging potatoes, cutting sugarcane, seining fish, herding cattle into a catch pen, shearing sheep, and milking cows. Logging, tapping trees for sap, and cutting Christmas trees are also forms of harvesting. Digging ornamental plants and cutting long-stem roses are two examples of harvesting horticulture crops.

Care must be used to assure a good product. Foods must be kept free of impurities and harmful organisms. Fibers must be kept dry and free of trash so their color isn't affected. Fruits and vegetables must be protected from wilting and injury.

26–16. Soybeans are being assembled from many farms at this elevator. (Courtesy, American Soybean Association)

Assembling is massing large quantities of products at a central location for the next marketing function. Assembling is important because of "economy of scale." This means that marketing people want to deal with large volumes. Some farms do not have large enough volumes unless they combine with other farms. Assembly locations include vegetable packing sheds, cattle sale barns, grain elevators, and cotton gins.

GRADING

Grading is the sorting of products for uniformity. It is often done during or after assembling. Standards are used in grading and are set by government agencies or producers' associations. Standards are intended to protect consumers from unsafe products and assure uniformity. Processors, supermarkets, restaurants, and others may have their own standards.

Grading often involves sorting based on several factors: size, color, blemishes, and varieties. Size is often a factor in the price a producer receives. For example, small cucumbers sell for more money than large cucumbers, and large watermelons sell for more money than small watermelons. Every product has standards of its own.

26–17. Eggs are being graded to provide uniform quality for the consumer. (Courtesy, Mississippi State University)

Color is a factor in some products. People want lettuce leaves that are green and eggs that have yellow yolks. They want cotton that is white and not yellowed by rain. They want oranges that are orange and limes that are green.

A ***blemish*** is a defect in a product. Potatoes with rotten spots are blemished. Hogs with injuries are discounted or rejected. Eggs that have dirty shells are undesirable. And, no one wants an apple that has a worm hole!

Products must be of uniform varieties or kinds. Fish that are mixed, such as tilapia and bass, are not desired and must be sorted. Red and white potatoes cannot be mixed. Yellow and white corn must be kept separate.

Inspection may be associated with grading. ***Inspection*** is an examination of a product to assure that it is free of disease or other defects. Products may be approved for shipment on the basis of inspection. Most inspection is used with live plants and animals that may transmit disease. Other inspection is to assure that quality standards are met in processing food products.

No 2240
S.C. STATE CROP PEST COMMISSION
CLEMSON, SC
Permit–Parcel Post Shipments Only
EXPIRES SEPT. 30,2008
Issued to Woodlanders, Inc.
1128 Colleton Ave., Aiken, SC 29801
whose nursery or premises has been duly inspected by an authorized inspector of the state of origin and found apparently free from injurious insects and plant diseases. This permit covers only stock growing in the said nursery at the time of inspection. The movement of stock other than that described herein is a violation of law.
O.J. DICKERSON, State Plant Pathologist
H.B. JACKSON, State Entomologist

26–18. A label on a box indicates that the plant materials being shipped have been inspected.

TRANSPORTING

Transporting is the moving of products from one place to another. Milk must be moved in refrigerated tanks from the dairy farm to a processing plant. Freshly picked beans must be quickly moved to a cannery (or packing shed for the fresh market). Sweet corn cannot be kept long without loss of the sugar content in kernels.

26–19. Export grain is being loaded on a ship at the Port of Charleston, South Carolina. (Courtesy, USDA)

Products are transported by trucks, barges, airplanes, wagons, and boxcars. Some, such as milk, are hauled in tanks; others, such as pulpwood, are hauled on open trailers. Some, such as hogs, cattle, and fish in tanks, must be hauled to keep them alive and in good condition.

PROCESSING

Processing is the preparation of a product for consumption. Many steps may be included that alter a raw product. Products are processed in different ways. Some processing is done to keep them from spoiling; other processing makes them easier to prepare. After or during processing, products are packaged.

PACKAGING

Packaging is the marketing function in which products are placed in containers. Many kinds of containers are used. Containers retain product quality and make handling easier.

Packaging materials include plastic, wood, paper, metal, and glass. Eggs and milk are often put in plastic cartons. Fruit may be put in wooden crates or pasteboard boxes. Strawberries and tomatoes may be put in paper packages. Asparagus and snap beans may be put in metal cans. Pickles and apple sauce may be put in glass jars.

Portioning

Containers allow products to be divided into amounts or sizes that consumers want. This is known as ***portioning***. Some containers are large, such as the can of beans used in a school cafeteria. Others are small, such as the jar of pimento peppers used at home when making pimento and cheese spread.

26–20. Many fresh vegetables in supermarkets are prepackaged for convenience of both the buyer and the seller.

26–21. Meat in a supermarket is packaged in consumer-size portions. (Courtesy, Education Images)

Crating

Crating is the placing of cans, jars, or fresh produce in crates (larger containers) for shipping. Many crates are pasteboard or paper boxes, though some may be plastic or wood. Crates make it easier to haul and move the products into and out of trucks, warehouses, and stores. Crates also protect the products from damage. Filler material may be added to a crate to keep the products from bumping together when moved.

26–22. Harvested lettuce is being moved from the field in large shipping crates.

Modern facilities use wooden pallets stacked with boxes or crates. Pallet contents may be held together with plastic wrap and moved with forklifts. All this saves human labor and reduces the cost of food and other products.

Labeling

Labeling is a step in packaging used to identify a product. The name of the product as well as other information, such as weight, date and place of processing, and nutritional value, may be given on the label. A label may also describe how to prepare the product, such as a microwavable dinner.

26–23. Apples may be individually labeled.

26–24. Interior view of a storage silo for pistachio nuts. (Courtesy, California Pistachio Commission)

STORING

Most products have to be stored—kept until needed. Some can be kept for a long time with little loss of quality, if stored properly. Others will deteriorate under even the best conditions.

Most marketing involves keeping the inventory (amount of product on hand) as small as possible to meet demand. This requires businesses to order and buy carefully. It reduces the amount of money a business invests in warehouses.

Products must be stored properly. Some must be in refrigerated coolers, such as fresh vegetables, eggs, meat, and milk. Others can be stored without refrigeration, as long as the temperature is kept reasonable, such as in a well-ventilated warehouse where canned foods are kept. Frozen foods must be kept frozen. Dried foods must be protected from moisture, insects, and rodents.

AgriScience Connection

ORANGES AND MARKETING TECHNOLOGY

Oranges are common in supermarkets across North America. They begin in a grove, where trees have been selected to produce citrus to meet consumer demand. The latest cultural practices are followed to assure quality fruit.

Once grown, the oranges are picked, placed in crates, shipped to a packing shed, graded, and placed in boxes. Trucks deliver them to warehouses and supermarkets. Of course, some are processed into juice and other products. (Courtesy, Florida Department of Citrus)

ADVERTISING AND PROMOTION

Advertising and promotion are used to motivate people to purchase a product. Agricultural marketing companies want to sell as much of a product as they can. They want people to buy what they have to sell.

Advertising

Advertising is any method of communicating information to people that is intended to persuade them to buy a product. Advertising may use radio and television, newspapers and magazines, billboards, direct mail, the Internet, and other means. The purpose is to make people aware of products and encourage them to buy.

Promoting

Promotion is something that promotes or advertises a product or service. Promotion may involve providing free samples of food in a supermarket. Special prices may be used to encourage people to buy a product who might not otherwise consider it. Reduced prices attract buyers and increase the amount sold.

Special events may be used. June is known as "dairy month." A variety of approaches are used to encourage the use of dairy products. Some places have strawberry festivals, apple harvest carnivals, and other events that promote products.

CAREER PROFILE

GRAIN GRADER

A grain grader evaluates and assigns a grade to grain. Samples of a grain are compared with standards set for the grain. Cracked kernels, miscolored kernels, damage by insects, diseased kernels, kernel size, and other factors are considered.

Grain graders need considerable knowledge of grain. They need practical experience as well as complete training in grading. A college degree is needed in an area related to agronomy.

Grain graders are employed by elevators, export companies, government agencies, grain buyers, and others involved in the grading of grain. This photo shows a grain grader at an export terminal. (Courtesy, South Carolina Farm Bureau)

26–25. A trade show exhibit may be used to promote products. (Courtesy, National Cotton Council)

SELLING

Selling is a transaction in which ownership changes from a seller to a buyer. Most people use money to buy products; however, sometimes things of equal value are exchanged.

Agricultural products may be sold several times during the marketing process. The producer may sell them to a processing plant. The processing plant may sell them to a warehouse company (known as a jobber). The warehouse may sell them to a retail store, which sells them to the consumer. All the people involved try to make a profit when they sell.

26–26. Bidding on cotton is underway at the New York Cotton Exchange. (Courtesy, New York Cotton Exchange)

Individuals working in this area may need the ability to make sales calls and use procedures that sell products. The skills are sometimes referred to as personal skills related to successful sales calls. Product knowledge is essential. Success in selling requires the seller to know the features, advantages, and benefits of the product. Good communication skills are also needed, including the ability to establish effective eye contact. Being honest and being a person who follows up on commitments promotes success in selling. Grammar and speaking skills are also needed for success in personal selling.

DISTRIBUTING

Distribution is the process of getting a product to the right place at the right time. The right amounts must arrive when needed. For example, if all the things needed to make hamburgers get to the restaurant but the buns, the restaurant cannot make and sell hamburgers.

Distribution requires careful planning. Once a product has been prepared, it must be moved to a place where the consumer can get it. This is much more than transporting, but transporting is an important activity. All activities must be carried out as planned. Orders must be placed and pro-

26–27. Container shipping is often used for import/export marketing. The containers readily stack on ships and semitrailers. (Courtesy, National Cotton Council)

Technology Connection

WHAT FAT CALIPERS SAY

Calling the device "fat calipers" is incorrect. The correct name is "body-fat skinfold calipers." The calipers are used to measure the thickness of a fold of skin along with its underlying layer of fat. With the measurement, an estimate of the total percentage of a person's body fat can be made. Since many people are conscious of their weight, the measurement can be used as a guide in choosing the foods consumed.

Each person is different. Percentages of bone, muscle, and organs vary. Physically active people have more muscle. People who exercise very little and spend a lot of time inactive tend to have a higher percentage of body fat. Food intake is also associated with body fat. Caliper measurements can be used as a guide in making choices about level of activity and about food intake.

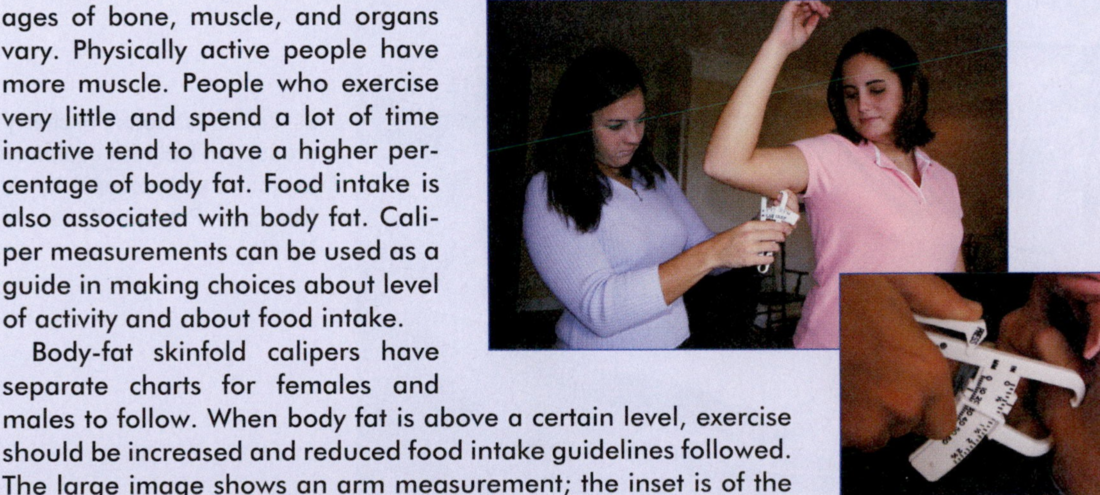

Body-fat skinfold calipers have separate charts for females and males to follow. When body fat is above a certain level, exercise should be increased and reduced food intake guidelines followed. The large image shows an arm measurement; the inset is of the waist area. (Courtesy, Education Images)

cessed. Trucks must make their runs and deliver what was ordered. All along the way, the product must be kept in a good, wholesome condition.

Getting products to the right places on time is a key to success. Restaurants and supermarkets without food have nothing to sell!

MARKETING INFRASTRUCTURE

Infrastructure is a big and important word in agricultural marketing. **Infrastructure** is all the facilities, services, and things that must be in place for marketing to occur. It includes production agriculture as well as the supplies and services producers need. Infrastructure definitely includes the marketing facilities for products.

Marketing infrastructure is those things that support marketing and make it possible. For example, a road is needed to haul a product to a processing plant. A truck with special equipment may be needed to do the hauling. The processing plant is infrastructure, because the product couldn't be processed without it.

HARVESTING INFRASTRUCTURE

Once plant or animal products have been produced, they must be harvested. A way to harvest them must be available.

26–28. Infrastructure in cherry production includes equipment to receive and process cherries. (Courtesy, Peterson Farms, Michigan)

Some products require expensive harvesting machinery. The machines may cost more than one producer can afford. An individual in a particular area may buy one machine and use it to harvest crops for several farmers. **Custom harvesting** is a harvesting service offered by one person who owns and operates equipment to harvest the crops of other producers. The owner charges a fee for the service. Custom harvesting is used with wheat, hay, cotton, and other crops.

Products should be produced only if there is a way to harvest them. For example, many farmers have built fish ponds, stocked them with fish, and fed out the fish without arranging for harvesting equipment. The equipment a farmer needs includes a seine, a seine hauler, a small boat to help in setting the seine, and haul trucks to take the fish to a processing plant. Never start a crop until harvesting equipment is available!

TRANSPORTATION INFRASTRUCTURE

Transportation is needed to move products. Roads, railroads, water ports, and airports are all used in marketing. Bridges and road surfaces must be strong enough to support the heavy weights of trucks.

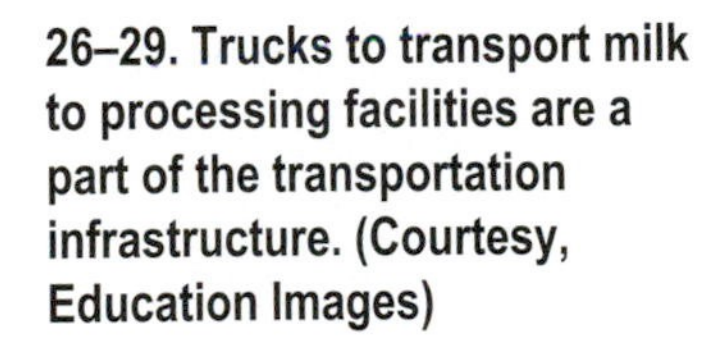

26–29. Trucks to transport milk to processing facilities are a part of the transportation infrastructure. (Courtesy, Education Images)

The right kinds of vehicles must be available. Trucks should be properly equipped to haul produce. If timber is being harvested, pulpwood, pole, or log trucks are needed. If the wood is chipped, trucks with chip trailers are needed. Milk must be hauled in refrigerated tank trucks, cattle hauled in cattle trucks, and cotton hauled in cotton wagons. Almost every plant or animal product has unique transportation requirements.

If the needed transportation is not available, marketing will be more costly and could be impossible.

ASSEMBLY AND PROCESSING FACILITIES

Packing sheds and processing plants need to be available. Hauling products long distances adds to the cost of production. Further, harvested products can deteriorate in quality while being hauled.

26–30. Onions are being unloaded at a processing plant.

Grain is a good example. Grain elevators or other outlets need to be convenient. Hauling grain long distances costs money. As the costs of hauling go up, profit goes down.

For many animals and plants, assembly and processing facilities are essential. They must be present before production begins. If not, careful thought should be given to producing a different product.

26–31. A milk sampler collects samples of milk for laboratory analysis from trucks as they arrive at the dairy processing facility. This individual is on top of a tank truck opening the hatch to take a sample. An employee such as this often also has other responsibilities related to receiving raw milk from dairy farms. (Courtesy, Education Images)

FINANCES

Producers must have money to pay the costs of producing crops and animals. Banks and other lending agencies are needed. If financial institutions are unavailable or none is willing to make a loan for a particular crop, the infrastructure of finance does not support marketing.

WORKERS WITH SKILLS

People who know how to perform certain work must be available. If the labor supply is short, the infrastructure will not support the production of certain crops. This is especially true when specific skills are needed.

Labor applies not only to production agriculture but also to all the marketing functions. People must be available to work in processing plants, drive trucks, and do other jobs.

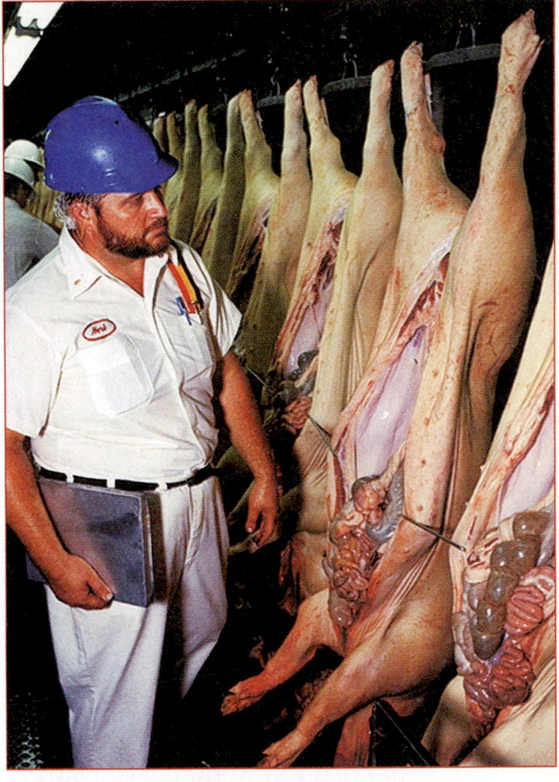

26–32. A meat inspector is examining pork carcasses with internal organs attached. (Courtesy, USDA)

REVIEWING

MAIN IDEAS

The setting for agricultural marketing is in the economic system. The free enterprise system is used in the United States. It allows individuals to make decisions and own production enterprises.

Marketing is the process of linking producers with consumers. Only products for which there is a market should be produced. A producer should never begin production without having a market.

Many marketing technologies are available and used. Each has advantages and disadvantages. The approaches used are often based on the preferences of the producer. The marketing channels are (1) roadside and retail markets, (2) central locations—packing sheds and elevators, (3) processing plants, (4) custom orders and vertical integration, and (5) niche markets.

Getting produce to consumers involves several marketing functions. These convert the raw produce of the farm or ranch into the product that consumers want. The functions are (1) harvesting and assembling, (2) grading, (3) transporting, (4) processing, (5) packaging, (6) storing, (7) advertising and promotion, (8) selling, and (9) distributing.

In order to market, certain facilities must be available. These are known as marketing infrastructure. Roads, vehicles, assembling and processing plants, finances, and labor are all necessary.

QUESTIONS

Answer the following questions, using complete sentences and correct spelling.

1. What are marketing and agricultural marketing technology?
2. Why are consumers important?
3. What methods are used in agricultural marketing? Briefly describe each method. Give an example of a plant or animal product that could be marketed using each method.
4. What functions are involved in marketing? Briefly describe each function. Give examples of what may be done to different products during the functions.
5. What is marketing infrastructure? What are the major areas of agricultural marketing infrastructure?
6. What is free enterprise? List major characteristics.
7. What are the ways of doing business in the United States? Briefly explain each.
8. What is the supply and demand curve? How does the curve relate to price?
9. What is management?
10. What is an economic system?

EVALUATING

Match each term with its correct definition.

a. roadside market
b. niche market
c. market function
d. blemish
e. grading
f. packaging
g. vertical integration
h. advertising
i. distribution
j. infrastructure

______1. The process of getting a product to the right place at the right time.

______2. Any method of communicating information to people that is intended to persuade them to buy a product.

______3. An on-the-farm retail facility to sell produce of the farm.

______4. A market that meets a unique need for a particular product.

______5. An activity or process that converts a product or performs a role in making the product readily available for consumers.

______6. A defect in a product.

______7. Facilities, services, and other things that must be in place.

______8. Placing products in containers.

______9. Sorting products for uniformity.

_____10. A process in which a nonfarm agribusiness owns the processing plant and all the sources of supplies for producing a plant or animal.

EXPLORING

1. How do producers market crops and livestock? Survey your community to determine how agricultural marketing is carried out. This might involve interviewing a few producers or agricultural professionals. The class may want to develop the questions to ask as a group activity.

2. Trace the marketing process of one of your favorite foods. If the food has more than one ingredient, trace all the ingredients. For example, if you like spaghetti with meat sauce, trace the origins of the raw materials and the processes involved.

3. Make a collection of the agricultural marketing information in the local newspaper. Gather the information for at least a week. Analyze price changes, and try to forecast future price levels.

4. Arrange to tour a local marketing facility, such as a packing shed, processing plant, or grain elevator. Have the manager explain how the facility operates and what it expects of producers.

CHAPTER 27

Processing Agricultural Products

OBJECTIVES

This chapter introduces important areas of agricultural processing to assure the desired food and fiber products. It has the following objectives:

1. Explain the meaning and importance of processing.
2. Explain spoilage in food.
3. List and describe methods of food preservation.
4. Describe methods in processing fiber products.
5. Describe methods in processing wood products.
6. Identify safety regulations in food processing.

TERMS

aseptic packaging
blanching
botulism
canning
classing
convenience preparation
cottonseed meal
curing
drying
fabricating
fermenting
fleece
food processing
freezing
ginning
HACCP
irradiation
lint cotton
log
olestra
pickling
planing
preserving
pulpwood
refrigeration
salmonellosis
saw timber
spoilage
synthetic food
textile mill
vacuum wrapping

27–1. Consumers expect foods in supermarkets to be nutritious and safe to eat.

PEOPLE expect things they buy to be safe and easy to use. They do not want to spend a lot of time preparing food or making clothing. Most people have fast-paced lives and want convenience and safety. This represents a marked change over the past when families were largely self-sufficient and provided for their own needs.

Many things we use today would be difficult to make at home from scratch. For example, imagine the work it would take to produce all the ingredients needed for your favorite pizza! Without the processing that has occurred, we would have to limit our pizzas to a few items that we could harvest from our gardens or in other ways.

Raw products, as harvested, must be prepared for use. Years ago, people did this in their homes. Today, we rely on others to produce and process our food and clothing. We also want good quality at an economical price. A worldwide network of activities helps meet consumer demand.

PROCESSING

27–2. Data are recorded and observed in food processing to guarantee that quality and safety standards are being met. (Courtesy, Mayfield Dairy Farms and Education Images)

Processing is all the steps in preparing products for people to use. Some products are made into final consumer forms; others require some preparation for consumption. Regardless, the goal is to provide quality, wholesome foods and other products in the desired forms for consumers.

Many people only think of food when they think about processing the produce of farms. Processing also includes fiber (such as cotton) and wood products.

PROCESSING FOOD

Food processing is all of the activities completed to prepare food for the consumer. The activities vary according to the food product and the preparation. Some items do not require much processing; others require a great deal of processing.

Processing includes converting foods into convenient forms. Some consumers want precooked food that they can make into a meal quickly. This may involve packaging a small can of tuna with crackers, pickle relish, and mayonnaise. Other times, precooked frozen meals that only need heating are produced.

27–3. Meat products, such as this pork roast that is being readied for cooking, are prepared in consumer-size portions. (Courtesy, Education Images)

Procedures in Processing

Food processing generally includes grading, fabricating, preserving, portioning, preparing convenient forms, and packaging. Some of these overlap with marketing technology. After all, agricultural marketing technology involves getting food, fiber, and shelter to people!

- **Grading**—Grading is the sorting of items for uniformity and quality. It is often done in marketing, but it may be needed in processing.

- **Fabricating**—***Fabricating*** is the activity or activities of making a product into a form that can be used. This varies with the product. For example, livestock must be slaughtered, and the carcasses must be cut into pieces that can be used. Humane practices must be used with live animals.
- **Preserving**—***Preserving*** is the process of treating food to keep it from spoiling. Many different procedures are involved. These are described in detail later in this chapter.

27–4. Hanging pork carcasses will be fabricated into desired cuts.

- **Portioning**—Portioning is the packing of a product into an amount or size that consumers want. Examples include putting two steaks in a package, placing 16 ounces of

AgriScience Connection

POTATOES INTO FRENCH FRIES

Huge amounts of potatoes are made into cut potato products each day. French fries are favorites with many people.

Potatoes are quickly washed, peeled, inspected, cut, frozen, and packaged for shipment. The top photo shows peeled potatoes being released by the peeler onto a conveyor for inspection. Following inspection, they are cut into the desired sizes and quick frozen. Very low temperatures are used in quick freezing—often in blasts of air at –25°F. After freezing, the potato products go into bags inside boxes where they are placed on a pallet and wrapped with plastic. (Courtesy, RDO Frozen, Park Rapids, Minnesota)

27–5. A package of frozen lima beans contains an amount (portion) appropriate for cooking.

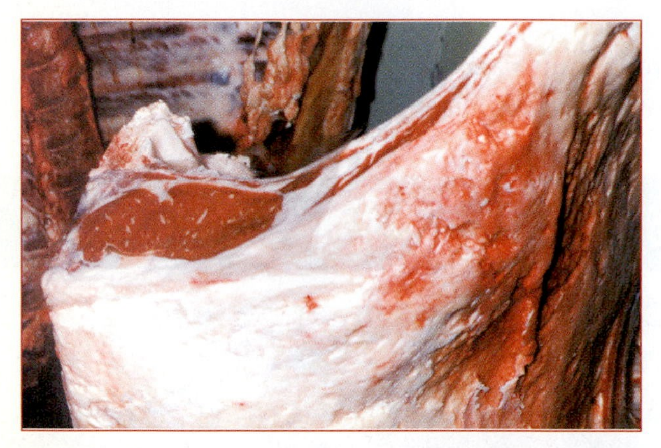

27–6. Beef loin is a tender and valuable part of the beef carcass.

27–7. Sausages are highly processed meat products.

vegetables in a can, and having the right combinations of vegetables, meat, and other items in a frozen dinner.

- **Preparing in convenient forms**—***Convenience preparation*** is the processing activity in which food is made easy for consumers to use. Sometimes the foods are precooked and only need to be heated. Some foods come in containers that can be used to heat and serve the food.

- **Packaging**—Packaging is the marketing function in which food is placed in a container. The container protects food from spoilage and keeps it in convenient portions. Packaging materials include glass, metal, plastic, mesh fabric, paper, and other materials.

Highly Processed Food Products

Foods may be made into a wide range of products. Some products require considerable processing. Some low-quality materials can be used to make high-quality products. In all cases, products that consumers want must be produced.

As examples:

- **Hams**—Hams may be prepared fully cooked and seasoned in various ways. Some are honey coated or injected with seasoning by using the blood vessels to distribute seasoning throughout the ham. The hams may be sliced using a spiral cutter before selling.

- **Boneless products**—Boneless products are prepared by removing the bones from fish fillets, turkey breasts, beef steaks, and sandwich meats. In some cases, the bones are carefully removed, and the meat is pressed back together. Many of these products are thoroughly cooked during processing. To be safe, always read the label to see if the product requires cooking before it can be eaten.

- **Oils**—Cooking oils are processed from plant and animal sources. Many are made from grains. Corn, soybeans, canola, and safflower grains as well as cottonseed are used to make oil, known as vegetable oil. A high-protein meal product remains. It is often used in animal feeds. Fatty tissue from hogs may be rendered (heated) to make a hog oil known as lard.

27–8. Soybean oil in a beaker and soybean meal in a scoop are surrounded by soybeans that have not been processed. (Courtesy, American Soybean Association)

PROCESSING FIBER

Processing fiber is converting raw material into cloth and other useful fabrics. Cotton, wool, and linen are three important natural fibers. Several artificial fibers are made.

Processing begins with cleaning and drying fiber to remove excess moisture and/or oil. The nature of processing varies with the kind of fiber. Cotton, for example, undergoes ginning, which is separating the seeds from the fibers (known as lint). Fibers are then woven into fabrics that are dyed or left their natural colors. Fabrics are usually sold as large rolls or bolts.

Fabrics are used in making clothing, furniture upholstery, carpet, towels, bed sheets, curtains, and other products. This occurs in cutting and sewing plants. Products that are knitted, such as socks and sweaters, are made in knitting mills.

The major use of fiber is in clothing manufacture. Clothing is made in garment plants. Automated cutting and sewing machines are used as much as possible in clothing construction.

More information is presented on processing fiber later in this chapter.

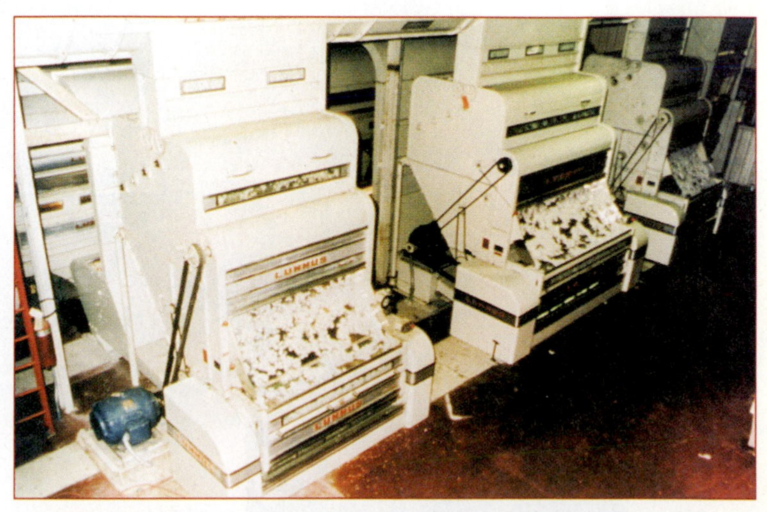

27–9. Gin stands have knives that separate the lint and the seed of cotton.

WOOD PROCESSING

Wood is produced and used to make many different products. When and how the wood is harvested are related to its later use.

The two kinds of wood are hardwood and softwood. Hardwood comes from oak, sweet gum, hickory, maple, and cherry trees. Softwood comes from pines, firs, and spruces. Softwood is typically used for lumber, paper, plywood, and similar products. In contrast, hardwood is typically used for furniture, flooring, and cabinets.

Pulpwood is harvested timber used in making paper, particle board, and other materials. The wood fibers are broken apart and pressed back together in a new form. Pulpwood trees are cut smaller and younger than saw timber.

Saw timber comes from the largest and oldest trees that are cut into logs. Saw timber is fabricated into lumber at a sawmill. Logs are cut into boards of different dimensions, such as 1, 2, or 4 inches thick and 4, 6, 8, 10, or 12 inches wide. Lumber is typically 8 to 20 feet long. Lumber has many uses in constructing homes and other buildings.

Specialty products are made from different kinds of wood. Examples are:

- Handles for hand tools from hickory wood
- Fine furniture from walnut
- Railroad crossties from various hardwoods
- Light wood structures, including floats on fishing lines, made from balsa wood

27–10. Logs are sawed into lumber at a sawmill.

SYNTHETIC FOOD MANUFACTURING

A ***synthetic food*** is an edible product made from nonfood raw materials or altered food materials. Synthetic foods achieve certain purposes in the human diet, such as no fat. One of the first synthetic food oils was ***olestra***—a fat substitute sometimes called "fake fat." (The trade name for olestra is Olean.)

In 1996, the U.S. Food and Drug Administration approved the use of olestra in making snack foods, such as potato chips, crackers, and tortilla chips. Since olestra adds no calories, potato chips, for example, have far fewer calories than those cooked in lard.

Olestra is a synthetic mixture of sugar and vegetable oil. It passes through the body without digestion or absorption. Some people who eat foods cooked in olestra may have side effects, such as upset stomachs and stomach pain. Some natural foods also have these same effects on people.

Synthetic foods should meet the same standards of wholesomeness as natural foods. They should be prepared in a clean environment and should be free of contamination.

SPOILAGE

Spoilage is a condition of food that makes it unsafe to eat. The flavor has usually changed. For example, sour milk has a definite change in flavor. Spoiled food is unfit for people to eat. Good, wholesome food can spoil if it is not properly handled and stored.

CAREER PROFILE

POULTRY INSPECTOR

A poultry inspector works in a processing plant and inspects birds during processing. Inspectors look for various traits of skin, flesh, bone structure, and body cavity. They are usually watching as birds move by on a conveyor. Birds with defects are removed from the line.

Poultry processors need specific training to perform their jobs. Post-secondary education in poultry science, meat science, or a similar area would be useful. Some poultry inspectors have baccalaureate degrees.

Jobs for poultry inspectors are available with state and federal agencies involved with inspection. Processing plants may employ individuals to assist inspectors. This photo shows inspectors observing freshly harvested birds for defects. (Courtesy, USDA)

27–11. Many foods are canned to prevent spoilage.

MICROORGANISMS AND SPOILAGE

Several kinds of microorganisms cause spoilage. Most spoilage is caused by bacteria and fungi, such as molds and yeast. Molds grow best at about 90°F (32°C) in places that are moist. For instance, the mold on bread develops when the bread is kept at this temperature.

Fresh vegetables and fruit will often begin to decay within a few days due to the presence of microorganisms. Storing fresh foods properly will slow the growth of these organisms.

Botulism and **salmonellosis** are probably the best-known food poisonings. The bacterium *Clostridium botulinum* causes botulism. Food contaminated by botulism may be fatal to people. Salmonellosis is caused by *Salmonella* types of bacteria, such as *Salmonella*

27–12. These oranges have completely spoiled and have no value. (Courtesy, Education Images)

27–13. A bulging can is a sign of spoiled food. Never eat food from such a can.

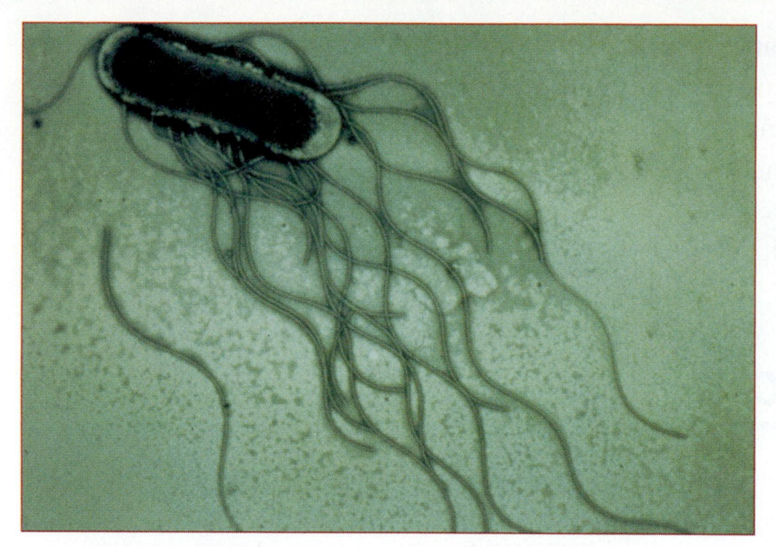

27–14. Greatly magnified *Salmonella enteriditis*—one cause of salmonellosis.

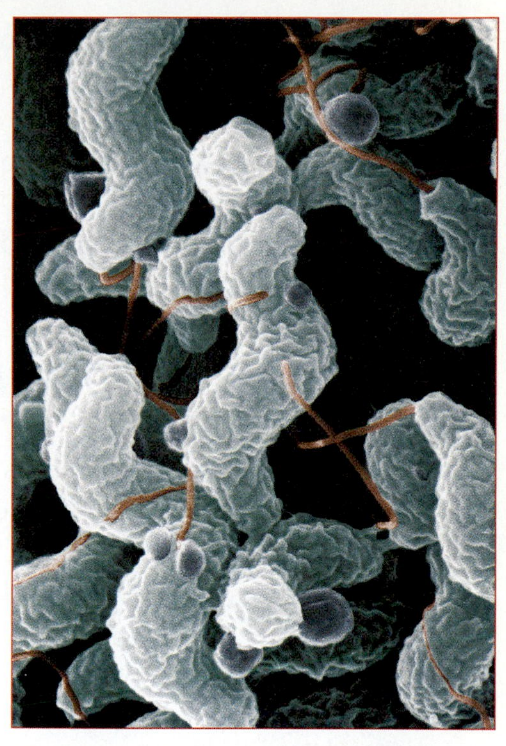

27–15. These *C. jejune* bacteria have been enlarged with a scanning electron microscope. (Courtesy, Agricultural Research Service, USDA)

enteriditis and S. *typhimurium.* These produce toxins in the intestines that cause upset stomachs. Salmonella are often found on raw meat, especially poultry, and inside eggs if the hens are infected.

The cause of more food-related gastrointestinal illness is the *Campylobacter* bacteria. It is a corkscrew-shaped organism visible with a scanning microscope. Campylobacter is the main cause of food-borne illness in developed countries. About a dozen species of Campylobacter have been implicated in causing food illness. The C. *jenni* and C. *coli* are the most common species. C. *fetus* is a cause of abortion in cattle and sheep. These bacteria enter the human body via contaminated food and water, contact with feces and other people, and raw meat consumption. Symptoms include stomach cramps, diarrhea, and, sometimes, fever. Good sanitation will help prevent contracting *Campylobacter* infections. Scientists are now working on the genetic sequencing of the organism to better understand how it enters the human body and causes illness.

27–16. Meats should be cooked thoroughly to destroy organisms that cause food poisoning.

Heat kills most microorganisms. They also do not survive in extreme cold, high-acid conditions, salt and sugar concentrations, and conditions without moisture. Chlorine bleach is often used in cleaning surfaces in food processing areas. The bleach kills nearly all microorganisms with which it comes into contact. The bleach is not used at full strength. It is diluted with water to achieve the desired goal.

All preserved food uses at least one method of controlling microorganism growth. Once the organisms are controlled, food can be sealed in airtight containers. Microorganisms must have oxygen to live.

RODENTS AND INSECTS

Rodents are small animals, such as rats and mice, that damage food and other products. They are more likely to cause problems with grain and similar products stored in bins and warehouses. The rodents eat some of the stored food, but they also contaminate the food with feces.

Insects can get into grain, fruit, vegetables, and other food products. Fruit flies may lay eggs on the fruit, and the eggs may hatch later. Weevils get into grain and eat the nutritious germ and starch. Laws dictate the acceptable level of insects and insect parts in human food. Most food producing standards have a very low tolerance level for insects and insect parts. No one wants flour that contains insects!

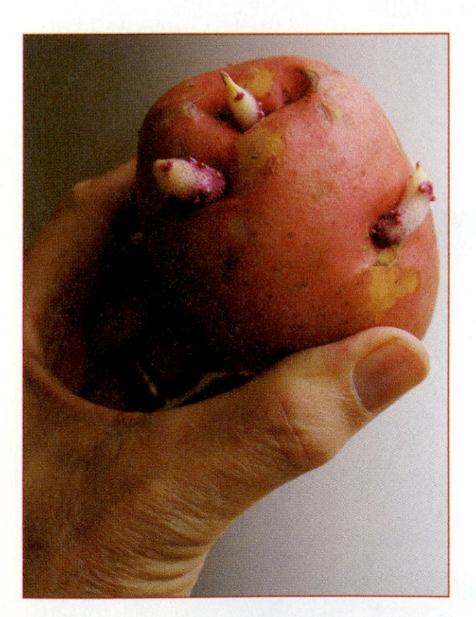

27–17. A potato is sprouting because of improper storage. (Courtesy, Education Images)

UNDESIRED GROWTH

Some food products will begin to grow after harvest if they are not stored properly. For example, an onion will sprout in warm storage. Beans and peas, as well as the seed in grapefruit and oranges, may also sprout. Sprouting is a particular problem with potatoes. Most people do not want to eat sprouts; however, some kinds of sprouts are used for food, such as alfalfa and bean sprouts.

CHEMICAL CONTAMINATION

Food products can be contaminated with chemical residues, such as pesticides. Harvested food crops should never be stored or hauled in areas where pesticides, fuels, or similar materials were or are stored. This also applies to feed crops for animals, such as feed for chickens and cattle.

FOREIGN OBJECTS

Undesirable objects in food make it unfit for human consumption. For example, metal and glass pieces should be kept out of food. Small rocks may be picked up accidently when crops are harvested, and rocks are difficult to remove from some crops. Wood and stem pieces are foreign materials, but they may not pose a serious danger to consumers. Pieces of rubber and plastic also contaminate food. As a result, food preparers must watch for rubber bands that hold some food in bunches.

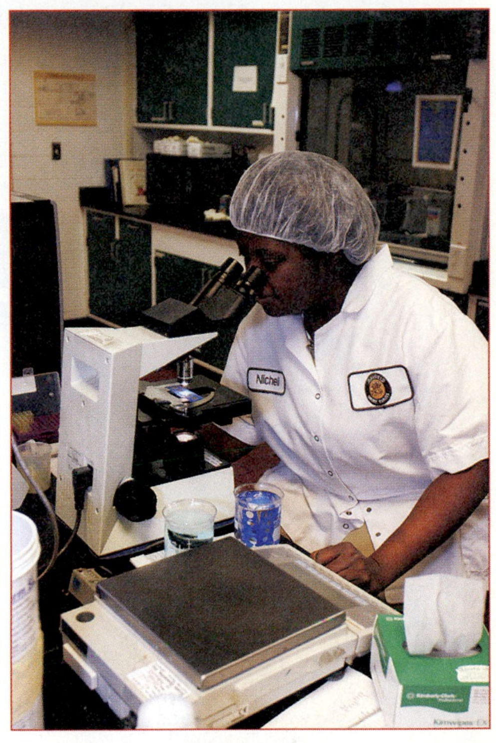

27–18. Purity and bacterial count are carefully determined before milk is processed. (Courtesy, Mayfield Dairy Farms and Education Images)

PURITY

Food products should be of uniform species. Beans and peas cannot be mixed. No one wants to open a can of tomatoes and find a kernel of corn! Beef and pork should be kept separate. Of course, various mixtures of foods are made when preparing soup, stew, and other products. Proper labeling is essential.

IMPROPER PRESERVATION

Foods will spoil if they are not properly preserved. Bulges in cans are signs of spoilage. Discoloration in food is also a sign that some spoilage is under way. Using the wrong additives can also make food unsafe to eat.

FOOD PRESERVATION

The main purpose of food preservation is to keep food from spoiling. Some foods will spoil quickly if they are not preserved; however, these same foods can often be kept for a long time without spoiling if preserved properly.

METHODS OF PRESERVATION

Food is preserved in different ways. Foods require methods of preservation suited to the chemistry of the raw product. Some foods have better tastes after preserving than they did fresh, such as cucumber pickles. Pickling preserves the cucumbers, and spice is added for flavor.

The acid levels of food products are important in food preservation. For example, tomatoes have higher acid than beans and corn. Microorganisms are less likely to grow in high-acid foods.

Methods of food preservation are presented in the following sections. Table 27–1 shows recommended methods of preserving selected food products.

Table 27–1. Recommended Methods of Preserving Selected Foods

Food	Method of Preserving		
	Canning	Freezing	Storing*
Vegetables:			
asparagus	xxxx	xxxx	
lima beans	xxx	xxxx	
snap beans	xxxx	xxx	
beets	xxx	xx	
broccoli		xxx	
carrots	xxxx	xxx	xxx
sweet corn	xxx	xxxx	
Irish potatoes			xxxx
sweet potatoes	x		xxxx
pumpkins	xx		xxx
squash	x	x	x
tomatoes	xxxx		
Fruit:			
apples	xx	xx	xxxx
berries	xxxx	xxxx	
grapes	xxxx	xx	
peaches	xxxx	xxxx	
Meats:			
beef	xx	xxxx	
pork	xx	xx	
chicken	xx	xxxx	

*Most foods can be stored temporarily in refrigeration. Some of the foods listed can be stored without refrigeration if the proper conditions are maintained.

Note: The number of x's indicates the rating of the method of preservation for the food, with more x's meaning the method is better. For example, canning asparagus (xxxx) is far better than canning sweet potatoes (x).

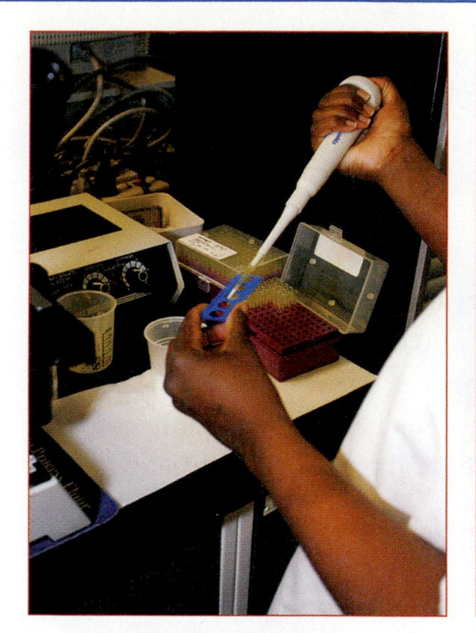

27–19. A milk sample is being carefully prepared for analysis. (Courtesy, Mayfield Dairy Farms and Education Images)

27–20. Unsealed cans are entering a steam tunnel as part of the canning process. (Courtesy, Education Images)

27–21. Cans are automatically sealed following the steam tunnel. (Courtesy, Education Images)

Canning

Canning is a preservation method in which food is placed in a container and heated to kill all microorganisms. The food containers are held at the right temperature for a period and then cooled. Products vary in the times and procedures used. Heating kills microorganisms that cause spoilage. An airtight seal on the container keeps other microorganisms from entering the container after it cools.

Most foods are canned in glass jars or metal cans. The cans are typically steel coated with a very thin layer of tin. Some cans are coated with a layer of enamel.

The seal is the final closure of the can or jar after the food has been put in it. The lid must be airtight. Seals that leak will result in spoilage.

Before food is placed in a jar or can, it should be prepared carefully. Only food free of decay and contamination should be canned. Some food will need trimming, cutting, or shelling. Also, food materials should be washed carefully.

Canning has long been used to preserve food. It is an economical way of keeping food from spoiling. After the canning process, most jars or cans can be stored at room temperature. Canned foods can be stored for months and, in some cases, years before use.

Proper equipment must be used in canning. For instance, large pressure cookers known as retorts are used to heat the cans of food. Heating sealed cans outside a pressurized container will result in the cans exploding. Improperly heating jars can result in the jars bursting and glass fragments cutting people.

27–22. Cans and jars are being lowered into a retort for pressure cooking. (Courtesy, Education Images)

Freezing

Freezing is a preservation method that uses low temperatures to freeze the water in food products. Freezing, which slows and prevents the growth of microorganisms, retains the nutrients in foods better than canning. However, some foods taste better if preserved in other ways.

27–23. Supermarkets usually have large frozen food areas.

As with canning, foods should be prepared carefully for freezing. Only food free of decay should be frozen. Most fruits and vegetables are blanched. ***Blanching*** is a process in which food is put in hot water, or exposed to steam, for a few minutes to loosen the skin and to destroy microorganisms that cause spoilage.

Fresh food is quickly frozen at temperatures well below freezing, often about –30°F (–34°C). After freezing, the food must be kept frozen until it is used.

Meats, vegetables, fruit, juices, and other foods are preserved by freezing.

27–24. A large facility that quick-freezes individual portions of fish.

Fermenting

Fermenting is a method of preservation that uses the controlled activity of certain bacteria, molds, and yeasts. It is used in making cheese, bread, vinegar, and other products. The action of the microorganisms must be regulated carefully to get the desired product. Yogurt, cheese, bread, and several other foods are made using a fermentation process.

27–25. In making cheese, milk passes through several steps in a controlled fermentation process. Steps in making cheddar cheese are: A—pasteurized milk forms curd that is cut into slabs (whey has been drained away); B—the curd slabs are milled; C—milled pieces are salted; D—salted milled cheese is placed in hoops for pressing; E—hoops are pressed under 40 psi for several hours; and F—the cheese is aged (to give flavor) in a refrigerated area for several weeks. (Courtesy, Education Images)

27–26. Raisins are being made by drying grapes on the ground near Fresno, California. (Courtesy, Education Images)

Drying

Drying is the removal of moisture from foods. Some drying is a natural process, such as sun-drying grapes to make raisins. Other drying is artificial and is known as dehydration. Products vary with the way they are dried or the additives that may be present while drying is underway. An example is grapes that are artificially dried in a chamber with a chemical additive to obtain golden raisins.

Many dehydrated foods need considerable processing to remove the moisture. Some dried foods can be made ready to eat (reconstituted) by adding water. Sometimes hot water is added, or the product is heated after water is added. Examples include dried milk, soup, instant coffee, and gelatin. Beef jerky is a dried food that does not need water added to it.

Technology Connection

ORGANISMS LIKE YOUR FOOD

Does the food you eat have microbes growing on it? The answer: most likely. Microbes are found on land, in air, in water, in refrigerators, and in about every other place our food may go. Some microbes may get on the crops as they are picked, transported, crated, and delivered to a local supermarket.

Scientists are trying to determine how to limit possible exposure. Everyone knows the value of sanitation and thoroughly cooking food. But, some foods are not cooked before they are eaten. Apples are often eaten raw, as is lettuce and nearly all other salad greens.

This shows a microbiologist spraying fresh-cut samples of iceberg lettuce with a substance that kills *E. coli* O157. The substance is a bacteriophage—a virus that invades and destroys certain bacteria. Obviously, the viruses are very small if they can invade bacteria invisible to the unaided human eye. (Courtesy, Agricultural Research Service, USDA)

Curing

Curing is a food preservation process in which substances are added to foods to prevent spoilage. Concentrations of salt, sugar, sodium nitrate, and other materials are used to prevent the growth of spoilage organisms. Since ancient times, meats—especially pork—have been preserved by salting. Smoke is sometimes used. Meat may be hung in a smokehouse or a liquid smoke flavoring may be added to the meat.

The natural circulatory system in meat may be used in curing. The ham is a good example. The curing solution may be injected under pressure into the blood vessels to assure that the curing solution reaches all parts of the ham.

Curing creates conditions where spoilage organisms do not grow well; salt and sugar prevent their growth.

Refrigeration

Refrigeration is a method of food preservation in which food products are kept at temperatures slightly above freezing. Refrigeration is also known as cold storage. Most fresh meats, milk, fruit, vegetables, and other fresh products high in moisture can be kept for several days with refrigeration because it slows the growth of microorganisms that cause food to spoil.

27–27. This supermarket has a large refrigerated fresh food section. (Courtesy, USDA)

Refrigerated products should not be allowed to freeze. Freezing might make them unacceptable by changing their texture and flavor. In some cases, freezing would spoil the food.

Nearly all homes have refrigerators. They are used to store fresh foods as well as cooked foods. However, most foods can be refrigerated for only a few days.

Irradiation

Irradiation is the preservation method of treating food with electrically charged particles, such as X-rays, electron beams, or gamma rays. Low amounts of radiation will kill spoilage organisms. Irradiation is used to keep potatoes from sprouting, to control insects in vegetables and grain, to control parasites in meat, and to slow the ripening processes in fruit and berries.

27–28. Vacuum wrapping a meat product.

Packaging

Aseptic packaging is a method of preservation in which the food and the container (plastic or foil) are sterilized. The amount of time the foods are heated while being processed is kept to a minimum. Long heating causes foods to lose their flavor. Foods in aseptic packaging can be stored for weeks or months at room temperature without spoilage.

Vacuum wrapping is a form of preservation packaging that removes oxygen inside the wrapper. Without oxygen, microorganisms cannot grow. Vacuum-wrapped products can be stored for quite some time at the proper temperature. Vacuum packing also prevents shrinkage and discoloration.

Pickling

Pickling is a method of preservation that prevents the growth of spoilage organisms by placing foods in acid solutions. Vinegar is commonly used. Cucumbers, cauliflower, okra, peppers, and other vegetables are pickled. Some meat products are pickled, such as pigs feet and eggs. Many pickling solutions have spices and other flavorings added to enhance the taste of the food product.

27–29. These rutabagas have been coated with wax to prevent spoilage.

Other Methods of Preservation

Other methods used to protect food from spoilage are:

- **Fumigants**—Fumigants may be used in grain bins and potato warehouses to kill insects and rodents.
- **Chemical additives**—Chemicals may be added to foods to prevent spoilage. For example, sulfur dioxide is added to dried fruit.
- **Waxing**—Some foods are coated with a thin layer of paraffin (wax). The wax holds moisture in and keeps organisms out. Apples and rutabagas are two examples.

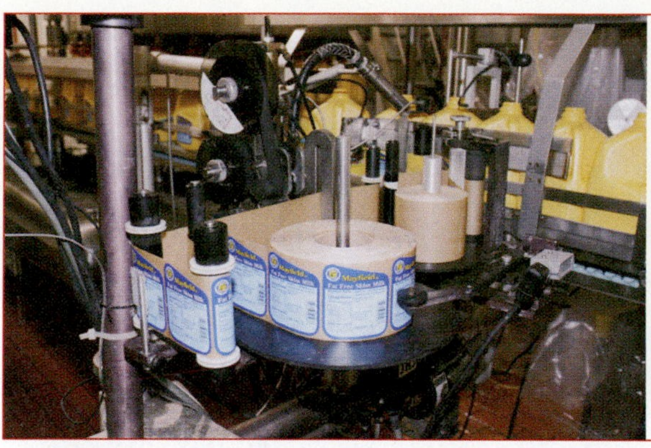

27–30. Gallon containers of milk are being labeled and filled in a modern dairy processing plant. (Courtesy, Mayfield Dairy Farms and Education Images)

- **Pasteurization**—Pasteurization is heating a food product to a certain temperature for enough time to kill microorganisms that may be present. Milk is the most widely used pasteurized food. All fresh milk should be pasteurized. In addition, dairy products should only be made with pasteurized milk. Fruit juices should also be pasteurized.
- **Jelly and jam making**—Fruit and berries are preserved with sugar to make jelly, jam, and preserves. These products are often used as spreads on breads.
- **Cooking**—Cooking stops the spoilage process. Cooked foods can be kept under refrigeration for several days.
- **Use of bases**—Bases are chemical compounds with a high pH. They are the opposites of acids. Examples are lye in making hominy from corn kernels and lime to keep potatoes from rotting.

ACADEMIC CONNECTION

NUTRITION

Relate your study of food processing and preparation to your studies in nutrition in a health or family and consumer science class. Explore the important role of nutritious, wholesome food as related to human health. Use your learning in agriscience class to support your mastery in nutrition.

PROCESSING FIBER PRODUCTS

Fibers can be classified as natural or artificial. Natural fibers are the most important in agriscience.

Natural fibers come from plants and animals. The major fibers from plants are cotton, linen, and jute. The major fibers from animals are wool, silk, and mohair. One natural fiber is made from a mineral—asbestos.

27–31. A bale of lint cotton is being compressed and made ready to go to a textile mill.

Artificial fibers may be manufactured into products similar to natural fibers. Nylon and rayon are examples.

COTTON PROCESSING

The first step in cotton processing is known as **ginning**. Several activities take place at the gin. The fiber and seed of the raw cotton, or seed cotton, are separated. The gin removes trash, dries the lint, and packages the lint cotton as bales. **Lint cotton** is cotton fiber without the seed. Cotton-producing areas have gins that are large processing plants.

Classing

After ginning, lint cotton is classed. **Classing** is judging cotton on the basis of length of fiber (staple), thickness of the fiber (micronaire), and grade, which is the presence of trash and color. Cotton should be white. Rain on open cotton bolls causes it to change to off-white. Human judgment in classing is being replaced with computer-based technology.

Weaving

Cotton is woven into cloth at a **textile mill**. Additional cleaning, spinning, and finishing are part of the process. Spinning is twisting the fibers into tiny threads. After weaving,

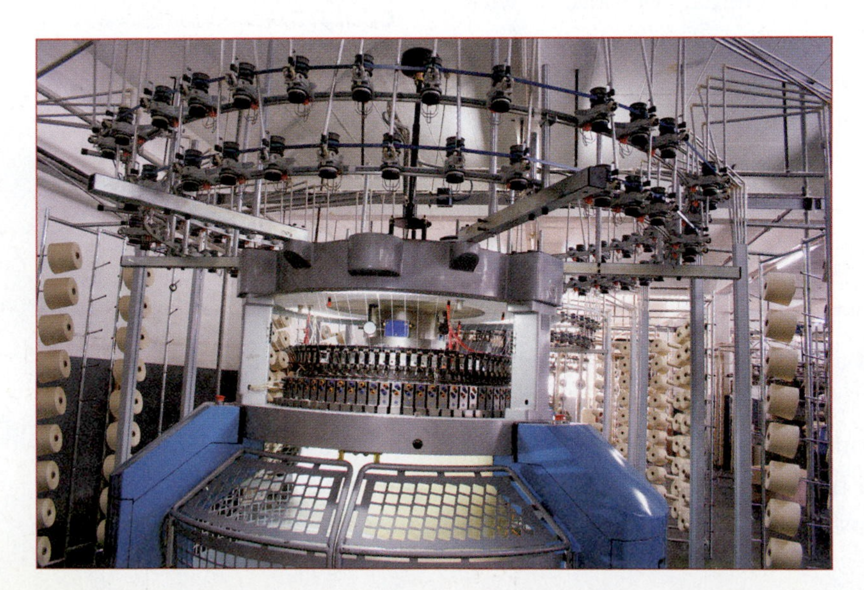

27–32. An industrial knitting machine that produces large amounts of cotton yarn. (Note the rolls of yarns on each side of the picture.)

the cloth may be dyed and treated with other chemicals to give it a particular finish. Some cloth moves through a machine that prints colored designs on it.

Garment Making

Cotton fabric is shipped from a textile mill to a garment (clothing) plant for manufacture. The fabric is cut into pieces, sewed, labeled, inspected, folded, packaged, and otherwise prepared for shipment.

Some cotton is made into yarn. This goes to a knitting mill for making knitted products, such as socks.

27–33. Manufacturing in a garment plant. (Courtesy, National Cotton Council of America, Memphis, Tennessee)

Cottonseeds

In the past, seeds removed from lint could be used to plant next year's cotton crop. This practice has declined because most cotton producers use improved seed. The genetically enhanced cotton varieties cannot be planted under current agreements between the seed company and farmer. These seeds are not thrown away; instead, they are used for making food and feed products.

Cottonseeds usually go to oil mills where they are made into vegetable oil and other products. A good example is oleomargarine—a butter substitute. The oil processing leaves a product known as ***cottonseed meal***. The meal is used as a high-protein concentrate to feed animals.

WOOL PROCESSING

Wool is used to make clothing, rugs, blankets, upholstery, and other fabric products. Various kinds of wool are produced by different species of animals. Most wool in North America is from medium-wool sheep.

Shearing

The wool on a sheep is known as ***fleece***. The fleece is sheared (removed) with clippers so the sheep is not injured. New procedures are being developed to remove wool chemically, but these are not widely used. Most sheep are sheared once a year—usually in the spring. This keeps the sheep cooler in the hot summer weather. A new coat is grown before winter.

Grading

After shearing, wool is sorted to remove trash and off-color fleece. It is then graded to determine its fineness, length, crimp, and color. Crimp refers to the straightness of the wool fibers. Fineness is determined by the diameter of the fibers.

Weaving

At the weaving plant, wool is cleaned with a detergent to remove excess oil and to prepare it for weaving. The fibers are arranged in parallel and woven into yarn or fabric. Different kinds of wool are used to make different fabrics. The wool is dyed to give it the desired color.

Manufacturing

After weaving, wool fabric is made into garments and other products. This requires cutting the fabric according to the pattern; sewing the pieces together; and adding buttons, zippers, and decorative trim to the final garment. Wool yarn is knitted into sweaters, hats, and other products that are known for warmth and durability.

AgriScience Connection

CLEAN FIELDS

Can food-borne infections get on crops while they are in the field? If they are present in the field, they can be on the harvested food crops. Producers can take steps to limit possible contamination.

A major goal of all producers is to limit the chance of food crops being contaminated while in the field. Cultural practices can help keep food products clean. Limiting human and wildlife access and other sources of field contamination will help. Some traffic is hard to control, such as birds that fly over a field. Using dirty irrigation water can also introduce unwanted microbes to a crop.

This shows a conservationist working with a producer to ensure the use of appropriate irrigation water and other practices with vegetable crops. (Courtesy, Natural Resources Conservation Service, USDA)

PROCESSING WOOD AND TIMBER PRODUCTS

Forestry products are made into many kinds of lumber, plywood, particle board, veneer, and paper. These products are used in many ways—shelter being the most important.

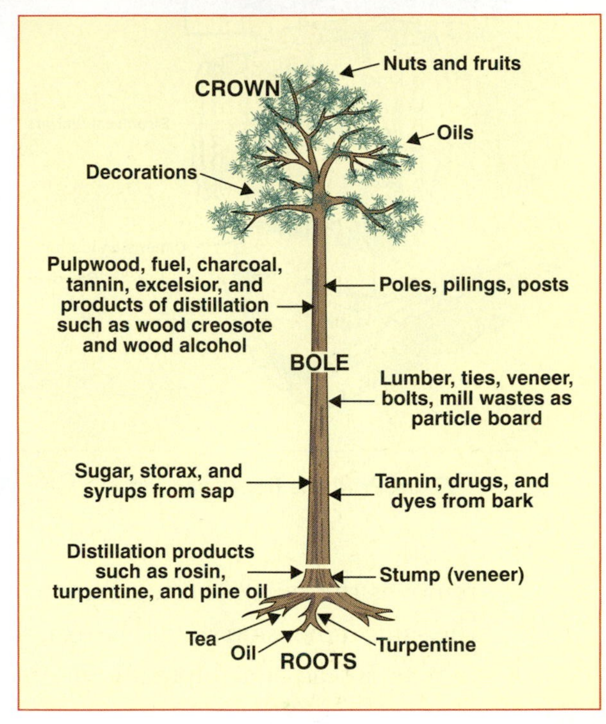

27–34. Examples of products made from different parts of trees.

PRODUCING LUMBER

Lumber is made by sawing logs into boards. A **log** is the large stem of a tree that required many years to grow. The trees are felled (cut down) and cut into log-lengths in the forest. Most logs are 12 to 20 feet (3.66 to 6.1 m) or longer. Logs must also have a minimum diameter that makes them useful in making wood products. The diameter varies with the kind of wood and the product to be made. It is obvious that a board 10 inches (25.4 cm) wide requires a log that is more than 10 inches in diameter at the smallest end.

Sawmilling

Logs are cut into boards at a sawmill. Large saws rip the logs open, creating sawdust where the cut is made. The boards are cut to specific sizes for lumber. As sawed, logs have a rough finish and are called rough lumber. After sawing, the lumber is graded. Important factors in grading are the presence of knots (where a limb grew on the stem), size, and presence of bark or defects, such as worm holes. Boards that grade high (No. 1) have few or no knots and all the bark has been removed.

27–35. A small, transportable sawmill. (Courtesy, Rita Lange)

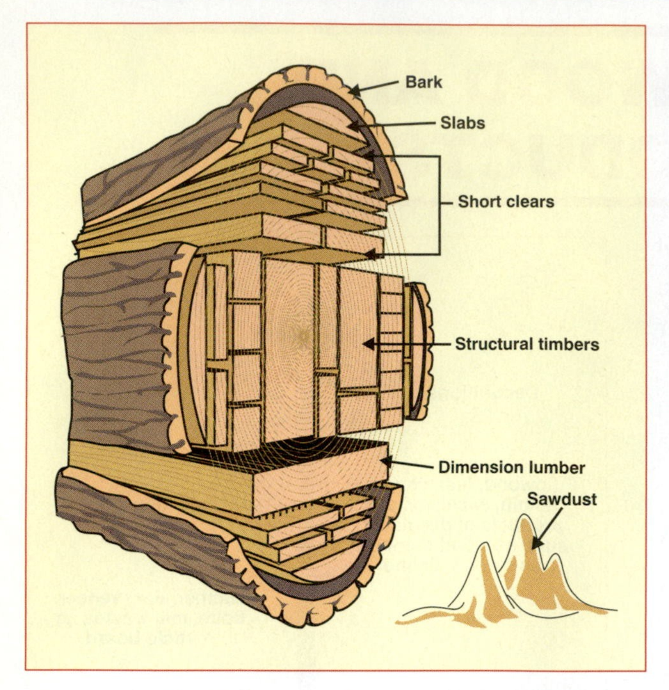

27–36. Products from a saw log.

Most logs are sawed into lumber while still green (sappy). The lumber cannot be used until the moisture has been reduced. Removing moisture from lumber is seasoning or drying. The lumber may be naturally or artificially dried. Natural drying is also known as air drying. The lumber is stacked so air moves between it in the lumberyard. This is a slow process, so artificial drying is more common. A dry kiln is used in artificial drying. The kiln is a specially constructed building that can be heated to speed up the drying process. Wood that is not dry will warp and will cause problems in building construction.

Planing

After drying, the lumber is planed. ***Planing*** is the smoothing and sizing of lumber. Planing removes the roughness resulting from sawing and carefully sizes the lumber. All boards of the same dimension should be exactly the size specified with no variation. After planing, the lumber is ready for shipment and use.

Treating

Some wood products are treated to make them resistant to rot and insect damage. Only approved chemicals should be used in treating wood. For instance, wood products treated with some preservatives cannot be used where people or animals will touch them. The chemicals are considered hazardous to health.

PRODUCING PAPER

Smaller trees are harvested as pulpwood and are used to make paper. They may be cut and hauled as long poles or made into pulpwood lengths of about 6 feet (2 m). Most go to a chipping mill where the bark is removed and the wood is chopped. The chips go to the paper mill.

Paper is made by breaking wood into pulp. The wood may be broken apart with chemicals and by a grinding action. Pulp is screened and washed to remove impurities. Large machines take the pulp and rapidly form it into continuous sheets of paper. The thin mixture of pulp

27–37. Poles await transportation to a paper mill. (Courtesy, Education Images)

and water is spread over a mesh wire that lets the water drain away. The remaining mat of fiber moves between large rollers that squeeze out more water. The paper goes through other rollers and dryers. The fibers bond together during the drying process.

OTHER WOOD PRODUCTS

Wood is used to make many products. These products often require the wood of certain kinds of trees.

Fine Wood

Furniture requires woods that are cut, milled, and seasoned. Most furniture is made from fine hardwoods, but some pines and cedars are used. Wood serves as the frame over which upholstery and cushions are fitted. Wood (e.g., cherry, black walnut, and mahogany) also makes beautiful table tops and other furniture with fine wood finishes.

Plywood

Both hardwood and softwood trees are used to make plywood. However, plywood made from softwood is more likely to be used in building construction. Plywood made from hardwood is typically used in cabinet and furniture manufacturing.

Veneer is a special kind of wood material. It is made by gluing a thin layer of valuable wood over thicker layers of lower-cost wood. For example, walnut may be put over gum or low-grade oak. Some veneers are used as paneling on the walls of homes and offices.

Plywood is a type of wood material made by slicing logs into thin sheets and gluing them together so the wood grain is perpendicular on adjoining layers. Plywood is quite strong and is a versatile material. The cutting process is somewhat like unrolling the log along its growth rings. The wood is "unwound" in a long, continuous piece. The wood is used while still green and is dried as a part of the plywood-making process. Plywood is made in various thicknesses, such as ¼ or ½ inch (0.63 or 1.27 cm). It is cut into sheets that are usually 4 feet (1.22 m) wide and 8 feet (2.44 m) long.

SAFETY AND REGULATIONS IN PROCESSING

Processing must follow certain standards. Many regulations have been made to protect consumers. Food processing is highly regulated by government agencies.

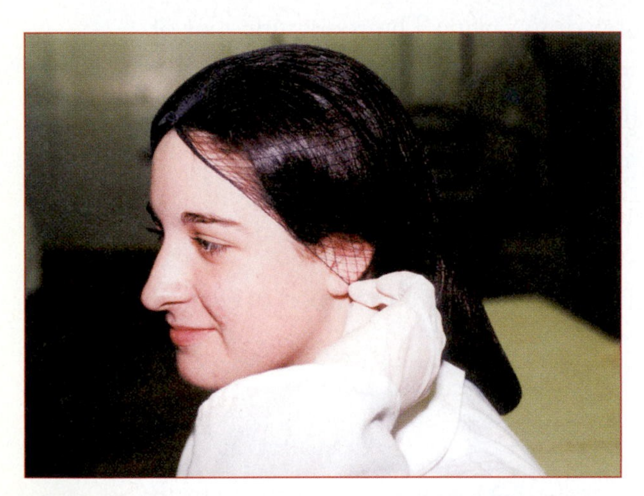

27–38. Hairnets, gloves, and clean clothing help prevent food contamination.

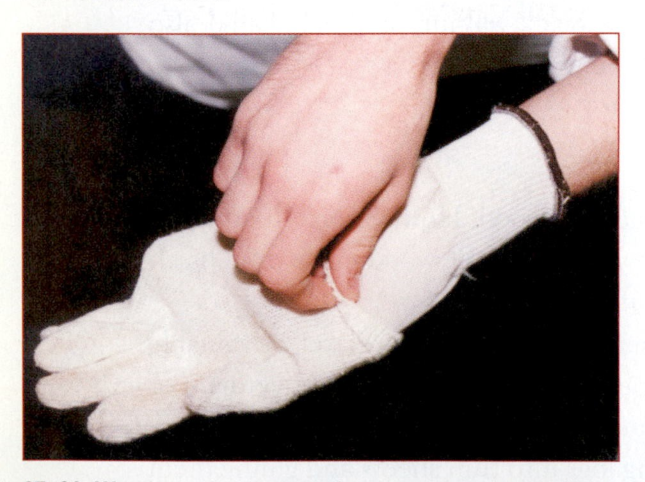

27–39. Wearing cut-resistant gloves protects hands from knife accidents in meat processing.

FOOD SAFETY

Food safety is following practices to ensure that foods are safe to eat. The food products are wholesome and provide important nutrients in the human diet.

A major effort in assuring food safety is HACCP: Hazard Analysis Critical Control Point. **HACCP** is a procedure used in a processing plant to identify potential points at which foods may be contaminated. HACCP is also used in taking steps to reduce hazards. Every commercial food processor develops a HACCP plan. HACCP is administered by the Food Safety and Inspection Service (FSIS) of the U.S. Department of Agriculture.

PERSONAL SAFETY

Processing often requires the use of machinery and equipment that could cause injury. People should know how to use equipment properly. Carelessness can result in injury.

Protective clothing should be worn. Boots and eye protection should be used. In noisy places, devices to protect hearing are needed. In some plants, gloves that protect workers from cuts or hazardous substances are worn.

SANITATION

Sanitation is the practice of keeping food processing areas clean. Regular washing and cleaning prevent the growth of microorganisms. Equipment, tables, floors, and other parts of the facility should be cleaned. Anything that the food touches must be cleaned.

27–40. Keeping food production clean promotes food safety. (This shows pastry products being prepared on an automated system.) (Courtesy, Education Images)

INSPECTION

A processing plant that produces food is inspected regularly for sanitation. Regulations must be met if the plant is to continue operating.

Inspectors may be hired by state or federal government agencies or by the processing plant. Their role is to assure wholesome food for consumers. With some food products, such as beef carcasses, inspectors observe each item processed. Other inspectors check on the physical facilities in which the processing is done. Such facilities must be clean and must provide the proper conditions for food safety.

Internet Topics

Selected topics for Internet discovery and reporting are listed here. Begin by searching for each topic. Next, read and learn about it. Conclude by preparing a brief report on your findings.

1. Botulism
2. HACCP
3. Olestra

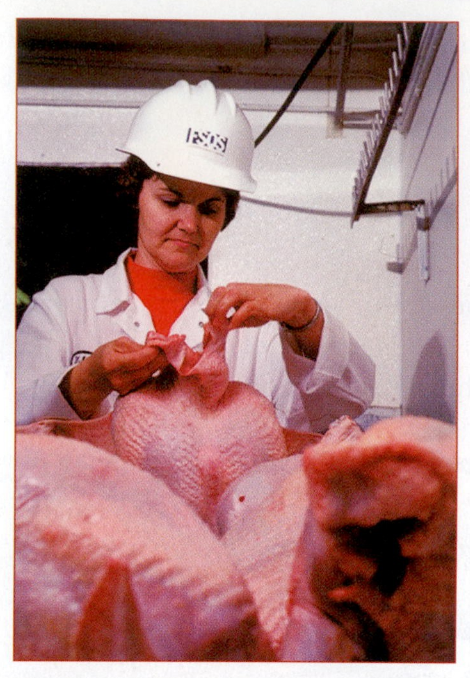

27–41. A U.S. Department of Agriculture inspector is assessing poultry carcasses. (Courtesy, USDA)

27–42. An inspection stamp indicates that meat has been inspected and passed.

WASTE DISPOSAL

Most processing creates waste materials, which must be disposed of properly. Water from processing cannot be released into streams without treatment. The general environmental rule is that water released into a stream cannot change the general condition of the stream.

27–43. Properly cleaning a food processing facility promotes safe food.

Some processing creates solid waste materials. These must be disposed of by moving them to landfills or by processing them into byproducts. Very little is actually thrown away at a processing plant. Most things can be used to produce other products. For example, some of the waste from meat processing is used to make protein concentrates for animal feed.

CONTROL OF RODENTS, INSECTS, AND OTHER ANIMALS

Food processing plants must control rats, mice, roaches, birds, and other animals. These should not be allowed in places where food is being processed. Birds should not fly around in canneries or other facilities. Pest management practices must be used inside and outside most food processing facilities. Pesticides and traps placed outside can often limit pest populations before they enter a food processing and/or storage facility.

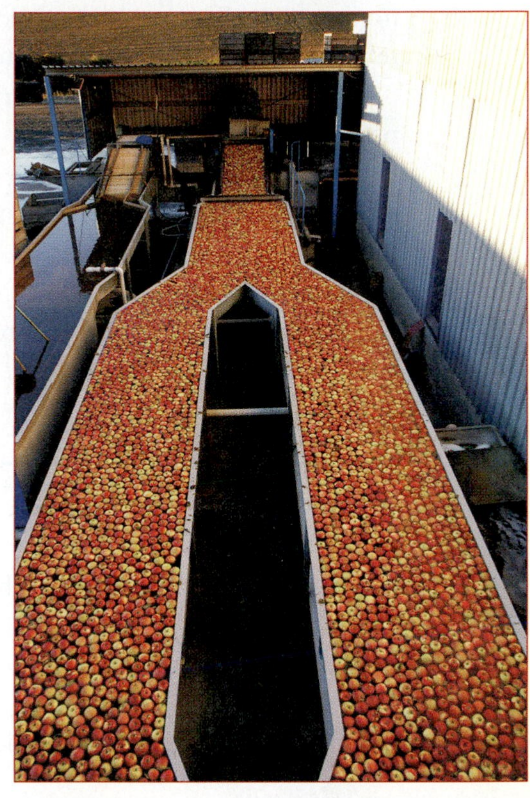

27–44. All processes in preparing food for the consumer should be carried out so contamination does not occur. (Contact with apples is minimal in this processing facility.)

PREPARATION

Food preparation requires attention to a wide range of safety practices. Food should be stored and prepared to minimize opportunities for spoilage and contamination. Proper heating to guarantee thorough cooking is essential. Many food preparers use cooking thermometers to be certain that the interior of foods, particularly meats, has reached an appropriate temperature.

REVIEWING

MAIN IDEAS

Processing is the act of changing crops and livestock into forms that consumers want. What is done varies with the kinds of products and the way in which they are prepared. The goal is to provide wholesome products that meet consumer demand.

Food and fiber processing is a part of the marketing process. The steps overlap and add value to what is produced. People are willing to pay more for products that are convenient to use.

Processing includes several functions: grading, fabricating, preserving, portioning, preparing in convenient forms, and packaging. All food products go through one or more of the processing steps.

Food preservation is an important area of processing. It involves preparing foods to keep them from spoiling. The methods of food preservation are canning, freezing, fermenting, drying, curing, refrigerating, irradiating, packaging, and pickling.

Fiber processing is converting the fiber into forms that people want. It includes weaving and spinning as well as manufacturing.

Wood products are made from softwood and hardwood timber. Lumber is made from larger trees, while paper is made from smaller trees. The manufacturing processes may be complex.

Safety practices and regulations must be followed in processing. Sanitation and other procedures are vital in food processing to make sure the consumer receives wholesome food products. USDA inspection is used in most food processing plants.

QUESTIONS

Answer the following questions, using complete sentences and correct spelling.

1. What is processing? Why is it important?
2. What are the five major steps in processing food?
3. What is food preservation? Why is it important?
4. What is spoilage? What causes food to spoil?
5. What major methods of food preservation are used? List and briefly describe each.
6. How are fibers processed? List and briefly describe the steps involved for cotton and wool.
7. How is lumber processed? What steps are involved?
8. How is paper made?
9. What wood products are made in addition to lumber and paper?
10. What safety and sanitation procedures should be followed in processing?

EVALUATING

Match each term with its correct definition.

a. HACCP
b. vacuum wrapping
c. preserving
d. fabricating
e. saw timber
f. lint cotton
g. spoilage
h. fermenting
i. pickling
j. salmonellosis

_____1. A type of food preservation that uses acid to prevent spoilage.

_____2. A program to identify and prevent food contamination.

_____3. Bacteria found on raw meat, like poultry, that cause food-borne illness.

_____4. Large trees harvested for sawing into lumber.

_____5. Using bacteria or other organisms to preserve foods.

_____6. A condition resulting from changes in food that make it unsafe to eat.

_____7. Cotton fiber without seeds.

_____8. Removing all the oxygen and placing foods in a tightly sealed plastic wrap.

_____9. Making products into the desired forms.

_____10. Treating food to keep it from spoiling.

EXPLORING

1. Arrange a tour to a local lumberyard. Note how the wood has been cut and the products that are available. Prepare a report on your findings.

2. Make a display of different kinds of food packaging materials. Use empty packages from your home, school cafeteria, or other places.

3. Preserve a food product using one of the preservation methods. Consult with a home economics teacher and use reference materials on food preservation.

4. Visit a clothing store, and observe the fabrics that are used. Make a list of the different articles of clothing and the kinds of fabrics used to make them. Determine if the fabrics are natural or artificial.

Appendixes

Appendix A. Common Weight Conversions and Equivalents Used in AgriScience

Customary System		
1 tablespoon (tbs)	=	3 teaspoons (tsp) or 0.5 fluid ounce (fl. oz)
1 cup	=	8 fl. oz
1 pint (pt)	=	2 cups or 16 fl. oz
1 quart (qt)	=	2 pt or 4 cups or 32 fl. oz
1 gallon (gal)	=	4 qt
1 bushel (bu)	=	8 gal or 32 qt
Metric System		
1 milliliter (ml)	=	1 cubic centimeter (cc) or 1 gram (g)
1 centiliter (cl)	=	10 ml
1 deciliter (dl)	=	10 cl or 100 ml
1 liter (L)	=	10 dl or 1,000 ml or 1 kilogram (kg)
1 dekaliter (dal)	=	10 L
1 hectoliter (hl)	=	10 dal or 100 L
1 kiloliter (kl)	=	100 hl or 1,000 L or 1,000 kg
Equivalents		
1 ounce	=	28.3495 grams
1 pound	=	453.59 grams or 0.45359 kilograms
1 kilogram	=	35.27 ounces or 2.2046 pounds
1 metric ton	=	2,204.6 pounds
1 liter (L)	=	1.057 quarts
1 liter (L)	=	0.2642 gallons
1 gallon (gal)	=	3.79 L
1 meter (m)	=	39.37 inches
1 acre	=	0.40468 hectares
1 inch	=	2.54 centimeters (cm)
1 centimeter (cm)	=	0.4 inch

Appendix B. Common Volume Measures Used in AgriScience

Customary Fluid Measures				
4 gills (gi)	=	1 pint (pt)	=	28.88 cubic inches
2 pints (pt)	=	1 quart (qt)	=	57.75 cubic inches
4 quarts (qt)	=	1 gallon (gal)	=	231 cubic inches
31.5 gallons (gal)	=	1 barrel (bbl)	=	4.21 cubic feet
2 barrels (bbl)	=	1 hogshead (hhd)	=	8.42 cubic feet
7.5 gallons	=	1 cubic foot	=	1,728 cubic inches
Customary Dry Measures				
2 pints	=	1 quart	=	67.2 cubic inches
8 quarts	=	1 peck (pk)	=	537.6 cubic inches
4 pecks (pk)	=	1 bushel (bu)	=	2,150.4 cubic inches or 1.244 cubic feet

Appendix C. Common Weight Measures Used in AgriScience

Customary Weights		
Avoirdupois Weight:		
437.5 grains (gr)	=	1 ounce (oz)
16 ounces	=	1 pound (lb)
100 pounds	=	1 hundredweight (cwt)
2,000 pounds	=	1 ton (T)
2,240 pounds	=	1 long ton
Troy Weight:		
24 grains	=	1 pennyweight (pwt)
20 pennyweight	=	1 ounce
12 ounces	=	1 pound
Apothecaries Weight:		
20 grains	=	1 scruple (sc)
3 scruples	=	1 dram (dr)
8 drams	=	1 ounce
12 ounces	=	1 pound
Metric Weights		
1 centigram (cg)	=	10 milligrams (mg)
1 decigram (dg)	=	10 centigrams and 100 mg
1 gram (g)	=	10 dg, 100 cg and 1,000 mg
1 kilogram (kg)	=	1,000 grams
1 metric ton (t)	=	1,000 kg

Appendix D. Measures and Weights of Selected Agricultural Commodities

Commodity	Unit	Customary Weight (pounds)
Alfalfa seed	bushel	60.00
Apples	northwest box	44.00
Apricots	lug	24.00
Asparagus	crate	30.00
Barley	bushel	48.00
Brussels sprouts	drums	25.00
Cabbage	open-mesh bag	50.00
Celery	crate	60.00
Coffee	bag	132.30
Corn		
ear (husked)	bushel	70.00
shelled	bushel	56.00
oil	gallon	7.70
Cotton (lint, no seed)	bale (gross)	500.00
Cream	gallon	11.84
Maple syrup	gallon	11.02
Milk	gallon	8.60
Oats	bushel	32.00
Onions (dry)	sack	50.00
Peanuts (unshelled)	bushel	17.00
Rice	bushel	45.00
Soybeans	bushel	58.00
Tobacco (Maryland)	hogshead	775.00
Tomatoes	crate	60.00
Wheat	bushel	61.00

Source: U.S. Department of Agriculture.

Appendix E. Determining Area of Common Shapes

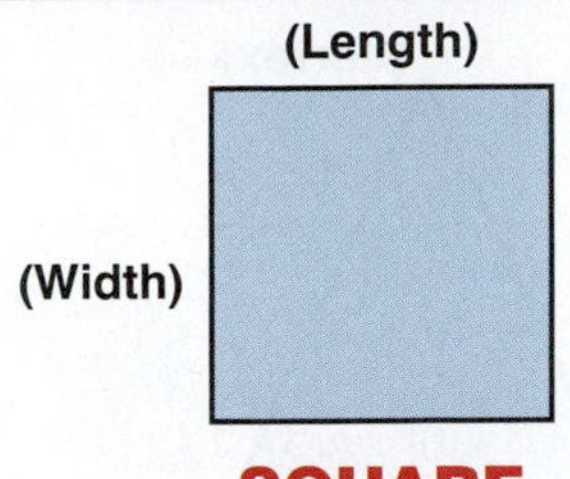

SQUARE

Definition: all sides equal and meet at right angles

Formula: Area = length × width

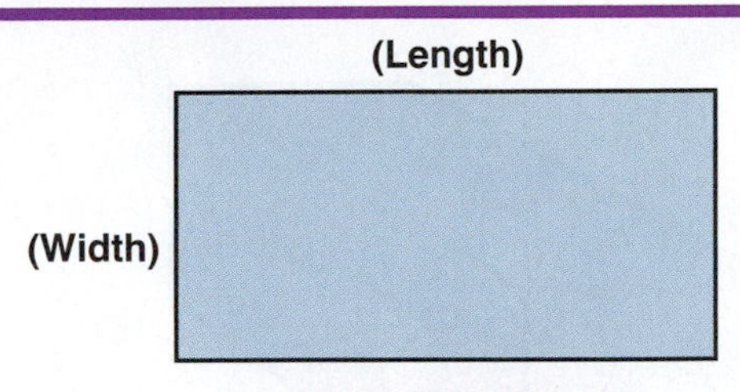

RECTANGLE

Definition: adjacent sides of equal length; sides meet at right angles

Formula: Area = length × width

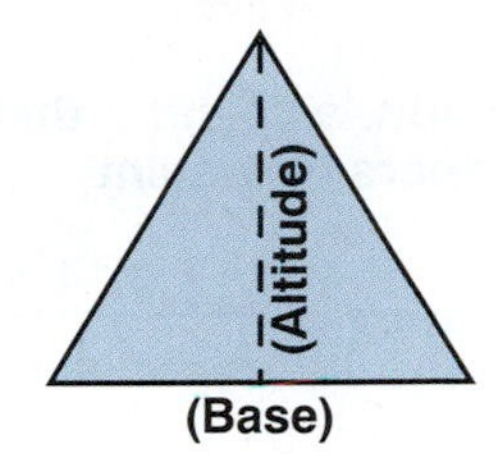

TRIANGLE

Definition: surface with three sides

Formula: $\text{Area} = \dfrac{\text{base} \times \text{altitude}}{2}$

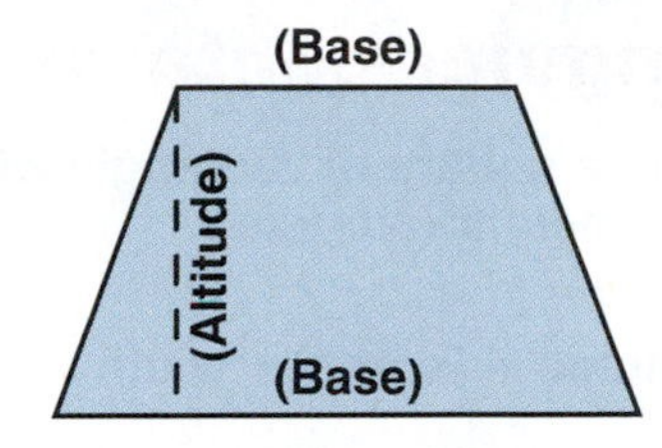

TRAPEZOID

Definition: four sided surface with two sides parallel

Formula: $\text{Area} = \dfrac{\text{base} + \text{base}}{2} \times \text{altitude}$

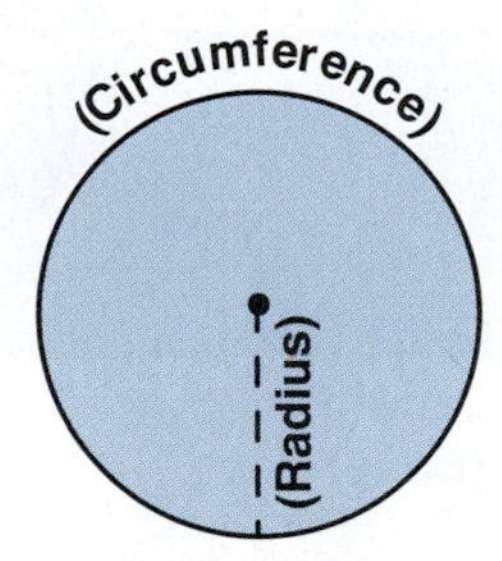

CIRCLE

Definition: surface that is round

Formula: $\text{Area} = \text{radius}^2 \times 3.14$

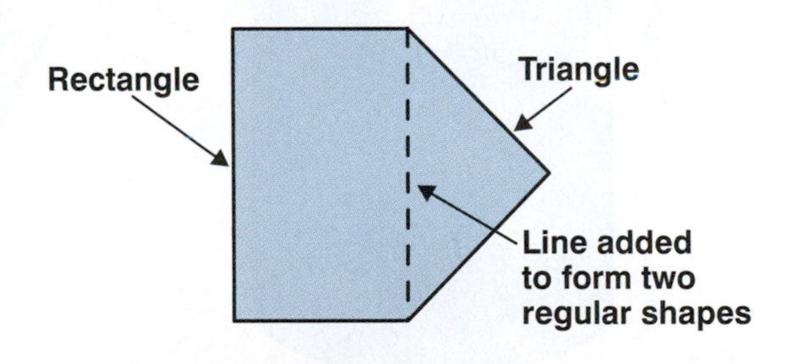

IRREGULAR

Definition: surface that is not a regular shape but can be divided into two or more regular shapes

Formula: use those for the shapes that are formed when the surface is divided

Appendix F. Determining Volume of Common Shapes

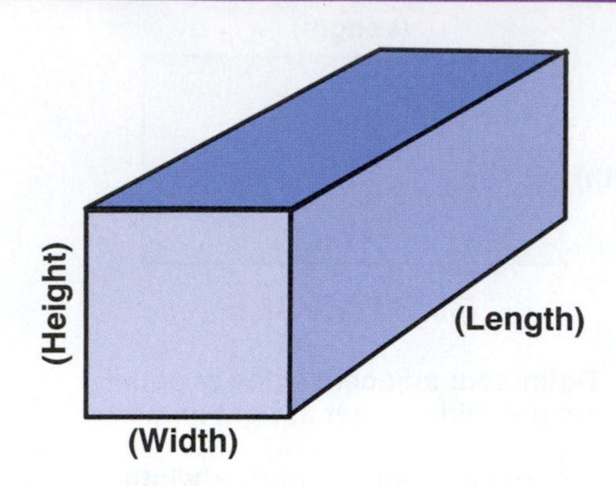

Square or rectangular containers

Definition: a square or rectangle with height (altitude)

Formula: Volume = length × width × height

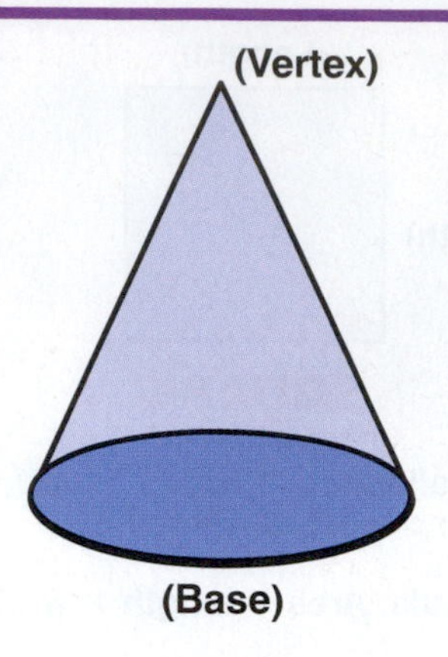

Cone

Definition: a cylinder that tapers to a point

Formula: Volume = $\frac{r^2 \times 3.14 \times \text{height}}{3}$

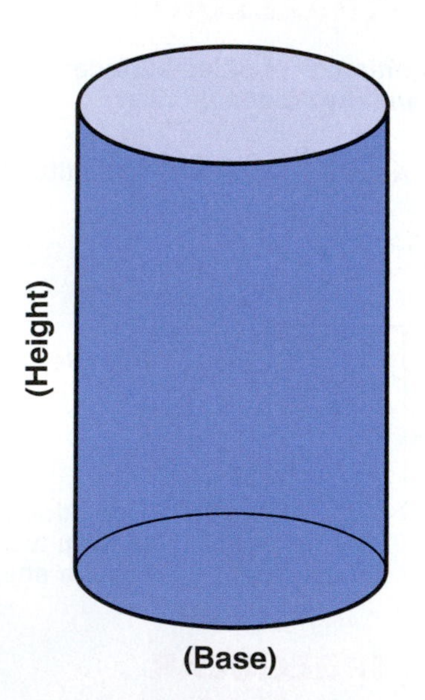

Cylinder

Definition: a circle with height (altitude)

Formula: Volume = $r^2 \times 3.14 \times$ height

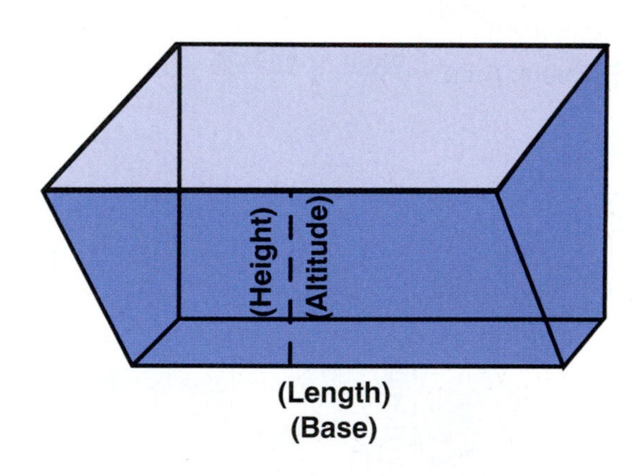

Irregular Shape

The volume of this shape is determined by obtaining an average area for the base and using the formula for a square or rectangular container

Appendix G. Common Forestry and Wood Measurements

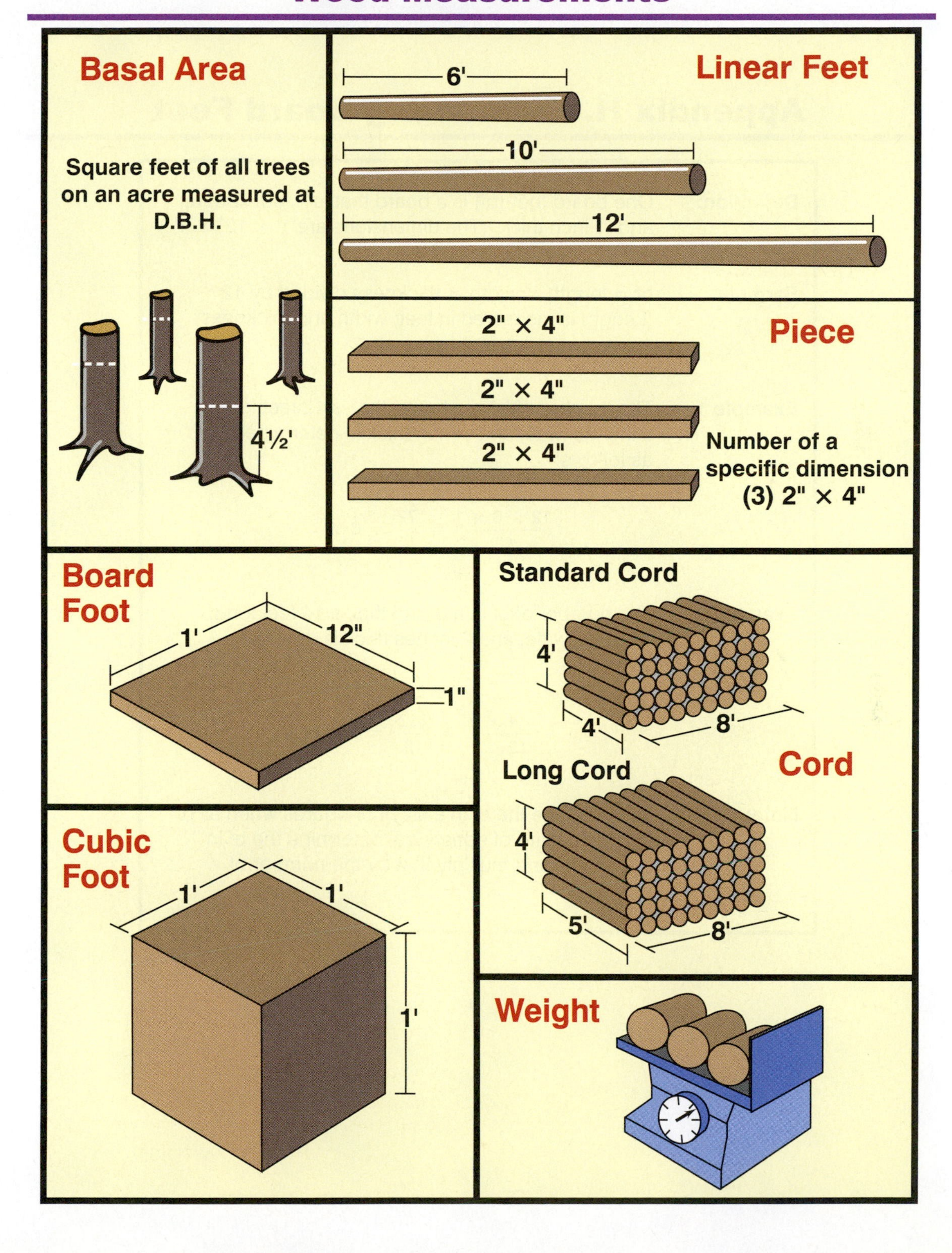

Appendix H. Calculating Board Feet

Definition: One board foot (bf) is a board that is 1 foot square and 1 inch thick. (The dimensions are 1' × 12" × 1".)

Formula: bf = length × width × thickness divided by 12 (Length is measured in feet; width and thickness are measured in inches.)

Example 1: The number of bf in a board that is 12 feet long, 6 inches wide, and 1 inch thick is determined as follows:

$$\frac{12 \times 6 \times 1}{12} = \frac{72}{12} = 6 \text{ bf}$$

Example 2: The number of bf in a board that is 14 feet long, 4 inches wide, and 2 inches thick is determined as follows:

$$\frac{14 \times 4 \times 2}{12} = \frac{112}{12} = 9.33 \text{ bf}$$

Note: To determine the bf in a stack of boards when all of the boards are of equal size, determine the bf in one board and multiply that by the number of boards.

Appendix I. Making Temperature Conversions

Converting Fahrenheit (F) to Celsius (C)

Formula: $C = {}^{5}/_{9}(F - 32)$

Steps:
1. Determine the F temperature.
2. Subtract 32 from the F temperature.
3. Multiply the difference found in step 2 by 5.
4. Divide the amount found in step 3 by 9.

(The result is the equivalent temperature in degrees C.)

Example: If the temperature is 100°F, what is the C temperature?

$100 - 32 = 68$
$68 \times 5 = 340$
$340 \div 9 = 37.7°C$

Converting Celsius (C) to Farenheit (F)

Formula: $F = {}^{9}/_{5}C + 32$

Steps:
1. Determine the C temperature.
2. Multiply the C temperature by 9.
3. Divide the amount found in step 2 by 5.
4. Add 32 to the amount found in step 3.

(The result is the equivalent temperature in degrees F.)

Example: If the temperature is 40°C, what is the F temperature?

$40 \times 9 = 360$
$360 \div 5 = 72$

Appendix J. Sample Format for a Bill of Material

Bill of Material

Name ______________________ Date ______________

to be used on ____________ project

Pieces	Description	Feet	Total Feet	Price	Amount

Total ________

Appendix K. Supplementary Technical Information

Safety Color Meanings of OSHA

The Occupational Safety & Health Administration of the U.S. Department of Labor has established a safety color code for marking physical hazards.

Red is the basic color for identifying fire protection equipment and apparatus. Safety cans or other portable containers of flammable liquids with a flashpoint at or below 80°F are to be painted red. Danger signs are to be painted red. Red lights are to be used at barricades and at temporary obstructions. Stop bars and buttons on hazardous machines are to be red.

Yellow is the basic color for designating caution and marking physical hazards. Hazardous situations include striking against, stumbling, falling, tripping and "caught in between." For more information, go to **www.osha.gov/** and search for safety color meanings.

Fire Safety and Extinguishers

Combustion (burning) is a chemical reaction between a fuel and an oxidant accompanied by the production of heat or heat and light. The three conditions necessary for combustion are (1) flammable material, (2) temperature to ignite the material (ignition point or above), and (3) oxygen to support combustion.

Fires are classified into five general categories. Fire extinguishers (devices used to control fires) are also classified into these categories. Different classes of fire extinguishers have geometric symbols on them that help identify the kinds of fires for which the extinguishers are intended. The classes are as follows:

- **Class A**—Fires in this class involve ordinary materials, such as paper, wood, and plastics; a green triangle is the geometric symbol.
- **Class B**—Fires in this class involve flammable or combustible liquids, such as gasoline, kerosene, and oil and grease; a red square is the geometric symbol.
- **Class C**—Fires in this class involve electrical equipment, such as switch boxes, power tools, and appliances; *never use water on a Class C file*; a blue circle is the geometric symbol.
- **Class D**—Fires in this class involve combustible metals, such as magnesium, potassium, and sodium; they usually occur in a chemical lab or similar facility; a yellow star is the geometric symbol.
- **Class K**—Fires in this class are kitchen fires that involve cooking oils and fats; a black hexagon is the geometric symbol.

Drawing Equipment for Scaled Agricultural Engineering Work

Preparation of a scaled agricultural engineering drawing requires certain drawing instruments. Though Computer-Aided Design (CAD) programs are now widely used, hand instruments are still useful with simple drawings. Examples of the equipment needed are as follows:

- drawing board
- T-square
- plastic triangle (30° × 60° × 90°)
- triangular scale
- sharp lead pencil
- eraser
- 12-inch scale (rule)
- compass
- protractor
- paper
- masking tape (for attaching paper to drawing board)

Tools Related to the Animal Science Industry

Tools that may be used in the animal science industry for veterinary medical care include the following: hypodermic needle and syringe, dose syringe, balling gun, hobbles, stethoscope, digital or other thermometer, scalpel blade and handle, otoscope, bandage scissors, forceps, and suture needles.

Woodlot and Forest Management Practices

Caring for the farm woodlot and forest requires certain basic practices. These small areas of wooded land often grow native species. Farm woodlot care may vary somewhat from the care provided on a tree farm because of the size of the operation and the management of the trees. Improvements in a woodlot are typically called timber stand improvement (TSI). Since trees require a number of years to reach harvest size, the practices are carried out over these years. Practices to include in caring for a farm woodlot include the following:

- **Thinning**—This practice is used to remove trees that are overstocked—too many are growing too closely together. Trees removed may have value for firewood and fence posts.

- **Culling**—This practice is cutting or killing trees that potentially have no saleable value as saw timber or pulpwood. Deformed and diseased trees are removed. Undesirable species (known as weed trees) are also removed. The removed trees may have value as fence posts, firewood, and pulpwood.
- **Intermediate cutting**—This practice is the removal of trees from a stand between establishment and final harvest cutting. It typically involves thinning and culling trees.
- **Harvest cutting**—This practice is done after the majority of the trees in the stand have reached harvest size. It is usually done so that the forest regenerates itself.
- **Maintenance**—This practice relates to improvements done in a farm woodlot. It includes managing insects and fire to prevent destruction or loss of the desired tree species.

Careers, Equipment, and Tools in the Forest Industry

The forest industry is an important part of the economic well-being of many local communities in the United States. Many people find jobs in the forest industry. Here are a few occupations: forester, forestry technician, consulting forester, timber harvester, sawmill manager, tree nursery supervisor, log truck driver, and tree planter. The Society of American Foresters lists some 14,000 employers with more than 700 occupations / job categories.

Individuals employed in forest occupations typically use a wide range of equipment and tools associated with the work. The kinds used may vary with the nature of the occupation.

Examples of equipment used in the forest industry are:

- mechanical tree planter—used to plant tree seedlings (place them in the soil).
- feller—large, powerful equipment that cuts trees and removes limbs on large forest harvest operations; may cut logs to length (known as bucking).
- skidder—pulls logs to a central point for loading onto trucks.
- loader (or knuckleboom loader)—lifts logs onto a truck.
- semi-truck—transports logs to sawmill, pulp chipper, or other location.
- chain saw—used to cut trees, particularly in smaller forest operations.
- firebreak tractor—used to plow around wooded areas to prevent spreading of fire.
- four-wheeler—used in forest management to ride unimproved roads and trails.

Examples of tools used in the forest industry are:

- dibble bar—used to hand plant trees on some farms.

- sharpshooter—used to hand plant trees on some farms.
- increment borer—used to obtain small wood samples (bores) for use in aging a tree.
- compass—used to measure direction in forestry land.
- GPS receiver with map—used to navigate in forest areas and cruise wood land.
- spot gun—used to mark trees for cutting or another purpose.
- odometer wheel—used to measure linear distances on forest land.
- clinometer—used to measure height, slope, and vertical angle; measures tree height.
- hypsometer—used to measure tree height.
- logger's tape—used to measure harvested logs.
- wood moisture meter—used to determine percent moisture in wood products.
- log metal detector—used to scan logs and locate metal.
- log tongs—used to hand-maneuver lightweight logs.
- peavy—used to rotate and position logs.

Careers in Wildlife and Environmental Management

Many careers are available in wildlife and environmental management. Some involve entrepreneurship; most involve employment with agencies or private businesses. Examples of occupations include the following:

- game warden—enforces wildlife laws (particularly hunting and fishing regulations), assesses wildlife population conditions, and provides educational programs.
- game farm operator—operates a farm to raise game animals, such as quail.
- fly fishing instructor—provides lessons on fly fishing.
- fishery biologist—studies a wide range of requirements related to fish and other aquatic species.
- park service ranger—oversees park use, provides educational programs, and promotes law enforcement; may collect user fees in some locations.
- naturalist—studies living and nonliving things in a park and provides educational programs; interprets nature to citizens.
- hatchery technician—performs a range of duties in the operation of a fish hatchery, such as collecting eggs, tending incubating eggs, and caring for fry (newly hatched fish).

- water quality technician—collects water samples for analysis and operates equipment to promote water quality.
- air quality technician—collects air samples and otherwise promotes air quality.

Tools and Materials Used in Plant Science Industries

A wide range of tools and materials are used in plant science industries. The kinds of tools and materials vary with the particular area of plant science. Those used in operating greenhouses are far different from those used in producing field crops, orchard crops, and vegetables.

In greenhouse and landscape work, the tools might include garden trowel, shears, leaf rake, shovel or spade, rotary spreader, wheelbarrow, bow saw, pruning snips, and rotary tiller. Materials used include seeds, potting media, fertilizers, and pest management materials (such as insecticides and fungicides).

In field crop, orchard, and vegetable production, the tools might include wrenches and other items to adjust and repair equipment. The use of hand tools in performing production practices is limited because of the high level of mechanization involving tractors, disc harrows, planters, applicators, and harvesters. Materials used include seed and/or seedlings, fertilizer, pesticides, fuel, and specialized materials, such as baling twine and crates.

Agricultural Entrepreneurship Considerations

Entrepreneurship is the owning and managing of a business in an effort to meet a market need and make a profit. The individual who starts and owns a business must carefully plan and manage the business to help assure success. In doing so, the entrepreneur assumes risk (possibility of a loss). The entrepreneur must obtain the necessary finances as well as establish a business that can prosper in the economic climate. Finances may be borrowed from a bank, obtained by selling stock, or acquired by other means, depending on how the business has been established. Of course, the owner receives the profit/reward, if any, for efforts in the entrepreneurship.

Soil Areas in States

Most states cover sufficient geographical area to assure a range of soil areas and types. Alabama, for example, has several major soil areas. The soils in an area were formed from materials with similar characteristics, such as limestone rock and flood plains. Soil surveys have been made to determine and map soils by series. A soil series is composed of soil with similar color, texture, depth of bedrock, and horizons.

Major soil areas or series in Alabama include limestone valleys and uplands, Appalachian plateau, Piedmont plateau, coastal plain, blackland prairie, major flood plains and terraces, and coastal marshes and beaches. More details and a map showing soil series locations can be found at **www.aces.edu/pubs/docs/A/ANR-0340/**.

Use of Crops as Alternative Fuel Sources

Increasing costs and declining supplies of fossil fuels (petroleum products) have resulted in considerable interest in the use of crops and native vegetation as sources of fuel. A great deal of research has been underway in recent years to develop fuels from crop materials. Some research and development has focused on using crop products (corn and soybeans, for example) to produce fuels, such as biodiesel. Other research and development has focused on using plant biomass (plant stalks and leaves, for example) that might otherwise be considered waste to produce fuels.

Here are examples of four crops that may be used as alternative fuel sources:

- **Sugar beet**—The sugar beet has been used to produce ethanol and holds promise as a useful source of this gasoline additive or substitute. With beet production of 20 tons per acre, yields of ethanol approximate 588 gallons per acre.
- **Soybean**—The soybean is a widely cultured crop in the eastern two-thirds of the United States. Oil from the bean is used for a wide range of food and nonfood products. Recent research has dealt with using the oil to make biodiesel. Scientists have found that 1 bushel of beans will yield 1.4 gallons of biodiesel.
- **Corn**—Corn is the single largest U.S. farm crop. Two approaches are being used with corn: grain for ethanol production and fodder biomass for fuel. Though some 1.3 billion bushels of corn were used for ethanol production in a recent year, some scientists feel that ethanol is relatively inefficient as a fuel source and the grain should be used for human food and animal feed.
- **Sunflower**—Some varieties of the sunflower produce seed with higher oil content. An example is the Peredovik sunflower, which produces a small, black seed high in oil and protein. The oil can be used to produce biofuels. Biomass from the plant may also hold fuel potential.

Aquaculture Enterprises

Aquaculture, the culture of aquatic species, was thought to hold considerable potential in the United States. Foreign competition has increased the need for efficiency in U.S. aquaculture. Some aquatic species are cultured in fresh water; others in salt water. Some require warm water environments; others prefer cold water. Decisions about producing aquatic crops must be carefully made and based on water quality and species requirements.

Aquatic species cultured in the southeastern United States include channel catfish, crawfish, shrimp, and tilapia. Catfish are often raised in ponds, where the water quality is carefully maintained. Crawfish are raised in marsh or flood plain areas, where water levels and quality can be managed. Shrimp may be raised in salt water or fresh water, depending on species water requirements. Tilapia is an imported fish that thrives in relatively harsh environments compared to other species.

Northern areas may produce trout, arctic char, salmon, and other species. Some species are in raceways or streams; others in cages or net pens in large lakes. Coastal areas may culture oysters, abalone, lobster, and the like.

Successful production of any species normally requires attention to its native needs. Feed must often be provided. In some cases, the young are artificially spawned. Diseases must be controlled. Quality of water is always critical in the success of an aquaculture enterprise.

Routine Maintenance of Small Internal-Combustion Engines

The service life provided by a small engine often hinges on the routine maintenance given the engine and on how it is operated. Always read and follow the operator's manual that accompanies an engine. The manual will indicate conditions for the performance of routine maintenance duties. Two critical maintenance activities with a small engine are changing the engine oil and cleaning the air filter.

An engine collects grit and other tiny abrasive materials that cause wear as the oil lubricates moving parts in the engine. The proper kind or SAE classification of oil should be used.

Cleaning the air filter allows the engine to take in clean air more efficiently. In some cases, the air filter can be washed, and in other cases, it must be thrown away and a new one installed.

In addition to these service jobs, always monitor an engine, and tighten loose bolts and screws. Keep the exterior of an engine clean.

Water and Sewer Service for Agricultural Structures

Many agricultural activities require a reliable supply of good water. Some farm operations, such as dairying, must have adequate water at all times. A cow will consume 4 to 5 pounds of water for each pound of milk produced. This means that a dairy cow will drink 35 to 50 gallons of water per day. In addition, other water must be used for cleaning and related uses. Water use on a dairy farm may reach 175 gallons per cow per day.

The water source may be a community or municipal system, a drilled well, a spring, or a stream or lake. The source must be clean and of adequate quality and free of contamination and microorganisms.

From the source, the supply system uses a main line, distribution lines, and branch lines. Distribution lines go from one building to another. Branch lines are within a building to get water to faucets and appliances. The size of pipe varies with water demand, with larger pipe needed to carry greater water volume. A 2-inch pipe may be installed to a greenhouse, with ¾- or 1-inch pipe used as branch lines inside the greenhouse. A pressure tank and conditioner may also be a part of the supply system. A greenhouse may include an irrigation line connected to a source of fertilizer or pesticides for application with the water.

Wastewater is handled in lagoons, ponds, septic tanks, or municipal waste systems. How it is handled depends on the contents of the wastewater. Rural residences may have septic tanks with leaching fields that dispose of wastewater. These systems would not be of sufficient size to handle wastewater from animal production facilities.

Always obtain needed approvals and permits for the installation of wastewater systems. The services of a licensed plumber may be required to install such systems.

Economic Impact of Agriculture in the United States

The agricultural industry in the United States has a major influence on the well-being of the nation. This includes work on farms as well as in agribusinesses that support farm production and that process and otherwise market the products of farms.

Providing an abundance of food and fiber meets a fundamental human need. Much of the food and fiber is consumed here in the United States, but considerable amounts of some products are exported. Such exports promote a more favorable balance of trade.

Jobs in agriculture provide employment for a few million people. Farming accounts for 1.5 percent of total employment. Add the nonfarm agricultural jobs, and the food and fiber system accounts for 16.7 percent of all employment. Food and fiber is about 12 percent of the U.S. GDP (gross domestic product). The value of food and fiber in the GDP is about $1.4 trillion. (Information obtained from the National Agricultural Statistics Service, USDA.)

Just as agriculture has a huge economic impact on the United States as a whole, it also has big impacts on the economies of many states. California and Texas are the largest in terms of agricultural impact. Other states have lesser but nevertheless important economic benefits from agriculture. An example is Louisiana. Sugar cane is the leading farm product in Louisiana. Other important crops are rice, soybeans, cotton, and corn. Beef cattle and calves produce 9 percent of the agricultural revenues in Louisiana. Sweet potatoes and tomatoes are the most important vegetable crops. Overall, the top five agricultural products in Louisiana are sugar, rice, cattle and calves, soybeans, and cotton.

Major Wildlife Species

Many areas of the United States have native wildlife adapted to the climate and other conditions that are found there. Some species have unique requirements and, therefore, cannot be relocated and survive. In general, wildlife includes all plants, animals, and other living things that have not been domesticated. Some are game species, which means that they are hunted for sport or food. Most states have agencies that enforce laws about the taking of wildlife by hunters and fishers. Regulations are intended to protect species and promote sustained enjoyment of wildlife. Hunters are well aware of the importance of bag limits. A bag limit is the maximum number of animals that a hunter, fisher, or trapper can take in a day or season. Some species are off limits and cannot be taken, particularly those whose populations are threatened, such as the grizzly bear and the gray wolf.

Game species may be aquatic or terrestrial or migratory. Arkansas is said to be the duck capital of the United States because of the abundance of ducks and other migratory birds that come south in the late fall. Farm land in the delta area becomes covered with water, providing good habitat for overwintering. (More duck stamps are sold in Arkansas than in any other state.)

A wide range of game wildlife is found in Arkansas, for example. Locations depend on the habitat available as required for wildlife to survive. Arkansas wildlife includes the alligator, whitetail deer, elk, brown bear, quail, rabbit, squirrel, and furbearers, including the badger, beaver, bobcat, coyote, gray fox, mink, muskrat, nutria, opossum, raccoon, red fox, river otter, spotted and stripped skunk, turkey, and weasel.

Meat Cuts

Harvested meat animals are prepared into retail cuts that provide portions for individuals, such as portion-size steaks, and for families, such as roasts. Some cuts are prepared for restaurants to meet unique needs in preparation. Beef, pork, lamb, veal, and poultry are often included as species that are fabricated into desired portions and products.

The nature of the meat industry has changed from shipping hanging carcasses to shipping boxes of meat as wholesale or retail cuts. Consumers demand convenience and want cuts to be in appropriate sizes to meet their needs. Skill in making cuts is essential. Making incorrect cuts can result in wasted high-value product. Individuals need the appropriate training and experience in meat fabrication. They need to know the names of cuts, where the cuts are from, and how to make the cuts. The National FFA Organization has a Meats Evaluation and Technology Career Development Event.

Here are two good sources of information on cuts of meat:

- The University of Nebraska maintains a meat identification Web page. You can access it at **http://animalscience.unl.edu/meats/id/**.
- Meals for You maintains an online "Guide to Meat Cuts." You can access it at **www.mealsforyou.com/cgi-bin/customize?meatcutsTOC.html**.

Glossary

A

abiotic factor—All the nonliving parts of an ecosystem.

acceleration—The increasing change in the speed of an object over time.

acetylene—A colorless gas that, with the addition of oxygen, can produce a hot flame for welding and cutting metal.

acid—A compound that gives up protons to water molecules to form hydronium ions.

acquired immunity—An immunity an animal develops by having a disease and developing immunity to the disease.

acre—An area measurement of land used in the United States; 43,560 square feet.

active membership—The FFA membership classification held by students 12 to 21 years of age, enrolled in agriculture class, and paying FFA dues.

acute disease—A disease that afflicts an animal for a short period and may be quite severe.

additive—A substance placed in feed during manufacture to preserve it and enhance animal growth.

ad hoc committee—A committee appointed to carry out a specific duty and dissolved when the duty is completed.

advertising—Any method of communicating information to people that is intended to persuade them to buy a product or service.

aerial stem—A stem that grows above ground.

agricultural drainage—A human-made system for moving excess water from land through open ditches and underground drain tiles or tubes.

agricultural education—Instruction in agriculture and related subjects, and usually considered at the secondary school level; a college major for a person wishing to pursue teaching agriculture as a vocation.

agricultural industry—All the processes involved in getting food, fiber, and shelter to the consumer.

agricultural marketing technology—All the processes in providing products of an agricultural origin to consumers in the forms they want, when they want; the movement of food, fiber and shelter materials from the producer to the consumer—the person who uses them.

agricultural mechanics—The use of machinery, tools, and associated devices and structures to perform agricultural jobs and manage resources.

agricultural services—The skills and knowledge of highly trained individuals who help agricultural producers.

agricultural supplies—The inputs (seed, feed, equipment, fertilizer, etc.) used to grow crops and livestock.

agriculture—The science of growing crops and raising livestock.

agriscience—The use of science principles in producing food, fiber, and shelter materials.

agronomy—The study of plants and how they relate to the soil.

air pressure—The force caused by the weight of the air on an object due to the earth's gravitational attraction.

alkane—A hydrocarbon that has single covalent bonds only.

allele—The different forms of genes.

alternating current (AC)—Electricity that regularly reverses, or alternates, its direction of flow.

alumni membership—An FFA membership category open to former members and others interested in supporting the FFA.

alveoli—Small structures in mammary glands that use nutrients from the blood to make milk.

ampere—A measure of the rate of flow of electrical current.

anatomy—The study of the form, shape, and appearance of an animal.

animal husbandry—The scientific management and control of animals to gain efficient growth and production.

animal science—The area of agriscience dealing with the production of animals for food.

animal unit—An animal measure used in pasture management; one animal unit is equivalent to a 1,000-pound beef cow.

animal well-being—The state of an animal's health and comfort.

Animalia—One of four kingdoms in the classification of organisms in the Domain Eucarya; Animalia organisms have many cells, can move about, and obtain food by ingestion.

annual—A plant species that completes its life cycle in one growing season.

annual growth ring—The result of secondary growth in a plant's lateral meristem.

antibiotic—A germ-killing substance made from bacteria or fungi; used to treat diseases in animals.

antibody—Proteins produced by organisms in response to an invasion of pathogens.

apomixis—The asexual reproduction of plants using unfertilized seed.

aquaculture—The science of water farming.

aquatic—Animals that live in water.

Archaea—In the classification of organisms, this domain is comprised of organisms that are very small, lack nuclei, and have cell walls. Archaea live in harsh environments, such as the bottom of lakes, and some produce methane gas.

area—The measurement of surfaces.

artificial erosion—Erosion caused by people disturbing the natural cover on land.

artificial insemination—The collection of semen from a male and depositing it in the reproductive tract of the female.

asepsis—The condition of being free of disease-causing germs.

aseptic packaging—A method of preservation in which the food and the container (plastic or foil) are sterilized.

asexual reproduction—The duplication or reproduction of a plant by using the vegetative parts of the plant.

assembling—The massing of large quantities of products at a central location for the next market function.

asset—An item of value that is owned.

association—The FFA organization at the state-level; chartered by the National FFA Organization.

atmosphere—The air that surrounds the earth.

atom—The smallest unit in which an element can exist.

atomic mass—The mass number of a chemical element; the total number of protons and neutrons in the element.

atomic number—The number of protons in the nucleus of the atom.

autotroph—An organism that captures energy from the sun and uses it to produce food.

auxin—A plant hormone that controls stem growth and regulates fruit development; causes the spurt of growth that plants have in the spring.

B

Bacteria—A domain in the classification of one-celled organisms that have cell membranes, a cell wall, and cytoplasm but without nuclei; corresponds to the kingdom Eubacteria.

bactericide—A chemical used to control bacteria; germicides.

balance sheet—An itemized statement of the assets and liabilities of a business.

balanced ration—An animal feed that provides all the nutrients in the right amount and proportion for a 24-hour period.

barometer—An instrument used to measure air pressure.

base—A compound that produces hydroxide ions in water.

base line—In land surveying, an east-west (horizontal) line.

basic human needs—The things that support human life worldwide—food, fiber, and shelter.

bathymetry—The measurement of the earth's crust beneath large bodies of water, such as an ocean, it includes the depth of the water.

beef-type cattle—Cattle with higher amounts of muscling in the loin and hind quarters.

beneficial insect—An insect that helps crops grow or provides other important benefits.

biennial—A plant that lives two seasons.

bill of material—An itemized list of the materials needed to construct a structure or other project.

binomial nomenclature—The system of assigning the two-part scientific name.

biodegradable—The term used to describe materials that will break down from the action of bacteria.

biogenesis—The theory that life comes from life.

biological pest management—The use of living organisms to limit pest populations; insects that eat other insects, bacteria and fungi that attack pests, and altering the reproductive processes of the pests.

biology—The study of living things; life science.

biosecurity—The use of approaches to manage risk and assure disease-free animals and other products.

biosphere—The area of Earth that supports life.

biotechnology—The application of science in the development of new products or processes that involve living things.

biotic community—A group of plants and animals that live together.

biotic factor—All the living parts of an ecosystem.

birding—The hobby of watching wild birds.

blanching—A process in which food is put in hot water, or exposed to steam, for a few minutes to loosen the skin and to destroy microorganisms that cause spoilage.

blemish—A defect in a product.

blood—The "liquid" in the circulatory system of animals; 90 percent water.

bloodline—A group of animals within a breed, all members of a bloodline tend to have one common ancestor.

board foot—A board that is 1 inch thick, 1 foot wide, and 1 foot long.

body system disease—A disease that attacks only certain systems of the body.

bonding—The action of joining two or more elements to form a compound.

bone—The hard tissue forming the skeleton of most full grown vertebrate animals.

bony fish—A class of vertebrates; Osteichthyes (trout, tilapia, salmon, and catfish).

botany—The science of plants.

botulism—A form of food poisoning caused by Clostridium botulinum.

boundary—The limit or line of land.

breathing—The process of air entering and leaving the lungs; respiration.

breed—Animals of the same species that have a definite identifying characteristic and a common origin.

breeding—The process of helping animals reproduce.

breeding system—The way animals are selected for mating to gain desired results in offspring.

broiler—A chicken 6 to 12 weeks old and weighing 2½ pounds or more.

browse—Leaves, twigs, and young shoots of trees or shrubs on which animals feed.

budding—A method of propagation in which a bud is taken from one plant and moved to another.

budget—A written statement of anticipated income and expenses for a time period, which is often one year.

bulb—An underground stem with layers similar to leaves; onion, tulip, and jonquil.

C

canning—A preservation method in which food is placed in a container and heated to kill all microorganisms.

capillary water—The thin layers of water that coat soil particles.

carbohydrate—Sugar, starch, and related substances that are formed in plants during photosynthesis.

career—The general direction of person's life in terms of employment; the sequence of jobs that a person holds to make a living.

career development event (CDE)—Competitive activities in the FFA that measure individuals and teams in the application of classroom-acquired knowledge; activities specifically designed to promote career skill acquisition in agricultural education.

career pathway—A grouping of careers based on similarities of duties, subjects, and skills; eight pathways in AFNR.

carrying capacity—The maximum stocking rate (number of animal units) on pasture land that is consistent with maintaining or improving vegetation on the land.

cartilage—A flexible material at the ends of bones; lubricates the joints and cushions shocks and protects bones from damage when they move against each other in a joint; tough, elastic, whitish animal tissue; the skeletons of embryos and young animals are composed largely of cartilage, most of which later turns to bone.

cash flow statement—A listing or report of the flow of cash into and out of a business.

cast iron—A popular, somewhat brittle, metal made of pig iron and containing considerable carbon and some impurities.

castration—Removing the testicles of a male animal.

cell—A tiny building block that has a definite structure.

cell cycle—The sequence of events in a cell from one mitosis to another.

cell division—The process of cells splitting.

cell membrane—A thin, flexible barrier that surrounds a cell and regulates the movement of materials into and out of the cell.

cell specialization—The development of a cell for a specific purpose or function.

cell structure—The general pattern of organization and relationship in a cell.

Celsius scale—The metric system of temperature measurement, in which water boils at 100°C and freezes at 0°C.

ceremony—A formal way to observe an accomplishment or carry out a FFA chapter activity.

change in state—The physical change of a substance from one form to another, such as water to ice.

chapter—The FFA organization at the local level, which is chartered by the state association.

chemical property—The way in which matter changes when it is combined with other matter.

chemical reaction—What occurs when a substance becomes another substance with different characteristics because of a chemical change.

chemical weathering—The alteration of rocks and minerals on the crust of the earth as a result of chemical reactions (interaction with carbon dioxide and oxygen).

chemosynthesis—The use by an organism of chemical energy to produce sugars and starches; performed by several types of bacteria.

chewing insect—An insect that bites off, chews, and swallows parts of plants.

chlorophyll—The green-colored substance in leaves and stems; chlorophyll must be present for photosynthesis to take place.

chromosome—The structure in a cell that carries the material that determines gender (male or female) and characteristics.

chronic disease—A disease that afflicts an animal for a long time.

circuit—The path electricity follows.

circulation—The movement of food nutrients, digested food, and other materials throughout an organism.

circulatory system—The complex blood-pumping system in an animal.

citizenship—Your conduct as a member of the population of the United States.

cladogram—A diagram or graphic depiction that shows the development of structures within clads (groups); it depicts how one organism branched from another over time.

classing—Judging cotton on the basis of length of fiber (staple), thickness of the fiber (micronaire), and grade, which is the presence of trash and color.

climate—The average of all the weather conditions for an area over time, usually a year or more.

climate zone—A large area of the earth's surface with similar temperatures.

clone—An offspring that is a genetic duplicate of the parent.

cloning—The process of asexually reproducing organisms—no union of male and female germ cells occurs.

cloud—A mass of condensed water vapor, droplets, and pieces of ice floating in the air.

colloid—A suspension in which the materials in a liquid can be fairly easily filtered out with parchment or other semipermeable membranes (semipermeable membranes let some substances through, but not others).

colostrum—The first milk after giving birth; it is high in antibodies and other substances that help the new animal survive.

committee—A small group or subgroup of a larger body charged with certain duties.

commodity organization—An association or group of shared interests developed around the production, marketing, or processing of a commodity or similar commodities.

common name—The plant and animal name used by non-scientists.

communication—The process of exchanging information.

companion animal—An animal that is kept as a pet.

compass—An instrument that is used to make direction measurements.

complete flower—A flower with all four principal parts: sepals, petals, stamens, and pistil.

composting—The promotion of the decomposition of organic matter.

compound—A substance made of two or more elements that have chemically combined.

compound leaf—A leaf that is divided into two or more leaflets; clover, rose, and locust tree.

computational science—The use of computers to solve problems in mathematics and science.

computer—A machine or electronic device that manipulates data; software provides the instructions.

computer simulation—The use of computer methods to depict or represent a situation.

concentrate—A feed low in fiber (less than 18 percent) and high in energy.

conductor—A substance that allows electrons to flow freely; copper, gold, and aluminum are good conductors of electricity.

cone—A cylinder that tapers to a point at one end.

conformation—The proportion of an animal's body features in relation to one another and to the whole body of an ideal of the breed or species.

consumer—A person who buys goods or services.

contact poison—A poison that must come into contact with the insect to be effective.

contagious disease—A disease spread by direct or indirect contact.

continental crust—The part of the earth's crust that is land: inorganic materials as well as materials from plants and animals.

contour plowing—The practice of plowing around or across a slope rather than "with" the slope.

controlled breeding—The selective act of mating plants or animals to achieve desired traits in the offspring.

convenience preparation—The processing activity in which food is made easy for consumers to use.

core—The center of the earth.

corm—An underground stem that is similar to a bulb, but it has thinner leaves and a thicker stem; taro, gladiolus, and garlic.

corrosion—A chemical reaction that destroys and weakens property; rusting, when a change in metal occurs through oxidation and reduction.

cottonseed meal—The product remaining after the extraction of oil from cottonseed; a high-protein animal feed concentrate.

country-of-origin labeling (COOL)—Label information required on products imported to the United States.

cow and calf production—A production system in which cows are kept to produce calves that will primarily be used for meat.

cracking and rolling—A method used to break the hard outer coatings on grain.

crating—The placing of cans, jars, etc., in larger containers or boxes for shipping.

crop rotation—The practice of alternately planting different crops that use different soil nutrients on the same land.

crossbreeding—The mating of animals of different breeds, but the same species.

crust—The outer layer of Earth's surface.

crustacean—Mostly aquatic animals with exoskeletons; arthropods, including shrimps, crabs, barnacles, crawfish, and lobsters, that usually live in the water and breathe through gills.

cultivar—A crop variety that is cultivated and retains its features when reproduced.

cultural pest management—A technique to keep pest populations at a low level in a crop; examples include rotating crops, roguing, trap cropping, burning, using resistant crop varieties, cleaning around fields, and cutting stalks.

cultural practice—An activity or treatment needed or used with a crop to gain the desired growth and product.

curing—A food preservation process in which substances are added to foods to prevent spoilage.

current electricity—The electricity caused by flowing electrons; the most important kind of electricity.

customary measurement system—The measurement system used in every day life in North America; English system: inch, foot, pound, pint, and other measures.

custom harvesting—A harvesting service offered by one person who owns and operates equipment to harvest the crops of other producers.

custom order marketing—A process in which buyers (processing plants and others) contract with producers to provide a specific crop or animal.

customs—Established ways of doing things.

cutting—A short section of plant stem used for propagation.

cytoplasm—The thick, semi-fluid material inside a cell but outside the nucleus.

D

dairy-type cattle—Cattle with the capacity to produce a large amount of milk.

debeaking—The removal of the tip of the beak to keep birds from attacking and pulling feathers out of each other.

deceleration—A decrease or slow down in the speed of an object over time.

decomposer—An organism that breaks down organic matter.

demand—A desire for a good or service and the ability to buy.

depreciation—An accounting process that spreads the cost of assets over the useful life of the asset.

detrivore—An organism that feeds on nonliving plant and animal remains or dead matter.

dew—A form of liquid precipitation that collects on outside surfaces at night when the weather conditions are conducive.

dew point—The temperature at which dew will begin to form.

diagnostic ultrasound—A noninvasive way of imaging soft tissues of the body; ultrasonography.

dicot—A plant with two seed leaves; most fruits, flowers, trees, and vegetables.

diet—The type and amount of feed and water an animal eats.

diffusion—The movement of a substance from an area of greater concentration to an area of lesser concentration.

digestible nutrient—The part of a feedstuff that can be digested, or broken down.

digestion—The process that changes food into simpler forms that can be absorbed.

digital safety—Using computer applications so that risk is reduced and safety is promoted.

diploid cell—A cell with all of the chromosomes characteristic of a species.

direct current (DC)—Electricity that always flows in the same direction.

disease—A condition of pain, injury, or the inability to function normally.

disinfection—The act of removing or lowering the germ population; the area is not sterile, but it has a low germ population.

distribution—The process of getting a product to the right place at the right time.

division—The duplication process of one cell splitting into two cells.

DNA (deoxyribonucleic acid)—The molecule in a cell that codes the genetic information of all living things; unique to a species and an individual.

DNA isolation—The process of extracting and separating DNA from other materials in a cell.

DNA profiling—The process used to identify individuals of a species and within a species based on DNA profiles; DNA fingerprinting.

domain—The broadest groups into which living organisms are classified; three domains are used: Bacteria, Archaea, and Eukarya.

domestication—Taming, or controlling, wild plant and animal species and producing them for specific purposes.

dominant trait—A trait caused by alleles that cover up or mask a recessive trait.

donor—A cow that provides ova.

donor cell—A cell that provides the new DNA material.

double helix—The structure that DNA forms.

draft—Resistance to movement, such as the resistance a plow has in the soil.

draft horse—A horse weighing upwards of 1,400 pounds that was developed to pull heavy loads (wagons, plows, or logs).

drawbar power—The power of a tractor in pulling a load attached to a drawbar.

drill—Seeding equipment that provides solid planting—seeds are not placed in rows.

dry fruit—Fruit that develops as a pod or in a hull.

drying—The removal of moisture from foods.

E

earth science—The study of the environment in which plants and animals grow.

economic system—The way goods are created, owned, and exchanged; varies among countries.

ecosystem—All the parts of a particular environment.

ectotherm—An animal whose body temperature adjusts to its environment.

effluent—Water that is discharged after use; discharged water from factories or farms.

electricity—The flow of electrons, which are small units of negative electrical charge.

element—A substance that cannot be broken down into simpler materials by ordinary means.

e-marketing—Marketing using the Internet.

embryo—An immature plant or animal that will grow if properly cared for.

embryo splitting—The asexual reproduction process in which an embryo is removed from a cow seven days after conception, cut in half, and each half is placed in the uterus of a recipient.

embryo transfer—The removal of an embryo from its mother (donor) and the placement of the embryo in a recipient to develop.

employer expectations—Those behaviors desired as employers seek and reward their employees; attributes employers want to see in their employees.

emulsion—A colloidal suspension made of two solutions.

endosperm—The tissue where food is stored for the embryo to use when the seed sprouts; the part of the grain that provides much of the nutrition in food and feed made from corn.

endotherm—An animal that maintains a certain body temperature.

energy—The power needed for body activity—both internal and external; in physics, the ability to do work.

energy level—An area (formerly orbital) in a chemical element in which electrons move varying distances.

entomology—The branch of zoology that deals with insects and related small animals.

entrepreneurship—The owning and managing of a business in an effort to meet a market need and make a profit; taking risk by owning an enterprise.

environment—All the factors that affect a living thing.

environmental disease—A disease caused by elements in the environment that are not appropriate and do not promote good health.

environmental science—The study of using and protecting natural resources.

equinox—The time when the sun is directly over the equator and day and night are of equal length.

erosion—The washing or wearing away of soil.

estrous cycle—The cycle in the female reproductive system that prepares it for reproduction.

estrus synchronization—The use of hormones to get several females to come in heat at the same time.

Eukarya—A classification domain for organisms that have cells with a nucleus; includes four kingdoms: Protista, Fungi, Plantae, and Animalia.

eukaryote—An organism whose cells have a nucleus.

euthanasia—The act of killing an animal to relieve it of suffering.

exoskeleton—A hard outer shell or covering that protects and supports the body of an arthropod..

experiment—A trial or test; used to discover something unknown or verify a procedure.

exploratory supervised experience—The foundation for other SE; an initial type of supervised experience that allows students to investigate agriculture opportunities and interests on a first-hand basis.

export—A good sold to another country.

extemporaneous speech—A speech that is not prepared in detail in advance, but notes may be used.

external parasite—A parasite that lives on the outside parts of an animal (see parasite).

external respiration—The exchange of gases in the lungs between the blood and the atmosphere.

eye contact—Looking at and into the faces of the audience or listener.

F

fabrication—The activity or activities of making a product into a form that can be used; constructing with material; cutting meat.

Fahrenheit scale—The customary measurement of temperature, in which water boils at 212°F and freezes at 32°F.

fair trade—a practice that assures that producers receive appropriate compensation for their labor.

fastener—A device used to hold two or more parts together.

fat—Concentrated source of energy; what excess energy is changed to in the body.

feed—A product containing nutrients; to provide as food something necessary for growth, development or existence.

feeder pig—A pig sold at a weight of 30 to 60 pounds to other producers to grow to a meat-size hog.

feedlot—A confinement finishing system whose aim is to get animals to grow and put on body fat.

feedstuff—A feed ingredient.

fermenting—A method of preservation that uses the controlled activity of certain bacteria, molds, and yeasts.

fertilization—The union of sperm (male sex cell) and ovum (female sex cell).

fertilizer—Any substance used to provide plant nutrients.

fertilizer analysis—A list of the amounts of nutrients in a fertilizer.

fetus—The unborn animal in the second half of the pregnancy.

FFA advisor—An agriculture teacher who oversees an FFA chapter in a school.

fiber—The material used to make clothing and shelter; what remains of feed after the sugar, protein, and other substances are removed.

fibrous root system—A root system in which many small roots spread out through the soil.

first aid—Providing initial care to an individual who has an injury or is ill.

fishing—The harvesting or taking of fish and other aquatic species form commercial and sport purposes.

fleece—The wool on a sheep.

fleshy fruit—Fruit that has large fibrous structures surrounding the seed.

floriculture—The production and use of flowers and plants with attractive foliage.

flower—The reproductive part of a flowering plant.

flushing—The use of a saline solution (water with some salt) in the uterus of the cow to remove developing embryos.

food—The solid and liquid material humans and other living things consume.

food chain—The ranking of species into successive levels, where each feeds on the one below.

food processing—All the activities to prepare food for human consumption.

food safety—The practices and procedures for keeping food safe to eat and preventing accidental contamination that would make it unsafe.

food security—The practices and procedures for keeping food safe by preventing the deliberate contamination of food with intent to cause harm or disruption.

food web—The feeding relationships among various organisms in an ecosystem that forms a network of interactions.

forage—Grasses, forbs, and other plants used as pasture or feeding areas for animals.

force—The push or pull on an object.

forestry—The science of growing and using trees.

free access—A feeding system that provides animals with access to the feed when they want it.

free enterprise—An economic system that is a modification of capitalism; used in the United States.

freezing—A preservation method that uses low temperatures to freeze the water in food products.

front—The boundary between two atmospheric air masses.

fruit—A mature ovary of a flowering plant.

fuel—Any material that provides energy, may be in a solid, liquid, or gas form.

fumigant—A pesticide in gas form that enters the respiratory system of organisms.

fungi—The kingdom that includes yeasts, mildews, and mushrooms.

fungicide—A chemical used to control disease caused by fungi.

G

galaxy—A system of stars, dust, gases, and other matter held together by gravitational forces.

game management—Practices used to protect and improve wildlife.

gender selection—Choosing the gender of offspring before birth or hatching.

gene—The specific determiner of heredity.

gene splicing—transferring genes from one plant, animal, or other species to another organism.

genetically modified organism (GMO)—An organism whose genetic material has been artificially altered by gene transfer; a transgenic crop.

genetic code—The specific order or arrangement of genes within the DNA.

genetic engineering—An advanced form of biotechnology.

genetic manipulation—The use of artificial ways of producing desirable traits in plants and animals.

genetic pest management—The use of genetics in the development of crops that are resistant to pests.

genetics—The study of heredity in plants and animals.

gene transfer—The act of moving a gene from one organism to another.

genome—The total genetic makeup of an animal.

genotype—The genes for a trait, represented by a combination of letters.

geology—The study of Earth's composition, structure, and history.

geotropism—The downward growth of roots; a response to gravity.

germination—The sprouting of a seed.

gestation—The time a female mammal carries a developing embryo in her uterus; period between fertilization and full development of the fetus.

gibberellic acid—A plant hormone that induces stem cell elongation and cell division; affects stem and leaf growth, fruit development, flowering, cell division, and other plant activities.

ginning—The first steps in processing cotton, which include the removal of seeds and trash, drying, and packaging.

glucose—Sugar made by photosynthesis in plants.

glycolysis—The process in which glucose is broken down to release energy.

gonadotropin—An ovulation-inducing hormone that is injected into a cow, which causes the cow to release 8 to 20 eggs during estrus (see superovulation).

good seed—Seed that will germinate, grow, and become the intended plant.

grading—The sorting of products for uniformity.

grafting—A method of propagation in which a section of a stem of one plant (scion) is placed onto another plant.

grain elevator—A marketing facility that receives, grades, dries, stores and sells grain to mills.

gravitational water—The water that fills cracks and air spaces between soil particles.

green lumber—Wood products, especially lumber, made from freshly sawed logs that contain sap and have not dried.

greenhouse—A specially-constructed enclosed building where plants can be cultured; usually constructed with a wood or metal frame covered with glass or plastic; climate-control equipment is used to gain the desired environment for plant growth.

grinding—The process that makes the particle size of feed smaller.

growing season—The period of time after the last frost in the spring and before the first frost in the fall; measured as the number of frost-free days.

growth—The process by which an organism increases in size by adding cells or by the cells getting larger.

H

habitat—The area where a plant or animal lives under natural conditions.

HACCP—Hazard Analysis Critical Control Points; a procedure used in a processing plant to identify potential points at which food may be contamination and taking steps to reduce the hazard.

haploid cell—A cell produced by meiosis; a cell with single chromosomes rather than homologous pairs of chromosomes.

hardpan—A layer of the B horizon, or subsoil, that is hard and compacted and restricts water percolation and plant root growth.

harvesting—The reaping, picking, or otherwise gathering of a product that is ready for marketing or other use; the first step after production in getting what has been produced ready for the consumer.

hay—The leaves and stems of plants that have been cut and dried for feed.

head squeeze—A mechanical device at the end of a chute that closes around an animal's neck; used in restraint.

health—The condition of the body and a measure of how well the functions of life are being performed.

heartwood—The innermost xylem of a tree.

heat—A form of energy that results from the motion of atoms and molecules.

heat engine—A type of engine that burns fuel (gasoline, diesel, etc.) to produce hot gases; an internal combustion engine.

herbaceous stem—A nonwoody stem that is soft and green stem and contains only a small amount of xylem.

herbicide—A chemical used to control weeds.

herbivore—An organism that receives its energy by eating only plants.

heredity—The passing on of the traits of parents to offspring.

heterotroph—An organism that consumes other organisms for its energy and food supply.

heterozygous—Different in structure; when the alleles are different in an organism.

homologous chromosome—The identical pairs of chromosomes in the nucleus of a diploid cell.

homozygous—The same in structure; when the alleles are the same in an organism.

horizon—A distinct layer of soil materials; one of the four layers of materials from the soil surface down several feet.

horsepower—The amount of power used to do 550 foot-pounds of work per second; 746 watts in SI; a way of stating power in engines and other power equipment.

horticulture—The science of growing plants for food, comfort, and beauty.

human need—An essential element or component that supports humans, with food, clothing, and shelter being most obvious.

humidity—The amount of moisture in the air.

humus—Well-decomposed organic matter.

hunting—The harvesting or taking of wildlife as game.

hydraulic power—Compression power that uses the force of a liquid or gas placed under pressure to do work inside a cylinder.

hydrocarbon—A simple organic compound made of two elements, hydrogen and carbon, which are commonly used for fuel.

hygroscopic water—A thin layer of water that sticks to soil particles and does not move about.

hydrosphere—All the water in all its forms on Earth.

I

immunity—The condition of being resistant to a disease.

imperfect flower—A flower that lacks either stamen or pistil; some imperfect flowers have the male sex organs; other imperfect flowers are female.

implant—A solid material placed under the skin to release substances over time.

import—a good brought into a country from another country.

impromptu speech—A type of speech that involves no specific preparation in advance.

inbreeding—The mating of closely related animals of the same breed.

incidental motion—A parliamentary procedure motion used to assure proper and fair treatment of all members.

inclined plane—A straight, sloping surface; a ramp.

incomplete flower—A flower that does not have all four principal parts; some flowers lack sepals and petals, such as the flowers on wheat and oats.

infectious disease—A disease caused by organisms that get inside the body.

infiltration—The action of water soaking into the soil after precipitation or irrigation.

information—Knowledge or news about an event, procedure, or other subject acquired in any manner.

infrastructure—All the facilities, services, roads, and other things that support and make marketing possible.

injection—The act of using a hypodermic needle and syringe to get a substance into the body system.

insect pest—A small, six-legged animal that damages plants in some way; characterized in the adult stage by division of the body into head, thorax, and abdomen with three pair of legs.

insecticide—A chemical that controls insects.

inspection—An examination of a product to assure that it is free of disease or other defects.

insulator—A material that does not conduct electricity; rubber, plastic, and glass are commonly used as insulation.

integrated pest management (IPM)—A planned process for limiting pests; the use of crop protection methods that have the best outcomes for the well-being of society.

internal drainage—The movement of water through the soil.

internal parasite—A parasite that lives inside an animal (see parasite).

internal respiration—The exchange of gases between the cells and the blood within the body.

international trade—The buying and selling of goods by two or more nations.

inventory—An itemized statement of the current assets of a business; something owned that has value.

invertebrate—An animal without a backbone.

in vitro—In glass (usually a test tube).

involuntary muscle—A muscle automatically controlled by a lower part of the brain; muscles that operate the heart, intestines, lungs, and other organs.

ion—An electrically charged atom.

irradiation—The preservation method of treating food with electrically charged particles, such as X-rays, electron beams, or gamma rays.

irrigation—The artificial application of water to meet plant growth needs.

irrigation-induced salinity—Salt in the soil saltiness from irrigation water.

isolation—The act of separating diseased and non-diseased animals.

isotope—An atom of the same element that has a different atomic mass.

K

kinetic energy—A result of an object's mass and motion.

kingdom—Within the Domain Eukarya, the broadest group into which organisms are classified; four kingdoms are used: Protista, Fungi, Plantae, and Animalia.

L

label—Printed information that provides detail about a product .

labeling—A step in packaging used to identify the product.

laboratory analysis—The process of scientifically observing samples or fluids or tissues for abnormalities.

lactation—The production of milk by female mammals.

lamb—The meat of a young sheep; a young sheep.

land capability classes—The classification of agricultural land on the basis of its productive capability.

land description—A written statement of the boundaries of land and its location in relation to other land.

land surveying—The process of measuring and marking real property (land).

landfill—A large earthen pit for waste disposal.

land forming—The group of activities in which the surface of the land is shaped so it is more productive.

landscaping—The use of plants and other materials to improve and create aesthetic qualities in the outdoor environment.

laser—A device that produces a special kind of light known as coherent light; electric eyes.

Law of Conservation of Mass—Matter can be changed from one form to another but it cannot be created or destroyed; matter is found in three states: solid, liquid and gas and can be changed from one form to another, but it cannot be created or destroyed.

Law of Definite Composition—All compounds have a definite composition by mass.

layering—A method of propagating in which roots grow from the nodes on the stem of a plant, while the stem is still attached to the parent plant.

leadership—A relationship among people in which influence is used to meet individual or group goals.

leadership development event (LDE)—Competitive activities that measure the personal skills of an FFA member, including public speaking, communicating, and relating to other people.

leaf—An organ on a plant stem that is usually green.

legume—A plant that stores nitrogen from the air in nodules that grow on its roots.

lever—A rod or bar that rotates on a fulcrum, or pivot point.

liability—An item with an obligation, such as a loan against a vehicle that is to be repaid.

life cycle—The sequence of changes the plant or animal goes through from its inception to reproductive maturity, includes the beginning, growth, and maturity stages of the life span.

life science—The study of living things; biology.

life span—The entire length of life.

ligation—The act of uniting or attaching two DNA fragments.

light horse—A horse weighing 900 to 1,400 pounds and is used for riding, driving, racing, and other purposes.

linear measurement—The distance between two points; length.

lint cotton—Cotton fiber without the seed.

liquid fertilizer—Fertilizer that is dissolved in water and applied by spraying on the soil.

living condition—The total of all the life processes.

loam—Soil that is nearly equal amounts of sand, silt, and clay.

locomotion—The ability to move from one place to another; movement.

lodge—When a plant falls over.

loess—The wind-blown parent material of soil.

log—The large main stem of a tree that required many years to grow; used for making lumber and other wood products.

lumber—Pieces of wood made by sawing logs, with cuts made to take advantage of the natural strength and attractiveness of wood.

lymphocyte—A type of white blood cell in the vertebral immune system.

M

machine—A device that helps transmit power, force, and motion.

macromineral—A mineral in larger amounts; major mineral.

main motion—The parliamentary procedure motion that brings business before an assembly (group).

maintenance—The level of nutrition needed to keep an animal from losing weight when not producing milk or other products.

major elements—Chemical nutrients that plants need in large amounts; macronutrients.

mammal—A class of warm-blooded, usually hairy vertebrates whose offspring are fed with milk secreted by the female mammary glands and reproduces by the mating of the male and female of the same species.

mammary gland—Milk-producing female gland by which the newborn of mammals are fed.

management—The use of resources to achieve the objectives of an enterprise; human effort in running an enterprise or business.

mantle—The layer between the earth's core and crust of the earth.

map—A drawing, photograph, or other representation of the features of Earth's crust.

marketing function—An activity or process that converts a product or performs a role in making products readily available for consumers.

market hog—A 240 to 260 pound hog ready for slaughter.

marketing—The part of the agricultural industry that moves crops and livestock from the producer to the consumer.

marketing infrastructure—Those things that support and make marketing possible.

mass—The amount of matter that an object contains; properties of matter determine mass.

mathematics—The science of numbers.

matter—Anything that has volume and mass.

meal—Ground feed.

measurement—A method of determining the number of units in something.

mechanical advantage—The way in which machines multiply the force that they receive.

mechanical pest management—The use of tools or equipment to remove or destroy a pest; plowing, mowing, mulching.

mechanical technology—The use of machines and equipment to do work.

mechanical weathering—The breaking of larger rocks into smaller ones.

mechanics—A division of physics that deals with forces exerted by objects when at rest or moving.

meeting—The assembly of a group of people for a particular purpose.

meiosis—Cell division in the sexual reproduction of organisms—both plants and animals; reduction division.

membership degrees—Levels in the active membership category based on individual achievement in the FFA.

meridian—In surveying, a true north-south line.

metal—An element with a brilliant appearance, known as metallic luster or shine.

metamorphosis—The series of changes from one form to another that an insect goes through in its life cycle.

metes and bounds—A system of describing land in which a known starting point is used to establish and indicate lines forming boundaries of property.

metric system—A measurement system based on a decimal system that increases or decreases numbers by 10s; International system of Units (SI)—grams, meters, liters, etc.

micromineral—A mineral needed in a smaller amounts; trace mineral.

mild steel—A relatively soft steel, it is made of pig iron and contains low carbon; popular with agricultural structures and machinery.

mineral—A chemical element or compound needed for maintenance, growth, reproduction, and other body functions.

mineral soil—Soil high in mineral content; soil with 6 to 12 percent organic matter.

minimum tillage—The cropping practice in which seed is planted with very little plowing.

minutes—The official written record of the business transacted in a meeting.

mission—A goal or task to be achieved.

mitosis—Cell division for growth and repair.

mixed feed—A feed made from a variety of ingredients.

mixture—A substance that has parts with different properties.

molecular biotechnology—The branch of biotechnology that deals with changing the structure and parts of cells; genetic engineering.

molecule—The smallest group of atoms that acts together to form a stable, independent substance.

mollusk—A number of invertebrate animals with soft bodies, most have hard shells for protection; hard shells: oysters, snails, clams, mussels, and abalone; those with no shell: slugs and octopuses.

monocot—A plant with one seed leaf; onions, corn, wheat, oats, and sorghum.

morphology—The form, structure, and configuration of an organism.

motion—The process of moving or changing position; a part of work and power.

motivational speech—A type of speech that is intended to arouse people and encourage them to take action.

MSDS (material safety data sheets)—Detailed information that accompanies chemical products; includes safe use and steps to take in case of an accident.

mulch—A cover on soil to hold the moisture; keeps down the growth of weeds. Common mulches are straw, tree bark and leaves, sheets of plastic, sawdust, and paper.

mulching—The practice of covering the soil with a layer of protective material.

multicellular organism—An organism composed of many cells; the cells are organized to form other structures, known as tissues, organs, and organ systems.

mutant—An offspring that has a trait genetically different from its parents.

mutation—A change that occurs naturally in the genetic material of an organism.

mutton—The meat of older sheep.

N

National FFA Advisor—The head of agricultural education in the U.S. Department of Education and who, by virtue of position, is chief of the National FFA Organization.

National FFA Center—The facility in Indianapolis that serves as the center of operations for the National FFA Organization.

National FFA Foundation—The fund-raising arm of the National FFA Organization.

natural immunity—An immunity that an animal develops without a vaccination.

natural insemination—The ejaculation of semen by the male in the vagina of the female during copulation.

natural resources—The things found in nature that support life, provide fuel, or are used in other ways by humans.

needle teeth—The eight sharp teeth of a newborn pig.

nematocide—A chemical used to control nematodes.

nematode—A tiny wormlike organism that lives in the soil and attacks the roots and stems of plants; sometimes, it causes knots on the roots of plants..

niche—The function or role of a living thing within its habitat.

niche market—A market that meets a unique need for a particular product.

nitrogen—The most important nutrient element in the growth of plants.

noncontagious disease—A disease caused by conditions or substances that are not transferred from one animal to another.

nonmetal—An element that is a poor conductor of heat, brittle, and not easily shaped.

nonpoint source pollution—Pollution from many sources that cannot be readily identified, such as all the engines that release exhaust.

nonrenewable natural resources—Resources that cannot be replaced once they are used.

nonruminant—An animal that does not chew a cud; its stomach has one compartment.

nonselective herbicide—A herbicide that will kill all plants upon contact; contact herbicide.

nonverbal communication—The exchange of information without spoken words or written symbols; communication through facial expressions, posture, and related means.

no-till cropping—The practice of growing crops without plowing the soil.

nuclear envelope—The double membrane layer that surrounds the nucleus.

nucleic acid—The substance in cells that directs all cellular structures and activities.

nucleolus—The dense region in the nucleus where the assembly of ribosomes begins.

nucleus—The storehouse of genetic information; central part of a cell that controls cell activity.

nutrient—The substances necessary for an organism to live and grow; anything nutritious for health and growth; chemical substances that support life processes.

O

Official FFA Manual—A book produced annually to provide official information about the FFA and its programs.

olericulture—The science of growing vegetables, such as snap beans and tomatoes.

olestra—A fat substitute; a synthetic nondigestible cooking oil; trade name is Olean.

omnivore—An organism (animal) that feeds on both plant and animal materials.

optics—The study of the use of light.

order of business—A list of what is to be accomplished in a meeting.

organ—A collection of tissues that work together to perform a specific function.

organic chemistry—The study of carbon.

organic matter—Rotting plants and animals.

organic soil—Soil high in organic matter; soil with more than 20 percent organic matter.

organism—Any living thing.

organismic biotechnology—The branch of biotechnology that deals with intact or complete organisms.

organ system—Several organs that work together to perform an activity.

ornamental horticulture—The use of plants and other materials for their decorative purposes.

osmosis—The movement of liquid from a greater concentration to a lesser concentration.

outcrossing—The mating of animals of the same breed, but of different families in the breed.

ovule—small egg; structure in plants that becomes a seed when fertilized.

ovum—The mature female sex cell; egg.

owner equity—The sum or amount remaining after liabilities have been subtracted from the assets.

ownership supervised experience—A type of supervised experience in which a student creates and/or owns an enterprise or business, such as a beef steer or greenhouse, which are operated on a financially sound basis.

oxidation—The chemical process between oxygen and other materials, such as when an atom of a metal gives up electrons; a process in animal respiration that produces chemical energy from digested food.

P

packaging—The marketing function in which products are placed in containers.

packing shed—A market outlet that receives, sorts, and combines produce from a number of different farms.

palatability—The way a feed feels in the mouth and tastes.

parasite—A multicelled animal organism that live in or on other animals.

parent material—The material from which soil develops.

parliamentary procedure—A method of conducting meetings in an orderly manner that gives all members the opportunity to participate in debating issues and making decisions.

particulate—The small, solid particles in the air.

parts per unit—For a given number of units, another set of units is added.

parturition—The process of giving birth.

PAS (National Postsecondary Agricultural Student Organization)—The student organization for those enrolled in agricultural education beyond the high school level; typically in the postsecondary or two-year college level.

pasture—Land where grasses and other plants grow for animals to graze.

pathogen—A microorganism that gets into the body and causes disease.

patriotism—Admiration for and loyalty to one's country.

pedigree—The names of the ancestors of an animal.

pellet—Feed pieces shaped by forcing ground material through holes in a heavy steel plate.

percent germination—The percentage of seeds that will sprout and grow.

percolation—The downward pull of gravity on water after infiltration.

perennial—A plant that lives for more than two seasons.

perfect flower—A flower that has the stamen and pistil in the same flower.

Periodic Law—The chemical and physical properties of the elements are periodic functions of their atomic numbers.

Periodic Table of the Elements—A method used to group and organize the elements for study.

permanent pasture—Land planted with grasses and legumes that will live and grow for years.

personal appearance—What other people see when they look at you.

pesticide—A chemical material used to control pests.

pH—a measure of the acidity or basicity of a substance; a scale of 0-14 is used to report pH.

phagocyte—Cells that surround and destroy disease-causing microorganisms.

phase change—The change of matter from one state to another.

phenotype—The appearance of an organism because of its genotype.

phloem—Tissue that moves food from where it is manufactured (usually in the leaves) to other parts of a plant.

phosphorus—A nutrient that plants need to store and transfer energy and to grow.

photoperiod—The length of the light period in a day.

photosynthesis—The process plants use to make food.

phototropism—The growth or turning of a plant in the direction of light.

physical property—A property that can be determined without changing the identity or composition of matter; color, odor, solubility (how much dissolves), melting and boiling points, hardness, density, and crystal formation.

physical science—The study of nonliving matter around us.

physics—The study of matter and energy and how the two relate to each other.

physiology—The study of the functions of the cells, tissue, organs and systems of the body; also the relationships of the systems to each other.

phytohormone—A naturally occurring hormone in plants that activates or regulates plant growth and development; also known as a plant hormone.

phytoplankton—Forms of algae that produces oxygen in water; used as food by some aquatic organisms.

pickling—A method of preservation that prevents the growth of spoilage organisms by placing foods in acid solutions.

placement supervised experience—The type of supervised experience in which the student has a planned program that involves working in a job for another individual or company for compensation.

placenta—A baglike structure in the uterus of the female that nourishes and protects the developing animal.

plan—In construction, an outline or drawing that describes the intended outcome.

planing—The smoothing and sizing of lumber.

planning—The process of identifying what is to be accomplished and how it is to be carried out.

plant disease—An abnormal condition in a living plant.

plant health—The condition of a plant related to growth and productivity; the condition of being free from disease.

planter—Seeding equipment that places seed in the soil with accuracy of rate, depth, and spacing.

plant hormones—A naturally occurring hormone that activates or regulates plant growth and development.

plant pest—Anything that causes injury or loss to a plant.

plant tissue culture—The propagation of plants using single cells or smaller groups of cells.

Plantae—One of four kingdoms in the Domain Eukarya; comprised of organisms that make their own food, cannot move about, and have cell walls containing cellulose.

planting depth—The depth at which, or how deeply, seeds are planted.

plasma—The liquid part of blood, which is 92 percent water; in chemistry, the high-temperature gaseous state of matter in which atoms lose most of their electrons.

platelet—Tiny cell fragment that is essential for blood to clot.

plywood—A type of wood material that is made by slicing logs into thin sheets and gluing them together so the wood grain is perpendicular on adjoining sheets; most plywood is in 4 × 8 ft. sheets.

pneumatic power—The use of compressed air to do work.

point source pollution—Pollution from a readily identifiable source, such as a factory.

polar climate zone—Those areas on earth where the average temperature for a year is below 50°F (10°C).

polarity—The condition of substances having different positive and negative charges.

polled—Naturally without horns.

pollen—The powdery substance that contains sperm, the male sex cell in plants; produced by the anther.

pollination—The transfer of pollen from an anther to a stigma of a flower of the same species.

pollution—The result of substances that spoil our environment; the presence of pollutants.

polymerase chain reaction (PCR)—The process of making copies of selected DNA molecules.

pomology—The science of growing fruits and nuts.

pony—A small horse weighing less than 900 pounds.

portioning—The division and packing of a product into an amount or size that consumers want.

potassium—A nutrient needed by plants for photosynthesis, moving sugar, and other functions.

poultry science—The study of birds used for food.

power—The rate at which work is done.

power take-off—A rotating shaft powered by the engine of a tractor.

PPE (personal protective equipment)—Devices that are used to protect the human body from injury, such as eye, hearing, and skin protection.

precipitation—Any form of water that falls to the ground.

precision farming—The use of cropping practices that improve yields based on the needs of the land.

preconditioning—The process of preparing an animal for stress; performing routine management well ahead of the time of stress.

pregnancy testing—One of several procedures used to determine if females are pregnant.

pregnant—The condition when a female has a developing animal in her reproductive tract.

prepared speech—The type of speech that is developed in detail and its delivery is practiced well ahead of time.

presentation—The position of the fetus during birth.

preserving—The process of treating food to keep it from spoiling; the many different procedures to protect food from decay.

presider—The person who runs or chairs a meeting.

pressure—The force on a solid or liquid.

primary defender—The skin and mucous membranes of the body; main defense of an organism against disease.

primary growth—The increase in the length of a plant; the plant is taller; linear growth.

primary root—The first root to grow from the seed.

primary succession—The development of a biotic community where none existed.

primary tillage—The cutting and shattering of the soil to prepare it for growing a crop.

principle of supply and demand—The price of a good is related to the interaction on a graph of the supply of the good and the demand for it.

prion—A tiny protein-type particle that is associated with certain diseases, such as mad cow.

privileged motion—Type of parliamentary procedure motion not related to the main question but one that helps a meeting smoothly carry out business.

processing—The act of changing crops and livestock into forms that people want; all of the changes made in a product that prepare it for consumption.

producer—An organism that makes its own food; an autotroph.

production agriculture—The farming part of the agricultural industry.

production system—The approach used in producing animals; focus on the outcome of production—the kind of product produced.

professional organization—An association or group focused around career pursuits of its members.

profit and loss statement—A detailed financial summary that provides information on the performance of a business in relationship to making a profit or a loss.

program—The software that provides instructions to a computer.

program of activities—An annual plan of the goals and procedures for an FFA chapter.

prokaryote—An organism with cells that do not have a nucleus or do not have a separate membrane enclosing the DNA.

promotion—Something that promotes or advertises a product or service; includes advertising, but goes further to attract new customers, reinforce the current customer base, or increase the amount sold.

propagation—The reproduction or increase in the number of plants.

protectant—A chemical used on disease-free plants to prevent disease from developing.

protein—The substance needed for maintenance, growth, reproduction, and other functions.

protein supplement—A feedstuff high in protein and TDN.

puberty—The age at which an animal is capable of reproduction.

public speaking—A type of communication that uses oral methods of conveying information; oral information may be enhanced with gestures and visuals.

pulley—A special kind of wheel that changes the direction of force using a belt or rope; a simple machine consisting of a kind of grooved wheel and a rope or belt that provides mechanical advantage in doing work.

pulpwood—Harvested timber (usually softwoods) used in making paper, particle board, and other materials; the wood fibers are broken apart and pressed back together in a new form.

pulse rate—The movement of the arteries as a result of the heart beat.

Punnett square—A diagram used in studying heredity to predict the outcome of breeding.

purebred—An animal registered with a breed association or eligible for registry.

purebred breeding—The mating of a purebred animal with another of the same breed.

Q

quality assurance—Promoting the production of quality products through good management practices.

quality-assurance program—An organized effort among producers with specific requirements promoted for compliance with production standards.

quorum—The minimum number of individuals or members that must be present for an organization to legally make decisions.

R

radiation—A form of energy that travels as waves.

rangeland—Land that grows forage plants and is used as pasture for animals.

ration—The total amount of feed an animal gets in a 24-hour period.

receptor—A receiver of information collected by various parts of the nervous system.

recessive trait—A trait that is covered up by dominant alleles.

recipient—The female animal that receives the split embryo.

recitation—A type of speech that is prepared and fully memorized ahead of the time of delivery.

recombinant DNA—The result of gene transfer in which a tiny amount of DNA is cut from one chromosome and moved to another; the DNA of two different organisms is combined.

recordkeeping—The act of recording the supervised experience activities, including financial, occupational, and personal competency development.

rectangular survey system—A method of describing land based on two fixed lines that are at right angles to each other; used in the United States.

recycling—The act of recovering and reusing materials instead of throwing them away.

red blood cell—The blood cell made in the red marrow of bones; contains a protein called hemoglobin that carries the oxygen in the blood.

refrigeration—A method of food preservation in which food products are kept at temperatures slightly above freezing; cold storage.

relative humidity—The amount of water in the air compared with the air's ability to hold water vapor.

remote sensing—The gathering and recording of data from a great distance.

renewable natural resources—Resources that can be replaced when they are used.

repair—The replacement of the worn or damaged parts of an organism; injuries are healed, worn cells may be replaced, etc.

reproduction—The process by which new organisms are produced.

research—The careful and diligent search for answers to problems.

research and development—The process of creating and implementing new technology or practices.

research supervised experience—The planned investigation of a topic in food, fiber, natural resources, or a related area.

respiration—The process an organism uses to provide its cells with oxygen so energy can be released from digested food.

respiration rate—The number of breaths taken in a minute; frequency of inhalation and exhalation of air.

restraint—The careful control of an animal so that it can be examined, transported, treated, groomed, or otherwise managed.

retail market—An off-the-farm retail facility; usually located near or in towns or cities and selling to consumers.

revolution—The movement of Earth in space around the Sun; one complete rotation.

rhizome—A long, underground stem that sends up shoots to start new plants, such as Johnson grass.

RNA (ribonucleic acid)—Similar to DNA, RNA consists of short segments and functions in translation, regulation, and processing related to gene expression.

roadside market—An on-the-farm retail facility; producer operating a retail market located on the farm in addition to producing what is sold.

rodent—A small mammal with two large front teeth; gnawing animal with four continuously growing incisors, including rats, mice, squirrels, and beaver.

roguing—The removal of infected plants from a field, forest, or orchard.

root—The part of a plant that usually grows underground.

root cap—The tip of the primary root that protects the root as it grows into the soil.

root hair—A small, hairlike growth that helps anchor the root; root that branches off primary and secondary roots that grow between soil particles.

rotation—The turning or spinning movement of Earth; revolving.

roughage—A feedstuff that is high in fiber and low in energy.

ruminant—An animal that chews a cud, such as a cow or goat; its stomach has three or four compartments.

S

safety—The condition of being free of harm and danger; preventing injury and loss.

salmonellosis—a type of food poisoning caused by the presence of bacteria in the Salmonella genus.

salt—A compound formed when a strong acid and base are combined.

sanitation—The practice of keeping areas clean.

sapwood—The xylem in a tree that is two to four years old; sapwood is lighter in color than heartwood.

saw timber—The largest and oldest trees harvested for making lumber.

scale—In drawing, preparing the details of an object proportionate to its actual size.

scheduled feeding—A feeding system that provides feed at certain times of the day.

science—Knowledge of the world around us arranged as facts or truths to show relationships.

scientific method—An organized way of asking questions and seeking answers; the five important steps in studying science.

scientific name—The two-word name in Latin of a species; it appears in italics or underlined.

scion—A section of stem or other plant part that is planted or grafted to another plant.

screw—A modified incline plane; a simple machine.

seasoning—Removing moisture from lumber; drying wood.

seaweed—Brown, red, and green algae forms cultivated in water for food or food ingredients.

secondary defender—The natural response within an organism to resist disease and stay healthy; certain antibodies and phagocytes.

secondary growth—The increase in the diameter of stems and roots in plants.

secondary root—A root that branches from the primary root.

secondary succession—The regrowth of living organisms in an area destroyed by a natural disaster or the actions of people

secondary tillage—The tillage that follows primary tillage to further prepare the seedbed or to cultivate crops and control weeds.

seed—A container of new plant life; formed in the ovary of a plant.

seed-borne pest—A plant pest that is carried by seed and goes where the seed goes.

seed certification—A method of classification that assures the seed is of good quality and measure up to expectation.

seed coat—the outer covering of a seed; also known as the testa; it protects the fragile embryo from injury and holds the seed parts together.

seed-germ—An embryo; an immature plant.

selective herbicide—A herbicide that kills only certain kinds of plants.

selling—The process of making a transaction between a buyer and a seller.

sexual reproduction—The reproduction or increase in the number of plants or animals; using seed to propagate plants or animals.

shelter—A building used for housing by humans and provided for animals.

shielded arc welding—Electric welding that uses a coated electrode.

side-dressing—The application of fertilizer along the sides of rows of growing plants; corn, cotton, and many vegetable crops are side-dressed. The fertilizer is placed 6 inches from the root area.

silage—A crop of green plants that are chopped into small pieces and placed in a silo to ferment.

simple leaf—A leaf that has only one blade; corn oak and elm trees, and wheat.

social behavior—The manner in which we relate to other people.

softwood—Wood that is light and easy to cut; from pine, fir, spruce, and similar trees.

soil—The outer layer of the earth's surface.

soil analysis—Tests used to learn what nutrients are present in soil.

soil pH—A way of expressing the acidity and alkalinity of soil.

soil profile—A vertical section of the soil at a particular place.

soil salinity—The amount of various salts in the soil.

soil sampling—The collecting of samples of soil for analysis.

soil structure—The arrangement of the soil particles into shapes or pieces in undisturbed soil.

soil texture—The proportion of sand, silt, and clay in the soil.

soil tilth—The physical condition of the soil.

solar system—A star and a group of large bodies, called planets, that revolve around it.

soluble—The characteristic of a solid material that dissolves in water or another fluid.

solute—The substance in a solution that is dissolved.

solution—A mixture of two or more substances.

solvent—The liquid material in a solution.

sound—The result of vibration through a medium, such as air or water, creating waves.

spay—A female animal that has been spayed or neutered; to remove the ovaries of a female.

species—The most specific division of classification; the seventh division in the scientific classification system of plants and animals where all organisms are very similar.

speed—The rate of motion.

sperm—The male sex cell.

spoilage—A condition of food that makes it unsafe to eat.

standard of living—The level of choice about both essential and nonessential goods and services that people can make based on what they can afford.

standing committee—A committee with ongoing activities.

steer—A calf castrated before sexual maturity.

stem—The plant structure that supports leaves and flowers, transports water and other material, and stores food.

stem cell—A cell that has the ability to divide for indefinite periods in the proper medium and give rise to specialized cells.

sterilization—The removal or destruction of all germs in an area.

stethoscope—An instrument used to hear and amplify sounds produced by the heart, lungs, and other internal organs.

stocker calf—A calf that is put in a feedlot or on pasture for additional growth and fattening.

stocking rate—The number of animal units that graze or otherwise use an acre of land.

stomach poison—A poison that must be eaten by the insect to be effective; toxic material that is eaten.

straightbreeding—The mating of animals of the same breed.

stratosphere—The second layer of the atmosphere above the earth extending out about 30 miles (50 km).

strip cropping—The practice of planting crops in strips, which slows down the speed of the wind and rate of water runoff.

subsidiary motion—A type of parliamentary procedure motion that is applied to the main motion to alter, dispose of, or stop debate.

subterranean stem—A part of a plant stem that grows below the ground.

succession—The pattern of growing and changing; the natural and continual replacement of the organisms.

sucking insect—An insect that sucks the sap from a plant.

superovulation—An increase in the number of eggs produced by a female.

supervised experience (SE)—The planned application of skills learned in agriculture classes; also known as supervised agricultural experience (SAE), or supervised occupational experience (SOE).

surfactant—A surface-active material that helps solutions spread over the waxy surfaces of leaves.

suspension—a solute containing a solvent that may settle out unless kept agitated.

sustainable agriculture—All the practices used to maintain our ability to produce food, fiber, and shelter.

symbiosis—A dependant relationship between organisms of different species.

symptom—The way a disease shows itself.

synthetic biology—The science of creating lifelike characteristics with chemicals and nonliving substances.

synthetic food—An edible product made from nonfood raw materials or altered food materials.

systemic medicine—A medicine that is absorbed into the bloodstream.

systemic poison—A poison that is absorbed by a plant through its roots.

T

taproot—One large root that grows downward.

tariff—A tax or fee placed on imports and/or exports.

taxonomy—The science of the classification of living things.

technology—The practical use of science.

temperate climate zone—A band around the earth between the tropical and polar zones where the average temperature is above 65°F (18°C) in the summer and below 50°F (10°C) in the winter.

temperature—The presence of heat as indicated by the activity of the molecules in the matter.

temporary pasture—Land planted for winter, summer, or semipermanent grazing.

terracing—The practice of building earthen ridges or embankments that slow the rate of water runoff.

terrestrial—An animal that grows on land.

textile mill—A facility where cotton is woven into cloth.

therapeutant—A chemical used on plants with disease present, therapeutic chemical.

thermal energy—Energy produced by heat.

thermometer—A device used to measure temperature.

thigmotropism—The growth of plants over and around solid objects, such as rocks.

tillage—The slicing, breaking, or cutting of the soil to prepare it for seeding and crop production; two types of tillage are primary and secondary.

tissue—A group of cells that are alike in activity and structure.

tissue analysis—The testing samples of tissue.

tissue culture—A method of propagation in which a plant is grown from a single cell or a group of cells.

tool—An implement used to perform a mechanical job.

top-dressing—The application of fertilizer after a crop is up and growing.

topical medicine—A medicine that is sprayed, dusted, or poured on an animal to control diseases or treat wounds; therapeutant applied to the skin or outer surface.

topography—The surface configuration of land areas; it includes elevation, streams, and other features.

toxin—A substance that contains poison or has the potential to poison.

trace elements—Chemical nutrients that are needed in small amounts for growth; micronutrients.

traceability—The identification of an animal or group of animals that can be traced back to the producer.

training agreement—A written statement that lists the terms or conditions of supervised experience (SE).

training plan—A list of the activities in supervised experience.

training station—The location where supervised experience is carried out.

transformation—The process of changing a cell through gene transfer.

transformed cells—A cell that receives new material through gene transfer.

transgenic organism—An organism whose genes have been artificially altered using genetic engineering (molecular) processes.

translocated herbicide—A herbicide absorbed into the plant and moved through the vascular system to all plant parts; growth regulators.

transpiration—A process through which the leaves of plants give off water.

transporting—The moving of products from one place to another.

trap cropping—Planting a small plot of the crop near the field about two weeks before the main crop is planted to trap pests.

tropical climate zone—A wide band around the earth on both sides of the equator where the average temperature is above 65°F (18°C).

tropism—Plant movement in response to its environment.

troposphere—The atmospheric layer closest to the earth.

trunk—The main stem of a tree.

tuber—An enlarged underground branch or stem of a plant, such as the potato.

U

ultrasonics—A science that uses high-frequency sound waves (ultrasound), which people cannot hear, to get an image of an organ or other feature.

unicellular organism—An organism that consists of one cell.

universe—All that exists in space—both known and unknown.

upgrading—Grading up to improve the quality of offspring—a purebred male is mated with a grade female with the intent to improve the quality of the offspring.

V

vacuum wrapping—A form of preservation packaging in which all air is removed and the wrapper is tightly sealed to keep air out.

variability—The difference between each animal in a species.

variety—A group of related organisms within a species.

vascular system—The circulatory system of a plant.

vector—A carrier of new DNA into a cell.

vegetative propagation—The asexual reproduction of a plant by getting a part of a plant to grow onto another plant; cloning that uses leaves, stems, or roots to grow the new plant.

veneer—A special kind of wood material that is made by gluing a thin layer of valuable or high-quality wood over thicker layers of low-quality wood.

verbal communication—The type of communication that uses words to express ideas.

vertebrate—An animal with an internal skeleton and backbone.

vertical integration—A process in which a nonfarm agribusiness owns the processing plant and all the sources of supplies for producing the plant or animal.

veterinary medicine—The branch of medicine that deals with animals; provided by individuals trained in the field and known as veterinarians.

veterinary technology—The science and art of providing professional support services to veterinarians in their practice.

vesicle—A tiny, sphere-shaped structure with cell-like characteristics; used in synthetic biology research.

viability—The ability of a seed to sprout and grow.

virus—A precellular group that is not classified into either Domain; viruses are tiny particles that contain genetic material and protein but are not complete organisms.

vital sign—An indication of the living condition; sign of life; breathing, pulse rate, and body temperature.

vitamin—An organic substance that performs an important function in an organism; needed for good health; important in regulating body functions, keeping the body health, and developing resistance to disease.

volt—The measurement of the potential energy of electricity.

volume—An amount of space measured in cubic units, such as cubic centimeters (cc) or cubic inches (in^3); the total size of an object.

voluntary muscle—A muscle controlled by the "thinking" part of the brain (cerebrum).

voting—How people make known their choices about issues and other matters being considered by a group.

W

waste—The solid and semisolid material that results from the activity of people and animals.

waste disposal—The process of getting rid of waste.

water—A colorless, transparent liquid; probably the most important of all nutrients.

water cycle—The circulation of water in the hydrosphere; also called the hydrologic cycle.

water table—The depth of the natural level of free water below the surface of the earth; point in the earth where all the spaces are filled and no more water can be held; natural level of free water below the surface of the earth.

watt—A measurement of the amount of electrical energy supplied or used.

weather—The general condition of the atmosphere as it pertains to temperature, air pressure, wind, moisture, clouds, and precipitation.

weathering—The effect of weather on rocks and minerals on the crust of the earth.

wedge—A solid piece of wood or metal that is thick at one end and thin at the other, much like an inclined plane; wedges have sloping surfaces that taper to a point.

weed—A plant that is growing where it isn't wanted.

weighing scale—A machine that measures weight.

weight—The heaviness of something.

welding—Using heat to fuse pieces of materials together; often used with metals and plastics.

wheel and axle—One of the simple machines; allows rotation (turning) and transfer of power.

white blood cell—The blood cells that help fight off disease.

wholesome—Fresh, healthy, and good to eat; promoting or conductive to good health.

wildlife—All animals, plants, and other living organisms that have not been domesticated.

wildlife management—All of the practices used in manipulating wildlife systems to achieve desired goals.

wilting—A plant condition in which the stems and leaves of a plant are no longer rigid and erect; may be due to excessive transpiration, low soil moisture, disease, or other factors..

wind—The movement of air.

woody stem—The stem of a tree, shrub, and other crop plants, such as cotton, that has rigid tissue known as xylem.

work—The moving of an object through a distance when there is some resistance to its movement.

working animal—An animal used for power, pleasure, and other purposes.

xylem—Tissue that conducts water and other nutrients throughout a plant.

Z

zoology—The science of animals.

zoonosis—Any disease that can be transmitted to humans, as well as to other animals.

Bibliography

Amor, D. *The E-Business Revolution.* Upper Saddle River, NJ: Pearson Prentice Hall, 2000.

Baker, M., and R. E. Mikesell. *Animal Science Biology & Technology,* 2nd ed. Upper Saddle River, NJ: Pearson Prentice Hall Interstate, 2005.

Ballard, B., and R. Cheek. *Exotic Animal Medicine for the Veterinary Technician.* Ames, IA: Blackwell Publishing, 2003.

Baumgartner, L. K., and N. R. Pace. "Current Taxonomy in Classroom Instruction," *The Science Teacher,* October 2007, Volume 74, Number 7.

Biondo, R. J., and J. S. Lee. *Introduction to Plant & Soil Science and Technology,* 2nd ed. Upper Saddle River, NJ: Pearson Prentice Hall Interstate, 2003.

Biondo, R. J., and C. B. Schroeder. *Introduction to Landscaping: Design, Construction, and Maintenance,* rev. 3rd ed. Upper Saddle River, NJ: Pearson Education, 2009.

Burton, L. D. *Agriscience and Technology.* Albany, NY: Delmar Publishers, Inc., 1992.

California Plant Health Association. *Western Fertilizer Handbook,* 9th ed. Upper Saddle River, NJ: Pearson Prentice Hall, 2002.

Campbell, N. A., J. B. Reece, M. R. Taylor, E. J. Simon, and J. L. Dickey. *Biology: Concepts and Connections,* 6th ed. San Francisco: Pearson Benjamin Cummings, 2009.

Damrosch, B., and R. Ryan. *The Garden Primer.* New York: Workman Publishing Company, Inc., 2008.

Davis, R., R. Frey, M. Sarquis, and J. L. Marquis. *Modern Chemistry.* Geneva, IL: Holt McDougal, Holt, Rinehart and Winston, 2006.

Eisenkraft. A. *Active Chemistry.* Armonk, NY: It's About Time, Inc., 2007.

Eisenkraft, A. *Active Physics.* Armonk, NY: It's About Time, Inc., 2000.

Faughn, J., and R. A. Seaway. *Holt Physics.* Geneva, IL: Holt McDougall, Holt, Rinehart and Winston, 2002.

Gibson, J. D., R. H. Usry, L. W. Hass, R. T. Liles, and G. E. Moore. *Agribusiness: Management, Marketing, Human Resource Development, Communication, and Technology.* Danville, IL: Interstate Publishers, Inc., 2001.

Hogan, K. A. *Stem Cells and Cloning.* San Francisco: Pearson Benjamin Cummings, 2009.

Hunsley, R. E. *Livestock Judging, Selection and Evaluation,* 5th ed. Danville, IL: Interstate Publishers, Inc., 2001.

Jackson, N. S., W. J. Greer, and J. K. Baker. *Animal Health,* 3rd ed. Danville, IL: Interstate Publishers, Inc., 2000.

Kahn, C. M., ed. *The Merck Veterinary Manual,* 9th ed. Whitehouse Station, NJ: Merck and Co., Inc., 2005.

Lawhead, J. B., and M. Baker. *Introduction to Veterinary Science.* Clifton Park, NY: Thomson Delmar Learning, 2005.

Laws, F., ed. *Delta Agricultural Digest.* Clarksdale, MS: Intertec Publishing, 2009.

Lee, J. S., G. J. Burtle, and M. E. Newman. *Aquaculture,* 3rd ed. Upper Saddle River, NJ: Pearson Prentice Hall Interstate, 2005.

Lee, J. S., J. Hutter, R. Rudd, L. Westrom, A. R. Patrick, and A. M. Bull. *Introduction to Livestock & Companion Animals,* rev. 3rd ed. Upper Saddle River, NJ: Pearson Education, 2009.

Lee, S. J., C. Mecey-Smith, E. M. Morgan, R. E. Chelewski, R. Hunewill, and J. S. Lee. *Biotechnology.* Danville, IL: Interstate Publishers, Inc., 2001.

Oster, D., and J. Walliser. *Grow Organic.* Pittsburgh, PA: St. Lynn's Press, 2007.

National Academy of Sciences. *Science, Evolution, and Creationism.* Washington, DC: The National Academies Press, 2008.

Pedigo, L. P., and M. E. Rice. *Entomology and Pest Management,* 6th ed. Upper Saddle River, NJ: Pearson Prentice Hall, 2009.

Phipps, L. J., G. M. Miller, and J. S. Lee. *Introduction to Agricultural Mechanics,* 2nd ed. Upper Saddle River, NJ: Pearson Prentice Hall Interstate, 2004.

Plaster, Edward J. *Soil Science and Management.* Albany, NY: Delmar Publishers, Inc., 1992.

Porter, L., J. S. Lee, D. L. Turner, and J. M. Hillan. *Environmental Science and Technology,* 2nd ed. Upper Saddle River, NJ: Pearson Prentice Hall Interstate, 2003.

Purdy, H. R., R. J. Dawes, and R. Hough. *Breeds of Cattle,* 2nd ed. Springfield, MO: TRS Publishing Corp., 2008.

Rolfe, G. L., J. M. Edgington, I. I. Holland, and G. C. Fortenberry. *Forests and Forestry,* 6th ed. Upper Saddle River, NJ: Pearson Prentice Hall Interstate, 2003.

Romans, J. R., W. J. Costello, C. W. Carlson, M. L. Greaser, and K. W. Jones. *The Meat We Eat,* 14th ed. Danville, IL: Interstate Publishers, Inc., 2001.

Schlock, D., ed. *Complete Guide to Vegetables, Fruits and Herbs.* Des Moines, IA: Meredith Books, 2008.

Schroeder, C. B., E. D. Seagle, L. M. Felton, J. M. Ruter, W. T. Kelley, and G. Krewer. *Introduction to Horticulture,* rev. 4th ed. Upper Saddle River, NJ: Pearson Education, 2009.

Seidman, L. A., and C. J. Moore. *Basic Laboratory Methods for Biotechnology,* 2nd ed. San Francisco: Pearson Benjamin Cummings, 2009.

Stewart, M., J. S. Lee, S. Hunter, B. Scheil, S. D. Fraze, and R. Terry, Jr. *Developing Leadership & Communication Skills,* 2nd ed. Upper Saddle River, NJ: Pearson Prentice Hall Interstate 2004.

Stutzenbaker, C. D., B. J. Scheil, M. K. Swan, J. S. Lee, and J. M. Omernik. *Wildlife Management: Science & Technology,* 2nd ed. Upper Saddle River, NJ: Pearson Prentice Hall Interstate, 2003.

Swiader, J. M., and G. W. Ware. *Producing Vegetable Crops,* 5th ed. Upper Saddle River, NJ: Pearson Prentice Hall, 2002.

Texas Cow-Calf and Stocker Beef Safety and Quality Assurance Handbook, The. Retrieved March 5, 2009, from http://www.texasbeefquality.com/handbook/Guides.htm.

Wagner, L. A., and G. J. Seperich. *Food Science & Technology.* Upper Saddle River, NJ: Pearson Prentice Hall Interstate, 2004.

Index

A

B

D

E

F

G

H

S

T

Z